# TELEVISION
# THE CRITICAL VIEW

# TELEVISION

## The Critical View

*Fifth Edition*

Edited by

HORACE NEWCOMB

New York   Oxford
OXFORD UNIVERSITY PRESS
1994

Oxford University Press

Oxford   New York   Toronto
Delhi   Bombay   Calcutta   Madras   Karachi
Kuala Lumpur   Singapore   Hong Kong   Tokyo
Nairobi   Dar es Salaam   Cape Town
Melbourne   Auckland   Madrid

and associated companies in
Berlin   Ibadan

Published by Oxford University Press, Inc.
200 Madison Avenue, New York, New York 10016

Oxford is a registered trademark of Oxford University Press

Library of Congress Cataloging-in-Publication Data
Television : the critical view /
edited by Horace Newcomb.—5th ed.
p.   cm.   ISBN 0-19-508528-0
1. Television broadcasting—United States.
I. Newcomb, Horace.
PN1992.3.U5T42   1994   791.45′0973—dc20   93-15634

2 4 6 8 9 7 5 3 2 1

Printed in the United States of America
on acid-free paper

*For Eric Michaels, 1948–1988*
*Among the Best of Critics*

*And for All the Students*
*With Whom I Have Been Privileged To Work*
*Since 1968*

# Preface

Television continues to change. Simple enough, and obvious. The changes seem more significant when one tries to keep pace with them, think about them, explain them, and present them in some understandable manner to others. To think quickly and precisely about the medium requires more than a regular reading of the trade press or even of scholarly journals. It seems even more difficult when one tries to make sense not only of the medium itself, but of the many efforts to understand and explain it. As I say below in the Introduction, Television Studies is now a flourishing academic enterprise. Thoughtful, detailed analysis, history, and research is more and more available. And this means that one must make a more and more concerted effort to find ways of organizing the work.

As always, this collection is such an attempt. The days when filling such a volume was a task in itself are long past. The assignment has now become one of surveying, sorting, and arranging.

Much of criticism rests in searches for patterns. The patterns found here will not be those chosen by others, but they are offered as guides and reference points. From here, the sorting becomes the responsibility of the user.

Special thanks for assistance in the preparation of this edition go to Dell Edwards, Michael Saenz, James Hay, and most especially, to Sara Newcomb.

*Austin*                                                                                   H. N.
*May 1993*

# Preface to the First Edition

The essays in this collection were selected because they view television in broad rather than narrow perspectives. Newspaper columns have not been included. This is not to say that newspaper criticism is excluded by definition from a breadth of vision, but simply that the pieces included here all develop their point of view in the single essay rather than over a period of time, as is the case with the columnist.

The essays in the first section all deal with specific program types. They serve as excellent models for practical television criticism because they show us that there is a great deal of difference between watching television and "seeing" it. They are, of course, involved with critical interpretation and assertion. Other analyses of the same programs may be offered by other critics, and the audience, as critic, must learn to make its own decisions. These essays will help in that learning process.

The second section is comprised of essays that attempt to go beyond the specific meanings of specific programs or program types. They suggest that television has meaning in the culture because it is not an isolated, unique entity. These writers want to know what television means, for its producers, its audiences, its culture.

The essays in the final section are concerned with what television is. They seek to define television in terms of itself, to determine how it is like and how it is different from other media.

All the essays are seeking connections, trying to place television in its own proper, enlarged critical climate. Consequently, many of them use similar examples, ask similar questions, and rest on shared assumptions. Some of the connections are obvious. Others will occur to the reader using the book. In this way the reader too becomes a critic and the printed comments may serve to stimulate a new beginning, a new and richer viewpoint regarding television.

I would like to express my thanks to John Wright of Oxford University Press for his initial interest and continued support for this book. His suggestions have strengthened it throughout. A special note of thanks must go to all my friends and colleagues who have made suggestions about the book and who, in some cases, have offered their own fine work for inclusion. Thanks, too, goes to my family for the supportive world in which I work.

*Baltimore*                                                              H. N.
*November 1975*

# Contents

# TELEVISION
# THE CRITICAL VIEW

# Television and the Present Climate of Criticism

## HORACE NEWCOMB

The purpose of this collection has always been the same, to contribute to what Moses Hadas many years ago referred to as the "climate of criticism." In defining that concept, he admonished "all who take education seriously in its larger sense" to "talk and write about television as they do about books." Earlier editions of the collection struggled with the fact that this was hardly the case, with the realization that television was rarely considered a prominent, significant, or special contributor to culture and society. It was seen as neither conduit for nor commentator on the aesthetic, political, or moral lives of citizens—except in the most negative manner. When television was "seriously" thought of by anyone who would take education seriously in its large sense, it was most often figured as intruder, as complicator, as rogue or pollutor. Many of these attempts to understand the medium must now be seen as incomplete, partial, narrow definitions supporting one limited perspective or another.

This is not to say that television should have been warmly and naively welcomed into society and home. Many aspects of the medium (less than fifty years old in its basic forms) were and are troublesome, threatening, oppressive, repressive, and obnoxious. So, too, are many books. The issue—the problem for criticism, in this collection and others, for teachers and students—has been the development of vocabularies sufficient to consider all these matters.

A second purpose of this anthology has been to explore the development of that vocabulary from the perspective of the "humanities" as broadly defined. That is to say, because television was early on considered as a social problem, many approaches and considerable terminology had already been developed to deal with the medium by the time scholars, critics, and thinkers who had traditionally focused their attention on books turned to this newer medium. Those first approaches, terms, and analytical paradigms were drawn largely from the realms of social psychology and sociology and had little use for strategies that would look closely at television as an expressive form.

This aspect of the "climate" of television criticism is the one that has altered most perceptibly in recent years, years encompassing previous editions of *Television: The Critical View* and now affording work for the present volume. And while this feature has changed, many others implied in Hadas's comments have remained essentially the same. This introduction focuses on the differences and similarities that can be marked in the present critical climate.

The most prominent change in our understanding and approach to television has occured within academic settings. Television Studies is now an established area of study in many universities. Although the concept, the term, and the designation are constantly under discussion and revision, it is clear that, from this academic perspective, Hadas's admonition is taken quite seriously. Interestingly enough, however, the development of Television Studies has modified our basic notions of what the medium is, what "humanities approaches" consist of, and what "academic" discourse surrounding television (or any other subject) has to do with "education in its larger sense."

Television Studies as an academic enterprise developed from four major backgrounds. In the United States, the first of these was literary studies that redirected critical analysis toward the study of popular entertainment forms: novels, material culture, magazines, radio programs, and so on. The study of popular films was also important in this enterprise, but developed through a somewhat different route, to be discussed below. The choice to examine these "inferior" or "unappreciated" forms was motivated by a number of concerns. Philosophically, scholars in this movement often felt the works they wished to examine were more indicative of larger cultural preferences, expressive of a more "democratic" relationship between works and audiences than the "elite" works selected, archived, and taught as the traditional canon of humanistically valued forms of expression.

Politically, these same impulses suggested that it was important to study these works precisely because their exclusion from canonical systems also excluded their audiences, devalued large numbers of citizens, or saddled them with inferior intellectual or aesthetic judgment. It is not coincidental that the study of popular entertainment developed momentum at the end of the 1960s, when many cultural categories were under question and when these questions made their way into educational institutions as questions concerning curricula, canonical content, and the value of "traditions."

For the most part, although these early studies were politically motivated and presented as part of far larger political movements and actions, they did not offer systematic ideological analyses of the works they examined. Their agenda was broadly cultural, examining popular forms for their contribution to ongoing discussions about various "meanings" within expressive culture. They did, however, open questions of how we might study nontraditional forms of cultural expression: games, events, designs, fashion. And it is necessary to remember that many of these analyses were carried out prior to the intense concern with theories of how meanings might be systematically imposed through cultural codes, social strategies, industrial organization, or other more embedded and socially grounded influences on expression. These early studies of popular culture were, rather, one of the final applications of critical analysis rooted in what is referred to in traditional literary studies as the New Criticism. And in this linkage, they represent the extension of New Critical questions to "profane" culture, thus contributing, in my view, to a radical modification and appropriation of that form of analysis. What remains of New Critical approaches, an immensely valuable contribution, is the close attention to textual detail, even when this "close analysis" is exercised in the examination of social constructs, cultural patterns, behaviors, and artifacts that far exceed earlier notions of "texts."

The second major influence on Television Studies emerged in Europe, its primary sources for American students coming from Great Britain. Cultural Studies, developing from the work of Raymond Williams, Richard Hoggart, and Stuart Hall, had already begun to examine similar artifacts, television among them, as American popular culture scholars, but had done so within a far more formidable analytical tradition. Where American approaches had acknowledged something about "politics" in their work, the British scholars systematically explored "ideology."

This work was profoundly influenced by continental Marxism and structural anthropology, two varying but in their early stages almost equally powerful forms of "structuralism." While Marxist theory analyzed structures grounded in the economic determination of social categories, structural anthropology searched (for and presumably in) "mental structures" that crossed specific social, geographic, and cultural boundaries. The argument that much of human experience, from "texts" to forms of social organization and action, was informed, ordered, and directed by "deeper" structures was immensely powerful. It allowed scholars and critics to bundle vast amounts of discreet artifacts and events into significant and manageable "patterns." The patterns, the formulas, the genres, the classes—of citizens or of texts—were the item to be analyzed. Complexities of contemporary life were thus more easily handled, described, and written about.

Work done at the Centre for Contemporary Cultural Studies at the University of Birmingham struggled with these varied and mixed approaches, sorted through theories and methods, and offered a number of powerful models for research on a number of topics. One of the primary syntheses accomplished centered on the work of Antonio Gramsci and the application of notions of hegemony to the study of contemporary culture. Mass media generally and television specifically occupied central positions of inquiry in these efforts.

When these approaches reached American universities, they reached into work already primed for more systematic analysis, and Television Studies, among other topics, eagerly appropriated some of the models, techniques, and styles of analysis. These American appropriations also distorted, in the view of some scholars and critics, the critical analysis of ideology that undergirded British work.

They did mesh well, however, with a third influence on the developing field of Television Studies. This work maintained the ideological focus and drew on the tradition of critical sociology associated with the Frankfurt School of sociological analysis. In the United States, analysis of television carried out in departments of sociology and philosophy was most involved with these sources. Academic critics working from this tradition were able to critique what they perceived to be a central weakness in the earlier "popular culture" approach, its reliance on a naive notion of "liberal pluralism" central to many expressive forms. The arrival of "British Cultural Studies" required and enabled some scholars working within that earlier tradition to sharpen their own critiques, to recognize weaknesses and gaps in their work, and to move to a more complex perspective on television and other aspects of popular expressive culture. Generally, the arrival and conflicted acceptance of Cultural Studies approaches in the United States overlaps heavily with the development of Television Studies, and ongoing debates have developed around the varying emphases on and definitions of ideology.

A fourth influence on Television Studies emerged from the growing body of film studies in the United States and abroad. Unlike television, film had been accepted, in some quarters since its earliest days, as a "fine art." In popular, general discussion, many of the same issues and concerns that would occupy critics of television—crass commercialism, "debased" moral attitudes, direct influence on viewer behavior, ideological agendas—were central to discussions of movies. Still, some early popular, and much academic, discussion of film took on a far more appreciative tone. And at times film occupied a decidedly noble position within public discourse. Put another way, the "climate of criticism" for film has, throughout its history, been more favorable than that surrounding television. In its most recent stages, film criticism, both for the general reader and for the academic community, developed sophisticated and complex forms and styles of analysis widely accepted as legitimate.

Even in these discussions, however, much of the work focused historically on its own canonical topics: films produced in Europe or other parts of the world, experimental or lyrical personal films created by individual artists in all countries, and early popular film that had taken on a valued status because of the announced "declines" (formal, social, aesthetic) that followed. Categories of analysis were initially quite close to those used for formal analysis of literary works by New Critics, but moved quickly to the appropriation and development of various forms of structuralism. And perhaps even more significantly, film studies explored an extraordinary range of applications of psychoanalytic theory to its objects.

The study of popular American film, the "Hollywood film," came later and with considerable opposition. But much of the recognition of the particular value of these works came from European critics and filmmakers who saw in Hollywood a form of personal and cultural expression with its own powerful creative techniques, its own intrinsic values. Gradually, the study of popular American film was accepted in the United States as a legitimate enterprise. At times this work was related to the broader study of popular culture, but because of the history of film studies, it remained, for the most part, separate.

When these approaches were infused by the same European theories of structuralism, Marxism, and Cultural Studies, another wave of work emerged; first slowly then more swiftly, film scholars turned their attention to television. In some cases this involved the straightforward application of film theory to the newer medium. And in many instances it also involved the modification or rejection of those theories because of the many significant differences between the two media.

Operating throughout this history, crossing all these influences, and so widespread as to be an influence on almost every field of study in the humanities and elsewhere, was and is feminist theory. I list it as a separate force because it is more far-reaching than Television Studies even while it is now central to that topic.

The recognition of gendered distinctions has been central to Television Studies in two ways. First, as in many other fields, it has called into question the theoretical base of many approaches from considerations of form to the organization of labor in the creation of television texts.

Second, and more problematic, television itself has been defined at times as a "feminine" medium. This designation rises on the one hand from the economic-social fact of television's "address" to women through its base in advertiser-supported, consumer-targeted content. On the other hand, it emerges from arguments that the form of television (again in advertiser-supported commercial systems such as American television), with its demand for open, unending narrative strategies, relates somehow to feminine experience. Discussion of such questions continues. Within either formulation, feminist theorists have been among the most active and perceptive in the study of television and all theoretical positions must take note of these approaches to the medium.

None of these influences, however, should be taken as fully definitive, fully encompassing the study of television. For one thing, their application to television was merely part of larger discussions of their usefulness in studying an array of cultural artifacts: literary, cinematic, pictorial, social. And they were applied to "high" as well as "low," or "popular," forms and behaviors.

Moreover, no sooner than one influence was fully absorbed, others were layered upon it. Much of the discussion in what might be termed the "theory wars" in the academic life of the seventies and eighties had to do with ways in which varying methods and systematic approaches might be applied in varying contexts or might be combined for more powerful forms of analysis.

These ongoing intellectual discussions and struggles are not conducted merely over the utility of methods and theories. They are also directed by more profound questions of epistemology, of how we know what we know. And they are legitimate attempts to come to terms with the role of expressive forms in the social and political lives of citizens as individuals and members of social groups. They are attempts to contribute to social and cultural improvement, to understand the role of expressive media in constructing or altering forms of inequality and oppression. Despite attempts to present these ongoing discussions as mere tempests in small teapots, or as the corrupt agenda of "radicals" in universities, these sometimes fierce exchanges, these often arcane presentations of ideas, have most often been good faith attempts to recognize and understand how social life, now dependent on mass media, on popular forms of expression and entertainment, on the far-reaching lines of information afforded by new technologies, can best be taught and understood, learned and used by all citizens.

Thus it is that various forms of structuralism have been challenged by what have now been defined as "poststructuralist" assumptions. From these perspectives the very notions of "unified" or "coherent" patterns of meaning, of social organization, or of individual psychology are called into question. How can invariant meanings be established by individuals who are themselves constructed of multiple and conflicting psychological, social, and cultural influences? How can mass audiences "receive" coherent meanings from ambiguous and conflicted texts, be they television programs, canonically approved "high culture," films, or rock 'n' roll? And if these varying, conflicting, multiple, unstable artifacts and users are constantly engaged in the process of "constructing" their own social meanings and behaviors, how can any coherent ideological effects be determined?

Added to poststructuralist problems are those suggested by a related category, "postmodernism." Here the issues of instability and potential incoherence are taken in still other directions, in some cases to still greater levels of challenge.

For some critics, commentators, and scholars, the notion of postmodernism means simply that all citizens are now both creator and critic, making up their own individual or group culture from the massive resources of our "cultural archives." In this view, the mixing of style and fashion, image and ideology, meaning and significance are evidence of a new, liberated consciousness.

For others, however, the consequences of postmodernism, are far more significant, more dire. The very possibility of meaningful discourse—about politics and ideology, about categories of thought and social organization, about any sense of "shared culture"—seems either impossible or now requires new forms of analysis, argument, and expression. This is especially the case if, as some arguments would have it, there is no longer any way to establish the "real" in human experience if all has dissolved into "representation."

Most important for our concerns here, television has often been seen as both cause and effect, source and symptom, agent and evidence of these newer social and cultural developments. This is especially the case when "television" is defined as or by "American commercial television." And because the form of American commercial television is shared throughout the world, either through the export of programming or through the collapse of various forms of public service broadcasting in the face of an advancing privatization of investment, "television" in any context often means "American TV."

American TV is defined most fundamentally in these discussions by its commercialism. And commercialism results in, indeed demanded, continuous programming flow, organized programming strategies matched to other forms of social organization, and the constant attempts to enforce existing power relationships. Thus, television secures conventional and dominant formations of race, class, gender, age, ethnicity, region, and style. So long as the fundamental purpose of the medium can be defined as the delivery of audiences to advertisers, or perhaps more significantly as uniting the world within an equation of television, consumerism, and "reality," using all possible strategies to accomplish this end, civilization itself must be thought of as sliding into the postmodern condition exemplified by television. This concern, fear, pessimistic conclusion has been heightened in recent years by vast changes in the technological and economic contexts of television. As we have moved, in American and world television, from a few central network presentations to the vast proliferation of televisual options, the postmodern model has taken on ever increasing power.

Again, however, our very notion of a "climate of criticism" should alert us to the fact that climates are changeable, always unstable, never the same in two places at once. The field of Television Studies is neither closely bounded or fixed. Questions implied by this brief survey of the winds that have crossed the field in the past ten years remain unsettled, unanswered. In many cases they are in constant states of reconsideration. Keeping this background in mind, then, I will now suggest some of the most pressing categories of questions that define the field at the present moment. These are, in some cases, the questions that have directed the reorganization of this collection, changing its format from that of previous editions. It is necessary in examining these questions to remember that none of the issues discussed above has gone away. Rather, new problems, new issues, are woven into others. Some of them are refined and some approaches are made stronger. Others are seen as less helpful and partially discarded. Still others are applied under new conditions, in response to changes in television itself. In all cases, criticism works with both the wisdom and the burden of its own pasts carried into the present. Moreover, it is equally important to remember that while questions may be grouped in patterns, answers remain more diverse. Even when several essays are presented here as focused on similar questions their approaches and techniques for understanding and answering those questions may differ widely.

Early television criticism focused on definitions of the television "text," its forms and conventions, and the meanings of these patterns. Often the analysis was conducted as a form of genre study, and the question of meanings circulated between individual instances of a genre—programs or series—and the larger group to which the instance belonged. Meaning was discovered in both the generic pattern and at the level of plot or story.

Another wave of critical discussion focused on issues of ideology, and here meaning, significance, and effect were discovered at the level of structure and organization. Clearly, genre remains central in such inquiry, but the primary shift was from meaning residing *in* genre and individual instance to the ways in which these patterns reorganized and reinforced existing social meanings.

This critical discussion of television and ideology continues to offer some of the most prominent questions for Television Studies, questions that inform many of the essays in this collection and are directly or indirectly addressed by all of them. In one sense this merely indicates that television criticism is part of a much larger critical enterprise, for the centrality of ideological issues is now common to almost all humanistic discourse. The remaining problem is, what are the unique features of television that demand critical explication?

Early explorations of television from this perspective often viewed television as little more than an "ideology machine," churning out replications of the most oppressive and repressive aspects of contemporary American society. Some of this work was grounded in film study and could not escape more widespread notions that television is merely a degenerate cinematic form. Other versions were far more sophisticated and sought to define television's specific economic, social, aesthetic, and cultural characteristics in order to show that it contributed to, rather than replaced, other forms of ideological discourse.

These views of television as ideologically monolithic, however, were quickly displaced by some of these same attempts. As the medium was analyzed and described more precisely and from different perspectives, the question shifted. Instead of asking how television performs this monolithic ideological function, questions came to center on whether or not television was so unified in its forms and effects. One way to formulate the problem at this moment is to grant that the medium (like all others) is varied and conflicted, but to ask whether or not it is more "open" or "closed," more rigid or flexible as a form of ideological expression. Even more precisely, we must seek to determine the circumstances in which either condition will define the case. Clearly, this means that all the questions of form, genre, content—broadly defined as the aesthetic questions—must still come into play. Many of the questions focused on these topics are addressed by essays in Part II of this collection, essays that focus on specific programs or program types. As some of these essays indicate, the questions must be examined in far richer contexts than in many early studies.

Defining contexts is the problem at the center of the second major focus of contemporary television criticism, and various attempts to establish these contexts now occupy a far more prominent place in Television Studies than before. Again, these developments lead to the reorganization of this collection.

Major contributions to this area of concern came from the introduction of Cultural Studies approaches to our critical repertoire for the reinsertion of expressive forms into social context lies at the heart of this enterprise. The two most significant contextual categories in contemporary Television Studies are broadly defined as the context of "production" and the context of "reception." Part I of the current collection focuses on various forms of production studies. The essays variously examine historical contexts of television production both social and industrial, broad applications of technique and design within television program types, and cases studies of the production of specific television programs.

Essays in Part III focus on the reception context. Here the studies are also informed by serious disagreements over the limits of audiences' interpretive power. Are meanings constructed in the process of "receiving" television? If so, are audiences able to "subvert" those meanings that interpretive critics might find repressive or regressive? In what instances is such power exercised? What other social factors might direct or limit this process of meaning making? These questions have become central to the current discussions of this medium and others.

The final section of the collection offers essays that attempt larger overviews of the entire medium of television. Some focus on aesthetic concerns, some on social. Many see the necessity of avoiding such easy dichotomies and seek various forms of synthesis. It is in this section that a clearer distinction between earlier approaches and later ones emerges. Overviews developed to explain television in "the network era" have a historical tone to them. Newer essays, more focused on television's social and technological proliferation, and informed by notions of poststructuralism and postmodernism, ask decidedly different questions. But the newer questions could not have been formulated without the older ones, and the resonances among these explanations are telling.

As stated above, almost all these developments in the burgeoning field of Television Studies have taken place within academic settings. There it is now commonplace for television to be taken as seriously as books by scholars and critics in various fields. This has not been the case, however, in writing offered to more general audiences. There the best writing about television remains essentially personal, individualized, subjective. While occasionally interesting, the commentary is most often repetitive, familiar, unwilling to consider the possibility that television presents itself in terms distinctive and definable. Questions remain much the same as over the past decades.

Answers to those questions that might be available from academic television criticism are overlooked or ignored. In worst cases, writers for general readers look on in bemused and thinly veiled contempt at the academic critics. The implication is that no one should take television *this* seriously.

There remains, then, another step, another stage in the history of television criticism. In that stage the serious histories, the detailed analyses, the studies of television reception that now remain, in every sense of the word, "academic" approaches to television, will be shared more widely with all audiences. This book is an attempt to assist in taking that next step. And it must be stated here that there are problems with presenting some academic criticism to general audiences. Academic critics and scholars bear responsibility for making their work clear, powerful, accessible, usable.

But an equal responsibility is borne by editors and writers for more generally distributed publication. One part of that responsibility rests in taking the academic criticism more seriously, learning from its insights, its hard historical work, its detailed case studies. Another part of the responsibility is grounded in the very lack of opportunity for anyone to write seriously about television for the public. Too many newspapers relegate television criticism to small columns on back pages. Too few magazines print television commentary of any sort other than the contemptuous or satirical.

There is no doubt that we are closer now to Hadas's preferred state of television commentary. For one thing, more and more students have the opportunity to make Television Studies a part of their general education. These opportunities do not replace their experiences with books, as some might fear. Rather, with books as their tutors, they learn to see television more clearly, more critically. And perhaps these students will come to demand more powerfully and precisely informed commentary about television in their wider experience as citizens.

More to the point, perhaps some of these students, the present users of this collection, will come to write the sort of criticism, the informed, precise, exploratory, and explanatory criticism that has been missing from our shared experiences of this medium in other times. Hadas's concern, finally, was with the making of critics, not merely with the making of criticism. That, too, is the primary purpose of this collection.

Nowhere are we made more aware of these matters than in Eric Michaels's astonishing essay, "For a Cultural Future." A portion of that essay closes this anthology. His case study of the making of television in an Australian Aboriginal community encompasses and capsulizes every issue touched on in this collection, for there all the issues are faced anew. More than in any other succinct single presentation, Michaels faces and analyzes the aesthetic and ideological, the cultural and social, the context of home and the context of nation, the application of practical politics and the analysis of ideology, apparently overwhelming issues of technological power and the recognition of personal resilience. As often stated throughout the essays that follow, it is television that gathers and focuses all these questions.

The intrusion—welcomed, feared, protested, and embraced—of this medium into collective human social experience is refigured in Yuendumu. Moreover, it is refigured in a world already altered by that intrusion in the worlds surrounding Yuendumu. It is the subtlety of these interactions that Michaels captures. We are constantly reminded that while we might generalize some of the comparisons of this encounter to the overall history of television, the significance of his analysis lies precisely in its individual case and context.

And the outcome of his analysis suggests that we must consider television as subtly as he does whenever we seek to discuss this medium in any form. For out of subtlety comes the extraordinary power of his conclusions. The political and moral authority with which Michaels speaks is a reminder that too often we waste words on trivial matters. Sometimes, as critics, we can justify our trivialities by arguing that from such minutia we build larger intellectual edifices. But such claims are often as much rationalization as justification.

This is not to argue that Michaels is correct, or true in every aspect of his argument, whether social, cultural, aesthetic, or political. Indeed, his work in Australia, closely tied to matters of actual practice and policy, remains much contested. Even for those of us removed from the actual locus of those debates, removed from all the complexities and details, however, the force of the arguments rings true. If we wish to disagree here we will have to do so in a manner equal to the arguments we confront. Michaels's work demands of us a higher level of critical exchange and discourse, and his too early loss from that exchange remains tragic in its implications.

But the work remains, powerfully present, distinctive in its voice, committed, engaged. It serves as a model for our attempts to come to terms with television both old and new. It reminds us of the power of the medium. But it reminds us, too, of the power of criticism. Those of us who have turned to the study of this medium have often done so because we experienced or sensed the kinds of links to lived experience limned by Michaels in his essay. Other essays throughout this collection are strong attempts to frame perspectives grounded in those connections. It remains now for users of the essays to do the same.

# I

# THE PRODUCTION CONTEXTS OF TELEVISION

Explorations of television's production contexts take many forms, but the central question for most of them can be constructed in a somewhat general fashion: How does this particular moment of television come to be? As simple as such a question might seem, it immediately involves us in consideration of topics ranging from the development of minute technological devices to inquiry into corporate strategy, national economic policy, agency and legislative regulatory decisions, and, at a still more expansive level, social and cultural history. It also involves us in the day-to-day practices of television producers, writers, and actors as they work to fashion "their" programs in all these other contexts.

The essays in this section offer several approaches to these questions and problems. As practical models for criticism they also pose some difficulties. On occasion each draws its materials, its evidence and examples from readily available sources: newspapers and magazines, public records archived in most university libraries, television programs still currently aired in original or rerun form. But each also depends on resources that might not be easily obtainable by students. Corporate records cannot generally be accessed, for example, without special permissions. Copies of scripts, proposals, and story conference revisions may not always be easy to come by, especially if the project has a historical component. The opportunity to spend time on the set of a television production requires that such time be available, that access be established, and that the researcher know, basically, what is going on when she arrives at the scene. Interviews with television makers are often somewhat more easily arranged. Students of television are often amazed at the availability of producers, writers, and other production personnel who will give their time for phone interviews. Network personnel are more difficult to reach, but can, on occasion, provide interesting material. Even in these more open situations, however, care must be taken for protocols, timing, and detailed preparation.

In spite of these difficulties, Television Studies can no longer proceed without these contextualizing explorations. Some of the most exciting, most precise, most fully developed work on television is now emerging in this area.

Lynn Spigel's superb, and essential study, *Make Room For TV,* from which her essay "Women's Work" is drawn, begins this book because it is one of the few major studies to remind us that television did "begin." Because television is so familiar at this point in our experience, we tend to overlook the fact that as cultures, as publics, as groups and individuals we had to learn to live with it. And there were major participants in the many worlds of television whose interests lay in "instructing" American society in that learning process. As Spigel indicates, from its earliest moments, television was presented as having a special relationship with women. The insertion of the receiver set into the home, the domestic space presumably defined by and for females, was a crucial part of our coming to terms with the medium. Spigel is the first scholar to trace the details of that process so that we may now better come to terms with our history.

Michael Curtin also demonstrates the power of social and ideological forces at work in early television, this time precisely focused on one aspect of the medium rather than on its large cultural patterns. In his discussion of the forces—industrial, social, political, personal—that shaped television documentaries, one of the most prestigious and prevalent early television forms, he requires that we reconsider our earlier judgments of that program category. His combination of careful contextual analysis with close attention to a specific example suggests how other television research might be conducted, particularly if we are to better understand, rather than generalize about, television history.

The sense of television as both domestic and ideological medium is also central to Christopher Anderson's analysis of the rise of the Disney Studios and that entity's appropriation of television for its own industrial strategy. This essay, too, is part of a larger study. There, Anderson points out that just as individuals, families, social groups—an entire culture—had to come to terms with television, so, too, did the industry to which television was most often compared. His work shows how early television makers, often based in film, learned to work with, indeed to invent, aspects of the newer medium.

With David Barker's discussion of production techniques we move from the historical to the industrial and aesthetic concerns that now inform many studies of individual television programs. In some ways this discussion shows that the processes of invention outlined by Anderson, continued at the microlevels of program production. Though Barker's examples are drawn from comedy in this instance, they have proven helpful for many other critics wishing to look more closely at production techniques. One of the most valuable contributions of this essay is that it offers us a vocabulary with which to discuss specific visual aspects of television productions. These elements also have a historical component, however, for we can compare these programs with others, before and after them.

Judine Mayerle focuses on a different aspect of the production process, the day-to-day work involved in bringing a specific program to air. Her ability to gain access to the *Roseanne* set, to observe, describe, and analyze the show's production practices, offers us far more detail than can be gained from textual studies alone. This is precisely the sort of grounding needed to anchor our interpretations and judgments of television's meanings. Increasingly it seems that neither sort of analysis is, alone, sufficient. In the case of *Roseanne* this is a particularly important observation, for the show has continued to change since Mayerle's study, changed as we have continued to watch and make it one of the most successful in the schedule. The show described by Mayerle is both the same as and different from the one we now view. And because the changes have often been widely discussed in both the trade and general press, we can match versions of the program as we read.

This newer form of analysis is illustrated in an especially strong and detailed manner in Rodney Buxton's essay on a specific espisode of *Midnight Caller*. Drawn from Buxton's massive and richly detailed study of the representation of AIDS on television, this piece describes one of the most widely discussed television treatments of the topic. Combining textual analysis with information drawn from interviews and reported details surrounding the case, Buxton provides a model for understanding television making as an active process. The activity in this instance involves the production company, the writers, and the community in a visible instance of personal, social, cultural, and ideological conflict. The controversy surrounding this television program exposed categories of cultural meaning and social power, demonstrating again the centrality of this medium in our individual and collective experience.

Taken together the essays in this section offer some of the most insightful newer forms of research, critical analysis, and cultural commentary surrounding television. This kind of work is requiring many students of the medium to reevaluate prior judgments based primarily on attention to content and form. Such work is indeed altering the climate in which many of us do our work.

Perhaps a full writing of television's history, a more precise examination of the choices and mechanisms surrounding its production could only come now, as the medium reaches a point of fundamental change. Realizing that what we have called "television" is, in fact, disappearing, changing as we view, those who have turned to a better understanding of its past and its processes offer us new and formidable ways of understanding what has intrigued us in this "first wave" of our experience with TV.

# Women's Work

## LYNN SPIGEL

The Western-Holly Company in 1952 marketed a new design in domestic technology, the TV-stove. The oven included a window through which the housewife could watch her chicken roast. Above the oven window was a TV screen that presented an even more spectacular sight. With the aid of this machine the housewife would be able to prepare her meal, but at the same time she could watch TV. Although it was clearly an odd object, the TV-stove was not simply a historical fluke. Rather, its invention should remind us of the concrete social, economic, and ideological conditions that made this contraption possible. Indeed, the TV-stove was a response to the conflation of labor and leisure time at home. If we now find it strange, this has as much to do with the way in which our society has conceptualized work and leisure as it does with the machine's bizarre technological form.[1]

Since the nineteenth century, middle-class ideals of domesticity had been predicated on divisions of leisure time and work time. The doctrine of two spheres represented human activity in spatial terms: the public world came to be conceived of as a place of productive labor, while the home was seen as a site of rejuvenation and consumption. By the 1920s, the public world was still a sphere of work, but it was also opened up to a host of commercial pleasures such as movies and amusement parks that were incorporated into middle-class life styles. The ideal home, however, remained a place of

Reprinted from *Make Room for TV: Television and the Family Ideal in Postwar America*, pp. 73–78, 206–211, by Lynn Spiegel, with permission of the publisher, University of Chicago Press, and the author.

revitalization and, with the expansion of convenience products that promised to reduce household chores, domesticity was even less associated with production.

As feminists have argued, this separation has justified the exploitation of the housewife whose work at home simply does not count. Along these lines, Nancy Folbre claims that classical economics considers women's work as voluntary labor and therefore outside the realm of exploitation. In addition, she argues, even Marxist critics neglect the issue of domestic exploitation since they assume that the labor theory of value can be applied only to efficiency-oriented production for the market and not to "inefficient" and "idiosyncratic" household chores.[2]

As feminist critics and historians have shown, however, the home is indeed a site of labor. Not only do women do physical chores, but also the basic relations of our economy and society are reproduced at home, including the literal reproduction of workers through childrearing labor. Once the home is considered a workplace, the divisions between public/work and domestic/leisure become less clear. The way in which work and leisure are connected, however, remains a complex question.

Henri Lefebvre's studies of everyday life offer ways to consider the general interrelations between work, leisure, and family life in modern society. In his foreword to the 1958 edition of *Critiquè de la Vie Quotidienne*, Lefebvre argues:

> Leisure . . . cannot be separated from work. It is the same man who, after work, rests or relaxes or does whatever he chooses. Every day, at the same time, the worker leaves the factory, and the employee, the office. Every week, Saturday and Sunday are spent on leisure activities, with the same regularity as that of the weekdays' work. Thus we must think in terms of the unity "work-leisure," because that unity exists, and everyone tries to program his own available time according to what his work is—and what it is not.[3]

While Lefebvre concentrated on the "working man," the case of the housewife presents an even more pronounced example of the integration of work and leisure in everyday life.

In recent years, media scholars have begun to demonstrate the impact that patterns of domestic leisure and labor have on television spectatorship. British ethnographic research has suggested that men and women tend to use television according to their specific position within the distribution of leisure and labor activities inside and outside the home.[4] In the American context, two of the most serious examinations come from Tania Modleski (1983) and Nick Browne (1984), who have both theorized the way TV watching fits into a general pattern of everyday life where work and leisure are intertwined. Modleski has suggested that the soap opera might be understood in terms of the "rhythms of reception," or the way women working at home relate to the text within a specific milieu of distraction—cleaning, cooking, childrearing, and so on.[5] Browne concentrates not on the individual text, but rather on the entire TV schedule, which he claims is ordered according to the logic of the workday of both men and women. "[T]he position of the programs in the television schedule reflects and is determined by the work-structured order of the real social world. The patterns of position and flow imply the question of who is home, and through complicated social relays and temporal mediations, link television to the modes, processes, and scheduling of production characteristic of the general population."[6]

The fluid interconnection between leisure and labor at home presents a context in which to understand representations of the female audience during the postwar years. Above all, women's leisure time was shown to be coterminous with their work time. Representations of television continually addressed women as housewives and presented them with a notion of spectatorship that was inextricably intertwined with their useful labor at home. Certainly, this model of female spectatorship was based on previous notions about radio listeners, and we can assume that women were able to adapt some of their listening habits to television viewing without much difficulty. However, the added impact of visual images ushered in new dilemmas that were the subject of profound concern, both within the broadcast industry and within the popular culture at large.

## The Industry's Ideal Viewer

The idea that female spectators were also workers in the home was, by the postwar period, a truism for broadcasting and advertising executives. For some twenty years, radio programmers had grappled with ways to address a group of spectators whose attention wasn't focused primarily on the medium (as in the cinema), but instead moved constantly between radio entertainment and a host of daily chores. As William Boddy has argued, early broadcasters were particularly reluctant to feature daytime radio shows, fearing that women's household work would be fundamentally incompatible with the medium.[7] Overcoming its initial reluctance, the industry successfully developed daytime radio in the 1930s, and by the 1940s housewives constituted a faithful audience for soap operas and advice programs.

During the postwar years, advertisers and networks once more viewed the daytime market with skepticism, fearing that their loyal radio audiences would not be able to make the transition to television. The industry assumed that, unlike radio, television might require the housewife's complete attention and thus disrupt her work in the home.[8] Indeed, while network prime-time schedules were well worked out in 1948, networks and national advertisers were reluctant to feature regular daytime programs. Thus, in the earliest years, morning and afternoon hours were typically left to the discretion of local stations, which filled the time with low budget versions of familiar radio formats and old Hollywood films.

The first network to offer a regular daytime schedule was DuMont, which began operations on its owned and operated station WABD in New

York in November of 1948. As a newly formed network which had severe problems competing with CBS and NBC, DuMont entered the daytime market to offset its economic losses in prime time at a time when even the major networks were losing money on television.[9] Explaining the economic strategy behind the move into daytime, one DuMont executive claimed, "WABD is starting daytime programming because it is not economically feasible to do otherwise. Night time programming alone could not support radio, nor can it support television."[10] Increasingly in 1949, DuMont offered daytime programming to its affiliate stations. By December, it was transmitting the first commercially sponsored, daytime network show, *Okay, Mother,* to three affiliates and also airing a two-hour afternoon program on a full network basis. DuMont director Commander Mortimer W. Loewi reasoned that the move into daytime would attract small ticket advertisers who wanted to buy "small segments of time at a low, daytime rate."[11]

DuMont's venture into the daytime market was a thorn in the side of the other networks. While CBS, NBC, and ABC had experimented with individual daytime television programs on their flagship stations, they were reluctant to feature full daytime schedules. With huge investments in daytime radio, they weren't likely to find the prospects of daytime television appealing, especially since they were using their radio profits to offset initial losses in prime-time programming. As *Variety* reported when DuMont began its broadcasts on WABD, the major networks "must protect their AM [radio] investment at all costs—and the infiltration of daytime TV may conceivably cut into daytime radio advertising."[12] In this context, DuMont's competition in the daytime market posed a particularly grave threat to advertising revenues. In response, the other networks gradually began expanding the daytime lineups for their flagship stations.[13]

It was in 1951 that CBS, NBC, and, to a lesser extent, ABC first aggressively attempted to colonize the housewife's workday with regularly scheduled network programs. One of the central reasons for the networks' move into daytime that year was the fact that prime-time hours were fully booked by advertisers and that, by this point, there was more demand for TV advertising in general. As the advertising agency BBDO claimed in a report on daytime TV in the fall of 1950, "To all intents and purposes, the opportunity to purchase good night-time periods of TV is almost a thing of the past and the advertiser hoping to enter television now . . . better start looking at Daytime TV while it is still here to look at."[14] Daytime might have been more risky than prime time, but it had the advantage of being available—and at a cheaper network cost. Confident of its move into daytime, CBS claimed, "We aren't risking our reputation by predicting that daytime television will be a solid sell-out a year from today . . . and that once again there will be some sad advertisers who didn't read the tea leaves right."[15] ABC vice president Alexander Stronach Jr. was just as certain about the daytime market, and having just taken the plunge with the *Frances Langford-Don Ameche Show* (a variety program budgeted at the then steep $40,000 a week), Stronach told *Newsweek,* "It's a good thing electric dishwashers and washing machines were invented. The housewives will need them." [16]

The networks' confidence carried through to advertisers who began to test the waters of the daytime schedule. In September of 1951, the trade journal *Televiser* reported that "forty-seven big advertisers have used daytime network television during the past season or are starting this Fall." Included were such well-known companies as American Home Products, Best Foods, Procter and Gamble, General Foods, Hazel Bishop Lipsticks, Minute Maid, Hotpoint, and the woman's magazine *Ladies' Home Journal*. [17]

Despite these inroads, the early daytime market remained highly unstable, and at least until 1955 the competition for sponsors was fierce. [18] Indeed, even while the aggregate size of the daytime audience rose in the early fifties, sponsors and broadcasters were uncertain about the extent to which housewives actually paid attention to the programs and advertisements. In response to such concerns, the industry aggressively tailored programs to fit the daily habits of the female audience. When it began operations in 1948, DuMont's WABD planned shows that could "be appreciated just as much from listening to them as from watching them." [19] Following this trend in 1950, Detroit's WXYX aired *Pat 'n' Johnny*, a program that solved the housework-TV conflict in less than subtle ways. At the beginning of the three-hour show, host Johnny Slagle instructed housewives, "Don't stop whatever you're doing. When we think we have something interesting I'll blow this whistle or Pat will ring her bell." [20]

The major networks were also intent upon designing programs to suit the content and organization of the housewife's day. The format that has received the most critical attention is the soap opera, which first came to network television in December of 1950. As Robert Allen has demonstrated, early soap opera producers like Irna Philips of *Guiding Light* were skeptical of moving their shows from radio to TV. However, by 1954 the Nielsen Company reported that soaps had a substantial following; *Search For Tomorrow* was the second most popular daytime show while *Guiding Light* was in fourth place. The early soaps, with their minimum of action and visual interest, allowed housewives to listen to dialogue while working in another room. Moreover, their segmented storylines (usually two a day), as well as their repetition and constant explanation of previous plots, allowed women to divide their attention between viewing and household work. [21]

Another popular solution to the daytime dilemma was the segmented variety show that allowed women to enter and exit the text according to its discrete narrative units. One of DuMont's first programs, for example, was a shopping show (alternatively called *At Your Service* and *Shoppers Matinee*) that consisted of twenty-one entertainment segments, all of which revolved around different types of "women's issues." For instance, the "Bite Shop" presented fashion tips while "Kitchen Fare" gave culinary advice. Interspersed with these segments were twelve one-minute "store bulletins" (news and service announcements) that could be replaced at individual stations by local commercials.[22] While DuMont's program was short-lived, the basic principles survived in the daytime shows at the major networks. Programs like *The Garry Moore Show* (CBS), *The Kate Smith Show* (NBC), and *The Arthur Godfrey Show* (CBS) catered to housewife audiences with their segmented variety of entertainment and advice.[23]

Indeed, the networks put enormous amounts of money and effort into variety shows when they first began to compose daytime program schedules. Daytime ratings continually confirmed the importance of the variety format, with hosts like Smith and Godfrey drawing big audiences. Since daytime stars were often taken from nighttime radio shows, the variety programs were immediately marked as being different from and more spectacular than daytime radio. *Variety* reported in October of 1951:

> The daytime television picture represents a radical departure from radio. The application of "nighttime thinking" into daytime TV in regards to big-league variety-slanted programs and projection of personalities becomes more and more important. If the housewife has a craving for visual soap operas, it is neither reflected in the present day Nielsens nor in the ambitious programming formulas being blueprinted by the video entrepreneurs. . . . The housewife with her multiple chores, it would seem, wants her TV distractions on a "catch as catch can" basis, and the single-minded concentration on sight-and-sound weepers doesn't jibe with her household schedule. . . . [Variety shows] are all geared to the "take it awhile leave it awhile" school of entertainment projection and practically all are reaping a bonanza for the networks.[24]

Television thus introduced itself to the housewife not only by repeating tried and true daytime radio formulas, but also by creating a distinct product tailored to what the industry assumed were the television audience's specific needs and desires.

Initially uncertain about the degree to which daytime programs from an audio medium would suit the housewife's routine, many television broadcasters turned their attention to the visual medium of the popular press. Variety shows often modeled themselves on print conventions, particularly borrowing narrative techniques from women's magazines and the women's pages. Much as housewives might flip through the pages of a magazine as they went about their daily chores, they could tune in and out of the magazine program without the kind of disorientation that they might experience when disrupted from a continuous drama. To ensure coherence, such programs included "women's editors" or "femcees" who provided a narrational thread for a series of "departments" on gardening, homemaking, fashion, and the like. These shows often went to extreme lengths to make the connection between print media and television programming foremost in the viewer's mind. *Women's Magazine of the Air,* a local program aired in Chicago on WGN, presented a "potpourri theme with magazine pages being turned to indicate new sections."[25] On its locally owned station, the *Seattle Post* presented *Women's Page,* starring *Post* book and music editor Suzanne

Martin. The networks also used the popular press as a model for daytime programs. As early as 1948, CBS's New York station aired *Vanity Fair,* a segmented format that was tied together by "managing editor" Dorothy Dean, an experienced newspaper reporter. By the end of 1949, *Vanity Fair* was boasting a large list of sponsors, and in the fifties it continued to be part of the daytime schedule. Nevertheless, despite its success with *Vanity Fair,* CBS still tended to rely more heavily on well-known radio stars and formats, adapting these to the television medium. Instead, it was NBC that developed the print media model most aggressively in the early fifties.

Faced with daytime ratings that were consistently behind those of CBS and troubled by severe sponsorship problems, NBC saw the variety/magazine format as a particularly apt vehicle for small ticket advertisers who could purchase brief participation spots between program segments for relatively low cost.[26] Under the direction of programming vice president Sylvester "Pat" Weaver (who became NBC president in 1953), the network developed its "magazine concept" of advertising. Unlike the single sponsor series, which was usually produced through the advertising agency, the magazine concept allowed the network to retain control and ownership of programs. Although this form of multiple sponsor participation had become a common daytime practice by the early 1950s, Weaver's scheme differed from other participation plans because it allowed sponsors to purchase segments on a one-shot basis, with no ongoing commitment to the series. Even if this meant greater financial risks at the outset, in the long run a successful program based on spot sales would garner large amounts of revenue for the network.[27]

Weaver applied the magazine concept to two of the most highly successful daytime programs, *Today* and *Home.* Aired between 7:00 and 9:00 A.M., *Today* was NBC's self-proclaimed "television newspaper, covering not only the latest news, weather and time signals, but special features on everything from fashions to the hydrogen bomb."[28] On its premier episode in January 1952, *Today* made the print media connections firm in viewers' minds by showing telephoto machines grinding out pictures and front page facsimiles of the *San Francisco Chronicle.*[29] Aimed at a family audience, the program attempted to lure men, women, and children with discrete program segments that addressed their different interests and meshed with their separate schedules. One NBC confidential report stated that, on the one hand, men rushing off to take a train would not be likely to watch fashion segments. On the other hand, it suggested, "men might be willing to catch the next train" if they included an "almost sexy gal as part of the show." This, the report concluded, would be like "subtle, early morning sex."[30]

Although it was aimed at the entire family, the lion's share of the audience was female. (In 1954, for example, the network calculated that the audience was composed of 52 percent women, 26 percent men, and 22 percent children.)[31] *Today* appealed to housewives with "women's pages" news stories such as Hollywood gossip segments, fashion shows, and humanistic features. In August 1952, NBC's New York outlet inserted "Today's Woman" into the program, a special women's magazine feature that was produced in cooperation with *Look* and *Quick* magazines.[32] Enthused with *Today*'s success, NBC developed *Home* with similar premises in mind, but this time aimed the program specifically at women. First aired in 1954 during the 11:00 A.M. to noon time slot, *Home* borrowed its narrative techniques from women's magazines, featuring segments on topics like gardening, child psychology, food, fashion, health, and interior decor. As *Newsweek* wrote, "The program is planned to do for women on the screen what the women's magazines have long done in print."[33]

In fashioning daytime shows on familiar models of the popular press, television executives and advertisers were guided by the implicit assumption that the female audience had much in common with the typical magazine reader. When promoting *Today* and *Home*, NBC used magazines such as *Ladies' Home Journal, Good Housekeeping,* and *Collier's* (which also had a large female readership) as major venues. When *Home* first appeared it even offered women copies of its own monthly magazine, *How To Do it*.[34] Magazine publishers also must have seen the potential profits in the cross-over audience; the first sponsor for *Today* was Kiplinger's magazine *Changing Times,* and *Life* and Curtis magazines were soon to follow.[35]

The fluid transactions between magazine publishers and daytime producers were based on widely held notions about the demographic composition of the female audience. In 1954, the same year that *Home* premiered, NBC hired W. R. Simmons and Associates to conduct the first nationwide qualitative survey of daytime viewers. In a promotional report based on the survey, Dr. Tom Coffin, manager of NBC research, told advertisers and manufacturers, "In analyzing the findings, we have felt a growing sense of excitement at the qualitative picture emerging: an audience with the *size* of a mass medium but the *quality* of a class medium." When compared to nonviewers, daytime viewers were at the "age of acquisition," with many in the twenty-five- to thirty-four-year-old category; their families were larger with more children under eighteen; they had higher incomes; and they lived in larger and "better" market areas. In addition, Coffin characterized the average viewer as a "modern active woman" with a kitchen full of "labor-saving devices," an interest in her house, clothes, and "the way she looks." She is "the kind of woman most advertisers are most interested in; she's a good customer."[36] Coffin's focus on the "class versus mass" audience bears striking resemblance to the readership statistics of middle-class women's magazines. Like the magazine reader, "Mrs. Daytime Consumer" was an upscale, if only moderately affluent, housewife whose daily life consisted not only of chores, but also, and perhaps even more importantly, shopping for her family.

With this picture of the housewife in mind, the media producer had one primary job—teaching her how to buy products. Again, the magazine format was perfect for this because each discrete narrative segment could portray an integrated sales message. Hollywood gossip columns gave way to motion picture endorsements; cooking segments sold sleek new ranges;

fashion shows promoted Macy's finest evening wear. By integrating sales messages with advice on housekeeping and luxury lifestyles, the magazine format skillfully suggested to housewives that their time spent viewing television was indeed part of their work time. In other words, the programs promised viewers not just entertainment, but also lessons on how to make consumer choices for their families. One production handbook claimed: "Women's daytime programs have tended toward the practical—providing shopping information, marketing tips, cooking, sewing, interior decoration, etc., with a dash of fashion and beauty hints. . . . The theory is that the housewife will be more likely to take time from her household duties if she feels that her television viewing will make her housekeeping more efficient and help her provide more gracious living for her family."[37] In the case of *Home,* this implicit integration of housework, consumerism, and TV entertainment materialized in the form of a circular stage that the network promoted as a "machine for selling."[38] The stage was equipped with a complete kitchen, a workshop area, and a small garden—all of which functioned as settings for different program segments and, of course, the different sponsor products that accompanied them. Thus, *Home*'s magazine format provided a unique arena for the presentation of a series of fragmented consumer fantasies that women might tune into and out of, according to the logic of their daily schedules.

Even if the structure of this narrative format was the ideal vehicle for "Mrs. Daytime Consumer," the content of the consumer fantasies still had to be carefully planned. Like the woman's magazine before it, the magazine show needed to maintain the subtle balance of its "class address." In order to appeal to the average middle-class housewife, it had to make its consumer fantasies fit with the more practical concerns of female viewers. The degree to which network executives attempted to strike this balance is well illustrated in the case of *Home.* After the program's first airing, NBC executive Charles Barry was particularly concerned about the amount of "polish" that it contained. Using "polish" as a euphemism for highbrow tastes, Barry went on to observe the problems with *Home*'s class address: "I hope you will keep in mind that the average gal looking at the show is either living in a small suburban house or in an apartment and is not very likely to have heard of Paul McCobb; she is more likely to be at a Macy's buying traditionally." After observing other episodes, Barry had similar complaints: the precocious stage children weren't "average" enough, the furniture segment featured impractical items, and the cooking segment showcased high-class foods such as vichyssoise and pot-de-crème. "Maybe you can improve tastes," Barry conceded, "but gosh would somebody please tell me how to cook corned beef and cabbage without any smell?"[39] The television producer could educate the housewife beyond her means, but only through mixing upper-class fantasy with tropes of averageness.

The figure of the female hostess was also fashioned to strike this delicate balance. In order to appeal to the typical housewife, the hostess would ideally speak on her level. As one producer argued, "Those who give an impression of superiority or 'talking down' to the audience, who treasure the manner of speaking over naturalness and meaningful communication . . . or who are overly formal in attire and manners, do not survive in the broadcasting industry. . . . The personality should fit right into your living room. The super-sophisticate or the squealing life of the party might be all right on occasion, but a daily association with this girl is apt to get a little tiresome."[40] In addition, the ideal hostess was decidedly not a glamour girl, but rather a pleasingly attractive, middle-aged woman—Hollywood's answer to the home economics teacher. When first planning *Home,* one NBC executive considered using the celebrity couple Van and Evie Johnson for hosts, claiming that Evie was "a sensible woman, not a glamor struck movie star's wife, but a wholesome girl from a wholesome background. . . . She works hard at being a housewife and Mother who runs a not elaborate household in Beverly Hills with *no swimming pool.*" Although Evie didn't get the part, her competitor, Arlene Francis, was clearly cut from the same cloth. In a 1957 fanzine, Francis highlighted her ordinariness when she admitted, "My nose is too long and I'm too skinny, but maybe that won't make any difference if I'm fun to be with."[41] Francis was also a calming mother figure who appealed to children. In a fan letter, one mother wrote that her little boy took a magazine to bed with him that had Arlene's picture on the cover.[42] Unlike the "almost sexy" fantasy woman on the *Today* show who was perfect for "morning sex," *Home*'s femcee appealed to less erotic instincts. Francis and other daytime hostesses were designed to provide a role model for ordinary housewives, educating them on the "good life," while still appearing down to earth.

In assuming the role of "consumer educator," the networks went beyond just teaching housewives how to buy advertisers' products. Much more crucially in this early period, the networks attempted to teach women and their families how to consume television itself. Indeed, the whole system pivoted on the singular problem of how to make the daytime audience watch more programming. Since it adapted itself to the family's daily routine, the magazine show was particularly suited for this purpose. When describing the habits of *Today*'s morning audience, Weaver acknowledged that the "show, of course, does not hold the same audience throughout the time period, but actually is a service fitting with the family's own habit pattern in the morning."[43] Importantly, however, NBC continually tried to channel the movements of the audience. Not merely content to fit its programming into the viewer's rhythms of reception, the network aggressively sought to change those rhythms by making the activity of television viewing into a new daily habit. One NBC report made this point quite explicit, suggesting that producers "establish definite show patterns at regular times; do everything you can to capitalize on the great habit of habit listening."[44] Proud of his accomplishments on this front, Weaver bragged about fan mail that demonstrated how *Today* changed viewers' daily routines. According to Weaver, one woman claimed, "My husband said I should put casters on the TV set so I can roll it around and see it from the kitchen." Another admitted, "I used to get all the dishes washed by 8:30—now I don't do a thing until 10 o'clock." Still another confessed, "My husband now dresses in the living room." Weaver boastfully promised, "We will change the habits of millions."[45]

The concept of habitual viewing also governed NBC's scheduling techniques. The network devised promotional strategies designed to maintain systems of flow, as each program ideally would form a "lead in" for the next, tailored to punctuate intervals of the family's daily routine. In 1954, for example, an NBC report on daytime stated that *Today* was perfect for the early morning time slot because it "has a family audience . . . and reaches them just before they go out to shop." With shopping done, mothers might return home to find *Ding Dong School*, "a nursery school on television" that allowed them to do housework while educator Frances Horwich helped raise the pre-schoolers. Daytime dramas were scheduled throughout the day, each lasting only fifteen minutes, probably because the network assumed that drama would require more of the housewife's attention than the segmented variety formats like *Home*. At 5 P.M., when mothers were likely to be preparing dinner, *The Pinky Lee Show* presented a mixed bag of musical acts, dance routines, parlor games, and talk aimed both at women and their children who were now home from school.[46]

NBC aggressively promoted this kind of routinized viewership, buying space in major market newspapers and national periodicals for advertisements that instructed women how to watch television while doing household chores. In 1955, *Ladies' Home Journal* and *Good Housekeeping* carried advertisements for NBC's daytime lineup that suggested that not only the programs, but also the scheduling of the programs, would suit the housewife's daily routine. The ads evoked a sense of fragmented leisure time and suggested that television viewing could be conducted in a state of distraction. This was not the kind of critical contemplative distraction that Walter Benjamin suggested in his seminal essay, "The Work of Art in the Age of Mechanical Reproduction."[47] Rather, the ads implied that the housewife could accomplish her chores in a state of "utopian forgetfulness" as she moved freely between her work and the act of watching television.

One advertisement, which is particularly striking in this regard, includes a sketch of a housewife and her little daughter at the top of the page. Below this, the graphic layout is divided into eight boxes composed of television screens, each representing a different program in NBC's daytime lineup. The caption functions as the housewife's testimony to her distracted state. She asks, "Where Did the Morning Go? The house is tidy . . . but it hasn't seemed like a terribly tiring morning. . . . I think I started ironing while I watched the *Sheila Graham Show*." The housewife goes on to register each detail of the programs, but she cannot with certainty account for her productive activities in the home. Furthermore, as the ad's layout suggests, the woman's daily activities are literally fragmented according to the pattern of the daytime television schedule, to the extent that her everyday experiences become imbricated in a kind of serial narrative. Significantly, her child pic-

tured at the top of the advertisement appears within the contours of a television screen so that the labor of childrearing is itself made part of the
narrative pleasures offered by the network's daytime lineup.[48]

### Negotiating with the Industry's Ideal Viewer

The program types, schedules, and promotional materials devised at the
networks were based upon ideal images of female viewers and, consequently, they were rooted in abstract conceptions about women's lives. These
ideals weren't always commensurate with the heterogeneous experiences and
situations of real women and, for this reason, industrial strategies didn't
always form a perfect fit with the audience's needs and desires. Although it
is impossible to reconstruct fully the actual activities of female viewers at
home, we can better understand their concerns and practices by examining
the ways in which their viewing experiences were explained to them at the
time. Popular media, particularly women's magazines, presented women with
opportunities to negotiate with the modes of spectatorship that the television industry tried to construct. It is in these texts that we see the gaps
and inconsistencies—the unexpected twists and turns—that were not fore-
seen by networks and advertisers. Indeed, it is in the magazines, rather than
in the highrise buildings of NBC, CBS, and ABC, where female audiences
were given the chance to enter into a popular dialogue about their own
relations to the medium.

In The Honeymooners, a working-class situation comedy, television's obstruction of household work was related to marital strife. The first episode

While the networks were busy attempting to tailor daytime programming to the patterns of domestic labor, popular media often completely
rejected the idea that television could be compatible with women's work
and showed instead how it would threaten the efficient functioning of the
household. The TV-addict housewife became a stock character during the
period, particularly in texts aimed at a general audience where the mode of
address was characterized by an implicit male narrator who clearly blamed
women—not television—for the untidy house. In 1950, for example, *The
New Yorker* ran a cartoon that showed a slovenly looking woman ironing a
shirt while blankly staring at the television screen. Unfortunately, in her
state of distraction, the woman burned a hole in the garment.[49] Women's
magazines also deliberated upon television's thoroughly negative effect on
household chores, but rather than poking fun at the housewife, they offered
sympathetic advice, usually suggesting that a careful management of domestic space might solve the problem. In 1950, *House Beautiful* warned of television: "It delivers about five times as much wallop as radio and requires in
return five times as much attention. . . . It's impossible to get anything
accomplished in the same room while it's on." The magazine offered a spatial solution, telling women "to get the darn thing out of the living room,"
and into the TV room, cellar, library, "or as a last resort stick it in the
dining room."[50]

In *The Honeymooners,* a working-class situation comedy, television's obstruction of household work was related to marital strife. The first episode

of the series, "TV or Not TV" (1955), revolves around the purchase of a television set and begins with an establishing shot of the sparsely decorated Kramden kitchen where a clothes basket filled with wet wash sits on the table. Entering from the bedroom in her hausfrau garb, Alice Kramden approaches the kitchen sink and puts a plunger over the drain, apparently attempting to unclog it. As pictured in this opening scene, Alice is, to say the least, a victim of household drudgery. Not surprisingly, Alice begs Ralph for a television set, hoping that it will make her life more pleasant.

In a later scene, after the Kramdens purchase their TV set, this situation changes, but not for the better. Ralph returns home from work while Alice sits before her television set. Here is the exchange between the couple:

> *Ralph:* Would you mind telling me where my supper is?
> *Alice:* I didn't make it yet. . . . I sat down to watch the four o'clock movie
> and I got so interested I . . . uh what time is it anyway?
> *Ralph:* I knew this would happen Alice. We've had that set three days
> now, and I haven't had a hot meal since we got it.

Thus, television is the source of a dispute between the couple, a dispute that arises from the housewife's inability to perform her productive function while enjoying an afternoon program.

A 1955 ad for Drano provided a solution to television's obstruction of household chores. Here the housewife is shown watching her afternoon soap opera, but this unproductive activity is sanctioned only insofar as her servant does the housework. As the maid exclaims, "Shucks, I'll never know if she gets her man 'cause this is the day of the week I put Drano in all the drains!" The Drano Company thus attempted to sell its product by giving women a glamorous vision of themselves enjoying an afternoon of television. But it could do so only by splitting the functions of relaxation and work across two representational figures—the lady of leisure and the domestic servant.[51]

If the domestic servant was a fantasy solution to the conflict between work and television, the women's magazines suggested more practical ways to manage the problem. *Better Homes and Gardens* advised in 1949 that the television set should be placed in an area where it could be viewed, "while you're doing things up in the kitchen." Similarly in 1954, *American Home* told readers to put the TV set in the kitchen so that "Mama sees her pet programs. . . ." Via such spatial remedies, labor would not be affected by the leisure of viewing nor would viewing be denied by household chores.[52] In fact, household labor and television were continually condensed into one space designed to accommodate both activities. In a 1955 issue of *American Home,* this labor–leisure viewing condensation provided the terms of a joke. A cartoon showed a housewife tediously hanging her laundry on the outdoor clothesline. The drudgery of this work is miraculously solved as the housewife brings her laundry into her home and sits before her television set while letting the laundry dry on the television antenna.[53]

The spatial condensation of labor and viewing was part of a well entrenched functionalist discourse. The home had to provide rooms that would allow for a practical orchestration of "modern living activities" that now included watching television. Functionalism was particularly useful for advertisers, who used it to promote not just one household item but an entire product line. In 1952, for example, the Crane Company displayed its kitchen appliance ensemble, complete with ironing, laundering, and cooking facilities. Here the housewife could do multiple tasks at once because all the fixtures were "matched together as a complete chore unit." One particularly attractive component of this "chore unit" was a television set built into the wall above the washer/dryer.[54]

While spatial condensations of labor and leisure helped to soothe tensions about television's obstruction of household chores, other problems still existed. The magazines suggested that television would cause increasing work loads. Considering the cleanliness of the living room, *House Beautiful* told its readers in 1948: "Then the men move in for boxing, wrestling, basketball, hockey. They get excited. Ashes on the floor. Pretzel crumbs. Beer stains." The remedy was again spatial: "Lots of sets after a few months have been moved into dens and recreation rooms."[55] In a slight twist of terms, the activity of eating was said to be moving out of the dining area and into the television-sitting area. Food stains soiling upholstery, floors, and other surfaces meant extra work for women. Vinyl upholstery, linoleum floors, tiling, and other spill-proof surfaces were recommended. Advertisers for all kinds of cleaning products found television especially useful in their sales pitches. In 1953, the Bissell Carpet Sweeper Company asked housewives, "What do you do when the TV crowd leaves popcorn and crumbs on your rug? You could leave the mess till morning—or drag out the vacuum. But if you're on the beam, you slick it up with a handy Bissell Sweeper."[56] In addition to the mess generated by television, the set itself called for maintenance. In 1955, *House Beautiful* asked if a "misty haze dims your TV screen" and recommended the use of "wipe-on liquids and impregnated wiping cloths to remedy the problem." The Drackett Company, producer of Windex Spray, quickly saw the advantage that television held for its product; in 1948 it advertised the cleaner as a perfect solution for a dirty screen.[57]

Besides the extra cleaning, television also kept housewives busy in the kitchen. The magazines showed women how to be gracious hostesses, always prepared to serve family and friends special TV treats. These snacktime chores created a lucrative market for manufacturers. For example, in 1952 *American Home* presented a special china collection for "Early Tea and Late TV," while other companies promoted TV snack trays and TV tables.[58] The most exaggerated manifestation appeared in 1954. The TV dinner was the perfect remedy for the extra work entailed by television, and it also allowed children to eat their toss-away meals while watching *Hopalong Cassidy*.

While magazines presented readers with a host of television-related tasks, they also suggested ways for housewives to ration their labor. Time-motion studies, which were integral to the discourses of feminism and domestic science since the Progressive era, were rigorously applied to the problem of increasing work loads. All unnecessary human movement that the television set might demand had to be minimized. Again, this called for a careful management of space. The magazines suggested that chairs and sofas be placed so that they need not be moved for watching television. Alternatively, furniture could be made mobile. By placing wheels on a couch, it was possible to exert minimal energy while converting a sitting space into a viewing space. Similarly, casters and lazy Susans could be placed on television sets so that housewives might easily move the screen to face the direction of the viewers.[59] More radically, space between rooms could be made continuous. In 1952, *House Beautiful* suggested a "continuity" of living, dining, and television areas wherein "a curved sofa and a folding screen mark off [the] television corner from the living and dining room." Via this carefully managed spatial continuum, "it takes no more than an extra ten steps or so to serve the TV fans."[60]

Continuous space was also a response to the more general problem of television and family relationships. Women's household work presented a dilemma for the twin ideals of family unity and social divisions, since housewives were ideally meant to perform their distinctive productive functions but, at the same time, take part in the family's leisure-time pursuits. This conflict between female isolation from and integration into the family group was rooted in Victorian domestic ideology with its elaborate social and spatial hierarchies; it became even more pronounced as twentieth-century lifestyles and housing contexts changed in ways that could no longer contain the formalized spatial distinctions of the Victorian ideal.

The problems became particularly significant in the early decades of the century when middle-class women found themselves increasingly isolated in their kitchens due to a radical reduction in the number of domestic servants. As Gwendolyn Wright has observed, women were now cut off from the family group as they worked in kitchens designed to resemble scientific laboratories, far removed from the family activities in the central areas of the home. Architects did little to respond to the problem of isolation, but continued instead to build kitchens fully separated from communal living spaces, suggesting that labor-saving kitchen appliances would solve the servant shortage.[61] In the postwar era when the continuous spaces of ranch-style architecture became a cultural ideal, the small suburban home placed a greater emphasis on interaction between family members. The "open plan" eliminated some of the walls between dining room, living room, and kitchen. However, even in the continuous ranch-style homes, the woman's work area was "zoned off" from the activity area, and the woman's role as homemaker still worked to separate her from the leisure activities of her family.

Women's magazines suggested intricately balanced spatial arrangements that would mediate the tensions between female integration and isolation. Television viewing became a special topic of consideration. In 1951, *House Beautiful* placed a television set in its remodeled kitchen, which combined "such varied functions as cooking, storage, laundry, flower arranging, dining and TV viewing." In this case, as elsewhere, the call for functionalism was related to the woman's ability to work among a group engaged in leisure activities. A graphic showed a television placed in a "special area" devoted to "eating" and "relaxing" which was "not shut off by a partition." In continuous space, "the worker . . . is always part of the group, can share in the conversation and fun while work is in progress."[62]

While this example presents a harmonious solution, often the ideals of integration and isolation resulted in highly contradictory representations of domestic life. Typically, illustrations that depicted continuous spaces showed the housewife to be oddly disconnected from the general flow of activities. In 1951, for example, *American Home* showed a woman in a continuous dining-living area who supposedly is allowed to accomplish her housework among a group of television viewers. However, rather than being integrated into the group, the woman is actually isolated from the television crowd as she sets the dining room table. The TV viewers are depicted in the background while the housewife stands to the extreme front-right border of the composition, far away from her family and friends. In fact, she is literally positioned off-frame, straddling between the photograph and the negative (or unused) space of the layout.[63]

The family circle motif was also riddled with contradictions of this sort. In particular, Sentinel's advertising campaign showed women who were spatially distanced from their families. In 1952, one ad depicted a housewife holding a tray of beverages and standing off to the side of her family, who were clustered around the television set. The following year, another ad showed a housewife cradling her baby in her arms and standing at a window far away from the rest of her family, who were gathered around the Sentinel console.[64] In a 1948 ad for Magnavox Television, the housewife's chores separated her from her circle of friends. The ad was organized around a U-shaped sofa that provided a quite literal manifestation of the semicircle visual cliché. A group of adult couples sat on the sofa watching the new Magnavox set, but the hostess stood at the kitchen door, holding a tray of snacks. Spatially removed from the television viewers, the housewife appeared to be sneaking a look at the set as she went about her hostess chores.[65]

This problem of female spatial isolation gave way to what can be called a "corrective cycle of commodity purchases." A 1949 article in *American Home* about the joys of the electric dishwasher is typical here. A picture of a family gathered around the living room console included the caption, "No martyr banished to kitchen, she never misses television programs. Lunch, dinner dishes are in an electric dishwasher." In 1950, an advertisement for Hotpoint dishwashers used the same discursive strategy. The illustration showed a wall of dishes that separated a housewife in the kitchen from her family, who sat huddled around the television set in the living room. The caption read, "Please . . . Let Your Wife Come Out Into the Livingroom! Don't let dirty dishes make your wife a kitchen exile! She loses the most precious hours of her life shut off from pleasures of the family circle by the never-ending chore of old-fashioned dishwashing!"[66]

This ideal version of female integration in a unified family space was contested by the competing discourse on divided spaces. Distinctions between work and leisure space remained an important principle of household efficiency. Here, room dividers presented a perfect balance of integration and isolation. In 1952, *Better Homes and Gardens* displayed a room divider that separated a kitchen work area from its dining area. The cutoff point was a television set built into the wall just to the right of the room divider. Thus, the room divider separated the woman's work space from the television space, but as a partial wall that still allowed for continuous space, it reached the perfect compromise between the housewife's isolation from and integration into the family. It was in the sense of this compromise that *American Home*'s "discrete" room divider separated a wife's work space from her husband's television space in a house that, nevertheless, was designed for "family living." As the magazine reported in 1954, "Mr. Peterson . . . retired behind his newspaper in the TV end of the living kitchen. Mrs. P. quietly made a great stack of sandwiches for us behind the discrete screen of greens in the efficient kitchen end of the same room."[67]

This bifurcation of sexual roles, of male (leisure) and female (productive) activities, served as an occasion for a full consideration of power dynamics among men and women in the home. Typically, the magazines extended their categories of feminine and masculine viewing practices into representations of the body. For men, television viewing was most often represented in terms of a posture of repose. Men were usually shown to be sprawled out on easy chairs as they watched the set. Remote controls allowed the father to watch in undisturbed passive comfort. In many ways, this representation of the male body was based on Victorian notions of rejuvenation for the working man. Relaxation was condoned for men because it served a revitalizing function, preparing them for the struggles for the workaday world. For women, the passive calm of television viewing was never so simple. As we have seen, even when women were shown watching television, they often appeared as productive workers.

Sometimes, representations of married couples became excessively literal about the gendered patterns of television leisure. In 1954, when the Cleavelander Company advertised its new "T-Vue" chair, it told consumers, "Once you sink into the softness of Cleavelander's cloud-like contours, cares seem to float away." Thus, not only the body, but also the spirit would be revitalized by the TV chair. But while the chair allowed Father "to stretch out with his feet on the ottoman," Mother's TV leisure was nevertheless productive. As the caption states, "Mother likes to gently rock as she sews."[68] Similarly, a 1952 advertisement for Airfoam furniture cushions showed a husband dozing in his foam rubber cushioned chair as he sits before a television set. Meanwhile, his wife clears away his TV snack. The text reads, "Man's pleasure is the body coddling comfort" of the cushioned chair while "Woman's treasure is a home lovely to look at, easy to keep perfectly tidy and neat" with cushioning that "never needs fluffing."[69] In such cases, the man's pleasure in television is associated with passive relaxation. The woman's pleasure, however, is derived from the aesthetics of a well-kept home and labor-saving devices that promise to minimize the extra household work that television brings to domestic space. In addition, the Airfoam ad is typical as it depicts a female body that finds no viewing pleasures of its own but instead functions to assist with the viewing comforts of others.

As numerous feminist film theorists have demonstrated, spectatorship and the pleasures entailed by it are culturally organized according to categories of sexual difference. In her groundbreaking article on the subject of Hollywood film, Laura Mulvey showed how narrative cinema (her examples were Von Sternberg and Hitchcock) is organized around voyeuristic and fetishistic scenarios in which women are the "to-be-looked-at" object of male desire.[70] In such a scheme, it becomes difficult to pinpoint how women can have subjective experiences in a cinema that systematically objectifies them. In the case of television, it seems clear that women's visual pleasure was associated with interior decor and not with viewing programs. In 1948, *House Beautiful* made this explicit when it claimed, "Most men want only an adequate screen. But women alone with the thing in the house all day have to eye it as a piece of furniture."[71] In addition, while these discussions of television were addressed to female readers, the woman's spectatorial pleasure was less associated with her enjoyment of the medium than it was with her own objectification, her desire to be looked at by the gaze of another.

On one level here, television was depicted as a threat to the visual appeal of the female body in domestic space. Specifically, there was something visually unpleasurable about the sight of a woman operating the technology of the receiver. In 1955, Sparton Television proclaimed that "the sight of a woman tuning a TV set with dials near the floor" was "most unattractive." The Sparton TV, with its tuning knob located at the top of the set, promised to maintain the visual appeal of the woman.[72] Beyond this specific case, there was a distinct set of aesthetic conventions formed in these years for male and female viewing postures. A 1953 advertisement for CBS-Columbia Television illustrates this well. Three alternative viewing postures are taken up by family members. A little boy stretches out on the floor, a father slumps in his easy chair, and the lower portion of a mother's outstretched body is gracefully lifted in a sleek modern chair with a seat that tilts upward. Here as elsewhere, masculine viewing is characterized by slovenly body posture. Conversely, feminine viewing posture takes on a certain visual appeal even as the female body passively reclines.[73]

As this advertisement indicates, the graphic representation of the female body viewing television had to be carefully controlled. It had to be made appealing to the eye of the observer, for in a fundamental sense, there was something taboo about the sight of a woman watching television. In fact, the housewife was almost never shown watching television by herself. Instead, she typically lounged on a chair (perhaps reading a book) while the television set remained turned off in the room. In 1952, *Better Homes and Gardens* stated one quite practical reason for the taboo. The article gave suggestions for methods of covering windows that would keep neighbors from peering into the home. It related this interest in privacy to women's work and television: "You should be able to have big, big windows to let in light and view, windows that let you watch the stars on a summer night without feeling exposed and naked. In good conscience, you should be able to leave the dinner dishes on the table while you catch a favorite TV or radio program, without sensing derogatory comments on your housekeeping."[74] Thus, for the housewife, being caught in the act of enjoying a broadcast is ultimately degrading because it threatens to reveal the signs of her slovenly behavior to the observer. More generally, we might say that the magazines showed women that their subjective pleasure in watching television was at odds with their own status as efficient and visually attractive housewives.

Although these representations are compatible with traditional gender roles, subtle reversals of power ran through the magazines as a whole. Even if there was a certain degree of privilege attached to the man's position of total relaxation—his right to rule from the easy chair throne—his power was in no way absolute, nor was it stable. Although such representations held to the standard conception of women as visually pleasing spectacles— as passive objects of male desire—these representations also contradicted such notions by presenting women as active producers in control of domestic affairs. For this reason, it seems that the most striking thing about this gendered representation of the body is that it inverted—or at least complicated—normative conceptions of masculinity and femininity. Whereas Western society associates activity with maleness, representations of television often attributed this trait to the woman. Conversely, the notion of feminine passivity was typically transferred over to the man of the house.[75] It could well be concluded that the cultural ideals that demanded women be shown as productive workers in the home also had the peculiar side effect of "feminizing" the father.

Perhaps for this reason, popular media presented tongue-in-cheek versions of the situation, showing how television had turned men into passive homebodies. In the last scene of *The Honeymooners'* episode "TV or Not TV," for example, the marital dispute between Alice and Ralph is inverted, with Alice apparently the "woman on top."[76] After Ralph scolds Alice about her delinquent housekeeping, Alice's TV addiction is transferred over to her husband and his friend Ed Norton, who quickly become passive viewers. Ralph sits before the television set with a smorgasbord of snacks, which he deliberately places within his reach so that he needn't move a muscle while watching his program. Norton's regressive state becomes the center of the comedic situation as he is turned into a child viewer addicted to a science-fiction serial. Wearing a club-member space helmet, Norton tunes into his favorite television host, Captain Video, and recites the space scout pledge. After arguing over program preferences, Ralph and Norton finally settle down for the *Late, Late, Late Show* and, exhausted, fall asleep in front of the set. Alice then enters the room and, with a look of motherly condescension, covers Ralph and Norton with a blanket, tucking them in for the night.

Men's magazines such as *Esquire* and *Popular Science* also presented wry commentary on male viewers. In 1951, for example, *Esquire* showed the stereotypical husband relaxing with his shoes off and a beer in his hand, smiling idiotically while seated before a television set. Two years later, the same magazine referred to television fans as "televidiots."[77] Nonetheless, while these magazines provided a humorous look at the man of leisure, they also presented men with alternatives. In very much the same way that Catharine Beecher attempted to elevate the woman by making her the center of domestic affairs, the men's magazines suggested that fathers could regain authority through increased participation in family life.

Indeed, the "masculine domesticity" that Margaret Marsh sees as central to Progressive era lifestyles also pervaded the popular advice disseminated to men in the 1950s. According to Marsh, masculine domesticity has historically provided men with a way to assert their dominion at home. Faced with their shrinking authority in the new corporate world of white-collar desk jobs, the middle-class men of the early 1900s turned inward to the home where their increased participation in and control over the family served to compensate for feelings of powerlessness in the public sphere. Moreover, Marsh argues that masculine domesticity actually undermined women's growing desire for equal rights because it contained that desire within the safe sphere of the home. In other words, while masculine domesticity presented a more "compassionate" model of marriage where men supposedly shared domestic responsibilities with women, it did nothing to encourage women's equal participation in the public sphere.[78]

Given such historical precedents, it is not surprising that the postwar advice to men on this account took on explicitly misogynistic tones. As early as 1940, Sydnie Greenbie called for the reinstitution of manhood in his book, *Leisure For Living*. Greenbie reasoned that the popular figure of the male "boob" could be counteracted if the father cultivated his mechanical skills. As he wrote, "At last man has found something more in keeping with his nature, the workshop, with its lathe and mechanical saws, something he has kept as yet his own against the predacious female. . . . And [it becomes] more natural . . . for the man to be a homemaker as well as the woman."[79]

After the war the reintegration of the father became a popular ideal. As *Esquire* told its male readers, "Your place, Mister, is in the home, too, and if you'll make a few thoughtful improvements to it, you'll build yourself a happier, more comfortable, less back breaking world. . . ."[80] From this perspective, the men's magazines suggested ways for fathers to take an active and productive attitude in relation to television. Even if men were passive spectators, when not watching they could learn to repair the set or else produce television carts, built-ins, and stylish cabinets.[81] Articles with step-by-step instructions circulated in *Popular Science,* and the *Home Craftsman* even had a special "TV: Improve Your Home Show" column featuring a husband and wife, Thelma and Vince, and their adventures in home repairs. *Popular Science* suggested hobbies through which men could use television in an active, productive way. The magazine ran several articles on a new fad—TV photography. Men were shown how to take still pictures off their television sets, and in 1950 the magazine even conducted a readership contest for prize winning photos that were published in the December issue.[82]

The gendered division of domestic labor and the complex relations of power entailed by it were thus shown to organize the experience of watching television. These popular representations begin to disclose the social construction of television as it was rooted in a mode of thought based on categories of sexual difference. Indeed, sexual difference, and the corresponding dynamics of domestic labor and leisure, framed television's introduction to the public in significant ways. The television industry struggled to produce programming forms that might appeal to what they assumed to be the typical housewife, and in so doing they drew an abstract portrait of "Mrs. Daytime Consumer." By tailoring programs to suit the content and organization of her day, the industry hoped to capture her divided attention. Through developing schedules that mimicked the pattern of her daily activities, network executives aspired to make television a routine habit. This "ideal" female spectator was thus the very foundation of the daytime programs the industry produced. But like all texts, these programs didn't simply turn viewers into ideal spectators; they didn't simply "affect" women. Instead, they were used and interpreted within the context of everyday life at home. It is this everyday context that women's magazines addressed, providing a cultural space through which housewives might negotiate their peculiar relationship to a new media form.

Women's magazines engaged their readers in a dialogue about the concrete problems that television posed for productive labor in the home. They depicted the subtle interplay between labor and leisure at home, and they offered women ways to deal with—or else resist—television in their daily lives. If our culture has systematically relegated domestic leisure to the realm of nonproduction, these discourses remind us of the tenuousness of such notions. Indeed, at least for the housewife, television was not represented as a passive activity; rather, it was incorporated into a pattern of everyday life where work is never done.

### Notes

1. This stove was mentioned in *Sponsor*, 4 June 1951, p. 19. It was also illustrated and discussed in *Popular Science*, May 1952, p. 132. The *Popular Science* reference is interesting because this men's magazine did not discuss the TV component of the stove as a vehicle for leisure, but rather showed how "a housewife can follow telecast cooking instructions step-by-step on the TV set built into this electric oven." Perhaps in this way, the magazine allayed men's fears that their wives would use the new technology for diversion as opposed to useful labor.

2. Nancy Folbre, "Exploitation Comes Home: A Critique of the Marxist

Theory of Family Labour," *Cambridge Journal of Economics* 6 (1982), pp. 317–29.

3. Henri Lefebvre, foreword, *Critique de la Vie Quotidienne* (Paris, L'Arche, 1958), reprinted in *Communication and Class Struggle*, ed. Armond Mattelart and Seth Siegelaub, trans. Mary C. Axtmann (New York: International General, 1979), p. 136.

4. See David Morley, *Family Television: Cultural Power and Domestic Leisure* (London: Comedia, 1986); and Ann Gray, "Behind Closed Doors: Video Recorders in the Home," *Boxed In: Women and Television*, ed. H. Baehr and G. Dyer (New York: Pandora, 1987), pp. 38–54.

5. Tania Modleski, "The Rhythms of Reception: Daytime Television and Women's Work," *Regarding Television: Critical Approaches*, ed. E. Ann Kaplan (Frederick, Md.: University Publications of America, 1983), pp. 67–75. See also the fourth chapter in Modleski, *Loving With A Vengeance: Mass-Produced Fantasies for Women* (New York: Methuen, 1984).

6. Nick Browne, "The Political Economy of the Television (Super) Text," *Quarterly Review of Film Studies* 9 (3) (Summer 1984), p. 176.

7. William Boddy, "The Rhetoric and Economic Roots of the American Broadcasting Industry," *Cinetracts* 6 (2) (Spring 1979), pp. 37–54.

8. William Boddy, "The Shining Centre of the Home: Ontologies of Television in the 'Golden Age'," *Television in Transition*, ed. Phillip Drummond and Richard Paterson (London: British Film Institute, 1985), pp. 125–33.

9. For a detailed analysis of the rise and fall of the DuMont Network, see Gary Newton Hess, *An Historical Study of the DuMont Television Network* (New York: Arno Press, 1979).

10. Cited in "DuMont Expansion Continues," *Radio Daily*, 12 April 1949, p. 23. See also "DuMont Skeds 7 A.M. to 11 P.M.," *Variety*, 22 September 1948, p. 34; "Daytime Tele As Profit Maker," *Variety*, 27 October 1948, pp. 25, 33; "Round-Clock Schedule Here to Stay As DuMont Programming Makes Good," *Variety*, 10 November 1948, pp. 29, 38.

11. Cited in "Daytime Video: DuMont Plans Afternoon Programming," *Broadcasting-Telecasting*, 28 November 1949, p. 3. See also "WTTG Gives Washington Regular Daytime Video with New Program Setup," *Variety*, 19 January 1949, p. 30; "Video Schedule on Coax Time," *Variety*, 12 January 1949, p. 27; "DuMont's 'Mother' Goes Network in Daytime Spread," *Variety*, 27 November 1949, p. 27.

12. "ABC, CBS, NBC Cold to Full Daytime Schedule; DuMont to Go It Alone," *Variety*, 6 October 1948, p. 27.

13. "CBS All-Day TV Programming," *Variety*, 26 January 1949, p. 34; "Video Schedule on Co-Ax Time," *Variety*, 12 January 1949, p. 27; "WNBT, N.Y., Swinging into Line as Daytime Video Airing Gains Momentum," *Variety*, 19 January 1949, p. 24; Bob Stahl, "WNBT Daytime Preem Has Hausfrau Pull but Is Otherwise Below Par," *Variety*, 9 February 1949, p. 34; "Full CBS Airing Soon," *Variety*, 2 March 1949, p. 29; "Kathi Norris Switch to WNBT Cues Daytime Expansion for Flagship," *Variety*, 1 March 1950, p. 31.

14. Cited in "Daytime TV," *Broadcasting-Telecasting*, 11 December 1950, p. 74.

15. *Sponsor*, 4 June 1951, p. 19.

16. *Newsweek*, 24 September 1951, p. 56.

17. *Televiser*, September 1951, p. 20.

18. In the early 1950s, many of the shows were sustaining vehicles—that is,

programs that were aired in order to attract and maintain audiences, but that had no sponsors.

19. "DuMont Skeds 7 A.M. to 11 P.M." *Variety,* 22 September 1948, p. 25.

20. "Pat 'N' Johnny," *Variety,* 1 March 1950, p. 35. This example bears interesting connections to Rick Altman's more general theoretical arguments about the aesthetics of sound on television. Altman argues that television uses sound to signal moments of interest, claiming that, "the sound track serves better than the image itself the parts of the image that are sufficiently spectacular to merit closer attention on the part of the intermittent viewer." See Altman, "Television/Sound," *Studies in Entertainment: Critical Approaches to Mass Culture,* ed. Tania Modleski (Bloomington and Indianapolis: Indiana University Press, 1986), p. 47.

21. Robert C. Allen, *Speaking of Soap Operas* (Chapel Hill: University of North Carolina Press, 1985).

22. See "Daytime Video: DuMont Plans Afternoon Program" and "DuMont Daytime 'Shoppers' Series Starts," *Broadcasting-Telecasting,* 12 December 1949, p. 5.

23. Some variety programs included fifteen minute sitcoms and soap operas.

24. "TV's 'Stars in the Afternoon'," *Variety,* 3 October 1951, p. 29.

25. "Women's Magazine of the Air," *Variety,* 9 March 1949, p. 33; "Women's Page," *Variety,* 1 June 1949, p. 34.

26. NBC had particular problems securing sponsors and, especially during 1951 and 1952, many of its shows were sustaining programs. So critical had this problem become that in fall of 1952 NBC temporarily cut back its schedule, giving afternoon hours back to affiliates. Affiliates, however, complained that this put them at a competive disadvantage with CBS affiliates. See "NBC-TV's 'What's the Use?' Slant May Give Daytime Back to Affiliates," *Variety,* 3 September 1952, p. 20; "Daytime TV—No. 1 Dilemma," *Variety,* 24 September 1952, pp. 1, 56; "NBC-TV to Focus Prime Attention on Daytime Schedule," *Variety,* 24 December 1952, p. 22; "NBC-TV Affiliates in Flareup," *Variety,* 6 May 1953, p. 23.

27. Weaver's concept was adopted by CBS executives who in 1952 instituted the "12 plan" that gave sponsors a discount for buying twelve participations during the daytime schedule. "Day TV Impact," *Broadcasting,* 3 November 1952, p. 73; Bob Stahl, "CBS-TV's Answer to 'Today,' " *Variety,* 12 November 1952, pp. 23, 58.

28. John H. Porter, memo to TV network salesmen, 11 June 1954, NBC Records, Box 183: Folder 5, Wisconsin Center Historical Archives, State Historical Society, Madison.

29. George Rosen, "Garroway 'Today' Off to Boff Start As Revolutionary News Concept," *Variety,* 16 January 1952, p. 29.

30. Joe Meyers and Bob Graff, cited in William R. McAndrew, confidential memo to John K. Herbert, 23 March 1953, NBC Records, Box 370: Folder 22, Wisconsin Center Historical Archives, State Historical Society, Madison.

31. *Daytime Availabilities: Program Descriptions and Estimates,* 1 June 1954, NBC Records, Box 183: Folder 5, Wisconsin Center Historical Archives, State Historical Society, Madison.

32. "Early Morning Inserts Get WNBT Dress-Up," *Variety,* 13 August 1952, p. 26.

33. "For the Girls at Home," *Newsweek,* 15 March 1954, p. 92. NBC's advertising campaign for *Home* was unprecedented for daytime programming promotion, costing $976,029.00 in print, on-air promotion, outdoor advertising, and novelty

gimmicks. See Jacob A. Evans, letter to Charles Barry, 28 January 1954, NBC Records, Box 369: Folder 5, Wisconsin Center Historical Archives, State Historical Society, Madison.

34. Jacob A. Evans, letter to Charles Barry, 28 January 1954, NBC Records, Box 369: Folder 5, Wisconsin Center Historical Archives, State Historical Society, Madison.

35. In a promotional report, NBC boasted that on *Today*'s first broadcast, Kiplinger received 20,000 requests for a free copy of the magazine. Matthew J. Culligan, sales letter, 27 January 1953, NBC Records, Box 378: Folder 9, Wisconsin Center Historical Archives, State Historical Society, Madison.

36. The report cited here was commentary for a slide presentation given by Coffin to about fifty researchers from ad agencies and manufacturing companies in the New York area. *Commentary for Television's Daytime Profile: Buying Habits and Characteristics of the Audience,* 10 June 1954, NBC Records, Box 183: Folder 5, Wisconsin Center Historical Archives, State Historical Society, Madison. For the actual survey, see W. R. Simmons and Associates Research, Inc., *Television's Daytime Profile: Buying Habits and Characteristics of the Audience,* 15 September 1954, NBC Records, Box 183: Folder 8, Wisconsin Center Historical Archives, State Historical Society, Madison. A short booklet reviewing the findings was sent to all prospective advertisers; *Television's Daytime Profile: An Intimate Portrait of the Ideal Market for Most Advertisers,* 1 September 1954, NBC Records, Box 183: Folder 5, Wisconsin Center Historical Archives, State Historical Society, Madison. For NBC's exploitation of the survey, see also Ed Vane, letter to Mr. Edward A. Antonili, 7 December 1954, NBC Records, Box 183: Folder 5, Wisconsin Center Historical Archives, State Historical Society, Madison; Hugh M. Bellville, Jr., letter to Robert Sarnoff, 27 July 1954, NBC Records, Box 183: Folder 5, Wisconsin Center Historical Archives, State Historical Society, Madison; Thomas Coffin, letter to H. M. Beville, Jr., 21 July 1954, NBC Records, Box 183: Folder 5, Wisconsin Center Historical Archives, State Historical Society, Madison. The survey also made headlines in numerous trade journals, newspapers, and magazines. For press coverage, see NBC's *clipping file,* NBC Records, Box 183: Folder 5, Wisconsin Center Historical Archives, State Historical Society, Madison.

37. Edward Stasheff, *The Television Program: Its Writing, Direction, and Production* (New York: A. A. Wyn, 1951), p. 47.

38. Consumer spectacles were further achieved through rear-screen projection, an "aerial" camera that captured action with a "telescoping arm," and mechanical devices such as a weather machine that adorned products in a mist of rain, fog, sleet, or hail. *Daytime Availabilities: Program Descriptions and Cost Estimates,* 1 June 1954, NBC Records, Box 183: Folder 5, Wisconsin Center Historical Archives, State Historical Society, Madison.

39. Charles C. Barry, memos to Richard Pinkham, 2 March 1954, 3 March 1954, and 4 March 1954, NBC Records, Box 369: Folder 5, Wisconsin Center Historical Archives, State Historical Society, Madison.

40. Franklin Sisson, *Thirty Television Talks* (New York, n.p., 1955), p. 144. Cited in Giraud Chester and Garnet R. Garrison, *Television and Radio* (New York: Appleton-Century-Crofts, Inc., 1956), p. 414.

41. Caroline Burke, memo to Ted Mills, 20 November 1953, NBC Records, Box 377: Folder 6, Wisconsin Center Historical Archives, State Historical Society, Madison; Arlene Francis, cited in Earl Wilson, *The NBC Book of Stars* (New York: Pocket Books, 1957), p. 92.

42. Cited in Wilson, *The NBC Book,* p. 94.

43. Sylvester L. Weaver, memo to Harry Bannister, 10 October 1952, NBC Records, Box 378: Folder 9, Wisconsin Center Historical Archives, State Historical Society, Madison.

44. Joe Meyers, cited in William R. McAndrew, confidential memo to John K. Herbert, 23 March 1953, NBC Records, Box 370: Folder 22, Wisconsin Center Historical Archives, State Historical Society, Madison.

45. A. A. Schechter, " 'Today' As An Experiment Bodes Encouraging Manana," *Variety,* 16 July 1952, p. 46. NBC also advertised *Today* by claiming that "people are actually changing their living habits to watch 'Today.' " See *Sponsor,* 25 February 1952, pp. 44–45.

46. *Daytime Availabilities: Program Descriptions and Cost Estimates,* 1 June 1954, NBC Records, Box 183: Folder 5, Wisconsin Center Historical Archives, State Historical Society, Madison.

47. Walter Benjamin, "The Work of Art in the Age of Mechanical Reproduction," in *Illuminations: Essays and Reflections,* ed. Hannah Arendt (New York: Schocken, 1969), pp. 217–51.

48. *Ladies' Home Journal,* April 1955, p. 130. See also *Ladies' Home Journal,* February 1955, p. 95; *Good Housekeeping,* July 1955, p. 135.

49. *The New Yorker,* 3 June 1950, p. 22.

50. Crosby, "What's Television Going to Do to Your Life?" *House Beautiful,* February 1950, p. 125.

51. *American Home,* October 1955, p. 14.

52. Walter Adams and E. A. Hungerford, Jr., "Television: Buying and Installing It Is Fun; These Ideas Will Help," *Better Homes and Gardens,* September 1949, p. 38; *American Home,* December 1954, p. 39.

53. *American Home,* May 1955, p. 138. The cartoon was part of an advertisement for the *Yellow Pages.*

54. *House Beautiful,* June 1952, p. 59.

55. W. W. Ward, "Is It Time To Buy Television?" *House Beautiful,* October 1948, p. 220.

56. *Ladies' Home Journal,* May 1953, p. 148.

57. "The Wonderful Anti-Statics," *House Beautiful,* January 1955, p. 89; *Ladies' Home Journal,* November 1948, p. 90.

58. Gertrude Brassard, "For Early Tea and Late TV," *American Home,* July 1952, p. 88.

59. In August 1949, for example, *House Beautiful* suggested that a swiveling cabinet would allow women to "move the screen, not the audience" (p. 69). Although portable sets were not heavily marketed in the early 1950s, they were sometimes presented as the ideal solution to the problem of moving the heavy console set.

60. *House Beautiful,* May 1952, p. 138.

61. Wright, *Building the Dream,* p. 172.

62. *House Beautiful,* June 1951, p. 121.

63. Vivian Grigsby Bender, "Please a Dining Room!" *American Home,* September 1951, p. 27.

64. *Better Homes and Gardens,* December 1952, p. 144; *Better Homes and Gardens,* February 1953, p. 169; see also *American Home,* September 1953, p. 102.

65. *House Beautiful,* November 1948, p. 5.

66. Edith Ramsay, "How to Stretch a Day," *American Home,* September 1949, p. 66; *House Beautiful,* December 1950, p. 77.

67. *American Home,* February 1954, p. 32.

68. *House Beautiful,* November 1954, p. 158. For additional examples, see *American Home,* November 1953, p. 60; *Better Homes and Gardens,* December 1951, p. 7; *TV Guide,* 18 December 1953, p. 18.

69. *Better Homes and Gardens,* October 1952, p. 177.

70. Laura Mulvey, "Visual Pleasure and Narrative Cinema," *Screen* 16 (3) (1975), pp. 6–18. Since the publication of Mulvey's article, numerous feminists—including Mulvey—have theorized ways that women might find subjective pleasures in classical cinema, and feminists have also challenged the idea that pleasure in the cinema is organized entirely around scenarios of "male" desire. For a bibliography on this literature and a forum on contemporary views on female spectatorship in the cinema, see *Camera Obscura* 20–21 (May–September 1989).

71. W. W. Ward, "Is It Time to Buy Television?" *House Beautiful,* October 1948, p. 172.

72. *House Beautiful,* May 1955, p. 131.

73. *Better Homes and Gardens,* October 1953, p. 151. There is one exception to this rule of male body posture, which I have found in the fashionable men's magazine *Esquire.* While *Esquire* depicted the slovenly male viewer, it also showed men how to watch television in fashion by wearing clothes tailored specifically for TV viewing. In these cases, the male body was relaxed, and the men still smoked and drank liquor, but they were posed in more aesthetically appealing ways. See "Town-Talk Tables and Television," *Esquire,* January 1951, pp. 92–93; and "Easy Does It Leisure Wear," *Esquire,* November 1953, p. 74. The figure of the fashionable male television viewer was taken up by at least one male clothing company, The Rose Brothers, who advertised their men's wear by showing well-dressed men watching television and by promising, "You Can Tele-Wise Man by His Surretwill Suit." See *Colliers,* 1 October 1949, p. 54.

74. Robert M. Jones, "Privacy Is Worth All That It Costs," *Better Homes and Gardens,* March 1952, p. 57.

75. This is not to say that television was the only domestic machine to disrupt representations of gender. Roland Marchand, for example, has argued that advertisements for radio sets and phonographs reversed traditional pictorial conventions for the depiction of men and women. Family-circle ads typically showed husbands seated while their wives were perched on the arm of the chair or sofa. In most of the ads for radios and phonographs in his sample, the opposite is true. Marchand argues that "in the presence of culturally uplifting music, the woman more often gained the right of reposed concentration while the (more technologically inclined) man stood prepared to change the records or adjust the radio dials." See *Advertising the American Dream,* pp. 252–53. When applied to television, Marchand's analysis of radio does not seem to adhere since men were often shown seated and blatantly unable to control the technology.

76. I am borrowing Natalie Zemon Davis's phrase with which she describes how women in preindustrial France were able to invert gender hierarchies during carnival festivities and even, at times, in everyday life. See "Women On Top," *Society and Culture in Early Modern France* (Stanford, Calif.: Stanford University Press, 1975), pp. 124–51.

77. *Popular Science,* May 1954, p. 177; *Esquire,* March 1951, p. 10; Jack O'Brien, "Offsides in Sports," *Esquire,* November 1953, p. 24.

78. Marsh, *Suburban Lives*, p. 82.

79. Greenbie, *Leisure for Living*, p. 210. Greenbie, in fact, presented a quite contradictory account of mechanization in the home, at times seeing it as the man's ally, at other times claiming that modern machines actually took away male authority.

80. "Home is for Husbands Too," *Esquire*, June 1951, p. 88.

81. In addition, companies that produced home-improvement products and workshop tools continually used television sets in their illustrations of remodeled rooms. Typically here, the Masonite Corporation promoted its do-it-yourself paneling in an advertisement that displayed a television set in a "male room" just for Dad. See *Better Homes and Gardens*, August 1951, p. 110. For similar ads, see *American Home*, June 1955, p. 3; *Better Homes and Gardens*, February 1953, p. 195; *American Home*, November 1952, p. 105. It should be noted that some of these ads also showed women doing the remodeling work.

82. "From Readers' Albums of Television Photos," *Popular Science*, December 1950, p. 166. See also "TV's Images Can Be Photographed," *Popular Science*, August 1950, pp. 184–85; R. P. Stevenson, "How You Can Photograph the Fights Via Television," *Popular Science*, February 1951, pp. 214–216.

# The Discourse of "Scientific Anti-Communism" in the "Golden Age" of Documentary

## MICHAEL CURTIN

Both media scholars and broadcast news veterans often characterize the early 1960s as the "golden age" of television documentary.[1] For it was during this era that documentary enjoyed a prominent place in prime-time network schedules and in public discussion about "quality" television. Furthermore, this golden age is remembered as an era of innovation. Albert Wasserman, one of the leading documentary producers at NBC during the early 1960s, recalls that the television networks developed the *journalistic* documentary as a conscious departure from the nonbroadcast heritage of the genre: "The whole history of the documentary film, which of course precedes television, was to a great extent a history of social indignation, allowing people who made films to express points of view about which the filmmaker felt strongly. With the evolution of the television documentary that is no longer appropriate. It is not appropriate for a television network to take a partisan position and to seem to be trying to force on an audience a predetermined editorial point of view."[2] Wasserman and others such as Fred Friendly and Irving Gitlin clearly distinguished their work from an earlier documentary tradition that included the inspirational efforts of John Grierson, the controversial docudramas of *The March of Time*, and the propaganda films of the Second World War. Unlike their predecessors, these "news professionals" were fond of noting that the power of the television documen-

From *Cinema Journal*, vol. 32, no. 1, Fall 1992, pp. 3–25. Reprinted with permission.

tary resided in its ability to offer a "mirror-like" reflection of social reality.[3]

However, careful analysis also suggests that the work of these documentarists was marked by contradiction. On the one hand, they contended that their programs were the product of professional, objective inquiry; on the other hand, the programs elaborated a worldview that can best be understood in relation to the Cold War. This contradiction was not exclusive to these documentaries. It was also found in many other aspects of American society at the time. Social scientists, policymakers, and American diplomats all endorsed the canons of objectivity and professionalism, but they were also stalwart proponents of "the American way of life" as a model for societies around the globe. Therefore, even though the early 1960s was an era that celebrated scientific expertise and professionalism, it was also a time that witnessed the appearance of a vigorous strategic, economic, and ideological campaign to secure American hegemony over much of the globe.

Previous criticism of these documentaries has failed to explore this contradiction. Erik Barnouw dismisses the programs as "a product of cold war and blacklists."[4] His historical survey of the documentary genre pays scant attention to these texts and fails to explain why network documentarists donned the mantle of journalistic professionalism and objectivity. Barnouw's analysis also leaves us wondering why so many scholars and news workers reflect back on this era as one of the pinnacles of journalistic accomplishment in the history of television. Furthermore, A. William Bluem, who has conducted the only extensive scholarly criticism of these programs, rejects Barnouw's conclusions. Bluem writes that "these documentaries achieved a force and authority which removed them from any possible accusation of 'slanting' or 'propaganda,' and dispelled once and for all the notion that the art of recording the great struggles of our society could destroy their journalistic validity."[5] Such disparate assessments as those of Barnouw and Bluem suggest a textual complexity that hinges on the relationship between the Cold War and the concept of objectivity.

The extensive scholarly literature regarding journalistic practice argues that pretensions to objectivity are at best an approximation of an ideal and at worst a strategic ritual that protects journalists from their critics.[6] Similarly, film scholars have painstakingly detailed the ways in which form, method, and narrative style undermine the truth-claims of documentarists.[7] Even the subgenre of cinema verite has been shown to rely on techniques that are modeled on conventions of the fiction film.[8] One of the things we have learned from this body of scholarship is that documentary claims to objectivity and truth are best understood by examining the ways in which such concepts are employed at particular historical moments.

In this essay, I argue that during the waning years of the Eisenhower administration, important corporate and political figures in the United States sought to mobilize public opinion behind an agenda for activist intervention along the frontiers of the "Free World." They promoted a dramatic military build-up in both nuclear and conventional weapons; advocated foreign aid and development programs for Third World countries; and embraced a partnership between government and business in order to ensure the expansion of American corporate activities at home and abroad. This was a marked departure from the priorities that dominated public policy during the Eisenhower administration. Not only did it involve a shift in political leadership, it also required a change in fundamental public assumptions about the nature of global relations and the U.S. role in the struggle against "monolithic Communism."

Documentary was one of the most important sites where this ideological work took place. In an era when television surpassed print as the American public's primary source of news and in which network newscasts were only fifteen minutes long, the television documentary played an important role in elaborating an analytical framework for understanding the significance of world events.[9] The genre seemed to transport the viewer to distant lands and created the illusion that one had direct access to information that would shape key foreign policy decisions. It is estimated that 90 percent of all households viewed network documentaries at least once a month and that the average ratings were somewhere in the low teens.[10] Although these audiences were small by the standards of prime-time entertainment, they far exceeded the audiences for cinema verite, a subgenre that has received far more scholarly attention.

My analysis suggests that the concept of objectivity was not simply a strategic ritual or the approximation of an ideal. Rather, objectivity played an important role in a discursive struggle whereby a transnational elite sought to reconfigure public attitudes regarding the Cold War. As we shall see, intellectuals, liberal politicans, and scientists—all of whom were largely excluded from positions of influence during the Eisenhower era—joined forces with corporate leaders during the latter part of the 1950s in a campaign to "get America moving again." The rhetoric of this group valorized the notion of American leadership in an active struggle against the forces of monolithic Communism throughout the globe. But just as important, it celebrated expertise, science, and professionalism as important weapons in the American arsenal. Thus, the dispassionate language of scientific method was married to the political rhetoric of superpower struggle, an articulation that I refer to as the discourse of scientific anti-Communism. In the wake of the Sputnik launch and in the midst of economic recession, this alliance of critics argued that the application of scientific technique to the problems of society was the only means by which the United States could meet the challenge from the East. Therefore, it is significant that this was the era in which all three television networks rapidly expanded their documentary production and shifted the exhibition of these programs from the "cultural ghetto" of Sunday afternoon to evening prime time. The networks contended that professional journalists, using an objective methodology, would render a public service through their examination of pressing issues at home and abroad.

Therefore, Barnouw and Bluem are both right and they are both wrong.

The network documentaries of this golden age were the product of the Cold War, but not in a reductive sense. Instead, they were part of a complex shift in public discourse and power relations during the late 1950s and early 1960s. Similarly, these programs were objective. But the concept of objectivity here must be understood as historically specific, as a key element in the discourse of scientific anti-Communism. Finally, the textual complexity of these programs suggests that the articulation of objective methodology with the hegemonic interests of transnational capital was neither natural nor eternal. Rather, embedded within this articulation were seeds of contradiction that resisted the closure that these programs sought to prefer.

This essay focuses its attention on the complex forces—economic, political, institutional, and discursive—that influenced the production of network documentaries during the early 1960s.[11] In order to present such an analysis, I begin by describing the rapid expansion of network investment in the documentary field, followed by an analysis of the thematic patterns within this body of programming. After a brief discussion of the narrative techniques at work in this genre, I then examine an exemplary text as a means of coming to grips with the complex and contradictory meanings that these programs produced. Finally, this leads to an analysis of the discourse of "scientific anti-Communism" and the social context in which it emerged.

## The Documentary Boom of the Early 1960s

We should begin by noting the distinctive nature of this moment in television history. Between 1960 and 1964 there were more documentaries produced and broadcast in network prime time than any other comparable period.[12] Indeed, during the peak season of 1962 there were six regularly scheduled prime-time documentary series as well as frequent specials like the *NBC White Paper* programs.[13] In all, the three networks produced 387 documentaries in 1962, more than ten times the number produced annually by the networks in recent years.[14]

Scholarly analysis of these programs has largely focused on the forces that fed this documentary boom. Raymond L. Carroll in his economic analysis of this phenomenon concludes that the rise and demise of this genre cannot be explained by market forces. Instead, he argues that other factors, such as politics, were most influential in promoting the rapid expansion of the documentary genre during this period.[15]

Other scholars have taken a similar position. For example, Erik Barnouw contends that documentary's golden age was a form of atonement for the transgressions of quiz show producers during the late 1950s. The networks, who were then leery of government intervention to clean up the industry, committed themselves to an expansion of public affairs programming in an effort to mend their tattered public image.[16]

Moreover, the industry's sensitivity to its public image had many aspects during this period, and the expansion of documentary programming served more than one purpose. James L. Baughman argues that, in 1959, CBS may have been less concerned about the quiz show controversy than it was about the revitalization of a then quiescent Federal Communications Commission. Baughman, in attempting to reconstruct the thinking of CBS executives, contends that the network premiered the *CBS Reports* series largely in response to a variety of criticisms leveled by government regulators and the public at large. By his account, CBS was less concerned about quiz programs than it was about government anti-trust inquiries.[17]

There were other reasons the networks were conscious of their public image as well. Fred Friendly and David Yellin, in their anecdotal recounting of this era, note the intense competition among the networks in the arena of news and public affairs. The so called "Friendly-Kintner wars" are often credited with the rapid expansion in both the number and prime-time scheduling of network documentary. They argue that all three major networks were vying for position in the field of "actuality" programming in an effort to establish a positive network identity in the public arena.[18]

Finally, the expansion of network documentary during this period was attributable to more than matters of imagery and prestige. Many have noted that documentary emerged as a compelling video genre at the moment when lightweight, sound-synchronized film cameras began to gain popular currency in network news organizations. The old, 35mm albatross started to give way to more portable 16mm equipment that not only allowed more fluid camera technique, but also permitted the camera to go places it had never gone before.[19]

Taken together, these explanations answer many questions about the network documentary boom in the early 1960s, but they do not explain other significant questions: What were these documentaries about? What role did they play in the circulation of certain meanings within society? Why did they circulate some meanings and not others? And how was this process of textual production related to other forces operating in society? These *political* questions remain a fascinating set of problems for a cultural history of the 1960s network documentary.[20]

## Cold War Discourse in Television Documentaries

A cursory examination of this body of programming makes clear that the most commonly shared documentary theme during this period was the struggle between East and West, between totalitarianism and democracy, between Communism and capitalism. The archival collections at the Library of Congress and the Museum of Television and Radio are well stocked with evidence to support this claim. In program after program, the viewer confronted the challenge of Communism on the battlefield, in the political arena, and in the classroom. The broadcasting trade press at the time also noted this phenomenon, arguing that the struggle to defend the "Free World" was the hottest topic in documentary television.[21]

To understand the scope of this tendency, we should begin by surveying the three television series that represented the most intensive commitment of network resources during this period: ABC's *Bell & Howell Close-Up*, *NBC White Paper*, and *CBS Reports*. These "flagship" series were considered the definitive expression of network commitment to the documentary form. They were pacesetters in style and content and, as a result, they attracted the greatest amount of public attention on the air, in the industry, and in the press. These programs therefore can be considered exemplars of the overall drift of documentary production at the major networks. Indeed, a sampling of programs from other documentary series during this period shows no significant disruption of the pattern that emerges from these "big three." As a result, these series are key objects for historical analysis.

A survey of these programs shows that during the first season of the *Bell & Howell Close-Up* series, ABC produced eighteen programs, of which six were explicitly about the Communist threat. These were primarily foreign policy programs such as "Ninety Miles to Communism," a program about Cuba; "The Red and the Black," an investigation of the red threat in Africa; and "Troubled Land," a profile of the peasant revolts in northeastern Brazil.[22] In each of these programs the central narrative conflict revolves around the Communist challenge to the Free World.

Another six programs during this premiere season of *Close-Up* engaged the red threat as a significant subtext. For instance, "The Flabby American" is a report on the physical fitness of the nation that is cast in the context of East–West relations.[23] Within the first minute the viewer sees Soviet children engaged in a hearty regimen of calisthenics while the narrator notes that young Soviets are trained to place their physical health at the disposal of the state. This is contrasted with reports of battle-weary GIs during the Korean conflict and concerns that the American soldier might not be tough enough to fight another war. Indeed, the opening scene of the documentary takes the viewer to a Marine boot camp in order to witness the sorry physical condition of newly inducted recruits. Although most of the program is centered on the fitness of the average, middle-class American, this fitness is framed as a Cold War issue.

Similarly, a documentary about racism, "Cast the First Stone," is contextualized by the concern that any systematically oppressed minority represents an inherent threat to the stability of the American political system.[24] Blacks are characterized as a restless segment of society that cannot be expected to endure the indignities of discrimination much longer. They are a potentially rebellious force, as are people of color in many parts of the developing world. A subtext that plays about the periphery of many race-related documentaries during this period is the tension between discrimination and American aspirations to leadership of the Free World. Many programs specifically ask, "How will the civil rights issue affect the U.S. image abroad?"

A similar thematic pattern can be found in the *NBC White Paper* series. Half of the twenty programs broadcast during its first four seasons explicitly revolve around the Communist challenge to the Free World, and three other programs draw on this theme as an important subtext. The Cold War is a subject that activates an important narrative tension in most of these NBC documentaries.

At *CBS Reports* there was less of an emphasis on foreign policy issues and, as a result, the Communist threat is not as explicitly pervasive. However, tropes of "freedom" are woven throughout the core of these documentaries, and freedom is cast in relation to its other: the often unspoken threat of Communism.[25] Thus, a much heralded documentary about "The Population Explosion" makes passing, but telling, reference to Communist takeover as a fate that threatens the nation of India. The program argues that liberal democracy cannot flourish in the midst of poverty and overpopulation. Furthermore, birth control among the poor in the United States is linked by implication to the problems confronting India. In both cases, traditional attitudes and religions are marked as obstacles to progress and modernity.[26] And, as we shall see, questions of modernity and reform are positioned within the larger context of superpower struggle.

### Documentary Narrative

Besides this thematic unity, it is also important to notice the ways in which these programs depart from the conventions of journalistic objectivity. Instead of providing a headline summary to frame the ensuing documentation, these programs often construct a narrative conflict and an enigma.[27] Unlike the conventions of broadcast news, which encourage the reporter to deliver the punchline at the very outset, many of these documentaries seek to draw the viewer into a narrative world by using some of the same techniques that one finds in classical Hollywood cinema.[28] Furthermore, this borrowing from Hollywood style is neither "natural" nor accidental. In fact, such techniques had been the subject of extensive discussion at all three networks for some time.

As early as 1955, Irving Gitlin, who would later be named executive producer of *NBC White Paper,* told a group of network reporters that television viewers would measure their documentary work by the same standards as fictional programming. From the viewer's perspective the equation is simple, said Gitlin: "I want a story; I want a beginning, middle and end; I want someone I can identify myself with, and most important of all, it has to be clear and I have to understand it."[29] Gitlin's approach was considered somewhat novel in 1955, but it was based on the fact that few of his documentaries at that time dealt with breaking news stories. As a result, he used narrative conflict as a vehicle to draw the audience into the program.

However, such techniques were still the exception, and most network documentaries were talky and pedantic. Even those who later were regarded as most sensitive to narrative considerations spent much of the 1950s under the sway of journalistic precepts. As producer Fred Freed later commented regarding the work of his colleague Albert Wasserman: "[Wasserman] didn't have the film sense that he later developed. He was very, very much still—as was almost everybody at CBS—under the influence of Fred Friendly and the radio approach: tell 'em what you're going to tell 'em; tell 'em; then tell 'em what you've told 'em. And make everything perfectly clear every step of the way."[30]

However, such didactic techniques were beginning to lose their preeminent position. In 1957, the CBS research department confirmed Gitlin's assessment of audience reactions to network documentaries. In a major study of thirty-two programs, based on responses by 2,500 adults, the network found that the most frequent complaints about CBS documentaries were: too much [voice-over] narration, not enough action, not entertaining, no unity or simplicity (too many scenes, characters, etc.), story gets under way slowly, too many issues and statistics. The documentaries that audiences found most satisfying were programs with "a strong unifying central character, a definite setting, and a strong unifying plot."[31]

Moreover, such network concerns about audience reception were linked specifically to questions of commercial viability. In 1958, CBS shifted the designation of its news division from a "staff operation" to an "operating division." Now, for the first time, CBS News was expected to generate a positive cash flow.[32] And in 1959, when Irving Gitlin was put in charge of the public affairs unit, his mandate was to develop programs with "sales potential."[33] Similar developments took place at NBC and ABC during this period. Audience appeal and sponsorship were a growing concern for top network executives. Indeed, the sponsor of ABC's *Close-Up* series, Bell & Howell, explicitly promoted the importance of entertainment values in documentary programming. In a letter to ABC executives, Peter G. Peterson, senior vice-president, wrote, "We felt that too many of these shows in the past might be characterized as newsmanship without showmanship."[34] Or, as Charles Percy, chairman of the board, put it, "Our objective is to prove that the times in which we live can be just as dramatic, just as exciting as those 'colorful, heroic days of the West.' "[35] Nor were ABC executives shy about adopting this logic. During a promotional premiere screening of the *Close-Up* series, ABC president Oliver Treyz declared, "We hope, we anticipate that [these documentaries] will be able to meet and compete with sheer entertainment shows on the level of audience appeal and will be seen by many more millions than customarily view so-called educational programs."[36]

Thus, by the early 1960s, narrative rather than journalistic structure had become the desideratum of the documentary producer. Although many of the programs were still rhetorically constructed, the most celebrated producers were invoking narrative strategies to deal with issues. "Anything I do I like to approach from a narrative rather than an interpretive position," commented NBC's Emmy-winning producer Reuven Frank. "Even great issues can be more successfully illuminated than expounded."[37]

That sentiment was echoed at ABC, where the network put John Secondari in charge of its nascent documentary unit. Secondari's appreciation of narrative techniques stemmed from his experience writing scripts for *The Alcoa Hour* and *Playhouse 90*. Furthermore, he came to ABC news with four novels to his credit, among them a book that was turned into the Hollywood feature film *Three Coins in the Fountain*.[38] Others at ABC shared Secondari's disposition as well. Producer Edgar Peterson, in charge of a historical documentary series about Winston Churchill during the war years, would liken the Churchill episodes to westerns with "a hero, a villain, and a chase."[39]

Thus, network documentarists consciously adopted narrative techniques as a way to organize information and attract audience attention. They were not pretending that the issues they addressed lent themselves naturally to narrative structure, nor were they arguing that events and relationships in the objective world would emerge in narrative form on their own. Instead, they were proposing that many of the conventions of fiction, particularly Hollywood cinema, should be applied to documentary treatment of pressing social issues. By the early 1960s, even Fred Friendly had come around to talking about documentary in narrative terms, "We hope each show will be just like reading through to the last page of a detective story to discover whether the butler did it. You won't know the outcome of any of our shows until you see it."[40] Therefore, by the dawn of the documentary's "golden age," narrative considerations had risen to the fore at all three networks, with Friendly in charge at *CBS Reports,* Secondari at ABC, and Gitlin at NBC. With documentary producers more sensitive to considerations of narrative, it follows that many of them sought to construct programs that revolved around a clearly defined conflict. As a result, the clash between East and West structures many of these programs. Thus, it is too simplistic to characterize these texts as products of "cold war and blacklists," or as the outcome of objective journalistic inquiry. Instead, the nature of commercial television was one of many influences that shaped these texts.

### *An Exemplary Text: "Panama: Danger Zone"*

However, skimming across the surface of these documentaries in search of thematic patterns and prevalent narrative strategies produces an inadequate explanation of the ways in which these programs produced and circulated meanings about the Cold War. To comprehend the complexity of this era of "actuality" programming, one must be willing to engage the intricate play of language and imagery within each of these documentaries. For the significance of these programs lies not only in their unqualified opposition to Communism, but in the ways they construct such opposition in relation to a whole range of other meanings.

Typical of this genre of programming was an *NBC White Paper,* entitled "Panama: Danger Zone," produced by Albert Wasserman, formerly of CBS.[41] The documentary begins with a tour of daily operations along the Panama Canal. Images of high technology, rationality, and stability are foregrounded here as we listen to a tour guide describe the canal as the "eighth wonder of the world." Chet Huntley's voice-over narration remarks that little has changed along the canal in more than four decades of operation. However, this state of normalcy is then ruptured by an edit that juxtaposes footage of the canal with footage of a street riot that broke out in Panama City one year earlier. As the visual imagery portrays chaotic violence and high emotion, Huntley's narration coolly informs us that, unlike the Canal Zone, things are changing rapidly in Panama due to increasing protests against the U.S. presence. Huntley says that American officials fear that, as the first anniversary of this riot approaches, violence could break out again.

Additionally, these contrasting images of the orderly operation of the canal and the disorderly nature of Panamanian politics are intercut with shots of a local calypso band throughout this opening sequence. Huntley informs us that the band is singing about last year's riot, which quickly earned a prominent place in Panamanian folklore. And although the song is being sung in Spanish, the passage of lyrics that is most intelligible to the casual viewer is the refrain, "Fidel, Fidel." In essence, the metaphoric clash between modern technology and Third World "unrest" provides the narrative frame for this documentary. It is as if North American technology and rationality have tamed the jungle but have failed to tame the darker side of Panamanian society. And it is this darker side that is being exploited by the forces of monolithic Communism.

After a commercial, we return to the program to find peasants seeking shelter as they scan the horizon in expectation of the daily storm, which, the narrator comments, "is almost inevitable." This motif of the gathering storm recurs throughout the program and suggests an impending deluge. This, of course, is not an objective observation but a visual metaphor. The fact of the matter is that the daily storms in Panama are usually brief and benign. Yet the documentary employs this metaphor in order to conjure up a narrative tension between contending forces. Thus, Wasserman distances himself from a more didactic journalistic approach in order to ground his analysis of the Communist threat in a specific location, at a particular moment of crisis, with a tightly structured plotline. There is a conflict, an enigma, and a cast of characters: the American expatriate community, the indigenous Panamanian society, and the global, monolithic Communist threat.

The documentary turns first to the expatriate community, which is referred to as "a tiny slice of America on the ninth parallel," and which is structured around the canal. Drawing upon historical footage of the construction and early operation of the waterway, the documentary shows how North American ingenuity subdued the jungle and converted it to a resource of great strategic value, not only to the United States, but to all nations of the Free World. Moreover, the canal is compared to a public utility, a nonprofit entity serving the commercial needs of all peace-loving nations. It is portrayed as the product of North American altruism, rather than as an early chapter in the expansion of Yankee imperial power.

However, the program also cautions that Panamanians have a different story to tell about the canal. They talk of discrimination against Panamanians who worked on the canal, arguing that white Americans were paid much higher wages than their local counterparts. Furthermore, the Panamanians claim that the United States receives most of the benefits from the canal's operation. And they claim that the canal treaty was never intended to grant the United States sovereignty over the Canal Zone.

These conflicting views have created tension between the two groups, and this is rendered visually in a number of scenes. For example, we observe a modern American elementary school where the day begins with the children singing the national anthem. In turn, this scene is crosscut with shots of an impoverished Panamanian school where the children sing their national anthem. Likewise, a visual profile of the prosperous, suburban-like trappings of the American community are juxtaposed with the wretched conditions in a poor, Panamanian barrio.

Furthermore, such contrasts are, like the opening sequence, positioned within the wider context of East–West struggle. As the viewer surveys specific visual documentation of Panamanian poverty—such as malnourished children playing in the midst of shantytown squalor—the narrator suggests a broader interpretation: "The setting is Panama, but this could just as easily be Caracas, Venezuela; or Rio, Tegucigalpa, Lima, Santiago, Havana. Panama in its basic problems is virtually indistinguishable from any other Latin land." Thus, as this segment draws to a close, the program reminds the viewer that Panama is a metonymic representation of the tensions at work throughout Latin America. Indeed, the voice-over narration works to erase the historical specificity of the conflict over the canal. Panama could be "any other Latin land." At this point, the documentary returns to the motif of the gathering storm and conjures up the specter of Fidel before it cuts to a commercial. This concluding strategy brings us back to the suggestion that the real conflict at work in this program is the clash between American interests and monolithic Communism.

Later in the program, a series of interviews with students and politicians make it appear that support for Castro is growing throughout Panama in response to the problems of poverty and Yankee imperialism. One student draws the link to Cuba by saying, "We have the same problems and the same enemy." Furthermore, a member of the Panamanian legislature points out that the Cuban revolution has been the most influential factor in spurring U.S. receptiveness to Panamanian concerns. For example, the U.S. State Department recently mounted "Operation Friendship," a sort of integrationist initiative aimed at encouraging social and cultural encounters between Panamanians and U.S. residents of the Canal Zone. According to the documentary, these measures may be enjoying some success. As the first anniversary of the riot draws near, the program builds to a moment of climactic tension, and yet the anniversary demonstration takes place without violence. Panamanians peacefully march through the streets demanding return of the Canal Zone. Despite this apparent moment of calm, the documentary remains skeptical, for the fundamental conflict between East and West has not been resolved.

Therefore, as the program draws to a close, the status of the structuring narrative is somewhat muddled. Neither the canal conflict nor the super-power struggle have been resolved. However, the documentary does provide its own form of closure in a scene that proposes a solution to the enigma alluded to during the opening segment. Here, the enigmatic Panamanian unrest is positioned within a field of meanings that strive to mediate the contradictions at work in the text. The camera cuts to a darkened street in Panama at night where there is some sort of festival going on. We see a candle-lit float garlanded with flowers and bearing a statue of a Catholic saint, probably the Virgin Mary. As we watch the float haltingly maneuver through the narrow streets on the shoulders of dark-skinned Panamanians, we hear them singing and then we hear the narrator commenting in a dramatic baritone as the scene unfolds: "The people of Panama, the people of Latin America have been slow to change. In their religious rituals they cling to an old Spanish march, two steps forward, one step back. But today, they're on the move, stirring from the darkness of their past. Throughout much of Latin America an explosive force is ready to erupt. They look to us for leadership or help. They no longer can be ignored."

As one of the documentary's concluding moments, this scene strives to mediate a number of tensions that were deployed throughout the hour-long program: primitive versus modern; low-tech versus high-tech; black versus white; darkness versus light; monolithic Communism versus the Free World. The documentary suggests that Panamanians wish to walk in light but are burdened by a cultural baggage that leads them down dark and narrow streets. These are not the broad boulevards of the urban planner; they are the crowded, rough-hewn barrios of the poor and uneducated. Political choices in these communities are steeped in passion, a passion that conjures up the specter of Fidel. This scene mediates a set of conflicting meanings in relation to the foreign policy discourse that emerged as part of the New Frontier. It constructs the problems confronting the Panamanian people as genuine but marks them as dysfunctions in the otherwise rational and stable system of the Free World. Moreover, it implies that while change is necessary, it can be contained and controlled from above: "They look to us for leadership or help. They no longer can be ignored." In essence, the program strives to affect a form of closure by offering a solution to the problem of Panama that integrates the rhetorics of liberal reform and staunch anti-Communism.

At the same time, this *NBC White Paper* produces a set of meanings by virtue of its "objective methodology." It does not speak the acrimonious metaphors of McCarthyism, isolationism, or brinkmanship. Rather this is a documentary constructed by an internationalist and a "nonideological" expert. Its call to action is predicated on rational evaluation of the empirical evidence and its narrator speaks with cool, factual authority. Nevertheless, this documentary draws conclusions that remain embedded in the politics of the Cold War. It operates within the boundaries of what might be referred to as the discourse of *scientific anti-Communism*.

## The Discourse of Scientific Anti-Communism

During the early 1960s, the network documentary presented itself as an unprejudiced examination of objective reality. Like social scientists, documentary journalists argued that the subject of their research was knowable through a series of procedures that were largely value free and subject to independent validation. Although neither group contended that their methodology was foolproof, both distinguished their work by reference to the concept of objectivity. In addition, both groups promoted the notion that technology, in the hands of experts, was capable of rendering accurate representations of social reality. While economists drew on the powers of the computer in order to profile and analyze aggregate commercial data, the journalist drew on the technologies of sound and light recording in order to profile and analyze issues of social conflict or concern. In essence, the television journalist, like the social scientist, operated according to certain codes of professional expertise.

Moreover, the television networks mobilized this professionalism toward practical ends. They used it to enhance their image during a period of scandal and public criticism. The television quiz show scandals of the late 1950s had brought into question the common supposition that "seeing is believing," and the networks sought to make amends for this breach of faith by expanding their commitment to a genre of programming produced by *professional* journalists. Thus, the documentary asked viewers to believe what they saw once again, for what they saw was presented as the product of trained expertise. Therefore, the persuasiveness of the documentary arose not only from its wide exposure, its critical acclaim, and its endorsement by elites, but from its claim to offer an unbiased view of social reality; *how* the programs presented themselves was as important as *what* they presented. One NBC press release touted Albert Wasserman, producer of the Panama documentary, for having "made it his career to capture on films true stories . . . exactly as they happen."[42]

This pursuit of "true stores" by the television eye emerged in the 1960s as one of the most important mandates of the networks. It was a pursuit that was no longer sequestered in the world of Sunday afternoon, nor was it the semiexclusive province of Mr. Murrow's newsmagazine. Rather, the documentary of the early 1960s was constituted as an objective, in-depth, prime-time examination of social reality. As such, it was just as important for the documentary to foreground its method as it was for it to draw conclusions. The method was "science," upon which rested documentary's ultimate claim to legitimacy. Even more, it was a form of social science that was inevitably linked to specific notions of political process.

NBC producer Fred Freed drew such a connection when discussing a

*White Paper* program about mainland China, a country that up to that time had been off-limits to the American broadcast media by order of the U.S. State Department: "We think the more information the public has, the more intelligently they can judge what our China policy should be. We ourselves don't intend to judge, just present as many of the objective facts as we can."[43] Freed argued that information is crucial to the political process and that democratic politics depend on the work of experts to deliver value-free facts so that various options might be weighed.[44]

Such a presumed linkage between objectivity and rational choice was not Freed's alone. Media scholar A. William Bluem places objectivity at the center of his analysis of the television news documentary published in 1965. "To argue that objectivity cannot exist is to pose the frightening possibility that we are hopelessly subject to all that is dark and irrational in human nature," wrote Bluem. "Without the concept of objectivity we abandon all outward meaning of life and events, even reason itself."[45] For Bluem, meaning was imbedded in social facts, and the news documentary was an assemblage of facts that provided the basis for rational choice. Thus, documentary expertise was, by implication, intimately bound up with democratic practice.

However, the genre enjoyed its heyday not only because it was the product of professionalism, but because that professionalism could be put to work by elites. It was put to work in the short term as a foil against public criticism of network indiscretions. But it was also put to work on behalf of a particular ideology. The network documentary genre was explicitly endorsed by a coalition of elites who would come to associate themselves with the foreign policy objectives of the New Frontier. Included among this circle of boosters were President John F. Kennedy; Newton Minow, chairman of the Federal Communications Commission; Edward R. Murrow, head of the U.S. Information Agency; Charles Percy, chairman of Bell & Howell and head of the 1960 Republican platform committee; and David Sarnoff, chairman of the Radio Corporation of America. Just as these leaders embraced the principles of internationalism and corporate liberalism, so too did they endorse the network documentary. Documentary was envisioned by these leaders as a vital link in their effort to build broad public support for key policies of the New Frontier. In order to understand why the documentary received such special attention, we must examine some of the larger social forces at work during this period.

### The Social Context of Scientific Anti-Communism

Changes in the global political economy throughout the decade of the fifties had a determinate effect on the network documentary. That is, even though the documentary was the product of human agency, professional codes, institutional imperatives, and new technological possibilities, it was also produced within a set of social relations that were delimited by a changing global economy. This is a fundamental reason why the object of documentary investigation and the strategies of documentary narration centered on the issue of East–West conflict.

In the first two decades after World War II, U.S. corporations and financial institutions stepped up their overseas activities at a dramatic rate. Manufacturers and advertisers turned their attention to growing markets in Europe and new markets in what was characterized as the developing world. Resource extraction from the latter also became important as sophisticated manufacturing processes grew increasingly dependent on a broad range of minerals that could only be found outside the United States. Furthermore, this growing web of corporate activity was stitched together by overseas banking operations that began to expand rapidly in the mid-1950s. Thus was constructed a Free World economy anchored by American technology, managerial practices, and investment dollars.[46]

The American media showed interest as well in overseas expansion during this period. Newswire services such as Associated Press and United Press International aggressively contested the dominance of European competitors in international markets. In addition, American broadcasting companies began to step up their overseas investments during the latter part of the 1950s. Network executives, who anticipated an increasingly saturated domestic market, began to look outward in search of further growth. They increased their involvement with the international marketing of television receivers and also expanded their investments in foreign stations. But most importantly, they projected a growing overseas syndication market as television audiences began to expand rapidly in countries around the world. Indeed, by 1962, revenues from overseas syndication equaled the revenues generated by domestic sales.[47] Thus, the networks, like other major American corporations during the late 1950s, were in the process of significantly enhancing the international scope of their operations. As a result, U.S. foreign policy objectives that were aimed at "containing" Communist aggression must be understood in relation to corporate ambitions to secure a sphere of operations for their growing and complex web of economic activity.[48]

However, such an explicit link between American economic prosperity and global hegemony was rarely drawn in public during the 1950s, and popular commitment to an interventionist foreign policy wavered throughout the decade. There was a strong isolationist sentiment among the public, as most Americans saw little reason for maintaining a military presence throughout the globe. Indeed, the major U.S. military intervention during this era was in Korea, where a muddled policy of containment crumbled when confronted by the prospect of full-scale war.[49] It was in this context that American voters selected a military hero for President in 1952 not because he would lead them to victory, but because he promised to lead them out of Korea. Such a lack of public resolve generated much discussion of a so-called Korean War syndrome. At the very core of this debate was a great deal of confusion about the nature of the Free World and the U.S. role in defending it against "the tide of Communist revolution."

Furthermore, the Eisenhower administration found itself severely hamstrung by a powerful isolationist element within the Republican party. Small business entrepreneurs and farmers had little to gain from an active and expensive international campaign to police the frontiers of the Third World. These businesspeople relied little on the resources from this area and were unlikely to market their products abroad. In fact, many of these small entrepreneurs felt that vast pools of cheap labor throughout the Third World were a threat to their operations. American manufacturers in such labor-intensive industries as shoemaking and apparel worried more about cheap imports than they did about Communist aggression in Asia. Thus, Eisenhower's foreign policy waffled through the fifties, constrained by public opinion and stiff opposition within the Republican Party.[50]

Meanwhile, among the leadership of capital-intensive firms in such industries as petroleum and electronics, there was a growing concern about stability in the developing markets of the Third World. Furthermore, many top executives at these firms understood that problems along the frontiers of the "Free World" not only raised questions of military strategy, but also questions of social reform. Thus, the competition between East and West was not only strategic; increasingly it took on economic and ideological dimensions as well.

By the late 1950s, studies published by the Ford Foundation and the Rockefeller Brothers Fund argued that American leadership of the Free World was in jeopardy. As the Soviet economy was racing ahead, the American economy was still sputtering along in the wake of the 1957 recession. Furthermore, Soviet preeminence in rocketry following the launch of Sputnik in 1957 had diminished American prestige and shifted the nuclear balance of power. The Soviets were suddenly capable of delivering nuclear warheads to North American cities within a matter of minutes. Images of American invulnerability were being challenged, and the elite think tank studies asserted that it was time to "get America moving again." This agenda was taken quite seriously by key leaders of transnational corporation, as many of them banded together behind Nelson Rockefeller's 1956 bid for the Republican Presidential nomination. Rockefeller's stinging rebuff by the isolationist faction within the Party encouraged many of these corporate leaders to redirect their campaign contributions to elements within the Democratic Party during the next Presidential campaign. John F. Kennedy happily appropriated the expansionist economic and foreign policies advocated by these major think tanks. Indeed, he even borrowed a leaf from Nelson Rockefeller's campaign rhetoric in his declaration of a "New Frontier."[51]

Kennedy's Presidential campaign rode the rising tide of crisis rhetoric that began with the 1957 Sputnik launch. This rhetoric was not only the product of corporate concern, but scientists, liberal politicians, and social critics—all of whom were on the "outs" during the Eisenhower era of homespun normalcy—also used Sputnik as a symbol of national decline. Thus, the Sputnik launch became a key moment of transition in American society not only because it elevated the level of American nuclear vulnerability, but because it elevated the visibility of social critics who yoked liberal politics to scientific technique in the global competition against monolithic Communism. Critics such as John Kenneth Galbraith and Arthur Schlesinger, Jr. enjoyed new celebrity and would soon join the inner councils of the Kennedy campaign. It was Schlesinger who argued that the future of the world rested at "the vital center" of the political spectrum and, in the wake of McCarthyism, his program for political change began to enjoy a renewed currency among corporate and political elites.[52] Inextricably linked to this centrism was a celebration of science and expertise as the fundamental material for political decision making. Wrote Schlesinger:

> American funds can buy out landlords; American methods of scientific farming and of land rehabilitation can increase production; American study of village sociology could help us to understand how we may most effectively release the energies so long pent up in the villages of Asia . . . .
>
> No people in the world approach the Americans in mastery of the new magic of science and technology. Our engineers can transform arid plains or poverty-stricken river valleys into wonderlands of vegetation and power. Our factories produce astonishing new machines, and the machines turn out a wondrous flow of tools and goods for every aspect of living. The Tennessee Valley Authority is a weapon which, if properly employed, might outbid all the social ruthlessness of the Communists for the support of the people of Asia.[53]

According to this perspective, scientific technique was a crucial tool in the struggle against monolithic Communism. Therefore, liberals like Schlesinger endowed it with a central role in their political program.

Furthermore, in the wake of the Sputnik launch, the image of the scientist, which had suffered ambivalent treatment in mass media throughout much of the 1950s, began to assume a new role.[54] By the end of the decade, liberal elites were actively promoting the notion that the only defense against Soviet missiles was a massive American counterstrike engineered by the scientific community. Therefore, it is not surprising that the scientific expert should emerge onto the high ground of public policy debates during the latter part of the decade. Nor is it surprising that the President who followed Eisenhower should be a proponent of "nonideological" pragmatism who would surround himself with a small inner circle of experts.

## The Articulation of Discourses

It was at such a conjuncture that the discourse of science was articulated with the discourse of anti-Communism. Documentary television was one of the important sites where this construction of meaning took place. These programs set out to explore events in places such as Panama by engaging a methodology that, like social scientific method, aspired to examine objectively the issues at hand. It was suggested that the facts would speak for themselves, and the documentary narrative would offer these facts a forum.

What made documentary so important as a site for the production of these meanings was its claim to provide the average citizen an unmediated experience with the distant realms of foreign policy.[55] The audience need not travel to Panama to see and experience that society directly. Documentary offered the illusion of being there, which enhanced both the subtlety and power of the stories that were being told. These programs presented themselves as "natural" representations of people and events. They spoke with factual authority and appeared to be documents that emerged unproblematically from the confusion of lived experience.

Many factors contributed to this nonideological characterization, but certainly one of the most important was the way in which the programs seemed to speak with a unified voice. This coherence grew from a conjuncture of practices—economic, political, institutional, ideological—that set the boundaries for documentary production. They range from the simple fact that documentaries were funded by television networks that were expanding their investments in international markets to more subtle factors such as the narrative conventions observed by television producers. It was on this terrain that the discourse of scientific anti-Communism operated.

Central to this discourse was the notion that the ultimate stakes in places like Panama were not simply economic or military; at stake was an entire conception of progress and politics. As a whole, network documentary promoted this conception by exploring "scientifically" the forces of darkness and irrationality. It pursued these forces at home and abroad. Thus, the battle against racial prejudice in Birmingham, Alabama, was not unlike the struggle against the passionate politics of Fidel in Latin America. "If people would only act rationally," the programs seemed to say, "we all could enjoy the fruits of prosperity." Documentary stood on the side of rationally organized societies and a rationally organized global economy. It stood on the side of a *Pax Americana*.

However, such an analysis should not be misread as an instrumentalist argument. This essay does not contend that a shift in economic and political forces unproblematically produced a shift in ideology or that network documentaries were simply tools for mobilizing public support behind the policies of an elite. I do not wish to suggest that news workers were simply ideological porters for the ruling class or that the process of documentary production was smooth and unconflicted.

Documentary, by its very nature, had to engage real people and events in order to tell its stories. Documentarists who mobilized the discourse of objectivity in the pursuit of their work were pressed to move among different cultures and subcultures in order to gather their narrative materials. Thus, Albert Wasserman's documentary about unrest in Panama includes a great deal of material not easily framed within the narrative of superpower struggle. Panamanian poverty is seriously examined in relation to the problem of political unrest, and the program works hard throughout the hour to reconcile a history of American exploitation with a policy of continued U.S. involvement. The narrative tries to tell the viewer that "we" can prevent a Castroite revolution in Panama, if "we" make our foreign policy more rational. But documented evidence of American abuses and the voices of the Panamanian opposition are also telling the viewer that something is wrong with a policy built on North American notions of a Free World. Thus, the television documentary during this period sought to construct a set of meanings around concepts of democratic process, scientific method, and American global leadership. Yet it could not contain within this dominant framework the excess of meanings produced by its own methodology.[56] As a result, both dominant and oppositional meanings can be found in these documentaries.

This tension reminds us of Stuart Hall's conception of a contest over the production of meaning. According to Hall, the moment of encoding results from a process of contestation and negotiation. Although dominant meanings do attempt to structure and contain subordinate ones, there is also an element of indeterminacy in the production of media messages, and this is further accentuated by indeterminacy at the moment of decoding.[57] Therefore, this essay is not arguing that documentary television led us down "the video road to Vietnam" in any instrumentalist sense.[58] Rather it is proposed that the complex and contradictory elements that these narratives brought together were assembled within a hierarchy of meanings that would most appropriately be described as the discourse of "scientific anti-Communism."

## Notes

I would like to thank the anonymous *Cinema Journal* referees and Melissa Owen Curtin for helpful comments on an earlier draft of this essay.

The following abbreviations appear in the notes below: BRTC, Billy Rose Theatre Collection, New York Public Library, and SHSW, Manuscript Collection, State Historical Society of Wisconsin, Madison.

1. For example, see panel discussion among sixties documentarists in Mary Ann Watson, "The Golden Age of the American Television Documentary," *Television Quarterly* 23, no. 2 (Summer 1988): 57–75; also, Bill Carter, "Whatever Happened to Documentaries?" *Washington Journalism Review,* June 1983, 43–46.

2. Interview with Albert Wasserman in Alan Rosenthal, ed., *The Documentary Conscience: A Casebook in Film Making* (Berkeley: University of California Press, 1980), 99.

3. Despite the fact that Edward R. Murrow was famous for his news analysis, he was fond of referring to the television news medium as a "mirror" of social reality. See Alexander Kendrick, *Prime Time: The Life of Edward R. Murrow* (Boston: Little, Brown and Company, 1969), 318. Friendly followed this lead in his characterization of the documentary; see, "CBS News Clinic, 1955," transcript, Sigfreid Mickelson papers, box 1, file 4, p. 64, SHSW.

4. Erik Barnouw, *Documentary: A History of the Non-Fiction Film* (New York: Oxford University Press, 1983), 226.

5. A. William Bluem, *Documentary in American Television: Form, Function, Method* (New York: Hastings House, 1975), 110.

6. Michael Schudson, *Discovering the News: A Social History of American News-*

*papers* (New York: Basic Books, 1978); Gaye Tuchman, "Objectivity as Strategic Ritual: An Examination of Newsmen's Notions of Objectivity," *American Journal of Sociology* 77, no. 4 (1972): 660–80; Herbert J. Gans, *Deciding What's News: A Study of CBS Evening News, NBC Nightly News, Newsweek, and Time* (New York: Vintage Books, 1980); Robert A. Hackett, "Decline of a Paradigm? Bias and Objectivity in News Media Studies," *Critical Studies in Mass Communication* 1, no. 3 September 1984): 229–59.

7. Bill Nichols, *Ideology and the Image: Social Representation in the Cinema and Other Media* (Bloomington: Indiana University Press, 1981), 170–284; E. Ann Kaplan, "The Documentary Form," *Jump Cut* 15 (1977): 11–12; Colin McArthur, *Television and History* (London: British Film Institute, 1980); Brian Winston, "Documentary: I Think We Are in Trouble," in *New Challenges for Documentary,* ed. Alan Rosenthal (Berkeley: University of California Press, 1988), 21–33.

8. Jeanne Hall, "Realism as a Style in Cinema Verite: A Critical Analysis of *Primary,*" *Cinema Journal* 30, no. 4 (Summer 1991): 24–50.

9. In the autumn of 1963, a public opinion poll by the Roper organization found, for the first time, that television had edged out print as the number one news source. *Broadcasting,* 27 January 1964, 70. Around the same time CBS premiered its expanded newscast of thirty minutes. *Variety,* 30 January 1963, 1.

10. *Broadcasting,* 12 September 1960, 27, and 5 March 1962, 52–53; *Business Week,* 9 June 1962, 50; *Printer's Ink,* 23 December 1960, 10; and *Sponsor,* 26 March 1962, 29–32.

11. Stuart Hall refers to this as the moment of encoding. Following Hall, my analysis pays attention to the multiple determinations at work in such institutional forms of expression as the network documentary. See Stuart Hall, "Encoding/Decoding," in Stuart Hall, Dorothy Hobson, Andrew Lowe, and Paul Willis, eds., *Culture, Media, Language* (London: Hutchinson, 1980). A related approach has been used by film scholars such as Robert C. Allen and Douglas Gomery, *Film History: Theory and Practice* (New York: Alfred A. Knopf, 1985). They argue for a Realist historiography that emphasizes an integrated analysis of the generative mechanisms that shape the system of cinema at any given historical moment. Allen also applies this methodology to broadcasting in *Speaking of Soap Operas* (Chapel Hill: University of North Carolina Press, 1985).

12. Raymond L. Carroll, "Economic Influences on Commercial Network Television Documentary Scheduling," *Journal of Broadcasting* 23, no. 4 (Fall 1979): 411–25.

13. Tim Brooks and Earle Marsh, *The Complete Directory to Prime Time Network TV Shows: 1946–Present* (New York: Ballantine, 1981), 868.

14. See Carroll, "Economic Influences," 415, and Christopher H. Sterling and John M. Kittross, *Stay Tuned: A Concise History of American Broadcasting,* 2d ed. (Belmont: Wadsworth, 1990), 502.

15. Carroll, "Economic Influences."

16. Erik Barnouw, *The Tube of Plenty: The Evolution of American Television* (New York: Oxford University Press, 1975), 247; others who take this position include Fred Friendly, *Due to Circumstances beyond Our Control . . .* (New York: Vintage, 1967), 99–113; and Thomas Whiteside, *The New Yorker,* 17 February 1962, 41.

17. James L. Baughman, " 'The Strange Birth of CBS Reports' Revisited," *Historical Journal of Film, Radio and Television* 2, no. 1 (1982): 28–38.

18. David G. Yellin, *Special: Fred Freed and the Television Documentary* (New

York: Macmillan, 1973); Friendly, *Due to Circumstances;* and Watson, "The Golden Age."

19. See discussions of technological developments and the television documentary in Robert C. Allen and Douglas Gomery, *Film History: Theory and Practice* (New York: Alfred A. Knopf, 1985), 220–23; Stephen Mamber, *Cinema Verite in America: Studies in Uncontrolled Documentary* (Cambridge, Mass.: MIT Press, 1974), 27–28 and passim; and Watson, "The Golden Age," 63, 65.

20. A. William Bluem has surveyed content as part of his larger project of tracing the development of the television documentary form; however, he does not address the politics of meaning in any substantive manner. See, Bluem, *Documentary.*

21. *Broadcasting,* 22 January 1962, 27–30.

22. "Ninety Miles to Communism," broadcast 18 April 1961; "The Red and the Black," 22 January 1961; "Troubled Land," 13 June 1961.

23. "The Flabby American," broadcast 30 May 1961.

24. "Cast the First Stone," broadcast 27 September 1960.

25. Between the premiere broadcast of *CBS Reports* on 27 October 1959 and the end of June 1964, the network produced ninety-six original programs in this series according to Daniel Einstein, *Special Edition: A Guide to Network Television Documentary Series and the Special News Reports, 1955–1979* (Metuchen, N.J.: Scarecrow Press, 1987). Of those, I would argue that thirty-six address themselves directly to superpower struggle as a major thematic text and sixteen others, such as "Eisenhower on the Presidency" and "Why Man in Space?" draw on this conflict as an important subtext.

26. Broadcast 11 November 1959.

27. See Michael Schudson, "The Politics of Narrative Form: The Emergence of News Conventions in Print and Television," *Daedalus* 111 (1982): 97–112.

28. Regarding the didactic style of television news reports see Justin Lewis, "Decoding Television News," in *Television in Transition: Papers from the First International Television Studies Conference,* ed. Phillip Drummond and Richard Patterson (London: British Film Institute, 1986), 205–34. As for the conventions of Hollywood style, see David Bordwell, Janet Staiger, and Kristin Thompson, *The Classical Hollywood Cinema: Film Style and Mode of Production to 1960* (New York: Columbia University Press, 1985). Of course, narrative conflict is not exclusive to Hollywood cinema. One finds it in other fictional forms as well. One also finds it in other cinema forms. Indeed, like Soviet narrative cinema, the network documentaries of the early 1960s often feature contending social forces rather than individual characters. Nevertheless, as I note below, Hollywood style was cited by the documentary producers themselves as having a more direct influence on their careers and their work.

29. Transcript of "CBS News Clinic, 1955," Sigfreid Mickelson papers, box 1, file 4, p. 91, SHSW; Gitlin makes a similar point in an interview in the *New York Times,* 28 January 1962, 15.

30. Yellin, *Special,* 81.

31. It also was noted that audiences preferred fewer and more developed scenes. See "What We Have Learned about Documentary and Educational Programs," CBS-TV Research Department, Program Analysis Division, Sept. 1957, Mickelson papers, box 1, file 27, p. 4, SHSW.

32. See "Draft, Budget," 1958 and confidential notes from budget meeting, 30 Jan. 1958, Mickelson papers, box 1, file 2, SHSW; also, *Report to the Stockholders of the Columbia Broadcasting System, 1959,* p. 8, BRTC.

33. Clipping, *New York Times,* 20 May 1960, Gitlin file, BRTC.

34. Letter from Peter G. Peterson to "Ollie," 1 November 1960, John Daly Papers, box 32, file 17, SHSW.

35. Clipping, *New York Herald Tribune,* 23 September 1960, Daly papers, box 32, file 3, SHSW.

36. Script from closed-circuit promotional screening of "Cast the First Stone," undated, Daly papers, box 31, file 17, SHSW.

37. *Variety,* 14 March 1962, clipping, Reuven Frank file, BRTC; see also, memo from Bob Saman (producer of *This Is NBC News*) to Correspondents, 13 September 1962, Saman papers, box 10, file 1, NBC collection, SHSW.

38. In 1963, Secondari received an Emmy award and was named Television Writer of the Year for producing documentary dramatizations of Columbus's voyage of discovery and the drafting of the Declaration of Independence. See the following clippings in John Secondari file, BRTC: *Variety,* 12 February 1975, *Newark Evening News,* 7 May 1961, *Current Biography,* April 1967.

39. *Variety,* 18 May 1960, 31. Bob Lang, one of the producers of CBS's *The Twentieth Century* noted that a similar formula had made that program a ratings success. Said Lang, "We built this show for 6:30 P.M. on Sundays, when kids control the sets. So we built in 'cops and robbers' (we called them Nazis and Poles and British and Americans) and the ratings were there. Here we have an outstanding example of thoughtful material, but . . . a review of the subjects treated might suggest certain of the controversial aspects of the subjects remained unexplored." *Sponsor,* 25 March 1963, 75.

40. *TV Guide,* 6 February 1960, 14.

41. Both Gitlin and Wasserman spent their early years in television documentary working for CBS, Wasserman as a producer and Gitlin as head of Special Projects. Gitlin left the network when he was passed over for the job as executive producer of *CBS Reports.* He was quickly picked up by NBC and put in charge of the *White Paper* series. One of his first actions in this capacity was to hire Wasserman and several other producers away from CBS.

42. "CBS Television Biography," CBS press release, 26 November 1958, Albert Wasserman file, BRTC. At the time this press release was written, Wasserman was working at CBS. In 1959 he moved to NBC. Also, I should point out that my discussion of objectivity should not be received as a sweeping or reductive argument. Both journalists and social scientists are well aware that the notion of complete objectivity is unattainable within their fields of endeavor. As Michael Schudson has argued, many journalists simply see objectivity as an ideal. However, during the early 1960s both journalists and social scientists were striving to perfect their methods so as to produce work that most closely approximated this ideal. Furthermore, as Gaye Tuchman has pointed out, when journalists are pressed to defend their work to superiors or outsiders, they tend to draw on the tropes of objectivity in order to defend their analysis. It is also common for outsiders to borrow selectively from the findings of journalists and social scientists as if their work provided objective data. Thus, politicians often borrow from newspaper accounts in citing "facts" to support their conclusions. Similarly, networks borrowed from the survey methods of social scientists in order to produce an objective audience that could be sold to advertisers. Therefore, the concept of objectivity circulated widely in American society during this period even though many expressed their doubts about the very possibility of objective analysis. For my purposes, I simply wish to argue that scientific methodology and objectivity played a powerful role in constituting claims to authenticity in

the network documentary. Discussions of the relationship between objectivity, science, and journalism can be found in Michael Schudson, *Discovering;* Gans, *Deciding;* and Dan Schiller, *Objectivity and the News: The Public and the Rise of Commercial Journalism* (Philadelphia: University of Pennsylvania Press, 1981).

43. *Variety,* 23 January 1962; Fred Freed file, BRTC.

44. It is interesting to compare this notion of documentary to the more unabashed advocacy style of Edward R. Murrow. The Murrow style emerges out of the World War II era when journalists, seeking to mobilize public support for the war effort, displayed little self-consciousness about offering opinion or commentary within the context of a news report. By the middle of the 1950s, CBS management began to show increasing concern about this advocacy style, particularly in relation to Murrow's work. See "Objectivity Study" and confidential memo, Sig Mickelson to Frank Stanton, 6 December 1955, Mickelson papers, box 1, file 11, SHSW.

45. Bluem, *Documentary in American Television,* 90–91.

46. Harry Magdoff, *The Age of Imperialism: The Economics of U.S. Foreign Policy* (New York: Modern Reader, 1966); Pierre Jalee, *Imperialism in the Seventies* (New York: Third Press, 1972); and Thomas J. McCormick, *America's Half Century: United States Foreign Policy in the Cold War* (Baltimore: Johns Hopkins University Press, 1989).

47. Wilson Dizard, "American Television's Foreign Markets," *Television Quarterly* (Summer 1964): 58; *Variety,* 21 November 1962, 23.

48. The alleged aggressor status of the Soviet Union has been thoroughly criticized by the work of such historians as William A. Williams, *The Tragedy of American Diplomacy* (New York: Dell, 1962); D. F. Fleming, *The Cold War and Its Origins* (Garden City: Doubleday, 1961); and Walter Le Feber, *America, Russia, and the Cold War, 1945–1966* (New York: Wiley, 1967).

49. In addition to the works cited above, my synopsis of this period draws upon such texts as Frederick F. Siegel, *Troubled Journey: From Pearl Harbor to Ronald Reagan* (New York: Hill and Wang, 1984); Eric Goldman, *The Crucial Decade and After: 1945–1960* (New York: Vintage, 1960); Stephen Ambrose, *Rise to Globalism: American Foreign Policy since 1938* (New York: Penguin, 1983); Godfrey Hodgson, *America in Our Time: From World War Two to Nixon, What Happened and Why* (New York: Vintage, 1978); Richard J. Barnet, *The Alliance: America, Europe, Japan, Makers of the Postwar World* (New York: Simon and Schuster, 1983); and McCormick, *America's Half-Century.*

50. Thomas Ferguson and Joel Rogers, *Right Turn: The Decline of the Democrats and the Future of American Politics* (New York: Hill and Wang, 1986), 46–57; Ambrose, *Rise to Globalism,* 186–244.

51. Ferguson and Rogers, *Right Turn,* 46–57; Siegel, *Troubled Journey,* 121.

52. Arthur M. Schlesinger, Jr., *The Vital Center: The Politics of Freedom* (1949; reprint, New York: Da Capo, 1988).

53. Schlesinger, *Vital Center,* 233.

54. Regarding the ambivalent treatment of science in popular film during the 1950s, see Peter Biskind, "Pods, Blobs, and Ideology in American Films of the Fifties," in *Shadows of the Magic Lamp: Fantasy and Science Fiction in Film,* ed. George Slusser and Eric S. Rabkin (Carbondale: Southern Illinois University Press, 1985), 58–72, and Michael Rogin, *Ronald Reagan, The Movie: And Other Episodes in Political Demonology* (Berkeley: University of California Press, 1988), 236–71.

55. Such pretensions to realism assume even greater significance in relation to agenda-setting research that indicates that media messages are most influential in

areas where the public has the least experience. J. P. Winter's examination of nearly twenty-five years of news coverage and public opinion trends found the media to be much more influential with issues such as foreign policy than issues such as inflation. J. P. Winter, "Differential Media Public Agenda-Setting Effects for Selected Issues, 1948–1976 " (Ph.D. dissertation, Syracuse University, Syracuse, N.Y., 1980).

56. Regarding the excess of meanings often produced by news texts, see John Fiske, *Television Culture* (New York: Methuen, 1987), 281–308. In this context, it is also interesting to note that Wasserman's Panama documentary was the subject of some controversy. Conservative elements in the U.S. Congress, led by Daniel Flood, a Republican from Pennsylvania, roundly denounced the program for suggesting that the U.S. was partly to blame for the unrest in Panama.

57. Stuart Hall, "Encoding/Decoding," 128–38, and "Notes on Deconstructing the Popular," in R. Samuel, ed., *People's History and Socialist Theory* (London: Routledge Kegan Paul, 1981), 227–40.

58. This is basically the position taken by J. Fred MacDonald, *Television and the Red Menace: The Video Road to Vietnam* (New York: Praeger, 1985).

# Disneyland

## CHRISTOPHER ANDERSON

The month of October 1954 marked a watershed for television production in Hollywood. Alongside those marginal movie industry figures who had labored to wring profits from television production during the late 1940s and early 1950s, there appeared a new breed of established producers attracted by television's explosive growth following the end of the FCC's station application freeze in 1952.[1] Early in the month, Columbia Pictures became the first major studio to produce episodic TV series when its TV subsidiary, Screen Gems, debuted *Father Knows Best* on CBS and *The Adventures of Rin Tin Tin* on ABC. Within three days in late October, two of the film industry's top independent producers, David O. Selznick and Walt Disney, joined the migration to prime time. Selznick, producer of *Gone With the Wind* (1939), made Hollywood's most auspicious debut with a program broadcast simultaneously on all four existing networks, a two-hour spectacular titled *Light's Diamond Jubilee*. Selznick was soon joined by fellow independent producer Walt Disney, whose premiere television series, *Disneyland*, entered ABC's regular Wednesday-night schedule on the twenty-seventh of October. Disney had forged a reputation as the cinema's maestro of family entertainment; now his *Disneyland* series promised to deliver what *Time* described as "the true touch of enchantment" to American homes.[2] Unlike their predecessors in television, these were established members of the movie industry who diversified into TV production without leaving movies behind. The first to link production for the two media, these producers sparked the full-scale integration of movie and TV production in Hollywood during the second half of the 1950s.

As the recipient of nearly two dozen Academy Awards for his studio's cartoon animation, Walt Disney was one of Hollywood's most acclaimed independent producers and certainly, along with Selznick, the most celebrated Hollywood producer to enter television by 1954. Disney possessed the independent producer's belief in television as an alternative to the movie industry's restrictive studio system, but his conception of television's role in a new Hollywood was more sweeping than that of colleagues who saw the electronic medium as nothing more than a new market for traditional film production. Unlike virtually every other telefilm producer in Hollywood, Disney harbored no illusions about dominating TV production; his modest production plans initially encompassed only the *Disneyland* series. Still, Disney was the first Hollywood executive during the 1950s to envision a future built on television's technical achievements—the scope of its signal, the access it provided to the American home. For Disney network television arrived as an invitation to reinvent the movie business, to explore horizons beyond the realm of filmmaking.

Disney later admitted that he was "never much interested" in radio, but television, with its ability to display the visual appeal of Disney products, was another matter entirely. The studio aired its first television program on NBC during December 1950. Sponsored by Coca-Cola, "One Hour in Wonderland" was set in a Disney Christmas party and featured excerpts promoting the studio's upcoming theatrical release, *Alice in Wonderland* (1951). In 1951 Disney produced its second hour-long program for NBC, a special sponsored by Johnson and Johnson. Disney's subsequent plans for a television series started with a seemingly outlandish demand: To obtain the first Disney TV series, a network would have to purchase the series and agree to invest at least $500,000 for a one-third share in the studio's most ambitious project, the Disneyland amusement park planned for construction in suburban Los Angeles. NBC and CBS balked at these terms, but ABC, mired in third place, decided to accept.[3] In uniting the TV program and the amusement park under a single name, Disney made one of the most influential commercial decisions in post-war American culture. Expanding upon the lucrative character merchandising market that the studio had joined in the early 1930s, Disney now planned to create an all-encompassing consumer environment that he described as "total merchandising." Products aimed at baby boom families and stamped with the Disney imprint—movies, amusement park rides, books, comic books, clothing, toys, TV programs, and more—would weave a vast, commercial web, a tangle of advertising and entertainment in which each Disney product—from the movie *Snow White* to a ride on Disneyland's Matterhorn—promoted all Disney products. And television was the beacon that would draw the American public to the domain of Disney. "We wanted to start off running," Walt later recalled. "The investment was going to be too big to wait for a slow buildup. We needed terrific initial impact and television seemed the answer."[4]

Television served a crucial role in Disney's plans for creating an economic and cultural phenomenon that exceeded the boundaries of any single communications medium. By raising capital through the ABC investment and raising consciousness through its depiction of the park's construction, television's figurative representation of Disneyland actually called the amusement park into existence, making it possible for the first time to unite the disparate realms of the Disney empire. With the home as its primary site of exhibition, television gave Disney unparalleled access to a family audience that he already had cultivated more effectively than any Hollywood producer in the studio era. As a result of the post-war baby boom, Disney's target audience of children between the ages of five and fourteen grew from 22 million in 1940 to 35 million in 1960.[5] Television provided the surest route to this lucrative market.

As a text, the *Disneyland* television program also marked a rite of passage for the Disney studio. Its broadcast signaled the studio's transition from the pre-war culture of motion pictures to a post-war culture in which Disney's movies were subsumed into an increasingly integrated leisure market that also included television, recorded music, theme parks, tourism, and consumer merchandise. By depicting the new amusement park as another of Walt's fantasies brought to life by the skilled craftsmen at the Disney studio, the *Disneyland* TV program gave a recognizable symbolic form to Disney's elaborate economic transformation, mediating it for the American public by defining it as another of the Disney studio's marvels. It is only a slight exaggeration, therefore, to claim that Disney mounted an entertainment empire on the cornerstone of this first television series.

Unlike many who groped for a response to the dramatic changes that swept the movie industry following World War II, Walt Disney and his brother Roy answered uncertainty with a calculated plan for diversification. Biographer Richard Schickel has suggested that the Disneys addressed the unstable post-war conditions more aggressively than other Hollywood leaders because their company had suffered misfortunes during the early 1940s, when virtually everyone else in Hollywood had prospered. During the late 1930s, Disney had stood for a moment at the pinnacle of the movie industry. Although an independent producer who worked outside the security of the major studios, Disney took extraordinary financial risks that ultimately paid off in the critical and financial success of *Snow White* (1937), which trailed only Selznick's *Gone With The Wind* as one of the two most profitable Hollywood movies of the 1930s.[6] But Disney's good fortunes lasted only briefly. Following *Snow White,* Disney nearly buried his studio beneath ambitious plans for expansion. With box-office disappointments like the costly animated feature *Fantasia* (1940), the closing of foreign markets due to the war, and over-investment in new studio facilities, Disney faced burdensome corporate debts that weighed even more heavily once banks shut off credit to the studio in 1940. Disney raised funds reluctantly by offering stock to the public, but only government contracts to produce educational cartoons kept the studio active during the war. "The only good thing about the situation," according to Schickel, "was that the problems that were later to plague the rest of the industry had been met by Disney at a time when the government could help out and when the general buoyancy of the industry could at least keep him afloat. The result, of course, was a head start in gathering know-how to meet the crisis that was coming—a head start in planning for diversification first of the company's motion picture products, then of its overall activities."[7]

Plagued by adversity during the 1940s, Walt and Roy Disney entered the 1950s with a plan to transform the Disney studio from an independent producer of feature films and cartoon short-subjects into a diversified leisure and entertainment corporation. Instead of retrenching, as others had, the Disneys fortified their company through a careful process of diversification. Beginning in 1953, the company implemented a series of changes designed to redefine its role in Hollywood. Disney established its own theatrical distribution subsidiary, Buena Vista, in order to end its reliance on major studio distribution. The studio also ceased production on its by-then unprofitable cartoon short subjects, cut back on expensive animated features, and began to concentrate on nature documentaries and live-action movies following the success of *Treasure Island* (1950) and *Robin Hood* (1952).[8] Blueprints for Disneyland and ideas about television production took shape during this period of corporate transition.

As Disney's schemes for expansion pointed toward television, the ABC-TV network eagerly cultivated ties with the motion picture industry.[9] Hollywood-produced television series became the cornerstone in ABC's plans for differentiating itself from NBC and CBS. As the third-place network, ABC elected to build its audience in direct opposition to the established networks. While the other networks touted their established stars or experimented with expensive spectaculars and the possibility of attracting viewers with unique video events, ABC remained committed to the traditional strategy of programming familiar weekly series that defined television viewing as a consistent feature in the family's domestic routine. Robert Weitman, the network's vice-president in charge of programming, emphasized the importance of habitual viewing in ABC's programming strategies. "The answer seems to be in established patterns of viewing," he explained. "People are annoyed when their favorite show is pre-empted, even for a super-special spectacular." Leonard Goldenson, who had spent decades in the business of movie distribution and exhibition, recognized the similarities between television viewing and the experience of moviegoing during the studio era. "The real strength and vitality of television," he claimed, "is in your regular week-in and week-out programs. The strength of motion pictures was always the habit of going to motion pictures on a regular basis, and that habit was, in part, taken away from motion pictures by television."[10] ABC's programming strategy was built on the belief that television's fundamental appeal was less its ability to deliver exotic events, than its promise of a familiar cultural experience.

As a result, ABC's regularly scheduled series would serve as the basis for network counter-programming, the principal tactic in the network's assault on CBS and NBC. Rather than compete against an established series or live event with a program of similar appeal, ABC hoped to offer alternative programming in order to attract segments of the audience not being served by the other networks. The network would construct and project a specific identity by treating its schedule as the expression of a unique relation to the broadcast audience. "Whatever the audience is not watching at any given time makes for new possibilities," Goldenson noted. "We are not trying to take away audiences from CBS and NBC. . . . We are trying to carve our own network character, to create new audiences."[11] This tactic was based on a related aspect of ABC's programming philosophy—its attention to audience demographics. Governed by the belief that "a network can't be all things to all people," ABC chose to target "the youthful families" with children, a section of the audience whose numbers had increased rapidly since World War II. "We're after a specific audience," claimed Goldenson, "the young housewife—one cut above the teenager—with two to four kids, who has to buy the clothing, the food, the soaps, the home remedies." As this statement implies, ABC chose to align itself with small-ticket advertisers, those selling the type of products that young families might be more likely to need and afford. Goldenson justified ABC's entire programming strategy when he remarked, "We're in the Woolworth's business, not in Tiffany's. Last year Tiffany made only $30,000."[12]

Anxious to acquire Hollywood programming that appealed to a family audience, ABC gambled on Disney by committing $2 million for a fifty-two-week series (with a seven-year renewal option) and by purchasing a 35 percent share in the park for $500,000. Without even a prospective format to present to advertisers, ABC invoked the Disney reputation alone to sell the program under a joint-sponsorship package to American Motors, the American Dairy Association, and Derby Food. Sponsorship of the season's twenty original episodes was sold at $65,000 per episode, and the network time was billed to advertisers at $70,000 per hour. During the late 1950s, when ABC's ratings and advertising revenue finally approached the levels of NBC and CBS, Leonard Goldenson consistently referred to the Disney deal as the network's "turning point." Indeed, *Disneyland* attracted nearly half of ABC's advertising billings during 1954, the final year during which the network operated at a loss.[13]

Although Walt Disney repeatedly assured the press that the *Disneyland* TV series would stand on its own terms as entertainment, the program served mainly to publicize Disney products. *Disneyland*'s identification of the amusement park and the TV series was confirmed during the first episode when Walt informed viewers that "Disneyland the place and Disneyland the show are all the same." Both the series and the park were divided into four familiar movie industry genres: Fantasyland (animated cartoons), Adventureland (exotic action-adventure), Frontierland (Westerns), and Tomorrowland (science-fiction). Introduced by Walt himself, each week's episode represented one of the park's imaginary lands through a compilation of sequences drawn from the studio's cartoon short-subjects, nature documentaries, animated and live-action features, or short films produced as outright promotions for Disney movies about to enter theatrical release. *Disneyland*'s format and pervasive self-promotion were unprecedented for television, but it had roots in the popular radio programs broadcast from Hollywood during the 1930s and 1940s. Hosted by actors, directors, or celebrity journalists, programs like *Hollywood Hotel* and *Lux Radio Theatre* offered musical performances or dramatizations of studio feature films, but their strongest lure was the glimpse they provided into the culture of Hollywood. Through informal chats with performers and other members of the industry, these radio programs perpetuated an image of Hollywood glamour while promoting recent studio releases.[14] Disney simply adapted this format for television. As the master of ceremonies, he turned himself into a media celebrity, much as director Cecil B. DeMille earlier had ridden *Lux Radio Theatre* to national fame.[15]

The actual production of *Disneyland* required a minimal financial investment by the Disney studio. At a time when the typical network series featured thirty-nine new episodes each season, Disney's contract with ABC called for only twenty original episodes, with each of them repeated once, and twelve broadcast a third time during the summer. Instead of producing twenty episodes of new television programming each season, Disney viewed the deal as an opportunity to capitalize on the studio's library of films dating back to the debut of Mickey Mouse in the late 1920s. The wisdom of this format, as Richard Schickel has noted, "was that it allowed the studio to participate in TV without surrendering control of its precious film library."[16] Long after many of the major studios had sold the TV rights to their films, the Disneys boasted that they still owned every film they ever made. Although it is not generally remembered, during the first three years of *Disneyland,* the studio produced only one narrative film made expressly for the series—the three-part "Davy Crockett" serial that took the nation by storm during that first season.[17] More typically, the *Disneyland* TV series introduced a new generation of children to the studio's storehouse of cartoons.

Even with a program that consisted largely of recycled material, the studio admitted that it would not turn a profit from its first year in television. There were production costs in preparing the theatrical product for broadcast (editing compilation episodes or filming Walt's introductory appearances) and in producing its limited amount of original programming. But these costs generally were defrayed throughout the studio's various operations. The three hour-long episodes of the "Davy Crockett" series, for instance, cost $600,000—more than three times the industry standard for telefilm production—and yet, during that year alone, the cost was spread over two separate network broadcasts and a theatrical release. By employing up to 80 percent of the studio's production staff, the television operation also enabled the Disney studio to meet the expense of remaining at full productivity. In addition, all costs not covered by the network's payments were charged to the studio's promotion budget—another indication of the program's primary purpose.[18]

Nearly one-third of each *Disneyland* episode was devoted directly to studio promotion, but the entire series blurred any distinction between publicity and entertainment. Indeed, *Disneyland* capitalized on the unspoken recognition that commercial broadcasting had made it virtually impossible to distinguish between entertainment and advertising. One episode, "Operation Undersea," provided a behind-the-scenes glimpse at the making of *20,000 Leagues Under the Sea* (1954) just one week before Disney released the film to theaters.[19] This episode was later followed by, "Monsters of the Deep," a nature documentary that provided another opportunity to plug the studio's most recent theatrical release. An episode titled, "A Story of Dogs," preceded the release of *Lady and the Tramp* (1955), Disney's second major feature distributed to theaters during the initial TV season.

Viewers didn't mind that *Disneyland* was simply a new form of Hollywood ballyhoo, because Disney framed the program within an educational discourse, reassuring viewers that they inhabited a position of privileged knowledge that was available only through television. Amidst paternalistic fears over the pernicious influence of television, comic books, and other forms of mass culture, Disney's middle-brow didacticism was disarming. In each episode, *Disneyland* rewarded its viewers with an encyclopedic array of general information borrowed loosely from the fields of history, science, and anthropology, while also sharing more specialized knowledge about the history of the Disney studio and its filmmaking procedures. Through this specialized knowledge about the Disney studio, the *Disneyland* TV series defined a particular relationship between television and movies, one in which television served an inchoate critical function by providing commentary on Disney movies. Though produced by the studio itself, *Disneyland* nevertheless contained elements of a critical discourse on the cinema. It educated viewers to perceive continuities among Disney films, to analyze certain aspects of the production process, and to recognize the studio's body of work as a unified product of Walt's authorial vision.

*Disneyland*'s most obvious strategy for educating viewers was its use of behind-the-scenes footage from the Disney studio. The first episode introduced the Disney studio through images of Kirk Douglas playing with his sons on the studio lot, James Mason fighting a man-made hurricane on the stage of *20,000 Leagues Under the Sea*, animators sketching models, and musicians recording the score for a cartoon. By representing the studio as an active, self-contained creative community bustling with activity, these scenes evoke impressions of studio-era Hollywood while masking the fact that historical conditions had rendered those very images obsolete. Disney also used behind-the-scenes footage to demonstrate the elaborate process of filmmaking, particularly the intricacies of animation. Although one might think that a filmmaker like Disney would be afraid of ruining the mystery of animation by revealing how its effects are achieved, Richard Schickel has observed that "Disney always enjoyed showing people around his studio and explaining to them exactly how the exotic process of creating an animated film proceeded." In fact, Disney originally planned for the amusement park to be located at the studio, with demonstrations of the filmmaking process as one of its major attractions.[20] In one feature film, *The Reluctant Dragon* (1941), Disney displayed the animation process by allowing Robert Benchley to lead moviegoers on a tour of the Disney studio. But this was a one-shot experiment that couldn't be repeated in other movies without becoming a distracting gimmick. Following in the tradition of the earlier Hollywood radio shows, therefore, Disney defined television as a companion medium to the cinema, an informational medium that could be used to reveal the process of filmmaking—since that impulse could not be indulged in the movies themselves. While Disney movies were presented as seamless narratives, television gave Disney the license to expose their seams.

Disney's willingness to display the process of filmmaking suggests that reflexivity in itself is not a radical impulse. More a disciple of Barnum than Brecht, Disney had no intention of distancing his audience from the illusion in his movies. Instead, he appealed to the audience's fascination with cinematic trickery. Disney exhibited what historian Neil Harris describes as an "American vernacular tradition" perhaps best exemplified by P. T. Barnum. Barnum's showmanship depended on his recognition that the public delights both in being fooled by a hoax and in discovering the mechanisms that make the hoax successful. Through his fanciful exhibitions, Barnum encouraged "an aesthetic of the operational, a delight in observing process and examining for literal truth."[21] Far from being hoodwinked by Barnum's artifice, the audiences that witnessed his exhibitions took pleasure in uncovering the process by which these hoaxes were perpetrated.

Inheriting Barnum's sense of showmanship, Disney developed his own "operational aesthetic" through television, enhancing his audience's pleasure—and anticipation—by offering precious glimpses of the filmmaking process. Of course, Disney's depiction of the production process was selective; it ignored the economics of filmmaking in favor of focusing on the studio's technical accomplishments. *Disneyland* never explored such issues as labor relations at the Disney studio or the economics of merchandising that sent the largest share of profits into Walt's pockets. Instead, in what has become a cliche of "behind-the-scenes" reporting on filmmaking, *Disneyland* treated each movie as a problem to be solved by the ingenuity of Disney craftsmen. This created a secondary narrative that accompanied the movie into theaters, a story of craftsmen overcoming obstacles to produce a masterful illusion. With this strategy, viewers were given an incentive to see the completed movie, because the movie itself provided the resolution to the story of the filmmaking process as depicted on *Disneyland*.

The program also educated viewers through its attention to Disney studio history. The determination to recycle the Disney library shaped the series during its early seasons, making *Disneyland* an electronic museum devoted to the studio's artistic achievements. Before the arrival of television, Hollywood's history was virtually inaccessible to the general public, available only sporadically through the unpredictable re-release of studio features and short subjects. Movies themselves may have been preserved in studio vaults, but for moviegoers accustomed to an ever-changing program at local theaters, the Hollywood cinema during the studio era was much like live television—an ephemeral cultural experience in which each text inevitably dissolved into memory, swept away in the endless flow of serial production. Although much of television in the mid-1950s traded on the immediacy of live broadcast, the sale of motion pictures to broadcasters meant that television also became the unofficial archive of the American cinema, in which Hollywood's past surfaced in bits and pieces, like fragments of a dream. One of the pleasures of *Disneyland* was the chance it offered to halt the flow of mass culture by remembering relics from the Disney vaults.

Although *Disneyland* may have struck a nostalgic chord for older viewers, the program's presentation of studio history was less sentimental than reverential. Cartoons nearly forgotten were resurrected with a solemnity normally reserved for the most venerable works of art. This attitude is apparent from the first episode when Walt announces that the end of each episode will be reserved for Mickey Mouse. After leading the viewer through an elaborate description of the proposed amusement park and other studio activities, Disney stands behind a lectern and turns the pages of a massive bound volume, an illustrated chronicle of Mickey's adventures. In spite of the flurry of changes at the studio, he explains, one should not lose sight of an eternal truth: "It all started with Mickey. . . . The story of Mickey is the story of Disneyland." As he continues, the scene segues into Mickey's first appearance in the cartoon "Plane Crazy," and then dissolves to one of his most famous appearances, as the Sorcerer's Apprentice in *Fantasia*. The tone of the scene—Disney's scholarly disposition, the sight of Mickey's history contained in a stately book—implies that the Disney studio's products are not the disposable commodities of pop culture, but artifacts worthy of remembrance. Walt's role as narrator is to reactivate forgotten cartoons in the public's cultural memory by demonstrating their canonical status within the artistic history of the Disney studio. As an electronic museum, *Disneyland* invoked the cultural memory of its audience mainly to publicize new Disney products. In spite of its commercial motives, however, the series also made it possible to conceive of Hollywood as having a history worthy of consideration.

"Monsters of the Deep," a typical episode from the first season of *Disneyland,* demonstrates the strategies for situating new Disney movies in the context of the studio's history and production practices. The episode introduces Walt in his studio's research department. Wearing a dark tweed jacket and surrounded by books and charts, he appears professorial. Inspired by knowledge, yet free from scholastic pretension, he is television's image of an intellectual, kindly and inviting. Speaking directly to the camera, he leads the viewer through a discussion of dinosaurs, using illustrations from enormous books to punctuate his presentation. This lecture seems motivated only by Disney's inquisitive character until the Disney sales pitch gradually seeps in. "We told the story of dinosaurs large and small in *Fantasia*," Disney reminds viewers as the screen dissolves to images from the animated feature. As Disney explains the habitat, feeding patterns, and behavior of dinosaurs, the footage from *Fantasia* becomes recontextualized, as though it were a segment from a nature documentary, a reminder that even Disney's most fantastic films have educational value. Disney segues into a report on sea monsters, asking whether giant squids have existed among the mysteries of the ocean, tracing the enigma through debates over the veracity of historical accounts. This query provides a transition to a discussion of the problems involved in creating a plausible giant squid for the Disney feature, *20,000 Leagues Under the Sea*. From the research department, the scene dissolves to a studio soundstage where star Kirk Douglas performs a song from the movie and then guides the television viewer through a behind-the-scenes glimpse of the special effects used to stage the movie's spectacular battle sequence featuring a giant squid. Afterwards, Disney draws a line of continuity through the studio's present and past accomplishments by introducing viewers to an extended sequence from the studio's most famous scene of undersea adventure, Pinocchio's escape from the whale, Monstro, in the 1940 feature, *Pinocchio*. Even the last two sequences, so clearly intended to advertise Disney products, carry the promise of edification as they define a limited and specialized knowledge—the Disney canon, the production of Disney movies—that is directed toward enhancing the experience of *20,000 Leagues Under the Sea*.

Because the *Disneyland* TV series delivered viewers like no program in ABC history, even the program's advertisers didn't mind subsidizing Disney's opportunity for self-promotion. *Disneyland* concluded the season as the first ABC program ever to appear among the year's ten highest-rated series. It was viewed weekly in nearly 40 percent of the nation's 26 million TV households.[22] The trade magazine, *Sponsor,* applauded Disney's skill at blending entertainment and salesmanship, quoting an unnamed ABC executive who quipped, "Never before have so many people made so little objection to so much selling."[23] Through its Emmy awards, the television industry affirmed its approval of Disney's venture, nominating Walt as TV's "Most Outstanding New Personality" and honoring "Operation Underseas"—an episode about the making of *20,000 Leagues Under the Sea*—as TV's Best Documentary.[24]

For the movie industry, the most telling detail in the entire Disney phenomenon was the surprising performance of the studio's feature films. By releasing its features through its own distribution company, Buena Vista, and by timing the release dates to coincide with simultaneous promotion on the television program, Disney emerged as the top-grossing independent production company of 1955. Undoubtedly aided by its exposure on the

TV series, *20,000 Leagues Under the Sea* grossed $8 million when it finally played in movie theaters—the largest sum ever reached by a Disney movie on its initial release. It finished the year as Hollywood's fourth highest-grossing movie and became the first Disney movie ever to crack the list of twenty all-time top-grossing films. In addition, Disney's new animated feature, *Lady and the Tramp,* pulled in $6.5 million—the highest figure for any of Disney's animated films since *Snow White.* Even its first feature-length True-Life Adventure, *The Vanishing Prairie* (1955), grossed a respectable $1.8 million.[25]

The most startling evidence of TV's marketing potential came from the studio's experience with Davy Crockett. Disney edited together the "Davy Crockett" episodes that already had aired twice on TV and released them as a feature film during the summer of 1955. *Davy Crockett: King of the Wild Frontier* may have been a typical "program oater," as *Variety* claimed, but it earned another $2 million at the box office because it had been transformed by television into a national phenomenon.[26] The accompanying Crockett merchandising craze gathered steam throughout the year, ultimately surpassing the Hopalong Cassidy boom of the early 1950s. By mid-1955, as "The Ballad of Davy Crockett" climbed the pop music charts, Crockett products—including jeans, pistols, powder horns, lunch boxes, the ubiquitous coonskin caps, and much more—accounted for nearly 10 percent of all consumer purchases for children, with sales figures for Crockett merchandise estimated to exceed $100 million by the end of the year.[27] Disney's apparent golden touch during 1954 and 1955 demonstrated to Hollywood that the studio had tapped into a rich promotional vein by integrating its various activities around television and the family audience.

The *Disneyland* TV program's most significant accomplishment, however, was the fanatical interest it generated in the Disneyland amusement park. Without the growth of national network television and the access it provided to the American family, Disney would not have gambled on the park. "I saw that if I was ever going to have my park," he explained, "here, at last, was a way to tell millions of people about it—with TV."[28] Disney needed television not simply to publicize the park, but to position it properly as a new type of suburban amusement, a bourgeois park designed to provide edifying adventures for baby-boom families instead of cheap thrills for the urban masses. To distinguish his park from such decaying relics as Luna Park at Coney Island, Disney assured the public that any amusement experienced in his park would be tempered by middle-class educational values. Disneyland wouldn't be another park trading in the temporal gratifications of the flesh, but a popular monument to human knowledge, a "permanent world's fair" built around familiar Disney characters and a number of unifying social goals, including educating the public about history and science.[29] The park was inextricably linked to television, because TV enabled Disney to redefine the traditional amusement park as a "theme park." With the assistance of the *Disneyland* TV series, Disney brought discipline to the unruly pleasures of the amusement park, organizing them around the unifying theme of Disney's authorial vision. By invoking cultural memories of Disney films, the TV series encouraged an impulse to re-experience texts that became one of the theme park's central attractions.

Just as it hooked American television viewers with the serialized story of Davy Crockett, the *Disneyland* TV series also bound up its audience in the ongoing story of what came to be mythologized as "Walt's dream." The seriality of *Disneyland*—and its direct relationship to the creation and continued development of the park—was crucial to the program's success. Before Disney, prime-time series were episodic; narrative conflicts were introduced and resolved in the course of a single episode. Open-ended serials were confined to daytime's soap opera genre. Disney certainly wasn't concerned about issues of TV narrative, but the *Disneyland* series demonstrated an incipient understanding of the appeal of serial narrative for network television. The success of the three-part "Davy Crockett" serial was attributable at least in part to its ability to engage viewers in an ongoing narrative. Similarly, with Walt as on-screen narrator, the *Disneyland* series, in effect, narrated the construction of Disney's amusement park, making the project a matter of continued concern for the show's viewers by creating a story out of the construction process and certifying it as the crowning achievement of an American entrepreneurial genius in a league with Thomas Edison and Henry Ford.

No less than three entire episodes, and portions of others, were devoted to the process of conceiving, building, and inaugurating the park. As the climax of the construction process, viewers witnessed the park's opening ceremonies on July 17, 1955, in a live, two-hour broadcast hosted by Art Linkletter, Robert Cummings, and Ronald Reagan. It was only appropriate that the first amusement park created by television should be introduced in a ceremony designed explicitly for television as a media event.[30] The first season of *Disneyland* was a unique type of television text, an open-ended series in which the episodes built toward a final resolution, staged as a television spectacular. By constructing a story around the events of the park's development, and by creating an analogy between the TV program and the park, the Disney organization provided a narrative framework for the experience of Disneyland.

The series represented the transition from the movie studio to the theme park by treating the park as the studio's most ambitious production. In the first episode, "The Disneyland Story," Walt introduces viewers to the park as an idea, shifting constantly between a huge map of the park, a scale-model replica, and stock footage that invokes each of the park's imaginary lands. The first season of *Disneyland,* he explains, will enable viewers "to see and share with us the experience of building this dream into a reality." The second construction episode, "A Progress Report," initiates the journey from the studio to the park as Walt takes a helicopter flight from his office to the new location. This episode also begins the process of identifying Disneyland with the culture of the automobile and the superhighway. Although the transition to the construction site could be managed by a straight cut or a dissolve, the helicopter flight instead laboriously tracks the highway that a typical traveler would follow to reach the park. With Walt providing commentary, the flight depicts both a literal and figurative passage from Hollywood to Disneyland, tracing a path from the Disney studio in Burbank, over the heart of Hollywood, down the Hollywood Freeway, connecting to the Santa Ana Freeway, and finally reaching the Disneyland exit in Anaheim—"a spot chosen by traffic experts as the most accessible spot in Southern California." Once at the construction site, the labor of construction is depicted through fast-motion photography. Accompanied by ragtime music, the scurrying workers driving bulldozers, digging ditches, and planting trees seem like animated figures; their labor takes on a cartoonish quality. In keeping with the tradition of the program's behind-the-scenes footage, Walt pauses to demonstrate how the technical feat of time-lapse photography works, but never addresses the actual labor of the workers whose activities are represented.

The third episode picks up the construction after the park's major structures have been built, as the various rides and special effects are being installed. Again, the series demonstrates how these devices, such as authentic-looking mechanical crocodiles, were designed and created at the Disney studio. This episode establishes a continuity between motion picture production and the creation of the park, demonstrating studio activities that have been reoriented to service the park. The underwater monorail employed to move the submarine in *20,000 Leagues Under the Sea* has become the basis for the Disneyland monorail train; the stage where Davy Crockett recently fought the battle of the Alamo is now the site of construction for the park's authentic Mississippi River steamboat; the sculptors and technicians who created the squid in *20,000 Leagues Under the Sea* are now making a mechanical zoo for the park. Once these devices are loaded onto trucks, they are transported to the park. As voice-over commentary reviews the route, viewers again follow the highway from the studio to the park, making the journey at ground level this time.

Television made the entire Disney operation more enticing by fashioning it as a narrative experience which the family TV audience could enhance—and actually perform—by visiting the park. Here again Disney shrewdly perceived television's ability to link diverse cultural practices that intersected in the domestic sphere of the home. In effect, Walt identified the program with the park in order to create an inhabitable text, one that would never be complete for a television-viewing family until they had taken full advantage of the postwar boom in automobile travel and tourism to make a pilgrimage to the park itself. A trip to Disneyland—using the conceptual map provided by the program—offered the family viewer a chance to perform in the Disneyland narrative, to provide unity and closure through personal experience, to witness the "aura" to which television's reproductive apparatus only could allude.

In a sense, Disney succeeded by exploiting the quest for authentic experience that has become central to the culture of modernity. In fact, tourism, as Dean MacCannell suggests, is based on the modern quest for authenticity, the belief that authentic experience exists somewhere outside the realm of daily experience in industrial society.[31] While Walter Benjamin predicted that mass reproduction would diminish the aura surrounding works of art, Disney seems to have recognized that the mass media instead only intensify the desire for authenticity by invoking a sublime, unmediated experience that is forever absent, just beyond the grasp of a hand reaching for the television dial. As a tourist attraction, Disneyland became the destination of an exotic journey anchored firmly by the family home, which served not only as the origin and terminus of the journey, but also as the site of the television set that would confirm the social meaning of the vacation experience. A father visiting the park expressed something of this sentiment. "Disneyland may be just another damned amusement park," he explained, "but to my kids it's the Taj Mahal, Niagara Falls, Sherwood Forest, and Davy Crockett all rolled into one. After years of sitting in front of the television set, the youngsters are sure it's a fairyland before they ever get there."[32] Television defined Disneyland as a national amusement park, not a park of local or regional interest like previous amusement parks, but a destination for a nation of television viewers. In the first six months alone, one million paying customers passed through the gates at Disneyland; 43 percent arrived from out of state. After the first full year of operation the park had grossed $10 million, one-third of the company's revenue for the year, and more than any Disney feature had ever grossed during its initial release.[33]

Disney's integration of television into the studio's expansive marketing schemes identified television as a worthy investment for Hollywood's major studios. Events leading up to Disney's debut may have suggested to executives of the major studios that they reconsider television production, but only Columbia, through its Screen Gems subsidiary, had acted decisively before Disney's triumph during the 1954–55 TV season. Disney's video success made it apparent that television had become the dominant national advertising medium by the mid-1950s. Providing a channel to an ever-expanding family audience, television could become the most effective marketing tool ever imagined by the movie industry. By following Disney's example and forming alliances with television networks, rather than with advertisers, the studios could ensure their access to the medium without surrendering autonomy to television's traditionally powerful sponsors.

Disney expanded his role in television during Fall 1955 with the premiere of *The Mickey Mouse Club* in ABC's weekday afternoon schedule. With this new program and the ongoing *Disneyland* series, Disney continued to use television mainly as an opportunity for studio publicity. Besides producing a sequel to "Davy Crockett," for instance, Disney created no original programming for *Disneyland* until the 1957–58 season. Disney's concept of "total merchandising" continued to shape the type of text that his company produced for television. Whereas traditional notions of textuality assume that a text is singular, unified, and autonomous, with a structure that draws the viewer inward, Disney's television texts were, from the outset, fragmented, propelled by a centrifugal force that guided the viewer away from the immediate textual experience toward a more pervasive sense of textuality, one that encouraged the consumption of further Disney texts, further Disney products, further Disney experiences. *Disneyland* drew the attention of viewers to the TV text only to disperse it outward, toward Disney products.[34]

Television made possible Disney's vision of "total merchandising" because it gave him the ability to integrate apparently isolated segments of the national commercial culture that developed after the war. In this sense, the entire Disneyland phenomenon may have been the first harbinger of Max Horkheimer and Theodor Adorno's prediction for the apotheosis of the television age, the moment when "the thinly veiled identity of all industrial culture products can come triumphantly out into the open, derisively fulfilling the Wagnerian dream of the *Gesamtkunstwerk*—the fusion of all the arts in one work."[35] By offering the first glimpse of a new Hollywood—in which television profitably obscured conventional distinctions among the media—Disney provided the impulse for the major studios to enter television and a blueprint for the future development of the media industries.

## Notes

1. For an account of the post-1952 boom in new television stations, television advertising revenue, and television set ownership, see J. Fred MacDonald, *One Nation Under Television: The Rise and Decline of Network TV* (New York: Pantheon, 1990), 59–62.

2. "This Week in Review," *Time,* 8 November 1954, 95.

3. Katherine and Richard Greene, *The Man Behind the Magic: The Story of Walt Disney* (New York: Viking, 1991), 119; Frank Orme, "Disney: 'How Old Is a Child?'" *Television* (December 1954): 37; "Disney 'Not Yet Ready' for TV," *Variety* (23 May 1951): 5; "Disney's 7-Year ABC-TV Deal," *Variety* (21 February 1954): 41.

4. "The Wide World of Walt Disney," *Newsweek* (31 December 1962): 49–51; "The Mouse That Turned to Gold," *Business Week* (9 July 1955): 74. The origins of Disney's character merchandising are described in "The Mighty Mouse," *Time* (25 October 1948): 96–98. For a more detailed discussion of the Disney corporation's use of character merchandising in relation to other TV producers of the early 1950s, see "He'll Double as a Top-Notch Salesman," *Business Week* (21 March 1953): 43–44.

5. John McDonald, "Now the Bankers Come to Disney," *Fortune* (May 1966): 141.

6. During its initial release release, *Snow White* grossed $8 million and became the first movie to exceed $5 million at the box-office. *Gone With the Wind* grossed over $20 million in its first year of release. Richard Schickel, *The Disney Version* (New York: Simon and Schuster, 1968), 229; Ronald Haver, *David O. Selznick's Hollywood* (New York: Bonanza Books, 1980), 309.

7. Schickel, *The Disney Version*, 28.

8. McDonald, "Now the Bankers Come to Disney," 141, 224; Schickel, *The*

*Disney Version,* 308–16; "Disney's Live-Action Profits," *Business Week* (24 July 1965): 78.

9. For a more detailed description of the history of ABC-TV's relations with the motion picture industry during this period, see Christopher Anderson, *Hollywood TV* (Austin: University of Texas Press, 1994).

10. "The Spectaculars: An Interim Report," *Sponsor* (15 November 1954): 31; "Twenty-Five Years Wiser About Show Business, Goldenson Finds TV the Biggest Star," *Broadcasting* (14 July 1958): 84.

11. Herman Land, "ABC: An Evaluation," *Television Magazine* (December 1957): 94.

12. Ibid., 93; "The abc of ABC," 17; "The TV Fan Who Runs a Network," *Sponsor* (15 June 1957): 45. It should be noted that ABC did not possess demographic ratings that would have enabled the network to determine the success of its programming strategy.

13. "Peaches and Cream at ABC-TV," *Variety,* 16 June 1954, 25; "The abc of ABC," 17; Klan, "ABC-Paramount Moves In," 242; Albert R. Kroeger, "Miracle Worker of West 66th Street," *Television* (February 1961): 66; "Corporate Health, Gains in Radio-TV Theme of AB-UPT Stockholders Meeting," *Broadcasting-Telecasting* (21 May 1956): 64; Frank Orme, "TV's Most Important Show," *Television* (June 1955): 32.

14. See Michele Hilmes, *Hollywood and Broadcasting: From Radio to Cable* (Champaign: University of Illinois Press, 1990), pp. 63–72; 78–112.

15. Walt also recognized that the Disney empire needed an identifiable author to crystallize the company's identity for the public, to "personify the product," as *Business Week* once noted. The naming of an author became an issue within the company as far back as the 1920s, when Walt convinced Roy to change the name of the company they had co-founded from Disney Brothers Productions to Walt Disney Productions. Consequently, as the studio expanded in 1953, Walt began to assume a more public persona, hosting the TV program and identifying himself with all things Disney, while diminishing Roy's identity. In 1953—against Roy's opposition—Walt formed Retlaw Enterprises (Walter spelled backwards), a private company which completely controlled merchandising rights to the name Walt Disney. In return for licensing the name to Walt Disney Productions, Retlaw received 5 percent of the income from all corporate merchandise. Since the Disney name was imprinted on everything associated with the company, Retlaw immediately generated enormous wealth for Walt. See John Taylor, *Storming the Magic Kingdom: Wall Street, the Raiders, and the Battle for Disney* (New York: Alfred A. Knopf, 1987), 7, 10.

16. Schickel, *The Disney Version,* 20.

17. "Disneyland Repeats Getting Bigger Audiences Than First Time Around," *Variety* (20 April 1955): 32. A complete filmography of Disney television programs through 1967 appears in Leonard Maltin, *The Disney Films,* second edition (New York: Crown Publishers, 1984), 321–26. For an examination of the Disneyland-inspired Davy Crockett phenomenon that swept through American culture beginning in 1954, see Margaret Jane King, *The Davy Crockett Craze: A Case Study in Popular Culture* (Unpublished Ph.D. Dissertation, University of Hawaii, 1976).

18. Orme, "How Old Is a Child?" 37, 72.

19. Critics within both the movie and television industries sarcastically referred to this episode as "The Long, Long Trailer," after the Lucille Ball–Desi Arnaz film

of the same title. See "A Wonderful World: Growing Impact of the Disney Art," *Newsweek* (18 April 1955): 62–63.

20. Schickel, *Disney Version,* 152; "Tinker Bell, Mary Poppins, Cold Cash," *Newsweek* (12 July 1965): 74.

21. Neil Harris, *Humbug: The Art of P. T. Barnum* (Chicago: University of Chicago Press, 1973), 79.

22. Tim Brooks and Earle Marsh, *The Complete Directory to Prime Time Network TV Shows,* Third Edition (New York: Ballantine Books, 1985), 1031. *Disneyland* remained among the top fifteen programs through 1957, and then fell from the top twenty until it shifted to NBC—and color broadcasts—in 1961.

23. Charles Sinclair, "Should Hollywood get it for free?" *Sponsor* (8 August 1955): 102.

24. Maltin, *The Disney Films,* 315.

25. "Disney Parlays Romp Home," *Variety* (30 November 1955): 3; "All-Time Top Grossing Films," *Variety* (4 January 1956): 84. At the time, *20,000 Leagues Under the Sea* was the nineteenth highest-grossing film of all time.

26. Ibid.

27. "The Wild Frontier," *Time* (23 May 1955): 92. Unfortunately for Disney, the studio could not control licensing of Crockett products, because it did not possess exclusive rights to the name or character of Davy Crockett. Since the mid-nineteenth century companies had used the Crockett name on products from chewing tobacco to whiskey. The Disney studio never again made this mistake. See also "U.S. Again Subdued by Davy," *Life* (25 April 1955): 27; "Mr. Crockett is a Dead Shot As a Salesman," *New York Times,* 1 June 1955, 38.

28. Schickel, *The Disney Version,* 313.

29. "Father Goose," *Time* (27 December 1954): 42; "Tinker Bell, Mary Poppins, Cold Cash," 74.

30. For an account of the opening ceremonies, see Bob Chandler, "Disneyland As 2-Headed Child of TV & Hollywood Shoots for $18 Mil B.O.," *Variety* (20 July 1955): 2. Chandler observes that the inauguration of Disneyland marked the "integration and interdependence of all phases of show biz."

31. Dean MacCannell, *The Tourist: A New Theory of the Leisure Class* (New York: Schocken Books, 1976), 159.

32. "How To Make a Buck," *Time* (29 July 1957): 76.

33. Ibid., Schickel, *The Disney Version,* 316.

34. Michele Hilmes describes the use of this strategy in the Hollywood-produced radio program, *Lux Radio Theater.* See Hilmes, *Hollywood and Broadcasting,* 108–10.

35. Max Horkheimer and Theodor W. Adorno, *Dialectic of Enlightenment* (New York: Continuum, 1987), 124.

# Television Production Techniques as Communication

## DAVID BARKER

Some scholars of mass communication have begun to approach the television message as a visual text and interpretation of this text as a process of
"decoding" (Fiske and Hartley, 1978; Silverstone, 1981). Perhaps the foremost proponent of this approach is Stuart Hall, who identifies this process
of "decoding" as a "determinate moment" in television discourse (1980:
129). Yet, Hall makes it quite clear that the process of "encoding" is equally
determinate:

> A "raw" historical event cannot, *in that form,* be transmitted by, say, a
> television newscast. Events can only be signified within the aural-visual
> forms of the television discourse. . . . The "message form" is the necessary
> "form of appearance" of the event in its passage from source to receiver.
> Thus, the transposition into and out of the "message form" is a determi-
> nate moment. (129)

Within the growing body of work based upon the encoding/decoding
model, however (e.g., Brunsdon and Morley, 1978; Morley, 1980; Wren-
Lewis, 1983), discussion has been restricted almost exclusively to only one
of these "determinate moments": decoding. Indeed, the process of encoding, despite its homologous position in the model, has by comparison, been
virtually ignored.[1]

From *Critical Studies in Mass Communication,* Barker, David. "Television Production
Techniques in Communication." September 1985. Pages 234–246. Copyright by
the Speech Communication Association. Reprinted by permission of the publisher.

The purpose of this study is to examine this process of encoding or, more specifically, the relationship between narrative structure and production techniques, as it is manifested in entertainment television. It is the thesis of this essay that the communicative ability of any television narrative is, in large part, a function of the production techniques utilized in its creation.[2]

The two programs chosen for this study were *All in the Family (AITF)* and *M\*A\*S\*H*. The decision to examine these particular programs was based on two factors. First, both programs were pivotal to their own specific narrative traditions. *AITF* represents the tradition of domestic situation comedies that revolved around a single axial character—in this case Archie Bunker. Archie was axial inasmuch as plot lines were usually built directly upon his character, and most other characters in the program were usually defined not so much by their own idiosyncrasies as by their relationship with him. We can find many of the seeds of *AITF* in *The Life of Riley, Make Room for Daddy, The Honeymooners,* and *Father Knows Best.*

*M\*A\*S\*H*, on the other hand, was pivotal to a tradition of what might be termed ensemble comedies in which each character had a persona of his or her own, distinct from the rest of the ensemble. Yet the interplay and conflict among such characters worked to strengthen the personalities of all concerned, making the ensemble comedy very much more than the sum of its parts. Foreshadowings of *M\*A\*S\*H* can be found in *Burns and Allen, The Addams Family, Gilligan's Island,* and *Hogan's Heroes.*

Prior to *AITF,* network situation comedy in general had become entrenched in what could be called "1960s telefilm values," which were characterized by a highly utilitarian approach to the communicative abilities of even the most basic production techniques. As exemplified by such programs as *My Three Sons* or *Leave it to Beaver,* this was a production style based on a highly repetitive and predictable shooting pattern: an exterior establishing shot (e.g., the Cleaver home as Ward pulls into the driveway) followed by sequences of alternating medium shots as the narrative progresses (e.g., Ward comes in the front door to be greeted by June who proceeds to inform him of the daily crises; Ward then decides on a course of action and initiates it). This pattern was occasionally punctuated with a tighter shot, such as a medium close-up (bust shot), but close-ups were virtually never used. Similarly, performer movement (blocking) was utilized only as a way to move characters into and out of an environment: Beaver walks into his bedroom and throws himself on the bed or Wally walks into the kitchen and stands by the sink. As Millerson (1979: 287) has pointed out, there is a great deal of meaning in such production variables as shot selection and performer blocking. Programs produced with these 1960s telefilm values suggest that reinforcing the structure of the program narrative (e.g., the ebb and flow of conflicts) with particular production techniques was not a fundamental concern for the producers.

With *AITF* and *M\*A\*S\*H,* however, close analysis of a number of episodes indicated that, for these two series, the relationship between narrative structure and production techniques was quite different. Thus, the second reason for choosing these two programs was that both were also products of conscious decisions by their respective creators, Norman Lear and Larry Gelbart, to utilize specific production techniques in specific ways.

In preparation for this study, twenty episodes selected at random of each series were analyzed for their use of particular production techniques, and it soon became clear that a number of these techniques—among them, control of screen space, lighting and set design, layers of action, and parallel editing—were manipulated significantly for narrative as well as aesthetic reasons. This analysis made it possible to draw some general observations about the role each of these techniques played in the communication of the particular narrative structures of *AITF* and *M\*A\*S\*H*. As a way of providing specific examples for these observations, one randomly chosen episode from each series was videotaped and subjected to a rigorous shot-by-shot analysis.[3]

Before moving to a discussion of the way these variables were manipulated in relation to the program narratives, it would be useful to explore the nature of the narratives themselves. In Archie Bunker, Lear had an axial character whose persona was often patriarchal in the sense he was myopic and tyrannical.[4] Lear has stated that Archie reminded him a great deal of his own father (Adler, 1979: xx) and that much of *AITF* was intended to portray the patriarch as buffoon.[5] Lear thus decided to shoot *AITF* in proscenium. This meant that the cameras (and thus the viewing audience) would maintain a distinct distance; they would not be allowed to move into the set—into Archie's domain—for reverse angles. This, and the fact that the program was produced live-on-tape in front of an audience, created a great sense of theatricality. There were practical economic reasons for such a decision (Adler, 1979: xxii), but there were equally important narrative reasons as well: shooting in proscenium helped maintain Archie not only as the axial character but as the buffoonish patriarch. By preventing the cameras from moving into the set for reverse angles, viewers were allowed only to look at Archie, not with him.

Lear's decision to shoot *AITF* in proscenium with multiple cameras pulled situation comedy back to its roots. He essentially employed an updated version of the Electronicam system, a multiple film camera system Jackie Gleason had employed successfully on *The Honeymooners* some twenty years earlier (Mitz, 1983:123). But if *AITF* pulled situation comedy back to its roots, *M\*A\*S\*H* pushed it forward into the somewhat more complicated aesthetic realm of the contemporary cinema. When Gelbart first saw the movie version of *M\*A\*S\*H,* he realized the intrinsic role director Robert Altman's reflexive shooting style played in the film's narrative (Gelbart, 1983). Altman was not dealing with the all-pervasive influence of a single axial character, but with the nuances of an ensemble of characters, and the apparent aimlessness of his camera movements, the sheer "busyness" of his shots, actually helped define the characters and their relationships.[6] Thus a shooting style similar to that employed by Altman would be important in maintaining the spirit of the ensemble: a single camera with multiple set-ups, unimpeded by spatial or psychological boundaries, able to capture visual patterns of great complexity. In short, one might say that a more "aggressive" use of the camera than had been the norm for situation comedies would be necessary; programs like *Leave it to Beaver* had employed a single camera, but little effort had been made to use the camera (or any production variable) to reinforce the narrative.

## *Theoretical Approach*

Effective control and manipulation of screen space is one of the most crucial elements of television aesthetics. While there are many ways of delineating such control (Arnheim, 1974; Burch, 1973; Heath, 1981; Zettl, 1973), for the purposes of this study I distinguished between two broad categories: camera space and performer space.

*Camera space* is composed of two elements: horizontal field of view and what I have termed "camera proximity." Horizontal field of view is the type of shot: CU (close-up, head and shoulders), MCU (medium close-up, taken at the bustline), MS (medium shot, taken at the waist), WS (wide shot, which encompasses the entire body or set), and so forth. The importance of field of view is underscored by Wurtzel and Dominick (1971) who argue that much of the perceived effectiveness of a television program depends on the way the director chooses the shots the audience is to see:

> The viewer, as represented by the camera, does not remain in one viewing position. He sees the action from both close-range and at a distance. (104)

The second component of camera space, camera proximity, is the location of the camera in relation to the performer—in front of them, behind them, and so forth. In *AITF,* camera proximity and, to a somewhat lesser extent, field of view, worked together to establish definite "geographical" boundaries for the cameras that were rarely violated. As will soon become evident, each performer in *AITF* had his or her own space that the other performers (to a greater or lesser degree) could move into and out of freely. But the camera could do so only at the risk of making the viewer uncomfortable. Such movement could be used for dramatic effect or to change the meaning of the narrative (e.g., the episode in which Archie and Mike were trapped in the basement and a camera was pulled into the set for a tight 2-shot as their usual antagonism gave way to a momentary rapport). In *M\*A\*S\*H,* on the other hand, geographical boundaries for the camera were loosely defined if at all. The camera was allowed to move at will without fear of the movement necessarily affecting a change in the narrative.

The second type of television space distinguished here is *performer space.* This is primarily a function of performer blocking (positioning and movement) along axes. These axes are defined in relation to the camera: the horizontal $x$-axis perpendicular to the camera's line of sight and the $z$-axis of depth toward and away from the camera or parallel with the camera's line of sight (Zettl, 1973: 174). Each of the cameras in multiple camera shooting or each new set-up of the camera in film style shooting has its own *x*- and *z*-axis. The distinction to be made concerns the way these axes are utilized.

Axis utilization is perhaps best described in terms of vectors. Zettl defines vectors as "directional forces [within the screen] which lead our eyes from one point to another within, or even outside of, the picture field" (1973: 140). Further, Zettl distinguishes among three types of vectors: graphic (created by stationary objects arranged in such a way that they lead the viewer's eyes in a particular direction (e.g., a row of smokestacks), index (something that points unquestionably in a specific direction, e.g., a person looking or pointing), and motion (created by an object that is actually moving or perceived as moving onscreen) (Zettl, 1973: 140–42). In discussing performer blocking the most important of these vectors is, of course, the motion vector.

## Comparative Analysis

In *AITF,* performers were blocked almost exclusively along the *x*-axis as they moved from the front door through the living and dining rooms to the kitchen and back again. The only time there was a real potential for movement toward or away from the cameras occurred when a character moved upstage to the stairway or downstage to the television set. *M\*A\*S\*H,* on the other hand, utilized a great deal of movement along *z*-axis motion vectors. Because the camera was free to move into the performers' space for shot-reverse shot sequences, performers could move toward and away from the camera as easily as they could move perpendicular to it. In the episode of *AITF,* the vast majority (92 percent) of performer blocking was along *x*-axes. In *M\*A\*S\*H,* the majority (78 percent) of performer blocking was along *z*-axes.

It is significant how the two components of space—camera and performer—worked together to reinforce the communicative ability of their respective narrative structures. In terms of space, the proscenium technique such as the one utilized in *AITF* would seem to favor an axial, somewhat patriarchal narrative structure in that it allows the viewer to only look at a character rather than with them. This is especially true when the patriarchal status of a character is the object of derision, as was the case with Archie. Such a status would be undercut somewhat if viewers were allowed to encroach upon that character's space proximally. Befitting a proscenium approach, the cameras thus became a fourth wall, allowing viewers to approach Archie's space from essentially only one direction.

Viewers could, and did, encroach upon Archie, however, through field of view: close-ups of Archie were often used for dramatic or comedic effect. This was especially true for reaction shots. In the particular episode under discussion here, there were six reaction shots, all were of Archie. Much of Archie's patriarchal stance, and the humor derived from making fun of such a stance, was a result not so much of what Archie said as the way he reacted to the words and actions of others. He appeared as the long-suffering father figure enduring the ignorance of the "children" about him (e.g., his reaction to Edith's news that she had promised their house for Florence and Herbert's wedding or his reaction to Edith's piano playing and singing during the wedding). Indeed, such reactions almost became narrative conventions within the series itself, aided considerably by the somewhat plastic quality of Carroll O'Connor's face and his remarkable ability to use it to show numerous inflections. Significantly, too, such reaction shots just as often showed Archie losing face, as he once again became the buffoon.

Reaction shots on *M*A*S*H*, however, were rare—the selected episode had none at all. Arguably, this was due to the fact that, in an ensemble comedy, the emotional reaction of any one character is in large part defined by the corporate reaction of the entire cast (witness the fact that in those episodes where a facial reaction of some sort was necessary the camera often panned across the faces of the entire cast).

The use of screen space to define a character in *AITF* was perhaps most conspicuous with regard to Archie and his chair. It was no accident that this particular chair occupied the point in the set where the *x*-axis vectors intersected the only real *z*-axis vector, that running from the television to the staircase. The action of *AITF* for the most part revolved around this throne; in fact, the space it usually occupied was the exact spot where Florence and Herbert said their vows. Normally, this was space Archie guarded zealously. Thus, the audience was never allowed to circle his chair or sneak up on it from behind, assuming Archie's place through a POV (point-of-view) shot.

The use of camera and performer space to reinforce the narrative was equally evident in *M*A*S*H*, where a large proportion of performer movement was employed as a transition device, to move the story from one subplot to another. In the sample episode, this was evident from the opening scene, which took place in the operating room (OR). In this scene, all three of the plotlines for the episode were laid out in dialogue between Hawkeye, Charles, B. J., and Potter. As they talked, Father Mulcahy paced back and forth, establishing a motion vector between them, asking questions that stimulated necessary plot information. The camera panned with Mulcahy as he moved, his movement acting not only as a transition between the surgeons but between the plotlines as well. More typically, however, movement-as-transition was achieved by having two characters moving along converging vectors meet, converse, then move on. The camera would pan or truck to follow them until they converged with another couplet standing still or moving in the opposite direction. The first couplet would "hand off" the story to the second and the camera would begin a new sequence with the second couplet.

This type of transition was seen when the Canadian leaves the *M*A*S*H* compound. Klinger escorts him to his truck and tells him goodbye. Klinger then continues across the compound along a *z*-axis until he converges with Hawkeye moving along the *z*-axis from the opposite direction. They stop, converse, and move on. Similar movement was used earlier in the episode when a nurse who had been talking to Hawkeye rises from their table and starts across the mess tent on the $x$-axis. The camera panned to follow her and came to rest on Klinger and the Canadian moving through the chow line on the $z$-axis.

This performer movement occurred in several planes simultaneously, thereby layering the action between foreground, middleground, and background. People were constantly moving in and out of the frame, past doorways and windows, between the camera and the subject of the shot, behind the subject, crowding the screen with visual information. To a very great extent, this layering was a result of the extensive use of $z$-axis motion vectors. As Zettl points out, "By placing objects and people along the $z$-axis in a specific way . . . we can make the viewer distinguish between foreground, middleground, and background rather readily . . ." (1973: 207). This can of course be compared with *AITF,* where extensive use of $x$-axis motion vectors resulted in the action usually being carried out on only one plane as opposed to three or four.

The use of performer movement as a transition device and the layering of that movement into multiple planes was obviously tailor-made for an ensemble comedy like *M\*A\*S\*H* which utilized interweaving characters and plots. The movement itself acted very much as a needle and thread, sewing the narrative together. But the multiple-plot narrative structure of *M\*A\*S\*H* also benefited from parallel editing (the intercutting of two activities occurring at roughly the same time in different locations) which was used to enhance the program narrative by playing the dialogue and actions in one subplot of the narrative off those in another. Put another way, parallel editing was used to help establish comedic relationships.

The comedic relationships in *M\*A\*S\*H* were sometimes obvious. For example, there was a cut from Hawkeye berating Klinger for paying off a five dollar debt with a bottle of wine to Charles begging Hawkeye to tell him where he got the wine. In another example, there was a cut from Potter and Klinger pouring Charles's wine into their stranded jeep to Hawkeye pulling the cork from a wine bottle in preparation for his tryst with the winner of his essay contest. Other times, however, the relationships were far more subtle, as when Charles, who has been talking to Hawkeye and B. J., turns his head towards the door of the Swamp and we cut to a moment later that day as Hawkeye and B. J. leave the Swamp and walk outside. In either case, however, it was the preciseness with which these words and actions were matched through editing (and judicious shot selection) that allowed the viewer to not only jump across space and time with no loss of orientation but to likewise make the comedic connection between a stranded Potter and Klinger and an amorous Hawkeye. These techniques also enabled the divergent strains of an ensemble comedy like *M\*A\*S\*H* to converge into a unified narrative structure rather than collapse into a series of isolated vignettes.

The establishment of comedic relationships also owed a great deal to timing. This was particularly true with regard to *AITF.* Inasmuch as *AITF* was shot live-on-tape with multiple cameras, most "editing" as such was done on the spot by switching between the cameras. Thus, postproduction editing, while usually necessary, was not the process of shot-by-shot assembly that an episode of *M*A*S*H* entailed. Nevertheless, in *AITF* the establishment of comedic relationships and the proper coordination of dialogue and movement were just as essential. Due to the factors of shooting in real time, comparatively fixed camera positions, and a lesser degree of postproduction editing, however, the comedy had to be played and the comedic relationships made clear by the performers themselves. This meant that split-second timing was paramount to the success of *AITF* (Lynch, 1973: 267–71).

By contrast, the single-camera film-style technique utilized in *M*A*S*H* did not require the performers to say their lines in real time to the extent the multiple-camera live-on-tape technique utilized in *AITF* did. In a shot/reverse-shot sequence, for instance, since only one camera was used, one member of the couplet would say all their lines for that camera position; then the camera would have to be moved for the lines of the second member of the couplet. Timing, so very important to the proper execution of comedy, was thus partially suspended, and only completely restored again when the editor assembled the film itself. Thus, performers in an ensemble comedy shot film-style must learn to set up their timing somewhat differently than those performers in a comedy like *AITF* shot in real time.[7]

Since the matter of timing was so crucial in *AITF*, every character had to know his or her blocking precisely, but this was especially true in Archie's case; it was through Archie's blocking that his role as the axial character within the narrative structure of *AITF* manifested itself so visibly. During the wedding ceremony for Florence and Herbert, Archie was constantly in motion, pulling one character after another about the set from one group and comedic situation to the next. It begins with Archie pulling a reluctant Herbert down the stairs and through the guests. Leaving him, he crosses the room to Edith and pulls her to the Priest in an effort to get the ceremony started. He then pulls Edith to the piano, pushes her onto the bench and tells her to start playing, only to return once again to tell her to stop. Archie then goes up the stairs and returns with Florence, depositing her next to Herbert. He crosses back to the piano, picks up Edith, and pulls her over to Florence and Herbert. The scene continues in a similar manner, with Archie orchestrating virtually every movement.

While this particular episode was exceptional in the number of people involved in this last scene and the degree of their movement, it was an exception that illustrates the point. In a scene this involved, precise execution of blocking and dialogue—both matters of timing—was essential. Any lapse in this execution would have prevented the director from getting the necessary shot. Further, the degree to which the timing in the scene depended on Archie's blocking was reflective of his axial status within the series as a whole.

The control of screen space through field of view, camera proximity, performer blocking, parallel editing, and timing all move a great distance towards explaining why *AITF* and *M\*A\*S\*H* "looked" the way they did. But much of the particular "look" of these two programs, especially with regard to the division of the visual field into planes and the articulation of depth, was also a function of lighting and set design.

## Lighting and Set Design

As Millerson (1982) points out, television is inherently a two-dimensional medium and the careful control of light and shadow is essential for creating the illusion of a third dimension. Zettl (1973) distinguishes two types of television lighting techniques. The first, chiaroscuro, is lighting for light-dark contrast. "The basic aim is to articulate space," writes Zettl, "that is, to clarify and intensify the three-dimensional property of things and the space that surrounds them, to give the scene an expressive quality" (38). The second type of lighting, Notan, "is lighting for simple visibility. Flat lighting has no particular aesthetic function; its basic function is that of illumination. Flat lighting is emotionally flat, too. It lacks drama" (44).

Notan lighting was obviously utilized on *AITF,* where the set was lit flatly and evenly. There was no regard for time of day—it was as bright inside the Bunker house at night as it was during the day. Similarly, there was no regard for light source—when it was day there was no appreciable difference in the amount of light coming through the windows than when it was night. Most importantly, shadows were virtually nonexistent. Thus, the fact that action occurred on only one plane was reinforced by a lighting design that, through the absence of shadow, helped to create an environment of only one plane.

*M\*A\*S\*H,* however, utilized chiaroscuro lighting, primarily in the form of source-directed lighting: during the day the sets were bright with "sunlight" streaming through windows while, at night, shadows increased markedly, the sets becoming dimmer, with darkened windows and light provided by lamps or overhead fixtures. Yet even during daylight hours, many depth clues were offered by shadows. As an example, in the mess tent, as Klinger glances at Hawkeye, there is a shot of Hawkeye sitting at his table. Even though there is nothing between the camera and Hawkeye to act as a point of reference, it is still obvious that Hawkeye is completely across the tent from Klinger. The fact that the shadows deepen as they move toward Hawkeye articulates the amount of space that exists between him and Klinger.

Set design can likewise articulate space and, through the use of depth clues, degrees of depth in the televisual image. Zettl (1973: 179) suggests a number of these clues, but three are of particular importance to the discussion here: overlapping planes, relative size, and height in the plane.

*M\*A\*S\*H* utilized sets of great depth, and these three depth clues were all conspicuously evident. Overlapping planes (in which one object partially covers another so that it appears to be lying in front) were used in numerous shots. For example, shooting across a bunk when Hawkeye and Charles were in the Swamp arguing over the wine or shooting through the jeep windshield as Potter and Klinger return with the curare. Relative size (guessing how large something is or how far away it is by the size of its screen image) was also used a great deal due to the placement of set pieces in relation to the set and the camera (e.g., in the mess tent, shooting across one table at Hawkeye sitting next to the window with Klinger and the Canadian sitting in the background) or movement in various planes (e.g., Hawkeye and B. J. leaving the Swamp and walking towards the camera with jeeps and trucks passing behind them and nurses walking between them and the camera). There were also several occasions when height in plane (the higher something is in the picture field the further away it is) was used (e.g., Klinger turning around as the Canadian leaves to see Hawkeye approaching from across the compound).

Another conspicuous characteristic of the set design on *M\*A\*S\*H* was the almost constant presence of the outside world. In the mess tent or in the "swamp," flaps were tied back, allowing viewers to watch the external as well as the internal workings of the camp: people walking by, jeeps passing, conversations being carried on, ballgames being played. Yet this presence in no way de-emphasized the importance of interiors. The more ensemble nature of *M\*A\*S\*H* necessarily required an environment of diversity, and despite the superficial similarity of green canvas and tent poles, each of the sets on *M\*A\*S\*H* maintained the distinct personality of its occupant. Character traits were conspicuously evident: the World War I memorabilia in Potter's office, the draped nylons and lace doilies in Margaret Houlihan's tent, the teddy bear tucked in Radar's bunk. In this regard, the most diverse habitat of all was the "swamp," where a still for making homemade hooch squatted side-by-side a phonograph and recordings of Rachmaninoff. The diversity here was of course due to the diversity of the occupants: at various times Hawkeye, Trapper John, Frank, B. J., and Charles.

The great use of depth in the set designs for *M\*A\*S\*H* stands in sharp contrast to the designs for *AITF,* where sets tended to be long and shallow. But because there was comparatively little movement toward or away from the camera, sets with any degree of real depth were unnecessary (this can also be seen on other situation comedies like *One Day At a Time, Maude, The Jeffersons,* and *Alice*). It should come as no surprise then, that the depth clues identified by Zettl were all missing from the particular episode of *AITF* under discussion here and were comparatively rare in any of the episodes analyzed in preparation for this essay. Indeed, the only time any of the depth clues even came close to utilization were the occasional instances of shooting across the television set.

In contrast to the efforts to include the outside world in *M\*A\*S\*H,* in *AITF* curtains were usually drawn over windows or, were they opened (e.g., the window in the Bunker's diningroom), the only thing visible through them was light, creating a sense of what Zettl calls "negative space" (1973: 177). Similarly, when front or back doors were opened, the audience saw painted backdrops and an occasional artificial bush or tree.

Up to a certain point, set design on *M\*A\*S\*H* and *AITF* was a function of performer blocking. As Alan Wurtzel reminds us, sets physically define the limits of performer blocking (1983: 424), and *M∗A∗S∗H*'s orientation towards *z*-axis blocking necessitated sets of great depth as much as *AITF*'s orientation towards *x*-axis blocking necessitated sets that were long and shallow. But, as Horace Newcomb has pointed out, one must also consider the degree to which sets reinforce the program narrative by "delineating a great deal of formulaic meaning" (1974: 28).

The two-dimensional treatment of the outside world in *AITF* led to little or no sense of space beyond the confines of the Bunker home. This sense was further compounded by the great degree of homogeneity from one *AITF* set to another. The living/dining area, the kitchen, Archie and Edith's bedroom, and Mike and Gloria's bedroom all looked very much alike. While movement from one of these environments to another sometimes occasioned rhetorical shifts in the narrative (e.g., Archie and Mike often seemed more vulnerable, less defensive when in their respective bedrooms), together these sets provided a dramatic gestalt, a sense of psychological closure.

Metallinos (1979) states that psychological closure is "one of the most crucial forces operating within the visual field" and recounts Zettl's definition of it as "the perceptual process by which we take a minimum number of visual or auditory cues and mentally fill in nonexisting information in order to arrive at an easily managed pattern" (211). The basically two-dimensional set design of *AITF*—drab and nondescript—and the minimal visual information provided concerning the outside world, encouraged the viewer to focus attention on those things inside the Bunker house that were three-dimensional: the characters and their confrontations.[8]

This focusing of attention had a narrative function as well as a physiognomic one, however. Inasmuch as *AITF* revolved about an axial character, the true essence of the program was that the world began and ended with Archie. It was essential, then, that *AITF* employ a narrative gestalt to a much greater degree than *M∗A∗S∗H,* and a great part of that narrative gestalt was the creation through specific set design and lighting techniques of a physical environment that was itself a complete, self-contained unit, apart from the outside world.

The need for narrative gestalt in *AITF* made it very much a drama of interiors. But in *M∗A∗S∗H,* as much of the outside world as possible was included. I would argue that this was due to the fact that, unlike those in *AITF,* the characters in *M∗A∗S∗H* were very closely tied to the outside world. So much of what happened in their lives was dictated by the ebb and flow of conflicts beyond their compound and beyond their control. Thus, a narrative heavily dependent on external realities was reflected in set designs that were open and emphasized the outside world.

## Conclusion

It has been the thesis of this essay that the communicative ability of any television narrative is, in large part, a function of the production techniques utilized in its creation. While I think it quite clear that *AITF* and *M∗A∗S∗H* support this thesis, one must be careful not to overemphasize the role of production techniques in the communication of entertainment television narratives based only upon the experience of two series. It would, however, seem appropriate to conclude that, at least in the case of *AITF* and *M*∗*A*∗*S*∗*H*, the creators of the two shows, faced with a number of options (including the standard "1960s telefilm values"), made some deliberate production choices that, while gambles of sorts, were obviously felicitous—witness the influence the two shows have had on subsequent programming. *AITF*'s proscenium style has dictated the course of situation comedy ever since, as the vast majority of sitcoms in the past decade have utilized multiple camera, live audience configurations. Similarly, *M*∗*A*∗*S*∗*H* has exerted a tremendous influence but, unlike *AITF,* not on programming within its own genre. The visually complex influence of *M*∗*A*∗*S*∗*H* can best be seen in recent hour-long comedy-dramas like *St. Elsewhere* and *Hill Street Blues,* an influence Robert Butler, who directed the pilot for *Hill Street Blues* called "making it look messy" (Gitlin, 1983: 293). Indeed, this influence can first be seen somewhat earlier in *Lou Grant,* when *M*∗*A*∗*S*∗*H* producer Gene Reynolds teamed up with James Brooks and Alan Burns of MTM, the production company later responsible for *St. Elsewhere* and *Hill Street Blues.*

This is not to say that other production techniques could not have been utilized on *AITF* or *M*∗*A*∗*S*∗*H* as, in fact, they were (e.g., the episodes where Mike and Archie were locked in the basement, where Gloria was molested, where the members of the 4077 were interviewed by a newsreel crew, where a subjective camera was used to show the 4077 from the point of view of a wounded soldier, etc.). These, however, were the rare exceptions rather than the rule and it is significant that such changes in production technique were not just the result of changes in narrative structure but were in large part responsible for the dramatic and emotional impact of these particular episodes.

In assigning meaning to specific production techniques, one runs the risk of overstating the importance of the television apparatus itself. Nonetheless, as Stuart Hall concedes, the way a message is encoded into televisual discourse has a great impact on what the message becomes and the way it is decoded. Indeed, the acknowledgement that the techniques of television production themselves have meaning questions the validity of looking at the production process as a given or, due to its often assembly-line, commercialized nature, as an endeavor unworthy of scholarly consideration. It argues, instead, that this process is an important link in human communication.

### Notes

1. There is a body of research that, while not working from the perspective of the encoding/decoding model, nonetheless deals with various production techniques. It has tended to fall into two categories. The first of these has dealt with television as a medium for instruction or information (e.g., McCain, Chilberg, and

Wakshlag, 1977; Schlater, 1969, 1970; Tiemens, 1970; Williams, 1965). The second category, on the other hand, has dealt more closely with television as a medium of entertainment, most often focusing on questions of aesthetics or semiotics (e.g., Herbener, Tubergen, and Whitlow, 1979; Metallinos, 1979; Metallinos and Tiemens, 1977; Porter, 1980, 1981, 1983). To varying degrees, the vast majority of this research—the current study included—owes a debt to the seminal work of Herbert Zettl, whose articulation of many of television's aesthetic tenets (1973, 1977, 1978) has provided it a foundation upon which to build.

2. In this context, I define "communicative ability" as the degree to which an encoded text determines its own decoding.

3. The episode of *AITF* concerned a conflict between Archie and Edith. Archie had planned a weekend fishing trip for himself and Edith along with Archie's friend Barney and his wife. Unbeknownst to Archie, Edith had agreed to have a wedding ceremony in their home for two octogenarians from the Sunshine Home, Florence and Herbert. The wedding was to take place the same day Archie wanted to leave on the fishing trip, the problem being how to schedule both to the detriment of neither.

The episode of *M*A*S*H* interwove three plots: procuring a drug, curare, used as a muscle relaxer prior to surgery; Klinger exchanging fruit cocktail with a Canadian M*A*S*H unit for several bottles of French wine; and Hawkeye holding an essay contest for the nurses with himself as prize.

4. My use of the term "patriarchal" in this context should be qualified, inasmuch as the term has gained a number of connotations, perhaps chief among them that of an historical system of male domination, though recently it has taken on a more Marxist bent, particularly in feminist film criticism. My use of it in reference to *AITF* is based upon the fact that while Archie is indeed a character axial to the narrative, his centrality has a blatantly oppressive, vituperative component to it (exemplified by his treatment of his family, particularly Edith) not necessarily associated with characters just because they are axial. Beyond this, however, further connotations of patriarchy are not intended.

5. During much of *AITF*'s first few seasons a debate raged as to whether or not the program endorsed bigotry or defused it by making it the object of ridicule. For a selection of literature from both sides, see Adler (1979).

6. Diane Jacobs (1977) calls this aspect of Altman's style "actualism." For further discussion of Altman's *mise en scène* and camera movement, see Rosenbaum (1975) and Tarantino (1975).

7. For an interesting discussion of playing comedy for a single camera versus playing it for multiple cameras, see Kelly (1981), pp. 28–35.

8. Indeed, Lear had originally intended to shoot *AITF* in black-and-white. When CBS balked, he compromised and made the sets a drab brown and as nondescript as possible.

## References

Adler, R. (1979) *All in the Family: A Critical Appraisal*. New York: Praeger.

Arnheim, R. (1974) *Art and Visual Perception*. Berkeley: University of California Press.

Brunsdon, C., and D. Morley (1978) *Everyday Television: "Nationwide."* London: British Film Institute.

Burch, N. (1973) *Theory of Film Practice*. Princeton: Princeton University Press.

Fiske, J., and J. Hartley (1978) *Reading Television*. London: Methuen.

Gelbart, L. (1983) "Its creator says hail and farewell to *M*A*S*H*." *New York Times*. February 27.

Gitlin, T. (1983) *Inside Prime Time*. New York: Pantheon.

Hall, S. (1980) "Encoding/decoding," pp. 128–38, in S. Hall, D. Hobson, A. Lowe, and P. Willis (eds.) *Culture, Media, Language*. London: Hutchinson.

Heath, S. (1981) *Questions of Cinema*. Bloomington: Indiana University Press.

Herbener, G., G. Tubergen, and S. Whitlow (1979) "Dynamics of the frame in visual composition." *Educational Technology and Communications Journal* 27 (Summer): 83–88.

Jacobs, D. (1977) *Hollywood Renaissance: Altman, Cassavetes, Coppola, Mazursky, Scorsese, and Others*. New York: A. S. Barnes.

Kelly, R. (1981) *The Andy Griffith Show*. Winston-Salem: John F. Blair.

Lynch, J. (1973) "Seven days with 'All in the Family': Case study of the taped TV drama." *Journal of Broadcasting* 17 (Summer): 259–74.

McCain, T., J. Chilberg, and J. Wakshlag (1977) "The effect of camera angle on source credibility and attraction." *Journal of Broadcasting* 20 (Winter): 35–46.

Metallinos, N. (1979) "Composition of the TV picture: Some hypotheses to test the forces operating within the television screen." *Educational Technology and Communications Journal* 27 (Fall): 205–14.

Metallinos, N., and R. Tiemens (1977) "Asymmetry of the screen: The effect of left versus right placement of television images." *Journal of Broadcasting* 20 (Winter): 21–33.

Millerson, G. (1979) *The Technique of Television Production* (10th ed.). London: Focal Press.

———. (1982) *The Technique of Lighting for Motion Pictures and Television* (2d ed.). London: Focal Press.

Mitz, R. (1983) *The Great TV Sitcom Book*. New York: Perigee Books.

Morley, D. (1980) *The 'Nationwide' Audience*. London: British Film Institute.

Newcomb, H. (1974) *TV: The Most Popular Art*. Garden City, N.Y.: Anchor Books.

Porter, M. (1980) "Two studies of Lou Grant: montage style and the dominance of dialogue." Unpublished paper.

———. (1981) "The montage structure of adventure and dramatic prime time programming." Unpublished paper.

———. (1983) "Applying semiotics to the study of selected prime time television programs." *Journal of Broadcasting* 27 (Spring): 63–69.

Rosenbaum, J. (1975) "Improvisations and interactions in Altmanville." *Sight and Sound*, 44 (Winter).

Schlater, R. (1969) "Effect of irrelevant visual cues on recall of television messages." *Journal of Broadcasting* 14 (Winter): 63–69.

———. (1970). "Effect of speed of presentation on recall of television messages." *Journal of Broadcasting* 14 (Spring): 207–14.

Silverstone, R. (1981) *The Message of Television: Myth and Narrative in Modern Society*. London: Heinemann.

Tarantino, M. (1975) "Movement as metaphor: The long goodbye." *Sight and Sound*, 44 (Spring).

# Roseanne—How Did You Get Inside My House? A Case Study of a Hit Blue-Collar Situation Comedy

## JUDINE MAYERLE

Roseanne *isn't television-ish. We have attempted to give the series the most real texture possible in terms of characters and setting. The conversations are the kind that go on in real life. The sets are designed and dressed to look as though the audience has just walked into someone's home. The washer and dryer are actually running when we shoot in the laundry room. The children in the series grow out of their clothes when money is tight, and everything's not beautiful in the Conner home all the time. We do fall into a bit of a trap because everybody always wants to look good, but we try to stick with realism the best we can.*[1]

Although American primetime entertainment television programming generally evolves out of its cultural milieu and, to varying degrees, reflects the tastes and concerns most prevalent among its audience, it is unusual for a series to do so as concretely as *Roseanne,* the breakout hit of the 1988–89 television season.[2] The blue-collar situation comedy achieves a gritty texture of character and setting that viewers recognize as similar to the fabric of their own lives (e.g., its lead characters sit at a messy table in their cluttered kitchen and argue about who left toast crumbs in the butter). Moreover, *Roseanne* reflects the growing influence of women in the production of primetime television programming, with women in such key positions as executive producer, producer, director, and writers for the series. Finally, although the multiple camera video production of *Roseanne* is typical of

From *Journal of Popular Culture,* Vol. 24, Spring 1991, pp. 71–78. Reprinted with permission.

contemporary situation comedy, the unusual production culture created by the often conflictive personalities of its production staff and ensemble cast has a significant effect on the overall tone of the series.[3]

The creative force behind *Roseanne* is The Carsey-Werner Company, an independent television production company founded by Marcy Carsey and Tom Werner in 1981. Their earlier collaboration at the ABC television network during the late 1970s resulted in such series as *Barney Miller, Soap, Taxi, Happy Days,* and *Mork and Mindy* helping to catapult ABC into the first place position. When ABC moved away from comedy development in 1980, Carsey left to begin her own company, and Werner joined her a year later. Although their first pilot, *Callahan,* did not make the 1982 fall schedule, their short-lived series *Oh, Madeline,* during the 1983–84 season was an indication of the comedic programming that was to become the company's hallmark. In 1983 Carsey and Werner joined forces with Bill Cosby in creating obstetrician Cliff Huxtable, along with Claire, his attorney wife, and their five children.[4] ABC passed on the series when it was pitched to them in 1984 because situation comedy was in a decline. However, NBC's Brandon Tartikoff bought the series and scheduled *The Cosby Show* on Thursday nights, an action largely responsible for moving NBC into a number one position in the ratings.

## *The* Roseanne *Concept*

Following the success of *The Cosby Show,* Carsey-Werner developed *A Different World* and, in mid-1987, listened to *Cosby* writer Matt Williams pitch his concept for a new situation comedy about three blue-collar working women, originally entitled *Life and Stuff.* "The idea was to take one married woman with kids, one divorced woman with a child, and another woman who is single, put them in a factory in Indiana—because that's my background—and explore all the things they confront and how they survive, in almost every case because of their terrific sense of humor," according to Williams.[5] While developing his concept, Williams went back to the Midwest, where he had grown up, to do research to create the series' fictional town of Langford, Illinois. His visits to Evansville, Indiana, and Elgin, Illinois, and his research-related conversations with approximately fifty women made him realize that "they're doing everything—raising kids, working full-time, keeping a house, supporting their other friends, dealing with family problems. And they did it with this stoic but fatalistic frame of mind of, 'Well, hell, it can't get any worse. So what are you going to do but laugh?' "[6]

While Williams was developing his series' concept, Carsey and Werner were having discussions with Roseanne Barr about the possibility of the stand-up comic starring in a series. Williams's concept seemed perfect for Barr, and when he viewed a special the comedienne had done for HBO, Williams agreed that not only was she very funny, but she also had something to say: "what Roseanne brought to the mix that I didn't was the strong feminist point of view." During her career as a stand-up comic, Barr had created a persona based on her own life, and she suggested that the focus of the show be a working woman with three children—a boy and two girls.

As the concept for *Roseanne* evolved, Williams knew he wanted to make both the stories and the set "worn and lived in," saying that the most important thing he learned from writing for *The Cosby Show* was that the key to a successful television series was audience identification. "I always remembered Bill saying he wanted the people at home to ask, 'How did you get inside my house?' "[7] To that end he worked with production designer Garvin Eddy to create sets that would allow the director to use four cameras to shoot the characters from a variety of perspectives (e.g., the bedroom, the bathroom, answering the door bell in the living room, standing at the kitchen stove), unlike the style of typical situation comedy which utilizes an end stage look. "I said, 'Let's invite the audience to come in and sit around the table.' That way it feels real because the whole point is that life in a household is continuous and rambunctious."[8] Williams asked the set designer to visit his (Williams's) grandmother's home in order to model the kitchen in *Roseanne* after hers, right down to the open shelving and louvered windows above the sink. Much of the set dressing, such as the living room couch and chairs, were bought out of the Sears catalogue, which is what the fictional Conner family (and many in the audience) would have done.

A number of story lines were developed for *Roseanne* once the concept was established and the series cast in March 1988. However, the actors did not know the characters very well nor did the writers know what the actors could bring to the screen. As a result, the production staff decided to let the series grow organically. For example, as first-season *Roseanne* producer Gayle Maffeo recalls, the relationship between Booker (the factory boss) and Jackie (Roseanne's sister) changed significantly once the two actors began to develop their characters. "It isn't just your everyday relationship, it's not a garden variety love affair. There's a love/hate kind of thing that's happening and it's fun to watch it happen. But until the chemistry of those two people met on stage, you couldn't conceive it. And it changes. So the direction changes."[9]

After passing on *The Cosby Show,* ABC's rueful Brandon Stoddard, given a second chance when offered *Roseanne,* scheduled the new series in one of the network's best time slots, between the popular *Who's the Boss* and *Moonlighting* on Tuesday nights. *Roseanne* initially received a six-episode order from ABC as a limited-run series in March 1988, but had only completed production on one episode before the beginning of the six-month strike by the Writers Guild of America. However, ABC executives were pleased with the episode and increased the order to thirteen episodes for the 1988–89 season. Twenty-four episodes were ultimately produced for the series' first season.

## The Production of Roseanne

*Roseanne* is at once brash and funny, compelling and moving, because of its integration of two comedic forms: presentational and representational. Audiences historically have enjoyed the stand-up comic who appears before the curtain and offers a monologue replete with jokes, anecdotes, and insults. From Milton Berle and Sid Caesar to Johnny Carson and David Letterman, the presentational style of the quick-tongued star daringly interacting with the audience has remained a favorite in American popular culture. The creators of *Roseanne* have kept the caustic dialogue of its stand-up comic stars but moved them back from the footlights and into the representational structure of the proscenium stage. There the easy improvisational style of Roseanne Barr and John Goodman, played within realistic sets in front of fluidly choreographed cameras, creates situations within family and friendships that the audience recognizes as similar to their own.

*Roseanne* is produced on Stage 2 of the CBS/MTM lot in Studio City, California, its 14,000 square feet making it one of the largest sound stages on the lot. This gives the set designers considerable latitude in building not only more sets for the series than are typical of situation comedy, but sets of different sizes as well as a "swing-set" which varies in type from week to week, depending on what is called for in the script. Producer Gayle Maffeo believes that the size of the sound stage "helps with the development of the show and makes it more interesting visually than situation comedies that are restricted to three small sets."[10] In addition to considerable flexibility in staging space and set design, the series' method of production (multiple-camera video) encourages a more improvisational style and allows for greater continuity of comedic performance.

An observation of the five days of the on-stage production process of a typical *Roseanne* episode reveals not only how the script evolves to fully utilize the talents of the ensemble cast, but also how interpersonal interaction, improvisational style, and creative conflicts create a tension on the set that contributes to the realistic dynamics of the cast when in character and ultimately affects the "look" of the series on the screen.[11]

Production number 208, entitled "Language Lessons," written by Laurie Gelman, was rehearsed and videotaped during the week of October 17–21, 1988. There are generally two interwoven stories in each episode of *Roseanne*. In the "A" story of "Language Lessons," Dan's growing despair at having been out of work for two weeks is aggravated by a visit from Roseanne's sister, Jackie, who suggests that he would not be so irritated by the amount of time she spends at the Conner home if he were not there all the time. The "B" story involves the efforts of younger daughter Darlene to create a castle out of popsicle sticks in order to earn a passing grade in her history class. Dan is so angered by Jackie's callous comments about how much time he spends at home that he refuses to speak to her; Jackie, for her part, was only teasing Dan, not knowing that he has just lost a bid for a dry-wall (construction) job. Young son D. J. accidentally breaks Darlene's castle, and she says it will be his fault if she fails history. Mediatrix Roseanne ultimately resolves the conflicts by suggesting that Jackie might be a bit more considerate of Dan, that Dan needs to remember that Jackie is her sister and has a right to visit the Conner home, and that Darlene's lack of studying during the regular school term necessitated additional work to raise her history grade.

## Day One / Monday

The weekly evolution of an episode of *Roseanne* (in this instance, "Language Lessons") begins with a production meeting on Monday morning at 9:30 A.M. when the production staff (e.g., producer, director, production manager, stage manager) and representatives of the ABC television network Program Practices Department gather around a long table on Stage 2 to discuss the script. The group reviews the stage directions indicated in the script regarding set changes, props, wardrobe, and so on, and discusses the production requirements necessary for the episode. Additional props, wardrobe, and even another set may be created during the first few days of the production week.

The cast members arrive at 10 A.M. and take their places at the table, while the production staff and network personnel sit behind them and listen to a first reading of the script. Roseanne Barr, a seven-year veteran of stand-up comedy, and John Goodman, comedian and motion picture actor (e.g., *Raising Arizona, Punchline, Sea of Love*), co-star as Roseanne and Dan Conner, with Laurie Metcalf in the supporting role of Roseanne's sister (Jackie Harris). Seven-year-old Michael Fishman (D. J. Conner) spends much of Monday in the classroom near Stage 2 with a welfare worker who is also a teacher; a stand-in takes his place on stage until it is imperative that he be there. Ninth-graders Sara Gilbert (Darlene Conner) and Lecy Goranson (Becky Conner) also attend classes in a room near the sound stage.[12]

Notes are taken during the reading and then discussed. Barr is harsh in her critique of the script because she created the "Roseanne" character and insists that the language and actions of the character be consistent with her conceptual vision. Series' creator Matt Williams envisioned the character as someone from his childhood in Indiana, albeit modified for Barr, while Barr believes her character would not do or say much of what is called for in the script. Although on the surface the discussion appears informal and relaxed, it is evident that the creative conflict and resultant tension will continue to build throughout the week and affect not only the ensemble players, but the production staff and writers as well. While this creates a production culture that is often difficult, the ultimate result will be a finely-honed script and performance by the time the episode is videotaped at the end of the week.[13]

Although the script has already gone through numerous revisions, the first draft (white) distributed on Monday will continue to go through revision up to the final runthrough on Friday before taping. After a lunch break, director Ellen Falcon begins to block and rehearse each scene with the actors from 1 to 6 P.M. Prior to the beginning of the first day's rehearsal, Falcon studied the script and visualized how the episode should "look." Although there is input from the actors during rehearsal, each scene is blocked according to Falcon's preconceived visualization. She may change that visualization during the next few days, sometimes moderately, sometimes drastically, depending on what it takes to make the comedic action flow. Although blocking during the first two days of rehearsal is done without cameras, the director blocks with specific shooting patterns in mind.[14]

Falcon is as fiercely protective of her prerogatives as director as Roseanne Barr is in her insistence that the series' concept and characters remain true to her vision. Both Falcon and Barr have strong personalities which are often in conflict during rehearsal, for example, Barr does not understand the direction in a scene and believes that her interpretation, rather than that of the director, is more appropriate for her character. The conflict between director and star will continue during subsequent episodes, until Falcon is compelled to leave the series during the winter (December) hiatus.

## *Day Two / Tuesday*

The cast members' return to Stage 2 on Tuesday brings a charged atmosphere to the set. The insults and disagreements found in virtually every episode of *Roseanne* have their origins as much in the production culture of the series as in the scripts. Although the insults are generally humorous and the disagreements generally mild, in a manner similar to the Conner family's interaction on screen, they can just as easily turn harsh and abrasive. Such conflict is intensified by the difficult progress of "Language Lessons" and the problems Barr has with her character's development.

A second reading of what is now a partial script is done at 10 A.M. on Tuesday. Act 2 did not work well on Monday and is being rewritten for distribution later that day. The yellow paged script indicates "Act 2 and Tag to come." Rehearsal of Act 1 is from 11 A.M. until 5 P.M., with a one-hour lunch break. Although Barr and Goodman follow the script during rehearsal, their improvisational comedic style surfaces as they interact with each other on the set. Their ad-libbed lines and comedic routines are often written into the script as it undergoes revision. A run-through of the rewritten episode is held from 5 to 6 P.M. for the executive producers, producer, network personnel, and writers, who make extensive notes regarding what problems are still evident in the episode. The writers revise the script that evening during what is traditionally known in the television industry as "rewrite night."

The tension that grows throughout the week is often dissipated by the antics of Michael Fishman (D. J.) who races about the stage, trying to escape a young stage manager, Barnaby Harris, whose major assignment is to take care of the little boy, entertain him, and see that he gets to rehearsal/taping and school classes on time. The pressure is also eased by the ministrations of craft services personnel who provide a variety of foodstuffs throughout every day of production: coffee and fruit, soft drinks and fruit juice, mineral water and peanuts, bagels and sweet rolls. Hot cereal is available in the morning, replaced at noon by Chinese food, lasagna, chili, and fruit. In the afternoon desserts cover the table: cake, pie, more fruit. One of the craft services personnel brings in a large box of pumpkins and announces a pumpkin carving contest. Members of the cast and crew pick up pumpkins with a promise to return them, creatively carved, on Halloween.

### Day Three / Wednesday

A revised script with orange pages is issued on Wednesday morning, as close to "fresh form" as will be seen between Wednesday and Friday. Although revision pages of different colors may be issued on Thursday and Friday, the result of Tuesday "rewrite night" is a fairly locked-in script. "Language Lessons" will continue to have problems, however, and new pages will be distributed right up to the final runthrough on Friday. For example, one issue is the manner in which Jackie and Dan, Darlene and D. J., resolve their difficulties and find the "language" to apologize to each other. Barr, Goodman, and Metcalf huddle with Falcon, while producer Maffeo watches anxiously from the sidelines. Although she has a split-screen monitor in her office in the Carsey-Werner administration building, with a direct video feed from Stage 2, the cameras will not be brought in until Thursday. Because she is in charge of the entire production of the episode, Maffeo often goes down to the stage to watch rehearsal, to listen to problems of the production staff, to occasionally offer suggestions to the director. When she cannot be there, Maffeo has one of her assistants monitor the episode's progress.

Despite the continuing problems with the script, the episode is "on its feet" and taking the shape that will be dramatized before the cameras on Friday. Barr and Goodman continue to improvise, not only during the blocking and rehearsal of scenes, but during breaks as well. Stage Manager Mark Samuels watches, bemused, and makes notes in his script. Falcon ignores their shenanigans until it gets in the way of rehearsal. Barr still does not know her lines, and this slows down rehearsal while Samuels prompts her or she refers to the script. The horseplay and improvisation of Barr, Goodman, and Metcalf during breaks in rehearsal spill over into performance and easily become part of a scene. Roseanne and Dan Conner's expressive faces, their comedic bits in the cluttered kitchen, the humorous nuances in ordinary dialogue are the result not only of what is called for in the script or requested by the director, but the culmination of the delight and often angry conflict the ensemble players experience during rehearsal.

Falcon watches the improvisation of the cast and listens to their jokes, even as she confers with her associate director during rehearsal breaks, noting moments that she may want to fall back on during the Friday night taping. A veteran television director, Falcon believes that her MA in film production from USC gave her the expertise to ultimately direct multiple camera video because she is able to block and edit "in her head."[15] Her approach to blocking during rehearsal is in terms of how she will edit the episode during the taping Friday night. Although all four cameras will tape simultaneously, Falcon's "first cut" will be done "live" as it is taped. Every scene of the episode is blocked, often out of order, in preparation for the cameras that will be brought in on Thursday.

The "creative differences" that Barr has with members of the production staff take their toll on Falcon, who is caught between the various factions. Although she tries to accommodate everyone, the struggle to move the episode forward makes her short-tempered. Despite the problems on stage, however, Falcon's manner is professional as she works with the script and technical crew in getting ready for the cameras that will be brought in the next day.

Falcon has studied the script and marked it with directions for the camera crew (e.g., camera movement, angles, which camera does what). She gives it to her associate director Wednesday night who copies all the notes into his script as well as into the technical director's script. This is in preparation for a meeting early Thursday morning between the director, associate director, technical director, and camera operators.

## Day Four / Thursday

The blocking and rehearsal on Tuesday and Wednesday have been "dry," that is, without technical crew and cameras present. Although Stage 2 has a production booth which could be used by the director, *Roseanne* utlizes a video truck that is parked outside Stage 2 by 7 A.M., Thursday and connected to the sound stage by the video and audio cables. Falcon will remain in the truck on Thursday and Friday, directing the rehearsal and taping from there, utilizing headsets to communicate with her stage manager on the sound stage. Falcon has a four-way split screen monitor which allows her to see all four cameras and decide which shot she will use in her "live" edited version. However, because all four cameras videotape simultaneously, she (and the producers) have four continuous tapings from which to later choose alternate shots.

When the camera operators have their equipment set up, Falcon talks them through each shot, telling them exactly what angle she wants, how close the camera should be to the actors, how wide the shot should be. The camera operators mark each of their shots on five-by-seven inch cards, following the numbering of shots in the script. Each camera operator thus knows which shot in each scene is his/hers and how it is to be done. The carefully pre-planned camera choreography on *Roseanne* results in a preliminarily edited episode at the end of Friday night's taping. Camera blocking with the cast runs from 9 A.M. to 4:30 P.M., with pink revision pages for Act 2, Scene 1 incorporated into the rehearsal.

Barr's ever-present acting coach takes notes, confers with makeup and wardrobe personnel, talks quietly to Barr who listens attentively. The star still has not learned all of her lines, and although Falcon is not present on stage, her irritation is evident from the reaction of stage manager Samuels who listens to her comments and instructions on his headset. A runthrough of "Language Lessons" begins at 6 P.M. By running the cameras under the light with the cast in partial wardrobe, the technical director is able to determine whether certain colors cause video flaring. Further refining of the episode continues as the network and production company personnel contribute their comments after the runthrough.

During a break in the afternoon rehearsal, Roseanne Barr flops onto the couch in the living room set, still unhappy with the revised script. She finds the overall quality of the series' scripts poor, she says, making it difficult for her to play her character because her lines and actions are not consistent with how she conceptualized Roseanne Conner.[16] Despite the efforts of her (then) husband, Bill Pentland (who is an executive consultant for the series as well as one of its writers), to calm her down, her agitation continues until it is time to resume rehearsal. Known for her acerbic tongue on and off stage, Barr will let her emotional force build until it spills over in performance on Friday, provoking a sense in the audience that she is not an actress playing a part, but a real person wearily or angrily confronting the mundane or difficult problems of everyday life. While this generally contributes to the realistic tone of the episode as videotaped, it also has a negative effect. For example, other members of the cast, such as John Goodman and Laurie Metcalf, not only have to be in character but must struggle to keep Barr on an even keel so a scene will work. This division and dissipation of creative energy is noticeable in the sometimes uneven comedic pace of an episode.

### *Day Five / Friday*

Although Friday is the day when the entire episode must come together in acting and technical performance, there is a playful mood on the set that belies the still increasing tension among the cast and production staff. The crew call for equipment set-up is from 11 A.M. to 12:30 P.M. The first makeup call for the cast is at 11:30 A.M. Yellow revision pages have been issued to replace dialogue that is still causing difficulty, and Falcon runs the cast through the changes. Cast notes are given from 12:30 to 1 P.M., followed by block and tape, which is essentially a dress rehearsal in full wardrobe, from 12:30 to 5:30 P.M. At the beginning of the season, the dress rehearsal for *Roseanne* was taped before a studio audience in the afternoon. However, because the children in the cast must be released by 10 P.M. on Friday, the production schedule was changed. Rather than having a studio audience for the afternoon dress rehearsal, that time is spent taping the segments involving the children and as much of the episode as possible.

What is known in the television industry as a "must-go" (required) dinner is catered for the cast, production staff, technical crew, and guests in an unused sound stage from 5:30 to 6:30 P.M. *A Different World* has its catered dinner on the other side of the stage, although the two groups from the Carsey-Werner shows do not mix. It is Michael Fishman's birthday (D. J.), and he is given a cake and gift by the series' cast and crew.

The studio audience, which has been waiting outside the studio lot, is brought in by pages (ushers) at 6:30 P.M. and seated in the bleachers. A standup comic "warms up" the audience at 6:45 P.M. by telling them about *Roseanne,* asking questions about the audience members, telling jokes, and so on. Although the audience warm-up is standard for multi-camera videotape and filmed situation comedies, the stand-up comic is chosen for his/her comedic style which typically matches the comedic tone of the show. In the case of *Roseanne,* the comic appears to deliberately try to get the audience into the "mood" of *Roseanne* by telling risque jokes, making ribald remarks, insulting specific audience members. The mood in the bleachers is thus similar to that of the Conner home or the plastics factory. It is imperative that the audience be primed for the show because their response to the evening's taping will be integrated into the soundtrack in postproduction sweetening. A condensed version of the "pilot" (broadcast on Tuesday night of that week) is shown on closed-circuit monitors mounted above the bleachers in order to familiarize the audience with the series. The cast has final notes and make-up touches from 6:30 to 7:00 P.M.

Taping begins at 7 P.M., with short breaks between scenes. The standup comic holds the audience's attention by jokes, running patter about the television industry, and comments on guests in the audience. Melissa Gilbert, star of *Little House on the Prairie* and older sister of Sara Gilbert (Darlene), is in the audience and is introduced. John Goodman comes up into the bleachers during a break in taping to talk with friends. Although asked for his autograph, he politely declines. Series' creator Matt Williams alternates between sitting with his guests in the bleachers and conferring with producer Maffeo on the stage. The stand-up comic continues to insult audience members who in general try to ignore him. However, the audience reacts angrily to his consistent singling out of an elderly gentleman. "Leave him alone," many shout. The gentleman gives as good as he gets, however, and before he leaves with his chartered bus group of senior citizens from San Diego, he sends a few zinging rejoinders to the astonished comic. The audience cheers.

The taping of Scene 3 of Act 2 is held up by problems that Goodman and Barr have with their lines. As the stand-up comic tries to hold the attention of the tired and restless studio audience, he is asked about Roseanne Barr's family. Barr, deep in conversation with her director and co-star on the other side of the sound stage, has been paying close attention to what is going on in the audience bleachers. She turns around, glares, and yells, "No, you don't answer that question!" Her personal world is private, she insists, and the audience has come to see her character, Roseanne Conner, not Roseanne Barr. Videotaping of scenes toward the end of the episode is often stopped and begun again because the acting is ragged, the cast weary from trying to remain in character despite the often emotional interruptions.

During one of the last scenes, the director's voice is heard over the speakers: "Cut!" Falcon comes on to the stage, and Barr leaves. For twenty minutes the audience sits, looking at the empty sets and the camera operators who talk quietly among themselves. One by one the cast members return to the set; no explanation is given for the delay. Then Barr returns to the set with her acting coach and takes her place. Taping continues. There are more interruptions, but finally the episode is completed. Each scene has been taped twice and the session has run long: it is 10:30 P.M. before the stage manager repeats the director's words, "that's a wrap." [17]

The audience wearily files down the bleacher steps and out to Radford Avenue, while the cast of *Roseanne* quickly disperses. The cameras are capped and pulled off-stage, the technical crew begins to take down the lights, and the swing-set is struck. Shortly afterwards the lights of Stage 2 are turned off. Although production usually begins on the next episode the following Monday morning, there is a week's hiatus built into the *Roseanne* production schedule every two or three weeks in order to give the writers time to polish first drafts of scripts. After a hiatus following completion of "Language Lessons," the cast and production staff reassemble on Monday morning, October 31, to begin rehearsal for the next episode entitled "Lover's Lane."

### *"How Did You Get Inside My House?"*

Although the conceptual evolution and production of *Roseanne* is typical of most primetime situation comedies, its distinct look—verbal as well as visual—is significantly different from its comedic counterparts. As a genre, situation comedy does not have as great an opportunity to be innovative as episodic series drama. The audience has certain expectations of situation comedy, in particular that whatever has upset the "situation" will be resolved at the end, and order will be restored. Moreover, situation comedy relies on its audience being distanced from the action, safe to laugh at the proverbial banana peel pratfall because it is not happening to them. In contrast, *Roseanne* has shrugged off the formulaic constraints and expectations of the genre and, in the words of Matt Williams, invited its audience to come in and sit around the sticky kitchen table.

Although the sets, set dressing, and wardrobe are realistic to the smallest detail—messy laundry room, living room couch with lumpy cushions, refrigerator covered with notes and children's drawings, kitchen counter cluttered with dirty dishes and groceries—the more significant realism lies in the interaction of the characters, the abruptly shifting storylines of episodes, the brash and funny dialogue. Matt Williams's research in developing the series' concept and Roseanne Barr's polishing of her character in comedy clubs throughout the country have created a familial culture more reflective of middle-class Americana than what is typically found on the television screen.

The creative team that developed *Roseanne* took a chance doing a series that was the flip side of the sanitized *Cosby Show*, but the audience seems to enjoy Roseanne Conner in a shabby sweat shirt as much as they admire Claire Huxtable in her designer sweater. While many in the audience may not say the things Roseanne Conner does, the popularity of the series indicates that they understand her language. Unlike Claire Huxtable who comes home to a house that seemingly cleans itself, to children who though occasionally difficult are fun to be with, to a husband who shares the household tasks, Roseanne's home is always in need of cleaning, her children are usually noisy and irritating, and her husband last dried the dishes ten years ago. Dan Conner dislikes his wife's sister, Jackie, and seldom tries to hide it. Roseanne tells her son D. J. that he's not stupid, just clumsy, like his dad. Darlene wonders whether she will still be a good softball pitcher now that she is developing breasts, and Becky sulks and refuses to speak to her mother when told that they cannot afford the expensive dress she wants for a school dance.

Roseanne Barr has suggested that *Roseanne* is in the tradition of *The Honeymooners* (with her own character modeled on Ralph Kramden and that of John Goodman a cross between Alice Kramden and Ed Norton) and in the more recent tradition of *All in the Family* with its outspoken dialogue.[18] At another level, however, *Roseanne* is similar to *M*A*S*H*, in that both series rely on humor to temper the reality of their characters' lives. For example, when Roseanne has a hard day at the factory and then an equally difficult evening at home, she leaves her husband and children to fend for themselves and walks until she finds a coffee shop that is just closing. She persuades the tired waitress to serve her a cup of coffee, and the two women, strangers, talk. The woman's husband has been deceased for some time, but she still misses him so much that even though she does not enjoy televised baseball and football games, she turns them on just to fill the empty house with the sounds she heard when he was alive. Moved, Roseanne returns home to discover that Dan has cleaned the house and sent the children to bed. Although the coffee shop conversation confronts Roseanne with a sobering thought unusual for comedy, it is tempered by a fiendishly grinning Dan who announces, "I saved the bathroom for you." The audience, sitting at the now-clean kitchen table, shares a laugh with Roseanne. "Well, hell, it can't get any worse. So what are you going to do but laugh?!!"

Although *Roseanne* is similar to *The Honeymooners, All in the Family,* and *M*A*S*H*, it has its roots in the stylistic blend of presentational and representational comedy of the variety shows of the 1950s. It also has its roots in the original domestic situation comedy form that was developed to dramatize the traditional nuclear family during the 1950s and early 1960s. Although far more realistic than the sanitized families headed by Ozzie and Harriet Nelson, Ward and June Cleaver, Jim and Margaret Anderson, like its comedic predecessors *Roseanne* succeeds by playing upon recognizable family situations.

During the 1970s and 1980s situation comedy broke away from the more traditional family ensemble in order to better emulate the changing structure of the American family. Comedies with single parents, extended families, and the surrogate family of the workplace and the school filled the primetime television environment. Although the advent of *The Cosby Show* in 1984 marked a return to the nuclear (albeit sanitized) family unit, it was *Roseanne* that moved the situation comedy more firmly into the reality of its viewers' lives, its realistic situations, dialogue, and sets appealing to a large and diverse audience. Although its blue-collar setting is not shared by all of its viewers, the problems of surviving in post-Reagan America (e.g., with a mortgage, three children, two wage-earners, and employment problems) has a broad appeal. Moreover, *Roseanne* also appeals directly to the greater majority of the thirty (or forty) something audience who are not yuppies. When the series lampoons what Roseanne and Dan Conner consider the pretensions of the stereotypical thirty-something crowd—gourmet dinners, fine wines, and mineral water, jogging and aerobics—it dramatizes the antithesis of yuppie culture. The angst that permeates the ABC dramatic series *thirtysomething* also permeates *Roseanne* but in a different way. Roseanne and Dan Conner, as well as others with whom they interact in the series, experience feelings of anxiety and depression in their lives, but they do not have the time or desire to really dwell on them. Their response is not to become more introspective, but to go bowling on a Friday night, to share a six-pack of beer with friends while pretending to fix the truck, to go to the mall even though they cannot afford to buy anything. *Roseanne* thus flies in the face of the pop culture establishment by its irreverent and candid putdowns, as well as in the face of the series' sponsors, most of whose products the Conner family cannot afford to buy.

### Conclusion

The success of *Roseanne* is the result of a unique confluence of factors. The Carsey-Werner Co. was negotiating a possible series with Roseanne Barr at the same time that Matt Williams was developing a concept for a new comedy series. Carsey-Werner veteran Matt Williams had learned the importance of audience identification from his writing on *Cosby,* and his intense research efforts lead him to make *Roseanne* as realistic as possible. Carsey-Werner, with its previous up-scale comedy hits *(The Cosby Show, A Different World),* was aware that there was another segment of the thirty-something generation with different characteristics than those usually attributed to the yuppies, a large audience segment that was not up-scale but blue collar.

There were other factors that helped to make *Roseanne* a breakout hit. Although the series could not go on the air in spring 1988, as originally scheduled, because of the strike by the Writers Guild of America, it had one episode completed before the strike and was able to go into production before most other series in the fall because it was a non-union show, thus gaining an early advantage on other series that had very late premieres.[19] Moreover, the writers' strike had forced network viewers to look to alternative entertainment, such as cable television programming. When *Roseanne* made its appearance in fall 1988, the high level of cable television penetration, with its programming of stand-up comics, had made the traditional network audience more receptive to the broad and often biting humor that typifies *Roseanne.*

The success of *Roseanne* has not been without problems, some of which continue to plague the series. The "creative conflict" on and off the set, precipitated in large part by its star, culminated in the midseason departure of the series' creator/executive producer, Matt Williams, and its director, Ellen Falcon, the result of irreconcilable differences with Barr. Although the production staff shake-up appeared to be a turning point for Barr, who realized she needed the supportive framework of her ensemble cast, crew, and production staff if she were to keep up the grueling pace of episodic series production, her insistence on controlling the series continued to the end of the first season when producer Gayle Maffeo also left the series. However, despite its problems, *Roseanne* continued to be in the Nielsen Top 5, often in first place, throughout its first two seasons.

Although the American people applauded *Roseanne,* as indicated by its consistently high ratings, the television industry did not. When the 1988–89 Emmy nominees were announced by the Academy of Television Arts and Sciences in August 1989, *Roseanne* and Roseanne Barr were virtually ignored. Although the series received four nominations, the only major nominee was Barr's co-star, John Goodman. Neither the star nor the breakout, top-rated series of 1988–89 was nominated for "Best Actress in a Comedy Series" or "Best Comedy Series." Speculation in the industry regarding why Barr was overlooked ran the gamut from "There are a lot of people in the profession who simply don't like the show. They feel it's too blue-collar and too broad. It's not a sophisticated comedy like *Cheers* or *Murphy Brown* (both Emmy nominees)" to "The snub is because Roseanne Barr is not a part of Hollywood. She made the [Nielsen] Top 5 right off the bat. You have to put in more time than that in this town, pay your dues."[20]

The creators of *Roseanne* put a new spin on an old genre by exploring terrain familiar to themselves and to their audience. "A lot of us grew up in the same circumstances as the Conner family," says Marcy Carsey, the daughter of a shipyard worker from Quincy, Massachusetts, and Matt Williams agrees. "If I hadn't gone to college on a football scholarship, I would have been a bricklayer."[21] *Roseanne* has gone back to its roots in mingling presentational and representational styles of comedy, in revitalizing the traditional nuclear family. It has broken new ground on the primetime television stage where the players can never be too rich or too thin, where children and relatives are charming, and given us, instead, characters who are just getting by financially, who are overweight and not concerned about dieting and exercise, whose children and relatives are not always pleasant. *Roseanne* has held up a mirror to the people and culture of the last decade of the twentieth-century, and the reflection is often bittersweetly accurate.

### Notes

1. Gayle Maffeo, producer, *Roseanne* (1988–89 season), interview with author, 19 October 1988, Studio City, California. Maffeo left the series at the end of the first season.

2. *Roseanne* ranked in the Nielsen Top 5 with its first episode and has remained in that group during its first two seasons, at times taking the first place position away from perennial frontrunner *The Cosby Show.*

3. "Production culture" refers to the work environment created by those involved in the production of *Roseanne,* for example, the interpersonal interaction between and among the cast, crew, and production staff, a collective (and sometimes conflictive) effort to achieve a shared production vision.

4. Cosby's original idea was for a comedy that would cast him as a janitor and his wife as a construction worker. Carsey and Werner disagreed, believing that he should be the head of an upper-middle-class family.

5. Nikki Finke, "The Blue-Collar Backgrounds Behind a Blue-Collar Hit," *Los Angeles Times* (January 26, 1989), section 6, p. 10.

6. Ibid.

7. Ibid.

8. Ibid.

9. Maffeo interview.

10. Ibid.

11. Discussion of a typical production week for an episode of *Roseanne* is based on the author's on-site observation of the series and conversations with members of the production staff, cast, and technical crew on Stage 2 of the MTM/CBS lot in Studio City, California, during the week of October 17–21, 1988.

12. California state law dictates the amount of time minor children must spend in school and how much time they may spend on the sound stage. The law also requires a welfare worker to be present whenever minor children are on the set to make sure they are handled correctly (e.g., that they do not exceed the number of hours they may be on the set, that they are not tired from too much rehearsal or too many takes).

13. Maffeo interview.

14. "Blocking" is the placement and movement of an actor or a group of actors in front of a camera, with the camera(s) following them from one position to another, according to a shot pattern dictated by a preconceived aesthetic idea developed by the director.

15. Ellen Falcon, interview with author, 19 October 1988, Studio City, Calif.

16. Roseanne Barr, interview with author, 20 October 1988, Studio City, Calif.

17. The week's rehearsal and performance, captured on videotape, is ready to be disassembled, reassembled, and polished through postproduction editing and audio enhancement. The twelve to fourteen hours of videotape from the Friday night's taping are sent to Compact Video on Monday morning. *Roseanne* editor Marco Zappia uses the director's script, as well as the script supervisor's book with all the timings and takes marked, to complete the day-and-a-half process of cutting and trimming the director's cut, using ¾-inch copies of the original master tapes. Director Falcon has chosen her cuts in the control room during the taping, so one completely edited version already exists. However, because each camera was isolated and recorded separately, the producers can re-cut the entire production if the director's cut is not to their satisfaction. For each cut, the editor enters the time code on the video tape into the computer, generating what is known as an "edit decision list" (EDL) that will be used to retrieve the original master videotape in making the on-line edited master.

The edited episode is returned to Carsey-Werner on Wednesday, and the producers decide what, if anything, needs to be changed. The editor will do as many versions as are necessary, but generally an episode of *Roseanne* goes through three cuts before it is finished. The producers then meet with the series' music composers and decide whether they will have an original score written for specific scenes, or whether they will use music for which Carsey-Werner has clearance from the American Federation of Television and Radio Artists (AFTRA). Once the episode has been edited and the music added, the show is considered "on-lined." Up to this point the episode has been on ¾-inch video cassettes. When the computer-generated "edit decision list" pulls all the chosen shots in correct order from the original videotape, the shots are dubbed down one generation to what is called the one-inch master tape. The master tape is taken to the sweetening department, where the audio is recorded from the videotape to a 24-track audio tape for mixing. This includes adding such sounds as door bells, interior factory noise, the studio audience's response to the taping, and all the other sound effects that contribute to the texture of the series. The mixer then refines and balances all of the sound elements and lays the mixed sound track back down on the master video tape, recording over the original sound. The episode is complete: all that remains is to make copies for distribution, including a master and a backup copy for the ABC television network.

18. Joy Horowitz, "June Cleaver Without the Pearls," *New York Times* (October 16, 1988), Section 2, p. 10.

19. *Roseanne* had other advantages when it went into production for its first year on the air which contributed to its becoming an instant hit. In October 1988, the primetime television environment was a virtual vacuum awaiting new episodes whose production had been significantly slowed by the six-month strike by the Writers' Guild of America. Most television production companies are signators to all of the television industry's unions and guilds and must abide by their rules. However, when The Carsey-Werner Company was founded, its creators had the choice of remaining a non-union company, although it is a signator to the Writers' Guild of America (WGA), the Directors' Guild of America (DGA), and the American Federation of Television and Radio Artists (AFTRA). Although the strike by the WGA from March to August 1988 shut down production of all signators to the Guild, Carsey-Werner and a few other smaller independent television production companies were offered a temporary contract and allowed to begin production. *Roseanne* thus went into production in early August, well ahead of most other series for the 1988–89 season. Because *Roseanne* is a non-union show, Carsey-Werner is able to be highly selective in hiring who they consider the very best for production staff and technical crew, even if some of these personnel are not members of the union. This cuts down on the cost of the production because union requirements call for a larger crew than the series would normally need. Moreover, although the series is produced on the CBS/MTM lot in Studio City, which historically is a union lot, Carsey-Werner is able to solicit bids for such work as the building of sets, designing and making of costumes, and so forth, and will accept a bid from CBS/MTM personnel only if the bid is the lowest received. Further, during the Teamsters' Strike in October 1988, although other series in production on the CBS/MTM lot were hindered by pickets and the inability to use transportation workers for many weeks at the beginning of production for the 1988–89 season (e.g., *thirtysomething*), Carsey-Werner had a restricted gate that gave their production company personnel and other workers entrance to the lot and was thus not affected by the strike.

20. Bill Bruns and Herma Rosenthal, "*Roseanne*'s Emmy Snub: The Reasons Why," *TV Guide*, August 19–25, 1989, pp. 33–34.

21. Finke, p. 10.

# "After It Happened . . . ": The Battle to Present AIDS in Television Drama

## RODNEY BUXTON

With the exception of one episode of the series *St. Elsewhere* in 1983, prime-time American network television programming did not address the topic of AIDS in the early 1980s, even as the disease was reaching epidemic proportions. Then, in the Fall 1985–86 broadcast season, several programs dealt with AIDS, paving the way for an increased discussion of the epidemic in mainstream American culture. NBC presented *An Early Frost*, a two-hour made-for-TV movie. CBS aired an episode of the medical series *Trapper John, M.D.* that examined the epidemic. ABC offered its own narrative in the series *Hotel*. Yet even as the decade drew to a close and a self-proclaimed kinder, gentler President prepared to take office, AIDS remained a topic that television could only address with some difficulty. Perhaps no incident better illustrates the societal struggles involved in contesting the ways in which representations of AIDS and people with the disease are created than the furor that surrounded the production of a single episode of NBC's television drama *Midnight Caller*. My research into this controversy suggests that the construction of mass-media representations occurs in an arena of cultural struggle where competing social forces—production companies, individual artists, political action groups, governmental and civic interests, as well as private citizens—vie to shape discussion of social issues like AIDS.

Reprinted from *The Velvet Light Trap*, Vol. 27 (Spring 1991), pp. 37–47. By permission of the author and the University of Texas Press.

### *Backstory*

When scriptwriters came off their six-month strike in the fall of 1988, production companies had to scramble to provide programming for the late-starting commercial television season. During the strike, relations between writers and production company representatives were at best strained and at worst antagonistic. Theoretically, no scriptwriter for any scheduled series was to have written during the time of the strike. Consequently, even after the dispute was settled, writers were under pressure to complete new scripts quickly. Production companies needed to minimize the unavoidable lag between the end of the strike and the airing of new programming. At this historic moment NBC added a new hour-long episodic drama, *Midnight Caller,* to its Tuesday night lineup and saw the script for an episode titled "After It Happened . . ." generate sharp controversy.

*Midnight Caller* made its broadcast debut October 25, 1988. Like most American television series, *Midnight Caller* incorporates a franchise as its narrative basis; that is, a narrative formula which provides a means of standardization and programming control for the networks and familiarity for program viewers. From the standpoint of narrative franchise, *Midnight Caller* has its roots firmly planted in the police procedural, a genre that pits legitimized social forces against antagonists who would undermine the legitimacy and stability of the social order. Despite such generic lineage, the series attempts to expand beyond these narrative constraints. John Perry, the line producer for the series, has stated: "This is not just a cops-and-robbers show. Every episode will leave you with a thought-provoking process" (Ford, "Script" 4). Each week *Midnight Caller* incorporates a different contemporary social issue as part of the series' franchise. As a focus for individual weekly installments, these social issues provide the basis for a self-contained, episodic narrative.

While *Midnight Caller* privileges an episodic narrative structure, the series does incorporate some cumulative narrative elements that provide series continuity from week to week. These cumulative elements are closely tied to the Jack Killian character. Killian, a white, middle-class male, is the axial center of the series. In the series' backstory, he is an ex-cop not entirely at peace with himself having accidentally killed his partner during a shoot-out. As part of *Midnight Caller*'s franchise, Killian must reconcile his overbearing moral standards with his morally reprehensible (at least from his perspective) involvement in his partner's death. Because of his intense morality, Killian had been unable to deal with his fellow officer's death and quit the police force. Whether his moral values were the result of his police work or his police work the result of his values, Killian is a flawed bearer of moral standards. The inability to accommodate his own behavior to his system of moral beliefs fuels the cumulative stakes that link individual episodes of the series together.

Since quitting the police force, a period of psychological self-flagellation has not healed Killian's psyche. As an economic measure of last resort, he accepts the offer made by Devon King, the owner of radio station KJCM, to host a phone-in, late-night talk show. The laid-back, subdued atmosphere of his new job provides a narrative backdrop whereby Killian can work through his moral conflicts and contradictions. In addition, the various individuals who call in to voice their opinions on the social issue of the week eventually prompt Killian back onto the street. Once Killian is there, his psychological state and the social issue are interwoven in the narrative fabric.

In one of the early episodes of the 1988 season, the producers of *Midnight Caller* decided to include a narrative about a bisexual man with AIDS who knowingly risks the infection of his sexual partners. In the earliest stages of script development for "After It Happened . . . ," members of the production company were not prepared for the controversy that was soon to engulf them. As Stephen Zito, scriptwriter for the episode, points out:

> As I said, we never saw this coming. When the first cast breakdown went out, which had a one-line description indicating what this episode was about, we began to get calls. We had agents say, "Boy, this is really very controversial." It was the first glimmer we had that this was more than your average television show. And from that point on, basically, we were playing catch up.[1]

Until this historical moment, producers and writers who had incorporated AIDS into programming content had to deal, for the most part, with the conservative attitudes of network standards and practices departments who did not want to provoke the ire of any social group. When advocacy pressure had been applied, it was generally directed at advertisers to pull out of programming already scheduled and produced. In these cases, advocacy pressure usually came from more reactionary social forces.

## The Confrontation

As explained by Zito, he initially conceived of "After It Happened . . ." as a hard-boiled detective story based loosely on the Gaetan Dugas profile in Randy Shilts's *And the Band Played On,* a journalistic chronicle of the AIDS epidemic until 1985. As I will argue later, Zito's use of Shilts's book as a reference would prove to be the basis for many criticisms of the script. Originally, the narrative centered upon the search for Mike Barnes, an HIV-positive bisexual man who was irresponsibly infecting other individuals with the virus. One of the people Barnes infects is Tina Cassidy, a former lover of Killian. In addition to Tina's positive HIV status, she is also pregnant from her sexual encounter with Barnes. Upon Tina's request, Jack Killian was to search through the dark streets of San Francisco to inform Barnes of his HIV status and of Tina's pregnancy. When Killian discovers that Barnes is knowingly infecting other individuals, his goal becomes getting Barnes out of the swinging singles circuit. In this first draft of the script, the episode is resolved when Kelly West, another of Barnes' sexual conquests, murders him on the street after Killian is unable to stop the man.

In a *San Francisco Sentinel* article, journalist Dave Ford stated that a copy of this draft of the script was leaked to a member of the San Francisco AIDS Coalition to Unleash Power (ACT UP). San Francisco ACT UP is the local chapter of a loosely affiliated national activist group campaigning for the civil rights and medical interests of persons with AIDS. Despite the national affiliation, each local chapter of the organization determines its own activist agenda and approach. Ford wrote that a member of the *Midnight Caller* crew (who was not involved in the "After It Happened . . ." episode) became frustrated over the producers' unwillingness to make changes in the initial shooting script, which the crew member felt to be inflammatory. The crew member then gave a copy of the script to a member of ACT UP ("Script" 4).

From the executive producer's perspective, Robert Singer stated that he was not approached before being confronted by members of ACT UP. According to him, the confrontation began when:

> The script got out because someone in our group, our crew leaked it. It was unfortunate because drafts go through rewriting. I'm not sure that script would have been the one we ended up shooting. There is constant rewriting that goes on with scripts. I felt betrayed and felt overwhelmed when these people came at me. No one in the crew approached me until after the snowball started rolling.[2]

While a difference of opinion exists as to when he was first approached, Singer did meet with Terry Sutton of ACT UP and the process of rewriting the shooting script began. In addition, as Ford reported, a further meeting between health department representatives, ACT UP members, and series producers was scheduled in an attempt to head off any further problems with the production ("Script" 4 –5). Even with such precautions, the production proved to be a volatile site of conflicting and contradictory ideological agendas. Economics, politics, medicine, creative freedom, and "realism" were the most obvious issues raised in the confrontation enveloping Midnight Caller Productions, AIDS activtists, gay media watchdogs, the mayor's office, and labor unions.

While the activists and Midnight Caller Productions were the major antagonists in this confrontation, other entities figured into it as well. The City of San Francisco, through Mayor Art Agnos and municipal politicians, were caught in the crossfire between Midnight Caller Productions and the activists. On the one hand, Ford points out that the city would have lost $10 million in revenues generated by the four-month, on-location production shooting schedule for the series ("Script" 5). The economic advantages of placating Midnight Caller Productions were obvious, but as Jim Harwood noted in *Variety,* the city officials also depended upon the support of the gay community in San Francisco (52). Alienating the gay community would not have been a politically prudent move. As it was, city officials were fairly impotent in the confrontation because they had contradictory political and economic agendas. When an injunction was served against the activists, little attempt was made by police officers to stop the protesters and no arrests were made when the activists failed to obey the injunction. In fact, according to Ford, the only power city officials would implement was the ability to facilitate the negotiation process between Midnight Caller Productions and the activists ("Melodrama" 3).

While the activists may not have had a primary economic interest in the production of the *Midnight Caller* series, the loss of potential wages for labor and trade unions generated by its production would not have been suffered without criticism from these groups (Harwood 52). Since the power in San Francisco's gay community stems, in part, from its inclusion within a political coalition that includes labor and trade unions, the economic interests of their political allies could not be entirely dismissed in the activists' quest to challenge Midnight Caller Productions.

Members of ACT UP obtained a copy of the first draft of the script during the week of October 9–14, 1988. Sutton met with Singer for an hour at the end of that week. As a result of that meeting, the script was revised over the weekend of October 15–16. In order to head off any further confrontations, the San Francisco Mayor's office set up another meeting between a coalition of AIDS activists and Midnight Caller Productions for Monday, October 17. Besides ACT UP, the coalition was comprised of members from the San Francisco AIDS Foundation, the Mobilization Against AIDS, the Community United Against Violence (CUAV), and to a lesser extent, the Gay and Lesbian Alliance Against Defamation (GLAAD) as well as the Alliance for Gay and Lesbian Artists in the Entertainment Industry (AGLA). During this meeting, the activists were allowed to read the revised script. The various groups agreed to meet the following day after everyone had a chance to thoroughly review the revisions. Despite those revisions, the Barnes character was still a bisexual man. However, as Singer pointed out, the script revision did reframe Barnes as an anomaly in the gay community. On Tuesday, October 18, the activists argued that there was no positive gay character to balance the negative image of the Barnes character. In addition, the activists were still incensed about the violence toward Barnes contained in the script. On Wednesday, October 19, Singer stated that the violent ending of the script would be revised. In changing the ending, Singer felt that the script no longer condoned violence as a reasonable solution to the AIDS epidemic. Despite the changes, AIDS activists, especially members of ACT UP who were not satisfied with the basic premise of the script, threatened to disrupt the filming of the *Midnight Caller* episode. Members of ACT UP felt the script exploited the AIDS epidemic for rating shares (Ford, "Script" 4).

From the perspective of the production company, Singer felt that the activists' agenda was continually shifting:

They [ACT UP] came in and their agenda was kind of scattered. . . . Their argument seemed to be disjointed. You couldn't really pinpoint what it was that they were trying to get to. We would make changes, and they would say, "Well, this is the most important thing, and this is what upsets us the most." Then we'd make that change and then they'd say, "Well, yeah, but what about this then." Then it got all muddled.

On Thursday, October 20, Midnight Caller Productions was filming on location at the intersection of Jones and Broadway. As reported by Nina Easton, a group of eighty protesters so disrupted the shooting by shouting and whistle-blowing that the production had to be shut down ("Protest"). In response to the activists' disruption, Midnight Caller Productions obtained a court injunction on Friday, October 21, that ordered the protesters to remain 100 feet from the set and out of camera shot (Harwood 52). In an attempt to head off even more disruptions, Singer and Perry met with gay filmmaker Rob Epstein (director of *The Times of Harvey Milk,* a documentary about the gay San Francisco City-Council member who was assassinated in 1978) and Patrick Mulcahey, a scriptwriter for *Santa Barbara* and a member of Project Inform, a clearinghouse for AIDS information. For Mulcahey, "The resourcefulness and humanity of San Francisco's response to the epidemic weren't taken into account." Even so, Mulcahey did not share the activists' viewpoint that commercial television could only exploit the AIDS epidemic for ratings. In contrast, Mulcahey stated, "I would like to see the overall silence of episodic TV on AIDS be broken" (Easton, "Protest" 8). All in all, eight major changes were made in the final draft of the script. Apparently, the changes were not enough to satisfy members of ACT UP. On Tuesday, October 25, 200 protesters violated the court order injunction as the production company attempted to film outside Lipps, a bar at the intersection of Ninth and Harrison streets. The protesters chanted and blew whistles to disrupt shooting for a second time (Ford, "Melodrama" 3).

The production of "After It Happened . . ." became a cultural arena whereby competing social forces struggled to shape a mainstream discussion of the AIDS epidemic. In the struggle to control the content of the narrative, many issues were raised by the various individuals and groups involved in the confrontation. Character definition and development, medical concerns, and appropriate responses to the epidemic were loosely gathered under the rubric of realism. Either *Midnight Caller* did not reflect social or medical reality or the program would shape social reality in a detrimental manner. The conflict also strained the efficiency of narrative and production economies, thereby creating tools that the activists could use in their struggle for control over the content. Finally, the confrontation raised the issue of freedom of speech and creative rights. In short, many systems of meaning were activated by all sides in the effort to shape the narrative of "After It Happened . . ."

## Social and Medical Realism

Not surprisingly, the representations of persons with AIDS and gay men from the perspective of "realism" were major issues for the AIDS activist coalition protesting the production of the episode. On one level, these groups felt that the script did not accurately reflect the historic reality of the San Francisco gay community in 1988. In contrast to the irresponsible sexual behavior of the narrative's antagonist (a bisexual man), the activists pointed out that it was the gay community which led the promotion of sexual responsibility and safe sex practices in San Francisco (Palermino). In order to separate the ideological link between gay men, irresponsible sexual behavior, and AIDS, the activists argued that Barnes' character be changed to that of a heterosexual man infected by a blood transfusion. In spite of the activists' protests, however, the Barnes character remained a bisexual in the finished episode. To offset the association between gay or bisexual men and AIDS, the script was instead changed to explicitly acknowledge that the disease is not limited only to this particular social group. While "After It Happened . . ." still maintained a link between gay or bisexual men and the epidemic, the strength of this ideological connection was undermined to some extent by the inclusion of a discussion during one of Killian's broadcasts about the irresponsible sexual behavior of both heterosexual and homosexual individuals. Also, another scene was changed so that the bartender of a gay bar outlines for Killian the shift to sexual responsibility as the dominant response of San Francisco's gay community to the AIDS epidemic.

Even so, the process of creating the Mike Barnes character illustrates the power of other discursive systems in shaping the content of television programming. Scriptwriter Zito explained his inspiration for the narrative, by saying, "I had read in Randy Shilts's *And the Band Played On* about Dugas and the willful behavior in which he engaged." Because it was the first mainstream book on the AIDS epidemic to gain wide public acceptance, *And the Band Played On* provides a powerful cultural model on which many individuals, including Zito, would, in part, base their understanding of the epidemic.

In many ways, *And the Band Played On* is a watershed book on AIDS. Even its detractors recognize that it chronicles the negligence of the federal government in even acknowledging that AIDS posed a serious health problem. In addition, Shilts argues quite persuasively that the problem was ignored because the majority of individuals inflicted with the disease were (and are) homosexual men. In spite of interjecting these issues into mainstream American consciousness, radical criticism of the book argues that its political benefits are overwhelmed by the construction of its historic narrative through the personal denial, defeats, and triumphs of various heroes and villains. Consequently, as noted by Donald Crimp, it is the individual heroes and villains in the book, rather than the insights into political bigotry, which continue to predominate in the swirl of American culture (241–42).

In particular, the Dugas profile, on which the Mike Barnes character was based, has come under attack from radical critics. In *And the Band Played On,* Shilts reports that as early as 1982, the Center for Disease Control in Atlanta began to notice that the French Canadian airline steward's name came up as a sexual partner in interviews with several gay men who were the earliest to be diagnosed with AIDS or, as it was then called, Gay Related Immune Deficiency. As it turned out, Dugas had sexual relations with 40 of the first 200 men diagnosed with the disease in the United States (Shilts 147). In addition, the book chronicles Dugas's sexual exploits even after he knew he was infected with the disease. Of all the other individuals, medical institutes, and political denials that the book encompasses, Crimp counters that Dugas is the entity that is most incorporated into mainstream discussions of the epidemic.

For one thing, Dugas provides a cultural talisman for heterosexuals who want to separate themselves from any indication that they may be at risk for getting AIDS. After all, Dugas was gay and promiscuous. For dominant, straight, American society, he was the perfect cultural icon to define the parameters of the epidemic and ease the psychological fears that heterosexuals could be at risk for infection as well. In addition, Dugas provided someone to blame for the epidemic. As part of the fabric of dominant American values, individualism—the ability to take charge of one's own destiny—tends to overshadow the constraints of social structures on human existence. In an ironic way, the focus upon Dugas defines the epidemic in terms of individualism because he was unable to change his sexual behavior. Consequently, he is a much easier target to point at as the cause of the epidemic rather than the ignorance and bigotry of the political, medical, and social systems.

Because the Barnes character in "After It Happened . . ." is based on the Dugas "character," it should come as no surprise that the AIDS activists were angered by both the script and its author for this similar ideological construction. Because of the powerful hierarchies at work in mainstream discursive practice, the criticisms of Shilts's book have not had the circulation and legitimation accorded to *And the Band Played On.* As a result, individuals who depend upon the book for their understanding of the AIDS epidemic are less likely to be aware of ideological problems within the book. When asked if he had knowledge of the criticism leveled at the book, Zito responded:

> What happened was we all came out of a six-month strike and we got thrown onto these shows real fast. And we turned out our scripts awfully fast. I don't think this excuses anything, just explains. There wasn't time to spend two or three weeks and find out everything we needed to know. I should have done that and didn't. I had no idea this was so controversial an issue. I certainly had no idea this would come back in our faces the way that it did. I was not aware that [Shilts] had been criticized for that.

The activists wanted a character who, instead of intentionally spreading the disease, at the very least, unknowingly infects others. As it was, the activists shifted their points of criticism about the narrative throughout the production of the episode, thereby making it harder to maintain narrative continuity. Redefining Barnes's character as a heterosexual man infected by a blood transfusion would have been economically costly, as Harwood points out, since it would have meant rewriting and reshooting not just a few scenes but the entire episode. Since the activists could not get the change in the Barnes character incorporated into the script, they suggested other changes which would be palatable. To be more acceptable and "realistic," Rene Durazzo, media spokesperson for the San Francisco AIDS Foundation, stated that the script should be about individuals who have "very real problems changing their behavior, out of fear, out of denial, out of a number of complex reasons—it's not just out of maliciousness" (Ford, "Melodrama" 3). This change was incorporated in the episode as it was eventually produced.

In addition, the AIDS activists disapproved of the fact that Barnes's ex-lover, Ross Parker, was represented as a victim and an outcast. As activist writer Tristano Palermino pointed out, "It would be a rare shame that someone like Ross could be so isolated. Thousands of volunteers, gay and straight, young and old, serve people with AIDS throughout our city" (Ford, "Melodrama" 6). For the activists, the initial conception of this episode of *Midnight Caller* reflected neither the reality of gay men nor of persons with AIDS in San Francisco, but rather the reality inscribed onto those social groups by dominant heterosexual discourse.

Medical discourse as a measure of realism, was also cited by the AIDS activists in their complaints. As a point of contention, the activists argued that *Midnight Caller* was distorting the development of the disease. For one thing, testing HIV-positive occurs at least six months after exposure to the virus; within this episode's narrative, Cassidy tests positive only two months after exposure.

Such compressions of medical time—that is, the reduced period for registering a changed HIV status and manifesting opportunistic infections—were unusual in light of previous demands by the networks (through their standards and practices divisions) upon the producers and writers of other programming dealing with AIDS. William Schwartz, who dealt with AIDS-related scripts while executive producer for the medical drama *Heartbeat,* stated:

> If there were any problems that we had, it was with standards and practices, which is, you know, the network form of—I don't want to say censorship—but it's their department that deals with accuracy. So the scripts, that particular script, was really gone over with a fine-tooth comb to make sure everything that we put in medically was correct.[3]

Deborah Dawson, story editor for the CBS hospital series *Trapper John, M.D.* offered similar comments:

> The medical stuff kept changing. Where, you know, day to day we were finding that, you know, there would be a new possible medicine. It was very hard because we wanted to be as positive as possible. We had a lot of flak from the CBS legal department and standards and practices to make sure that we didn't offer any false hope.[4]

In fictional television programming, the privileged status of the medical profession has shaped and continues to shape the representation of AIDS. Incorporation of the epidemic into narrative content has to adhere strictly to the empirical methodology of the medical profession. Information and facts which cannot be verified by medical empiricism cannot be used—except in one case: the misinformed character as a straw man/woman. In such a narrative strategy, the character who is misinformed is chastised and given the correct information by another character. From the network perspective, medical accuracy is imperative if fictional programs are to address AIDS.

Depending upon the particular issue at hand, medical accuracy was just as important for the activists as it was for the offices of broadcast standards and practices. As a basis of ideological struggle, medical realism as well as social realism provided much of the critical argument for the activist position.

However, in response to activists' criticisms about medical distortion, Zito contended that the medical profession, in spite of its aura of empiricism, does not yet have its own consensus about the epidemic:

> Since we're talking about medical facts here—everybody's got their own facts. So even though we had checked out and cleared things, when we ran into certain areas we were told that we were wrong, that we were medically incorrect. It just depends on which expert you want to talk to. . . . We used the Center for Population Options . . . and I used a doctor here [in Los Angeles].

From his different sources, Zito found that professional opinions varied about the ability to determine an individual's HIV status. As it turned out, the range extended from three months to ten years. With such a wide margin of variability, Zito did not understand why he received so much criticism in the way he depicted Cassidy's HIV diagnosis. For him, the scenario was realistically plausible.

While the activists agreed with the standards-and-practices perspective on accuracy in depicting the onset of the AIDS virus, the inevitable outcome of HIV infection was highly contested. Once Cassidy finds out her HIV status, she interprets her antibody condition as an inevitable death sentence. Within medical discourse, professionals have ascertained that the development of the disease can take up to ten years before opportunistic infections manifest themselves and even longer before they completely debilitate an individual.

Addressing the issue of inevitability, Mulcahey, scriptwriter for the NBC soap opera *Santa Barbara* and liaison between the activists and the *Midnight Caller* production executives, said, "That is unknown. By the time she might get sick, there might be all sorts of treatments to stall infection" (Ford, "Melodrama" 3). According to Nina Easton, "the strategy of informed San Francisco HIV-positive virus carriers is to keep themselves alive long enough to see . . . new therapies come into currency" ("Unrealistic" C10). As the media spokesperson for the San Francisco AIDS Foundation, Rene Durazzo added concerns about Ross Parker as well: "The gay character is portrayed as an individual with no hope, who has given in to the disease. . . . That is not the case today. There are individuals who are determined to live their lives with deep quality" (Ford, "Melodrama" 3).

The activists wanted to replace the dominant portrayal of individuals infected with AIDS as hopeless victims with a positive image of empowerment and hope. However, for NBC's Broadcast Standards Department accuracy was the important factor. No cure had yet been found for AIDS. Consequently, any incorporation of hope within the script needed to be highly qualified. While not disavowing empowerment for HIV-positive individuals, Zito's intent was to persuade people to act sexually responsible. For him, overemphasizing hope undermined the focus on sexual responsibility:

> So my argument was: "We're giving out the wrong information to people," because any chance that we have to be educational disappears when you start watering down this stuff. I mean you tell people they're going to die if they get AIDS, that's one thing; you tell people they might die, or might not—that's another thing. And you know, people are always looking for some little out when they don't want to change their behavior.

Realism, as a critical stance for the activists, centered upon the reflection of the social and medical reality of gay men, persons with AIDS, and the disease itself. Another point of contention in constructing "reality" proved to be even more troubling. Violence, as a means of confronting and stopping Barnes's activities, fueled even more anger in the activist coalition (Ford, "Script" 5). Within the narrative, Killian's character enacts a pursuit of Barnes which, in the first confrontation between the two, results in Killian's physical antagonism toward Barnes. While this was inflammatory enough on its own merit for the activists, the original ending of the episode angered them even more. In Zito's first draft, the narrative was to end with the assassination of Barnes by Kelly West, a vengeful former lover.

Far more than reflecting reality, the activists argued that this narrative contributed to an ideological climate in which violence against gay men and persons with AIDS could be sanctioned as a viable solution to the AIDS epidemic (Ford, "Melodrama"). The activists' fears were not without foundation. During the 1980s, both the increased social visibility of the gay community within American society and the emergence of the AIDS epidemic tended to goad conservative social forces. Concurrently, the amount of physical violence directed against gay men and lesbians increased in the 1980s. According to the National Gay and Lesbian Task Force, which monitors hate crimes against gay men and lesbians, the rate of such crimes rose from a reported 2,042 incidents in 1985 to 7,248 incidents in 1988 as reported in "The Scale of Hate." Given that social climate, it is not surprising that the AIDS activists protested so vehemently about the original resolution of this *Midnight Caller* episode and tried to stop the production of it.

To his credit, Zito acknowledged that his own social experience may have blinded him to the potential implications of his script. According to his interpretation of the activists' concerns:

> It was going to create anger against, and possibly even violence against, the gay community because they would be perceived as irresponsible cruisers. . . . I never had to deal with this the other way around. I've never been a subject to this kind of prejudice. I'm straight. So, I don't know what it's like to have people dislike you because you're gay, or think bad thoughts about you. Apparently, people have, and apparently, they're very sensitive to these things.

In spite of the changes in the Barnes character and the revamped ending that resulted from early conferences, many activists felt the revisions did not go far enough. As a response to what they felt were inadequate [revisions], the activists engaged in civil disobedience at sites of location shooting, disrupting production of the episode.

## Production and Narrative Economy

Like all television series, *Midnight Caller* depends upon the elements that define its narrative franchise to eliminate the need to explain the basic premise of the series week after week. From the standpoint of narrative economy, the franchise elements provide a narrative shorthand that can be translated into production economy as well. With little need for lengthy scenes of exposition, the narrative can move quickly into the dramatic conflict and hopefully engage the viewer so that she or he does not change channels. In U.S. commercial television, viewer engagement translates into advertising revenues for the networks and continued profits from licensing fees for the production company. In addition, the final shooting script is a blueprint that cuts down on waste in time and materials in the production process.

When the AIDS activists engaged in civil disobedience, they were able to undermine both narrative and production economies for the *Midnight Caller* series. Because *Midnight Caller* is quite dependent upon location shooting in San Francisco, it was relatively easy for the activists to interrupt production. Within the series, location shooting provides *Midnight Caller* with a look no other show has. The rolling hills, winding streets, and fog-drenched atmosphere provide the series with a narrative ambience that a series shot in Los Angeles would not have. As a result, location shooting provided the activists with an effective pressure tool to influence the production company.

While the ideological ramifications of representation fueled the agenda of the AIDS activists, one major concern for Midnight Caller Productions was the escalation of economic costs when the production was shut down. Addressing the costs of the shooting disruption, Singer offers the following estimate:

> The cost of production was probably $50,000 a day, but actually [we] never experienced a day in which we were totally shut down. We lost some time because of the disruptions. . . . The first night that things kind of exploded, we had to leave the street, because things were getting out of hand. But overall, it probably cost us a day and a half of shooting. And so, that was probably $70,000 or so. And the legal fees were kinda high.

After the injunction against the San Francisco AIDS Foundation and ACT UP failed to curb the disruptions, Harwood reported in *Variety* that Midnight Caller executives "hinted" that the benefits of location shooting in San Francisco were being outweighed by the production costs encumbered when production was shut down.

While moving the production of *Midnight Caller* back to Los Angeles was a feasible economic option, it would have created problems in narrative economy that would have impacted the production economy as well. At a narrative level, the pilot and succeeding installments of *Midnight Caller* had already established the franchise elements of the series which included San Francisco as its geographical backdrop. Because exterior shooting is an integral element of *Midnight Caller,* moving the fictional location of the series would have undermined the narrative economy that is gained through series continuity. To maintain continuity, at least one or two episodes of the series would have had to address the impetus behind any geographical change. Such a move would not have been impossible, but given the cramped production schedules for television series following the 1988 writers' strike, this was not an option that would have provided much breathing space.

Eliminating external location shooting and restricting production to internal sets would have provided yet another option in producing the series, but would have meant changing the ongoing narrative agenda for Killian's character. Within the series franchise, much of the ongoing narrative tension that links different episodes of *Midnight Caller* stems from Killian's alienation from the external world. Conventionalized within the series, this psychological dilemma is visually depicted as an opposition between the controlled security of the radio studio interior and the threatening forces that lurk in exterior San Francisco locations such as the waterfront, the Tenderloin, and North Beach. If the series format had to eliminate exterior shooting, the overarching conflict driving Killian's character week in and week out would lose much of its narrative strength. More likely, such production changes would have redirected the cumulative subtext of the series. Because the character of San Francisco is so important to the *Midnight Caller* franchise, the series could not afford a generic urban simulation to denote the city. As a means of dealing with the activists, Midnight Caller Productions' long-term interests would not have benefited by moving to Los Angeles or retooling the usual series format.

### *Freedom of Speech and Creative Rights*

While both narrative and production economies were major concerns for Midnight Caller Productions, they were not the only interests for the company. From the perspective of the episode's scriptwriter, creative issues were at stake as well. Zito argued:

> There were creative rights issues. There were First Amendment free speech issues. Basically, we were denied that by the fact that we were being shut down on city streets. You know, if somebody went out and burned a bunch of books, everybody would be up in arms, but if somebody goes down and shuts down a television show, people don't see the issue being the same thing.

As a tactical maneuver, civil rights issues would tend to strike a resonant chord with the activists. After all, the Gay Liberation Movement is centered upon the civil rights of gays and lesbians. Also, civil rights is part of the agenda for AIDS activists who seek to eliminate discrimination against individuals infected with the disease. However, some activists gave little credence to the production company's argument. From their perspective, the ideological structuring of this narrative did not promote civil rights, rather it eroded them. As Ford reported, Waiyde Palmer of ACT UP contended:

> We're not trying, at any time, to not allow freedom of speech. We only have objections when freedom of speech can cause genocide, murder, prejudice or violence against any community. ("Script" 5)

While the First Amendment is a pertinent issue for commercial broadcasting generally, and for this confrontation specifically, neither Midnight Caller Productions nor the AIDS activists fully articulated the limitations that constrain freedom of speech in American television. For one thing, a difference does exist between freedom of speech on a street corner and freedom of speech within the broadcasting system. From an economic standpoint, everyone has access to a street corner in order to address important social issues. In light of the institutional structure of the American broadcast system, access to freedom of speech on television for everyone is implausible as the economic cost of disruptions for Midnight Caller Productions illustrates.

In addition, the distinction between television as entertainment and television as an outlet for dogmatic presentations permeates the broadcast industry. Echoing other producers, story editors, and scriptwriters, Singer, in his assessment of the changes incorporated into "After It Happened . . . ," stated, "they [the activists] did not get everything that I think they would have wanted to, which would have become a polemic and not a television show, really, anymore." From the perspective of both the networks and the creative community, fictional television programming fulfills the function of entertainment. Because this conception of programming is so deeply embedded in the cultural practice of program production, overt political stances are deemed inappropriate for entertainment content.

As Zito pointed out, the personnel at NBC were not enthralled with the focus on AIDS in his script, even in its initial form:

> When the network read the draft of the script, they were aware that it was going to be controversial. No one had any idea how controversial. But when they read it, what they said was, "AIDS is a turn-off. And we don't want it mentioned before we even get to the first commercial, because TV sets all over America are going to turn off when they hear AIDS."

For the networks, maintaining the distinction between entertainment and polemics is, at an abstract level, an important defense mechanism to circumvent potential criticism about programming. From the perspective of network programmers, television programming is, after all, only entertainment. In spite of being positioned as mere entertainment, however, a television program is a discursive universe that constructs a hierarchy of meaning. This hierarchy privileges some ideological positions over others. From the perspective of discursive analysis, the gap between entertainment and polemics is more a matter of degree than of antithetical difference. Indeed, conventional dramatic entertainment requires conflict. Conflicts in traditional fictional programming necessarily involve the collision of value systems embodied by the protagonist and antagonist. Because of their conventional narrative structure, television dramas always exploit some level of polemicization.

As a result, the pragmatics of program production are more contradictory. Despite his distinction between entertainment and polemics, Singer does not entirely dismiss the power of fictional television to construct meaning and shape people's belief systems. In explaining his shift, he stated:

> I found myself in the middle of the night really thinking about this a lot and thinking, "Well, God. If just one person misconstrued what we're trying to do here, that might have some negative effect." That would be a bad thing to happen. I sort of searched my soul on that and decided, "Well, better err on the side of caution on this one."

Just because entertainment is not overtly polemical does not mean it has no ideological agenda. Almost every fictional television narrative dealing with AIDS provides, to some degree, "infotainment" about the transmission of AIDS and the people infected by it. "After It Happened . . ." is no different. While it provides this information, it does so within its own unique ideological arena structured by the parameters of *Midnight Caller*'s narrative franchise.

As it turned out, the activists and Midnight Caller Productions employed inverse strategies in their confrontation with each other. Although the activists had a political agenda, they used economic pressure to challenge the production. In addition, they were not under the constraints of broadcast production. The activists had little to lose in either time or money since their agenda was political. In this case, the capitalist mode of popular culture production worked against the production company. Because the production of an hour-long series is expensive and hectic—especially in the wake of the writers' strike—Midnight Caller Productions had less of a bargaining position in this ideological struggle. At least in part, the production company's economic interest was best served by accommodating some of the protesters' demands.

While economic and creative concerns were important issues for them, neither Singer nor Zito were indifferent to the concerns of the activists. While Singer did not agree with the activists on the effective potential of the script to promote violence, he did change his perspective about AIDS after the conference between the involved parties. As activist Palmer put it, "Bob Singer's consciousness has been raised" (Ford, "Protest" 4). As a result of the conferences, Singer acknowledged that he was exposed to issues associated with AIDS of which he was unaware:

> Now, in ACT UP's defense, one thing they did do was open up this can of worms which really allowed for some very meaningful dialogue, and resulted in a less charged situation ultimately, and something that really helped us and AIDS, too.

From a creative standpoint, the changes in the script were more painful for Zito because he had written it. Evoking First Amendment rights, he felt that the production company had given in to too many demands of the activists and made too many changes in the script. In spite of these criticisms, he acknowledged that some positive changes came out of the confrontation:

> You know, it's a funny thing, because I don't think it's so much a line or a fact as it is that the episode that aired was far more compassionate than the first or the working draft . . . we had in fact been moved to certainly a more humanistic, or a more sensitive approach to what is clearly a difficult and complicated subject.

As a result of the various rewrites, the final version of "After It Happened . . ." (broadcast on December 13, 1988) presents a different ideological landscape than that of the first draft of the episode. Most obviously, the ending was rewritten so that Barnes, the antagonist, is not killed. Indeed, the ending is so ambiguous that one cannot be sure if Barnes has decided to change his sexual behavior or not. After his close brush with death, he walks into the darkness of the San Francisco night.

In addition to this major change, other modifications were implemented as well. Some of these changes incorporated the concerns of the activists. For one thing, Barnes is no longer a social malcontent; instead, he is a person whose sexual irresponsibility is the result of fear and denial. The protagonist, Killian, while still having problems with Barnes's behavior, is much more compassionate in his attempt to fathom the impact of AIDS upon an individual's life. Overall, the focus of many scenes was changed from an emphasis on physical aggression and confrontation to an expanded discussion of issues associated with AIDS. In the initial draft, for example, King, Killian's boss, was concerned with the commercial links among AIDS as a controversial topic, a ratings booster, and a profit generator at the radio station. Because of the rewrites, she is no longer concerned with profits but the impact of Killian's broadcast upon those most affected by the epidemic. In the final draft, violence is criticized as an improper response to either the epidemic or individuals infected with the virus.

While not all of the activists' demands were met by the changes in the final script, the ideological thrust of the story is not the reproduction of conservative values it had been initially. By diffusing the violence and providing only weak narrative closure, this episode of *Midnight Caller* fails to provide any cohesive ideological construction of the AIDS epidemic. However, as it was finally produced, "After It Happened . . ." entered into the discursive arena as a more progressive, humanistic discussion of AIDS, especially in comparison to the reactionary responses of traditionalist politicians and journalists. In a society where allocated funding for AIDS education and research is bridled by the interests of heterosexual monogamy and fundamentalist religious fervor (as evidenced by Senator Jesse Helms's amendment to a crucial 1987 congressional bill),[5] this episode offered an alternative and, at times, oppositional voice against these dominating social factions. Without the struggle and negotiation between the AIDS activists and Midnight Caller Productions, "After It Happened . . ." would not have presented this alternative cultural perspective as a response to the AIDS crisis.

## Notes

1. The author conducted a telephone interview with scriptwriter Zito on May 7, 1990. Remarks by Zito appearing in this paper were taken from this interview. In addition, the author used Zito's first and final drafts of the "After It Happened . . ." episode to compare the versions before and after Midnight Caller Productions met with the AIDS activists.

2. The author conducted a telephone interview with Robert Singer, the executive producer of *Midnight Caller,* on May 7, 1990. Remarks by Singer appearing in this paper were taken from this interview.

3. The author conducted a telephone interview with William Schwartz, executive producer of *Heartbeat,* on September 28, 1988.

4. The author conducted a telephone interview with Deborah Dawson, story editor for *Trapper John, M.D.,* on August 17, 1988.

5. *Congressional Record,* 14 October 1987: S14217. Washington DC: GPO, 1987. Restrictions were placed on federal funding for AIDS by Amendment No. 964 during the 100th Congress's first session, demonstrating the power of reaction-

ary social forces to define the appropriate response to the AIDS epidemic. In this amendment, archconservative Senator Jesse Helms (R-North Carolina) was able to constrain AIDS education through a reactionary agenda. According to the wording of the amendment, federal AIDS funding cannot promote or condone homosexual behavior between males. Consequently, discussion of gay safe-sex practices and even compassion toward those infected with the disease are eliminated from the political arena. Conservative social factions have been able to structure a response that ignores the impact of the AIDS epidemic upon the lives of those most affected by it.

## *Works Cited*

*Congressional Record* 14 October 1987: S14217.

Crimp, Douglas. "How to Have Promiscuity in an Epidemic." *AIDS: Cultural Analysis, Cultural Activism*. Ed. Douglas Crimp. Cambridge, Mass., and London: MIT Press, 1988. . Pp. 241–42.

Dawson, Deborah. Telephone interview. 17 August 1988.

Easton, Nina J. "Gays Protest *Midnight Caller* Episode." *Los Angeles Times* 25 October 1988: VI, 1: 2.

———— "*Midnight Caller* AIDS Episode Called Unrealistic." *Austin American-Statesman* 13 December 1988: C10.

Ford, David. "*Midnight Caller* Script Provokes Gay Activists' Ire." *San Francisco Sentinel* 21 October 1988: 4.

———— "TV Melodrama Script Continues to Spark Controversy." *San Francisco Sentinel* 28 October 1988: 3.

Harwood, Jim. "*Midnight Caller* AIDS Episode Wrapped: Shaky Truce with Gays." *Variety* 22 November 1988: 52.

Palermino, Tristano. "Tap Dancing Past Midnite." *Coming Up* [San Francisco] January 1989: 6.

"The Scale of Hate." *Austin American-Statesman* 5 May 1990: A6.

Schwartz, William. Telephone interview. 28 September 1988.

Shilts, Randy. *And the Band Played On*. New York: St. Martin's Press, 1987.

Singer, Robert. Telephone interview. 7 May 1990.

Zito, Stephen. "After It Happened . . ." First Draft. 28 September 1988.

———— "After It Happened . . ." Final Draft. 12 October 1988.

———— Telephone interview. 7 May 1990.

# II

---

# TELEVISION TEXTS

Readings in this section strongly reflect changes in the television industry and consequent changes in programming strategies of American television in the 1980s and 1990s. That is to say, they reflect the passing of "the network era." Audiences accustomed to familiar, regular, genre-defined and genre-driven television of earlier periods found in recent years that the medium was undergoing substantial change. As suggested earlier, even the term "television" seemed to mean something new.

These changes were defined and directed primarily by economic and technological shifts. Wider availability and increased subscription to cable services gave millions of viewers increased access to more and more television programming. Easy and apparently delighted use of remote control devices made selecting from more channels—or at least surveying them in the "graze" or "zapping" mode—part of the viewing experience.

Offerings from local access channels to high-dollar subscription services now competed with network offerings for viewers' attention. The growing number of independent stations made it possible for FOX Broadcasting to create a new competitor modeled on the network style. But these stations also created a demand for and display of older television shows in reruns. The old programs were often redefined by their new contexts, very self-consciously framed for marketing purposes as "family" or "hip" or "women's" television.

Economic conditions in the network television production context were more and more constrained, forcing both networks and studios to rethink their offerings to one another and to the public. Increased reliance on video cassette recorders (primarily for watching theatrically released movies rather than for time-shifting network offerings) diminished the overall size of the audience for network TV.

Taken together, these forces changed television as it had come to be known through a forty-year history. Audiences "tutored" in specific viewing strategies quickly altered those patterns, learned, and created new ones.

One of the primary results for the television industry was a shifting pattern of program offerings. Once thought "dead," the half-hour situation comedy returned with the enormous success of *The Cosby Show*. Costs of one-hour, dramatic programming—prime-time soaps, melodrama of all sorts, police and detective programs—were deemed excessive because these programs drew less and less interest in the "after market" of syndication, location of television's richest financial rewards. By contrast, more and more stations and cable services were eager to pay for and to program half-hour comedies.

New programing types emerged to fill schedule gaps caused by the diminished number of one-hour programs. The eighties and nineties witnessed a proliferation of talk shows, "reality" programming, and one-time, "Movie of the Week" performances. "Prestige informational" programming from network news divisions competed with the newer tabloid versions of "reality." Music videos influenced programs and repaid their stylistic and technical debt to commercials by offering even more styles for commercials to reappropriate. Self-defined audiences were elaborated by a new wave of "demographic" planning and new markets were designated, sometimes in the blending of a network with an age group (FOX's specific creation of its youth orientation), sometimes with a gendered group (*Lifetime*'s appeal to women), sometimes with a "concept" (*Nick at Night*'s "television for the television generation").

Partly as a result of these shifts, partly as a result of individual interest and the altering of theoretical concerns mentioned in the introduction, the essays in this section focused on television texts, on the programming itself, are more varied than ever. Some of the essays are still primarily concerned with patterns of generic meaning. Others are still concerned with specific topics, such as race and gender and the negotiation of other social issues within individual episodes of particular programs. But many of the essays discuss their topics in terms of the contexts of television's new forms, new mixes, new strategies.

Betsy Williams's essay on *Northern Exposure* focuses on the show's relation to television history. In so doing it can be readily and usefully compared with essays in Part I, offering a contemporary instance of industry responses to changing conditions. The first relationship Williams develops draws from theories of "quality television" developed in academic contexts and focused on at MTM Productions. Williams relates *Northern Exposure* to that history in terms of its creative personnel, its style, and its construction of particular social realities. But she also points out how the show is, quite significantly, a product of the current period of economic retrenchment and narrowcast programming strategy.

Laurie Schulze explores one of the forms that has come to fill the "void" left by reduced production of continuing television drama, the Made-for-TV-Movie. Because these forms can be produced quickly, on relatively low budgets, with topics that seem to have immediate appeal, they are used to fill programming gaps. Moreover, they can often be sold to cable programmers or, in some cases, to international theatrical and television programmers. In discussing the industrial conditions and strategies that led to the rise of this form, Schulze, like Williams, leads us back to the production context. But she also analyzes the cultural operations of these movies and suggests that their use and popularity cannot be totally explained by those industrial decisions.

Herman Gray's discussion of television's representation of African American experience in comedic form is not context driven by the same sense of industry–text interaction. But it does indicate how television's search for and construction of markets and images is constantly intertwined. More importantly his analysis makes clear how forms of fiction are directly linked to other forms of social knowledge and practice. At the same time, his exploration of a multi vocal, conflicted text helps explain why so much discussion has surrounded a phenomenon such as *The Cosby Show*.

Jimmie Reeves's essay on *The Wonder Years* suggests that such developments are possible, at least in the context of dominant, mainstream images. He carefully charts the differences of this program, differences that occur within old formal contexts. As part of eighties television's demographic lurch toward the baby boom generation, *The Wonder Years* seems a particularly perceptive meshing of the old and new in TV. This essay, too, relates directly to Part I, focusing on notions of "authorship" as central to the production context.

"New" is also available in *Roseanne*, perhaps the most popular success of the decade. This success is especially telling in light of its significance for female star power, and Kathleen Rowe examines this element as it functions both industrially and narratively. *Roseanne*'s and Roseanne's "unruliness" cuts across the production process and the process of representation, and Rowe's careful linkage of these issues to historical instances of the same cultural categories enriches our perspective. When paired with Mayerle's discussion of the actual production processes of this series, a far more complete and powerful analysis of the show is possible. This pairing is indicative of newer impulses in Television Studies to establish more detailed accounts of the determining forces in our relationships with the medium.

Another sort of comparison emerges by reading Rowe's discussion of *Roseanne* beside Denise Kervin's analysis of *Married . . . with Children* (MWC). This series, one of the FOX network's first major hits, enables us to see shades of cultural process at work and raises major questions. Is *MWC* stretching boundaries in our notions of family, offering a critique of widely accepted ways of thinking about ourselves? If so, is it able to offer such a critique only because of its venue, or because audiences of many sorts are now "prepared" for such images and stories by a larger sense of shifting social reality? Or is *MWC* actually reinforcing older patterns of social power, offering us the illusion of critique in order to "sell" in a particular context of manufactured "difference?" Placed side by side, do *Roseanne* and *MWC* set new models in place, call old ones into question, or merely reproduce dominant ideologies by serving as a distraction from the true realities of television?

Are those true realities to be found in the commercials? Hal Himmelstein's analysis of a Kodak advertising campaign suggests that the images of family, of other aspects of American social and cultural life are more powerfully appropriated and rewritten here than in the programs they surround. Certainly, the advertisement's linkage of such fundamental mythologies to those found in the programs warrants continued critical work on this topic.

The analysis of commercials moves not only back into our understanding of various television programs, but forward, as well, into Lisa Lewis's discussion of music videos. While the Kodak commercials may not exhibit as much formal resemblance to MTV as some, the issue of mythologies remains central to both forms. Lewis argues that music videos offer female performers new opportunities for the presentation and representation of both personal and larger, gender-defined concerns. Comparing these videos both to commercials and the programs about women, about families, increases our sense of the way each stands alone. But perhaps more importantly, it also increases our sense of the ways in which television is able to assume (and potentially consume) all of them within its larger patterns and relations.

Both Bernard Timberg and Mimi White provide extended commentary on forms of television talk. Timberg provides the first thorough examination of the television talk show as a distinct genre. These programs, so dependent on the distinctiveness of individual host-stars, are often considered radically different from one another. Timberg shows how they are actually rooted in formal similarity. In doing so he offers us new tools with which to explore the expanding number of such shows and explains some of the reasons for this proliferation.

White's discussion of television therapy, focused on *Good Sex with Dr. Ruth,* is part of a larger discussion of the therapeutic functions of television. In her longer study White includes discussions of other television forms, situation comedies, prime-time melodrama, and so on. Breaking the reliance on genre studies in television, she suggests that the relationships formed between audiences and programs far exceed notions of cultural myth and narrative.

But it is precisely to myth and narrative that we return with Richard Campbell's study of the formulas of *60 Minutes.* Turning the analytical power of genre theory to this most popular of news shows serves two purposes. It offers us a detailed and extended examination of the program and helps explain its cultural power and popularity. But it also enables us to consider more carefully the mythologies, the narrative strategies of other types of news and reality offerings. If the present trend continues, if more and more such offerings are made, Campbell's work can serve as one map for weaving our way through the varied versions of our world.

Even the tabloid television programs, however, are, like *60 Minutes,* somehow related to "old" television, to network television. The newer versions remind us that television is changing, but they do not suggest that the various forms of fragmentation actually constitute a disappearance, of sorts, of a medium as we have known it.

Elihu Katz and Daniel Dayan offer a rather different model. If indeed the centralizing qualities of television are dissolving, what remains to draw the mass audience to a shared experience? If television has now become far more like a book store or a big-city newsstand, offering specialized information and entertainment to ever more precisely defined groups, what sort of content, what program, has the power to initiate the kind of massive viewing once taken for granted? Katz and Dayan argue that there are specific types of "media events" that can call audiences from their special concerns to a collective experience of emotional, political, or cultural significance. If they are right, they may have seen in certain specific events from the past a significant aspect of the future of television.

Changing technological factors make possible another future, however, one even more specialized than current multiple choices. William Boddy outlines the programming strategies and practices of Paper Tiger Television, a production and programming collective offering truly alternative visions via video. Paper Tiger may then be a model that moves toward a far more "personalized" televisual form. And with this more precise focus, this more direct sense of audience and interaction, the alternatives are social and political more than formal. Boddy, and Paper Tiger, remind us forcefully that what we have known as "television" has little to do with the inherent qualities of video technology and everything to do with the deployment of social, economic, and political power. "Television" is the construction of a particular set of social relations in particular historical conditions, and as many of these essays indicate, one of the most exciting aspects of the critical task is its self-reorientation as it watches its object shift.

Personal television of the Paper Tiger sort may become central in the new technological and economic contexts. Combined with the possibility for globally acknowledged media events that have a potential to address the population of the earth "as audience," we may find ourselves oscillating from one room in our media library to another, from the quietness of our study carrel to the auditorium where we meet with others. On the way, we may pass through the shelves of fiction, therapy, information, and advertisement. Little that we know of television will go away. Much is in the process of being "recatalogued." The essays that follow are among our instruction manuals.

# "North to the Future": *Northern Exposure* and Quality Television

## BETSY WILLIAMS

"Everyone knows quality [television] when they see it," suggests a November 1991 *Broadcasting* article aptly entitled "Quality TV: Hollywood's Elusive Illusions."[1] Not surprisingly, the article falls short of a strict definition of the term, and instead names series that exemplify "quality": among them *Cheers, Murphy Brown, Brooklyn Bridge, Northern Exposure, Homefront,* and *The Simpsons.* As this diverse list suggests, "quality television" represents many things to many people in the industry. For instance, "one rule is that the biggest hits on television are usually considered quality series," although *America's Funniest Home Videos* would not make the above inventory.[2] Another somewhat contradictory rule would suggest that a series "will last longer with low ratings than a show without quality."[3] In industry press, the term "quality" is bandied about frequently and imprecisely, and furthermore is often used interchangeably with other terms such as "name-brand television" ("a Bochco series"), "hit TV" (a *Murphy Brown*), and "boutique television" (a *Twin Peaks*).[4]

Despite the confusion, there seems to be one show which meets the industry's wide-ranging qualifications for "quality." "What ultimately works and what the networks need to remember is quality. . . . The usual comparison is with *Northern Exposure,*" comments an NBC affiliate board member in 1991.[5] A kinder, gentler *Twin Peaks, Northern Exposure* is an "outback" drama/comedy set in the "Alaskan Riviera," as the show's astronaut

From *The Spectator* (USC School of Film and Television) forthcoming, 1993. Reprinted with permission.

cum entrepreneur dubs the region. At the center of a remarkable ensemble cast is inveterate Manhattanite Dr. Joel Fleischman, reluctantly working off his medical school loans in remote Cicely. The show, which could often be called "The Sentimental Education of Joel Fleischman," explores a range of cultural oppositions related to Joel: east versus west, frontier versus civilization, science versus mysticism, male versus female. The complex ensemble enables a wide range of perspectives to be expressed on the show and is certainly one reason that the show is often invoked these days as an exemplar of "quality" television. How, then, does this Emmy award winning series[6] fit into the paradigm of quality television in both industrial terms and in the discourses of contemporary television criticism? What are the economic and aesthetic implications of that designation? Moreover, what role does quality television serve the networks, both historically and in the present moment?

As discussed by Jane Feuer et al. in *MTM: "Quality Television,"* the term quality refers most generally to television's ongoing negotiation of the tension between economics and aesthetics, or as Feuer puts it, "the relationship between textual production and commodity production."[7] Feuer's view of quality television comes from an analysis of Grant Tinker's independent production company, MTM Enterprises, Inc., which began doing business in the early 1970s. As discussed by Feuer, "quality TV" refers to those programs produced at an independent house (in this case MTM Enterprises, Inc.) which exhibit certain industrial and stylistic features, as follows: these shows (*The Mary Tyler Moore Show, Lou Grant, The Bob Newhart Show, Hill Street Blues, St. Elsewhere,* etc.) maintain a fairly consistent demographic profile and display an authorial style evinced at the textual level by two key characteristics: self-reflexivity and liberal humanism. Perhaps the most important aspect of Feuer's definition is her assertion that MTM Enterprises was "in the business of exchanging quality TV for quality demographics."[8] In other words, advertisers were willing to pay top dollar for time on a lower rated "quality" series such as *Hill Street Blues* because the audience, although smaller than that of a top 10 rated series, was composed of urban-based, active consumers. Thus, the relevance of Feuer's quality paradigm to a discussion of current television series is precisely because it focuses on this "thorny" relationship between commodity production and textual production, thereby enabling an analysis that "links specific historical and institutional conditions of production to the [textual] discourses thus produced," as Michelle Hilmes puts it in her discussion of *Cheers.*[9]

Like the creative team behind *Cheers* (Glen and Les Charles, Charles Burrows), the producers of *Northern Exposure* are MTM alumni *(St. Elsewhere).* The prevalence of MTM alumni working within the industry at present raises an interesting question, as posed by Susan Boyd-Bowman: "What features of quality television carry over into the output of other production companies working in the same industrial conjuncture, often employing personnel who are MTM alumni?"[10]

Boyd-Bowman's comment suggests the need to recontextualize the concept of quality television given industrial, economic, and technological de-velopments of the decade since MTM was an industry force. This [essay] attempts such a recontextualization, at least in terms of *Northern Exposure,* and will use the paradigm of quality television as it circulates both in con-temporary criticism and in industry discourse to historicize the show and to help identify and explain its narrative strategies. The series represents a par-ticularly rich site of analysis for the simple reason that it is clearly inflected by the tradition of quality television from whence it springs, yet it is also quite different from other quality series on television in its rigorous nego-tiation of social, sexual, and spiritual issues. That is, *Northern Exposure* evades the typical quality characterization as represented by a direct throwback to the flagship quality series (the domesticated workplace comedy *The Mary Tyler Moore Show*) such as *Murphy Brown,* yet it also veers away from the "boutiquey" elite status of a *Twin Peaks*. In fact, *Northern Exposure* has been referred to as a kinder, gentler *Twin Peaks* both in terms of its narrative strategies and its popularity.

The comparison to *Twin Peaks* is pertinent because *Northern Exposure* has succeeded where *Twin Peaks* failed, in garnering both critical acclaim and consistently solid (top 15) ratings. That is a difficult balance for even a quality series to achieve but one which reflects the medium's ongoing ne-gotiation between economics and aesthetics.

As Michelle Hilmes suggests, this essential tension between aesthetics and economics, or high art versus popular art, results from the contradic-tion between what has been, until recently, a government sanctioned mo-nopolistic broadcasting system that uses a public resource, the electromag-netic spectrum, for "free" and in return provides programming that is in "the public interest, convenience, or necessity."[11] Historically, that tension has played itself out in many ways—in 1950s debates over live versus filmed programming, hour-long drama versus half-hour situation comedies, quiz show scandals, or in the network responses to regulatory criticism, such as FCC chairman Newton Minnow's 1961 characterization of the medium as "a vast wasteland." The turn to "quality television" in the 1970s and early 1980s, particularly with its emphasis on the hour-long drama epitomized by *Hill Street Blues, Lou Grant,* and *St. Elsewhere,* renegotiated this essential tension in a moment of industry crisis. The instability was a result of a combination of several factors: the FCC's Cigarette Ad Ban cut the net-works' revenue base by 12 percent and the FCC's Syndication and Financial Interest Rules further reduced the networks' profits from syndication, while creating opportunities for independent producers such as Grant Tinker, Larry Gelbart, and Norman Lear.[12] More importantly, as Paul Kerr points out, the A. C. Nielsen Company began measuring audiences demographically in response to advertiser pressures.[13] "Quality demographics" took on extraor-dinary significance as vice president in charge of audience measurement Paul Klein at NBC successfully campaigned for "better" numbers over bigger numbers. CBS's "rube-shucking" response was particularly dramatic—the network cancelled six extremely popular and long-running series in the 1970–

71 season, most of which were "hayseed" comedies reaching a bipolar—old and young—audience concentrated more in rural areas away from urban centers of consumption.[14] The need to "think demographically," as CBS president Robert Wood put it to his new head of programming, Fred Silverman, meant that the empty slots would be filled with shows like *The Mary Tyler Moore Show* aimed at baby boomers concentrated in urban areas with access to consumer goods and services.[15] The attention to quality demographics also enabled the networks to redress the aesthetic imbalance indicated by Minow's characterization of the period as "a vast wasteland." However, with audience shares totaling 90 percent in the early 1970s the networks could afford to "disenfranchise one section of the audience to the benefit of another," as Paul Kerr puts it.[16]

Now, of course, the networks contend with radically increased competition and a correspondingly decreased audience share overall, skewing the "class for mass" formula underlying quality series of the seventies and eighties. Thus it is that that *TV Guide* calls *Northern Exposure* "a surprise smash hit," even though the show averages a 14/25 rating/share.[17] A show that pulls down those moderate numbers could only be called a "smash hit" at a time when the networks' share of the audience, not including Fox, is only 65 percent.[18] There is no longer a margin for quality, yet quality television seems to be more pervasive than ever. Thus, although the specific economic and historic circumstances which facilitated the emergence of quality television have changed, the networks' obligation to simultaneously serve the public and deliver them to advertisers remains strongly in force. "Advertisers yearn for viewers for whom the tube is a periodic activity," writes Mark Schapiro in a recent interview with John Falsey and Joshua Brand.[18] Moreover, "*Northern Exposure* is the type of show that gives people not predisposed to television a reason to watch," suggests Peter Tortorici, executive vice president of CBS Entertainment.[20] Thus, there is tremendous commitment on the part of the networks to "name brand" writers whose shows carry the quality label in terms of both aesthetic features and conditions of production, and there is increased commitment to the hour-long drama, which dominated prime-time television in the early 1980s.

*Northern Exposure* was born into and is succeeding in an industrial climate whose watchword in the early 1990s is "stop the slide"—as NBC Entertainment president Warren Littlefield puts it—of viewership to alternative delivery systems.[21] The show is "quality" in terms of industrial lineage and its structural elements, but it also further refines the MTM style in four important ways: (1) its aggressively hybrid nature; (2) the composition and configuration of its ensemble; (3) its unprecedented and completely innovative use of backstory to structure episodes; and (4) its self-consciously "bardic" voice. Thus, bearing in mind Hilmes's goal of linking "specific historical and institutional conditions of production to the [textual] discourses thus produced," the rest of this essay will consist of three sections: the first provides a brief background on MTM alumni and quality television, the second discusses current industry conditions that bear on the making of *Northern Exposure,* while the third looks at the text of the show itself.

The original MTM Enterprises Inc. "family" (itself something of a domesticated workplace ensemble) consisted of its president, Grant Tinker and his then wife, Mary Tyler Moore, writer/producers such as James Brooks, Allen Burns, Ed Weinberger (who would later be pivotal in creating *Cosby*), Stan Daniels, and the creative offspring who have since become luminaries in their own right—Stephen Bochco *(Hill Street Blues, Bay City Blues, LA Law, CopRock, Civil Wars); Glen Charles, Les Charles, and Charles Burrows (Taxi, Cheers);* and of course, John Falsey and Joshua Brand *(St. Elsewhere, A Year in the Life, Northern Exposure, I'll Fly Away,* and *Going To Extremes).* The company's history is very nearly synonymous with the major programming developments of recent decades, that is, the CBS sitcom renaissance of the early seventies and the move to hour-long drama in the early eighties, signaled by the change in Lou Grant's role as Mary's boss at WJN-TV to his role as city editor at the *Los Angeles Tribune.*

Whether comedies or hour-long dramas, Thomas Schatz suggests that "the essential feature of all MTM's series has been the ensemble itself."[22] And while MTM Enterprises is less active now than formerly, its alumni have been successful implementing and adapting the strategies that so clearly demarcated the best MTM products from the rest of television. Schatz lists them as follows: "an ensemble cast, domesticated workplace, multiple plots in a semi-serial format, aggressive cinematic technique, and 'quality' viewer demographics."[23] These characteristics permeated not only the work of MTM alumni in the mid-to-late eighties but the industry in general, and were/are clear features of such shows as *thirtysomething, China Beach, The Wonder Years, Homefront, I'll Fly Away, Brooklyn Bridge, The Simpsons,* and of course, *Northern Exposure.*

According to Feuer, however, the MTM style is above all a function of the creative independence of the production company from the network.[24] Given that, there are several parallels between the current situation at CBS and the two most significant moments in the development of quality television—at CBS in the early seventies and NBC in the early eighties—that point to the desirability of this admittedly tense relationship. CBS, the third place, fuddy-duddy network for much of the eighties, is now being hailed as fertile ground for creative and inventive talent, just as it was in the early seventies when Brooks's *The Mary Tyler Moore Show,* Lear's *All in the Family,* and Gelbart's *M\*A\*S\*H* constituted what critic Gary Deeb calls the "most soul-satisfying comedy block in television history."[25] An August 12, 1991, *Broadcasting* article suggests that "CBS has been enjoying a resurgence of its reputation among the Hollywood creative community after years of rejection from many writers and producers."[26] Both Gary David Goldberg *(Brooklyn Bridge)* and Barney Rosenzweig *(Trials of Rosie O'Neill)* commend CBS for its supportive, hands-off management style.[27] Another studio executive calls CBS "clearly the most attractive place to shop."[28] CBS network executives, independent producers, and industry observers alike attribute the change to the arrival of entertainment president Jeff Sagansky in January 1990. Sagansky was at NBC in the early eighties, along with Brandon Tartikoff and Grant Tinker, when that network fostered the talents of Stephen Bochco, Gary David Goldberg, and Joshua Brand and John Falsey. In fact, Tinker recently suggested that Sagansky is following a strategy for developing exceptional series that Tinker himself instituted at MTM Enterprises in the early seventies: "One of the things I believed in was that you got the best people and then got out of their way, and Jeff is the same way. He gets involved, but he doesn't confuse himself with being the producer."[29] As Rosenzweig points out, "the current situation at CBS is a good sign for the future; showmanship and minimal involvement. . . . It's bound to reflect in the type of product that winds up on television."[30] In other words, the legacy of quality television—of CBS in the early seventies and NBC in the early eighties—is not simply limited to formal attributes and audience demographics but to a certain kind of relationship between producer and network that, if somewhat adverserial, is nevertheless characterized by maximum support and "minimal involvement."

Still, the current relationship between quality producer and network seems troubled, despite reports of a creative resurgence at CBS. For instance, Grant Tinker says that execs now face greater pressures from the top than ever before,[31] which might account for producer Barney Rosenzweig's perception that networks are failing to act as a buffer between producers and the advertising community.[32] Similarly, according to Stephen Bochco, many writers feel that in the 1991–92 season "there's a knee-jerk response of economic fear that seems to have made creative decision making at the networks more conservative this year."[33] In the same article, Bruce Paltrow refers to the situation as the "whitebreading of America." The impact of competitive delivery systems and a nationwide recession makes the networks skittish and even desperate (consider, for example, CBS's purchase of ABC-produced *Davis Rules* or ABC's unprecedented airing of competitor MTV's tenth anniversary celebration on November 27, 1991[34]). "The business is changing so much that the old rules just don't apply anymore," says Ted Harbert, executive vice president of ABC Entertainment.[35]

Whether the networks are acting out of desperation or financial savvy (or both), their programming strategies seem to vary from one season to the next as they look for the combination that will "stop the slide." For instance, in the 1990–91 season the networks implemented a spate of high-profile innovative programs to heighten visibility and compete with cable, Fox, and VCR use. Commenting on the fall 1990 schedule, *Variety* wrote: "the search for innovative programming is no longer arbitrary or intermittent, it's an agenda."[36] In the same article, producer Stephen J. Cannell notes what he considers to be a seismic change in network management behavior: "They no longer insist that you come up with an idea for a mass audience. They want to know if you can develop something for a specific demographic."[37] The practice of "narrowmarketing" (for want of a better term) to an upscale albeit slender slice of the audience is the philosophy that launched *Twin Peaks* and other risky series, such as *Shannon's Deal,* and was the banner under which *Northern Exposure* was conceived and first produced. Along these lines, producer Linda Bloodworth-Thomason *(Designing Women, Evening Shade)* says that "the question of franchise is looser" and she "credits the pioneering efforts of comedy kings Jim Brooks, Norman Lear, and Larry Gelbart for making it possible now for writers with track records to call their shots when pitching to the networks."[38] Thus, the very real threat of increased competition seemed to goad the networks into adopting bolder programming strategies, granting greater creative license, to a degree, dispensing with the idea of always capturing a mass audience.

However, 1991–92 witnessed the cancellation of many of those risky shows. Despite the accolades that greeted Sagansky's first year at CBS, his 1991 strategy was to program "traditional shows with a new spin."[39] A top TV agent pegs Sagansky as "without doubt the most mainstream-minded president since Fred Silverman."[40] *Broadcasting* dubbed Sagansky's plans: "The same ol' song and danceky" and writes that "his programming instincts are middle-American, middle-brow, and more middle-aged than yuppie."[41] Accordingly, CBS cancelled its riskier series and programmed new shows such as Stephen J. Cannell's *Palace Guard* and *Royal Family* ("shows that would not have been out of place in 1975"[42]). The network's new traditionalism was further signaled by two weekends of nostalgia-based programming, one in February 1991 and another in November 1991. The latter consisted of *The Best of Ed Sullivan, The Best of M\*A\*S\*H* (which, by the way, finished fourth in the ratings), and *The Best of Bob Newhart.*

What does all this mean for the fate of quality television? As economic pressures increase, venturous series have less of a chance, given that they often require time to build an audience. However, commenting on the networks' skittishness regarding innovative series in the 1991–92 season, Paul Schulman, president of the Paul Schulman Co., asks: "Are they getting away from certain shows? I don't think so. *The Wonder Years* for example, I haven't seen writing like that in years. There's also *Northern Exposure.* These are shows with strong writing and mass appeal."[43]

Despite *Northern Exposure*'s "mass appeal," the show barely made it on the air, taking "an even tougher route than normal to make it,"[44] although ultimately this route accommodated both the network—being less risky— and the show itself—allowing it to build an audience slowly. CBS Entertainment's executive vice president Peter Tortorici bluntly admits, "when *Northern Exposure* got on the air it did so because it was able to be produced on a lower than normal budget for its initial summer run."[45] Not only did the producers have to make do with a production budget one-third less than that of most hour-long primetime shows, but it was scheduled opposite *LA Law* on Thursday nights. Former CBS Entertainment president Kim Lemasters had given the original commitment to Falsey and Brand, which was somewhat backhandedly honored by Sagansky's offer of a summer (1990) replacement run. Although some analysts perceive CBS's practice of trying out riskier series during the summer as "bold," Brand thinks that "the expectation on the network's part was that it wouldn't continue [beyond summer]. I don't think they thought it would catch on, so I think it kind of surprised them."[46] Fortunately, *Northern Exposure* played well enough with audiences to induce CBS to order an additional eight episodes to air beginning in April of 1991. CBS then reran the first eight episodes in the summer of 1991, pulling in a more than respectable 23 percent share, and bringing new viewers up to speed before the fall season began.[47] Now the network seems firmly behind its surprise smash hit and has placed it in the coveted Monday night slot following *Murphy Brown* and *Designing Women*.

Thus, the show that was almost certainly conceived of as a kinder, gentler *Twin Peaks*—that is, an offbeat, narrowly targeted ensemble "outback" series—is very popular with what passes nowadays for a "mass" audience. I qualify that term because the idea of a single mass audience is somewhat obsolete, along with network dominance of the industry. It is important to reconceptualize the audience as an amalgam of specialized reading formations because such a view underscores the complexity of both the show and its audiences. For instance, a commentary by Scott Sherman in the now defunct radical queer journal *Outweek* celebrates *Northern Exposure* as a "must see; a really great TV show." Sherman waxes euphoric over an episode featuring a gay couple:

> Unlike so much television, it's not gay people who are depicted as having the problem. . . . Thank the producers of *Northern Exposure* for this wonderfully fair and humorous show. Encourage them to bring back the gay couple as ongoing characters. . . . Let's make sure this popular program continues presenting positive gay images.[48]

It is likely that this reaction is shared by many groups somewhat marginalized by much of network fare, such as Native Americans, who figure prominently within this unusual ensemble. While further reception study on this show is warranted, I wish to concentrate on the narrative strategies that enable *Northern Exposure*'s diverse appeal. For example, the show takes place on the edge of the civilized world, in Cicely, Alaska, a frontier town founded by two gay women seeking sexual freedom in the 1890s. As might be inferred from this small but significant piece of backstory provided at the outset of the premiere episode, the discourse of the show has to do not only with the town's foundation myth as a site of social and sexual freedom, but with the narrative's negotiation of a variety of issues pertaining to our own foundation myths, myths currently undergoing revision in a climate of multiculturalism and accompanying changes in the ways we see and (re)write history(s).

Moreover, the show's generic recombination and its penchant for self-reflexivity enable it to engage a range of issues almost unprecedented on primetime. Thus while *Northern Exposure* is usually billed as a drama (and has received several Emmy nominations as such), it functions more as an hour-long ensemble comedy with an a slight nod to the medical franchise, another to primetime melodrama, another to the fish-out-of-water sitcom, and still another to the sixties' "magicom" (a sitcom employing magical or fantasy elements). While generic recombination and self-reflexivity is standard operating procedure on most quality shows, *Northern Exposure* references genres and texts in a more organic way than that of any previous quality show. What we have here is neither pastiche nor parody but a conscious mining of its historical antecedents, which include Freud and Mother Goose as well as *St. Elsewhere* and *Marcus Welby.*

Superficially, *Northern Exposure* is about Joel Fleischman, who owes four years of doctoring to the state of Alaska, whose taxpayers financed his top-notch medical education. To Joel's utter dismay, he ends up in Cicely, population 839, rather than Anchorage, as he had expected.[49] However, Joel is not the focal character of this "round" ensemble, nor is his East Coast perspective at all dominant. Rather, each character of this remarkably complex ensemble represents a different discursive strategy. The ensemble, which makes up what Falsey and Brand call a "nonjudgemental universe of unconventional characters in a clash of cultures,"[50] consists of Native Americans, a (closet) descendant of Louis XIV, an ex-astronaut, a former Miss Northwest Passage, a Grosse Point socialite turned bush pilot, a West Virginian disc jockey/priest/Jungian/ex-con, and an Upper West Side yuppie physician. This varied ensemble facilitates the show's multi-perspectival approach to the vagaries and complexities of modern life and emphasizes an exploration of "the cultures rather than focusing on the clashes,"[51] according to Falsey and Brand. Put another way by Thomas Schatz, the show exists outside of "the war zone"—nuclear families, and so forth—that characterizes the narrative universes of most television series.

In addition, the show mines its characters' past for backstory in a way that is particularly distinctive among quality series. One aspect of the MTM style was the development of a quasi-serial strategy, which allowed for the ongoing reactivation of long-term stakes through irresolveable conflict (such as the relationship between Sam and Diane on *Cheers,* or that between Captain Furillo and Joyce Davenport on *Hill Street Blues*) but within the confines of a more plot-based episodic franchise that inevitably achieved closure. In contrast, *Northern Exposure* "remembers" pieces of its characters' pre-series lives and uses them as episodic fodder, reaching some kind of closure that yet resonates serially. Generally, the serial aspects of most quality shows had to do with static features—Mary Tyler Moore's single girl status, Sam and Diane's incompatible backgrounds, or Furillo and Davenport's irresolveable conflicts over the legal system—whereas the episodic plots were a function of series franchise and setting (police precinct, bar, etc.). In *Northern Exposure,* the episodic plots are more organic, being character based, and steadily add to a storehouse of information we know about a given character and which is woven into future episodes.

For instance, Chris Stevens turns out to have a symbiotic black half-brother—Bernard—whom he didn't know about until Bernard showed up one day in town. The two share thoughts, dreams, and have the same vision of a sculpture representing the Aurora Borealis before they discover they share the same rambling father. Bernard returns to Cicely in a subsequent episode, when Chris is having dreams of Africa that turn out to belong to Bernard. Similarly, Joel receives a ghostly visitation from a former occupant of his cabin whose ultimately suicidal misanthropic ways bear an uncanny resemblance to Joel's own emotional insularity. Learning of his predecessor's suicide due to loneliness, Joel renegotiates his own standoffish relations to the inhabitants of Cicely, a narrative move that yet resonates deeply with Joel's past actions within the community and bodes well for his future as a more engaged member. In another example, when Maggie's conservative, buttoned-down father visits from Grosse Point, she cajoles Joel into acting like her fiance, because that is what she feels her father has always expected from her, explaining in part the marital ambivalence that has been an ongoing source of plot material (five boyfriends dead of natural, albeit mysterious, causes). Even the town has an important backstory, as mentioned above, which has cropped up in several episodes, one of the most important of which is the season finale of 1992 depicting Cicely and Rosamunde's transformation of the town from a squalid frontier town to a cultivated utopia. The show's unique balance of serial and episodic stakes is narratively pleasurable; it also fosters viewer identification and loyalty, and will make the show a more likely prospect for syndication. *Northern Exposure* is perhaps even more episodic than it is serial, since our sense of seriality comes largely from just knowing the characters.

Related to this, *Northern Exposure* has a very unusual, even anomalous voice. That is quite a claim, given the multiplicity of voices emitting from the black box. However, one can't listen to all of them and it is important to make judgments based on personal politics and taste in order to determine "the relation between ideologies and aesthetics, between the progressive and the pleasurable," as Susan Boyd-Bowman puts it in her meditation on the merits of quality television.[52] I find the show progressive and inclusive because its overarching discourse centers the relationship of narrative to lived experience; that is, it interrogates the role of stories (and hence ideology) in the construction/formation of individuals. Any conclusions are always tentative and function to bring the dominant discourse into an arena of negotiation, which may be seen on one level as Cicely, Alaska, and on another as that cultural forum sometimes known as television. The show seems to have an awareness of itself as an example of the kind of activity that can transpire within the cultural forum, where, as Newcomb and Hirsch point out:

> Our most traditional views, those that are repressive and reactionary, as well as those that are subversive and emancipatory, are upheld, examined, maintained and transformed. The emphasis is on process rather than product, on fusion rather than indoctrination, on contradiction and confusion rather than coherence.[53]

In other words, the show is self-consciously "bardic"—aware of its role in the "transmission of culture and mediation of language" that, according to Fiske and Hartley, characterizes the medium.[54] I use the term bardic television with reservations because, technologically and culturally speaking (for starters), television no longer presents the same kind of "unified" or dominant voice it did when they constructed this model. The metaphor still pertains, though nowadays the bard's voice is more inclusive, polymorphic, and fragmented. *Northern Exposure* neatly exemplifies this postmodern bard.

Most episodes employ the voice of Chris Stevens, the disc jockey of K-BEHR, to both organize and comment upon each episode's day-in-the-life-of-Cicely structure, in a fashion reminiscent of *Hill Street Blues*'s morning "roll call." Chris's eloquent musings draw upon a range of cultural sources, from Native American writers to Jung, Thoreau, Proust, Albert Einstein, Maurice Sendak, and Chicken Little, among others. Chris is this community's philosophical troubadour: he narrates the town and its stories, and by extension, ours, invoking centuries-old traditions in which culture is transmitted orally and thereby continually recontextualized, a tradition in which television itself now plays a part.

Often Chris spearheads the show's rigorous negotiation of the transmission of culture(s).[55] In a fall 1991 episode, the discovery of a 176-year-old body preserved in ice sets off a debate about the roles of truth and myth in the construction of history. The body is identified as one Pierre Le Moulin who was apparently traveling with none other than the Emperor Napoleon. Napoleon stayed after Pierre left to return to France, founding a mixed tribe of very short people. Chris plays a central role at a town meeting to discuss whether the truth regarding Napoleon's absence at the battle of Waterloo should be shared with the rest of the world. Arguing against, Chris suggests that history serves a mythical function; national and personal identities are built upon its "fragile edifice." Joel—self-proclaimed man of science—wants to let the facts speak for themselves. At the end of this meeting, three members of the Telekutan tribe supposedly fathered by Napoleon and Pierre enter and demand the body, which is sacred to them. Maurice refuses—he sees in the body the economic future of Cicely, whereas the Telekutans see in it their past. Pierre has become a contested battleground in a war about whose stories dominate, who owns them, who has the right to suppress them, to worship them, or to commodify them. In the end, it is the spiritual meaning of Pierre that prevails, for the Telekutans spirit the body away from its temporary resting place in the Brick's (a local watering hole) freezer in the dawn. The final shot tracks their canoe down the river in a series of point-of-view shots motivated by Marilyn, Joel's Native American nurse. It was Marilyn who, at the beginning of the episode, suggested that Pierre might belong to the Telekutans in the first place. She is what Douglas Sirk would call the "secret owner" of this episode; it is her unheeded suggestion that speaks "truth" about Pierre and his ultimate fate—restoration to the tribe whose stories are constructed around him. Over the final shot we hear Chris, bidding farewell to Pierre and reading, appropriately enough, from the closing passage of *Remembrance of Things Past,* in which Proust refers to the construction of the past as "the vast structure of recollection."

The constructedness of culture(s) and the inscription of humans within those fragile edifices are renegotiated week after week in *Northern Exposure* to a degree unprecedented on television, even within the self-reflexive and humanist tradition of quality TV. Can we hope to see more network fare like this? According to Peter Tortorici, "television has never been better because it's never been as challenged as it is today—there have never been more alternatives."[56] His optimism raises an important point. The number of alternatives has fragmented the audience, to the point where the networks realize that a very popular show will only bring in so many households. Whether the present moment represents one of challenge or one of crisis for the networks, the implications are the same. The quality series of the seventies and eighties didn't have to reach a mass audience; the quality series of the nineties simply cannot reach a mass audience. As one analyst puts it, the networks are acting more and more like cable webs all the time, meaning they show a greater propensity to target specific audience groups.[57] Thus a quirky, introspective show like *Northern Exposure* has greater chances for survival. Perhaps the narrowly targeted but polyvocal series is the way of the future for quality network television, and if so, *Northern Exposure* is a vital signpost beckoning . . . north.

## Notes

1. Steve Coe, "Quality TV: Hollywood's Elusive Illusions," *Broadcasting,* November 18, 1991, p. 3.

2. Ibid.

3. Paul Schulman, president of the Paul Schulman Co., quoted in ibid.

4. Bill Carter, "Wooing the Real Stars Behind the Hit Series," *New York Times,* February 23, 1992, pp. H 33–34.

5. Steve Coe, "Networks Limit Risk for Reward of Season Win," *Broadcasting,* 5.

6. *Northern Exposure* swept the 1992 Emmy Awards, receiving six in all including Best Drama series (*USA Today,* September 30, 1992, p. D1).

7. Jane Feuer, "The MTM Style," in *MTM: Quality Television,* ed. Jane Feuer, Paul Kerr, and Tise Vahimagi (London: British Film Institute, 1984), p. 38.

8. Ibid., p. 34.

9. Michelle Hilmes, "Where Everybody Knows Your Name: *Cheers* and the Mediation of Cultures," *Wide Angle* 12, no. 2 (1990): 64–73.

10. Susan Boyd-Bowman, "The MTM Phenomenon," *Screen* 26, no. 6 (1985): 87.

11. Hilmes, "Where Everybody Knows Your Name," p. 66.

12. Paul Kerr, "The Making of (the) MTM (Show)," in *MTM: Quality Television,* ed. Feuer, Kerr, and Vahimagi, p. 66.

13. Ibid., p. 63.

14. Ibid. Cancelled shows included *Mayberry R.F.D., The Beverly Hillbillies, Hee Haw, Green Acres, The Glen Campbell Goodtime Hour,* and *Family Affair.*

15. Ibid., 64.

16. Ibid., 68.

17. Neil Hickey, "Review of *Northern Exposure*," *TV Guide,* September 7, 1991, p. 4. (At the outset of the series's third season (fall 1992), the rating/share for the weeks 8/31–9/6 and 9/6–13 were, respectively, 15.9/29 and 12.4/21, *Broadcasting* (September 14, 1992, p. 28, and September 21, 1992, p. 32).

18. This is the figure as determined by *Broadcasting*'s Ratings Week March 2–8 (*Broadcasting,* March 16, 1992, p. 18).

19. Mark Shapiro, "Trademarks for Offbeat," *Austin American Statesman,* March 29, 1992, p. 5.

20. Tortorici, quoted in David Kissinger, ibid.

21. Littlefield, David Kissinger quoted in "Nets' Plans: Better Safe than Sorry," *Broadcasting,* August 26, 1991, p. 39.

22. Thomas Schatz, "*St. Elsewhere* and the Evolution of the Ensemble Series," in *Television: The Critical View,* 4th ed., ed. Horace Newcomb (New York: Oxford University Press, 1987), p. 88.

23. Ibid., p. 90.

24. Feuer, "MTM Style," p. 34.

25. Gary Deeb, "The Man Who Destroyed Television," *Playboy,* February 1980, p. 220.

26. Steve Coe, "CBS's Stock Rises in Hollywood," *Broadcasting,* August 12, 1991, p. 36.

27. Ibid.

28. Ibid.

29. Ibid.

30. Ibid.

31. Coe, "Quality TV," p. 17.

32. Ibid., p. 1.

33. David Kissinger, "Scribes Say Network Wimps Crimp Their Style," *Broadcasting,* July 1, 1991, p. 19.

34. Bill Carter, "Rivalries Give Way to Ratings," *New York Times,* November 25, 1991, p. C1.

35. Ibid.

36. Elizabeth Guider, "Off the Wall Gets You in the Door, if Not on the Air," *Variety,* August 15, 1990, p. 85.

37. Ibid.

38. Ibid.

39. David Kissinger, "Same 'Ol Song and Danceky from CBS Chief Sagansky," *Broadcasting,* August 5, 1991, p. 25.

40. Ibid.

41. Ibid., p. 27.

42. Ibid.

43. Steve Coe, "Networks Limit Risk for Reward of Season Win," *Broadcasting,* June 3, 1991, p. 35.

44. Steve Coe, "Second Chance for *Northern Exposure*," *Broadcasting,* February 11, 1991, p. 53.

45. Ibid.

46. Ibid.

47. Ibid.

48. Scott Sherman, "Glaad Tidings: CBS's *Northern Exposure*," *Outweek*, June 26, 1991, p. 22.

49. This number was designated because $839,000 was the budget per episode of the initial summer run. Production budget has increased by about one-third since then. Interestingly, the show is shot on location in a Washington state burg that was a creative haven for an alternative film and video community during the 1970s, which resonates with Cicely's own backstory as a frontier town offering alternative lifestyles, etc.

50. Schapiro, "Trademarks for Offbeat," p. 5.

51. Ibid.

52. Susan Boyd-Bowman, "The MTM Phenomenon," *Screen* 26, no. 6 (1985): 87.

53. Horace Newcomb and Paul Hirsch, "Television as a Cultural Forum," in *Television: The Critical View,* ed. Newcomb, p. 459.

54. John Fiske and John Hartley, *Reading Television* (London: Methuen, 1978), p. 86.

55. I want to note that Chris's bardic function has evolved. He did not begin the series with such a preeminent role in the ensemble, but audience reaction to him has been overwhelmingly positive and his role has been expanded accordingly. In conjunction, as reaction to Joel has been tepid, his centrality has shifted to make way for Chris.

56. Coe, "Quality TV," p. 3.

57. J. Max Robins, "Fall Skeds Trade Class For Mass," *Variety,* May 27, 1991, p. 1.

# The Made-for-TV Movie:
# Industrial Practice,
# Cultural Form,
# Popular Reception

## LAURIE SCHULZE

The made-for-TV movie's initial reception by popular criticism was marked by a particularly vehement hostility. Despite the fact that NBC's 1966 *Fame Is the Name of the Game* (usually credited with being the first made-for-TV movie) captured 40 percent of the viewing audience, *TV Guide*'s Richard K. Doan labeled *Fame* a "non-movie," describing it as "Grade-B melodrama . . . given the promotional trappings of a 'world premiere' because neither moviegoers nor TV audiences had seen it before."[1]

Variously referred to in popular critical discourse as "non-movies," "quasi-movies," and "quickies," made-for-TV movies were immediately positioned by reviewers as objects unworthy of serious consideration. The inevitable point of comparison was the theatrical feature, and made-for-TV movies were designated the "film industry's stepchildren." As television critic Judith Crist wrote in 1969, the TV movie "wouldn't . . . earn a B rating on any theatrical meter bill."

Crist went on to say that "what separates the television movie from the theatrical film is, basically, production values—cheap ones—not quite casts and hiccuping plots" that lurch "from climax to climax in time for the cluster of commercials at set intervals, allowing no subtle developments of character or story." Crist condemned TV movies for their formulaic character-istics, and included a quote from a network executive admitting that TV movies are "all the same . . . and they get the same ratings so who cares?"

From *Hollywood in the Age of Television*, Balio, ed., Routledge 1990. Reprinted by permission of the author and the publisher.

Early popular criticism like Crist's almost always justified its dismissal of the TV movie by pointing to the movie's mode of production—its limited budget, tight shooting schedule, and the sheer amount of text generated implicates the TV movie as the product of the "factory," not the "atelier." "Popularity," Crist concluded, "is small proof of quality."[2]

If the meaning constructed for the TV movie in early popular criticism might best be summed up as "junk," eventually reviewers attempted to make a space in which "exceptional" TV movies could be singled out and positioned against the background of the form. Articles like Dwight Whitney's "Cinema's Stepchild Grows Up" (*TV Guide,* July 20, 1974) situated themselves as an apologetic for made-for-TV movies, while acknowledging that most of them were "shoddy."

Whitney's attempt to recuperate some TV movies begins by pushing the made-for-TV movies as "mere potboiling trash" back into history: "in the beginning, the made-for-TV movie was . . . a repository for the deposed stars of former TV series." A few years later, Alvin H. Marill and Patrick McGilligan used similar arguments, Marill referring to "those escapist pieces of the early days" and McGilligan echoing Whitney's opening gambit: "in the beginning, most of the made-for-television movies were run-of-the-mill exploitation pictures."[3] This clears a space for the "good" TV movie by relegating the "bad" TV movie to the formative stages. That space is then filled by gesturing at the serious topics taken up by recent TV movies and making a case for the existence of TV movie auteurs.

Marill refers to the "more mature themes" (homosexuality, alcoholism, rape) dealt with by the contemporary TV movie, the kind approvingly labeled the "sociological film" by Whitney and the "public service drama" by McGilligan. The implication is that TV movies that deal with serious social problems themselves deserve to be taken seriously, much like David Thorburn's argument concerning television melodrama, which urges that melodrama deserves critical consideration because it acts as an "arena" for dealing with "disturbing" social and moral problems. These attempts to valorize the social-issue TV movie also emphasize its public service or informational value: Whitney singled out *The Morning After* because it "showed what it is like to be afflicted with alcoholism." If the public service seriousness of the TV movie was the basis of one of the earlier critical strategies used to make a case for the "good" TV movie, however, it seems to have become a point of potential dissatisfaction with the form. In a review of *The Burning Bed* (NBC, 1984), Richard Zoglin refers to social-problem TV movies as "TV's issue-of-the-week parade," arguing that they are so wrapped up in teaching viewers about social issues that they become very bad drama.[4]

If serious themes are no longer necessary and sufficient conditions for elevating the TV movie to a place in the popular canon, pointing to the identifiable presence of TV movie authors has been maintained as a critical strategy for legitimating some made-for-TV movies.[5] Some popular television critics, following the lines of the auteurist project in film studies, argue that the made-for-TV movie can be shown to be popular art because it can be shown to have authors. Those TV movies that can be linked with an authorial presence (and issuing source of meaning) are marked off as deserving serious critical study and critical acclaim. This strategy is implicit in Whitney's earlier article; those films that he singles out for attention are labeled "Christiansen and Hosenberg's *The Autobiography of Miss Jane Pittman,* or "Paul Junger Witt's *Brian's Song,*" or "Roger Gimbal's *I Heard the Owl Call My Name,*' " Later "auteurist" critical work draws heavily on directors who began in TV movies and then became recognized auteurs in film. Steven Spielberg is most often cited as an identifiable artist in the TV movie genre *(Something Evil, Duel, Savage)*. Since Spielberg is now an acknowledged cinematic (and thus, from the point of view of traditional criticism, more prestigious) director, the invocation of his name in connection with the TV movie adds weight to the claims of critics who attempt to canonize a TV movie by attributing it to an author. Patrick McGilligan extends the visible presence of a single auteur to cover the entire field of TV movies, claiming that "it took a director like Steven Spielberg to legitimize the made-for-television movie."[6] At the least, the TV movie can be pointed to as a place where "real" auteurs serve their apprenticeship and graduate to "real" movies.

The auteurist project for TV movies, however, is limited to a few movies by a few TV movie directors. Tom Allen, for example, attempts to identify patterns of meaning and style marking the TV movies of John Sargent, Steven Spielberg, John Badham, Lamont Johnson, John Korty, and William A. Graham as authored movies that therefore command critical attention. But while he valorizes the TV movies of the TV auteurs, he sets them against the "great bulk" of TV movies as a whole, which, he writes, is "impossible to imagine . . . anywhere but on subsidized, time-killing commercial TV." If the TV movie can be demonstrated to have "some golden talents," Allen claims, it has not produced a "golden age." Some TV movies are special because golden talents have managed to resist the institutional constraints of the television system, inscribing a personal vision in their movies that can be extracted from them by the critic. But the TV movie as a whole is characterized by Allen as "a common heap of dross."[7]

Faced with the immense popularity of the TV movie, which seems to fly in the face of reviewers' low estimations of the genre, critics often offer up the audience as a scapegoat. Sometimes the popular audience is conceptualized as merely unsophisticated, helpless to resist the machinations of the powerful and persuasive television system. Herbert Gold, for instance, speaks of the promotion of the TV movie as an "attack" on audiences, and cites the unquestioned ability of the TV movie's dramatic construction to "hold" and "grip" viewers in spite of themselves.[8] A recent essay in *Time* on the "dangers" of the made-for-TV movie based on real events worries that the public may not be capable of understanding that "a network may have one standard of fidelity to fact in its 7 P.M. newscast, and another an hour later in its docudramas."[9]

More often, however, traditional criticism holds the audience itself responsible for its low sensibilities. Television critic Tom Shales *(The Washington Post)* bemoans the fact that TV movies like ABC's "awesomely stupid" *Lace* and "ludicrously smarmy" *My Mother's Secret Life* received much higher ratings than the "wholesome movie hits" *Star Wars* and *Chariots of Fire*. Shales quotes an independent producer who charges that "the public deserves some of the blame for these programs. They always flock to the trashy things." The trashy things Shales is concerned about trade in "stereotypes" that are "most demeaning to women." Significantly, Shales goes on to argue that the female audience is at fault. "Such programs," Shales claims, "would not be on the air if they didn't appeal to women." This implicates the female audience in either feeblemindedness, reactionary antifeminism, or a kind of masochistic aesthetic, all of which Shales appears to imply. The moral turpitude Shales attributes to the TV movie is traced to some deficiency on the part of its female viewership, setting up a kind of "us" and "them" opposition. Made-for-TV movie "trash" is something that "they" (presumably uneducated, ideologically defective, or self-destructive women) want to watch. "We" (the implied reader of Shale's essay) would like to see what Shales approvingly labels "serious, intelligent, unsmutty TV movies," which won't be made because the (female) audience wants trash. (The "we" is presumably educated, liberal, and male.) [10]

Although popular criticism does not seem to explicitly categorize the TV movie as a women's genre, it lurks around the edges of the common complaint about the TV movie's reliance on melodrama, a genre historically linked with female viewership. When critics like Richard Corliss comment that "for a cathartic sob one must go to TV for a Movie of the Week" instead of to the cinema, a connection is clearly drawn between the TV movie and the woman's film—the "tearjerker" or the "weepie." [11] There are also frequent allusions to the similarities between TV movies and the soap opera, a narrative strongly connected with the female audience in popular and academic discourse. The TV movie, with its reliance on the family, melodrama and the romance, its tendency to take up domestic issues, and its penchant for female protagonists and female stars, may indeed lean toward what has come to be called a feminine narrative form. The point is that a female audience for the TV movie is taken to task by popular criticism to strike another blow against what is perceived to be a nonaesthetic and morally defective form of popular culture. [12]

Incapable of constituting the typical TV movie as an aesthetic object and thus incapable of attributing any positive aesthetic or social pleasures to the viewing experience, popular criticism pushes the TV movie even closer to the edges of its discourse by positing an audience that is also marked as "different," and in some way as aesthetically and morally bankrupt and as incomprehensible as the TV movie is thought to be. Traditional aesthetic criticism refuses to make a space for the possibility of legitimate meanings or pleasures on the part of the typical made-for-TV movie audience, except for the sophisticated pleasures of the "schlock buff," whose superior sensibilities can legitimately be trained on trash. But if reviewers, on the whole, have not taken to the TV movie, the popular audience has, and the history of the made-for-TV movie is largely the history of a television form that has been increasingly profitable for networks and that has taken on a significant role in competitive programming strategy.

In the 1950s, Hollywood's major motion picture companies released their pre-1948 feature films to local television stations for broadcast. The major Hollywood studios were already involved in producing programs for television, but until 1956, the only theatrical features shown on television came from foreign studios or from B-movie American producers. Their "cheap production values" notwithstanding, these theatrical features became, according to Douglas Gomery, a "mainstay of local television programming practice" by 1955. In 1956, C & C Television, which had acquired the rights to the RKO features, released the first theatrical titles from a major Hollywood studio to local markets and within a year pulled in an estimated $25 million. The other major studios soon followed, and by 1958, all of the majors' pre-1948 features were in heavy circulation on local stations. Yet the networks, although occasionally including a feature film in their prime-time schedules, did not take up Hollywood films until the 1960s, when the consistently high local ratings for pre-1948 films convinced them that Hollywood movies on network prime time might generate an even more impressive share of the audience. In the 1960s, television successfully negotiated for recent feature films, and by the early 1970s the three networks were broadcasting ten prime-time "Movie Nights" each week. The ratings were high, and the recent Hollywood feature became, as Gomery puts it, "one of the strongest weapons" the networks could deploy in the ratings wars, especially during the rating period called "sweeps weeks," which determine the rates local stations and networks will charge advertisers for commercial time.[13]

This programming practice, saturating prime time with Hollywood films, rapidly led to a shortage of available and appropriate features: television broadcast them faster than Hollywood could produce them. In May 1966, *Television Magazine,* describing the "seller's market" created by the movie shortage, reported that the stations were so desperate they were buying "anything with sprocket holes." Opportunistic studios used the impressively high ratings commanded by their features to exact even higher prices from the networks for the right to broadcast their more popular films. When *The Bridge on the River Kwai* went to ABC for $2 million in 1966, the television industry began to realize that there might be no limit to what studios could demand for their product.[14]

With the increasing costs of Hollywood features cutting away at their profits, the networks began to commission the production of films exclusively for television. MCA, Inc. and its subsidiary, Universal City Studios, were influential players in the development of TV movies. Lew R. Wasserman, chairman of MCA, the top talent agency in the 1940s, had long been interested in getting into film production, but Screen Actors Guild regulations prevented a talent agency from moving into the movie business. When television came along, however, the union allowed MCA to produce filmed material for television, and MCA's Revue Productions, with programming chief Jennings Lang, was turning out popular programs like *Alfred Hitchcock Presents, General Electric Theater,* and *Bachelor Father* by the late 1950s. In 1962, Wasserman bought the Universal Pictures lot, modernized the run-down studio, and then Wasserman, Lang, and MCA began pitching the concept of one-shot movies made just for television to the networks. CBS and ABC were skeptical, but NBC was more optimistic. In 1965, NBC contracted with MCA for more than 30 "World Premiere" movies to be produced over a several year period. The first, *Fame Is The Name of the Game* (Universal, 1966), turned out to be a "back-door pilot" for a subsequent series, as well.[15] What critic Patrick McGilligan terms a "declaration of independence" from the Hollywood theatrical feature by the networks had begun.[16]

For the next three seasons, movies made for television appeared intermittently in the networks' prime-time programming, but they were, according to McGilligan, "infrequent" and considered "risky business." In 1969, ABC introduced its *Movie of the Week,* a regular series of films made exclusively for television. No one, according to Barry Diller, who created the "Movie of the Week" format, expected the genre to be that popular. But the made-for-TV movie demonstrated that it could command extremely large audiences, sometimes even overstripping the ratings power of popular theatrical features.[17] In addition to its rating potential, the made-for-TV movie was also more economical to produce than the theatrical feature and often cost less than the rights to broadcast popular Hollywood films (in the early 1970s, production costs were typically under $1 million). By 1972, the popular and profitable made-for-TV movie, was, as Gomery puts it, "established as a force on network television," with all three major networks placing made-for-TV movies in their prime-time programming. ABC's *Brian's Song* (1971), a docudrama about the friendship between Chicago Bears football players Gale Sayers and Brian Piccolo (Piccolo, tragically, died of cancer in 1970), made the top ten list of all movies (theatricals and made-for's) ever shown on television, winning Emmies and critical acclaim, as well as the highest rating yet achieved by a made-for-TV movie.[18] TV movie production more than doubled between 1970–71 and 1971–72.

In the mid-1970s, producers and networks expanded the made-for-TV movie form to include novels for television and miniseries. *Rich Man, Poor Man* in 1975, *Roots* in 1977, and *Holocaust* in 1978 proved to the industry that the long-form television movie could produce spectacular ratings. The long-form movies became what McGilligan calls the "command items" of the genre. These special events, however, have equally special production costs compared with the average TV movie. Currently, a typical made-for-TV movie costs about $2 million to produce, while the CBS miniseries *Space* (1985) reportedly cost over $32 million and the recent ratings hit *Lonesome Dove* (1989) over $20 million. Since the long-form movies need high ratings to recoup their high production costs, they represent more of a gamble than the ordinary movie of the week, and the two-hour made-for-TV movie continues to be the backbone of the genre. Not that the average made-for-TV movie is uncompetitive. In 1979, ABC's *Elvis* (a biography of Elvis Presley) captured the largest share of the audience, against *Gone With the Wind* on CBS and *One Flew Over the Cuckoo's Nest* on NBC. In the 1983–84 season, two ABC made-for-TV movies, *The Day After* (a drama about a nuclear strike on Lawrence, Kansas, starring Jason Robards and JoBeth Williams) and *Something About Amelia* (a made-for about incest, starring Ted Danson and Glenn Close), ranked second and fourth in the ratings for all television shows of the season. NBC's *Adam* (missing children) and *Policewoman Centerfold* (a police officer's career is at stake after she poses for a centerfold) both out-performed ABC's *Monday Night Football,* while blockbuster theatricals like *Star Wars* and *Chariots of Fire* failed to win their time slots.

Since its entry into broadcast television, the made-for-TV movie has become a major part of network programming practice. The three networks together commission, on average, over a hundred movies each year, sometimes outstripping domestic feature film production. The made-for-TV movie occupies about 20 percent of network prime time.[19] Since the growth of cable and pay TV, made-for-TV movies may be more important to the networks than ever, because cable and pay TV are getting the first-run deals for theatrical features. In 1983, Lew Erlicht, president of ABC Entertainment, reported that the first-run exposure of theatricals on cable and pay TV cost the eventual network run of the same theatricals at least 10 to 15 share points. Erlicht said, "there's no reason we should be buying 24-share films, why should we?" The TV movie, with real first-run status, represents a network programming alternative to overexposed theatricals. ABC chose to open the 1983–84 season with *The Making of a Male Model,* getting a 33 share. Steve Mills, CBS vice president/motion pictures for TV, claims that CBS's TV movies perform well in "defensive" programming, going up against the competition's top-rated shows and coming away winners. Made-fors also have a highly desirable flexibility. Functioning as television's "switch-hitters and designated schedulers" in the programming game, TV movies can come off the bench to undercut the other networks' feature films, special events, or series.[20] CBS, currently running third behind ABC and NBC in prime time, recently announced that since series are not pulling viewers to the CBS schedule, their programming strategy for the 1989–90 season would focus on "high concept" made-for-TV movies to attract the audience.[21]

The TV movie is undeniably a mass-produced text. The networks underwrite the production of over two hundred hours of made-for-TV movies every year, and the economic constraints on made-for-TV movie budgets as well as the demand for quantity limit the actual shooting schedule of the average TV movie to about three weeks, although there may be months in which to develop a concept and write a script. Months spent in preproduction and three weeks in production might seem a rather leisurely pace in comparison with the production schedules of television's other narrative forms: an episode of a soap opera, for example, must be shot in a single day.

In critical discourse, however, made-for-TV movies were invariably compared to the "real" movies, and the comparison resulted in a characterization of made-for-TV movies as "quickies." Judith Crist, perhaps the most influential television critic, labeled them "nonmovies." Crist, like many other critics, claimed that the mode of production of the TV movie consigns it (with rare exceptions) to the trash heap of formulaic mass-produced narratives. Made-for-TV movies, Crist wrote, "don't bear detailing" because they are "all the same," identical products of the "assembly-belt system" of the television "factory."[22]

Popular aesthetic discourse on the made-for-TV movies as mass-produced culture does have a point: made-for-TV movies are organized according to an industrial system of production, have limited production funds, and restricted production schedules. But leveling all made-for-TV movies to the same narrative text and deducing the way the text will be read and its aesthetic value from the mode of production obscures a necessary difference between mass-produced consumer goods and "mass-produced" texts. As Robert Allen points out, the analogy breaks down at a crucial juncture, often lost on traditional critics.

> The absolute standardization required for the mass production of consumer items is inapplicable to the production of narratives. The consumer expects each bar of Ivory soap to be exactly like the last one purchased, but he or she expects each new movie or episode of a television program to bear marks of difference.[23]

While the mode of production and the institutional function assumed by television texts—the establishment of a regular audience habituated to a particular program—does demand that programs exhibit marks of similarity so that viewer familiarity will ensure repeated and predictable return to them, they must also exhibit marks of difference. Television's fictive narratives must be both the same and yet not the same. Their differences, ignored by much traditional criticism, are as essential to their popularity as their similarities. Meaningful differences among TV movies may include generic variations: there are TV movie versions of hard-boiled detective genre (*Calendar Girl Murders*), the screwball comedy (*Maid in America*), science-fiction (*V*), the family melodrama (*Family Secrets*), the docudrama (*Flight #90: Disaster on the Potomac*), the biography (*The Jesse Owens Story*), the historical romance (*Mistress of Paradise*), the suspense-thriller (*Through Naked Eyes*). TV movies take up different social issues, from alcoholism (*The Boy Who Drank Too Much*) to union activism (*Heart of Steel*). Some TV movies experiment with visual style (NBC's *Special Bulletin*, a 1983 made-for about protestors threatening to detonate nuclear warheads in Charleston, South Carolina, was shot on videotape as a simulated newscast).

Alongside their differences, made-for-TV movies have also developed identifiable similarities that have solidified into "codes" that distinguish them from, for instance, the theatrical feature. Critics and audiences both exhibit a sense for what those distinguishing characteristics of the TV movie genre are, although they may not necessarily articulate them. But almost everyone would understand what is meant when, for example, a film reviewer remarks that the 1988 theatrical release *The Good Mother* (starring Diane Keaton and Liam Neeson, about a divorced woman who risks losing custody of her daughter when her exhusband charges that the six-year-old's exposure to the relationship between Keaton and her lover is damaging the child) "looks like a standard Issue-of-the-Week TV-movie that somehow wandered onto the big screen."[24]

Shooting for the box rather than the big screen makes a difference in the visual style of the TV movie. Because the television image is much smaller, and because the image has relatively low resolution and permits less detail than the cinematic image, the medium shot and the close-up dominate TV movie practice, combined with a tendency to use relatively shallow focus: compared to the cinematic image, television's *mise en scène* is stripped down. Limited budgets also restrict the TV movie's scope: huge crowd scenes and on-location spectacles cost money, money that may not be available to the typical movie-of-the-week. But the TV movie has fashioned an aesthetic from its limitations. Horace Newcomb identifies "intimacy" as one of television's aesthetic principles, pointing out that stylistic intimacy is appropriate for television's reduced visual scale and for TV's normative viewing conditions: television is watched in private spaces, in the home. TV movies extend the principle of intimacy into their narrative material as well, concentrating heavily on the personal story.[25] (The TV movie's preference for melodrama as the generic vehicle for intimate human interest stories will be discussed at greater length below, as will the TV movie's reliance on hot social issues for topical material. Both "melodramatic" and "topical" are practically synonymous with the cultural connotations indentified with the TV movie genre.)

Television's viewing situation is generally a distracted one, and viewers' attention to the small screen is often intermittant. John Ellis argues that watching TV is organized by a "regime of the glance," rather than the concentrated gaze at the image encouraged by the theatrical film viewing situation. Television sound is audible, however, even when the viewer may not be looking directly at the screen. So TV places a greater emphasis on sound to attract the viewer's glance and to carry the message.[26] Thus, TV movies tend to rely on dialogue more than their theatrical counterparts.

Unlike theatrical films, TV movies are interrupted by commercials. Made-fors must structure their narratives around commercial breaks, resulting in what David Thorburn terms "segmented dramatic structure," each act achieving a "localized vividness," with dramatic mini-climaxes occurring just before the commercial breaks, so viewers will stick with the story across the interruptions. Thorburn goes on to point out that one key strategy functioning to achieve this segmented vividness is an aesthetic of performance:

the actors must be "intense" and "energetic," so that their highly emotional performances give each segment "independent weight and interest."[27] The effects of segmentation on narrative structure and the actor's performance is perhaps most clearly evident in the opening act of the TV movie. As Todd Gitlin puts it, "all salient [narrative] elements have to be established with breathtaking haste" and characters' traits must "leap out of the screen," or the TV movie risks losing its audience.[28]

TV movie acting seems to have produced, as one critic observes, its own "galaxy of stars," with names that seldom appear in theatrical features but that are almost guaranteed to win high ratings for made-fors. Lindsay Wagner, Elizabeth Montgomery, Mariette Hartley, Angie Dickinson, Richard Thomas, Robert Wagner, and Richard Crenna, for example, have or have had TV movie star power comparable to the box office clout of a Meryl Streep or Sylvester Stallone. While movie stars occasionally appear in made-fors, the TV movie has its own independent star system, marking the made-for off from theatrical films. Both the aesthetic of the close-up and narrative segmentation, as well as the reliance on melodrama, invest the TV movie heavily in the actor's performance, and the TV movie star is a significant aspect of the genre's popularity.

Most made-for-TV movies are the product of independent producers. The networks are the buyers. According to independent producer Frank von Zerneck *(Katie: Portrait of a Centerfold),* the initial determination on the production of a made-for-TV movie is the development of a "concept" that will interest a network. Network interest is predicated on generating the maximum audience of the right demographic kind to deliver to advertisers. Networks will make the decision to underwrite TV movie projects based on whether or not they think the movie can be marketed in such a way as to ensure a sizable share of the audience.[29]

Todd Gitlin notes that the case of the made-for-TV movie presents the networks with a rather unique problem of promotability. Since TV movies anchor the relatively closed end of television's spectrum of narrative closure/openness, and since they are seen only once—sometimes twice if rerun in the summer—their audience cannot be developed by word of mouth or critical acclaim, as can the audience for a theatrical feature or a television series or serial. The TV movie cannot constitute, in effect, it own advertisement, as a series or serial or soap opera can.[30]

The marketing strategies available to the networks to bring an audience to a made-for-TV movie are limited to a few sentences in *TV Guide* and a brief trailer broadcast the week before the movie is scheduled to air. Von Zerneck says, "the audience [for the made-for-TV movie] is fresh, it's new each time. You have to convince them that there's something in this movie they want to see."[31] Being able to persuade an audience that "there's something in this movie they want to see" with a short series of clips and a sentence in *TV Guide* practically demands what one former network vice-president in charge of TV movies refers to as "hot concept" movies, already invested with "high promotability."[32]

In the absence of continuing characters and situations, which build an audience for the series and serial forms, the TV movie must depend on a brief, condensed one-shot narrative image to solicit its audience. As Gitlin puts it, to be marketable to networks in the first place, the concept for the TV movie has to be "sensational."[33] The TV movie therefore depends on a high degree of responsiveness to whatever issues are currently in heavy cultural circulation for its economic survival and effectiveness in commanding an audience. It must provide curiosity by the promise of the unusual or the scandalous and immediately mark itself off as different. Yet it must be familiar at the same time, and reassure by its reference to the instantly recognizable. If the concept proposed to the network by the independent producer cannot be condensed into this brief sensational/familiar narrative image, the project stands little chance of being developed. Von Zerneck sums up the situation from the network point of view: "No matter how good or bad a movie is, for television, if you can't summarize it in a sentence that will appear in *TV Guide,* and if you can't describe it in a paragraph, then you'll have a great deal of difficulty selling the project."[34]

If a network feels that the concept will be promotable—sometimes a random sample of viewers are surveyed to find out if they would watch a TV movie on the basis of a brief plot summary like the one that would appear in *TV Guide*—the independent producer will hire a writer to construct a treatment. If the treatment is approved by the network, a writer will proceed with a script. The producer, according to von Zerneck, usually pays for the script, but is reimbursed by the network if the network wants to fund the project. (Von Zerneck's company must develop twenty to twenty-five scripts for the networks to produce three or four made-for-TV movies per year. The development of a script does not guarantee that the network will eventually decide to bankroll the movie.) If the network accepts the script, it pays the producer a license fee, which will finance the production of the movie. This license fee gives the network the right to broadcast the finished movie twice.

The advantage to the independent producer is that after the network broadcasts the movie, the rights then revert back to the producer, who may sell it into syndication, release it on videocassette or theatrically in foreign markets, or sell it back to the networks for late-night programming. The producer also has the option of entering into partnership with a studio that provides production and postproduction facilities and the services of its marketing organizations to sell the movie into syndication or foreign release after the network has given the movie back to the producer. The studio exacts an overhead, built into production costs, which means that the producer must expect more of a delay before profits from syndication or foreign or videocassette sales are returned. The studio might also make it more difficult for the producer to go over budget or over schedule.

The greater determination, however, is exerted by the network, which first decides, in effect, whether or not the project will be produced at all. Von Zerneck admits, "they control the money and the network. They also have a point of view about what they want on their network, what kind of movie they want. So you're essentially writing it to order. The script is custom made for them." The bottom line for the networks, von Zerneck claims, is, "is it going to be a movie they can exploit? Will it get an audience? Will it have footage that they'll be able to put into a trailer to entice people to watch it?"[35]

The economic basis of the commercial broadcast television system is delivering the largest, most desirable audience to advertisers. So the networks tend to underwrite movie projects with a high degree of exploitability, projects for which the popular audience can be presumed to have the necessary "cultural capital," in Pierre Bourdieu's useful term, required for its narrative image to be immediately salient and relevant. The network notion of where popular sensibilities are to be monitored is summed up by a former network vice-president in charge of TV movies: "I look at TV commercials to look for a trend. Or places where money is spent, not free stuff like television. What's on the covers of magazines? What are the advertisers using to sell soap? What are they saying about what's going on in the country?" Hot concepts are mined from national magazine covers or from television's reality programming, talk show issues and tabloid television scandals that provide the narrative skeletons that can be fleshed out into what Gitlin rather sarcastically calls "little personal stories that executives think mass audiences will take as relations of the contemporary."[36]

The made-for-TV movie, perhaps because its story world begins and ends in two hours, because its narrative situations and characters do not come into the American home week in and week out, appears to be capable of pushing at the limits of the controversial without losing its audience. As Les Brown observes, when NBC broadcast *My Sweet Charlie* in 1970, a TV movie about an intimate relationship between a black man and a white Southern girl, it received a 53 share in the ratings; very few viewers complained that the movie was offensive.[37] Almost twenty years later, interracial sexual relationships between continuing characters on series, serials, or daytime soap operas seem to be very difficult for the networks to risk. Television's practice implies the assumption that the mass audience is more willing to tolerate sensational or controversial subjects for a single evening than on a continuing basis. Nevertheless, networks (in search of ratings) are reluctant to chance deeply offending even the TV movie audience. *TV Guide* reports that NBC "toned down" the recent *Roe vs. Wade* (1989), a TV movie about the again controversial Supreme Court decision legalizing abortion, to avoid criticism by antiabortion groups, who felt that the docudrama might be too sympathetic to "Jane Roe" (the woman denied an abortion under Texas law whose case eventually came before the Supreme Court). According to a senior vice-president in charge of TV movies, changes were made in the script to make *Roe vs. Wade* less "pro-choice."[38]

In broadcast television practice, the made-for-TV movie emerged as a privileged site for the negotiation of problematic social issues. What has been variously termed the issue-of-the-week movie, the public service drama, and the social problem TV movie by popular critics became the primary form of the TV movie genre. The issues taken up by the TV movie, however, tend toward the hot concept problems that can be transformed into salacious, highly promotable trailers: housewives turned afternoon prostitutes, lesbian mothers, gay fathers, venereal disease, rape, white slavery, addiction, abortion, domestic violence, incest, adultery, bisexuality, child pornography. If some critics complain that TV movies "take a spicy topic . . . and publicize it with a lot of juicy, lead in advertising," they also argue, as Richard Zoglin does in a review of *The Burning Bed* (NBC, 1984), a TV movie about domestic violence starring Farrah Fawcett, that "one by one, TV movies take up a topic, adorn it with stars, and promote it as another prime-time break through. As drama, these TV crusades have such familiar faults—too simplistic, too preachy, too ponderously 'educational'—that a good one can easily get lost in the shuffle."[39]

Whether the controversial issues the made-for-TV movie appropriates are easily marketed as "provocative" (teenage hitchhikers, male strippers) or more "serious" social problems (missing children, teenage suicide, toxic waste, the homeless, union activism, racism), both criticisms found in popular reviews—too sensational, too educational—point to interesting functions the TV movie has assumed in prime-time broadcast television. The issue-of-the-week movie, the center of the genre, already constitutes a response to points of social struggle. It opens up the site of an immense and intense ideological negotiation, limning, as it does, the more salient and disturbing phenomena on the social agenda.

That the topics "torn from the headlines" (a promotional slogan used to market *The Burning Bed*) are capable of being sensationalized in the movie's narrative image can be traced to the function any television text must assume in the economic system of broadcast television—getting the maximum audience of the demographic kind attractive to advertisers—and to the unique problem of promotability attached to the TV movie. That movies made for television also take on what I will call a pedagogical function with respect to the issues they organize—educating the audience by having characters cite the latest statistics on whatever subject is taken up, describing a social problem with presumed accuracy (if within a fictional context), and indicating the solution to the issue—brings up what Stuart Hall has called the "framing" function of popular television.[40]

Todd Gitlin, among other academic critics, argues that the made-for-TV movie, while taking up sensational and disturbing problems, strongly tends to organize or frame these issues in a way that domesticates them. Gitlin writes that "if the networks like a dollop of controversy now and then, they usually want it manageable: social significance with a lifted face." The TV movie's social issues are "routinely depoliticized" when turned into "little personal stories," and the made-for-TV movie ends up as "mass culture's equivalent of the squarish, hard-skinned, tasteless tomato grown for quick, reliable, low-cost machine harvesting, untouched by human hands."[41] Douglas Gomery, in his analysis of *Brian's Song* (ABC, 1971), concludes that made-for-TV movies simplify complex social issues at best and more often than not, through narrative structure and mythologizing stereotypes, manage to make the social issues they invoke "non-controversial."[42] Gitlin's analysis points to the effects that framing social issues in the context of the melodramatic personal story might have on a TV movie's ideological edge. The pedagogical stance frequently adopted by the made-for-TV movie might have a similar domesticating function.

*Something About Amelia,* for instance, spends the balance of its narrative energy in depicting the steps that need to be taken in a case of incest: believing the child, alerting the proper authorities, removing the father from the home and temporarily placing the child in a treatment center for abused children, counseling with qualified professionals to enable the family to deal with the consequences in ways that will lead to "healing" and the reformation of the family. All the facts and statistics about incest are oddly comforting, as if by simply knowing them we gain control over the problem. And the movie seems to reassure us, in a teaching sort of way, that existing social institutions and professionals already in place can and do provide solutions, that the problem is being handled.

Some TV movies bring in extra-textual material to educate the audience about issues and suggest ways in which they might be prevented or managed. *Adam* (NBC, 1983) was followed by a broadcast of photographs and descriptions of missing children, with a public appeal to call a special telephone number with any information that might help locate them. More than thirty missing children were found, and the solution indicated in the movie—the establishment of a nationwide network for finding missing children—was eventually put in place on the strength of public awareness generated by the TV movie. (Adam's real-life father, John Walsh, went on to host *America's Most Wanted*.)

*Surviving* (ABC, 1985) was broadcast accompanied by a number of extra-textual programs on the issue of teenage suicide, including a thirty-minute documentary on the problem, local and national news features, and educational material made available for use in schools in conjunction with the movie. The movie itself was bracketed with direct appeals to teenagers considering suicide, encouraging them to seek help. One network executive described the movie as "preventive medicine" for teenage suicide.[43] Whether it brings in extra-textual informational programming, or whether the shape of the problem and its solutions are left to the movie alone, the TV movie's pedagogical strategies do potentially work to pull the disturbing issue it takes up toward socially manageable limits. The problem can be understood, statistics cited, fictive representatives of professionals consulted, solutions spelled out, the issue reassuringly negotiated.

It is also always the case that articles in national news magazines and newspapers, stories on local and national news programs, and educational materials operate as part of a marketing strategy as well, putting a specific made-for-TV movie on the agenda, promoting it as a socially significant television event. For example, although the full-page ad in *TV Guide* reads

"She had only just begun . . . and suddenly her world was falling apart. The life and death of a superstar" to promote *The Karen Carpenter Story* (CBS, 1989), the close-up emphasizes that Karen had a "serious health problem," anorexia nervosa, a "life-threatening eating disorder." Immediately following the close-up, in the Denver edition of *TV Guide,* is a full-page advertisement for the Rader Institute for the treatment of all eating disorders. The caption, above a drawing of an emaciated young woman staring into a mirror filled with an overweight reflection, reads "The Agony of Anorexia." The young woman is saying, "If only I were thinner, then I'd be happy." The advertisement gives the institute's address and telephone number, under the phrases "It's not your fault . . . you're not alone." In the same *TV Guide,* an article by Karen's brother, Richard, discusses her anorexia, and an article by staff writer Susan Littwin offers an analysis of the eating disorder by a psychiatrist specializing in the treatment of anorexia. The 5:00 P.M. news on the local CBS affiliate (KMGH, channel 7) broadcast a story on anorexia and bulimia the day the movie aired. The movie was also a "Read More About It" book project selection, and in this case "it" was eating disorders.[44] Whatever informational aspects all this material surrounding *The Karen Carpenter Story* might have had, it also certainly functioned to add another layer to a hot concept and to promote the TV movie as a social issue must-see. And in fact, as of mid-March 1989, *The Karen Carpenter Story,* with a 26.4 rating and a 41 share, was the top-rated made-for-TV movie of the season.

The TV movie, then, typically activates what might be termed referential codes—codes that tie an issue depicted in a fictive narrative to the shape the problem and its possible solutions assume in social reality. However, the made-for-TV movie also depends on certain fictional codes. The made-for-TV movie, like much of fictional television, relies heavily on the family melodrama as a generic code that pulls whatever issue is taken up into familiar terrain.

Melodrama, in particular, is a preferred fictional context for addressing disturbing social materials. Peter Brooks, in his study of melodrama, defines the genre in terms of its tendency toward "excess," the unmasking of things that might otherwise be repressed. As Brooks writes, "the genre's very existence is bound to [the] possibility, and necessity, of saying everything."[45] If melodrama involves itself with the excessive, its function consists, many critics have argued, in invoking desires or anxieties only to put them back into the box again. John Cawelti describes melodramas as narratives whose worlds "seem to be governed by some benevolent moral principle," presenting a "moral fantasy" . . . showing forth the essential 'rightness' of the world order," an order that "bears[s] out the audience's traditional patterns of right and wrong, good and evil."[46]

While there is something to be said for conceptualizing melodrama as a "fantasy of reassurance," some critics step back from Cawelti's definition of the genre. David Thorburn, for instance, in his definition of television melodrama, argues that while the genre's "reassurance-structure," its "moral simplification," and its "topicality" are among its generic conventions, TV melodrama is not escapist fantasy that sensationalizes the controversial only to neutralize it by arbitrarily invoking the established moral and political order in the end. TV melodrama, Thorburn claims, functions as a "forum" or "arena" in which "forbidden or deeply disturbing materials" may be addressed; melodrama is "not an escape into blindness . . . but an instrument for seeing."[47]

That the made-for-TV movie has a distinct preference for melodramatic codings of the issues it takes up, especially those issues that center around the family, has not escaped the attention of popular critics. Crist, for example, complains that made-for-TV movies are sunk in the "melodrama rut," a rut that mires the genre in predictability, exaggerated sentimentality, and shallow characters.[48] While Crist is quite right to point out that the TV movie does indeed almost invariably use melodrama, and the family melodrama in particular, as its central fictional code, what Crist calls a rut I would call a generic convention. It is, however, a generic convention not without possible consequences.

Ellen Seiter, in her analysis of television's family melodramas, identified some of the key characteristics of the genre: "conventionally, the structure of melodrama limits it to the presentation of only those conflicts which can be resolved within the family." Further, the family melodrama presents its problems as privatized ones that can be managed by privatized solutions, domesticating what are in fact profound social conflicts by pulling them down to the personal level of the individual character.[49] Todd Gitlin's critique of the made-for-TV movie centers largely around this point: The "iron embrace of the little personal story" smothers the social issues the TV movie would like to think it is dealing with.[50]

Not surprisingly, the made-for-TV movie, concentrating as it does on the family and domestic issues, frequently features women centrally in its narratives. From *Sybil* to *Money on the Side,* from *The Autobiography of Miss Jane Pittman* to *Katie: Portrait of a Centerfold,* TV movies are one of the few prime-time programming forms that permit a woman's story to be the story being told. Prime-time series and serial melodramas—*Dynasty* and *Dallas,* for example—tend to place a community of male characters more centrally in their narratives. The made-for-TV movie may be filling a gap left by the "woman's film" of the Hollywood studio system. A statement concerning the affinity of the TV movie for the woman's story has come, interestingly enough, from television producer Aaron Spelling *(Charlie's Angels),* who claims that TV movies are one of the few prime-time contexts in which women can be stars. Spelling remarked, in an interview, "tell me one woman who has carried a regular dramatic series since the days of Loretta Young and Barbara Stanwyck. Women like to watch women and they seldom got a chance on prime time before the TV movie."[51]

That the made-for-TV movie appears to address itself to the female audience is not without its advantages to the broadcast television system. The audience most attractive to most television advertisers are women between the ages of eighteen and fifty-four, since they make the balance of consumer decisions, purchasing goods not only for themselves but for all members of the household. There may be, however, a certain paradox in the way in which the family melodrama places women at the center of its narratives and in the way it addresses the female audience.

Ellen Seiter argues that the "obsessive questioning of the characters in terms of their family relationships and a delineation of what is appropriate for mothers and fathers, wives and husbands, daughters and sons" is one of the main preoccupations of the genre. The "examination" to which women in melodrama are subjected typically reveals that the burden placed on them is an "impossible" one, for which they must nonetheless bear the responsibility. Caught in contradiction, the women in melodrama are "never let off the hook," either loving too much or not enough, caring too much about the home or too much about a career. Male characters in television's melodramas, on the other hand, are "treated with a remarkable degree of permissiveness." If they are cold and ambitious, the narrative suggests that these qualities are necessary masculine strategies for survival. Underneath the veneer men in the family melodrama are really sensitive and vulnerable, and the responsibility for bringing out those qualities rests with the female characters, who must understand their fathers, husbands, or lovers and win their trust if this essential male tenderness is to emerge. The history of the family melodrama is heavily weighted down with patriarchal values that locate a woman's place in the traditional roles of "nurturing, patient, forgiving . . . wives, mothers and daughters." Even when a melodrama puts women at the center of the narrative, men, Seiter claims, are "the heart" of the genre.[52]

In an earlier article, I explored some of these issues with respect to *Getting Physical* (CBS, 1984), a made-for-TV movie about female bodybuilders (a hot topic at the time). And, although I would hesitate to extend the analysis into a general claim about melodrama and the TV movie, it does appear that one way of reading *Getting Physical* reveals that its use of the family melodrama exerts considerable influence on narrative and characters. For example, *Getting Physical* at first seems as if it will be, perhaps, a feminine version of *Rocky,* the story of a female underdog who, with hard work and strong will, transforms herself into a winning athlete. However, the movie quickly turns into a story about how Kendall's commitment to bodybuilding upsets her family in general and her father in particular, and her romantic relationship with her boyfriend, Mickey. Kendall's father charges her with being a rebellious, inconsiderate, and unfeminine daughter, who has disrupted the family's unity, while Mickey forces a break-up because Kendall is ignoring him to concentrate on her sport. Most of the movie's narrative energy is spent on the problems Kendall's bodybuilding causes in terms of her position as daughter in the nuclear family and as woman in the heterosexual romance. And, in the end, Kendall's choice is rephrased into compliance with her father's prime directive, to "pick one thing to be good at and stick to it." Kendall's father, realizing that Kendall has been a good daughter, comes to see her compete in a contest, and it turns out that he is really a benevolent and caring father, after all. Mickey, in a similar move, decides to take Kendall back, and supports her dedication to the sport. *Getting Physical,* using the family melodrama and the romance, transcribes Kendall's story into a story that concentrates on her relationships with men, a story that not so subtly insists that her choices are valid only if they are compatible with being the kind of daughter her father wants her to be and the kind of girlfriend Mickey wants her to be.[53]

If popular critics are less than happy with the TV movie's investment in melodrama because of what they take to be melodrama's aesthetic defects, academic critics tend to be skeptical of melodrama's ideological tendencies and the effects that they have on the TV movie's politics, perhaps its gender politics in particular. However, we cannot underestimate the possibility that melodrama may, in some contexts, work to expose the problems that some critics argue it works to repress. As Geoffrey Nowell-Smith argues, "the importance of melodrama lies in its ideological failure. Because it cannot accommodate its problems either in a real present or an ideal future, but lays them open in their contradictoriness, it opens a space which most . . . films have studiously closed off."[54] For their female audiences, melodramatic TV movies may very well lay bare more conflicts and contradictions in the culture's ideologies of masculinity and femininity, and of the family, than they lay to rest. Further work around the made-for-TV movie audience needs to be done, to explore this possibility.

The made-for-TV movie, then, in current broadcast television practice, assumes a certain general function and a certain ideological shape. Within the television system, made-for-TV movies are both popular and profitable. Relatively inexpensive, quickly produced, they are financed by the networks, which brings institutional and economic pressures to bear on their form. Because the economic point of programming is to deliver the largest, most profitable audience to advertisers, the conditions of its production dictate that the TV movie possess a certain saliency. It must be sensational, because its limited promotion must provide enough interest to command a sizable share of the audience. Yet it must not be offensive enough to provoke viewers into changing the channel or never tuning in at all. Typically shown on the networks only once, the made-for-TV movie anchors the closed end of broadcast television's spectrum of narrative closure/openness. In addition to its topicality, the TV movie's narrative form is also used as a promotional device. With respect to broadcast television's other narrative forms—the soap opera, the serial, and the series—the TV movie assumes a sort of urgency. It is a "special event," a "tonight-only world premiere."

Made-for-TV movies have merged as a privileged site for acknowledging and negotiating some of the most contested issues in American society. The social issue TV movie, the backbone of the genre, has taken on a pedagogical function with respect to the problems it organizes, educating the audience about the issue and indicating ways of managing the conflict. Melodrama, the preferred fictional code of the TV movie, which focuses on familial and domestic issues, functions to enable the expression of disturbing material, while working to contain it within traditional frames of meaning.

Nevertheless, the TV movie's politics cannot be taken for granted. The made-for-TV movie, responding as it does to the most salient points of sociocultural strain, opens itself up to the possibility of alternative or subversive readings on the part of its vast audience. In its search for ratings through the controversial, the TV movie frequently brings the socially marginalized—women, people of color, gays and lesbians, the working class, the homeless and unemployed, the victims—onto popular terrain. Despite critical charges that it does so only to domesticate or depoliticize social issues or emerging ideologies, the TV movie may very well, for some audiences, make a space for progressive or even radical perceptions of the conflicts and fault-lines in American culture. Many critics of the made-for-TV movie seem to underestimate both the essential contradictoriness of the TV movie and the audience's role in actively making meanings and pleasures from popular texts. Unless we are to suppose that popular audiences are indeed simply helpless, stupid, and capable of being duped into watching (and taking pleasure from) television that is not in the best interests, we must suppose that even the typical TV movie is more than escapist trash or depoliticized fluff. As Richard Dyer argues in "Entertainment and Utopia," popular entertainment is "escapist" in the sense that it "offers the image of 'something better' to escape into, or something we want deeply that our day-to-day lives don't provide." And, as Dyer goes on to say, the way in which popular entertainment points out the gaps between what we actually have and what we want, between what society promises us and what it gives us, plays with "ideological fire."[55] The popular TV movie, in terms of its possible meanings, pleasures, and politics for the audience, deserves a second look.

### Notes

1. Richard K. Doan, "The Name of the Game Is Quickies," *TV Guide,* December 10, 1966, p. A1.

2. Judith Crist, "Tailored for Television Movies," *TV Guide,* August 30, 1969, pp. 6–9; Douglas Stone, "TV Movies and How They Get That Way: An Interview with TV Movie Makers Frank von Zerneck and Robert Greenwald," *Journal of Popular Film and Television* 7 (1979): 146.

3. Dwight Whitney, "Cinema's Stepchild Grows Up," *TV Guide,* July 20, 1974, p. 21; Alvin H. Marill, *Movies Made for Television* (New York: DaCapo Press, 1980), pp. 21, 26; Patrick McGilligan, "Movies Are Getting Better Than Ever on Television," *American Film* 5 (1980): 52.

4. Richard Zoglin, "A Domestic Reign of Terror," *Time,* October 8, 1984, p. 85.

5. See Tony Bennett, "Text and Social Process: The Case of James Bond," *Screen Education* 41 (1982): 3–14, for an analysis of the relationship between the concept of the "author" and the process of canon-formation.

6. McGilligan, "Movies Are Getting Better," p. 52.

7. Tom Allen, "The Semi-Precious Age of TV Movies," *Film Comment* 15 (1979): 22–23.

8. Herbert Gold, "Television's Little Dramas," *Harper's* (March 1977), p. 88.

9. William A. Henry III, "The Dangers of Docudrama," *Time,* February 25, 1985, p. 95.

10. Tom Shales, "Networks Refuse to Acknowledge Sexual Revolution," *The Sunday Oregonian,* March 18, 1984, p. 41.

11. Richard Corliss, "The Revenge of the Male Weepie," *Time,* April 20, 1985, p. 65.

12. See Terry Lovell, "The Social Relations of Cultural Production," in *One Dimensional Marxism,* by Simon Clarke, Victor Jelenieski Seidler, Kevin McDonnell, and Kevin Robbins, and Terry Lovell (London: Allison & Busby, 1980), pp. 232–56, for an interesting discussion of how women and children are inevitably positioned as the social groups most susceptible to the supposed ill-effects of mass culture. See also Robert C. Allen, *Speaking of Soap Operas* (Chapel Hill: University of North Carolina Press, 1985), esp. pp. 25–29, for how popular criticism and academic research has constructed the (female) soap opera audience as somehow "abnormal," and pp. 11–18 for a clear explanation of why "traditional aesthetics" discourse can find little good to say about popular television.

13. Douglas Gomery, "Television, Hollywood, and the Development of Movies Made for Television," in *Regarding Television,* ed. E. Ann Kaplan (Frederick, Md.: University Publications of America, 1983), pp. 124–25; Douglas Gomery, 'Brian's Song': Television, Hollywood, and the Evolution of the Movie Made For Television," in *Television: The Critical View,* 4th ed., ed. Horace Newcomb (New York: Oxford University Press, 1987), pp. 197–200.

14. Bruce Edwards, "Co-production: Ready When You Are TV," *Television Magazine* (May 1966), p. 56.

15. Peter J. Schuyten, "How MCA Rediscovered Movieland's Golden Lode," *Fortune* (November 1976), pp. 122–23.

16. McGilligan, "Movies Are Getting Better," p. 126.

17. McGilligan, "Movies Are Getting Better," p. 52; Gomery, "Television," p. 126.

18. Gomery, "Television," p. 126; Gomery, " 'Brian's Song,' " pp. 197–98.

19. Todd Gitlin, *Inside Prime Time* (New York: Pantheon, 1983), p. 157.

20. Fred Silverman, "Made-For-TV Movies: Seen Playing Greater Role in Web Strategy," *Television/Radio Age,* November 7, 1983, pp. 34, 104.

21. Jeff Kaye, "CBS Plans 'High Concept' TV-Movies," *TV Guide,* December 31, 1988, p. A-1.

22. Crist, "Tailored for Television," pp. 6–9.

23. Robert C. Allen, "Speaking of Soap Operas," p. 46.

24. Art Durbano, "This Week's Movies," *TV Guide,* September 23, 1989, p. 51.

25. Horace Newcomb, "Toward a Television Aesthetic," in *Television: The Critical View,* ed. Newcomb, pp. 614–20.

26. John Ellis, *Visible Fictions* (London: Routledge and Kegan Paul, 1981), pp. 126, 143.

27. David Thorburn, "Television Melodrama," in *Television: The Critical View,* ed. Newcomb, pp. 633–34.

28. Gitlin, *Inside Prime Time,* pp. 161–62.

29. Stone, "TV Movies," pp. 147–48.

30. Gitlin, *Inside Prime Time,* pp. 158–59.

31. Stone, "TV Movies," p. 148.

32. Albert Auster, "If You Can't Get 'Em Into the Tent, You'll Never Have a Circus: An Interview with Len Hill," *Journal of Popular Film and Television* 8 (1981): 13.

33. Gitlin, *Inside Prime Time,* p. 161.

34. Stone, "TV Movies," p. 148.

35. Stone, "TV Movies," pp. 150–51.

36. Gitlin, *Inside Prime Time,* pp. 161, 163–64.

37. Les Brown, *Television: The Business Behind the Box* (New York: Harcourt, Brace, Jovanovich, 1971), p. 30.

38. Joanna Elm, "NBC Tones Down 'Roe vs. Wade' TV-Movie to Avoid Angering Abortion Pressure Groups," *TV Guide,* May 6, 1989, pp. 49–50.

39. Auster, "If You Can't Get 'Em," p. 10; Zoglin, "A Domestic Reign of Terror," p. 85.

40. See Stuart Hall, "Culture, the Media and the 'Ideological Effect,' " in *Mass Communication and Society,* ed. James Curran, Michael Gurevitch, and Janet Woollacott (Beverly Hills: Sage Publications, 1979), pp. 315–48. See also Todd Gitlin, "Prime Time Ideology: The Hegemonic Process in Television Entertainment," in *Television: The Critical View,* ed. Newcomb, pp. 507–32.

41. Gitlin, *Inside Prime Time,* pp. 163, 194.

42. Gomery, " 'Brian's Song,' " pp. 213–16.

43. Richard Zoglin, "Troubles on the Home Front," *Time,* January 28, 1985, p. 65.

44. *TV Guide* (Denver ed.), December 31, 1988, pp. A-53, A-54, A-58; pp. 26–29.

45. Peter Brooks, *The Melodramatic Imagination* (New York: Yale University Press, 1976), p. 42.

46. John G. Cawelti, *Adventure, Mystery, and Romance* (Chicago: University of Chicago Press, 1976), p. 45.

47. Thorburn, "Television Melodrama," p. 630.

48. Crist, "Tailored for Television," p. 8.

49. Ellen Seiter, "Men, Sex and Money in Recent Family Melodrama," *Journal of the University Film and Video Association* 35 (Winter 1983): 19.

50. Gitlin, *Inside Prime Time,* p. 179.

51. Whitney, "Cinema's Stepchild," p. 26.

52. Seiter, "Men, Sex and Money," pp. 23–26.

53. See Laurie Schulze, " 'Getting Physical': Text/Context/Reading and the Made-for-Television Movie," *Cinema Journal* 25 (Winter 1986): 35–50, for a more detailed analysis of the relationship between melodrama and ideology in this particular made-for-TV movie.

54. Geoffrey Nowell-Smith, "Minelli and Melodrama," in *Home Is Where the Heart Is: Studies in Melodrama and the Woman's Film,* ed. Christine Gledhill (London: BFI, 1987), p. 74.

55. Richard Dyer, "Entertainment and Utopia," in *Genre: The Musical: A Reader,* ed. Rick Altman (London: Routledge and Kegan Paul, 1981), p. 177.

# Television,
# Black Americans, and
# the American Dream

## HERMAN GRAY

William F. Buckley Jr. has observed, "it is simply not correct . . . that race prejudice is increasing in America. How does one know this? Simple, by the ratings of Bill Cosby's television show and the sales of his books. A nation simply does not idolize members of a race which that nation despises" (Demeter, 1986, p. 67). Buckley seems to suggest that if racial prejudice exists at all in the United States it does not figure significantly in the nature of American society, nor does it explain very much about social inequality based on race and characterized by racial discrimination, racial violence, economic dislocation, and social isolation. Still, what is perhaps most interesting about Buckley's observation is his reliance on Bill Cosby's successful media presence as a barometer of American racial equality.

An open class structure, racial tolerance, economic mobility, the sanctity of individualism, and the availability of the American dream for black Americans are represented in a wide range of media. Representations of such success are available in *The Cosby Show,* the box office power of Eddie Murphy, the international popularity of Michael Jackson, and the visibility of Oprah Winfry. Equally important to the contemporary ideology of American racial openness, however, are representations of deprivation and poverty such as those shown on network newscasts and documentaries. In media reports of urban crime, prisons overcrowded with black men, increased violence associated with drugs, and the growing ranks of the homeless are drawn the lines of success and failure.

As Buckley's observations demonstrate, the meanings of these representations are not given; rather, viewers define and use the representations differently and for different reasons. One message of these representations of success and failure is that middle class blacks (and whites) succeed because they take advantage of available opportunities while poor blacks and other marginal members of our society fail because they do not (Glasgow, 1981; Lewis, 1984). These representations operate not just in terms of their relationship to the empirical realities of black life in America but also in relationship to other popular media constructions about black life. My interest here is in the relationship between representations of black life in fictional and nonfictional television and the ideological meanings of these representations when television is viewed as a complete ideological field (Fiske, 1987a). In the following section, I theoretically situate the problem. I then turn to a discussion of black failure as represented in the CBS News documentary *The Crisis of Black America: The Vanishing Family* and the representation of upper middle class black affluence in *The Cosby Show*.

### Theoretical Context

In order to describe how television representations about race communicate and to examine their ideological meanings, I draw on Gramsci's notion of ideological hegemony (Gramsci, 1971; Hall, 1982). Media representations of black life (especially middle class success and under class failure) are routinely fractured, selectively assembled, and subsequently become a part of the storehouse of American racial memory. The social and racial meanings that result from these processes appear in the media as natural and given rather than as social and constructed. In *Ideology and the Image* (1981, p. 1) Bill Nichols states that "ideology uses the fabrication of images and processes of representation to persuade us that how things are is how they ought to be and that the place provided for us is the place we ought to have." I use "hegemony" to specify the material and symbolic processes by which these racial representations and understandings are produced and naturalized (Fiske, 1987a; Hall, 1982).

Media representations of black success and failure and the processes that produce them are ideological to the extent that the assumptions that organize the media discourses shift our understanding of racial inequality away from structured social processes to matters of individual choice. Such ideological representations appear natural and universal rather than as the result of social and political struggles over power.

The process of media selection and appropriation, however, is only one part of the play of hegemony. Mass media and popular culture are, according to Stuart Hall (1980), sites where struggles over meaning and the power to represent it are waged. Thus, even as the media and popular cultural forms present representations of race and racial (in)equality, the power of these meanings to register with the experiences (common sense) of different segments of the population remains problematic. Meanings constantly shift and are available for negotiation. It is in this process of negotiation that different, alternative, even oppositional readings are possible (Fiske, 1987a; Hall, 1980). Because of this constantly shifting terrain of meaning and struggle, the representations of race and racial interaction in fictional and nonfictional television reveal both the elements of the dominant racial ideology as well as the limits to that ideology.

Within this broad struggle over meaning, Fredric Jameson (1979) shows how popular cultural forms such as film and television work symbolically to establish preferred, even dominant ideological meanings. In popular culture, ideology is secured through the psychological appeal to utopian values and aspirations and a simultaneous repression and displacement of critical sensibilities that identify the social and economic organization of American society as the source of inequality. In television representations of blacks, the historical realities of slavery, discrimination, and racism or the persistent struggles against domination are displaced and translated into celebrations of black middle class visibility and achievement. In this context, successful and highly visible stars like Bill Cosby and Michael Jackson confirm the openness and pluralism of American society.

The commercial culture industry presents idealized representations of racial justice, social equality, and economic success. Idealized middle class black Americans increasingly populate fictional television. They confirm a middle class utopian imagination of racial pluralism (Gray, 1986). These idealized representations remain before us, driven, in the case of television, by the constant search for stable audiences and the centrality of advertising revenue as the basis for profits (Cantor, 1980; Gitlin, 1983).

As Jameson further notes, however, utopian possibilities are secured against the backdrop of reified nonfictional (and fictional) representations. In the case of racial representations, the black under class appears as menace and a source of social disorganization in news accounts of black urban crime, gang violence, drug use, teenage pregnancy, riots, homelessness, and general aimlessness. In news accounts (and in Hollywood films such as *Colors*), poor blacks (and Hispanics) signify a social menace that must be contained. Poor urban blacks help to mark the boundaries of appropriate middle class behavior as well as the acceptable routes to success. As a unity, these representations of black middle class success and under class failure are ideological because they are mutually reinforcing and their fractured and selective status allows them to be continuously renewed and secured. Furthermore, the meanings operate within a frame that privileges representations of middle class racial pluralism while marginalizing those of racial inequality. This constant quest for legitimacy and the need to quell and displace fears at the same time as it calls them forth are part of the complex ideological work that takes place in television representations of race.

The representations of black American success and failure in both fictional and nonfictional television, and the assumptions that organize them, are socially constructed according to commercial, professional, and aesthetic conventions that guide producers and consumers of television (Gray, 1986). These conventions guide personnel in the selection and presentation of images to ensure that they are aesthetically appealing, culturally meaningful, politically legitimate, and economically profitable.

Although fictional and nonfictional representations of blacks emanate from separate generic quarters of television, they activate meanings for viewers across these boundaries. That is, the representations make sense in terms of their intertextuality between and within programs (Fiske, 1987a; Fiske and Hartley, 1978; Williams, 1974). Television representations of black life in the late 1980s cannot be read in isolation but rather should be read in terms of their relationship to other television texts.

The meanings that these representations express and activate are also significant in terms of the broad social and historical context in which they operate. Fictional and nonfictional representations of black life appear at a time when political and intellectual debate continues over the role of the state in helping the black urban poor and whether or not affirmative action ought to remain an active component of public policy. Within the black political and activist community, sharp differences remain over the role of the black middle class and the efficacy of black generated self-help programs to battle problems facing black communities. Increased racial violence and antagonisms (including those on college campuses), economic dislocation, a changing industrial base, ethnic and racial shifts in the demographic composition of the population, and the reelection of a conservative national administration help set the social context within which television representations of black life take on meaning.

A myriad of community, institutional, social, political, and economic forces shape the broad public discourse on the conditions of blacks in contemporary American society. In the absence of effective social movements such as those for civil rights, students, women, and against the war, which, at the very least, helped ground and mediate media representations, these representations take on greater authority and find easier access to our common sense (Winston, 1983, p. 178). Under these conditions, the ideological potency of media representations remains quite strong.

Media representations of black success and failure occur within a kind of gerrymandered framework. Through production conventions, political sensibilities, commercial pressures, and requirements for social organization and efficiency, television news and entertainment selectively construct the boundaries within which representations about black life occur. The primacy of individual effort over collective possibilities, the centrality of individual values, morality, and initiative, and a benign (if not invisible) social structure are the key social terms that define television discourses about black success and failure.

### Reification and the Under Class

To explore the reification side of the Jameson formulation, I begin with a discussion of the CBS News report about the black urban under class. The special report which aired in January 1985 is titled *The Vanishing Family: Crisis In Black America*. CBS senior correspondent Bill Moyers hosted the ninety-minute documentary which was filmed in Newark, New Jersey. Through interviews and narration by Moyers, the report examines the lives of unwed mothers and fathers, detailing their education, employment, welfare history (especially across generations), hopes, frustrations, and disappointments.

The appearance of the terms "vanishing family" and "crisis" in the title of the program implicitly suggests the normalcy of everyday life when defined by stable nuclear families (Fearer, 1986; Fishe, 1987a). Missing is recognition that families and communities throughout the country are in the midst of significant transformation. Instead, the program title suggests an abnormal condition that must be recognized and addressed.

In the report's opening segment, visual representations also help frame the ideological terms of the report. Medium and long camera shots are used to establish perspective on the daily life in the community. Mothers are shown shopping for food and caring for children; groups of boys and young men appear standing on street corners, playing basketball, listening to music, and working out at the gym. Welfare lines, couples arguing, the police, housing projects, and the streets are also common images.

These shots tie the specific issues addressed in the story into a broader discourse about race in America. Shots of black men and youth standing on corners or blacks arrested for crime are conventionally used in newscasts to signify abnormalities and social problems. These images operate at multiple levels, so even though they explicitly work to frame the documentary, they also draw on and evoke images of crime, drugs, riots, menace, and social problems. People and communities who appear in these representations are labeled as problematic and undesirable.

The documentary's four segments are organized around three major themes, with each segment profiling unmarried couples. By the end of the four segments, the dominant message of the report is evident: self-help, individual responsibility, and community accountability are required to survive the crisis. This conclusion is anticipated early in the report with a promotional tease from a black social worker. In a thirty-second sound bite, the social worker notes that the problem in the black community is not racism or unemployment but the corruption of values, the absence of moral authority, and the lack of individual motivation. This dominant message is also reinforced in the introduction to the report by correspondent Moyers:

> A lot of white families are in trouble too. Single parent families are twice as common in America today as they were twenty years ago. But for the majority of white children, family still means a mother and a father. This is not true for most black children. For them things are getting worse. Today black teenagers have the highest pregnancy rate in the industrialized world and in the black inner city, practically no teenage mother gets married. That's no racist comment. What's happening goes far beyond race.

Since blacks dominate the visual representations that evoke images of crime, drugs, and social problems, little in the internal logic and organization of the documentary supports this contention. Even when voice-over data is used to address these issues among whites, it competes with rather than complements the dominance of the visual representations. Moyers' comment is also muted because the issues are examined primarily at the dramatic and personal level.

For example, the first segment considers the experiences of urban single parent families from the viewpoint of women. The opening piece profiles Clarinda and Darren, both young and poorly prepared emotionally or financially to care for an infant. Clarinda supports the baby with welfare and is also the baby's primary source of emotional nurturance. Darren occasionally sees his baby but takes little economic or emotional responsibility for her. On camera he appears distant and frustrated.

The second segment focuses on Alice, 23, and Timothy, 26. They are older but financially no more prepared to raise a family than Clarinda and Darren. Unlike Darren, Timothy is emotionally available to Alice. (On camera they confess their love for each other, and Timothy is present at a birthday party for one child and the delivery of another.) In the interview Alice freely shows her frustration with Timothy, especially his lack of work and unwillingness to take responsibility for his family.

Timothy on the other hand lives in a world of male sexual myths and a code that celebrates male sexual conquest and virility (Glasgow, 1981). Although he confesses love for Alice and his kids, he avoids economic and parental responsibility for them, especially when his own pleasures and sexual conquests are concerned.

The mothers in these segments are caring, responsible, and conscientious; they raise the children and provide for them. They are the social, economic, and emotional centers of their children's lives. As suggested in the interviews and visual footage, the fathers are absent, immature, selfish, irresponsible, and exploitive. Where women are shown at home with the children, the men are shown on street corners with other men. Where women talk of their children's futures, men speak in individual terms about their present frustrations and unrealistic aspirations.

The dramatic and personal tone of these representations makes them compelling and helps draw in the viewer. These strategies of organization and presentation also help personalize the story and, to a limited extent, give the people texture and dimension. Nevertheless, these representations are also mediated by a broader set of racial and class codes that continue to construct the people in the documentary as deviant and criminal, hence marginal. The members of the community are contained by these broader codes. They remain curious but distant "others."

The third segment features Bernard, a fifteen-year-old single male who still lives at home with Brenda, his thirty-year-old single mother of three. This segment tells the story of life in this community from the young male point of view. The male voice takes on resonance and, in contrast to Darren and Timothy, we learn that the men in this community have feelings and hopes too. The segment shows Bernard's struggle to avoid the obstacles (drugs, educational failure, unemployment, homicide, jail) to his future. From Brenda's boyfriend (and role model for Bernard) we learn about the generational persistence of these obstacles to young male futures.

In each of these segments the dramatic dominates the analytic, the personal dominates the public, and the individual dominates the social. Individual mobility, character, and responsibility provide powerful explanations for the failures presented in the story. Indeed, by the final segment of the report the theme of moral irresponsibility and individual behavior as explanations for the crisis of the under class is fully developed. Moyers introduces the segment this way:

> There are successful strong black families in America. Families that affirm parental authority and the values of discipline, work, and achievement. But you won't find many who live around here. Still, not every girl in the inner city ends up a teenage mother, not every young man goes into crime. There are people who have stayed here. They're outnumbered by the con artists and pushers. It's not an even match, but they stand for morality and authority and give some of these kids a dose of unsentimental love.

As a major "actor" in the structure of this report, Moyers is central to the way that the preferred meanings of the report are conveyed. As an economically and professionally successful white male, Moyers' political and moral authority establishes the framework for identifying the conditions as trouble, for articulating the interests of the dominant society, and for demonstrating that in the continued openness of the social order there is hope. Through Moyers' position as a journalist, this report confirms the American dream even as it identifies casualties of the dream.

Moyers' authority in this story stems also from his position as an adult. During his interviews and stand-ups Moyers represents adult common sense, disbelief, and concern. This adult authority remains throughout the report and is reinforced (and activated) later in the story when we hear from caring (and successful) black adults of the community who claim that the problems facing the community stem from poor motivation, unclear and unsound values, and the lack of personal discipline. Like Moyers, these adults—two social workers, a psychologist, and a police officer—do not identify complex social forces like racism, social organization, economic dislocation, unemployment, the changing economy, or the welfare state as the causes of the crisis in their community. They blame members of the black community for the erosion of values, morality, and authority. This is how Mrs. Wallace, the social worker, puts it:

> We are destroying ourselves. Now it [the crisis] might have been motivated and plotted and seeded with racism, but we are content to be in this well now. We're just content to be in this mud and we need to get out of it. There are not any great white people running around this block tearing up stuff. It's us. We've got to stop doing that.

When combined with the personal tone of the documentary and Moyers' professional (and adult) authority, this comment, coming as it does from an adult member of the community, legitimates the emphasis on personal attributes and a benign social structure.

At the ideological level of what Stuart Hall (1980) calls preferred readings, each segment of the documentary emphasizes individual personalities, aspirations, and struggles for improvement. These assumptions and analytic strategies are consistently privileged over social explanations, and they provide a compelling vantage point from which to read the documentary. This displacement of the social by the personal and the complex by the dramatic both draws viewers into the report and takes them away from explanations that criticize the social system. Viewers question individual coping mechanisms rather than the structural and political circumstances that create and sustain racial inequalities.

## Middle Class Utopia

I consider the utopian side of the Jameson formulation by exploring the theme that media representations of black success and failure are ideological, precisely to the extent that they provide a way of seeing under class failure through representations of middle class success. Implicitly operating in this way of viewing the under class (and the middle class) is the assumption that since America is an open racial and class order, then people who succeed (and fail) do so because of their individual abilities rather than their position in the social structure (Lewis, 1984).

In contrast to the blacks in the CBS documentary, successful blacks who populate prime time television are charming, unique, and attractive individuals who, we assume, reached their stations in life through hard work, skill, talent, discipline, and determination. Their very presence in formats from talk shows (Bryant Gumbel, Arsenio Hall, Oprah Winfrey) to situation comedy (Bill Cosby) confirms the American value of individual success and mobility.

In the genre of situation comedy, programs like *The Cosby Show, 227, Frank's Place,* and *Amen* all show successful middle class black Americans who have effectively negotiated their way through benign social institutions and environments (Gray, 1986). Their family-centered lives take place in attractive homes and offices. Rarely if ever do these characters venture into settings or interact with people like those in the CBS documentary. As doctors, lawyers, restaurateurs, ministers, contractors, and housewives, these are representations of black Americans who have surely realized the American dream. They are pleasant and competent social actors whose racial and cultural experiences are, for the most part, insignificant. Although black, their class position (signified by their occupations, tastes, language, and setting) distances them from the codes of crime, drugs, and social problems activated by the urban under class. With the exception of the short-lived *Frank's Place,* the characters are never presented in situations where their racial identity matters. This representation of racial encounters further appeals to the utopian desire in blacks and whites for racial oneness and equality while displacing the persistent reality of racism and racial inequality or the kinds of social struggles and cooperation required to eliminate them. At the level of the show's dominant meanings, this strategy accounts in part for the success of *The Cosby Show* among blacks and whites.

In virtually any episode of *The Cosby Show,* the Huxtable children— Sandra, Denise, Vanessa, Theo, and Rudi—are given appropriate lessons in what appear to be universal values such as individual responsibility, parental trust, honesty, the value of money, the importance of family and tradition, peer group pressure, the value of education, the need for independence, and other important guides to successful living in America.

In contrast to the experience of the young men in the CBS documentary, *Cosby*'s Theo learns and accepts lessons of responsibility, maintaining a household, the dangers of drugs, the value of money, and respect for women through the guidance of supportive parents. In Theo's relationship to his family, especially his father Cliff, the lessons of fatherhood and manhood are made explicit. Theo and his male peers talk about their aspirations and fears. They even exchange exaggerated tales of adolescent male conquest. Because similar discussions among the young men in the documentary are embedded within a larger set of codes about the urban black male menace, this kind of talk from Timothy, Darren, and Bernard signals their incompetence and irresponsibility at male roles. In the middle class setting of *The Cosby Show,* for Theo and his peers this same talk represents the ritual of adolescent male maturation. Together, these very opposite representations suggest a contemporary version of the culture of poverty thesis which attributes black male incompetence and irresponsibility to the absence of male role models, weak personal values, and a deficient cultural environment.

The strategy of imparting explicit lessons of responsibility to Theo (and to young black male viewers) is deliberate on the part of *Cosby*. This is not surprising since the show has enjoyed its greatest commercial success in the midst of increasing gang violence and epidemic teen pregnancy in urban black communities. The show's strategy illustrates its attempt to speak to a number of different audiences at a number of different levels (Fiske, 1987a; Hall, 1980).

Shows about middle class black Americans revolve around specific characters, settings, and situations (Gitlin, 1983; Gray, 1986). The personal dimension of social life is privileged over, and in many cases displaces, broader social and structural factors. In singling out *The Cosby Show,* my aim is not to diminish the unique qualities, hard work, and sacrifices that these personal representations stress. Nevertheless, I do want to insist that the assumptions and framework that structure these representations often displace representations that would enable viewers to see that many individuals trapped in the under class have the very same qualities but lack the options and opportunities to realize them. And in the world of television news and entertainment, where production conventions, ratings wars, and cautious political sensibilities guide the aesthetic and journalistic decisions of networks, the hegemony of the personal and personable rules. Whether it is Bill Cosby, Alicia Rashad, Darren, Alice, or Bill Moyers, the representation is of either deficient or gifted individuals.

Against fictional television representations of gifted and successful individuals, members of the urban under class are deficient. They are unemployed, unskilled, menacing, unmotivated, ruthless, and irresponsible. They live differently and operate with different attitudes and moral codes from everyone else; they are set apart. Again, at television's preferred level of meaning, these assumptions—like the images they organize and legitimate—occupy our common sense understandings of American racial inequality.

## Conclusions

The assumptions that organize our understandings of black middle class success and under class failure are expressed and reinforced in the formal organization of television programming. Formally, where representations of the under class are presented in the routine structure of network news programming, it is usually in relationship to extraordinary offenses such as drugs, homicide, and crime. In contrast, middle class blacks are very much integrated into the programming mainstream of television. Successful shows about black life inhabit a format and genre that has a long tradition in television entertainment—the situation comedy. The rhythm, texture, and form of this type of show are comfortable and familiar to most viewers. Moreover, these programs are coupled with others that are similar. Thus, for instance, the Thursday evening schedule is built around *The Cosby Show* and *A Different World*. *227* fits snugly into the Saturday evening programming flow with *Golden Girls* and *Amen*. Still, even though representations of under class and middle class life are presented in the "bracketed" space of the news documentary and the situation comedy, at the level of decoding, the meanings of these shows circulate in the programming flow across programs and genres.

Surely, then, the failure of blacks in the urban under class, as Mrs. Wallace suggested in the CBS documentary, is their own since they live in an isolated world where contemporary racism is no longer a significant factor in their lives. The success of blacks in the television middle class suggests as much. In the world of the urban under class, unemployment, industrial relocation, ineffective social policies, power inequalities, and racism do not explain failure, just as affirmative action policies, political organization, collective social and cultural challenges to specific forms of racial domination, and the civil rights movement do not help explain the growth of the black middle class.

The nonfictional representations of the under class and the fictionalized treatment of the middle class are significant in other ways. Contemporary television shows in general and shows about black life in particular have reclaimed the family; they are either set in the nuclear family of *The Cosby Show* and *227* or the work place family of *Frank's Place* and *Amen* (Feuer, 1987; Taylor, 1988). The idealized representations of family presented in these shows maintain the hope and possibility of a stable and rewarding family life. At the same time, this idealization displaces (but does not eliminate) possibilities for critical examination of the social roots of crisis in the American family (Jameson, 1979).

Family stresses such as alienation, estrangement, violence, divorce, and latch key kids are typically ignored. When addressed in the television representations of black middle class families, they are presented as the subject of periodic and temporary disagreements rather than as expressions of the social stresses and disruptive impulses that originate in the social organization of society and the conflicting ideologies that shape our understanding of the family as a social institution.

At the negotiated level of meaning (Hall, 1980), *The Cosby Show* effectively incorporates many progressive moments and impulses from recent social movements. The show presents Claire's independence, autonomy, and authority in the family without resorting to exaggeration and trivialization (Downing, 1988). Again, this utopian impulse is one of the reasons for the show's popular appeal. And yet it is also one of the ways the explicit critical possibilities of the show are contained and subverted. Claire's independence and autonomy are expressions of her own individual character; they are confined to the family and put in the service of running a smoother household. This claim on the family and the affirmation of female independence are especially appealing when seen against the crisis of the family dissolution, female-headed households, and teenage pregnancy presented in the CBS documentary. Ironically, this celebration of Claire's independence and agency within the family has its counterpart in the CBS documentary. In each case, black women are assertive and responsible within the contexts of their various households. Thus, even within the constraints of under class poverty this moment can be read as an appeal to the utopian ideal of strong and liberated black women.

Ideologically, representations of under class failure still appeal and contribute to the notion of the black poor as menacing and threatening, especially to members of the white middle class. Such a menace must, of course, be contained, and through weekly visits to black middle class homes and experiences, whites (and middle class blacks) are reasonably assured that the middle class blacks with whom they interact are safe (Miller, 1986). Whites can take comfort in the fact that they have more in common with the Huxtables than with those representations of the family in crisis—Timothy, Clorinda, Darren, and Alice.

The twin representations of fictional and nonfictional television have become part of the public discourse about American race relations. While, no doubt, both the fictional and nonfictional representations of blacks are real, like all ideology, the realities are selected, partial, and incomplete. Where the television lens is trained, how wide, which angle, how long, and with whose voice shapes much of what we see and how we understand it. As these fictional and nonfictional television representations indicate, television helps shape our understandings about racial (in)equality in America.

## *References*

Cantor, M. (1980). *Prime-time television: Content and control.* Beverly Hills: Sage.

Demeter, J. (1986). Notes on the media and race. *Radical America,* 20(5): 63–71.

Downing, J. (1988). "The Cosby Show" and American racial discourse. In G. Smitherman-Donaldson and T. A. van Dijk (eds.), *Discourse and discrimination* (pp. 46–74). Detroit, Mich.: Wayne State University Press.

Feuer, J. (1986). Narrative form in American television. In C. MacCabe (ed.), *High theory/low culture: Analyzing popular television and film* (pp. 101–15). New York: St. Martin's Press.

Feuer, J. (1987). Genre study and television. In R. Allen (Ed), *Channels of discourse* (pp. 113–34). Chapel Hill: University of North Carolina Press.

Fiske, J. (1987a). *Television culture.* London: Methuen.

Fiske, J. (1987b). British cultural studies and television. In R. Allen (Ed.), *Channels of discourse* (pp. 254–91). Chapel Hill: University of North Carolina Press.

Fiske, J., and Hartley, J. (1978). *Reading television.* London: Methuen.

Gitlin, T. (1983). *Inside prime time.* New York: Pantheon.

Glasgow, D. (1981). *The black underclass.* New York: Vintage.

Gramsci, A. (1971). *Selections from the prison notebooks.* New York: International Publishers.

Gray, H. (1986). Television and the new black man: Black male images in prime-time situation comedy. *Media, culture, and society,* 8: 223–42.

Hall, S. (1980). Encoding/decoding. In S. Hall, A. Lowe, and P. Willis (eds.), *Culture, media, language* (pp. 128–39). London: Hutchinson.

Hall, S. (1982). The rediscovery of ideology: Return of the repressed in media studies. In M. Gurevitch, T. Bennett, J. Curran, and J. Woollocott (eds.), *Culture, society, and the media* (pp. 56–91). London: Methuen.

Jameson, F. (1979). Reification and utopia in mass culture. *Social Text,* 1: 130–48.

Lewis, M. (1984). *The culture of inequality.* New York: New American Library.

Miller, M. C. (1986). Deride and conquer. In T. Gitlin (Ed.), *Watching television* (pp. 183–229). New York: Pantheon.

Nichols, B. (1981). *Ideology and the image.* Bloomington: University of Indiana Press.

Taylor, E. (1987, October 5). TV families: Three generations of packaged dreams. *Boston Review of Books,* p. 5.

Williams, R. (1974). *Television: Technology and cultural form.* New York: Oxford University Press.

Winston, M. (1983). Racial consciousness and the evolution of mass communication in the United States. *Deadalus,* 111: 171–83.

# Rewriting Culture:
# A Dialogic View of
# Television Authorship

## JIMMIE L. REEVES

*And no man putteth new wine into old wineskins; else the new wine will burst the skins, and be spilled, and the skins will perish.*

*But new wine must be put into fresh wineskins; and both are preserved.*

*No man also having drunk old wine straightway desireth new; for he saith, The old is better.*[1]

Luke 5:37–39

In some circles, these three verses are known as the Parable of the Wineskins. Authorship of the parable is, of course, attributed to Jesus of Nazareth. But, as in all texts, the meaning of this allegory is dependent on contextual factors. Within the parameters of Luke's biography of Jesus, the parable is clearly meant to be interpreted as Jesus' response to harsh criticism from an influential group of religious authorities known as the Pharisees. Disturbed by Jesus' practice of mingling with social outcasts, the Pharisees publicly confronted the Nazarene with the loaded question, "Why do you eat and drink with publicans and sinners?" (Luke 5: 30). As the parable suggests, the very people whom the Pharisees despised and ostracized were, for Jesus, receptive vessels able to accommodate the new wine of his radical message. In the context of the New Testament, then, the Parable of the Wineskins was meant to illustrate how a revolutionary theology could not be contained by the attitudes and prejudices of orthodoxy.

This essay, though, will place the parable into another context—the context of a general critique of traditionalism. Situated in this broader frame,

From *Making Television Authorship and the Production Process,* Robert J. Thompson and Gary Burns, eds., New York: Praeger 1990, pp. 147–160, an imprint of Greenwood Publishing Group, Inc., Westport, CT. Reprinted with permission.

the parable provides metaphoric commentary on the struggle that accompanies the institution of any new order of experience. In symbolizing this struggle, "old wineskins" represent the inflexibility of traditional views of the world; the image of "new wine" expresses rapidly changing cultural and material conditions; and "fresh wineskins" speak of the innovation and vitality of emerging modes of thought that are better able to accommodate the ferment of precipitous change.

Taken as a metaphor for a crisis in traditionalism, the parable provides decisive insights into the history of critical inquiry directed toward the study of television. As Booth (1987, p. 414) suggests, television criticism is unique in that it is the only body of criticism that practices *de*-preciation more than appreciation. And one of the chief grievances leveled at television echoes the Pharisees' rebuke of Jesus: The tired complaint "Why does TV have to cater to the lowest common denominator" expresses the same elitist sentiments of "Why do you eat and drink with publicans and sinners?" Clearly, traditional assumptions and attitudes regarding art and the critical enterprise are challenged by the crass commercialism and brute popularity of American television. Notions about artistic expression being the province of an individual author's personal vision are not relevant to television's collaborative production process. And standard interpretive methods that treat the isolated work as the object of critical evaluation are not able to accommodate the turbulence of television's fragmented form, nor the fundamentally derivative character of television's intertextual storytelling dynamic.

It's not at all surprising, then, that critics trained to appreciate the old wine of the high arts are almost unanimous in dismissing the aesthetic possibilities and the ideological power of television with the nostalgic lament, "The old is better." As the final line of the parable observes, "No man also having drunk old wine straightway desireth new."

## Rupturing an Old Wineskin

In attempting to fashion a new critical wineskin that is flexible enough to embrace the ferment of television, this essay will adopt few of the assumptions underlying traditional views of authorship. In the first place, such views are based on a disavowal of the commercial that masks the economic dimensions of all taste cultures and expressive practices. As Bourdieu (1986, p. 133) forcefully argues, "the ideology of creation, which makes the author the first and last source of the value of his work" conceals the enterprise of the "cultural businessman." According to Bourdieu, art dealers, gallery owners, and publishers do much more than merely exploit "the labor of the 'creator' by trading in the 'sacred'. . . ." In fact Bourdieu considers the marketing, publishing, and staging of a work of art to be nothing less than *acts of consecration*. In this consecration, the art trader anoints a product he has "discovered" with the oil of his reputation as a man of "distinction." And, in Bourdieu's words, "the more consecrated he [the art trader] personally is, the more strongly he consecrates the work."

By indulging in the prestige and posturing of economic "disinterested-ness," the legitimate artist is also implicated in the ongoing hypocrisies of the legitimate art business. Such feigned disinterestedness is only plausible because, after all, the art trader acts as a mediating agent who both links the artist to and screens the artist from the market. According to Bourdieu (1986, p. 136) the pretenses of this producer-trader relationship are, in part, sustained by a curious double bind: artists "cannot even denounce the exploitation they suffer without confessing their own self-interested motives." Ultimately then, Bourdieu sees the makers and marketers of works of art as "adversaries in collusion":

> [E]ach abide by the same law which demands the repression of direct man-
> ifestations of personal interest, at least in its overtly "economic" form, and
> which has every appearance of transcendence although it is only the prod-
> uct of the cross-censorship weighing more or less equally on each of those
> who impose it on all the others. (1986, p. 136)

Blind to and blinded by this "cross-censorship," the legitimate art estab-lishment perpetuates the affectations of what Newcomb and Alley (1983, p. 37) term a "historically grounded class-bias." According to Newcomb and Alley, this bias tends to value the work of art "not only for its innova-tive vision, but for its sheer difference":

> With other Western societies we have, since the Renaissance, distinctly
> valorized the innovative, the daring, the non-traditional, the "unique." The
> presence of these qualities rests, in turn, on our willingness to grant the
> artist something like "freedom" or "individuality." The autonomy of the
> artist is viewed as the source of creativity, and without it, "true" art is
> deemed impossible. (1983, p. 35)

Of course, "individuality" and artistic "freedom," like "sovereign ego" and "private property," are relatively new concepts in the history of ideas—con-cepts that are closely linked to the rise of individualism and capitalism.

However, a more ancient notion of art is expressed in the ornaments adorning the Pharaoh's sarcophagus, or the psalms guiding the thoughts of the Israelites, or the totem poles ordering the gods of the Haida Indians. Here, the aesthetic object is understood to project the ideals and values of the community, not the vision and aspiration of an individual. In indivi-dualizing authorship, traditional criticism fails to recognize that even the most inaccessible piece of modernist fiction has some of the same adorning, guiding, and ordering properties of pre-Renaissance and pre-industrial art. For, as Bourdieu demonstrates, high art and the aesthetic sense essentially operate as *class definers* in contemporary technocracies:

> The most intolerable thing for those who regard themselves as possessors
> of legitimate culture is the sacrilegious reuniting of tastes which taste dic-
> tates shall be separated. This means that the games of artists and aesthetes
> and their struggles for the monopoly of artistic legitimacy are less innocent
> than they seem. At stake in every struggle over art there is also an imposi-

tion of an art of living, i.e., the transmutation of an arbitrary way of living into the legitimate way of life which casts every other way of living into arbitrariness. (1984, pp. 56, 57)

Ultimately, in legitimating an "arbitrary way of living," traditional views of authorship adopt what Allen and Gomery (1985) identify as the "great man" theory of history—"the belief that history is made by the inspired acts of outstanding individuals, whose genius transcends the normal constraints of historical context." Of course, in the West, aesthetic histories based on the "great man" theory are principally written by privileged white men, tend to celebrate "masterpieces" made by privileged white men, and usually uphold the "arbitrary ways of living" valued by privileged white men.

Based on a denial of the economic and contaminated by an elitist bias, the "great man" view of authorship has dire interpretive consequences. Like Judge Robert Bork's interpretive approach to the U.S. Constitution, traditional notions of authorship tend to reduce meaning to a matter of "original intent." In privileging authorial intention, such approaches endorse a static view of the communication process that isolates the message-product from the flow of history and treats meaning as a frozen asset that is somehow deposited in the work by its creator. However, as Bakhtin (1981), Newcomb (1984), Fiske (1987a, 1987b), and other critical scholars argue, human communication is much more complicated than a series of monologues articulated by self-determined individuals. Rather than approach meaning solely in terms of original intent, these scholars see the text as an arena for symbolic and political contestation. And in this contestation, meaning emerges from a multidimensional struggle: between writing and reading; between text and context; between "Us" and "Other"; between said and unsaid; between known and unknown; between domination and resistance; between revulsion and relevance; between estrangement and pleasure.

### Fashioning a New Wineskin

In trying to make sense of what Allen (1987b, p. 4) calls "the circumscribed role of the author," I hope to elaborate a theory of cultural production that emphasizes the dynamic operation of "rewriting" in television's storytelling process. This emphasis is, in part, an acknowledgment of the logistical, intellectual, and creative demands of TV's assembly-line storytelling. The brutal time constraints of this assembly line force the creative staff of most TV series to work on several episodes, in various stages of completion, at the same time. In other words, in series television, rewriting is an overlapping cyclic process that requires the precise coordination of a community of creators in an efficient routine. And, as Anderson observes:

[I]t is no wonder that these stories develop through formulaic repetition and the invocation of references, stereotypes and cliches. . . . This is necessarily the way in which popular culture works. Meaning develops according to a delicate operation of similarity and difference. In this process, a single story gains significance both through its identity with the stories that precede it and through its disruption of these stories. (1987, p. 119)

In this "delicate operation of similarity and difference," rewriting acts as a *generative ritual* that insures each episode of a series is faithful to narrative patterns established in previous episodes. From the earliest moments of scripting, the rewriting conforms to the age-old imperative that the show, indeed, must go on. And elsewhere (Reeves, 1988), in an analysis of preproduction documents associated with the making of a single episode of *Newhart,* I have demonstrated how the necessity to rewrite the series activates, informs, directs, and inhibits every stage of creative collaboration.

But, here, the accent is placed on a different level of rewriting—the rewriting of generic conventions into the chain of texts collectively known as the "TV series." Following Newcomb and Alley, this essay proposes that it is at the level of series creation that questions of authorship become a pertinent critical concern. Of course, at this level of discourse, the dominant creative force is the television producer.

Because of the scale of series production, television producers can ill afford to indulge in the pretenses of the "legitimate" art business. Rather than wear the "legitimate" artist's mask of "disinterestedness," rather than engage in a ritual disavowal of economic interest, television producers combine the role of the narrative artist with that of the trader in cultural goods. And as storytellers who cut deals, watch the bottom line, and live and die by the ratings, television producers are actively engaged as interpreters of the culture. When devising a new series or revamping an old one, they take their place beside advertising agents and fashion designers as what Sahlins calls "hucksters of the symbol."[2] In Sahlins' words:

> In the nervous system of the American economy, theirs is the synaptic function. It is their role to be sensitive to the latent correspondences in the cultural order whose conjunction in a product symbol may spell mercantile success. (1976, p. 217)

This [essay] proposes that the spirit of rewriting provides the synaptic spark that brings these "latent correspondences in the cultural order" to life. For, as hucksters of the symbol, television producers are engaged in an ongoing revision of the American way—a revision that translates the conflicts and contradictions of modern life into terms that are comprehensible and forms that are meaningful to a vast, heterogeneous audience.

In sketching out a "dialogic view of authorship," I aim to underscore the *open orientation* of this rewriting of culture. For the work of television producers is not simply directed toward the series at hand. Instead, they must always take into account the force of other texts (i.e., previous texts, competing texts, future texts) and other contexts (i.e., the production budget, the legal climate, the ratings). On television, this open orientation is responsible for both the conventional and the innovative.

Of course, the sector of conventional storytelling demands a central place in television studies. Like burial charms, sacred psalms, or totem poles, conventional series deserve to be interpreted as documents that give concrete form to widely held values and beliefs. In fact, some of the most provocative scholarly television criticism has centered around the careful analysis of such conventional programming as *The Love Boat* (Schwichtenberg, 1987), *Guiding Light* (Allen, 1987a), and *The A-Team* (Fiske, 1987a, 1987b).

To account for changes in the operation and evolution of television's system of stories, studies investigating narrative innovation have generally placed authorship in a more prominent position on the critical agenda. In the first decade of television, this innovation was directed toward adapting popular story formulas from film and radio to the formal, economic, scheduling, and ideological constraints of a new and evolving storytelling system. During this formative stage, Jack Webb's work on *Dragnet* serves as a particularly cogent example of early generic adaptation that would have a lasting influence on the future of television. In Marc's words:

> *Dragnet* was not merely a hit. It was an ideology, a "look," and an object of satire that made it a household word even in households that did not necessarily tune into it. It was TV's first big crimeshow money-maker, drawing serious network attention—and cash—to the genre. . . . Webb had moved the TV crimeshow from the thin artifice of tiny New York studios (most were hastily converted radio facilities) to the cavernous sound stages and spacious boulevards of Hollywood. Webb's purchase of an old Republic Pictures studio lot in North Hollywood, where he constructed a new TV production complex, was a sign of the times. Cop, car, criminal, and camera would not be separated again. (1984, p. 74)

As Marc, Newcomb (1974), and other television scholars have demonstrated, this early generic translation/adaptation/innovation is also apparent in such classic TV fare as *I Love Lucy, Gunsmoke,* and *Perry Mason.*

Later, the innovation would primarily be directed toward refining the formulas which survived and thrived on the new medium. Consider, here, the corporate authorship of MTM. With series like *The Mary Tyler Moore Show, Rhoda,* and *The Bob Newhart Show,* MTM literally rewrote the situation comedy formula to accommodate more complex characters. As Feuer (1987, p. 55) puts it:

> "Character ensembles," "motivation," "a set of little epiphanies," have transformed the problem/solution format of the sitcom into a far more psychological and episodic formula in which—in the hand of MTM—the situation itself becomes the pretext for the revelation of character.

This refinement of genre is also evident in the work of Richard Levinson and William Link *(Columbo),* Larry Gelbart *(M\*A\*S\*H),* and Danny Arnold *(Barney Miller).*

In the last decade—in part because of television's maturation as a storytelling system and the parallel maturation of the "television generation"—innovation has often resulted in talented producers exploring and blurring generic boundaries. Steven Bochco and Michael Kozoll did this with *Hill*

*Street Blues,* enriching the cop show formula with the humor of the dark comedy and the tangled relationships of the soap opera. But more recently, this generic blurring is evident in a group of shows that labor under the label of "dramedy": *Frank's Place, The Days and Nights of Molly Dodd, The "Slap" Maxwell Story,* and *The Wonder Years.*

Although the shows are profoundly different from each other, they do share certain common features: all, admittedly, integrate moments of poignance into a comedic framework; all are shot film style; none use a laugh track. And in *Electronic Media's* semi-annual poll of newspaper critics, all of the so-called "dramedies" placed in the top ten (1988, p. 18).[3] The audience, though, has not been as enthusiastic. Because of low ratings, *Molly Dodd, "Slap" Maxwell,* and *Frank's Place* vanished from prime-time network television after only one season.

*The Wonder Years,* though, is another story. ABC gave *The Wonder Years* an early boost by running it as a mid-season replacement that premiered immediately following the network's broadcast of the 1988 Superbowl. This auspicious beginning was certainly a contributing factor in making *The Wonder Years* the only one of the four celebrated "dramedies" to attract a consistently large audience. Primarily because of this commercial success, I will use *The Wonder Years* to further illustrate the dialogic character of television authorship.

### Analyzing the Ferment of New Wine

To understand how a show like *The Wonder Years* gets developed, we first have to consider the economic context of the major broadcast networks in the 1980s. Thanks to the most extreme shift in the structure of the American communication complex since the 1950s, the dominance of the three-network oligopoly that television inherited from radio has been severely and swiftly undermined by new forces in the marketplace. Just over ten years ago, 91 percent of TV's prime-time audience watched programs broadcast on the three national networks (Stevenson, 1985). But at that time, only a little over 17 percent of America's TV homes were wired into a cable system (Sterling and Kittross, 1978, p. 536) and the domestic VCR was still too expensive for most medium-income families. By 1985, cable had penetrated 46 percent of America's households, 30 percent owned VCRs (Miller, 1986, p. 16), and the major networks' combined share of the prime-time audience had slipped to 73 percent. Ironically, during this period of decline for the networks, Americans devoted even more time to television—the average family's weekly consumption increased from forty-seven hours and forty-seven minutes in 1982 to forty-nine hours and fifty-eight minutes in 1984 (Stevenson, 1985).

To grab a share of the growing mass and shrinking particular audience, ABC has instituted a programming strategy that involves targeting specific age and generational groups. And one of the prime groups that is especially attractive to both ABC and its advertisers is the first television generation—

the infamous baby boomers. *Moonlighting, thirtysomething,* and *The Wonder Years* are all shows that form the programming base for ABC's strategy.

But, where yuppie sagas like *thirtysomething* look at baby boomers booming in the Reagan-Bush era, *The Wonder Years* takes a different perspective. Significantly, the show's coproducers (Neal Marlens and Carol Black) are very self-conscious about this difference. Their self-consciousness was, in fact, the subject of a newspaper article appearing on the front page of the *Los Angeles Times:*

> Even though some in the TV industry have jokingly called "The Wonder Years" "twelve-something" or "The Little Chill," Marlens said the show views the world with 12-year-old wonder rather than the disillusionment of radicals who have reluctantly entered the Establishment.
>
> "We saw that a lot of the shows that seemed to be made by yuppies, baby boomers, people in their mid-30s to mid-40s, seemed to be so directly and literally about contemporary life," Marlens said. "I think that there's some value in that, but it's also real dangerous. It's like looking in a mirror, which is not necessarily the best way to get a perspective on yourself." (Haithman, 1988, pp. 1, 24)

Beyond being made for and about the television generation, *The Wonder Years* is also made by members of that generation: Marlens and Black are both in their early thirties.[4] And their familiarity with the television experience is evidenced by the savvy references to television history that enrich the series. Indeed, in combining a two-syllable, emotionally charged adjective (Wonder) with a one-syllable time designator (Years), the very title of the series acknowledges the influence of *Happy Days*. Where *Happy Days* examined the banal problems of middle-class family life in the 1950s, *The Wonder Years* centers on a suburban family caught up in the turmoil of the 1960s.

And *The Wonder Years* invokes the history of situation comedy in other meaningful ways. Like *Leave it to Beaver,* the organizing sensibility in the series is the youngest member of the family, a twelve-year-old named Kevin Arnold (Fred Savage). And his older brother, Wayne (Jason Hervey), is Wally Cleaver with Eddie Haskell's horns. Providing much of the humor of the series, Kevin and Wayne's sibling rivalry is established in the early moments of the pilot when the voice-over narration of an adult Kevin tells us: ". . . years later, when we were both adults, we finally got to know and understand each other, and Wayne explained that, basically, he just deeply regretted the fact that I was born, and he wanted me to feel the same way."

Voice-over narration, of course, is not a standard feature of the sitcom. And, the voice of the adult Kevin speaks to the show's status as a generic hybrid. In addition to triggering memories of other situation comedies, *The Wonder Years* also resonates with *The Waltons,* a family drama set in the 1930s. The narration of John Boy Walton reminiscing about those bygone days contributed to what Newcomb (1987) has appraised as *The Waltons's* profound sense of "intimacy, continuity, and history." While the voice-over narration on *The Wonder Years* is often used for comedic effect, providing ironic commentary on the action, it also frequently accomplishes the aesthetic effects valued by Newcomb.

The sense of history aroused by the voice-over is further stimulated by the show's frequent use of pop music of the era. In an early episode, for instance, "Black Bird"—the Beatles' acoustical dirge—enhances the pathos of one of Kevin's many disappointments on the path to adulthood. And Marlens and Black did not commission a theme song for *The Wonder Years*. Instead, a hit tune like Joe Cocker's rendition of "I Get By With a Little Help from My Friends" plays as the opening credits appear under nostalgia-inducing "home movies" of *The Wonder Years*'s family.

For me (because of my dual status as a baby-boomer and a son), the most moving episode of the first season explored Kevin's relationship with his father. In my analysis of this episode, I intend to demonstrate how Marlens and Black rely on viewer familiarity with the television experience—and how this reliance, in turn, enables them to fashion a relatively compelling commentary on contemporary suburban life.

Interestingly, television plays a prominent and humorous role in disclosing the premise of the story. In fact, the episode opens with a close up of an ancient, black-and-white set exhibiting a nature documentary. As a large gorilla moves through a forest, the documentary's narrator explains, "The male enters a hostile environment to find sustenance. He returns after an unsuccessful foray, aggressive and unpredictable." At this time, a wide shot reveals that the TV is located in the Arnold kitchen and is being watched by Kevin, Wayne, and their mother, Norma (Alley Mills). We continue to hear the documentary's narrator, but now he seems to be describing the action of the Arnold family. Offscreen, a barking dog and screeching tires signal the homecoming of Jack Arnold (Dan Lauria). As Norma rushes to the kitchen window, the docu-narrator continues with bland scientific detachment: "Notice the reaction of the startled mother and her offspring as they begin to sense the presence of the male." Norma, on reading Jack's mood, issues a warning to the brothers: "Your father's had a bad day at work, so no noise." Kevin and Wayne exchange knowing looks as the despondent father enters through the kitchen door. Norma greets him with a neutral, "Hi, hon. How's work?" Without acknowledging the existence of Kevin and Wayne, Jack moves quickly through the kitchen, grunting "Work's work" as he exits into the living room. The close-up of the TV reappears, showing young gorillas scurrying into the undergrowth as the docu-narrator observes: "The irritable male gives out unmistakable signals that tell the young to keep their distance." Cutting back to a wide-shot of the kitchen, the screen shows Norma and her boys react with fear to Jack's offscreen slamming of a door. Finally, after the boys flee the house to go play catch, the introductory scene ends with another shot of the television, now showing an angry gorilla beating its chest.

The clever construction of this opening sequence speaks not only to the wit of *The Wonder Years,* but also to its deviation from the norms of domestic comedy. Jack Arnold represents a significant rewriting of the benign TV father regulating such conventional shows as *Leave it to Beaver, Happy Days,* and *The Cosby Show.* As a stern disciplinarian who has trouble communicating with his spouse and children, Jack is one TV father who, very often, doesn't "know best." Unlike the fatherly lawyers, obstetrician-gynecologists, and briefcase-carrying businessmen of other domestic comedies, Jack is a mid-level manager who is frustrated by the pressure and drudgery of his job.

Viewers discover, in voice-over narration immediately following the opening scene, that Jack copes with this frustration in two ways:

> When my father had a bad day at work, he'd just sit in the dark by himself and watch TV. We learned early on that this was a danger signal and we adapted our behavior accordingly. And when he had a really bad day, I'm talking about a very not-good day, he had this telescope and he'd go out in the back yard and just look through it for hours.

Ultimately, the star gazing will be a device that helps soften and humanize Jack's character.

Jack's television viewing, though, is presented as something that separates him from his family. For instance, a turning point in the story occurs when Kevin interrupts Jack while he watches the sports news. Curious about Jack's work and not understanding what it means to manage "distribution and product support services," Kevin asks Jack what he "does all day." Miffed because Kevin causes him to lose track of the day's baseball scores, Jack snarls, "Shovel other people's crap so you kids can eat." Feeling rejected by his father, Kevin retreats to his room to brood. And Norma, predictably, intervenes on Kevin's behalf:

> *Norma:* You gotta relax a little, Jack.
> *Jack:* Dammit, Norma, don't tell me to relax. I mean what does he want to know. About the seven *S-14* forms I've got to fill out every time I turn around? About the whining customers? About the incompetent jackasses in Shipping and Receiving?
> *Norma:* Yeah. Yes, that's exactly what he wants to know. He wants to know more about you, Jack. I don't know why that's so hard for you to understand.

The intensity of this wife–husband interchange, of course, runs counter to the cheerful spousal collusion typical of most television comedy. Marriage, here, is not a merry adventure. In fact, as we discover, Jack partially blames his marriage and family responsibilities for his miserable existence.

Even so, to soothe Kevin's injured pride, Jack agrees to take him to work the next day. In a humorous sequence, Kevin meets the boring people who work in Jack's office, observes Jack handling a crisis involving the incompetence of a subordinate, and fantasizes about wielding the power and authority of a manager. But the highlight of Kevin's excursion into the world of corporate work comes when he accompanies his father on a coffee break. Away from the hectic atmosphere of the office, Kevin learns that when Jack was his age he wanted to be the captain of a ship:

> *Jack:* Yeah. You know, one of those big ocean liners, or a freighter, or an oil tanker. Be out there in the ocean in the middle of the night . . . navigating by the stars. Of course, they use instruments for all that now, but I didn't know that. Yeah, I thought that'd be the greatest thing in the world.
> *Kevin:* How come you didn't do it?
> *Jack:* How come? Well, you know, one thing leads to another. Went off to college, met your mom, next summer I got a job on a loading dock here at Norcom. The rest is history.
> *Kevin:* You'd have made a great ship's captain, dad.
> *Jack:* Nah. Probably not. Probably get sea sick. You know, Kevin, you can't do every silly thing you want to in life. You have to make your choices. You have to try to be happy with them. . . .

The tranquil mood of this intimate moment is shattered when father and son return to the office. Jack's supervisor ambushes him there. And Kevin watches, in shock, as Jack suffers a humiliating chewing out.

As I hope this analysis demonstrates, this touching treatment of the father–son relationship also investigates what it means to be male in our society, the alienation of corporate life, the naive dreams of childhood, the sobering realities of adulthood, and the illusory character of freedom of choice. What is lost in this analysis is much of the historical texture of this story. Miniskirts, the military-industrial complex, the Vietnam War, the generation gap, the Washington Senators, and the "Buckle-Up-for-Safety" campaign are all invoked at various moments during this episode.

Admittedly, at one level *The Wonder Years* is an apologia that is often militant in its defense of middle-class life in the suburbs. This militancy is, in fact, even brought to the fore in the series itself. Consider, here, for instance, the closing narration of the pilot episode:

> I think about the events of that day again and again . . . whenever some blowhard starts talking about the anonymity of the suburbs or the mindlessness of the TV generation. Because we know that inside each of those identical boxes with its Dodge parked out front and its white bread on the table and its TV set glowing blue in the falling dusk, there were characters and stories, there were families bound together in the pain and the struggle of love, there were moments that made us cry with laughter, and there were moments, like that one, of sorrow and wonder.

Yet, in telling the story of a middle-class family "bound together in the pain and the struggle of love," *The Wonder Years* also continually unmasks and explores central contradictions of the mainstream American experience. At times, these contradictions involve crises in the patriarchal order; at other times, they concern the alienation of corporate life; at still other times, they deal with the economic and social inequities of a stratified class system. Although the resolutions to most episodes generally confirm centrist American values—especially those associated with individualism—the show rarely offers complete closure. In other words, the force of the cultural contradictions represented on *The Wonder Years* tends to overwhelm the contrivances of the reassuring endings.

In fact, the contradictions often persist because, unlike in the conventional sitcom, the characters in *The Wonder Years* are allowed to change. For instance, after Kevin's excursion into the world of work in the episode analyzed above, the boy's relationship with his father enters a new stage. The final scene of the episode begins with Kevin watching through a window while Jack, once again, gazes at the stars through his telescope. As Kevin hesitantly steps out onto the back porch we hear the pensive narration of his adult voice: "That night, my father stood there looking up at the sky the way he always did. But suddenly I realized I wasn't afraid of him in quite the same way anymore. The funny thing is, I felt like I'd lost something." As the narration concludes, Jack sadly looks over, sees Kevin, and beckons for him to come join him. The episode then ends on a bittersweet note with Kevin peering through the telescope as his father tells him: "That's Polaris, the North Star. That's how the sailors used to find their way home."

At the close of the episode, then, Marlens and Black do not rescue the fictional father from the contradictions unmasked by the story. Jack Arnold is still very much a lost soul trapped by his job, his family responsibilities, and his place in the patriarchy.

## Preserving the Wine of Authorship

Obviously, Marlens and Black have mastered the recombinant style that is much maligned by Gitlin (1985). However, we are not only unfair, but we are also irresponsible, when we depreciate the generic innovation of such perceptive cultural interpreters simply because it is popular—that is, because it appeals to modern-day "publicans and sinners" and it doesn't conform to an elitist definition of art.

Therefore, while I obviously agree with much of Bourdieu's stinging critiques (1984; 1986) of the ideology of charisma underlying traditional notions of authorship, I am not at all comfortable in completely writing off authorship as a dead issue. In acknowledging that all producers are inspired by self-interest, that they are intent on manufacturing a "product symbol [that] may spell mercantile success," we are still faced with the challenge of making distinctions between the good steward and the bad, between the friend and the panderer, between benefaction and exploitation. Although TV's creative artist is circumscribed by multitudinous forces—both intertextual and extratextual—we must continue to recognize that she still plays a significant, and sometimes decisive, role in the struggle for meaning. In discarding the old skin of traditional criticism, then, we must take care to preserve a portion of the wine of authorship—for to siphon out the community of creators from the blend of culture is to deny human beings any responsibility for history, and to surrender to the despair of some reductive form of psychological, economic, or technological determinism.

Of course, we can, and indeed should, reject such despair. If we consider *reality* to be a social construction and *thought* to be a public enterprise, then ideologies, and economies, and institutions, and technologies are all human products—and human intervention and struggle can still change history.[5] But in sharing responsibility for history, we must also recognize the need to hold people accountable for their contributions to maintaining or transforming the world. This chapter proposes that a dialogic view of authorship is both geared to the realities of contemporary cultural production and crucial to enforcing such artistic accountability.

### Notes

The author thanks Brian Nienhaus for championing the work of Bourdieu. The author also expresses appreciation to Richard Campbell for reading and suggesting changes in earlier drafts of this essay.

1. Here, Jesus of Nazareth intended to translate a furious religious controversy into concrete terms that were both accessible and comprehensible to his followers. Because of the technological limitations of that period, wine was stored in animal skins before the fermentation process was complete. It was common knowledge that the inflexible leather of old skins could not withstand the internal pressures generated by this fermentation. The common folks listening to the parable realized that new wine required containers fashioned from fresh leather.

2. Newcomb and Alley also incorporate Sahlins's work into their view of television producers: "These creators are, as Sahlins suggests, true readers, true analysts of the cultures in which they live and work, the society in which they must seek and create an audience. They must be sensitive to many sorts of cultural change, technological as well as sociological, cognitive, or political" (1983, p. 32).

3. *The Wonder Years* and *Frank's Place* came in second and third behind the critic's favorite, *L.A. Law*. *"Slap" Maxwell* came in eighth, and *Molly Dodd* tenth.

4. As of this writing, the six existing episodes of the series have all been co-written by Marlens and Black.

5. This conception of reality and thought is based on what Carey (1975) identifies as a "ritual view of communication."

### References

Allen, R. (1987a). *The Guiding Light:* Soap opera as economic product and cultural document. In H. Newcomb, ed., *Television: The critical View,* 4th ed., pp. 141–63. New York: Oxford University Press.

Allen, R., ed. (1987b). Talking about television. In R. Allen, ed. *Channels of discourse: Television and contemporary criticism,* pp. 1–16. Chapel Hill: University of North Carolina Press.

Allen, R., and D. Gomery. (1985). *Film history: Theory and practice.* New York: Knopf.

Anderson, C. (1987). Reflections on *Magnum, P.I.* In H. Newcomb, ed., *Television: The critical view,* 4th ed., pp. 112–25. New York: Oxford University Press.

Bakhtin, M. (1981). *The dialogic imagination: Four essays.* Ed. M. Holquist. Trans. C. Emerson and M. Holquist. Austin: University of Texas Press.

Booth, W. (1987). The company we keep: Self-making in imaginative art, old and new. In H. Newcomb, ed., *Television: The critical view,* 4th ed., pp. 382–418. New York: Oxford University Press.

Bourdieu, P. (1984). *Distinction: The social critique of the judgment of taste.* Trans. R. Nice. Cambridge: Harvard University Press.

———. (1986). The production of belief: Contribution to an economy of symbolic goods. Trans. by R. Nice. In R. Collins, J. Curran, N. Granham, P. Scannell, P. Schlesinger, and C. Sparks, eds., *Media, culture and society: A critical reader,* pp. 131–63. Beverly Hills: Sage.

Carey, J. (1975). A cultural approach to communication. *Communication* 2: 1–22.

*Electronic Media* critics poll. (1988, May 2). *Electronic Media,* p. 18.

Feuer, J. (1987). The MTM style. In H. Newcomb, ed., *Television: The critical view,* 4th ed., pp. 52–84. New York: Oxford University Press.

Fiske, J. (1987a). British cultural studies and television. In R. Allen, ed., *Channels of discourse: Television and contemporary criticism,* pp. 255–90. Chapel Hill: University of North Carolina Press.

———. (1987b). *Television culture.* London: Metheun.

Gitlin, T. (1985). *Inside prime time.* New York: Pantheon Books.

Haithman, D. (1988, April 8). *"Wonder Years"* Pays its Respects to '60s suburbia. *Los Angeles Times,* Part 6, pp. 1, 24.

Marc, D. (1984). *Demographic vistas: Television in American culture.* Philadelphia: University of Pennsylvania Press.

Miller, J. (1986). International roundup: The global picture. *Channels of Communications 1986 Field Guide,* pp. 16–18.

Newcomb, H. (1974). *TV: The most popular art.* New York: Anchor Books.

———. (1984). On the dialogic aspects of mass communications. *Critical Studies in Mass Communication,* 1, 1: 34–50.

———. (1987). Toward a television aesthetic. In H. Newcomb, ed., *Television: The critical view,* 4th ed., pp. 613–27. New York: Oxford University Press.

Newcomb, H. and R. Alley. (1983). *The producer's medium: Conversations with creators of American TV.* New York: Oxford University Press.

Reeves, J. (1988). Rewriting *Newhart:* A dialogic analysis. *Wide Angle,* 10, 1: 76–91.

Sahlins, M. (1976). *Culture and practical reason.* Chicago: University of Chicago Press.

Schwichtenberg, C. (1987). *The Love Boat:* The packaging and selling of love, heterosexual romance, and family. In H. Newcomb, ed., *Television: The critical view,* 4th ed., pp. 126–40. New York: Oxford University Press.

Sterling, C., and J. Kittross. *Stay tuned: A concise history of American broadcasting.* Belmont, Calif.: Wadsworth.

Stevenson, R. (1985, October 20). The networks and advertisers try to recapture our attention. *New York Times,* Section F, p. 8.

# Roseanne: Unruly Woman as Domestic Goddess

## KATHLEEN K. ROWE

*Sometime after I was born in Salt Lake City, Utah, all the little babies were sleeping soundly in the nursery except for me, who would scream at the top of my lungs, trying to shove my whole fist into my mouth, wearing all the skin off on the end of my nose. I was put in a tiny restraining jacket. . . . My mother is fond of this story because to her it illustrates what she regards as my gargantuan appetites and excess anger. I think I was probably just bored.*

Roseanne: My Life as a Woman[1]

Questions about television celebrities often center on a comparison with cinematic stars—on whether television turns celebrities into what various critics have called "degenerate symbols" who are "slouching toward stardom" and engaging in "dialogues of the living dead."[2] This [essay] examines Roseanne Barr, a television celebrity who has not only slouched but whined, wisecracked, munched, mooned, and sprawled her way to a curious and contradictory status in our culture explained only partially by the concept of stardom, either televisual or cinematic. Indeed, the metaphor of decay such critics invoke, while consistent with a strain of the grotesque associated with Barr, seems inappropriate to her equally compelling vitality and *jouissance*. In this essay, I shall be using the name "Roseanne" to refer to Roseanne Barr-as-sign, a person we know only through her various roles and performances in the popular discourse. My use follows Barr's lead in effacing the lines among her roles: Her show, after all, bears her name and in interviews she describes her "act" as "who she is."

Nearing the end of its second season, her sitcom securely replaced *The Cosby Show* at the top of the ratings. The readers of *People Weekly* identified her as their favorite female television star and she took similar prizes in the

From Kathleen Rowe, "Roseanne: Unruly Woman as Domestic Goddess," *Screen* 31:4, 1990, pp. 408–419. Reprinted with permission.

People's Choice award show this spring. Yet "Roseanne," both person and show, has been snubbed by the Emmies, condescended to by media critics, and trashed by the tabloids (never mind the establishment press). Consider *Esquire*'s solution of how to contain Roseanne. In an issue on its favorite (and least favorite) women, it ran two stories by two men, side by side—one called "Roseanne—Yay," the other "Roseanne—Nay." And consider this from *Star*: "ROSEANNE'S SHOTGUN 'WEDDING FROM HELL'—"Dad refuses to give pregnant bride away—'Don't wed that druggie bum!' ";
"Maids of honor are lesbians—best man is groom's detox pal"; "Ex-hubby makes last-ditch bid to block ceremony"; "Rosie and Tom wolf two out of three tiers of wedding cake" (6 February 1990). Granted that tabloids are *about* excess, there's often an edge of cruelty to that excess in Roseanne's case, and an effort to wrest her definition of herself from the comic to the melodramatic.

Such ambivalence is the product of several phenomena. Richard Dyer might explain it in terms of the ideological contradictions Roseanne plays upon—how, for example, the body of Roseanne-as-star magically reconciles the conflict women experience in a society that says "consume" but look as if you don't. Janet Woollacott might discuss the clash of discourses inherent in situation comedy—how our pleasure in Roseanne's show arises not so much from narrative suspense about her actions as hero, nor from her one-liners, but from the economy or wit by which the show brings together two discourses on family life: one based on traditional liberalism and the other on feminism and social class. Patricia Mellencamp might apply Freud's analysis of wit to Roseanne as she did to Lucille Ball and Gracie Allen, suggesting that Roseanne ventures farther than her comic foremothers into the masculine terrain of the tendentious joke.[3]

All of these explanations would be apt, but none would fully explain the ambivalence surrounding Roseanne. Such an explanation demands a closer look at gender and at the historical representations of female figures similar to Roseanne. These figures, I believe, can be found in the tradition of the "unruly woman," a *topos* of female outrageousness and transgression from literary and social history. Roseanne uses a "semiotics of the unruly" to expose the gap she sees between the ideals of the New Left and the Women's Movement of the late sixties and early seventies on the one hand, and the realities of working class family life two decades later on the other.

Because female unruliness carries a strongly ambivalent charge, Roseanne's use of it both intensifies and undermines her popularity. Perhaps her greatest unruliness lies in the presentation of herself as *author* rather than actor and, indeed, as author of a self over which she claims control. Her insistence on her "authority" to create and control the meaning of *Roseanne* is an unruly act par excellence, triggering derision or dismissal much like Jane Fonda's earlier attempts to "write" her self (but in the genre of melodrama rather than comedy). I will explain this in three parts: the first takes a brief look at the tradition of the unruly woman; the second, at the unruly qualities of *excess* and *looseness* Roseanne embodies; and the third, at an episode of her sitcom which dramatizes the conflict between female unruliness and the ideology of "true womanhood."

## The Unruly Woman

The unruly woman is often associated with sexual inversion—"the woman on top," according to social historian Natalie Zemon Davis, who fifteen years ago first identified her in her book *Society and Culture in Early Modern France*. The sexual inversion she represents, Davis writes, is less about gender confusion than about larger issues of social and political order that come into play when what belongs "below" (either women themselves, or their images appropriated by men in drag) usurps the position of what belongs "above." This *topos* isn't limited to Early Modern Europe, but reverberates whenever women, especially women's bodies, are considered excessive—too fat, too mouthy, too old, too dirty, too pregnant, too sexual (or not sexual enough) for the norms of conventional gender representation. For women, excessive fatness carries associations with excessive wilfulness and excessive speech ("fat texts," as Patricia Parker explains in *Literary Fat Ladies,* a study of rhetoric, gender, and property that traces literary examples of this connection from the Old Testament to the Twentieth Century).[4] Through body and speech, the unruly woman violates the unspoken feminine sanction against "making a spectacle" of herself. I see the unruly woman as prototype of woman as subject—transgressive above all when she lays claim to her own desire.

The unruly woman is multivalent, her social power unclear. She has reinforced traditional structures, as Natalie Davis acknowledges.[5] But she has also helped sanction political disobedience for men and women alike by making such disobedience thinkable. She can signify the radical utopianism of undoing all hierarchy. She can also signify pollution (dirt or "matter out of place," as Mary Douglas might explain). As such she becomes a source of danger for threatening the conceptual categories which organize our lives. For these reasons—for the power she derives from her liminality, her associations with boundaries and taboo—she evokes not only delight but disgust and fear. Her ambivalence, which is the source of her oppositional power, is usually contained within the license accorded to the comic and the carnivalesque. But not always.

The unruly woman has gossiped and cackled in the margins of history for millenia, from Sarah of the Old Testament who laughed at God (and figures in Roseanne's tribute to her grandmother in her autobiography), to the obstinate and garrulous Mrs. Noah of the medieval Miracle Plays (who would not board the Ark until she was good and ready), to the folk figure "Mère Folle" and the subject of Erasmus's *The Praise of Folly*. Her more recent incarnations include such figures as the screwball heroine of the 1930s film, Miss Piggy, and a pantheon of current female grotesques and sacred monsters: Tammy Faye Bakker, Leona Helmsley, Imelda Marcos, and Zsa Zsa Gabor. The media discourse around these women reveals the same mixed bag of emotions I see attached to Roseanne, the same cruelty and tendency to carnivalize by pushing them into parodies of melodrama, a genre which, unlike much comedy, punishes the unruly woman for asserting her desire. Such parodies of melodrama make the unruly woman the target of *our* laughter, while denying her the power and pleasure of her own.

The disruptive power of these women—carnivalesque and carnival-ized—contains much potential for feminist appropriation. Such an appro-priation could enable us to problematize two areas critical to feminist the-ories of spectatorship and the subject: the social and cultural norms of femininity, and our understanding of how we are constructed as gendered subjects in the language of spectacle and the visual. In her essay "Female Grotesques," Mary Russo asks: "In what sense can women really produce or make spectacles out of themselves? . . . The figure of female transgressor as public spectacle is still powerfully resonant, and the possibilities of rede-ploying this representation as a demystifying or utopian model have not been exhausted."[6] She suggests that the parodic excesses of the unruly woman and the comic conventions surrounding her provide a space to act out the dilemmas of femininity, to *make visible* and *laughable* what Mary Ann Doane describes as the "tropes of femininity."

Such a sense of spectacle differs from Laura Mulvey's. It accepts the relation between power and visual pleasure but argues for an understanding of that relation as more historically determined, its terms more mutable. More Foucaldian than Freudian, it suggests that visual power flows in mul-tiple directions and that the position of spectacle isn't entirely one of weak-ness. Because public power is predicated largely on visibility, men have tra-ditionally understood the need to secure their power not only by looking but by being seen—or rather, by fashioning, as author, a spectacle of them-selves. Already bound in a web of visual power, women might begin to renegotiate its terms. Such a move would be similar to what Teresa de Lauretis advocates when she calls for the strategic use of narrative to "con-struct other forms of coherence, to shift the terms of representation, to produce the conditions of representability of another—and gendered—so-cial subject."[7] By returning the male gaze, we might expose (make a spec-tacle of) the gazer. And by utilizing the power already invested in us as image, we might begin to negate our own "invisibility" in the public sphere.

## Roseanne as Spectacle

The spectacle Roseanne creates is *for* herself, produced *by* herself from a consciously developed perspective on ethnicity, gender, and social class. This spectacle derives much of its power from her construction of it as her "self"—an entity which, in turn, she has knowingly fashioned through interviews, public performances, and perhaps most unambiguously her autobiography. This book, by its very existence, enhances the potency of Roseanne-as-sign because it grants a historicity to her "self" and a materiality to her claims for authorship. The autobiography describes key moments in the development of "Roseanne"—how she learned about female strength when for the first time in her life she saw a woman (her grandmother) stand up to a man, her father; how she learned about marginality and fear from her childhood as a Jew in Utah under the shadow of the Holocaust, and from her own experience of madness and institutionalization. Madness is a leitmotif both in her autobiography and in the tabloid talk about her.[8] Roseanne's eventual discovery of feminism and counter-culture politics led to disillusionment when the women's movement was taken over by women unlike her, "handpicked," she writes, to be acceptable to the establishment.

Co-existing with the pain of her childhood and early adulthood was a love of laughter, the bizarre, a good joke. She always wanted to be a writer, not an actor. Performance, however, was the only "place" where she felt safe. And because, since her childhood, she could always say what she wanted to as long as it was funny, *comic* performance allowed her to be a writer, to "write" herself. While her decision to be a comedian was hampered by a difficulty in finding a female tradition in which to locate her own voice, she discovered her stance (or "attitude") when she realized that she could take up the issue of female oppression by adopting its language. Helen Andelin's *Fascinating Womanhood* [1965] was one of the most popular manuals of femininity for the women of her mother's generation. It taught women to manipulate men by becoming "domestic goddesses." Yet, Roseanne discovered, such terms might also be used for "self-definition, rebellion, truth-telling," for telling a truth that in her case is both ironic and affirmative. And so she built her act and her success on an exposure of the "tropes of femininity" (the ideology of "true womanhood," the perfect wife and mother) by cultivating the opposite (an image of the unruly woman).

Roseanne's disruptiveness is more clearly paradigmatic than syntagmatic, less visible in the stories her series dramatizes than in the image cultivated around her body: Roseanne-the-person who tattooed her buttocks and mooned her fans. Roseanne-the-character for whom farting and nose-picking are as much a reality as dirty dishes and obnoxious boy bosses. Both in body and speech, Roseanne is defined by *excess* and by *looseness*—qualities that mark her in opposition to bourgeois and feminine standards of decorum.

Of all of Roseanne's excesses, none seems more potent than her weight. Indeed, the very appearance of a 200-plus-pound woman in a weekly prime-time sitcom is significant in itself. Her body epitomizes the grotesque body of Bakhtin, the body which exaggerates its processes, its bulges and orifices, rather than concealing them as the monumental, static "classical" or "bourgeois" body does. Implicit in Bakhtin's analysis is the privileging of the female body—above all the *maternal* body which, through pregnancy and childbirth, participates uniquely in the carnivalesque drama of inside-out and outside-in, death-in-life and life-in-death. Roseanne's affinity with the grotesque body is evident in the first paragraph of *Roseanne: My Life as a Woman*, where her description of her "gargantuan appetites" even as a newborn brings to mind Bakhtin's study of Rabelais.[9] Roseanne compounds her fatness with a "looseness" of body language and speech—she sprawls, slouches, flops on furniture. Her speech—even apart from its content—is loose (in its "sloppy" enunciation and grammar) and excessive (in tone and volume). She laughs loudly, screams shrilly, and speaks in a nasal whine.

In our culture, both fatness and looseness are violations of codes of feminine posture and behavior. Women of "ill-repute" are described as loose, their bodies, especially their sexuality, seen as out of control. Fatness, of course, is an especially significant issue for women, and perhaps patriarchy nowhere inscribes itself more insidiously and viciously on female bodies than in the cult of thinness. Fat females are stigmatized as unfeminine, rebellious, and sexually deviant (under-or over-sexed). Women who are too fat or move too loosely appropriate too much space, and femininity is gauged by how little space women take up.[10] It is also gauged by the intrusiveness of women's utterances. As Henley notes, voices in any culture that are not meant to be heard are perceived as loud when they do speak, regardless of their decibel level ("shrill" feminists, for example). Farting, belching, and nose-picking likewise betray a failure to restrain the body. Such "extreme looseness of body-focused functions" is generally not available to women as an avenue of revolt but, as Nancy Henley suggests, "if it should ever come into women's repertoire, it will carry great power."[11]

Expanding that repertoire is entirely consistent with Roseanne's professed mission.[12] She writes of wanting "to break every social norm . . . and see that it is laughed at. I chuckle with glee if I know I have offended someone, because the people I intend to insult offend me horribly."[13] In an interview in *People Weekly,* Roseanne describes how Matt Williams, a former producer on her show, tried to get her fired: "He compiled a list of every offensive thing I did. And I do offensive things. . . . *That's who I am. That's my act.* So Matt was in his office making a list of how gross I was, how many times I farted and belched—taking it to the network to show I was out of control" [my emphasis]. Of course she was out of control—*his* control. He wanted to base the show on castration jokes, she says, recasting it from the point of view of the little boy. She wanted something else—something different from what she sees as the norm of television: a "male point of view coming out of women's mouths . . . particularly around families."[14]

Roseanne's ease with her body, signified by her looseness, triggers much of the *unease* surrounding her. Such ease reveals what Pierre Bourdieu describes as "a sort of indifference to the objectifying gaze of others which neutralizes its powers" and "appropriates its appropriation."[15] It marks Roseanne's rebellion against not only the codes of gender but of class, for ease with one's body is the prerogative of the upper classes. For the working classes, the body is more likely to be a source of embarrassment, timidity, and alienation, because the norms of the "legitimate" body—beauty, fitness, and so on—are accepted across class boundaries while the ability to achieve them is not. In a culture which defines nature negatively as "sloppiness," physical beauty bears value that is not only aesthetic but moral, reinforcing a sense of superiority in those who put some effort into enhancing their "natural" beauty (p. 206).

Roseanne's indifference to conventional readings of her body exposes the ideology underlying those readings. Concerning her fatness, she resists the culture's efforts to define and judge her by her weight. Publicly celebrating the libidinal pleasure of food, she argues that women need to take up more space in the world, not less. And her comments about menstruation similarly attack the "legitimate" female body, which does not menstruate in public. On an award show she announced that she had "cramps that could kill a horse." She startled Oprah Winfrey on her talk show by describing the special pleasure she took from the fact that she and her sister were "on their period"—unclean, according to Orthodox law—when they were allowed to bear their grandmother's coffin. And in her autobiography she writes about putting a woman (her) in the White House: "My campaign motto will be 'Let's vote for Rosie and put some new blood in the White House—every 28 days'" (p. 117). Rather than accepting the barrage of ads that tell women they can never be young, thin, or beautiful enough and that their houses—an extension of their bodies—can never be immaculate enough, she rejects the "pollution taboos" that foster silence, shame, and self-hatred in women by urging them to keep their genitals, like their kitchen appliances, deodorized, anticepticized, and "April fresh." Instead she reveals the social causes of female fatness, irritability, and messiness in the strains of working class family life, where junk food late at night may be a sensible choice for comfort after a day punching out plastic forks on an assembly line.

## Demonic Desires

The episode I'm going to talk about (7 November 1989) is in some ways atypical because of its stylistic excess and reflexivity. Yet I've chosen it because it so clearly defines female unruliness and its opposite, the ideology of the self-sacrificing wife and mother. It does so by drawing on and juxtaposing three styles: a realist sitcom style for the arena of ideology in the world of the working class wife and mother; a surreal dream sequence for female unruliness; and a musical sequence within the dream to reconcile the "real" with the unruly. Dream sequences invariably signal the eruption of unconscious desire. In this episode, the dream is linked clearly with the eruption of *female* desire, the defining mark of the unruly woman.

The episode begins as the show does every week, in the normal world of broken plumbing, incessant demands, job troubles. Roseanne wants ten minutes alone in a hot bath after what she describes as "the worst week in her life" (she just quit her job at the Wellman factory). But between her husband Dan and her kids, she can't get into the bathroom. She falls asleep while she's waiting. At this point all the marks of the sitcom disappear. The music and lighting signal "dream." Roseanne walks into her bathroom, but it's been transformed into an opulent, Romanesque pleasure spa where she is pampered by two bare-chested male attendants ("the pec twins," as Dan later calls them). She's become a glamorous redhead.

Even within this dream, however, she's haunted by her family and the institution that stands most firmly behind it—the law. One by one, her family appears and continues to nag her for attention and interfere with her bath. And one by one, without hesitation, she kills them off with tidy and appropriate means. (In one instance, she twitches her nose before working her magic, alluding to the unruly women of the late sixties/early seventies sitcom *Bewitched*). Revenge and revenge fantasies are of course a staple in the feminist imagination (Marleen Gorris's *A Question of Silence* [1982], Nelly Kaplan's *A Very Curious Girl* [1969], Cecilia Condit's *Possibly in Michigan* [1985], Karen Arthur's *Lady Beware* [1987]). In this case, however, Roseanne doesn't murder for revenge but for a bath.

Roseanne's unruliness is further challenged, ideology reasserts itself, and the dream threatens to become a nightmare when she is arrested for murder and brought to court. Her family really *isn't* dead, and with her friends they testify against her, implying that because of her shortcomings as a wife and mother she's been murdering them all along. Her friend Crystal says: "She's loud, she's bossy, she talks with her mouth full. She feeds her kids frozen fish sticks and high calorie sodas. She doesn't have proper grooming habits." And she doesn't treat her husband right even though, as Roseanne explains, "The only way to keep a man happy is to treat him like dirt once in a while." The trial, like the dream itself, dramatizes a struggle over interpretation of the frame story that preceded it: the court judges her desire for the bath as narcissistic and hedonistic, and her barely suppressed frustration as murderous. Such desires are taboo for good self-sacrificing mothers. For Roseanne, the bath (and the "murders" it *requires*) are quite pleasurable for reasons both sensuous and righteous. Everyone gets what they deserve. Coincidentally, ABC was running ads during this episode for the docudrama *Small Sacrifices* (12–14 November 1989), about a real mother, Diane Downs, who murdered one of her children.

Barely into the trial, it becomes apparent that Roseanne severely strains the court's power to impose its order on her. The rigid oppositions it tries to enforce begin to blur, and alliances shift. Roseanne defends her kids when the judge—Judge Wapner from *People's Court*—yells at them. Roseanne, defended by her sister, turns the tables on the kids and they repent for the pain they've caused her. With Dan's abrupt change from prosecutor to crooner and character witness, the courtroom becomes the stage for a musical. He breaks into song, and soon the judge, jury, and entire cast are dancing and singing Roseanne's praises in a bizarre production number. Female desire *isn't* monstrous; acting on it "ain't misbehavin'," her friend Vanda sings. This celebration of Roseanne in effect vindicates her, although the judge remains unconvinced, finding her not only guilty but in contempt of court. Dreamwork done, she awakens, the sound of the judge's gavel becoming Dan's hammer on the plumbing. Dan's job is over too, but the kids still want her attention. Dan jokes that there's no place like home but Roseanne answers "Bull." On her way, at last, to her bath, she closes the door to the bathroom to the strains of the chorus singing "We Love Roseanne."

The requirements for bringing this fantasy to an end are important. First, what ultimately satisfies Roseanne isn't an escape from her family but an acknowledgment from them of *her* needs and an expression of their feeling for her—"We love you, Roseanne." I am not suggesting that Roseanne's series miraculously transcends the limitations of prime-time television. To a certain degree this ending does represent a sentimental coopting of her power, a shift from the potentially radical to the liberal. But it also indicates a refusal to flatten contradictions. Much of Roseanne's appeal lies in the delicate balance she maintains between individual and institution and in the impersonal nature of her anger and humor, which are targeted not so much at the people she lives with as at what makes them the way they are. What Roseanne *really* murders here is the ideology of "perfect wife and mother" which she reveals to be murderous in itself.

The structuring—and limits—of Roseanne's vindication are also important. Although the law is made ludicrous, it retains its power and remains ultimately indifferent and immovable. Roseanne's "contempt" seems her greatest crime. More important, whatever vindication Roseanne does enjoy can happen only within a dream. It cannot be sustained in real life. The realism of the frame story inevitably reasserts itself. And even within the dream, the reconciliation between unruly fantasy and ideology can be brought about only deploying the heavy artillery of the musical and its conventions. As Rick Altman has shown, few forms embody the utopian impulse of popular culture more insistently than the musical, and within musicals, contradictions difficult to resolve otherwise are acted out in production numbers. That is what happens here. The production number gives a fleeting resolution to the problem Roseanne typically plays with: representing the unrepresentable. A fat woman who is also sexual; a sloppy housewife who's a good mother; a "loose" woman who is also tidy, who hates matrimony but loves her husband, who hates the ideology of "true womanhood" yet considers herself a domestic goddess.

There is much more to be said about Roseanne and the unruly woman: about her fights to maintain authorial control over (and credit for) her show; her use of the grotesque in the film *She Devil* (1989); her performance as a standup comic; the nature of her humor, which she calls "funny womanness"; her identity as a Jew and the suppression of ethnicity in her series; the series' move toward melodrama and its treatment of social class. A more sweeping look at the unruly woman would find much of interest in the Hollywood screwball comedy as well as feminist avant-garde film and video. It would take up questions about the relation between gender, anger, and Medusan laughter—about the links Hélène Cixous establishes between laughing, writing, and the body and their implications for theories of female spectatorship. And while this [essay] has emphasized the oppositional potential of female unruliness, it is equally important to expose its misogynistic uses, as in, for example, the Fox sitcom *Married . . . With Children*

(1988). Unlike Roseanne, who uses female unruliness to push at the limits of acceptable female behavior, Peg inhabits the unruly woman stereotype with little distance, embodying the "male point of view" Roseanne sees in so much television about family.

Roseanne points to alternatives. Just as "domestic goddess" can become a term of self-definition and rebellion, so can spectacle-making—when used to seize the visibility that is, after all, a precondition for existence in the public sphere. The ambivalence I've tried to explain regarding Roseanne is evoked above all, perhaps, because she demonstrates how the enormous apparatus of televisual star-making can be put to such a use.

## Notes

With thanks to Ellen Seiter for her helpful comments on an earlier draft of this [essay].

1. Roseanne Barr, *Roseanne: My Life as a Woman* (New York: Harper and Row, 1989), p. 3.

2. The phrase "slouching toward stardom" is Jeremy Butler's.

3. Janet Woollacott, "Fictions and ideologies: The case of the situation comedy," in Tony Bennett, Colin Mercer, and Janet Woollacott, *Popular Culture and Social Relations* (Philadelphia: Open University Press, 1986), pp. 196–218; Patricia Mellencamp, "Situation comedy, feminism and Freud," in Tania Modleski (ed.), *Studies in Entertainment* (Bloomington: Indiana University Press, 1986), pp. 80–95.

4. Patricia Parker, *Literary Fat Ladies: Rhetoric, Gender, Property* (New York: Methuen, 1987).

5. Natalie Zemon Davis, *Society and Culture in Early Modern France* (Stanford: Stanford University Press, 1975), pp. 124–51.

6. Mary Russo, "Female grotesques," in Teresa de Lauretis (ed.), *Feminist Studies, Critical Studies* (Bloomington: Indiana University Press, 1986), p. 217.

7. Teresa de Lauretis, *Technologies of Gender* (Bloomington: Indiana University Press, 1987), p. 109.

8. For example "Roseanne goes nuts," in the *Enquirer,* 9 April 1989, and "My insane year," in *People Weekly,* 9 October 1989, pp. 85–86. Like other labels of deviancy, madness is often attached to the unruly woman.

9. Mikhail Bakhtin, *Rabelais and His World,* trans. Helene Iswolsky (Bloomington: Indiana University Press, 1984).

10. Nancy M. Henley, *Body Politics: Power, Sex and Non-verbal Communication* (Englewood Cliffs, N.J.: Prentice-Hall, 1977), p. 38.

11. Ibid., p. 91.

12. In "What am I anyway, a Zoo?" *New York Times,* 31 July 1989, she enumerates the ways people have interpreted what she stands for—the regular housewife, the mother, the postfeminist, the "Little Guy," fat people, the "Queen of Tabloid America," "the body politic," sex, "angry womankind herself," the notorious and sensationalistic La Luna madness of an ovulating Abzugienne woman run wild," etc.

13. *Roseanne,* p. 51.

14. *People Weekly,* pp. 85–86.

15. Pierre Bourdieu, *Distinction: A Social Critique of the Judgement of Taste,* trans. Richard Nice (Cambridge: Harvard University Press, 1984), p. 208.

# Ambivalent Pleasure from
# *Married . . . with Children*

## DENISE J. KERVIN

While widely considered to be coarse and controversial since its debut in April 1987, *Married . . . with Children* has been the Fox Broadcasting Company's most watched program (Polskin 2). The new network originally presented itself as looking for programming that was somewhat different from the major networks' offerings and specifically aimed at attracting the eighteen to thirty-four and eighteen to forty-nine-year-old middle to upper middle-class urban segments of the television audience, the market prized by advertisers for making major product purchases (Grover, "Rupert Murdoch" 50). While many of Fox's offerings duplicated conventional fare from the Big Three networks, *Married . . . with Children,* produced by Ron Leavitt and Michael Moye, is different. It is a comedy focusing on Al Bundy, "a shoe salesman who hates his life," and his interactions with his wife, children, and liberal neighbors he sees as responsible for his misery (Grover, "Fox's New" 43). Ratings information indicates that the Bundys are watched by the desired target audience and other audience segments as well. While Nielson ratings indicate the show does not pull the numbers the major networks do for their successful programs, *Married . . . with Children* does well by Fox's standards. A major network series is usually doomed unless it wins ratings of at least 15 points, with one ratings point equal to approximately 874,000 homes with television; Fox, however, expected ratings lower than 15 (Block). According to Fox's research department, *Married . . . with*

From Denise Kervin, "Ambivalent Pleasure from *Married . . . with Children,*" *Journal of Film and Video,* vol. 42, no. 2 (June 1990). Reprinted with permission.

*Children* averages a 10 to 11 rating for men and women eighteen to thirty-four years old and an 8 to 9 rating for all viewers eighteen to forty-nine-years old; ratings are also slightly higher in urban markets.[1]

Given that the series is being watched by the target audience, *Married . . . with Children* is interesting for the ambivalent viewing position it offers to members of that group. *Married . . . with Children* creates this ambivalence through its characters and comic situations as they present contradictory ideas about family, class, and in particular, gender. While television programs generally contain contradictions, most work hard to naturalize the conflicts inherent within these social divisions. *Married . . . with Children,* though, seems to use them self-consciously for humor, with the result of creating a potentially uncomfortable viewing position for members of the target audience: eighteen to forty-nine-years old, upscale, likely to be married, and possibly having children. The series does this by addressing different aspects of these viewers' ideological subject positions—the beliefs and experiences resulting from economic and social status, gender, education, politics, that is, virtually all the intersections between person and world. By creating humor from the clash of subject positions represented by the Bundys and the Rhodes, their neighbors, *Married . . . with Children* invites this audience to accept a viewing position of ambivalent pleasure.

*Married . . . with Children* is not the first television comedy series to focus on conflicts and contradictions arising from families, class positions, and gender roles. Other shows that focused on one or more of these social divisions include *All in the Family, Soap, The Honeymooners,* and *Buffalo Bill.* The primary factor that seems to differentiate *Married . . . with Children* from these earlier comedies is the way in which it might be read by members of the target audience. More specifically, *Married . . . with Children* seems to be constructing an ambivalent viewing position for viewers who were raised in lower to upper middle-class families during the 1950s and taught the dominant values and beliefs of that time. As Fiske argues, a person's subject position is already constructed, a product of his or her place in history and society. A childhood in the 1950s, therefore, would help to create for members of this audience segment a subject position that incorporates current hegemonic beliefs, such as the centrality of the nuclear family, attainment of success through hard work, and natural superiority of males over females. However, as Fiske goes on to argue, future experiences may be contradictory to the original self, resulting in the formation of new, potentially oppositional subject positions ("Television: Polysemy" 404). These "more recently acquired and less deeply rooted subject positions can and do conflict with the original, given one" (405). For viewers born in the 1950s, but growing to adulthood in the 1960s and 1970s, the possibility of taking on beliefs oppositional to their "original, given" subject position is great. These are the viewers who began life with dominantly defined, traditional beliefs, but through the ensuing years took on more "liberal" views, such as that men and women are equals and that household tasks should be shared.

*Married . . . with Children* seems to address these audience members' contradictory subject positions, most obviously within the representations of the Bundys and the Rhodes: the former "speaking to" viewers' dominantly defined childhoods, the latter to viewers' more recent, "oppositional" beliefs. The pleasure of *Married . . . with Children* for this audience resembles what Fiske calls "the pleasure of recognition," a recognition of one's contradictory beliefs, one that works to create an ambivalent pleasure (*Television Culture* 51). It is, of course, impossible without empirical research to say exactly how such viewers actually read *Married . . . with Children*. It is also true that the viewing position offered by the program will be inhabited differently by these viewers based on their individual situations. There will always be a range of decodings, even within a group of individuals occupying similar social positions. However, it is possible to examine the viewing position constructed by the show in order to determine how certain textual strategies speak to the conflicts inherent in the members of the target audience's original and more recent subject positions.

A major source of contradiction and humor in *Married . . . with Children* comes from the use of intertextual references, particularly to television and popular culture of the last thirty years. For example, in one episode Bud "helps" Kelly with a book report on Daniel Defoe's *Robinson Crusoe* by supplying her with the premise from *Gilligan's Island* complete with theme song supposedly sung by Crusoe and Friday. The same episode's main plot line revolves around retrieving Marcy's Barbie doll, sold by Steve and Al in order to buy classic Chicago Cubs baseball cards. The plotlines invite members of the audience to draw on their past and present attitudes towards these cultural artifacts, ones perhaps collected in youth, and even collected by their children today. The humor evoked may create some ambivalence for viewers, a result of both remembered fondness for the things of childhood and a critical attitude learned during adulthood, perhaps seeing these artifacts as problematic inculcators of dominant standards of female beauty and male physical power.

An even more fundamental use of intertextual references occurs in the series' presentation of location, set, and characters, all three drawing on and satirizing the conventions of American television's domestic comedy genre. Newcomb describes the prototypical domestic comedy as taking place in established neighborhoods with large houses and big lawns (43–48). The conventional set is comfortable, clean, and gendered: Mom in her kitchen, Dad in the den or living room, the kids in their bedrooms. The characters form a family, whether the traditional nuclear family, or one defined by the workplace. "The real basis for domestic comedy is a sense of deep personal love among the members of the family . . . a sense of groupness, of interdependence" (48).

*Married . . . with Children* takes these conventions and uses them ironically. This irony is evident within each episode and even within the series' opening credit sequence during its first two seasons. Over shots introducing the show's location, set, and characters, Frank Sinatra sings "Love and Marriage," the familiar song extolling the "inevitable" movement from romance to its associated social institution. Meanwhile, the consequences of such a combination are ironically commented on by the visuals. The first shots show Chicago locations, a setting that is referred to within the program at times and which serves to implicitly place the Bundys' economic and social position within the urban lower middle class. Most of the series' action takes place in the Bundy home, shown under the credits as a "modest" split-level ranch. The main set is then introduced: the Bundy living room, located next to the dining room and kitchen, which essentially creates one large room. The design of the set allows both some privacy within the individual areas and interaction between them. As in conventional domestic comedies, this space is gendered. The female characters often converse in the kitchen or dining room, while the male characters interact in the living room; the openness of the entire area allows the sexes to "face off."

The Bundy house and its furnishings do not fit the usual presentation of what Gitlin terms "consumer happiness" provided by most television programs (228). Instead, the sets are reminiscent of those in *All in the Family*—rather old, worn, and never particularly attractive. In the flow from program to advertising, the Bundy home may cause the world of commercials to appear more attractive than usual and, perhaps simultaneously, comment ironically on the perfection of that world of consumer goods. The contrast and potential irony implicit in the juxtaposition of the program and commercials is evident each week as the sponsor's product is superimposed between the embattled Al and Peg sitting at either end of their sofa.

After establishing the location and set of *Married . . . with Children,* the credit sequence introduces the Bundy family, continuing to play off the cultural knowledge and expectations of the conventional television family. Instead of happy parents with cute, happy children, the audience sees Al—husband, father, and service worker—sitting on the sofa mechanically handing out money to each of the other family members (including the dog) as they pass by him. He does this with a resigned and glassy stare, but it is unclear whether he is enduring yet another reminder of what he sees as his primary role in life (money machine to his ungrateful family), or if he is attempting to watch television. Al is presented as an average-looking guy in his early forties, usually dressed in shirt and slacks for work. His informal speech, nasal twang, and less than perfect enunciation connote the Midwest middle class.

Peggy, wife and mother, stands behind the sofa saying goodbye to their children. Bud, in his early teens, often offers a more mature point of view than his parents, when he is not trying to get his sister in trouble. Kelly, a little older, has always been more strongly stereotyped as a lower middle-class female, with her bleached blonde hair and punk fashion. Both children take money from Dad, seemingly a habitual activity. After the children leave, Peg sits down next to Al, who then mechanically hands money over to her. Peg is the most stereotyped and exaggerated of the show's characters. Her clothing harkens back to the 1950s and 1960s, with tight calf-length pants, tight tops, four-inch high heels, and teased hair worn to her shoulders. Peg walks with small, mincing steps, usually holding her arms stiffly with the elbows bent and close to her body. It is a very mannered characterization, connoting a frivolousness associated primarily with the stereotyped female.

The other continuing characters on *Married . . . with Children* are Marcy and Steve Rhodes, the Bundys' neighbors. Newly married and childless when the series began, the Rhodes provide a contrast to the Bundys in terms of class, education, profession, and gender roles. Marcy and Steve belong to the group that Fox is particularly interested in attracting: solidly middle class, upwardly mobile, college educated, and in white-collar careers. Their relationship is depicted as liberal and "enlightened" so, for example, Steve does not (usually) mind that Marcy is an executive at her bank, while he is a teller, or that she makes more money than he does. For certain members of the target audience, the Rhodes present a set of familiar ideas and values, while the Bundys appear so stereotyped they might be labeled the "Hen-Pecked Husband" and the "Ditzy Housewife." As Marc argues in relation to *The Beverly Hillbillies,* such two-dimensional characters as the Bundys would seem to forestall audience identification, and instead encourage viewers to turn their attention to the characters' "outlooks, beliefs, and methods of coping with the world and evaluate these in terms of our own—and official—wisdom" (54). However, for the audience members who have taken on a subject position similar to the Rhodes', but who began life with beliefs closer to the Bundys', the series creates an ambivalent viewing position.

As the above description of the credit sequence implies, the irony of *Married . . . with Children* relies on the viewer's knowledge of conventional television families, a familiarity with characters like the Cleavers and the Andersons. From this perspective, the series can be seen as satirizing the depiction of the conventional happy family in television's domestic comedies. *Married . . . with Children* also focuses on the nuclear family, but as a source of conflict; this family provides little comfort for its members, implicitly questioning the naturalized belief that family love is a given. The first episode of *Married . . . with Children,* for example, set up the way in which the program would continue to treat the discourse on family, as well as class and gender. In the episode, Kelly sits on the sofa as we see a large plant behind it moving towards her. Bud lunges from behind its cover to grab Kelly's head and pull it back by her hair. He holds a knife (obviously a toy) to her neck and says, "Die, Commie Bimbo," thereupon drawing the blade across her neck. Peg ends the confrontation by mildly chiding Bud and complacently sending the children off to school. As they leave Kelly tells Bud she hates him; Bud replies, "Good." Al then enters from the upstairs bedroom carrying a small, potted cactus, asking if it is Peg's. She had placed it where the alarm clock previously was located, but forgot to tell Al, resulting in his bandaged hand. Al says he'll forgive her forgetfulness, as he stopped the bleeding with her slip.

These interactions between brother and sister, parents and children, and husband and wife are part of the Bundy family's "normal" daily life. In its outline, this is the nuclear family, the norm of past domestic comedies;

assumed to be one of the strengths of our society, its "breakdown" is mourned in the media. *Married . . . with Children,* however, problematizes the naturalized belief that family members love and support one another, even if they sometimes fight. The program even suggests that the Bundy family and, by implication, the family in society, is held together largely through economic necessity rather than love.⌈To a certain extent, then, *Married with Children* critiques the traditional family structure by rarely bringing the Bundys into complete harmony, instead continually drawing on their conflicts for humor. This is the reverse of what Feuer argues is the conventional structure of situation comedies, where the narrative traditionally works to reintegrate the "ruptured" family (113); integration appears to be impossible for the permanently "ruptured" Bundy family. *Married . . . with Children* thus introduces contradictions into the discourse of the family, contrasting naturalized notions of love and nurturance with the daily life of the Bundy family.[2] Closely related to the articulation of family within *Married . . . with Children* is the representation of social class through the differential positioning of the Bundys and the Rhodes within the socio-economic system. Again, the program contradicts traditional domestic comedies, those with the implicitly white-collar father who is never seen at the office, the efficient mother who is the careful manager of the home, and the children who, perhaps grudgingly at times, appreciate the life their parents give them. In *Married . . . with Children* the Bundys regularly bemoan their financial and social standing, pointing up their lower middle-class position and rejecting the belief that they will inevitably move up the socio-economic ladder. The show's humor involves the Bundys' subversion of such naturalized beliefs as the virtue of hard work and the selflessness of parents toward their children.

This subversion addresses significant issues within the representation of class as discussed by Steeves and Smith, including issues of control within the workplace, and who possesses education, social skills, and the "proper" cultural values (45–46). For example, in the initial episode, Al describes his job as a shoe salesman as "minimum wage-paying slow death." His lack of pleasure or interest in his job is a continuing element in the program, often juxtaposed to the feelings held by the Rhodes, especially Marcy. Al's ongoing comments concerning work make clear his sense of powerlessness in the workplace, a situation duplicated at home, leading Al to question another naturalized belief: that he is, and should be, working for his family, creating a home for his wife and children. This belief is held firmly by the other Bundys; Al works so they can consume. Again, this is in contrast to the two-career Rhodes, who earn enough to afford luxuries unavailable to the Bundys. This also contrasts with a current trend within television programs, such as *thirtysomething,* which *celebrate* the joys and frustrations of men supposedly involved equally in both home and work.

The Bundys and Rhodes are also juxtaposed according to other indicators of class mentioned above: education, social skills, and cultural tastes. The Rhodes are represented as college-educated, well-mannered, and knowledgable about culture. Their "good" up-bringing, as defined by dominant social values, creates conflicts with the high-school educated, boorish, and culturally illiterate Bundys. For some members of the target audience—upscale, educated, eighteen to forty-nine-years old—a potentially ambivalent viewing position is created, however. For those who have moved from a dominantly-defined subject position to one of a more oppositional nature, the Rhodes' liberal ideas and life-style may seem familiar, inviting identification. Besides offering a humorous contrast in attitudes to the Bundys, Steve and Marcy can be seen as figures of identification for this audience, their reactions to the Bundys perhaps felt to some extent by many of the prime viewers addressed by Fox.

If these viewers do identify with the Rhodes, perhaps enjoying with the couple a slight feeling of liberal superiority over the Bundys, the program soon works to call that superiority into question. While the Bundys come off as uncouth and uncultured compared to the Rhodes, the latter couple often is presented as naive, snobbish, and lacking in knowledge about how the world really works. The viewers who identify with Steve and Marcy may end up uncomfortable with the Rhodes', and their own, assumed superior position because the couple is often made to appear as, in fact, less conscious and critical than the Bundys of how the economic and social systems in which they are enmeshed really operate. The Bundys display a realization of the inevitability of their socio-economic position, rejecting the American dream of success through hard work. Their representation also contains an awareness of the dynamics between family units and the economic system, pointing to consumerism as perhaps the main source of familial cohesiveness today. The show thus presents the Bundys as openly aware of and dissatisfied with their position, although they still desire the "rewards" promised by consumer capitalism. In contrast, the more solidly middle-class Rhodes can literally and figuratively afford to "buy into" that system.

Thus far it has been argued that *Married . . . with Children* "speaks to" the subject position of some members of its target audience in terms of childhood memories of television and popular culture, using that intertextual knowledge ironically in the construction of the show's characters and narratives. The potential for a stable subject position is offered in the characterization of the Rhodes; however, this position is consistently undermined and replaced by a certain ambivalence. Even though the couple is upwardly mobile and well-educated, conveying generally positive connotations, they still often end up looking less attractive within the series' articulation of the discourse of class. In the show's contrast of the Bundys' and Rhodes' social attitudes towards male and female roles, however, the class differences established *between* the households become secondary to gender differences *within* the couples. As class divisions give way to gender divisions, the Rhodes gradually display a similarity to Peg and Al; the "enlightened" couple displays a dominantly defined core.

The representation of gender within the characters of the Bundys draws on dominant stereotypes associated with males and females within our society. This stereotyping is blatant, especially in the representation of Kelly and Peg. As Steeves and Smith found in studying gender and class in prime-time television, there is the association of lower-class females with overt and promiscuous sexuality (51), a continuing source of humor in the depiction of Kelly's "easy" virtue. Likewise, *Married . . . with Children* uses the stereotyped assumption about the lack of value of women's work within the home. A continuing intersection between the discourses of family, class, and gender concerns Peg as the inverse of the conventional television wife and mother, that nurturing, selfless angel, first mate to her husband's captain, guide to their children in the proper ways of society. Instead, Peg's character draws from a different, negative store of stereotypes associated with wives and mothers: sarcastic, lazy, manipulative, vindictive, emasculating. In the initial episode Peg is shown lying on the sofa, eating candy, reading a magazine, smoking, and watching a television talk show examining an Amazon tribe in which the women devour the men after the mating season. (Emasculation is worldwide.) The program's continuing commentary on women's work (or lack thereof) within the home is represented here as Peg, hearing Al arrive, quickly starts to clean up evidence of her inactivity.

While Peg is presented as a failure in the conventional role of the "good" housewife, she is adept at understanding and using the system of "tricks" taught implicitly to females, enabling them to get some of what they want despite a lack of access to male-dominated forms of power. For example, Peg uses consumerism as therapy and as a form of coercion. In the initial episode, Al tells Peg that he plans to go to a basketball game that evening instead of meeting the Rhodes for the first time. Peg replies: "No you're not." Al: "I work all day. And when somebody works all day, they need to have some fun at night. Now, I don't actually expect you to understand any of this, but trust me, I'm your husband. I know best." Peg: "The bank book is in both of our names. The credit cards are in both of our names. And the stores are still open." Peg's characterization combines the traditional placing of the female, particularly within marriage, with the accommodations and small resistances that women make within subservient positions.

Slightly less blatantly stereotypical, the representation of Al and Bud still goes to the heart of masculine representation—power—in the use of negative characteristics associated with males. A continuing source of humor involves Bud's trouble attracting girls, a great embarrassment to him but happily pointed out by Kelly. Al is presented as the beleaguered and essentially emasculated male, both at work and at home. The cause of this emasculation is mainly women, specifically Peg at home and the stereotyped (overweight, overbearing) female customers Al deals with at the shoe store. In addition, the traditional power attached to being the provider, having the "real" (outside the home) job, is undercut by a family that takes such provision for granted and does not think very highly of Al's earning power.

The program's humor in part depends on Al's attempts to regain and exhibit power through his demands on Peg and resistance at work; even if sometimes successful, the show implies these efforts are only temporary.

The representation of the Rhodes as liberal, especially in their definition of gender roles, provides a continuing source of friction between the couples. In the first episode Marcy and Steve are presented as a "progressive" couple through their (especially Marcy's) rather ostentatious comments about their mutually defined, equal relationship. Al's response is to conclude that this is yet another example of a man allowing his naturally dominant position to be undermined by a woman. Worse yet, when Al asks Steve which team he favors winning the NBA championship, Steve explains that Marcy does not like sports and they decided to do only those things they both enjoyed; ergo, no sports. Marcy describes how sports glorify violence and competition, continuing: "When we have a child we don't want it to grow up with the 'winning is everything' attitude. A child is better off not being exposed to sports." Al's reply: "You gonna neuter him too?" Both Al and Peg, ideological counterparts, find Al's remark very funny, while Steve and Marcy look at one another with appalled expressions. The audience is thus treated to two essentially oppositional views of "proper" gender roles and married life, each exaggerated for humor, but each recognizable as present within society. Audience members occupying the viewing position that finds points of similarity with the Rhodes may share in that couple's astonishment at the sexism and amazingly poor social skills exhibited by the Bundys. However, for the Rhodes and perhaps for the viewer, that feeling of moral and political superiority may not last long.

While this exchange has initially separated the couples along class lines, the Bundys decide individually to share the benefit of their wisdom with the Rhodes, shifting the episode's conflict from one of class differences to one of gender differences. The females go to the kitchen together to make coffee, while the males stay in the living room. Al begins by complaining about how women always ask what men are thinking. Al's reply: "If I wanted *you* to know I'd be talking." Meanwhile, in the kitchen Peg is ladling tablespoons of instant coffee into cups, explaining that the way to get men to take you out is not allow them to enjoy eating and drinking at home too much. Peg goes on to question Marcy about whether Steve is staying up after Marcy has gone to bed. Discovering that he is, Peg warns Marcy that she is "letting him slip away" because this is a step toward allowing Steve to have fun *alone,* without her. Al, meanwhile, is telling Steve that without sports his son will grow up to be a "sissy-Mary," prompting Steve to nostalgically remember how he "used to love sports." Al: "Of course you did, you're a man."

The result of these interactions is to bring out the previously hidden, dominantly defined core within Steve and Marcy, tearing away their liberal veneer and moving them towards agreement with Al and Peg. This shift is marked visually through gestures associated earlier in the episode with the Bundys: Peg's unconscious leg kicking becomes a sign of Marcy's anger at

Steve; Al's habit of putting his hands down his pants when relaxing in front of the television becomes a sign of Steve's rebellion from his no-sports agreement. As Marcy demands to know what Steve does after she goes to bed, the previously "enlightened" pair gradually sound more and more like the Bundys. After the Rhodes storm home to "redefine their relationship," the Bundys comment on how it will be hard for the young couple, but then agree that though it was hard for them too, they made it. Sitting close together on the sofa, Al asks Peg if she wants to "go upstairs." As they climb the stairs together, Al pats Peg's derriere. Thus, as the Bundys draw out the hidden core of dominant ideas present under the Rhodes's superficial equality, Peg and Al are reconstituted as a couple. Having reasserted stereotyped gender roles, the Bundys share a rare moment of agreement and affection.

While the series has changed somewhat since its inception, the movement of Steve and Marcy from representing liberal ideas and gender equality to being Al and Peg's ideological counterparts is an ongoing source of humor within *Married . . . with Children.* This transformation is usually helped along by the Bundys, as in the first episode, although the Rhodes always return the following week with their liberal views restored. The characters of Marcy and Steve seem to exist partially to clash with and draw attention to the Bundys dominantly structured positions, but then to reveal their own hegemonic heart of hearts. They may act like liberals (their position becomes no more oppositional than that), but their roots are in the fertile ground of the dominant ideology. As the program "speaks to" the subject position held by members of the target audience, the ambivalent viewing position constructed seems to pivot on this use of the Rhodes. Initially, viewers' newer, more liberal social practices, beliefs, and values may encourage identification with the Rhodes. However, *Married . . . with Children* also addresses these viewers in terms of where they have come from through the beliefs and values held as true during their dominantly defined childhood, beliefs and values ironically represented by the Bundys. For such viewers, identifying with the Rhodes creates an ambivalent viewing position. As Steve and Marcy's liberalism is stripped away, displaying their original subject position formed within dominant ideology, the inherent contradictions between both characters' and audience members' "layers" of beliefs emerge.

How individual viewers respond to *Married . . . with Children* and its negotiation of subject positions requires further investigation. It seems that the series works towards producing an ambivalent pleasure for members of the target audience. Relying on familiarity with conventional television characters and storylines, and with reference to current discourses on gender, class, and the family, this segment of the show's viewers may read the show as satirizing hegemonic beliefs. The program could be understood as focusing on contradictions within definitions of masculinity and femininity, socio-economic positions, and the traditional nuclear family, a process in keeping with liberal ideals. For viewers holding such ideals, the program may offer both humor and what might seem like an oppositional social consciousness. However, while *Married . . . with Children* seems to offer ideological satire, liberal beliefs are also used for humor, especially as they are recuperated into the dominant ideology through the characters of the Rhodes. Viewers may therefore find themselves "caught" between their past and present subject positions, between the comfortable ideas learned as a child and more recent, perhaps less ingrained ideas acquired as adults. The program may allow these viewers, if willing, to examine the contradictions within themselves)

## Notes

1. Information on Fox's performance and audience comes from a telephone conversation with a representative of the Fox Broadcasting Company's research department on 11 May 1988, as well as ratings information provided by the Nielsen Rating Service.

2. It should be mentioned, however, that every so often *Married . . . with Children* does acknowledge that there are emotional ties that bind the Bundys together, though this acknowledgement is usually emphasized as out of the ordinary. For example, Fox invited viewers on the 1988 Valentine's Day episode to decide whether Al would tell Peg he loved her. That dilemma was left unresolved within the program itself; the voted-on scene was shown later the same evening, during another Fox program. Such a ploy was a clever device to encourage viewers to continue watching Fox programming and an interesting chance to see if the audience would vote for continuing the "normal" combative relationship between Al and Peg or vote them into a more conventionally loving relationship. The majority of viewers calling in chose to have Al express his affection; unfortunately, no tally of votes was given, so it was impossible to gauge the strength of the potentially oppositional "con" votes. In general, though, such an acknowledgement of affection is a rarity within the Bundy family.

## Works Cited

Block, Alex. " 'Hey Gang, Let's Put On a Show.' " *Forbes* 6 April 1987: 140+.

Feuer, Jane. "Narrative Form in American Network Television." *High Theory/Low Culture: Analyzing Popular Television and Film,* Ed. Colin MacCabe. New York: St. Martin's Press, 1986. Pp. 101–14.

Fiske, John. *Television Culture.* New York: Methuen, 1987.

———. "Television: Polysemy and Popularity." *Critical Studies in Mass Communication* 3.4 (1986): 391–408.

Gitlin, Todd. "TV's Screens: Hegemony in Transition." *Cultural and Economic Reproduction in Education.* Ed. Michael Apple. Boston: Routledge, 1982. Pp. 202–46.

Grover, Ronald. "Fox's New Network Goes After the Baby Boomers." *Business Week* 6 April 1987: 41+.

———. "Rupert Murdoch, Can We Talk?" *Business Week* 1 June 1987: 50.

Marc, David. *Demographic Vistas: Television in American Culture.* Philadelphia: University of Pennsylvania Press, 1984.

Newcomb, Horace. *TV: The Most Popular Art*. New York: Anchor Books, 1974.

Polskin, Howard. "Does *Married . . . with Children* Go Too Far?" *TV Guide* 29 July 1989: 2–5.

Steeves, H. Leslie, and Marilyn Smith. "Class and Gender in Prime-Time Television Entertainment: Observations From A Socialist Feminist Perspective." *Journal of Communication Inquiry* 10.3 (1987): 43–63.

# Kodak's "America": Images from the American Eden

## HAL HIMMELSTEIN

*"The illiterate of the future," it has been said, "will not be the man who cannot read the alphabet, but the one who cannot take a photograph." But must we not also count as illiterate the photographer who cannot read his own pictures? Will not the caption become the most important component of the shot?*

Walter Benjamin, "A Short History of Photography" (25)

*Anything can be associated with anything else for a viewing subject who is structured by the rhetoric of the commercial.*

Mark Poster, "Foucault, the Present and History" (120)

During 1984 and 1985, the J. Walter Thompson U.S.A. advertising agency conceived and produced the "Because Time Goes By"[1] national television advertising campaign for the Eastman Kodak Company. J. Walter Thompson has serviced the Kodak account, its second oldest account, since 1930.

The campaign comprised nine spots. Five were produced in 1984: "Reunion," "Music Makers," "Baseball," "Olympics," and "The Gift." Four others were produced in 1985: "America," "Henry, My Best Friend," "Old Lovers," and "Summer Love."[2] The campaign was "transitional" for the client. It marked a return to Kodak's tradition of warm, often lyrical "emotional appeals" commercials without completely abandoning the "product benefits" focus of recent campaigns.[3] Among those featured in the new spots were: young lovers on the beach sharing their memories in photographs as their summer of love comes to an end; elderly couples expressing their continued love; a little boy whose puppy quickly grows to a giant dog while

From *Journal of Film and Video*, vol. 41, no. 2 (Summer 1989). Reprinted by permission of the author.

the boys grows very little; and a Christmas commercial featuring a young child giving a gift to an elderly black man.

The following discussion focuses on one of these spots, "America," which was first aired in the Spring and Summer of 1985 in both a sixty-second and a thirty-second version. "America" was chosen for study because it represents a dominant strain in television commercials from the period which employed themes of American patriotism and restoration. It was also chosen for study because the apparent rupture between its seamless aesthetic and its socially provocative imagery opens up an important discourse on the process and power of symbolization and mythification in our public imagery.

The resulting analysis is grounded in both a close reading of the text of "America," and a series of telephone interviews conducted in 1987 with key individuals in the agency's creative department and the client's Consumer Products Division's Marketing Communications Department who were intimately involved in the creation of the spot.[4] The interviewees were very open in their responses to all questions regarding the creative and administrative aspects of the commercial's production. At Kodak's request, financial information regarding the cost to produce the commercial was not forthcoming.[5] While the interviews provide important insights into the "mind" of the commercial and the minds of those who conceived it, one must be careful to separate creator intentionality and the text itself as viewed.

### *The Social Context of "America"*

The "America" spot was part of an advertising trend. The advertising industry's success is dependent in large measure on its ability to interpret a "complex, multi-layered, fast-changing society," and to follow the society's emotional mood as quickly and accurately as possible (Stevenson D4). This is especially true for commercials seeking a heterogeneous national audience. When an ascending mood is perceived, agencies and their clients jump on the bandwagon; when it passes, they jump off. Themes and symbols representative of those themes are appropriated from the store of existing public imagery to tap the perceived mood. The confluence of two national events in 1984—the Summer Olympic Games in Los Angeles and Ronald Reagan's presidential re-election campaign—provided advertisers a clear signal that a patriotic mood was sweeping the country.

The swell of national pride surrounding the 1984 Los Angeles Games was an extension of the patriotic fervor generated by the underdog American ice hockey team's unexpected victory over the Soviet team four years before in the 1980 Winter Olympic Games in Lake Placid. Olympic competition, always a source of nationalism among competing countries, has for years been a forum for Cold War ideological warfare between the United States and the Soviet Union. The U.S. boycott of the 1980 Summer Olympics in Moscow, following the Soviet invasion of Afghanistan, and the Soviets' reprisal boycott of the Los Angeles Games are clear examples of the intrusion of international politics into sports.

Motivated by national ideology, the Carter administration in 1980 took the moral high ground of a nation of "free" people supporting the cause of a nation oppressed by a totalitarian regime. The 1984 Los Angeles Games provided Americans another opportunity to promote the American way of life in relation to the Soviets. In this case the ideological frame was economic. Rather than massive government support for the construction of sport facilities for the Games, American private enterprise contributed large sums of money for such construction. Corporate public relations considerations aside, the privatization of the Los Angeles Games provided a convenient symbol of the contrast between the Soviet athletic program, with its well-financed yet subservient athletes carrying out the goals of the state on the playing fields, rinks, and arenas, and the American program in which athletes are autonomous "free spirits" nurtured by free enterprise. This contrast became a subtext of many products' advertising campaigns saluting America's Olympic athletes.

The Reagan political spots drew heavily on the theme of national revitalization reflected in the American entrepreneurial spirit. The campaign's optimism seemed myopic to many, given evidence to the contrary of America's waning economic empire. Americans continued to lose jobs in the heavy manufacturing and high-technology electronics sectors of the economy to overseas competitors, foremost among them the Japanese; and, ironically, the farm crisis in America's heartland was reaching devastating proportions just as the Reagan commercial campaign was presenting images glorifying rural American life. Compounding the individual despair caused by unemployment in the industrial and high-technology sectors, and the increasing loss of family farms, was a general perception of America's loss of international prestige. Viewed in the context of the American military defeat in Vietnam and its embarrassment in the Iranian hostage debacle, America's economic decline took on added psychological importance. The central theme of the Reagan advertising campaign, that "Life is better . . . America's back . . . and people have a sense of pride they thought they'd never feel again" (Stevenson D4), became the anthem for a nation gripped by malaise and apprehensive of its future.

Suddenly product campaigns extolling patriotic values cluttered the electronic landscape. Notable among them were Chrysler's "The pride is back, born in America," and Miller Beer's "Made the American way, born and bred in the U.S.A.," both of which drew directly on the Reagan campaign theme. Kodak's "America," while eschewing the overt symbolism of Chrysler's giant American flag and the equally overt lyrics of both the Chrysler and Miller spots, nonetheless drew heavily on the Reagan message, and for good reason, as we shall see.

## Conceiving the Spot

In 1984, according to a J. Walter Thompson account supervisor, Eastman Kodak "was feeling somewhat vulnerable" because 10 percent of its film business in the United States had been taken away by Fuji Photo Film USA Inc., a Japanese film and videotape manufacturer that had entered the U.S. market in 1970.[6] Fuji had developed a film stock with color reproduction characteristics different from Kodak's. While Kodak had maintained its emphasis on the faithful rendition of skin tones, Fuji emphasized the bright colors (some people at Kodak called them "garish") to which the viewers of color television had become accustomed. Fuji had also taken the lead in product marketing, introducing multi-roll packages of film and huge, brightly colored point-of-sale advertising displays.

There was a difference of opinion in two camps within Kodak's Marketing Communications Department as to whether a new campaign should stress product benefits or brand image.[7] Kodak's top management historically "had a clear prejudice for brand image maintenance. It was proud of its image."[8] Central to Kodak's decision to stress brand image in the campaign was the reality that Fuji had a more successful product on the market—"Quite honestly, Kodak wasn't in a position to make grand product performance claims because Fuji, at the time, had surpassed Kodak in product quality"—and the agency's belief that "nine out of ten Americans shoot film to save their memories, not to worry about the industrial quality of film stock."[9]

Kodak decided to take the high ground, a smart move since its major competition was Japanese and the campaign coincided with the political window during which a patriotic theme would be well received. The "America" spot would "ride the tide of the new patriotism, a place that only Kodak could be."[10] The patriotic approach would allow Kodak to capitalize on the "heritage" of the brand.

Market research indicated that viewers had come to expect emotional appeal advertising from Kodak, spots with themes such as a mother's love for her son. At the same time, Kodak was trying to avoid the "trap of being [perceived as] a sentimental brand." It was committed to "moving its image forward without becoming high-tech or punk."[11]

Kodak's headquarters are located in Rochester, New York, a "stable environment out of the cultural 'firing line' of New York City," and this may account in part for its conservative mindset. According to a former advertising agency executive and long-time observer of the industry, Kodak has a "deeper sense of responsibility about the country as a whole" than a company headquartered in New York City or Los Angeles, which tends to be "more cynical." "A company's personality," he added, "is the key to the way commercials are [ideologically and aesthetically] constructed."[12] An account director for J. Walter Thompson, who was assigned the Kodak account after "America" had been shot, described Kodak's corporate philosophy:

The company has a Midwestern mentality, a general outlook on life and conception of American society that is safer than most New York City-based companies, Kodak takes a more conservative approach. It is willing to be a little more provocative in terms of shooting style, but not in terms of politics. It reminds me of my Dow Chemical account in Indianapolis. It's not New York City.[13]

Kodak's Director of Marketing Communications for Consumer Products agreed with this assessment, saying Kodak "goes out of its way not to be controversial." If controversy arises, he added, it is "by accident."[14]

Not only does Kodak eschew the high-tech look which has become a signature of many contemporary brand image commercials; it also does not want to be perceived as a "funny" company. A recent example offers proof of this mindset at work. In 1985, J. Walter Thompson shot a spot called "Ostriches" as part of a campaign for Kodak's 400 film. The spot started on a close-up shot of jockeys lining up for the start of a race. As the race began, the shot widened to reveal the jockeys astride ostriches, not horses. While the spot post-tested very well, Kodak's Director of Marketing Communications for Consumer Products insisted that "humor is not an emotion" in Kodak's lexicon. The spot never aired.[15]

Kodak's cautious corporate philosophy is reflected in its formal, highly-structured commercial approval process. It was through this process that the "Because Time Goes By" campaign and the "America" spots were transformed from ideas to completed commercials. Kodak's formal approval process, while rare in the advertising milieu of the 1970s,[16] has become far more common in the 1980s where, with increasing conglomeration, there is "a less clear [corporate] attitude toward larger visions of America" and an increased bottom-line consciousness.[17] In Kodak's case, television advertising is "so important for presenting an image of the company" that all the various levels of Kodak's management are involved in deciding the thrust of a campaign.[18]

In developing television spots for its film products, Kodak's first decision-making level is the Marketing Communications Director for Film,[19] who oversees the advertising production process for commercials, print, and outdoor advertising of all film products. Generally in attendance at this initial meeting between client and agency staff are, for the client, the Marketing Communications Director for Film and "communications specialists" who will oversee the actual production of the spots and monitor costs, and for the agency, several key members of the creative team. This meeting is intended to narrow down a number of ideas for the proposed spots to those that have special merit.

The Marketing Communications Director for Film makes a recommendation to Kodak's Director of Marketing Communications for Consumer Products (including, but not limited to film products). This stage is "the most important buy-in. Management above [this level] will usually buy the Marketing Communications Director's judgment."[20] It was at this level that the debate between brand image and product-benefits approaches for the campaign took place, with brand image carrying the day.

The next level of decision-making resides with Kodak's Marketing General Manager for Film. Here we move from Marketing Communications to overall marketing management. Storyboards and related creative materials are presented by the agency. A "go-ahead" is given for the shoots. The succeeding stages of Kodak management decision-making are generally "for review only." These include Kodak's General Manager for the Consumer Products Division, and the Group Executive Vice-President for the Photographic Products Group. Once the spot is completed, it is presented to Kodak's Chairman, Vice-Chairman, and President for *pro forma* approval. This is often done the day before the spot is scheduled to premiere. The purpose of the review is to familiarize the corporate officers with the spot or spots so that they can intelligently respond to any feedback they might receive from viewers. Of course, "a look from the Chairman" is worth a thousand words. If he doesn't like it, it won't air.[21]

The team assembled by the agency to produce a spot generally includes: a Vice President and Creative Supervisor, primarily an administrator, who assigns creative people to the account and acts as liason with the client; an Account Supervisor; a Group Creative Director who develops ideas for the spot (in the case of "America," the Group Creative Director was selected to direct the shooting of the spot; J. Walter Thompson often goes outside the agency to hire commercial directors); another Group Creative Director who writes music and lyrics for the spot; an Executive Producer who supervises the production, hires a production house to do the shooting, and monitors production costs (on the "America" shoot, the Executive Producer functioned as the Second Unit Director and also shot stills that were used in the spot); and an Art Director who does most of the casting for the spot.

It is difficult to credit any individual in a group creative effort, such as that which resulted in "America," as the sole originator of an idea or theme. We know that in a collective endeavor, ideas are generated by numerous informal dialogues among key participants as well as in more formal meetings. This is especially true in the creative departments of advertising agencies. According to one observer with many years of first-hand agency experience as an executive of one of the country's leading agencies:

> One probably won't get an answer as to the exact process in which an idea is constructed, because decisions get made in a non-decisional way at most agencies. The crystallization of the decision-making process is announced in a very brief memo from the agency's account executive: "We've decided to go forward with this idea." No one knows exactly who "we" was.[22]

Recognizing this difficulty, the author conducted telephone interviews, ranging in length from twenty minutes to an hour-and-a-half, with seven key players involved in the creation of "America" in an attempt to develop as clear a picture as possible of the gestation of the images, music, and lyrics that constitute the finished product. These interviews occurred two years after the spot's completion and may suffer to some extent from the problem of participant recall of specific facts and the exact sequence of events. In addition, as with any highly successful group creative effort, some participants may claim credit for others' ideas. The author sought to avoid this potential problem by cross-checking participant accounts and asking follow-up questions. In each instance of an unresolved discrepancy, and there were but three such cases, the author accepted as the accurate account that on which two or more participants concurred. Through this process a clear portrait of the spot's development emerged.

The J. Walter Thompson creative team was in California working on another project when word came from New York to develop an idea with a "patriotic theme" for the "America" spot. According to Greg Weinschenker, who had been assigned to direct "America," "they sent back an idea with people saluting the American flag." Weinschenker found that idea "boring." He suggested instead a motorcycle journey across the United States. His suggestion was autobiographical. In 1970 Weinschenker had travelled across the country on a motorcycle from New York City to Albuquerque, New Mexico, with his dog. Unlike "America," Weinschenker's journey "was not that romantic. In fact it was physically painful. It took two or three weeks."[23] According to Weinschenker, Kodak wanted the spot to show more color than the previous spots in the campaign, most of which were shot indoors, so the idea of an outdoors shoot was appealing.

Kodak's Marketing Communications Director for Film at the time, Bruce Wilson, was clear from the outset of discussions with agency staff as to the "concept and feeling" Kodak sought to achieve with "America," as well as its style. It was to be a "vignette" spot. Rather than a coherent narrative, a vignette commercial is a sequence of "slightly interlocking little scenes and situations" (Arlen 7).

The agency's creative team went to work developing a storyboard based on Weinschenker's and Kodak's conceptions. Weinschenker sat down with Linda Kaplan, an agency Group Creative Director and lyricist, to flesh out the idea. Kaplan and Weinschenker had worked closely on many projects prior to "America." Kaplan proposed the idea of a Vietnam draft evader returning from Canada following years of exile to see his family and reunite with his old girlfriend. Not only did Weinschenker find this idea too controversial, but he said "the rider would be too old." He wanted the rider to be "youthful, not jaded." According to Weinschenker, "the character was to have no pre-conceived notions about anything."[24] Had the agency proposed the draft evader idea to Kodak, it would never have entertained such a storyline, according to Kodak's Bruce Wilson.[25]

The "board" that emerged showed a young motorcycle rider going across the country, meeting different people. One original concept had the cyclist coming into a big city, Los Angeles, at the end of his journey. In contrast to the warmth and friendliness of the "country" he had seen, Los Angeles was a "cold place." Sitting on Hollywood Boulevard with tears in his eyes, the motorcycle rider internalizes the contrast.[26]

The precise storyboard for the spot was not that important to Kodak in giving the go-ahead to begin production. Director Weinschenker's "loose" shooting style, developed in many successful Kodak spots prior to "America," was well known to the client, which voiced no objection to such an approach. Kodak allowed the agency creative group sufficient latitude to pick up shots on location as opportunities for unplanned scenes became available. Because "America" was part of a larger "marketing package"—stills from the shoot were to be used for print advertisements in magazines in addition to their use in the commercial—the precise budget for the shoot was not of critical concern to Kodak.[27]

### *Shooting the Spot*

Director Greg Weinschenker emphasized seeking "real people" rather than actors for the vignettes that would comprise the spot. The motorcycle rider protagonist would be a professional actor. The people he encountered on his journey across America would for the most part be cast from non-actors who lived in the places he visited.

Casting began. Greg Blanchard, the actor chosen to play the cyclist, was cast in Los Angeles. He "had the right spirit" according to Weinschenker. Blanchard's youthful, clean-cut appearance was ideally suited to the 'depoliticized" image both agency and client sought. The "clean cyclist" would maximize the viewer's identification and provide a functional hero, an ego-ideal. Conversely, the viewer's potential anxiety and defensive response to the iconography of the cyclist-as-anarchist, cultivated through decades of B-grade motorcycle gang movies and culminating in the anti-establishment, pot-smoking anti-heroes of *Easy Rider,* would be minimized (Steinbock 64–66). A professional motorcycle rider was cast as Blanchard's double. He would appear in panorama shots picked up by the Second Unit, headed by Executive Producer Sid Horn.

Sites for shooting were selected by agency Art Directors Kathy McMahon and Marisa Acocella. Primary sites were Oregon, Arizona, and Los Angeles. The rural locations would provide the palette of outdoor colors that Kodak desired—the lush greens and golds of the Oregon countryside, and the earthy browns and reds of the Arizona desert and mesas. Los Angeles would provide the contrasting image of big city "coldness."

These West Coast locations offered additional economic benefits. Agency Executive Producer Sid Horn had hired Ron Dexter, of Dexter, Dreyer and Lai, a Los Angeles-based film production house, as Director of Photography for "America." Art Directors McMahon and Acocella went out into the field two weeks before the crew to do the initial pre-casting of the "real people" Weinschenker sought. Weinschenker asked them to "just find the right people and the right look" for the scenes he had planned. Most of the final casting decisions were made by the director on site.

Weinschenker and Dexter headed the First Unit, which shot all the scenes featuring interactions between Greg Blanchard and the characters he encountered in Oregon, Arizona, and Los Angeles. Kodak Marketing Communications Specialist Ann Winkler, the client's representative on the shoot, worked with the First Unit. Executive Producer Sid Horn headed the Second Unit, which travelled south from Oregon to Monument Valley, Utah, picking up shots of Blanchard's double riding through the countryside. In addition to the film footage shot for the spot, Weinschenker, Horn, and a Kodak photographer shot stills, which were subsequently used in the spot and in magazine advertisements.

The persons appearing in the spot, and those who were cast and shot, but did not appear in the final version (the "out-grades"), were paid between $300 and $600 a day, depending on the length of the shooting day. Because the spot was so successful, airing over the course of many months in the national market, those appearing in major roles in the spot's final version made between $15,000 and $20,000 with residuals.[28]

The motorcyclist's first human encounter in the spot, with the elderly woman, was not storyboarded. While the crew was shooting the Oregon school bus scene (which became the spot's concluding vignette) the elderly woman wandered out of her house to see what was happening, then disappeared back into the house. Weinschenker spotted her, stopped the shoot, and ran after her. Her daughter, who lived in the house, talked with the director and got her mother to play in the spot. The elderly woman was totally deaf. As it turned out, the elderly woman's great-granddaughter was one of the children getting off the school bus. Weinschenker set up the two-shot of Blanchard and the elderly woman at the rural mail box. Blanchard began to point, and the elderly woman, picking up the cue, also pointed. The shot was in the can. According to Weinschenker, the elderly woman "never did know what was going on."

The scene of the motorcycle rider sitting with a bearded man on a bench in front of a country store and reaching out to shake the collie's paw was shot in an old Arizona town fallen on hard times. The town, once thriving because of its proximity to the mining industry, was now a virtual ghost town. The shot had the look of the "old west" we have come to expect in our popular imagery of the area.

The pick-up truck scene with Blanchard and the two black men was storyboarded before the shoot began. It was shot in Oregon. The original concept for this vignette had the three men sitting in the truck bed filled with hay. One of the black men was playing a harmonica, while Blanchard played guitar. This was shot, but Kodak's Winkler did not like it. She felt that "portraying blacks as musicians was a cliche."[29] Kodak was very careful not to represent groups in ways that some might consider stereotyped or demeaning. The scene was reconceived at the site. Weinschenker had the three men begin to unload the hay, which quickly turned into the playful hay-tossing scene used in the completed spot. Another version of the scene was shot, but not used. In this version, some black teen-agers had gotten their pick-up truck stuck in the mud. Blanchard stopped to help them free the truck. The consensus among the crew was that this shot didn't work.

When asked about the appropriateness of casting blacks in such a setting, Weinschenker replied "We had to have blacks in the commercial. If you have six people in a spot, one has to be black."[30] This statement should not be taken literally as evidence that racial quotas are employed by the agencies or clients. It does, however, generally reflect agency and client sensitivity to demographics. One longtime agency insider noted that "everyone in the ad business is very familiar with and sensitive to demographics," and that there is a "racial mindset" at the agencies.[31]

The question of the portrayal of blacks resurfaced a year later at Kodak. J. Walter Thompson had cast a black saxaphone musician in one of the spots in Kodak's "The Color of Life" campaign. Before giving the go-ahead for the shoot, Kodak insisted that the role be recast with a white musician in the role, arguing, as had Winkler on the "America" shoot, that the black musician was a cliché.[32]

The scene of the cowboy greeting the motorcycle rider along the roadside was shot in Arizona. The cowboy was a real cowboy cast at the site. Similarly, three Vietnam veterans were lined up at a veterans hospital in Portland, Oregon, as possible actors in the scene where the motorcyclist offers a friendly handshake to a disabled veteran confined to a wheelchair. (None of the veterans who were actually physically disabled would agree to appear on camera in a wheelchair.) It rained constantly in Oregon during the time scheduled to shoot the scene, so the veteran's shoot was moved to Arizona.

The First Unit found Richard, the veteran who appeared in the scene, in a Phoenix, Arizona, veterans clinic. He was the only veteran who wanted to do the shot. While not physically disabled, he did have emotional problems. Like the other veterans, Richard didn't want to appear in a wheelchair but, said Weinschenker, "he needed the money." Weinschenker added, "he was a nice guy who couldn't cope with society. He had done drugs." Winkler was not comfortable with placing a veteran who was not physically disabled in a wheelchair (the contradiction between the artifice of such a representation and Weinschenker's professed desire to use "real people" in the spot was all too apparent—the veteran was "real," the nature of his injury was not). Winkler suggested the scene be shot two ways, one in the wheelchair and one not, and that the decision on which to use be made in editing. Weinschenker did a number of takes of Richard in the wheelchair, but before he could get the second shot without the wheelchair it started to rain, so that shot was never done. Winkler said, partly in jest, partly with conviction, "Greg must have planned it that way."[33] During the shooting of the scene, Weinschenker said, the veteran started to cry, although this version is not the one used in the completed spot.[34]

The Native American was cast in Arizona. McMahon wanted to use Will

Sampson, the powerful Indian who starred in the film *One Flew Over the Cukoo's Nest,* but he was unavailable. Instead, she found an Indian actor friend of Sampson's. The Indian was the head of an entertainment family who travelled the world doing tribal dances. He also worked with Indians with alcohol problems. During the shoot, Weinschenker wanted the Indian to hug the motorcyclist, but he discovered that Indians "don't like to hug." Instead, the scene as shot became a lighthearted conversation between the two men seated on a grassy hilltop.

The scene of the farmer on the old tractor following the cyclist down the two-lane rural highway was shot in Oregon. The final scene in the spot is a shot of the motorcyclist stopped on a country road as children alight from a yellow school bus and cross in front of him. The last child off, a small blond-haired boy carrying a lunch pail, stops in the middle of the road and turns toward the cyclist, who gives the boy a military salute. This scene, like one featuring the elderly woman, was not storyboarded. The idea for the scene was generated by Ann Winkler, Kodak's Marketing Communications Specialist, during the shoot. At that point in the shoot, Winkler pointed out to Weinschenker, no kids had been cast for the spot. Winkler and Kodak "wanted kids and dogs."[35] Weinschenker and the crew hated Winkler's suggestion, but shot the scene anyway. Weinschenker set up the scene with the children crossing in front of Greg Blanchard, the motorcycle rider. While they were shooting, one little boy stopped on his own to look at the cyclist. Weinschenker liked that image, and re-created the scene, directing Blanchard to salute. At the time of the shoot, this scene did not particularly stand out from the others, and it was not being considered as an ending. According to Kodak's Bruce Wilson, the agency creative people "felt the image was 'hoakey,' but during editing it turned out to be a very powerful ending."[36]

Two alternative endings were shot, but neither was used. Weinschenker felt that the Hollywood Boulevard ending described earlier was poignant, but not right for the spot as it developed in editing, where the decision was made to confine the vignettes to the "country" with the possibility that the city footage shot, including the Hollywood Boulevard scene, might comprise a totally separate spot. (This, however, never materialized.) The other ending, which was not storyboarded, was shot in the same Arizona "ghost town" as the scene featuring the bearded man and collie. In it, the protagonist-rider meets an aging biker. After circling one another, Blanchard catches up with the aging biker and the two exchange greetings. According to Executive Producer Sid Horn and Kodak's Ann Winkler, although the scene was shot in numerous takes, upon viewing the dailies it was obvious that the scene "did not click."[37] It was discarded.

Lyricist Linda Kaplan, working separately from Weinschenker while he was on the shoot, wrote the lyrics for "America," then waited for the footage to see if the lyrics and images worked well together. If they didn't, Kaplan would adjust the lyrics to fit the images. According to Kaplan, they ended up matching well. Kaplan chose to establish a "Christopher Cross" feeling in the music, rejecting rock and roll.[38]

The First and Second Units together shot "between 40,000 and 50,000 feet of film" for the spot.[39] This resulted in a shooting ratio of about 450 feet shot for every foot actually used—a very high ratio by normal nationally-distributed commercial standards.

The respective skills of director Weinschenker and director of photography Ron Dexter, who led the First Unit, complemented one another. Weinschenker, who generated the ideas for most of the scenes shot, was particularly adept at working with actors. Dexter, who had an excellent sense of color and composition, was very good at picking up the perfect shot, and brought the "look" of the spot together.[40] Sid Horn's Second Unit brought in many very effective panorama shots.

Editing was done in New York City, with Weinschenker in charge. Kodak's Ann Winkler watched the editing for a few days. She had become emotionally involved in the spot, as had the other participants on the crew, and was reluctant to "let it go" at that time. At various points Weinschenker asked her for "suggestions" as to which scenarios she preferred, but she stayed out of the decision-making process.[41]

The edited spot was post-tested. According to Kodak's Bruce Wilson, Kodak "was trying to be a little more contemporary than in previous campaigns." "America" appeared to achieve this. The spot "tested more positive among younger people than older people, yet it did not test as more 'emotional'." According to the research, the youth and patriotism theme of "America" was relevant to younger, single viewers who saw the spot. "They could relate," added Wilson.[42]

The spot began airing during Kodak's peak advertising period, the time of year most Americans take their vacations, Spring and Summer 1985. "America" won numerous creative awards, including a 1986 Clio award for Best Original Music with Lyrics; a 1986 *Advertising Age* award for Best Commercial, "Leisure Entertainment" category; Mobius awards for Best Direction, Best Music, and Best Cinematography; an Art Directors Club Certificate of Merit; and a Silver award for Best Original Music with Lyrics at the 1985 International Film and TV Festival.

### *Viewer Responses*

According to Kodak's Bruce Wilson, "the motorcycle community was absolutely ecstatic" about the portrayal of the motorcyclist. Kodak "received hundreds of letters in appreciation of the spot."[43] Ann Winkler noted that Kodak received "less than five negative letters."[44] The response received at the agency was similar. According to Vice President and Creative Supervisor Geraldine Killeen, more letters were received on "America" than for any spot she can remember. Especially praiseworthy were motorcycle riders and motorcycle magazines, who noted the clean-cut image of the cyclist which "reflected the majority of weekend motorcycle enthusiasts."[45]

While Bruce Wilson indicated Kodak did not receive "a single letter regarding the Vietnam veteran image," agency lyricist Kaplan noted that the agency did receive a few letters expressing the feeling that viewers did not want to be reminded of the War.[46]

## Interpretive Contexts

The "story" embedded in the images, lyrics and music of "America" may appear at first glance to be seamless. Its lush greens and highly-saturated golds and reds, its majestic panoramas, slow-motion action and slow lap-dissolves, its "kids and dogs" and other warm, friendly, compassionate, fun-loving denizens of rural America, all produce an almost hypnotic, transcendental vision of an American Eden in which the proud "American family" celebrates the sacrosanct American value of individual freedom, represented by the cyclist's journey of personal and geographic discovery. Nonetheless, as John Berger wrote:

> No story is like a wheeled vehicle whose contact with the road is continuous. Stories walk, like animals and men. And their steps are not only between narrated events but between each sentence, sometimes each word. Every step is a stride over something not said. (284–85)

While the "America" director Greg Weinschenker insisted that there was "no story" in the spot, and that the motorcycle rider "was to have no preconceived notions about anything," it is clear from the creators' own descriptions of the spot's gestation that narrative "strides" were taken; specific, conscious decisions were made regarding what to include in the spot and what to withhold. On one level, these decisions in practice appear to be more "negative reactive" than active. For example, the alternative ending, with the motorcycle rider sitting on Hollywood Boulevard, tears welling in his eyes as the stark contrast between the purity of the country and the coldness of the big city comes to consciousness, was rejected in favor of the theme of the celebration of the heartland.

The conceptualization process in commercial-making is essentially inductive, with the creators trouble-shooting individual scenes, rejecting representations that might offend sub-groups within the commercial's target audience. On another level, however, the decisions can be seen as ideologically grounded. For example, Weinschenker rejected the possible draft evader scenario at the outset. However, he subsequently made reference to Vietnam in the scene, filled with pathos, in which the cyclist meets and shakes hands with the disabled Vietnam veteran. In what was to become the concluding vignette in the spot, the scene Kodak's Bruce Wilson called the spot's "central image," Weinschenker directed the motorcyclist to throw a military salute to the young boy, last off the school bus, who stopped to stare at the rider.

The inclusion of the Vietnam veteran vignette and the image of the cyclist's military salute in the final version of the spot is significant when considered in the context of the experiences of the two men most responsible for the spot's creation: director Weinschenker and Kodak's Marketing Communications Director for Film, Bruce Wilson. Weinschenker characterized himself as a "protester during Vietnam." He did not go to war; his draft lottery number never came up. A friend of Jeff Miller, one of the students killed at Kent State, he had "animosity toward the establishment." He felt his 1970 motorcycle journey across America, the basis for his idea for the spot's theme, produced a change in him, resulting in "a great love for the country and its people."[47] Wilson, who felt the veteran image was "powerful and appropriate," and was concerned only that the handicapped person was presented in a "natural way," served in the Army for four-and-a-half years, although not in Vietnam, and indicated he was in favor of the war.[48]

The "things not said" are part of the story of "America." So too are those things which, having been said, are then wrenched from the hands of their creators and entered into the larger cultural discourse. "Cultures are dramatic conversations about things that matter to their participants, [arguments] about the meaning of the destiny its members share" (Bellah et al. 153). The creators never have the final word. Their work is alive in the culture, part of its "constituted narrative," appropriated by professional critics and regular viewers alike. It is therefore both intensely personal and political.

One of the most significant applications of themes and symbols by the advertising and public relations storytellers is "the power to create support for ideas and institutions by personalizing and humanizing them" (Fleischman and Cutler 144). Advertising, like all fictional narrative,

> is a mode constantly susceptible to transformation into myth. The creation of a narrative almost always entails the shaping of awkward materials into a smooth, closed structure. And this is the essence of myth. Like bourgeosis ideology most narrative . . . denies history, denies material reality as contradictory, and denies the fact of its own production. (Davies 70)

The apparent seamlessness of "America" may be interpreted in such a mythic context.

Shortly after "America" began its run in the national television market, one critic, reviewing the spot in *Advertising Age,* the industry's most respected trade publication, undertook to demythologize the spot. The review begins by comparing "America" with a scene from the film *Cabaret,* in which a blond-haired German youth, wearing a Nazi uniform, sings the song "Tomorrow Belongs To Me." The review continues:

> What distinguishes the Kodak campaign, entitled "Because Time Goes By," [from other patriotic commercials] is that it manipulates the concept of time as a means of political propaganda. . . .
> If the definition of a reactionary is one who tries to reconstruct a

romanticized past for an uncertain future, then the Kodak campaign is the most ultra-conservative selling job of the '80s. . . .

"America" . . . shows a young white man on a motorcycle "setting off to find America." Set in the present, the spot shows little but rustic surroundings and human symbols of a sanitized past.

It's bad enough to show our grinning Everyman encountering a fluffy collie, a weatherbeaten cowboy and a little old lady at a roadside mailbox. But when he starts shaking hands with a crippled Vietnam veteran or yukking it up with a deliriously happy Indian, the spot crosses the line from mere nostalgia to pure demagoguery.

The implication that all past ugliness can be whitewashed for a brighter future is driven home by the final image: As the cyclist stops for a little boy carrying a lunch box, he raises his hand to his forehead and flashes the boy a military salute. It's as if he is saying to his diminutive comrade: "Tomorrow belongs to us."

. . . freezing moments in time has less to do with sentimental glory than precisely documenting the past. Tomorrow has never belonged to those who misremember yesterday. (McWilliams 52)

The critic, Michael McWilliams, had been reviewing spots on a free-lance basis for *Advertising Age* for about a year prior to his review of "America." McWilliams was a conscientious objector during the Vietnam War. He had been drafted and performed two years of alternative service. The issue, argued McWilliams, is "simply that fifty-five thousand men died there. And that's what's important. Nothing else."[49]

The *Advertising Age* review prompted a strong rebuttal in the "Letters" column of *Advertising Age* from Stephen G. Bowen Jr., at the time Executive Vice President and General Manager, New York, J. Walter Thompson U.S.A. (Bowen was appointed President of J. Walter Thompson U.S.A. in May 1987.) Bowen wrote that McWilliams "has an extraordinarily fertile imagination reminiscent of the defeatists' laments of the national Democratic party during the past two elections." Bowen implied that McWilliams was out of touch with American values that helped elect Ronald Reagan, who campaigned on a platform of optimism and "for a healing of the wounds that kept our society from living up to its promise in the 60s and 70s." Bowen concluded that if McWilliams became enlightened and had an insight, "he should be sure to get it to Ted Kennedy first. He needs it . . ." (20).

Kodak would have preferred that Bowen not respond to the review. Kodak's own public position was not to respond, a company policy regarding any public debate in reference to one of its commercials. Internally, "no one [at Kodak] understood" how McWilliams could interpret "America" in this way. The Marketing Communications staff thought McWilliams was "a crackpot."[50] Director Greg Weinschenker expressed anger at the review, indicating he could not see how the spot could be interpreted as "Nazi propaganda."[51]

According to McWilliams, "Not a word or a comma was ever changed in any of my *Ad Age* reviews. My editor, Fred Danzig, who had been with

*Ad Age* for many years, stood behind me in the whole affair."[52] Shortly after the appearance of his "America" review, McWilliams found his work was no longer being accepted for publication by *Advertising Age,* and he began writing about television for *Rolling Stone.*

The deeply personal nature of the varied responses to the images of the Vietnam veteran and the cyclist's military salute in the spot's final scene were brought into clearer focus by a colleague to whom the author showed the sixty-second version of "America." The colleague, himself able to avoid military service through a student deferment, found the imagery in the spot "profoundly troubling, calling into question one's own actions and decisions made during the war. It revives unresolved conflicts about personal ethics."[53]

These varied responses to the Vietnam War in particular, and to militarism in general, point to the power of public imagery to evoke cultural discourse which is profound, contentious, and psychologically complex. In a commercial, as in a photograph,

> The *seen,* the revealed, is the child of both appearances and the search. . . . appearances . . . go beyond, they insinuate further than the discrete phenomena they present, and yet their insinuations are rarely sufficient to make any more comprehensive reading indisputable. The precise meaning . . . depends on the quest or need of the [viewer]. . . . The one who looks is essential to the meaning found, *and yet can be surpassed by it.* (Berger and Mohr 118)

While multiple readings must be acknowledged and respected as representing differences in individual perspective, one must nonetheless, in the words of Stuart Hall, guard against an undue "emphasis of difference—on the plurality of discourses, on the perpetual slippage of meaning . . . on the continuous slippage away from any conceivable conjuncture" ("Signification" 92–93). As a communicative form, "America" posits, through its codes, an organized system of core values and mythic constructs that circumscribes an American ethos. The mythic constructs of "America" are grounded in ideology. As Hall wrote, "Ideologies are the frameworks of thinking and calculation about the world—the 'ideas' which people use to figure out how the social world works [and] what their place is in it . . ." (Signification" 99). Myth transforms the common sense of ideology into the society's "sacred social discourse." Myth, a living tradition, is suspended in time; it is both "primordial" and "indefinitely recoverable" (Eliade 5–6). Myth is, above all, discourse about power, about founding and maintaining a way of life, about a fundamental order of being.[54]

The mythology in "America" surpasses not only the cursory readings of McWilliams and Bowen, but also the intentions of the spot's creators. As Hall noted, "Just as the myth-teller may be unaware of the basic elements out of which his particular version of the myth is generated," so may he be unaware that "the frameworks and classifications . . . [he is] drawing on reproduced the ideological inventories of [his] society" ("Rediscovery" 72).

A closer examination of the signification, representation, and ideology embedded in the mythic constructs of "America" will reveal the discursive practices in the text, its "ideological work."

Kodak is "about" photography. "Photographed images do not seem to be statements about the world so much as pieces of it, miniatures of reality that anyone can make or acquire" (Sontag 4). Kodak's vision, to create a nation of casual snapshooters and "serious amateur" photographers, relies heavily on this distinction between "making statements" and the capturing and preserving of raw "reality." The photograph "seems to have a more innocent, and *therefore more accurate,* relation to visible reality than do other mimetic objects like painting and drawings" (Sontag 5–6; emphasis added). Beyond this appearance of the accurate reporting of reality, "photographs actively promote nostalgia. Photography is an elegiac art, a twilight art. Most subjects photographed are, by virtue of being photographed, touched with bathos" (Sontag 15).

The combination of the seemingly real and the sentimental mystifies that which is photographed. The camera "can bestow authenticity upon any set of appearances, however false. The camera does not lie even when it is used to quote a lie. And so this makes the lie *appear* more truthful" (Berger and Mohr 96–97). In advertising "the lie is constructed before the camera" (Berger and Mohr 96).

The mystifying power of the photograph is ideally suited to the construction of mythology. The elegiac quality of "America," its bathos, is firmly rooted in the mythology of the American Eden. Kodak celebrates the Jeffersonian ideal of the honest ordinariness of the rural American, whose "good" labor and desire for community are morally superior to the vices of idleness, diversion, and intemperance of his urban brothers. Jefferson's ideal citizen was the diligent democratic husbandman, who could earn an independent living and participate in the civic life of a community of self-governing relative equals. Like the Puritans, Jefferson feared and abhorred the utilitarian individualism associated with the economic man of cities and industrialization and characterized by the "spirit of enterprise and the right to amass wealth and power for oneself" (Bellah et al. 28).

The mythic hero in American culture "must leave society [the Metropolis], alone or with one or a few others, in order to realize the moral good in the wilderness, at sea, or on the margins of settled society" (Bellah et al. 144). While Kodak's hero, the cyclist-wanderer, sets off to find America, and to "realize the moral good in the wilderness," he comes from nowhere. Existing outside history, he does not bear the taint of the Metropolis, of the American technological and military hegemony which was used for class subordination in the industrial city, projection of class war outward into racial war against Native Americans on the borders, and subsequently for creation of a vast international economic and political empire following World War II (Slotkin 51–52). References to the city as a "cold place" with its implicit socio-political implications of unbridled commercial power, class and race warfare, and lack of community were removed from the spot by eliminating the Hollywood Boulevard ending.

The history of the nineteenth-century conquest of Native Americans and the mythic ideological frame that rationalized it is glaringly absent from the discourse of "America." This complex mythic ideological system, consisting of two dominant readings of history, agrarianism and progressivism, provided an intellectual justification for American expansionism. The agrarian ideology of the Jeffersonians was "an antidote to the class antipathies generated by industrialization," substituting "the cultivation of the land, the interaction of man with pure and inanimate nature, for the human conflict of Indian dispossession" (Slotkin 52). The accompanying literary mythology of agrarianism saw the brutal Indian wars of the eighteenth and nineteenth centuries as an unpleasant prelude to the story of clearing and cultivating the soil by democratic farmers. The literary mythology of the Progressives, on the other hand, saw "the naturalness and inescapability of violence arising when two countries or races compete for the same territory" (Slotkin 52).

In either case, Native Americans' claims to the vast frontier and wilderness lands were invalidated in the name of progress and "manifest destiny." True, Jefferson promised Native American leaders in 1809 that:

> we will never do an unjust act toward you. On the contrary, we wish you to live in peace, to increase in numbers . . . and furnish food for your increasing numbers. . . . We wish to see you possessed of property, and protecting it by regular laws. . . . all our people . . . look upon you as brethren, born in the same land and having the same interests. (1267)

But there was a price. Native Americans would have to shift from hunting to farming, thereby reducing the land holdings they would need to support their own increasing population while opening frontier lands to accomodate the increasing white settler population.[55] Jefferson's desire to accomplish this land transfer justly and nonviolently ignored both Native Americans' sacred ties to the land and the drives of the Metropolitan machinery of capitalism for continual resource exploitation.

Native Americans' status as marginal people living on the social and cultural periphery of Anglo America continues to the present. In the mid-1950s, under a U.S. government policy of "relocation," Native Americans began moving en masse from the reservations to selected cities, among them Los Angeles, Seattle, San Francisco, and Minneapolis, in anticipation of receiving education and job training. Many of those who signed up for these programs were placed in run-down Army barracks and given little or no training. Many didn't speak English. The problems of the reservations— poverty, poor health, alcoholism, and suicide—followed Native Americans to the cities. Today, over half of the 1.4 million Native Americans live in metropolitan areas, where they have assumed the status of an urban underclass. There are great fears in the Native American community that these displaced persons are losing the "spiritual path" as they become assimilated into urban culture.

The nineteenth-century visual representations of Native Americans portrayed them as the colorfully primitive, gloriously doomed uncivilized (Dorris 27, 36).[56] Kodak's "America," in its own mystifying, portrays the Native American as an ideal Jeffersonian citizen living in Edenic bliss, unproblematically anchored in the vast space of the American wilderness.

"America" is equally mystifying in its representation of the American farm. Images of lush farmland are combined with a shot of a farmer sitting high atop an anachronistic tractor, following the cyclist down a winding two-lane country road. The images evoke memories of the family farm and a simple agrarian life. These images belie not only the contemporary economic and political struggles of family farmers, but also the social history of American agriculture. In fact, the images of the farm and farmer offered by the spot accomplish their ideological work by mythologizing a pre-existing myth.

The continued expansion of the "corporate farm" and its subsumption of the small family farm, excluded from Kodak's vision of the heartland, is not unique to the twentieth century. Nineteenth-century American bourgeois development belied the extant mythology linking national prosperity with the pioneer farmer. In reality, according to historian Richard Slotkin, "the special environment of some frontiers . . . gave . . . agricultural enterprises a particularly monopolistic and tyrannical form that was inimical to the 'individualism' of entrepreneurs" (43–44). This was especially true of farmers in regions dominated by railroad land companies and small ranchers on range land desired by land-owning "oligarchies." Not only did the railroads sell land at vastly inflated rates to individual settlers or organized colonies, they also controlled the best land of the plains region and kept farm holders continually in their debt through manipulation of freight rates. In the South, the main cash crop, cotton, was instrumental in the organization of farming into a system of large plantations. The work of harvesting cotton was accomplished through the institution of slavery. Today's rural working class, among them tenant farmers and migrant farmworkers, is largely ignored in contemporary accounts of the "farm crisis." Also ignored is the exploitative relationship between migrant workers and the family farmers glorified in populist mythology. This relationship was exposed more than a quarter-century ago in the *CBS Reports* documentary *Harvest of Shame*.

In the process of producing the mythology of the American Eden, Kodak's "America" falls victim to "the general impulse toward . . . malignant possessiveness [which] shows signs of being stronger than ever in American life." According to psychiatrist Robert Jay Lifton, this impulse

is populist in its rural, common-man, anti-cosmopolitan tones, nativist in its easy rage toward whatever is "foreign" and "alien," chauvinistic in its blindly "patriotic" distinction between "us" and "them." It is an impulse that not only runs deep in the American grain but in the universal grain as well. For it is associated with a broader image of restoration—an urge, often violent, to recover a past that never was, a golden age of perfect harmony during which all lived in loving simplicity and beauty. ("Introduction" 4)

The need for restoration is grounded in "symbolizations around national virtue and military honor" (Lifton, *Home* 132). One of the most powerful of these is that of the "warrior as hero" who functions as "a repository of broad social guilt. Sharing in his heroic mission could serve as a cleansing experience of collective relief from whatever guilt had been experienced over distant killing, or *from the need to feel any guilt whatsoever*" (132; emphasis added). The insistence on "the continuing purity and guiltlessness of American warriors" (132–33) is transferred to American society as a whole.

The American ethos has always contained a strong strain of self-righteousness. "In matters of war and national destiny," Lifton wrote, "Americans have always felt themselves to be a 'blessed' or 'chosen people' " (*Home* 158). America's post-World War II economic, technological, and military power rendered unpalatable our defeat in Vietnam at the hands of a "third rate" military power. Those who called into question the American warrior-hero's "purity of mission" were considered by many if not most Americans as disloyal to the American vision.

The draft-evader scenario rejected by Weinschenker at the outset of the conceptualization process would have raised troubling questions regarding the war, reawakening nagging questions of collective guilt. The scenario which did evolve substitutes a recruit for warrior-hero symbolization. This symbolization functions in the spot on both levels described by Lifton. The first level—the cleansing experience of collective relief from the guilt that has been experienced—is contained within the scene featuring the cyclist and the Vietnam veteran in the wheelchair. This scene presents the cyclist's mission as one of reconciliation. He becomes the agent of adjustment, restoring harmony to a country torn apart by the war, welcoming the tainted veteran back into company of "good men."[57] This scene is placed at the temporal center of the sixty-second spot (0:27–0:31). The second level—the cleansing experience of collective release from the need to feel any guilt whatsoever—is contained in the spot's final image, the freeze-frame of the cyclist's military salute to the young school boy. Is this young boy another echo of the "warrior-hero-to-be," carrying the American purity of mission into the future?

The warrior images in "America" are a far cry from the narrative of revenge exploited in mainstream theatrical films such as *Rambo,* in which the soldier-hero's retributive actions are decidedly individualistic and ultimately anti-social (Hallin 22). Nonetheless, in its quiet way the iconography of redemption in "America" seeks to build ideological consensus around the notion of the retrieval of America's greatness, a legacy of power and control re-established. By implication, war in general, and the Vietnam War in particular, are legitimized.

The warrior-hero iconography is a distinctly male iconography. Pure, rational, and strong, the male is the culture's source of stability. His journey is that of the "straight path." "America" is a commercial about the actions of men. The only women in the spot, the elderly woman and the young girls in the school bus scene, signalize the spot's main action, but are not integral to it. The four young school girls who run past the cyclist prior to the little boy's entrance in the frame do not acknowledge the cyclist's presence, and therefore cannot bond with him. On the other hand, the school girls' presence in the shot serves to announce and frame the subsequent militaristic bonding that does occur between the young boy and the cyclist. The elderly woman's pointing gesture directs the cyclist to the various men with whom he subsequently connects in the remainder of the spot.

The restorative impulse which motivates "America" is inscribed, in the "Because Time Goes By" caption, over the spot's final image. Benjamin's epigraph, quoted at the beginning of this essay, focuses on the power of captions to direct meaning and thereby to liberate the photographic object from its aura. For me, the "Because Time Goes By" caption works *against* the mystifying and depoliticizing practices inherent in the commercial while appearing at first reading to reinforce them. For a critical viewer in post-Vietnam America, the images in this commercial are not innocent ones.[58] This is so precisely because in spite of the passage of time, the real social relations and human connections evoked in the mythology of this spot are haunting. They refuse to be turned entirely into "art." Benjamin views history as a process of mourning, not in the elegiac sense, but rather as the history of the oppressed brought to present consciousness. Here the dialectic of official or dominant ideology and the "other" history of oppression becomes the source of emergent oppositional ideology. The photographer, a soothsayer predicting the future from remnants of the past, uncovers guilt and names the guilty in his pictures. "America," in spite of its massive, corporate attempt to cover and sanitize the past, ultimately succeeds, for a critical viewer, in doing just the opposite. The restorative impulse in "America" runs counter to the model of "transformation" proposed by Lifton: a model according to which we *confront,* through sustained questioning, the values and symbols that lead to acts of violence; *reorder* our values; and *renew* ourselves through the creative exploration of alternative social forms (*Home* 388–406). These "animating" principles both expose and refute the denial associated with the restorationist impulse.

## Notes

The author wishes to thank Bernard Timberg for his many helpful editorial suggestions.

1. The "Because Time Goes By" theme is an appropriation of a memorable and still resonant older popular culture coding, the nostalgic "As Time Goes By" lyric in the film *Casablanca* (1942). The lyric evokes the memory of a brief, passionate love affair between Rick Blaine (Humphrey Bogart) and Ilsa Lund (Ingrid Berg-

man) in Paris just prior to the Nazi occupation. As Prefect of Police, Capitaine L. Renault reminded Rick, "under that cynical shell, you're at heart a sentimentalist." A similar appropriation, but with a sardonic twist, is the theme song of the 1970s social comedy *All in the Family,* "Those Were the Days."

2. Conversation with Sid Horn, former Executive Producer, J. Walter Thompson U.S.A., and Second Unit Director, "America," October 8, 1987. (Horn left the agency at the end of 1987.)

3. Conversation with Ann Winkler, Marketing Communications Specialist, Eastman Kodak (1980–86), October 14, 1987. Since these spots were introducing Kodak's Kodacolor VR film, the product-benefits approach could not be completely ignored.

4. The only key person directly involved in the spot's creation who declined an interview was the Director of Photography, Ron Dexter. Dexter indicated his busy production schedule prevented his participation in the study.

5. "America" is clearly a very high cost commercial by contemporary standards. Its six-week shooting schedule and 450:1 shooting ratio lead the author to estimate the cost to produce the spot in excess of $500,000.

6. Conversation with Warren Milich, Account Supervisor on "America," J. Walter Thompson U.S.A., May 8, 1987.

7. Bruce Wilson, at the time Kodak's Marketing Communications Director for Film, believed that the political climate in the country was right for an emotional approach. His immediate supervisor, Jack Powers, at the time the Director of Marketing Communications for Consumer Products, felt on the other hand that a product-benefits approach would be more successful. Wilson's view prevailed in the ensuing discussions. The spots would be produced to appeal more to emotions, with less emphasis on touting the specific benefits of Kodak films.

8. Conversation with Warren Milich.

9. Conversation with Warren Milich.

10. Conversation with Warren Milich.

11. Conversation with Warren Milich.

12. Conversation with Loomis Irish, Professor of Television and Radio, Brooklyn College of The City University of New York, and an executive for many years with Batten, Barton, Durstein and Osborne, April 29, 1987.

13. Conversation with Mitchell Brooks, Account Director on the Eastman Kodak account, J. Walter Thompson U.S.A., May 6, 1987.

14. Conversation with Bruce Wilson, Director Marketing Communications and Support Services for Consumer Products, Eastman Kodak, May 8, 1987. Kodak's marketing philosophy should not be generalized to all "heartland" companies, however. Some practitioners note that certain heartland companies, such as 3M and B. F. Goodrich, have supported controversial programs on Vietnam and China, while more "urban" East Coast companies such as General Electric and DuPont have scrupulously avoided controversy.

15. Conversations with Sid Horn and Ann Winkler.

16. Traditionally in the advertising business an idea on how to proceed with a campaign oftentimes was agreed upon by agency executives and clients over a game of golf or on a fishing trip. This was particularly true with family-owned clients and publicly-owned companies with a very secure senior management which had a clearly-articulated company philosophy on the perception of the American "reality."

17. Conversation with Loomis Irish.

18. Conversation with Bruce Wilson.

19. Kodak's organization of its marketing division is highly unusual for a large corporation. The division is divided into two distinct units—marketing and marketing communications. Marketing's primary functions include pricing of products, developing sales programs, and financial planning. Marketing communications, on the other hand, employs "communication specialists" working solely on advertising for all "paid communication." These communication specialists are involved in overseeing the advertising production process.

20. Conversation with Bruce Wilson.

21. Ibid. In *Thirty Seconds,* an eyewitness account of the "Reach Out" campaign for American Telephone & Telegraph, author Michael J. Arlen indicates final approval of the spots was given by an AT&T corporate vice-president in charge of public relations. Although the lines of decision-making authority may differ from corporation to corporation, it is clear that the ultimate decision to air or not air a spot is made at the highest level of the client management structure—in Kodak's case, at the very top.

22. Conversation with Loomis Irish.

23. Conversation with Greg Weinschenker, Director, "America," J. Walter Thompson U.S.A., April 11, 1987.

24. Conversation with Greg Weinschenker.

25. Conversation with Bruce Wilson. In all fairness, it is hard to imagine that any client would accept such a draft-evader scenario.

26. Conversation with Sid Horn.

27. Conversations with Bruce Wilson and Ann Winkler. The client pays the cost to shoot the commercial. Kodak Marketing Communications Specialist Ann Winkler was on the shoot and monitored production costs along with J. Walter Thompson Executive Producer Sid Horn. Horn is highly-regarded by Kodak as being "very budget-conscious."

28. Conversation with Sid Horn.

29. Conversation with Ann Winkler.

30. Conversation with Greg Weinschenker.

31. Conversation with Loomis Irish.

32. Conversation with Sid Horn.

33. Conversation with Ann Winkler.

34. Conversation with Greg Weinschenker.

35. Conversation with Ann Winkler.

36. Conversation with Bruce Wilson.

37. Conversations with Sid Horn and Ann Winkler.

38. Conversation with Linda Kaplan, Group Creative Director and Lyricist, J. Walter Thompson U.S.A., March 4, 1987.

39. Conversation with Greg Weinschenker.

40. Conversation with Sid Horn.

41. Conversation with Ann Winkler.

42. Conversation with Bruce Wilson.

43. Conversation with Bruce Wilson.

44. Conversation with Ann Winkler.

45. Conversation with Geraldine Killeen, Vice President and Creative Supervisor, J. Walter Thompson U.S.A., April 13, 1987.

46. Conversations with Bruce Wilson and Linda Kaplan.

47. Conversation with Greg Weinschenker.

48. Conversation with Bruce Wilson.

49. Conversation with Michael McWilliams, April 29, 1987.

50. Conversation with Ann Winkler.

51. Conversation with Greg Weinschenker.

52. Conversation with Michael McWilliams.

53. Conversation with Drewery McDaniel, Professor of Telecommunications, Ohio University, October 10, 1986.

54. Sheldon Wolin, "The Modern Political System: Myth Without Ritual." International Conference on the Presence of Myth in Contemporary Life. New York City, October 12, 1983.

55. Advocating the westward expansion of white settlers, and recognizing its inevitability, Jefferson nevertheless vowed to protect the Native American against the exploitation of unscrupulous whites, who made a practice of plying Indians with alcohol and getting them to sign over their land. Unlike the Puritans, who saw armed conflict with Native Americans as the inevitable struggle between God-fearing Christians and immoral heathens according to divine order, Jefferson, echoing French policy, envisioned the Indian and Caucasian intermarrying and forming a single community of noble farmers.

56. See also Slotkin, *The Fatal Environment*. Slotkin explores the "industrial and imperial version of the Frontier Myth whose categories still inform our political rhetoric of pioneering progress, world mission, and eternal strife with the forces of darkness and barbarism" (12).

57. The symbolization of reconciliation is firmly rooted in American history. In *The Fatal Environment*, Slotkin describes the Fourth of July celebration held in conjunction with the 1876 Centennial Exposition in Philadelphia, which "carried these themes of [national] growth and reconciliation into the realm of civic ritual. . . . The great parade . . . included a prominent contingent of former soldiers and officers of the Southern Confederacy—a display that was meant to symbolize the binding up of the wounds from the terrible Civil War that had torn the nation apart in four years of battle and twice that many of rancorous and uneasy peace. . . . But the imagery was a mask, the oratory hollow. The United States in 1876 was in the midst of the worst economic depression in its history, and of a crisis of cultural morale as well. The reality outside the fairgrounds put the Exposition's triumphant pageantry in a context that was corrosively ironic" (5).

58. My own reading is informed by my experience of six years in the U.S. Army Reserve during the period of the Vietnam War. That experience was marked by my disapproval of the war, a critical awareness of the contradictions of my role in the army, and the resultant feelings of guilt that I had not taken alternative action, such as evading the draft by going to Canada or claiming conscientious objector status and thereby refusing to participate in any manner, no matter how tangential, in the machinery of killing.

## Works Cited

Arlen, Michael J. *Thirty Seconds*. New York: Farrar, Straus & Giroux, 1980.

Bellah, Robert N., et al. *Habits of the Heart: Individualism and Commitment in American Life*. Berkeley: University of California Press, 1985.

Benjamin, Walter. "A Short History of Photography." *Screen* 13:1 (1972): 5–26.

Berger, John, and Jean Mohr. *Another Way of Telling*. New York: Pantheon, 1982.

Bowen, Stephen G. Letter to the Editor. *Advertising Age,* 23 September 1985: 20.

Davies, Gil. "Teaching About Narrative." *Screen Education* 29 (Winter 1978–79): 56–76.

Dorris, Michael. "Mythmaking in the Old West." *New York Times,* 21 September 1986: 27, 36.

Eliade, Mircea. *Myth and Reality.* New York: Harper & Row, 1963.

Fleischman, Doris E., and Howard Walden Cutler. "Themes and Symbols." *The Engineering of Consent.* Ed. Edward L. Bernays. Norman: University of Oklahoma Press, 1955. Pp. 138–55.

Hall, Stuart. "The Rediscovery of 'Ideology': Return of the Repressed in Media Studies." *Culture, Society and the Media.* Ed. Michael Gurevitch et al. London: Methuen, 1982. Pp. 56–90.

———. "Signification, Representation, Ideology: Althusser and the Post-Structuralist Debates." *Critical Studies in Mass Communication* 2.2 (1985): 91–114.

Hallin, Daniel C. "Network News: We Keep America on Top of the World." *Watching Television.* Ed. Todd Gitlin. New York: Pantheon, 1986. Pp. 9–41.

Jefferson, Thomas. "Speech to the Chiefs of Various Indian Tribes." *Modern Eloquence: Political Oratory.* Vol. 13. Ed. Thomas B. Reed. Philadelphia: John D. Morris, 1903.

Lifton, Robert Jay. "Introduction." *America and the Asian Revolutions.* Ed. Robert Jay Lifton. New York: Aldine, 1970. Pp. 1–9.

———. *Home From the War: Vietnam Veterans—Neither Victims nor Executioners.* New York: Simon and Schuster, 1973.

McWilliams, Michael. "Kodak Ads Strain Credulity." *Advertising Age,* 26 August 1985: 52.

Poster, Mark. "Foucault, the Present and History." *Cultural Critique* 8 (Winter 1987–88): 105–21.

Slotkin, Richard. *The Fatal Environment: The Myth of the Frontier in the Age of Industrialization, 1800–1890.* Middleton, Conn.: Wesleyan University Press, 1985.

Sontag, Susan. *On Photography.* New York: Delta, 1977.

Steinbock, Dan. *Television and Screen Transference.* Helsinki: Finnish Broadcasting Company, 1986.

Stevenson, Richard W. "Red, White and Blue is Out." *New York Times,* 16 March 1987: D1, D4.

# Form and Female Authorship in Music Video

## LISA A. LEWIS

"Sexist and violent against women." That's the reputation that MTV has acquired since its inception in 1981 as America's twenty-four-hour music video channel.[1] What with parents buying lock boxes to prevent music video viewing, and senators' wives on the trail of pornographic rock lyrics,[2] it's small wonder that women critics have been reluctant to speak up in its defense. Austin, my adopted home, has a proud and progressive musical heritage traditionally supportive of female musicians, yet when the local magazine that covers the music scene asked writer Brenda Sommer if she'd like to do a regular column on music video, she turned it down. *The Austin Chronicle* printed her explanation (Sommer, 1985: 9):

> You asked me to tell you why I couldn't review music videos for you. And I told you that I realized I couldn't because I didn't have anything nice to say about them.

It's true music video does bring together two cultural forms that have notorious histories as promulgators of female objectification—rock music and television imagery. And specific examples of women in chains, in caged boxes, and strewn across sets in skimpy leather outfits can certainly be called upon to justify such claims. But focusing on the sexist representations present in many male videos overshadows an aggregate of videos produced for

From Lisa Lewis, "Form and Female Authorship in Music Video," *Communication*, Vol.9 pp. 355–377. © 1987 Gordon and Breach Science Publishers S.A. Reprinted with permission.

songs sung by female musicians and their popularity with female fans. Several news magazines and television news broadcasts have featured stories on "rock's new women," acknowledging the fact that female rock musicians have never been so popular. *Ms.* (January 1985) magazine, the mouthpiece for liberal American feminism, has helped give musician Cyndi Lauper the feminist recognition she deserves by placing her on the cover as a "Woman of the Year." Still, music video's role in popularizing female musicians and in serving as the newest terrain for the negotiation of gender politics has been largely ignored, particularly by the academic press.[3]

"Girls Just Want To Have Fun," "Love Is A Battlefield," "She Works Hard for the Money," and "What's Love Got To Do With It?" are among the music videos that explicitly work toward overturning sexist representations of women, addressing a female audience as women through forceful references to female experience and desire. They are examples of what I call "woman-identified" videos after Adrienne Rich's (1983) characterization of "woman-identification" as "a source of energy, a potential springhead of female power" (p. 199). It isn't necessary to read the videos "against the grain"[4] to recoup specific sequences in the name of the female spectator, or to ignore a narrative closure that "claws back"[5] to patriarchal relations—familiar recourses for feminist critics interested in retrieving a female voice from dominant media. Much of the feminist content in woman-identified videos is readily accessible, although its processing by the audience is indisputably complex.

Each of the videos contains images in which women are shown appropriating space that is culturally the privilege of men. Each one includes scenes prominently displaying female subjectivity and solidarity. And each makes symbolic references to modes of patriarchial power which oppress women. The on-screen women's roles in these videos are expansive, even omnipotent. The generic form itself structurally contributes to female musicians' control over the soundtrack and visual action. It is the forceful presence of the female musicians in these videos that presents a struggle to regain (if ever it were truly lost) woman's image on the screen for women, and makes music video into a major site for the contestation of gender inequality.

Crucial to the creation of woman-identified music videos are the agency of the female musician, the formal conventions of the music video genre, and ultimately the audience's interpretation. Exposure on MTV has and is contributing to an upsurge in female rock and roll musicianship. Female musicians are actively participating in making the music video form work in their interest, to assert their authority as producers of culture and to air their views on female genderhood. The generic emphasis in music video on using the song as a soundtrack, together with the centrality of the musician's image in the video, formally support the construction of female authorship. The result is a body of video texts that refer to an explicitly female experience of life, addressing a gendered spectatorship even as they maintain a generic consistency that secures their broad appeal.

First, consider the importance of MTV in providing female rock musicians the opportunity to gain the audience recognition and industry backing that women interested in music have historically been denied.[6] In the years leading up to the start of music video programming, female rock musicians were struggling for recognition both as vocalists (the traditional female niche), and as instrumentalists and composers. The contemporary women's movement in the late 1960s and early 1970s provided momentum for change, but the early punk movement in Britain at the end of the 1970s was equally, if not more, important to female musicians. Although punk emerged essentially as a working-class male subculture, Hebdige (1983: 83–85) makes the point that punk included a minority of female participants who aggressively tried to carve out a specifically female form of expression, a sharp contrast to the usual subsuming of women by subcultural phallocentrism:

> Punk propelled girls onto the stage and once there, as musicians and singers, they systematically transgressed the codes governing female performance. . . . These performers have opened up a new space for women as active participators in the production of popular music.

Punk's advocacy of "defiant amateurism" (Swartley, 1982: 28) undermined the devalued status of the amateur musician, granting women unprecedented access to musical information and audiences.

Under the capitalist economic system that operates rock and roll as a commercial enterprise, commercial distribution commands the largest audiences and the financial backing to produce music. Indeed the aspirations of most rock musicians, women included, lie with commercial distribution. But in 1979, just when new female musicians were preparing to break into the music scene, the U.S. recording industry went into a tailspin as the combined effects of a sluggish economy, home-taping, and the diversification of the home entertainment market began to be felt. Any individual or group without a proven track record, and this especially applied to women musicians, was hard pressed to win a record company contract, an essential step in the quest for a large audience. That began to change, however, in the summer of 1981 with the introduction of music video programming.[7]

Six weeks after MTV went on the air in selected test markets like Tulsa, Wichita, Peoria, Syracuse, Grand Rapids, and Houston, record sales rose for certain musical artists getting heavy play on the channel. Retailers in these areas received requests for music that was not even getting airplay on the radio in their communities. A Nielsen survey, commissioned in 1983 by MTV owner Warner-Amex, showed MTV to be influencing 63 percent of its viewers to buy certain albums. For every nine albums bought by MTV viewers, four purchases could be directly attributed to the viewing of the record company-produced music videos. Lee Epand, vice-president of Polygram Records, one of the companies originally reluctant to turn over free copies of music video tapes to MTV, finally admitted that the cable channel had proved to be "the most powerful selling tool we've ever had" (Levy, 1983: 78).

MTV initiated an upward spiral for many new and unknown bands and vocalists, women musicians included. As album and singles sales began to rise to all-time highs, in some cases even surpassing industry sales records, financing and promotion brought new female faces to MTV. A national audience rewarded their favorites by buying more records, thereby catapulting the new and unknowns to star status virtually overnight.

In 1982, the Go-Gos became the only all-female vocal and instrumental group ever to make the Top 10. Their first album, *Beauty and the Beat,* was also the first album by an all-women rock band to hit number one on the charts. That same year, *Ms.* magazine ran an article entitled "At Last . . . Enough Women Rockers to Pick and Choose" (Brandt, 1982). Although many women gained recognition as instrumentalists, the real success story was in the musical category where women had traditionally excelled, that of vocalists. But seldom has success come so big and so fast as it did on MTV.

The 1985 winner of the top Grammy award, Tina Turner, was a woman without a record deal one year, and with a top hit single the next. Cyndi Lauper's debut album, *She's So Unusual,* remained in the Top 30 for more than sixty weeks, having sold close to 4 million copies in the United States alone. Lauper's album produced four Top 5 hit singles, a new record for a female singer. In a February 1985 *Rolling Stone* Readers' Poll, Lauper was ranked first in the category for "New Artist," second to Tina Turner in the "Female Vocalist" category, and while a distant third to Bruce Springsteen and Prince as "Artist of the Year," she outranked Michael Jackson. Madonna, the newest female vocalist success story, sold 3.5 million copies of her album *Like a Virgin* in just fourteen weeks. The album was "triple platinum" before its artist had even set foot on a touring stage, the principal promotion device before the age of music video. Female musicians such as Pat Benatar, Chaka Khan, and the Pointer Sisters have all reached a million in sales with recent albums.[8]

There is a dialectic operating with respect to female musicianship, MTV distribution of their videos, female fan viewing, and the creation of woman-identified videos. Music videos of female vocalists shown on MTV give those women the exposure they need to become popular music stars. The popularity of the musicians and their videos among fans provides clout to musicians looking to use woman-identified videos as vehicles of expression and breaks ground for fellow female musicians. The generic conventions of form and content that have developed to define the MTV product (and music video in general) structurally enable female musicians to promote a woman-identified image. Music video form and female authorship work hand in hand to make woman-identified videos possible.

Conventionally, music video is organized by the use of a pre-recorded popular music song (which usually contains lyrics) as the soundtrack of the video, and by the use of the musician in key on-screen roles. The use of a musical track as the video soundtrack involves two distinct modes of production, the production of the song and the subsequent production of the video. Female musician authorship of music videos, therefore, takes shape along these two axes.

The relations of production of the song tend to assign authorship to the vocalist, particularly the responsibility for the lyrical content, despite the fact that the division of labor in the production of pop songs often means the vocalist has neither written the song nor its instrumentation. Expanding on this tendency for vocalists to be handed authorship on a platter, female musicians have employed several tactics to insure that their authorship takes a woman-identified form, to counter-balance the fact that they must often work with male writers' lyrics. Neither Tina Turner nor Cyndi Lauper, for example, wrote the songs that have made them famous, yet each has won female authorship through their vocalization style, their rewriting of selected lyrical phrases, and their manipulation of their images in the promotional press.

In constructing her authorship of the song "What's Love Got To Do With It?" Tina Turner concentrated on vocalization style. She self-consciously describes in the fan magazine *Record* how she made the award winning song her own by reworking its musical rendition:

> The song was this sweet, little thing. Can you imagine me singing like Diana Ross or Barbra Streisand, trying to sound velvety and smooth? I really fought. Eventually, we roughened it out instrumentally and I added some (rock) phrasing, and we changed the song's attitude and got a hit. I have input, not just in song selection but in treatment too. I'll never be a musician, but I know what's right. (Mehler, 1984: 20)

Turner's self-deprecating assessment of herself as a non-musician at the end of the statement is in line with what Rieger (1985) suggests is a historically-constructed myth that positions performance outside of creative practice. The false distinction between performance and composition as it relates to creative contributions is in effect responsible for the disproportionately high number of female vocalists compared to female lyricists and composers: "Women have always only been allowed a first foothold in those areas where creativity was considered to be of secondary importance" (p. 136). In fact, the example provided by Turner reveals musical rendition to be not only a creative process but a politically empowered one.

A promotional blitz, timed to coincide with the release of Turner's album, *Private Dancer* (on which "What's Love Got To Do With It?" appears), publicized the previously hidden details of her years as a battered wife and personal slave to her ex-music partner and husband, Ike Turner. The biographical information functions to turn the lyrics of her songs into autobiographical statements. Stanzas of "What's Love Got To Do With It?" may not have been written by Turner, but her authorial voice is created as a consequence of the fan's attempt to reconcile the text with the extratextual information about the singer's personal history:

> It may seem to you,
> that I'm acting confused
> when you're close to me.
> If I tend to look dazed,

> I read it someplace,
> I've got cause to be.[9]

Quotes by Turner in the press saying things such as "I've never sung anything I couldn't relate to" (Mehler, 1984, p. 21) lend credibility to autobiographical readings of the songs she sings. In fact, Turner's appeal to women seems largely based on the notion of experience. Her ability to survive and come out on top after many years of male harassment and lack of professional recognition makes her a fitting female hero. Her courageous image is only underscored for women by the rumble of reactionary male attraction to counter-statements by Ike Turner such as the feature article in the new music magazine, *Spin* entitled "Ike's Story": "Yeah, I hit her, but I didn't hit her more than the average guy beats his wife" (Kiersh, 1985: 41).

The appearance of Cyndi Lauper's mother in the video, "Girls Just Want To Have Fun," works to turn song lyrics that specifically address a maternal figure into a reference to the singer's own life experience in much the same way that autobiographical notes in the press serve Tina Turner's association with the songs she sings. Lauper also picked the authorship strategy of selectively rewriting lyrics to make the song "Girls Just Want to Have Fun" better match the woman-identified image she was constructing. Songwriter Robert Hazard's first version of the song is fashioned as an inflated male fantasy of female desire:

> My father says, "My son,
> what do they want from your life?"
> I say, "Father, dear, we are the
> fortunate ones.
> Girls just want to have fun."

But Lauper's alteration of the song's lyric text, as suggested by producer Rick Shertoff, results in a custom-made vehicle for the expression of her views on female inequality:

> My mother says, "When you gonna
> live your life right?"
> "Oh, Mother, dear, we're not the
> fortunate ones.
> And girls just want to have fun."

It is the conventional usage of the song as soundtrack in music video that formally extends female musician authorship to the video text. In videos that feature a narrative scenario, the soundtrack provided by a female vocalist can operate like a narrator's omnipotent voiceover guiding the visual action. Sometimes she manages to literally put words in the mouths of other characters (sometimes male) through the use of a common music video device whereby a selected lyrical phrase is lip-synched as if it were dialogue. In "Girls Just Want To Have Fun," the burly ex-wrestler Lou Albano, as Cyndi Lauper's father, lip-synchs Cyndi's lyric, "What you gonna do with your life?" as she is shown pinning his arm behind his back in a self-reflexive wrestling manuever. The technique in this scene, by replacing the father's scolding voice with the daughter's, parodies and undermines the authority of the father, and by symbolic extension, patriarchy itself.

In terms of the videos' conventional visual representation, the generically-mandated appearance of the musician becomes another structure operative in the authorship construction process. Musicians actually appear in one of two roles in music video. Usually they are formally distinct, but sometimes they are so convergent that separation becomes difficult: (1) the role of musician, and (2) the role of actor in the video's narrative scenario, a role that is realized through the interaction of the musician in performance contexts and the musician as protagonist within the narrative line constructed by the video.

"Girls Just Want To Have Fun" and "Love Is A Battlefield," for instance, both maintain the musician in the role of protagonist in narratives that remain intact, undisrupted by the intercutting of performance footage. "She Works Hard for the Money," on the other hand, features Donna Summer primarily in performance shots that periodically interrupt a narrative which unconventionally uses an unknown actor instead of the musician in its central role. The lines between the two categories blur in "What's Love Got To Do With It?" The video presents itself essentially as a performance video in which city streets appear as backdrop for Turner as musician: yet as the video unfolds, her interactions in "mini-dramas" with the various street characters she encounters create an impression of her as a character.

Vocalists are privileged actors in the videos because they play lead roles and lip-synch most of the lyrics. Feminist critics, in their desire for female musician parity with male musicians in the broad spectrum of divisions of labor in music, have sometimes pointed to and criticized the assignment of female musicians to the category of vocalist. In music video, however, the prerogative rests most squarely on the shoulders of the vocalist, making female musicians' traditional role into an asset.

The visual appearance of the musician in the video affords a greater range for performance than that offered on the concert stage. Eye contact and facial gestures available to only a few concertgoers are equally accessible to all video viewers. Role-playing, limited to costuming changes and the use of props on the stage, can be intricately elaborated in music video through location shooting, the use of sets, and interactions with other actors. In other words, the full gamut of devices available to television productions is opened up to musicians in music video.

Many female musicians have proved to be quite adept at manipulating elements of visual performance in their video act, thereby utilizing music video as an additional authorship tool. In "What's Love Got To Do With It?" the gestures, eye contact with the camera and with other characters, and the walking style of Tina Turner add up to a powerful and aggressive on-screen presence. Her miniskirt, show of leg, and spike heels could operate to code her as a spectacle of male desire. Instead, the image she projects struggles for a different signification. It's easier to imagine the spikes as an offensive weapon than as a sexual lure or allusion to her vulnerability. Turner's control over her own body and interactions with others in the video, particularly with men, encourages a revaluation of her revealing clothes and high heels from indices of her objectification to signs of her own pleasure in herself.[10] Similarly, in "Girls Just Want To Have Fun," Cyndi Lauper's kinetic body movement mediates against the voyeur's gaze. Almost constantly in motion, Lauper's choreographed performance fills the frame, her gyrating arms and legs stealing space away from men on city streets in location footage.

While the formal device of using the musician as video persona is useful as an instrument to empower female musicians textually, the genre's thematic motifs provide the pegs on which woman-identified content may hang. Certain political articulations are, of course, also constrained within the confines of genre, but the dominant motifs of adolescence—life in the street and time spent in leisure practice—provide rich routes to woman-identified content by focusing on themes that culturally relate to gender contradictions.

The thematic motifs, in general, are targeted to an adolescent audience although they have a wide ranging salience for younger and older viewers. Barbara Hudson (1984) describes adolescence as a professionally defined and publicly accepted and reinforced discourse which maintains the existence of a lifestage between childhood and adulthood, a kind of liminal space associated with troubled and tumultuous behavior. Adolescence is unproblematic as a "masculine' construct," but becomes the source of contradictory expectations when applied to female teenagers because it must coexist alongside gender discourses such as femininity (p. 35). Hudson's fieldwork in schools and social service settings revealed that the major authorities in contact with female adolescents were asking them "to develop 'masculine' characteristics of independence, political and career interests, whilst at the same time expecting them to develop a personality style of caring for others, looking after children, being gentle and unassertive" (p. 42). Hudson finds that the two discourses of adolescence and femininity are mutually subversive of one another "and, in particular, that adolescence is subversive of femininity" (p. 31).

Music video reproduces adolescence as a male-defined discourse through its use of characteristic themes and symbolic representations. The city street is a central symbol of freedom from parental authority which allows young viewers to participate vicariously in an independent, public existence outside the limitations that school and home life impose on middle-class youth. The endless array of images of video characters loitering on sidewalks, strolling the avenues, dancing in the street, and traveling in cars also celebrates the specifically male youth attachment to the street as a site of sociability and escape, of subcultural formation and male bonding. At the same time, the images valorize leisure activity, the socially-extended right of middle-class adolescents and the socially-induced fate of lower-class and minority teen-agers. It is in leisure activity that adolescent boys carve out their own domain in the world. As McRobbie (1980) suggests, it represents a privileged "space" for the "sowing of wild oats" and for experimentation with roles and dangers before a lifetime of work.

But herein lies the problem for adolescent girls, for they are subject not only to expectations based on their age, but overwhelmingly by those assigned to their gender. The female gender experiences streets as dangerous and fearful places. What Reiter (1975) described as a "sexual geography" in her study of male and female use of public space in southern France applies to the United States as well. Women are expected to use streets as the route between two interior spaces, be they places of employment or consumption activity. The social consequence of street loitering or strolling is the label "prostitute" and the coding of one's body as available for male pursuit. Women's level of comfort on city streets is tenuous at best; rape and harassment are constant threats structuring their behavior. Girls learn early the gestures of deference: to avoid making eye contact, to cause one's body to "shrink" so as to take up as little space as possible.

Leisure time and leisure practices are also subject to a division along gender lines. Middle- and lower-class girls in the United States grow up in a culture in which women's work in the home is a constant, yet devalued, activity and where work outside the home is still underpaid and limited in form. In the many instances where women work outside the home, domestic laboring becomes an even more relentless form of double duty. As teenagers, girls encounter the expectation that they will assume the role of invisible worker at home, even when they are encouraged to seek a higher education and a career. Leisure practices engaged in by boys are, as well, often deemed inappropriate for girls. Subcultural youth groupings, often allied in part by their musical preferences, are usually off limits to girls. McRobbie (1980) confirms this point in "Settling Accounts with Subcultures," saying that subcultural use of the street as a primary site of activity tends to preclude female participation (p. 120). Girls' leisure takes different forms as a consequence, or perhaps more aptly stated as a result, of female resourcefulness and the will to resist subordination.[11]

In music video, representations of leisure make up the symbolic arena where gender battle lines are drawn. Woman-identified videos rely heavily on images that signal both the appropriation of male leisure forms and practices, and the celebration of specifically female ones. In this way, motifs based on street symbolism and leisure themes can become powerful social commentaries for female audiences. This is the context in which feminists have made the refrain of "Girls Just Want to Have Fun" into a slogan. The song gets its hard edge from the audiences' reading of the music video's image content and the quality of Cyndi Lauper's look and performance.

Shots of a mother (played by Lauper's own mother) at her morning duties in the kitchen contrast with the Lauper character's role as her adolescent daughter who comes bounding into the kitchen after a wild night out.

The mother, to express her distress over the daughter's disregard of appropriate feminine behavior, breaks an egg over her heart. A montage sequence of the daughter chattering on the phone with her many girlfriends celebrates a girl leisure practice which is usually ridiculed in the public media. The bouncing Lauper then leads her band of girlfriends through New York City streets in a frenzied snake dance, a carnivalesque display that turns women's experience of the street upsidedown. Their arms reaching out for more and more space, the women push through a group of male construction workers who function as symbols of female harassment on the street. The image of men cowering in the wake of the women's dance epitomizes female fantasies of streets without danger or fear, and women's desire for an unmitigated release from socially-imposed restrictions on female bodily expression. The men, their threatening status overturned, are brought back to the daughter's home to experience female fun: dancing with wild abandon in one's bedroom.

Similar sequences in both "Love Is A Battlefield" and "She Works Hard for the Money" turn the street motif into a celebration of sisterhood, openly presenting a symbolic takeover of male patriarchal privilege. Benatar, like Lauper, plays a teenager on the outs with her parents, particularly her father. Cast out of the house by him, she settles back into a fantasy sequence which speaks directly to the experience of women in public space, the state of relationships between the sexes, and the desire for closer contact among women. She is shown as a runaway on city streets at night, braving the threat of male attack. Ending up in a bar, she is one of many women in the room slow dancing like a zombie, her head turned away from her various male partners in utter boredom, her absolute lack of pleasure apparent. The scene constructs male/female relations as strictly reduced to a prostitute/trick/pimp configuration. The turning point of the video is marked by the scream of a woman whose voice, laid unconventionally over the musical soundtrack, shrieks, "Leave Me Alone!" The image of a pimp physically reprimanding her stirs the Benatar character and her female counterparts to join together in action. A new dance succeeds. As lyrics sung by Benatar are heard—"we are strong, no one can tell us we're wrong"—the women's aggressive chest thrusts and kicks force the pimp back against the bar. The scene culminates with Benatar splashing a drink in his face, a moment given formal priority by the addition of a sound effect. Unwilling or unable to retaliate, the pimp is successfully subdued, and Benatar and her female comrades retreat to the street and display a private language made up of gestures to demonstrate their solidarity and celebrate their defeat of the male ego. Benatar turns and saunters down the street, at last its rightful owner.

The bar sequence is reminiscent of an account by Ardener (1975: 30) of a ritual performed among the Bakweri women in Africa when a culturally-specific insult against women is perpetrated by a male member of the tribe:

> Converging again upon the offender, [the women] demand immediate recantation and a recompense. . . . The culprit will be brought forward, and

the charges laid. . . . The women then surround him and sing songs accompanied by obscene gestures. . . . Finally the women retire victoriously to divide [a] pig among themselves.

The mythical association with female tribalness and male street gang activity that the video establishes presents a potent "grudge fantasy" for female viewers.

The music video "She Works Hard for the Money" traces in its narrative the daily routine of a single mother as she struggles through four jobs—a scrub woman at an office building, a waitress, a factory seamstress, and a mother of two children. The zealous repetition of female labor on the screen is framed by representations of the protagonist's fantasy about being a dancer. McRobbie's (1984: 132–33, 145) essay "Dance and Social Fantasy" addresses the special significance that dance holds for girls and women:

> Dance has always offered a channel, albeit a limited one, for bodily self-expression and control; it has also been a source of pleasure and sensuality. . . . This is especially important because it is the one pleasurable arena where women have some control and know what is going on in relation to physical sensuality and to their own bodies.

It is the video's use of the dance motif to explore issues of female oppression and resistance, rather than exploiting it to make a spectacle for the male gaze, that makes "She Works Hard for the Money" so subversive.

The video opens with a soft focus, slow motion shot of the swirling protagonist, an image promptly revealed as a dream when she is awakened by an alarm clock calling her to work. A photograph of the protagonist as a younger woman in dance pose is meant to illustrate her unfulfilled ambition of becoming a dancer. Its strategic placement in a scene with her two unruly children contextualizes the failure to meet her goal as the familiar female sacrifice for love and family. In response to her day of "hard work," she allows herself to return to the fantasy in a daydream, a device that allows the video a return to the street motif to formulate its conclusion.

In this sequence, the protagonist takes to a city boulevard as the dancer of her dreams and is quickly joined by a cadre of dancing women dressed in uniforms representing a variety of work classifications. A zoom angle shot shows the women literally taking up a city block, all the while stretching out their arms for more space. The choreography, like that of "Love Is A Battlefield," includes aggressive chest and hip thrusts which mock the male spectator like the tribal dance of insult.

The juxtaposition of work and leisure, of reality and fantasy, throughout "She Works Hard for the Money" and the explosive outcome at its conclusion speak more directly to a female viewer. Holdstein (1985) argues in *Jump Cut* that musician Donna Summer, by not playing the protagonist and appearing mainly as the glamorous musician in interspersed shots, undermine the video's political impact. I would argue that Summer's presence adds yet another contradiction for the viewer to unravel, another source of tension based on notions of what serves and what impedes female solidarity.

The smattering of data related to female fan reception of woman-identified videos reveals the extent to which the videos are being actively integrated as tactics for coping with conflicts posed by gender constructs. One of the most visible signs of female audience impact lies in the modeling behavior exhibited by the "dress-like" and "wanna-be" fans. Hudson (1984) discusses at length how what one wears and what personality characteristics one exhibits are tied to cultural definitions of femininity. Both Cyndi Lauper and Madonna are key figures of style imitation because of the ways their dress and use of the body violate the discourse of femininity.

Lauper's unconventional hair and dress style have won her credit with feminists. In awarding her the "Woman of the Year" title, *Ms.* cited her "for taking feminism beyond conformity to individuality, rebellion, and freedom, both in her personal philosophy and her style as a performer" (Homeday, 1985: 47). By tipping the balance in the direction of an adolescent presentation, Lauper, and those who mimic her, challenges the core assumptions of what Hudson calls "the master discourse" (Hudson, 1984: 51) of femininity.

Madonna's image, on the other hand, is more often than not ridiculed by feminists, although her popularity among young women is strong. Admonished for making a sexual spectacle of herself, her "slut" presentation is, in actuality, marred by the indifference she projects toward men and the self-assurance she displays as an image of her own creation. Her style combines the very trappings of the discourse of femininity—wedding dress, undergarments—with its counterimage—material desires, blasphemous use of religious symbols, sexual lures. *Time* (Skow, 1985) quotes the owner of the New York video club "Private Eyes" on the subject of the Madonna wanna-bes who frequent his club: "The guys are scared of these girls. 'What do I do?' they ask. The girls come on so strong, dressed in their mothers' best fake jewelry, saying 'Don't touch me, I'm the material girl, spend money on me' (p. 77)." One female teenage fan clearly interprets Madonna's image in a feminist way: "She gives us ideas. It's really women's lib, not being afraid of what guys think" (p. 77).

At least part of the appeal of Madonna's sexual, "slut" image for adolescent girls lies with the way it can be used to counter feminine ideals of dependency and reserve. Empirical studies of girls in British secondary schools found that many females rely on an extreme, affected, sexual self-presentation. Hudson (1984) interprets this as a strategy to counter the discourse of femininity, a schoolgirl tactic which carries substantial subversive potential. Whereas behaviors associated with the discourse of adolescence may be accepted as a "phase" or "status a teenager is moving out of," signs that feminine values are not being incorporated are far more threatening since "femininity is what a girl is supposed to be acquiring" (p. 44).

This analysis could be applied to the Julie Brown video "The Homecoming Queen's Got A Gun," temporarily banned from MTV in 1985. An account printed in *People Weekly* (Donahue and Gold, 1985: 57) will serve to describe the video:

The video is a spoof of the blood-soaked finale of *Carrie*. Brown portrays a gun-crazed homecoming queen who, at the zenith of her teen life, goes berserk. From her float she picks off cheerleaders "one by one," the whole glee club ("no big loss") and "half the class," until she is ordered by the police to throw down her gun and her tiara.

The threat of "The Homecoming Queen's Got A Gun" lies not so much in the way it violates the bounds of acceptable adolescent behavior but for the way its portrayal of aggressiveness in women denigrates the ideological essence of the femininity construct. Ostensibly a victim of MTV's enforced crackdown on violence, the video was allowed back on the air after the female in-joke "Are you having a really bad period?" was cut.[12]

A final issue related to female authorship in music video pertains to the more traditional view of authorship as directorial input. I have argued that a combination of generic textual practices in music video—the female musician singing "her" song on the video soundtrack, her on-screen performance as musician and actor, the use of street and leisure themes in the video's narrative—can create a context of female authorship even when videos are directed by men. But the musician's direct control over the means of video production is still an important aspect of the struggle for the redefinition of female screen presentation.

The amount of directorial input that musicians are allowed in the creation of their videos' concept has not yet been fully standardized and is inconsistent. Music video director Julia Heyward, who has directed male and female musicians' videos, tells of conflict she has experienced because of the difference between what the bands she had worked with wanted and what the record company wanted (Zuckerman, 1985). Nevertheless, female musicians have managed to exert considerable control over their videos in selected cases. Jerome (1984) reports that the final scene of "Love Is A Battlefield" was reshot after Benatar expressed her discontent at an ending that showed her character returning to the home she had left. As a result, the video's last image features Benatar alone in the street. But perhaps nowhere is there a better example of female musician input that the case of Cyndi Lauper's involvement in the directing and shooting of "Girls Just Want To Have Fun."

Rolling Stone Press's *Rock Video* (1984) by Michael Shore provides evidence of Lauper's directorial voice in its day-by-day account of the production of "Girls Just Want To Have Fun." Lauper's name appears over and over in the account as a contributor at virtually every stage of production. It is Lauper who picks the video's producer, Ken Walz, and director, Ed Griles, based on her work with them on videos for her previous musical group Blue Angel. Lauper suggests the video's concept, picks location sites in New York City, brings in choreographer Mary Ellen Strom, and finds extras to appear in the video. The construction workers that serve as pivotal symbols in the video's snake dance sequence were actual workers who Lauper coaxed into the on-camera action. Speaking of Lauper's coaching of other passers-by drawn into the scene, Shore (1984: 171) says:

> Cyndi, who appears to be doing as much directing as Griles or anyone else, runs them through their paces several times while waiting for the new chorus-line members to return to the location.

Lauper also suggests the antique boutiques where campy items used in the creation of interior sets were purchased. She herself spends hours splatter-painting furniture for the bedroom scene. Her input even extends to post-production work, as she screens rushes, approves the rough cut, and checks in on the progress of the time-consuming special effect that appears midway through the video.

In his diary-like chronicle of the shoot, Shore (1984: 167) is attentive to Lauper's many initiatives and interventions, and even includes snatches of interviews that allow her to voice her intentions.

> Finally, there is the artist herself: Cyndi is not just a pretty face on stage, a pretty voice on record. She's an experienced actress as well. . . . Cyndi plays an unusually large creative role in the conceptualizing and staging of the video itself, from start to finish . . . says Cyndi . . . "I know what I want and don't want—I don't want to be portrayed as just another sex symbol."

The multiplicity of authorship opportunities for female musicians and the strong woman-identified statements that MTV and the music video genre have made possible should by now be apparent. This is no small accomplishment in a mass medium largely closed to personal expressions from a female perspective and to portrayals of female fantasies of the overthrow of male domination and the forming of alliances among women, and where creative control over production by women is severely limited. Far from being the absolute bastion of male desire, as some critics argue, MTV is providing a unique space for the articulation of gender politics by female artists and audiences.

### Notes

Special thanks to Jane Marcus, Horace Newcomb, and Robert Sabal for their editorial contributions and to John Fiske for encouragement.

1. MTV is an advertiser-supported cable television channel dedicated to the programming of record company-produced music video clips for a target audience aged twelve to thirty-four. It premiered on August 1, 1981. Its owner/operator is the Warner Amex Satellite Entertainment Company, a joint venture between Warner Communications and American Express. In 1985, MTV was sold to Viacom, a television syndication company. Examples of popular criticism that describe music video as sexist and violent include Levy (1983) and Barol (1985).

2. The co-founders of the Parents Music Resource Center, organized in protest of "pornographic" rock music lyrics, includes Susan Baker, wife of Treasury Secretary James Baker; Tipper Gore, wife of Senator Albert Gore (D-Tenn.); and Ethe-

lynn Stuckey, wife of Williamson Stuckey, a former representative from Florida. Hoping eventually to see the enactment of a system for rating records similar to the one used for rating movies, the Center won a lesser concession in August 1985 when nineteen top record companies agreed to start printing warnings of sexually explicit lyrics on album and music video packaging. Although male musicians, such as Prince and Twisted Sister, are most under fire, Cyndi Lauper's song "She Bop" has been targeted because of its reference to female autoeroticism. For more information see *Broadcasting* (July 15, 1985).

3. Frith and McRobbie (1978/79) provide an early exploration of gender and rock music. McRobbie's work on girl subcultures (1976, 1980) and dance (1984) is pertinent to the analysis of music video and female audiences although it has not been used as such. Steward and Garratt (1984) trace the history of female musician involvement in rock and roll and provocatively suggest ways that pop's success depends on female fan support. Brown and Campbell (1984) focus on gender in their content analysis of music video, but without attention to female musicians or woman-identified videos. Holdstein (1985) provides a textual reading of Donna Summer's "She Works Hard for the Money," arguing against a feminist interpretation. Kaplan has explored female representation in music video in several papers (1985a, 1985b, 1985c, and 1985d).

4. Here the notion of "reading against the grain" refers to the interpretive practice of reading "through" a dominant male discourse to locate a subtextual female one. The practice is linked historically to theoretical debate within feminist film criticism. By challenging the concentration of Mulvey (1975) and others on film as the domain of voyeuristic male pleasure, the practice of "reading against the grain" raises issues related to the complexity of the interpretive process and to female pleasure in film texts. Still, as the practice involves conceptualization, it assumes that the female voice in a text exists as a secondary and embedded texture or quality. I am arguing against the appropriateness of this model for the analysis of woman-identified music videos although my concern with female reception is consistent with this approach. See Gaines (1985) for a survey and critique of theories pertaining to female representation on the screen, and de Lauretis (1984) for a discussion of modes of analysis that foreground female pleasure.

5. "Claw back" is a term used by Fiske and Hartley (1978) to describe the way television functions to maintain a socio-central position, thereby mediating against ideologically aberrant readings by audiences.

6. Rieger (1985) dates the institutional exclusion of women from musical composition and performance back to the beginning of these institutions themselves. Churches in the middle ages made it an official practice to bar females from participation in liturgical rites, effectively creating a gender boundary to "high music" culture. Early educational institutions reserved musical training and opportunities primarily for their male students. Women's music-making was forced into popular culture forms, and, with respect to the formation of the bourgeoisie in the eighteenth century, into domestic space. Female piano-playing and singing were designed as appropriate forms of musical expression for women and incorporated into the bourgeois woman's role in the family: "It was important to a man's prestige that his wife could entertain his guests with music, and of course a musical education for his daughter served as a good investment for an advantageous marriage" (p. 141). Music performed by women was conceived as a service provided for fathers, husbands, and children, not as a source of pleasure for themselves, or as a career direction, a means for making money. Prior to the influx of women, men were accus-

tomed to performing music in the home. But as music in a domestic setting became associated with bourgeois female roles, men responded by establishing professional standards and devaluing the amateur status. The legacy of too little institutional support and the ideological attitude toward the suitability of musical expression for women form the basis for male-dominated musical forms today, including rock and roll.

7. The following sources enabled me to trace the decline of the record industry and to feel justified in crediting the start of music video cable distribution with its subsequent turnaround: Henke (1982), Hickling (1981), Kirkeby (1980), Loder and Pond (1982), Pond (1982), Sutherland (1980), and Wallace (1980).

8. Information about sales and rankings of female musicians were constructed from the following sources: Brandt (1982), Loder (1984), Miller et al. (1985), *Rolling Stone* (February 28, 1985), and Swartley (1982).

9. These are selected lyrics from "What's Love Got To Do With It?" Written by Terry Britten and G. Lyle, the song was recorded under the Capitol Records label. Copyright © 1984 by Myaxe Music Ltd. & Good Single Music Ltd. Myaxe Music Ltd. published in U.S.A. by Chappell & Co., Inc. International Copyright Secured. All rights reserved. Used by permission.

10. The term "revaluation" is borrowed from the work of anthropologist Marshall Sahlins (1981). In this regard, I have also found useful the discussions of "active" signs in Volosinov (1973) and de Lauretis (1984). Another example of Tina Turner's attempt to revalue signs that have acquired an association with male desire is her adoption of the song "Legs," popularized by a ZZ Top video, for her concert stage act.

11. See Brake (1980) and McRobbie (1976) for reviews of literature pertaining to subcultures and feminist critiques of this literature; see Griffin (1985) for a discussion of leisure as it relates specifically to girls.

12. According to Donahue and Gold (1985), a reference to drugs in the lyric "The whole school was totally coked!" was also cut to enable "The Homecoming Queen's Got A Gun" to air on MTV. Lyrics are by Julie Brown, Terrence McNally, and Charles Coffey.

## *References*

Ardener, Shirley. "Sexual insult and female militancy." In Shirley Ardener (ed.), *Perceiving Women*. New York: Wiley, 1975.

Barol, Bill. "Women in a video cage." *Newsweek,* March 4, 1985, p. 54.

Berrand, Judy. "Sound, image and the media: Rock video and social reconstruction." *Parachute,* December 1985.

Brake, Mike. "The invisible girl: The culture of femininity versus masculinism." Chapter 5 in *The Sociology of Youth Culture and Youth Subculture*. London: Routledge & Kegan Paul, 1980.

Brandt, Pam. "At last . . . Enough women rockers to pick and choose." *Ms.,* September 1982, pp. 110–16.

*Broadcasting* "The women behind the movement." Vol. 109, no. 42, July 15, 1985.

Brown, Jane D., and Kenneth C. Campbell. "The same beat but a different drummer: Race and gender in music videos." Paper presented at the University Film and Video Association Conference, Harrisonburg, Va., January 1984.

Cretcher, Jeff. "Hard up was hard to do: The production of a rock video." *American Cinematographer,* September 1983, pp. 56–59, 108–13.

Cubitt, Sean. "Box pop." *Screen* 26 (1985): 84–86.

de Lauretis, Teresa. *Alice Doesn't: Feminism, Semiotics, and Cinema.* Bloomington: Indiana University Press, 1984.

Donahue, Deirdre, and Todd Gold. " 'Val gal get your gun' Julie Brown blasts her way onto MTV," *People Weekly,* May 20, 1985, pp. 56–58.

Dyer, Richard. *Stars.* London: British Film Institute, 1979.

Fiske, John and John Hartley. *Reading Television.* London and New York: Methuen, 1978.

Frith, Simon. *Sound Effects: Youth, Leisure, and the Politics of Rock 'n' Roll.* New York: Pantheon Books, 1981.

Frith, Simon, and Angela McRobbie. "Rock and sexuality." *Screen Education* (1978/79): 3–19.

Gaines, Jane. "Women and representation." *Jump Cut* 29 (1985): 25–26.

Griffin, Christine. "Leisure: Deffing out and having a laugh." Chapter 5 in *Typical Girls? Young Women from School to the Job Market.* London: Routledge & Kegan Paul, 1985.

Hebdige, Dick. "Posing threats, striking poses: Youth, surveillance, and display." *SubStance* 37/38 (1983): 68–88.

Henke, James. "1981: Another bad year for the record industry." *Rolling Stone,* March 4, 1981, p. 51.

Henley, Nancy. *Body Politics.* Englewood Cliffs, N.J.: Prentice-Hall, 1977.

Hickling, Mark. "Record sales hold steady with last year's." *Rolling Stone,* October 15, 1985, p. 52.

Holdstein, Deborah H. "Music video: Messages and structures." *Jump Cut* 29, no. 1 (1985): 13–14.

Hornaday, Ann. "Cyndi Lauper." *Ms.,* January 1985, p. 47.

Hudson, Barbara. "Femininity and adolescence." In Angela McRobbie and Mica Nava (eds.), *Gender and Generation.* London: Macmillan, 1984.

Hustwitt, Mark. "Rocker boy blues." *Screen* 25 (1984): 89–98.

Jerome, Jim. "They film the songs." *Gentlemen's Quarterly,* March, 1984, pp. 90, 98, 100.

Kaplan, E. Ann. "A postmodern play of the signifier? Advertising, pastiche and schizophrenia in music television." In Richard Collins et al., (eds.), *Proceedings of the International Television Conference.* London: British Film Institute, 1985a.

———. "History, the historical spectator and gender address in music television." Paper presented at the Yale Conference on History and Spectatorship, March 1985b.

———. "The representation of women in rock videos." Paper presented at Lafayette College, April 1985c.

———. "Sexual difference, visual pleasure and the construction of the spectator in rock videos on MTV." Paper presented at the Conference on Sexual Difference, Southampton University, July 1985d.

Katz, Cynthia. "The video music mix." *Videography,* May 1982, pp. 28–35.

Kiersh, Edward. "Ike's story." *Spin,* August 1985, pp. 39–43, 71.

Kinder, Marsha. "Music video and the spectator: Television, ideology and dream." *Film Quarterly* 38 (1984): 2–15.

Kirkeby, Marc. "The pleasures of home taping." *Rolling Stone,* October 1980, pp. 2, 62–64.

Laing, Dave. "Music video: Industrial product, cultural form." *Screen* 26 (1985): 78–83.

Levy, Steven. "Ad nauseam: How MTV sells out rock and roll." *Rolling Stone*, December 8, 1984, pp. 30–37, 74–79.

Loder, Kurt. "Sole survivor." *Rolling Stone*, October 11, 1984, pp. 19–20, 57–60.

Loder, Kurt, and Steve Pond. "Record industry nervous as sales drop fifty percent." *Rolling Stone*, September 30, 1982, pp. 69, 78–79.

Marchetti, Gina. "Documenting punk: A subcultural investigation." *Film Reader* 5 (1982): 269–284.

McRobbie, Angela. "Girls and subcultures." In Stuart Hall and Tony Jefferson (eds.), *Resistance Through Rituals: Youth Subcultures in Post War Britain*. London: Hutchinson, 1976.

———. "Settling accounts with subcultures: A feminist critique." *Screen Education* 34 (1980): 37–49.

———. "Dance and social fantasy." In Angela McRobbie and Mica Nava (eds.), *Gender and Generation*. London: Macmillan, 1984.

Mehler, Mark. "Tina Turner's still shaking that thing." *Record*, December 1984, pp. 17–21.

Miller, Jim, et al. "Rock's new women." *Newsweek*, March 4, 1985, pp. 48–57.

Mulvey, Laura. "Visual pleasure and narrative cinema." *Screen* 16 (1975): 6–18.

Pond, Steve. "Record rental stores booming in U.S." *Rolling Stone*, September 2, 1982, pp. 37, 42–43.

Reiter, Rayna R. "Men and women in the south of France: Public and private domains." In Rayna Reiter (ed.), *Toward An Anthropology of Women*. New York: Monthly Review Press, 1975.

Rich, Adrienne. "Compulsory heterosexuality." In Ann Snitow et al. (eds.), *Powers of Desire*. New York: Monthly Review Press, 1983.

Rieger, Eva. " 'Dolce semplice'? On the changing role of women in music." In Gisela Ecker (ed.), *Feminist Aesthetics*. London: Women's Press, 1985.

*Rolling Stone* "The winners: Readers' poll/Critics' poll." February 28, 1985, pp. 26–27.

Sahlins, Marshall. *Historical Metaphors and Mythical Realities*. Ann Arbor: Unviersity of Michigan Press, 1981.

Shore, Michael. *The Rolling Stone Book of Rock Video*. New York: Rolling Stone Press, 1984.

Skow, John [with Cathy Booth and Denise Worrel]. "Madonna rocks the land." *Time*, May 27, 1985, pp. 74–77.

Sommer, Brenda. "At home with video." *Austin Chronicle*, February 22, 1985, p. 9.

Stern, Bert. "Let the mascara run." *Vanity Fair*, August 1985, p. 71.

Steward, Sue and Sheryl Garratt. *Signed, Sealed, and Delivered: True Life Stories of Women in Pop*. London: Pluto Press, 1984.

Sutherland, Sam. "Record business: The end of an era." *Hi Fidelity* May 1980, p. 96.

Swartley, Ariel. "Girls! Live! On stage!" *Mother Jones*, June 1982, pp. 25–31.

Volosinov, V. N. *Marxism and the Philosophy of Language*. New York: Seminar Press, 1973.

Wallace, Robert. "Crisis? What crisis?" *Rolling Stone*, May 29, 1980, pp. 17, 28, 30–31.

Worrel, Denise. "Now: Madonna on Madonna." *Time*, May 27, 1985, pp. 78–83.

Zuckerman, Faye. "Heyward longs for creative freedom." *Billboard*, March 23, 1985, p. 30.

## *Video Credits*

Pat Benatar, "Love is a Battlefield" (Chrysalis Records: Mary Ensign, producer; Bob Giraldi, director, 1984).

Julie Brown, "Homecoming queen's got a gun" (Rhino Records: Terrence McNally, producer; Tom Daley, director, 1984).

Cyndi Lauper, "Girls just want to have fun" (Epic Records: Ken Walz, producer; Ed Griles, director, 1983).

Donna Summer, "She works hard for the money" (Polygram Records: Chryssie Smith, producer; Brian Grant, director, 1983).

Tina Turner, "What's love got to do with it?" (Capitol Records: John Caldwell, producer; Mark Robinson, director, 1984).

# The Unspoken Rules of Talk Television

## BERNARD TIMBERG

The talk show, like the daily newspaper, is often considered a disposable form. When *Tonight* show host Jack Paar mentioned to director Hal Gurnee in the early 1960s that he was going to throw away the taped masters from the first years of his show, Gurnee's response was quick: "You're not going to throw away the hubs, are you?" (The aluminum hubs of the two-inch video masters were worth $90 at the time.) [1] The first ten years of Johnny Carson's *Tonight* shows were erased by NBC without any thought to future use. The producers of television talk had no idea that what they were sending out over the airwaves might have future value.

The academic industry has similarly neglected talk shows. While there has been an explosion of interest in television on the part of scholars and critics in the 1980's, most of that interest had been directed toward news and dramatic programming. With news and drama, critics from the arts, humanities, and social sciences had a place to start: historical and sociological studies of news production and a tradition of narrative studies in literature and film that could be transferred to the small screen.

Talk shows came out of a series of less well defined traditions—Chatauqua, vaudeville, newspaper opinion and gossip columns, radio panel and information shows, variety and musical theater. Talk shows also crossed wildly uneven informational terrain, from entertainment talk shows with

From *Television Talk*, The University of Texas Press (forthcoming 1994). Reprinted with permission.

stars and celebrities to talk shows that feature headline news and current events to specialized talk shows on things like home repair and cooking.

Talk shows of one kind or another made up 24 percent of all radio programming from 1927 to 1956, with general variety talk, audience participation, human interest, and panel shows comprising as much as 40 to 60 percent of the daytime schedule.[7] Network television from 1949 to 1973 consistently filled over half its daytime program hours with talk programming of one kind or another, and devoted 15 to 20 percent of the evening schedule to talk programming.[3] As the networks went into decline in the 1980s, talk shows continued to expand. How was scholar or critic to make sense of such a massive array of material? What exactly was the worth of this incessant flow of words?

We are beginning to learn that yesterday's trash is today's history. Over time, through Peabody and Emmy awards, books, articles, and critical acclaim, the talk show has begun to receive recognition for its role in defining social, cultural, and political life in America over the past four decades. Politicians have always known the power of television. Adlai Stevenson was probably the last Presidential candidate to disdain its use. Since the media-created campaign of Ross Perot and his announcement of candidacy on *Larry King Live* in 1992, Presidential candidates have made television talk shows central to their campaigns.

In circumstances such as these the hosts of television talk shows have become increasingly powerful. They speak to cultural ideas and ideals as forcefully as politicians or educators, and their impact may at times be greater.[4] National talk show hosts become surrogates for the citizen. Interrogators on the news or clown princes and jesters on entertainment talk shows, major television hosts have the license to question and mock—as long as they play within the rules.

Some names stand out in any account of television talk—Steve Allen, Ernie Kovacs, Jack Paar, Johnny Carson, David Letterman in entertainment talk; Edward R. Murrow,[5] Ted Koppel, Bill Moyers, Phil Donahue, and Oprah Winfrey in news and public affairs. These hosts have become part of the history and folklore of television. They are the titans of talk.

Just as the titans of television talk shape the talk show as it emerges in the network era,[6] so these titans are themselves shaped by previous talk show traditions—by sponsors, network officials, professional writers, producers and directors trained on other shows, and by generations of Americans who have enjoyed the form. An investigation of the television talk show must look at the landmark figures who have established the talk show as a form, but it also must describe the intricate web of forces, the people and passions, that put a talk show on the air.

Though hosts and shows change over time, the principles remain the same. The talk show is host-centered, forged in the present tense, spontaneous but structured, churned out within the strict formulas and timed segments of costly network time, and designed to play to the hottest topics of the day. Whoever may be the host, these are the operating principles of the television talk show.]

If you had been watching the Letterman show the evening of October 15, 1986, you would have heard Letterman conclude his opening remarks by asking director Hal Gurnee: "The last we heard the ballgame between the Astros and the Mets is tied 3–3 in the bottom of the tenth? Is that right?" Looking up to an undefined space in the sky, Letterman continues, "Hal, oh Hal—this will be our director Hal Gertner—can you keep us posted if there are any developments in that ball game?"

Letterman is talking about the National League playoff game between the New York Mets and the Houston Astros, a hard fought contest that has gone into extra innings. The Mets have not been in a World Series since 1973 and have not won a pennant since the "Miracle Mets" of 1969.

Letterman's voice at this moment is reminiscent of Jack Benny's invocations of his "valet" Rochester. Gurnee, or "Gertner," as Letterman calls him, is, like Rochester, the formally obedient servant who is not the least cowed by his master's voice. "Yeah, Dave," the off-screen Gurnee replies, "by the way, it's 3–3 in the tenth."

Letterman shoots back: "I just announced that Hal." The director calmly replies that he just wanted to be sure everyone knew. A suspicion forming on his brow, Letterman asks, "Are you watching the game in there?" The director's response is instantaneous, "Sure we are!" The studio audience breaks into laughter, and Letterman himself begins to laugh at the thought of his director, feet on the console, paying more attention to the Mets game than his star.

"But that shouldn't affect the quality of tonight's show, should it?" Letterman asks.

"Not at all, Dave. Craig Reynolds is up, by the way."

"OK, OK," Dave attempts to break off and go back to the show, but he can't get his vagrant director off his mind. "Do you guys have snacks in there?"

"They're on their way," Gurnee replies, followed by another explosion of laughter from the studio audience. Letterman himself cackles at this latest effrontery. "Well, just let us know what happens," he trails off.

When Letterman announces his first guest, NBC news anchor Tom Brokaw, Gurnee/Gertner interrupts again. His voice is quiet, polite, assured. "Excuse me, Dave. He fouled one off." (A new roar of laughter from the studio audience.)

Letterman waits for the laughter to die down. "All right, fine, Hal." Then, his voice dripping with sarcasm, "let us know if there are any broken bats, too, all right?"

Things run smoothly through the first guest interview. Then, just as Letterman comes back from the commercial break, the director intercedes again, "Excuse me, Dave."

"That's our director," Letterman explains to his viewers. A look of uneasy amusement crosses his face.

"Keith Hernandez is up now, Dave."

"OK," Letterman replies quickly, seeking to go on.

"I'm sorry, he just stepped out of the box."

"OK, Hal." The emphasis on the word "OK" telegraphs Letterman's impatience. Gurnee is not deterred and continues, "Looks like he has something in his eye."

"Uh-huh." Letterman says. A scowl crosses his face. He is clearly no longer amused. "*OK, Hal.*"

This brings a moment of silence, and though it appears that the host has finally shaken a director more concerned with the baseball game on a competing channel than directing his show, just as Letterman is about to launch his discussion with his second guest, a familiar voice is heard over the intercom.

"By the way, Dave, I thought you might want to know, it's still 3–3."

"3–3, what inning?" Letterman asks.

"Bottom of the eleventh," the director replies.

"Bottom of the eleventh," the host repeats. Mechanically consulting his blue sheet of notes, Letterman turns to the next item of business, "OK. Tomorrow on the program we have—"

Before he can get the next words out of his mouth, the familiar voice returns, "Excuse me, Dave. Billy Hatcher's up."

"Uh?" Letterman is startled. (A small laugh builds in the studio audience.)

"He just stepped out," Gurnee continues. "He's knocking some dirt off his spikes."

"I'll knock some dirt off your spikes, Hal." Letterman mutters under his breath. (A small titter rises in the audience.)

"Excuse me?" the director replies. This act of chutzpah elicits a huge, four-second roll of laughter from the audience.

Letterman, caught off guard, retreats. "Nothing, Hal. You go back to your corn dogs and beer. Don't let *us* interrupt anything out there."

The entire exchange has taken one minute and forty-five seconds of air time.

### *Highlighting the Rules by Breaking Them*

This scenario is not just amusing. It illustrates, by challenging but ultimately reaffirming them, a set of talk show rules and principles that govern virtually all television talk shows. These are the principles of "fresh talk"[7]— talk that is not scripted or canned but generated moment to moment.

What distinguishes the fresh talk of one talk show from another is the personality of the host, the topics addressed, and the talk traditions within which the host operates. Certain hosts have redefined what talk shows do and have been enormously successful at it. They are the titans of television talk, and they have almost always come from news and entertainment shows. We'll take a quick look at those rules now, and then take a closer look at them in the following pages.

Letterman took the entertainment talk show on television in new directions by "problematizing" it—questioning its rules and playfully subverting them on the air in each show. In the incident above, five basic unspoken rules of television talk are brought into question.

The first is the host's centrality and control of the show. The primary rule of television talk is that the host is the master (or mistress) of ceremonies. From the beginning, emcees and femcees have centered television talk.

The second rule brought into question concerns the present-tense flow of a television talk show and the unspoken understanding that no one is entitled to stop it. It is this present-tense flow, orchestrated skillfully and unobtrusively by the host, that is so abruptly halted by the director's interventions. The viewer is made uncomfortably aware of how little it takes to stop the smooth technological operation of a television talk show. When Gurnee intercedes and threatens to redirect the show at his own whim, he pointedly breaks its flow.

Gurnee also willfully ignores Letterman's special role, as talk show host, of speaking to and amusing millions as if engaged in a private conversation with them. The private conversations that take place on the show are of course destined for a wide viewing audience—some three or four million viewers each night. That unspoken compact between talk show host and viewer is violated by a purely private intervention—what the director and his crew choose to do at the moment.

Fourth, the commodity function of the show is interrupted. This is the most important function of television talk to its network management (in this case NBC). It is the commercial imperative to reach the biggest possible audience in the show's time slot with the commercials that accompany the program. This purpose is clearly subverted here by a technical team that prefers to watch a competing show (with its commercials).

Finally, the conscious structuring and crafting of what seems "spontaneous" in television talk is brought to light in the director's actions. For the seamless functioning of a talk show, the invisible hands and eyes of the technical crew must be unquestioningly trained to their tasks. Wandering off to see another show, even if it only occupies part of the technical crew's time and attention, is not acceptable—treasonable even, when the crew's job is to fashion the sound and picture images that put the show out over the air.

Indeed, this comedy bit on the Letterman show has called into question the unspoken rules that distinguish television talk from the kinds of talk that occur in the street, on the playground, or in the house, and are silently assumed in the creation of all television talk shows.

While some talk show hosts are innovators who, like Letterman and Donahue, take the television talk show in new directions, others, like Johnny

Carson, are synthesizers of previous talk show traditions. Taking over the *Tonight* show from Steve Allen, Ernie Kovacs, and Jack Paar (all significant innovators) in 1963, Carson's long reign as "king of late night talk" institutionalized the late night comedy/variety talk show. He set a standard for his competitors as well as successors in late night talk.

When Carson appeared on the air for the last time on May 22, 1992, fifty million people watched the show, a talk show record. Carson's retirement was front page news across the country, with publicity that built steadily in the week before his final appearance. In the week of his retirement after thirty years on the air, Carson was hailed as a cultural institution. Though Letterman rose to prominence in the 1980s as the host who took late night comedy talk in new directions, Carson was the host who more than any other consolidated its traditions. When we look at the unspoken rules of television talk in depth in the next section, we will be referring back to Carson as a standard.

When I speak here of "television talk" rather than "television talk shows," I am referring to a special kind of talk that occurs in front of the camera. It is often directed to an individual in the studio (host, guest, or sidekick) but also, simultaneously, to a national television viewing audience of millions. On entertainment talk shows and on entertainment/public affairs shows like *Donahue,* and occasionally on shows oriented toward hard news like Ted Koppel's electronic town meetings, a studio audience serves to varying degrees as a surrogate for the viewing audience at home.

Television talk is always present tense, though it is produced in varying degrees of "liveness": truly live, taped as-if-live, or filmed or taped and edited afterwards. TV talk is also produced, and played back, with different degrees of frequency. There are daily shows, weekly shows, and talk shows recorded at irregular intervals, like Bill Moyers *Conversations with . . .* series. However frequently produced, talk shows become a regular viewing habit at the times they are broadcast. It is common, for instance, for a weekly or monthly show to be "stripped" as a daily show in syndication.

In practice, however blurred these differences are in a viewer's experience of television talk, they play a part in how viewers form relationships with talk show hosts. Letterman, for instance, was taped as-if-live daily at 5:30 PM for a 12:30 EST broadcast and was careful never to stay on the air too long with reruns. Whatever the degree of "liveness" or frequency of a talk show on the air, the basic rules or principles remain the same.

## Unspoken Rules

### Rule No. 1: Television Talk Is Host-Centered

Even though much television talk is planned or scripted by an off-screen team (indeed, as we will later see, Gurnee's interruptions were planned in advance), television talk invariably centers, in one way or another, on the host. In every talk show explored [here] the host turns out to have a high degree of control over the show and over the production team that builds the show. From a production point of view, the host is frequently the managing editor of the show. From a marketing point of view, the host is the label that sells the product. A successful talk show host also becomes the fulcrum of the show's power in negotiations with the network. He or she is the one irreplaceable part. Without the "brand-name" host, the show may continue (as, for instance, the *Tonight* show continued when Leno replaced Carson), but network executives are fully aware that the success of the show hinges on the host.

Early talk had an array of formidable hosts. Eleanor Roosevelt had a show in New York from 1950 to 1951 called *Mrs. Roosevelt Meets the Public.*[8] Albert Einstein was her first guest. Women were frequent hosts and among television's first prominent talk personalities beginning in 1948. Then came a shakedown in the industry. The highly verbal panel and game shows that early television had inherited from radio days gave way to entertainment and variety talk. During the 1960s and 1970s the networks grew in power, higher rates were charged and talk time became more costly. Television talk was institutionalized.

Johnny Carson rose to power during this time. The *Tonight* show was inexpensive to produce, guests being paid little or nothing to appear on the show, and *Tonight* became a steady profit center for the network. Such consistency was a rare in an industry where booms and busts in program cycles were the norm. Carson also set a standard by his astute management of the show. Through well-publicized walkouts and thorny negotiations with the networks when his contracts came up, Carson gradually established better conditions for himself, leading to his famous four- and three-day work weeks. With each contract he assumed greater degrees of and control and ownership over his show. Carson set a standard. By the 1980s, major news talk and entertainment talk show hosts, aided by personal managers and agents, commanded multi-million dollar salaries and significant degrees of control over their shows.

Along with making business decisions about their shows, successful talk show hosts work closely with their producers to select and manage their writer/producer teams. From the earliest days, talk show hosts had writers and producers who worked for them. Steve Allen generally had two writers on his staff—one of them, Herb Sargent, going on to write with a much younger team of writers twenty-five years later on *Saturday Night Live.* Carson listed eight writers on his final *Tonight* show. Letterman expanded his writing staff from nine to twelve in his first ten years on the air. There was never any question for whom the writers wrote or to whom they were responsible.

This was true of talk show producers by and large as well. Jack Paar went through three producers in his first two years. Letterman switched producers after five years on the air and then added another one (Peter LaSally from the *Tonight* show) when Carson went off the air. Carson had two producers before Fred de Cordova's long stewardship in the 1970s and 1980s. De Cordova, a veteran producer-director who had worked with such major show business figures as Fred MacMurray, Jack Benny, and George Burns, always took his orders directly from Carson. Carson oversaw every detail of the show and his control over his writers was absolute. Carson's writers were always on six-week renewable contracts.

Johnny Carson followed time-honored talk show traditions here. Talk show writers are almost always anonymous and are almost always on short-term contracts. Letterman varied that tradition by allowing his writers to be pictured with him in publicity shots and assume speaking roles in skits,[9] but he too left most of his writers' work invisible. During Carson's final show, which stood to some extent as a commemoration of everyone who had worked on the show for thirty years, none of the writers was mentioned by name. Only in the final credits at the very end could the viewer catch the names of Carson's writers, a form of recognition that did not occur on the nightly show.

Television talk begins and ends with the host. Talk shows are, to use the words of one producer, "he" or "she" shows.[10] The producers and talent coordinators line up guests and topics for "him" or "her"; the writers write lines that will work for "him" or "her," to be spoken as he or she would speak them. The tone and pacing of everything that happens within the show is set by the host.

## Rule No. 2: Television Talk Is Produced and Experienced in the Present Tense

Videotaped, live, or taped as-if-live, television talk shows take place in the present tense. By 1992 most were videotaped before they went on the air. Talk shows go to reruns from time to time, notably when hosts go on vacation breaks. Some shows, like Johnny Carson's *Best of* series, have been packaged as syndication reruns. Yet even "reruns" participate in the illusion of the talk show present. Talk shows are conducted, and viewers participate in them, as if talk show host and guest and viewer *occupy the same moment*—the moment the show goes on the air.

This should not be surprising. Television is a present-tense medium. In the early 1950s viewers were thrilled to see the first coast-to-coast cable transmission as both the Pacific and Atlantic Oceans appeared together on Edward R. Murrow's *See It Now*. Television's most exciting moments in news, politics, and sports still take place in an electronically instantaneous present. This kind of present-tense immediacy distinguishes television from film and photography, which capture images and relay them back later. Film and photography render a frozen present tense, bringing a present-tense moment back to us at a later time. News, sports, and talk shows alone require viewers to "suspend disbelief" and think of the host as speaking directly to them.

## *Rule No. 3: Talk Show Hosts Speak Intimately to Millions*

Talk show hosts sustain a sense of history, continuity, and intimacy with millions simultaneously. On his last show in May 1992, Carson was clearly aware of the history and continuity of his relationship with his audience. Before launching his last monologue, Carson brought a stool on stage. This was the stool with which he had begun the show in 1962, inherited from Jack Paar before him. "We'll end this show as we began," he said, surrounded by "the same shabby set and props." Several times his voice choked with emotion as he spoke. This indeed was a form of present tense immediacy through the camera, a private conversation to a very public audience.

## *Rule No. 4: Words = Dollars*

Johnny Carson's monologue on his last show included a joke about the proliferation of talk shows on television in the early 1990s. "When we started this show," he said, "the total population of the earth was 3 billion, 100 million. This summer it is 5 billion, 500 million people, which is a net increase of 2 billion, 400 million people. . . . A more amazing statistic is that half of those 2 billion, 400 million people will soon have their own late night talk show."

Carson's joke about multiplying talk shows suggested that even on the eve of his retirement he had his eye on the competition. Talk shows were, as they had always been, relatively cheap to produce in 1992, often costing less than $100,000 a show compared to upwards of a million dollars or more for a prime time drama. That, combined with the insatiable demands of the entertainment industry to promote its products and a larger number of new outlets through satellite and cable TV, had given rise to the new cycle of talk shows to which Carson referred.

With all that was new, however, commercials still funded most TV talk shows. Carson referred to this too on his final broadcast, catching himself as he reminisced with sidekicks Ed McMahon and Doc Severinsen and almost forgot a commercial break. Should he go to a commercial, he asked his old staff members, and then, shaking his head as if that were a silly question, he said, "Of course, 'the dollar.'" As McMahon and Severinsen nodded assent, Carson muttered "some things never change" and went dutifully into the commercial.

Some things do not change on commercial television. Television talk is a commodity. Carson's joke on the proliferation of talk shows, which was comprised of seventy-five words and ran only thirty seconds on the screen, was a $150,000 joke—the cost to advertisers of a thirty-second "spot" on Carson's last show. Each word of that joke cost approximately $2,000.[11]

Although the rates on Carson's last show were particularly high that evening,[12] commercial time on television is invariably costly. An industry of network and station "reps," time buyers and sellers, manages the fluctuations of advertising time in the talk show market.

Hosts are commodities, too. Their worth to networks and advertisers is reflected in their yearly salaries. In 1991 NBC paid Carson approximately $30 million.[13] The only talk show host to exceed Carson's salary was Oprah Winfrey, whose syndication contracts brought her an annual $42 million. In 1991 Arsenio Hall received $12 million dollars for his role as host, Phil Donahue $8 million, David Letterman, $6 million (jumping to $14 million in his deal with CBS the next year), Jay Leno, $3 million.[14] These figures were supplemented by income from host-owned production companies that produced films and other television shows, and from real estate and other investments. Oprah Winfrey's net worth in 1991 was an estimated $250 million. That made her, along with Carson, one of America's wealthiest 400 people, according to *Forbes* magazine's annual survey. Television talk makes money for those who own it and produce it. Money is the fuel that drives the machine.

### Rule No. 5: Television Talk Structures Spontaneity

Talk on television is by necessity highly planned. It must fit the format, commercial imperatives, and time limits of commercial television. Though it can be entertaining, even "outrageous," it must never permanently alienate advertisers or viewers.

A good example of the limits of television talk can be seen in the career of talk show host Morton Downey, Jr. Confrontational and abrasive, an "in-your-face" kind of host, Downey had a loyal following for his 1980s syndicated show out of New Jersey. Though critics were frequently repelled by Downey as a personality, they also found him provocative and a welcome change from the blandness of most television. When Downey blew cigarette smoke into the camera and leered challenges at viewers and guests, "blue chip" advertisers did not want their products associated with his show and it was taken off the air.[15]

Downey represents an extreme example, but all television talk shows are regulated by invisible rules of acceptability. Guests are carefully chosen, questions prescreened. In the comedian Richard Pryor's appearance on the Letterman show in 1986, as much as 80 percent of the interview was set up in advance. This was standard procedure in the preparation of guest interviews. The questions, the order of the questions and Pryor's possible responses were discussed by Letterman with segment producer Robert Morton, who prepared the interview. Morton's notes were in front of Letterman throughout and structured his replies.[16]

The give and take on a show like Carson's or Letterman's is structured by scores of such invisible hands.[Carson's final show lists two executive producers, a line producer, a director, a co-producer, two writing supervisors, six writers, a music supervisor, a musical conductor and assistant musical conductor, three talent coordinators, a commercial administrator, two production managers, an associate director, an art director, a production assistant, two stage managers, a technical director, a lighting director, two

## Major National Talk Show Hosts on Television (1948–92)
### By Robert Erler and Bernard Timberg

| Talk Show Host | Years of National Talk Prominence |
|---|---|
| Faye Emerson (1948–60) | 12 |
| Arthur Godfrey (1948–61) | 13 |
| Arlene Francis (1949–75) | 26 |
| Dave Garroway (1949–61) | 12 |
| Garry Moore (1950–77) | 27 |
| Art Linkletter (1950–70) | 20 |
| Steve Allen (1950–84) | 34 |
| Ernie Kovacs (1951–61) | 10 |
| Mike Wallace (1951– ) | 41 |
| Merv Griffin (1951–86) | 35 |
| Hugh Downs (1951– ) | 41 |
| Edward R. Murrow (1951–59) | 8 |
| Dinah Shore (1951–62, 1970–80) | 21 |
| Jack Paar (1951–65, 1973) | 15 |
| Mike Douglas (1953–82) | 29 |
| Johnny Carson (1954–92) | 38 |
| David Susskind (1956–86) | 30 |
| Barbara Walters (1963– ) | 29 |
| William Buckley (1966– ) | 26 |
| Dick Cavett (1966– ) | 26 |
| Joan Rivers (1969, 1983– ) | 10 |
| Phil Donahue (1970– ) | 22 |
| Bill Moyers (1971– ) | 21 |
| Tom Snyder (1973–82) | 9 |
| Geraldo Rivera (1978– ) | 14 |
| Robert McNeil and Jim Lehrer (1979– ) | 13 |
| Ted Koppel (1979– ) | 13 |
| David Letterman (1980– ) | 12 |
| John McLaughlin (1982– ) | 10 |
| Larry King (1983– ) | 9 |
| Oprah Winfrey (1986– ) | 6 |
| Arsenio Hall (1987– ) | 5 |

Note: Hosts associated exclusively with game shows, even when those shows have centered around clever forms of talk (for example, Groucho Marx in *You Bet Your Life* or John Daly and the panelists of *What's My Line?*), have not been included in the list. Neither have we included long-standing hosts of talk shows that have centered on special topics, like Oral Roberts and Pat Robertson, who had successful religious talk shows; Dick Clark, who has been on the air almost continuously since the 1950s with his shows on music; or Ruth Westheimer with her highly successful shows on sexual topics.[17]

audio technicians, a video engineer, three videotape editors, a head carpenter, a prop master, a head electrician, a production staff of five, a makeup person, a hairdresser, a graphic arts department, a wardrobe supervisor, and various assistants to the producers. Fifty-five names appeared in all, and this represented only the employees of Carson Productions, not the NBC camera crews and studio technicians in Burbank who taped the show. Nor do we see in the credits the scores of network officials who oversaw, budgeted, did legal work, and publicized the show. Over a hundred talk show professionals put a show like *Tonight* on the air each evening.

The rules of television talk are clear. They have been formulated over forty years of television practice, an inheritance of radio and performance traditions that preceded television by many years. Host-centered, a commodity in competition with other commodities, topical and spontaneous within professionally prescribed limits, these are the unspoken rules that guide television talk.

## Notes

1. Hal Gurnee made these remarks in a seminar co-sponsored by the Museum of Television and Radio and the Directors Guild at the Museum of Television and Radio on May 26, 1992. Gurnee was the person responsible for my unusual access to the Letterman show 1983–86.

2. These figures come from Christopher H. Sterling and John M. Kitross, *Stay Tuned: A Concise History of American Broadcasting,* Belmont, Calif.: Wadsworth Publishing, 1978, Programming Charts: pp. 519–26.

3. Ibid., pp. 528–32.

4. See Frank Mankiewicz, "From Lippmann to Letterman: The 10 Most Powerful Voices," *Gannett Center Journal* 3:2 (Spring 1989), p. 81. In Mankiewicz' decade by decade summary, television hosts and news anchors have progressively assumed the role once played by nationally syndicated newspaper columnists, cartoonists, and radio commentators as influencers of public opinion.

5. With his celebrity interview show *Person to Person* from 1951–61, Edward R. Murrow was a founder of entertainment talk as well as news talk. Murrow resisted the celebrity orientation of *Person to Person* at first. One of his early shows featured a mail carrier in Harlem, for instance. But non-celebrity interviews did not do well in the ratings and were abandoned. (Joseph Persico, *Edward R. Murrow: An American Original,* New York: McGraw-Hill, 1988, p. 345.) *Person to Person* was a forerunner of the Barbara Walters specials and other talk shows based on one-on-one celebrity visits and the work of the so-called lower Murrow established some of the earliest traditions of entertainment talk on television.

6. Thomas Schatz uses de Saussure's terms *parole* and *langue* to describe the impact of certain Hollywood films on the "language structure" of Hollywood film genres. The same could be said of the impact of certain talk show hosts. Their shows redirect the development of the "language" of the television talk show.

7. The term is borrowed from Erving Goffman's *Forms of Talk,* Philadelphia: University of Pennsylvania Press, 1981. The different kinds of fresh talk on television correspond in interesting ways to the four "aims" of discourse theorists have posited for social speech: aims of information, entertainment, persuasion, and, more

recently, self-expression. Television talk exhibits, in one form or another, all four of these aims. As is the case in written or spoken language, these aims frequently overlap. Each of the major traditions of television talk—that of the news, the entertainment talk/variety show, the reframed event, and the sales pitch—privileges one or another of these four major aims of speech. For a detailed discussion of the discourse aims of television talk shows, see *Titans of Talk,* Chapter 2, "The Range and History of Television Talk Show," Austin: University of Texas Press, 1994. For more detailed general discussion of the aims of discourse and their history in written and spoken speech, see James L. Kinneavy, *A Theory of Discourse,* Englewood Cliff, N.J.: Prentice-Hall, 1971.

8. *Mrs. Roosevelt Meets the Public,* originally titled *Today with Mrs. Roosevelt,* ran from February 12, 1950, to July 15, 1951.

9. He also allowed his writers to do outside work, and many went on to other writing jobs after two or three years with the show.

10. Tony Geiss, who was an associate producer on *Late Night with David Letterman* in its first year on the air.

11. Figures supplied July 28, 1992, by Gene Cunningham, research director for Blair Television, from sources in the NBC Sales Department. Admittedly, the cost of advertising time was especially high during Carson's final week. The NBC sales department hiked the rate considerably; normally a thirty-second spot would cost from $30,000 to 35,000 and this would have been a $30,000 joke.

12. The show had a record audience that night, with approximately 50 million viewers tuning in.

13. These figures are from a 1992 *Forbes* magazine article.

14. Ibid.

15. The source of this information is Fred Gold, Research Director of WWOR-TV at the time of the Downey show.

16. Chapter 9 of *Titans of Talk,* in the part of the book that goes behind the scenes to show how a talk show is produced from the control room and the studio floor, describes the detailed preparation that goes into a show of this kind.

17. Other prominent hosts of special topic talk shows include Julia Childs with her cooking shows and Bob Vila of *This Old House* on PBS. Among the hosts who upped the ante of confrontation on television talk shows were Joe Pyne, Alan Burke, Les Crane, and Morton Downey, Jr. The confrontational talk show hosts tended to have only a regional impact or did not last for long as national shows (sometimes because of sponsor decisions).

The decision to exclude some of the most prominent hosts of game shows was a particularly difficult one for us, since game show hosts sometimes represented some of the most entertaining and original forms of fresh talk on the air at the time. Certainly, the talk of Groucho Marx on *You Bet Your Life* was as important as the game, and, as David Marc and Robert Thompson point out in *Prime Time, Prime Movers* (Boston: Little, Brown, 1992), *What's My Line?* was as much a talk show as a game show, featuring such well-known raconteurs as book publisher Bennett Cerf, syndicated columnist Dorthy Kilgallen, and poet Louis Untermeyer. Ultimately, however, we decided that the talk of the game show revolves around the game that is central to the show's format, and the ratio of talk to game, as well as the role of the host or master of ceremonies of a game show, should be left to a book that centers on that genre. Similarly, though hosts of special topic talk shows do take on national prominence and certainly influence the form of television talk—Dr. Ruth

Westheimer and her appearance as a guest on national talk shows signalled a new openness and legitimacy to discussions of sexual topics, and Pat Robertson became a national figure through his discussions of political and cultural issues as well as religion—these hosts are primarily identified with the topics that are their special concern rather than general news and entertainment programming.

# Tell Me More:
# Television as Therapy

## MIMI WHITE

"The Shows That'll Make You Feel Better," by Dr. Joyce Brothers; "A Psychiatrist Looks at Prime Time," by John P. Docherty, M.D.; "Prime Time on the Couch: A Psychiatrist Wonders What's Happening to Romantic Love," by Willard Gaylin, M.D.; "Your Therapy Could be Watching *Dallas* or *Dynasty*," by Teresa Kochmar Crout; "Can TV Cause Divorce?" by David Hellerstein; "Dr. Ruth to the Golden Girls: How to Keep That Spice and Sparkle in Your Lives," by Dr. Ruth Westheimer—these are only some of the articles that were published in *TV Guide* over the course of the 1980s.[1] They are all notable for promoting the idea that television functions therapeutically within a familial and interpersonal context. Watching television can help or hinder your relationship with your spouse or children. Television can speak a therapeutic discourse and is equally open to being addressed from a therapeutic perspective. On more than one occasion in the pages of *TV Guide,* Dr. Ruth Westheimer gives advice about their behavior to television's fictional characters, while Dr. Joyce Brothers tells readers which programs to watch in order to set a romantic, humorous, or cheerful mood at home.

These articles coincide with the rise of the therapy or counseling show on television in the 1980s, a hybrid subgenre of reality programs and talk shows.[2] A number of these shows gained national prominence, including two programs featuring Dr. Ruth Westheimer, first *Good Sex! with Dr. Ruth Westheimer* on cable and then *Ask Dr. Ruth* in syndication.[3] Throughout the *TV Guide* articles and the counseling programs, television is promoted as both therapeutic subject and object from a variety of positions and perspectives. The medium functions as a therapist; it sometimes needs the advice of a therapist; it may promote symptoms among its viewers that will lead them to need therapy from an external source; or, indeed, it may lead viewers to the therapists who have programs on television.

The *TV Guide* articles are significant because they repeatedly associate television and therapeutic discourses in the largest-circulation magazine published in the United States. Even if people who subscribe to *TV Guide* do not actually read the articles, at a minimum the headlines, often featured on the cover (and therefore on display at checkout counters in countless grocery and drug stores), promote the idea that therapeutic issues are relevant to any consideration of the process of watching and understanding television. *TV Guide* is a magazine that basically helps people watch television. It includes a detailed weekly schedule of what shows will air when (often including plot summaries), rates movies that are appearing on cable and broadcast stations, provides information on what is happening at various networks and stations, previews forthcoming programs and specials, profiles television personalities, and more generally addresses current issues in television programming (the article titles cited above fall into this latter category).

With this array of information, viewers who read *TV Guide* participate in current institutional and critical discourses regarding television. In this light, it is important that the magazine has regularly promoted recognition of therapeutic discourses on television. Popular consciousness of the therapeutic thereby becomes an integral dimension of television viewing. This is not, strictly speaking, limited to *TV Guide;* rather, I am using the magazine as an exemplary instance of popular media that promote particular ways of watching, and understanding, television programs, personalities, and institutions.[4] In *TV Guide* awareness of the therapeutic is evinced in articles that highlight the counseling programs of the 1980s as an emergent genre, but it is also evident in more sociologically oriented analyses of how television might have an impact on our lives and in therapeutic analyses of fictional television characters.

Taken together, these articles suggest a pervasive and encompassing view of therapy as a proper, even natural, discourse to conjoin with television. In other words, the idea that television functions as a therapeutic apparatus, and should be explored in these terms, is an integral part of the everyday discourses and practices of regular television viewers. In this [essay] I examine these articles, with an emphasis on the ways in which the viewer is invited, or positioned, to participate in television's therapeutic discourses. I then look at one of the more prominent therapy programs of the 1980s, *Good Sex! with Dr. Ruth Westheimer,* as an example of the most literal (and public) generic manifestation of television's therapeutic practice.

## TV Guide: *Watching Television Therapeutically*

The variety and range of articles in *TV Guide* that invoke therapy make it clear that therapeutic and psychological issues pervade the medium. Characters on television are subject to being analyzed; watching television might have beneficial or detrimental effects on your familial or romantic relations; you might learn something by watching television therapy shows, or even more by participating in one; it might even be appropriate to watch television as part of your own therapy. One article details a psychiatrist's use of television as a part of therapy:

> Dr. Young, a Dayton, Ohio, psychiatrist, has been prescribing television to help patients increase self-awareness and develop more effective ways of coping with problems since 1963. What's more, he routinely discusses the behavior of TV characters at therapy sessions, often underscoring his points by using instant replay—a viewing room complete with TV set, VCR and a collection of selectively recorded tapes occupies a prominent place on the second floor of Young's spacious office.[5]

As the doctor explains, "We can learn a lot about ourselves by watching our TV twin."[6]

The professional imprimatur for all of *TV Guide*'s perspectives on therapy and television derives from the stature of the authors—frequently identified as psychologists, psychiatrists, or M.D.'s—or from references in articles to professional experts and associations, such as the Association for Media Psychology, the American Psychological Association, or the American Psychiatric Association. However silly the positions expressed in the articles might seem at first glance, the reader is given the necessary cues to recognize the position of authority that sanctions them. The article on the psychiatrist who uses television as an integral part of therapy falls into this category.

As the article states, the psychiatrist in question has been using television successfully in therapy with patients for almost thirty years. Yet toward the end of the article a negative note is sounded by some comments on the potential pitfalls of Dr. Young's strategy that seem especially to underscore the frivolous dimensions of the whole idea of using television in a clinical context. "Sometimes," the article reports, "Dr. Young's prescriptions do go awry. Like the time a patient was following Pam and Cliff's mother [Rebecca] on *Dallas*. 'All was progressing well,' recalls Young, 'until Rebecca's untimely death.' And, of course, dozens of male patients needed new prescriptions following the unfortunate demise of Bobby Ewing."[7] (The reader familiar with *Dallas* will no doubt wonder what happened to those patients when the character of Bobby was revived on the series a year later!) Still, as if to assure readers that they should take this seriously, the article ends with an authoritative statement to support the linking of fiction and therapeutic practice. "I don't see any problem with it," says psychiatrist Jerome Logan, referring to Dr. Young's somewhat unorthodox approach. "It's another tool to stimulate thought on the part of the patient and facilitate interaction between patient and therapist. And if a patient is willing to follow through, it speeds up the therapy." Fiction and dramatization are often used to describe and promote understanding of behavior, adds Logan. "We talk about the Peter Pan Syndrome and the Cinderella Complex. And even Freud depended on Oedipus to get his point across."[8]

Overall, *TV Guide*'s perspectives on television and therapy are structured by contradictory arguments and points of view. This is systematic in their approach to the subject. No decisive opinion or point of view regarding the issue is ultimately allowed to hold sway. In relation to television's therapy programs the terms of the debate are relatively obvious: Are the programs helpful and educational or harmful and distorting? Do they promote more openness and understanding about emotional problems or turn real human suffering into television spectacle for the sake of profits? One article explains the dilemma as follows: "Critics worry that the advice given on TV may be hazardous to a viewer's health. . . . And while TV may demystify the counseling session, it may also distort it. The problem that a patient presents, after all, generally disguises a much deeper psychological question, and it is this that a therapist, with skill and sensitivity, seeks to uncover."[9] Yet in the same article the author writes, "TV can provide current information on common problems. It can, while respecting privacy, encourage the discussion of feelings. Whether the present state of the art is helpful or harmful, however, remains a matter of heated debate."[10]

What is less obvious are the ways in which, over time, the magazine publishes articles by experts representing more distinctive positions and points of view that nonetheless implicitly contradict one another. This may occur over the course of weeks, months, or even years. Thus, an article about how watching television can enhance your romantic life may appear at one time, and an article about how television has contributed to the decline of romantic love may be featured at another.[11] Similarly, one article by a Ph.D. psychology professor advises watching one of the couples on *L.A. Law* to develop successful intimacy, while another by a psychiatrist suggests that television has a profound and primarily negative effect on marriages.[12]

Even within articles by experts there are often subtle, and sometimes blatant, contradictions. Psychiatrist David Hellerstein's article "Can TV Cause Divorce?" for example, begins with the strong implication that the effect of television on couples can only be negative, but by the end of the article the author suggests that a number of shows make beneficial contributions to marriage. And the same program may be referred to by experts as exemplifying both the positive and negative impact of television on relationships. In "Can TV Cause Divorce?" Bill Cosby is cited as a potentially negative influence because he creates a character that real-life husbands cannot match. "To a large degree," the author states, "the couples I talked to gauged the success of their marriages by what they saw on TV. In one couple, the wife is a devotee of *The Cosby Show*. She told me, 'I always compare how my husband deals with a situation with what Bill Cosby would do.' When her

husband is crabby or nasty, or can't defuse an argument with a warm Cosby-
esque joke, she ends up feeling angry, even cheated."[13] Yet, in the same
article, *The Cosby Show* is also mentioned as an example of new, positive
tactics in the depiction of couples on television, with beneficial influence.
Hellerstein explains: "Take *Cosby*, for example. Dr. Huxtable is funny, warm,
and strong, as is Clair, his wife. When they fight, they generally end up
solving their problems—with style and humor. Granted, *The Cosby Show* is
sentimental at times, and it doesn't deal with many of the complex, irresolv-
able problems that real life often poses—but still, it represents a step forward for television."[14]

Another article, which deals with the dangers and value of television
for lonely people, is no less equivocal. First the author argues that viewers
develop an unreal sense of community with figures on television that substitutes for actual human relationships. "Some viewers actually go to bed
feeling someone out there is connected to them," she complains.[15] Then
later in the same article the author proposes that television can be a benefit
to lonely people by providing them with a sense of community.

> There are at least two ways, however, in which television provides com-
> panionship not only to the very lonely, but to the rest of us as well. One
> of the feelings that plague lonely people borders on self-pity—"I'm all alone
> in the world. Nobody knows the trouble I've seen." Some television shows
> address and assuage this sense of being a pariah. . . . To help people feel,
> "I am not alone in this; others have similar experiences"—whether or not
> solutions are offered—is a tremendously reassuring aspect of the best of
> television's talk shows.
>     Even more important is the sense of community that television can
> provide. Since lonely people feel they are not part of anything larger than
> themselves and have no sense of belonging, television can give them a
> common language and a common set of references at times. . . . Watching
> serious coverage on television addresses the need to gather at the river; it
> is the modern version of an ancient ritual.[16]

The author concludes the article by neatly summarizing this equivocation.
"Television—a box with a picture tube—neither causes nor cures loneliness,
but can be *used* as a pal or as a trap for the lonely."[17] In other words, the
article is structured to pose a question, as it does explicitly in its title, "Is
TV a Pal—or a Danger—for Lonely People?" It proceeds to answer the
question by presenting conflicting and contradictory arguments, and it concludes by indicating that in the final analysis there is no definitive answer to
this question.

Through the course of these articles, the psychiatric and psychological
perspectives presented emphasize television's therapeutic function. But the
precise or final value of this function is indeterminate, or at least not gen-
eralizable. This equivocation is typical of American television. The articles
explore and give voice to "both sides of the issue," following the traditional
American journalistic formula for objectivity. Pro or con, good or bad, right

or wrong—on any issue, be sure both of these perspectives are incorporated; to address only one is to editorialize (and thereby potentially alienate readers, advertisers, and broadcasters). Moreover, the very format of *TV Guide* not only follows this rhetorical model of objectivity but is also highly formulaic, including as one of its primary principles the importance of brevity in presentation. Therefore, the nuances of an argument and complexities in an author's position may be lost, emerging instead as blatant forms of contradiction. And clearly the titles of the articles are designed to attract attention, no matter what the specific arguments of the author may be.

Yet it is crucial that within the general contours of proper journalistic objectivity, these articles—and others like them—persistently assert the importance of watching television in the terms of a therapeutic relation. The articles also suggest that television does have social effects and therefore warrants being taken seriously by the likes of psychiatrists and medical doctors. But since the exact nature and degree of these effects are ultimately equivocal, all viewers as individuals have to take the medium seriously by continuing to watch. Indeed, following these articles, watching television is as likely as not to have positive effects. Psychiatrists watch it and write articles in *TV Guide* for people like you and me; and one in Ohio goes so far as to have his patients watch television during their therapy sessions. Even if you find these positions and opinions suspect, you have to watch some television to confirm its status as a serious social apparatus.

Thus the articles not only encourage a particular posture of viewing—the therapeutic relation—but also promote watching in general. In addition, from the therapeutic perspective advanced in these articles, it is mainly television's fiction programs that are considered important, rather than the news and documentary programs that are more typically considered to fulfill the serious social mission of the medium. As a result, the association of therapeutic discourse with television programs and television viewing also functions as a way of making the medium's entertainment programming a matter of serious consideration and a worthwhile object of viewing activity.

At the same time, by promoting a conflict of opinion about the precise value and meaning of the therapeutic transactions at stake in television, the articles encourage readers (and viewers of television) to make their own assessments. The issue of therapy is set in place for the public from a perspective of authority, an authority premised on the very pervasiveness of the issue as a topic of investigation. But, once the issue is set, readers/viewers are implicitly enjoined to make their own judgments about it. Here journalistic objectivity works to support freedom of opinion among that portion of the population that reads *TV Guide* and watches television. The terms of the debate are set through discourses of television and bolstered by reference to professional authority. However, one cannot contest the existence of therapeutic discourse, since that provides the a priori ground for all these articles. By the same token, the freedom to dispute the value of therapeutic discourse promotes a certain skepticism regarding the professional authorities themselves; they do not have the final word. This too serves to facilitate further viewing since you have to watch in order to assess the situation for yourself, to exercise your freedom of opinion.

Readers/viewers are positioned to judge for themselves when and whether they take the discourse of therapy seriously, even while they are trained to see it widely deployed through the medium. To further complicate the situation, the articles specifically dealing with therapy shows note that although various psychological and psychiatric associations recognize "media therapy" as a practice, they include guidelines that discourage specific diagnosis or treatment on the air. The educational value and impetus behind these shows is identified with the fact of televising general discussion about problems that may be common to many people and encouraging (confessional) talk about one's feelings. In other words, the precise value and meaning of therapy programs is indeterminate; and the imprimatur of professional associations really only applies to programs that do not actually do therapy—at least in conventional terms—in the first place. However, a viewer of the programs would also note—as some of the articles do—that television therapists often engage in diagnoses and directive individual advice that is presumably not allowed by the very professional associations that nonetheless endorse a more generic deployment of therapeutic and psychological issues on television.

These articles, and the programs and viewing strategies they describe, further suggest that conventional therapies are losing their force in contemporary culture, especially in a culture dominated by television consciousness. "TV is not what would happen in an office," says Dr. David Viscott, host of the syndicated therapy show *Getting in Touch.* "But you also have to be a realist. Most people watching television have about as much attention as a morning glory. You are competing with all this 't and a' and also the religious stations."[18] What we are offered is a therapeutic relation that is skeptical of too much authority yet attracted to the celebrity aura that accrues to people on television (as therapist or patient), that introduces more voices in a less hierarchical therapeutic transaction yet infuses all relations with the commodity and consumer effects of the television apparatus (at least as it is constituted in the United States).

If we move beyond the counseling program strictly speaking, we find manifestations of therapeutic discourse permeating the medium. Therapy in all its manifestations and permutations is thus situated as a strategy of engagement. It is a way to attract viewers or a posture a viewer may (perhaps even should) assume in relation to a wide range of programs. The therapeutic is hereby recognized as a discursive strategy, establishing relations between television and its viewers in two directions. Indeed, in the terms of the therapeutic relation, the medium does not simply exercise social effects; it is itself the object of diagnoses that come from viewers. Therapeutic discursive strategies are thus represented as having the potential to overcome the technological limits of one-way communication dominated by the box, as viewers are enjoined to treat television and the people who populate it as patients. In *TV Guide* this posture is modeled in articles by Dr. Ruth Westheimer and Dr. Joyce Brothers, two therapists whose careers and celebrity are intimately bound up with television and radio. Dr. Ruth lets us share her professional perspective on Maddie and David in *Moonlighting* as well as her advice on healthy sex lives for television's Golden Girls, while Dr. Joyce suggests programs that will make us feel better if we watch them.

In *TV Guide*'s promotion of therapy as a constitutive component of television viewing, soap operas and game shows are often mentioned as the generic precedent and functional equivalent of television's therapeutic vocation. The previously mentioned article on lonely people and television notes that soap operas are particularly popular among those who are lonely: "They have pseudo-relationships with the people on the soaps that are very much like real-life relationships—you know all the characters' secrets, how their houses are furnished and what's inside their refrigerators." [19] In another article the program *Crisis Counselor* is specifically described in the terms of a soap: "In true *General Hospital* style, every counseling session seems to contain a startling confession." [20] The impulse that leads people to participate in and watch television counseling programs is compared to interest in the kinds of daytime couples shows that will be discussed in the following chapter. "Down deep in that part of the national psyche to which we seldom admit," one *TV Guide* article explains, "that impulse that has fed for twenty years upon shows like *Divorce Court* and *The Love Connection* still craves something more. For the unsatiated, real people and their real soap-opera lives may be the only things left. Not only for viewers, but for those who so desperately want to use television as their personal confessional booth." [21]

In context, these comparisons are meant to be vaguely derisory. Television's therapeutic discourses are implicitly associated with a specific, gendered audience, sharing with the soap operas and game shows of daytime television a predominantly female audience. This is especially the case with counseling and therapy shows, most of which have existed as syndicated, cable, or local programs airing during the daytime or as fringe programming. In the 1980s the Lifetime Channel was one of the main places where one could find different therapy and counseling programs, including *Couples,* with psychiatrist and family therapist Dr. Walter Brackelmanns . . . and Dr. Ruth Westheimer's first television show, *Good Sex!*. Lifetime started out as a medical and health-oriented cable station. During the course of the 1980s it evolved into a "woman's" cable station, showing a range of talk shows, counseling shows, and reruns of situation comedies and dramatic series, as well as featuring some original fiction series,[22] all considered of primary interest to women viewers (while reserving Sunday for more serious medical-issues talk shows).

Generic associations notwithstanding, the therapeutic dimensions of television are by no means limited to daytime programming. In *TV Guide* articles, not only soap operas and game shows but also prime-time serials, prime-time dramatic series, situation comedies, and made-for-TV movies are included as grist for the therapeutic and confessional mill. The therapeutic mode may be most strongly associated with women's genres and female viewers, but it is also popularly associated with the medium as a whole, or at least with the full range of daytime programs and prime-time fiction.[23]

## Good Sex! with Dr. Ruth Westheimer

Dr. Ruth Westheimer is a celebrity sexologist who rose to national prominence in the 1980s. Her television show on Lifetime was an offshoot of a radio sex-counseling program and was developed for television in the wake of a series of successful appearances by Westheimer on TV talk shows including *The Tonight Show with Johnny Carson, Late Night with David Letterman,* and *Donahue. Good Sex!* combines elements of the radio call-in show and television talk show formats. During the course of the show, Dr. Ruth takes phone calls from around the country, responds to letters on the air, engages in dramatized therapy sessions, and has a celebrity guest with whom she talks (usually about romance, family life, and sex) and who stays with her while she continues to take phone calls from viewers. The guests are usually entertainment figures or authors, not other psychologists, therapists, or sexologists. During segments where Dr. Ruth responds to phone calls and letters, she shares the set with co-host Larry Angelo, who helps with introductions and with transitions from one segment to another or between program segments and advertisements, reads letters aloud, and chats with Dr. Ruth.[24]

Each episode begins with a printed verbal title which is also heard on the soundtrack, establishing Dr. Ruth's credentials and the sexual purport of the show. The double coding—verbal title and voice-over reading—assures that the program is addressed not only to people actually watching but also to those who are within earshot of the television. Nightly, as the program opens, the viewer is told:

> The following program was pre-recorded.
> It includes explicit sexual references and may not be suitable for all viewers.
> The program is hosted by Dr. Ruth Westheimer, noted sex therapist and adjunct associate professor in the sex therapy teaching program at New York Hospital–Cornell Medical Center.

The implication that the program is not appropriate for all viewers—that there is potentially something improper about it—is immediately counterbalanced by the clarification of Dr. Ruth's institutional standing and reputation. However you may feel about explicit sexual references, the ones that you may hear on this show are clearly authorized. But, interestingly, the program concludes with another title that casts this authority in more relative terms. "The opinions expressed in the program represent the opinions of Dr. Ruth Westheimer and may not represent those of other qualified professionals."

In part these disclaimers and qualifications are a function of television's normal standards and practices and of the legal apparatus to which the medium is subject. Subscribers to a cable service carrying the show might be shocked to discover the expression of sexually explicit material that they find unsuitable for themselves and their family. The program might otherwise be liable to lawsuits from unhappy viewers whose problems are exacerbated, rather than solved, by Dr. Ruth's advice. However, the tension between qualified and unqualified promotion of the star-host's credentials is emblematic of the tensions that pervade the show as a whole. This includes, notably, the tension between Dr. Ruth's status as a serious medical authority on the one hand and her status as a popular celebrity in the world of mass media on the other, and the concomitant possibility of confusion as to who speaks with what authority at all on the program, especially regarding the physical, emotional, and psychological problems and issues at stake in the investigation of sexual matters.

Sexual literacy and mature, informed choice are a recurrent concern of the show, and Dr. Ruth is represented as an authoritative educator, eager to discuss and promote information that will help viewers as sexed and sexually active individuals. In this capacity she excels as the therapist, enjoining individuals to speak freely by reassuring them that their questions and concerns are important and intelligent and that they are not alone in their concerns, problems, or lack of knowledge. Issues discussed in this vein include masturbation, oral sex, orgasm, loss of desire, need for affection, virginity, and impotence. The information and advice dispensed to callers is tempered, however, by Dr. Ruth's intermittent avowals that she cannot "do therapy" over the telephone. This cautionary tactic follows the previously noted guidelines of professional organizations regarding media psychology and therapy. It also puts in place a division between formal clinical therapy and the more diffused social therapy practiced by the show. Here, Dr. Ruth's program exemplifies the tension discussed earlier in relation to *TV Guide* articles about whether what we see on television is or is not therapy—or what kind of therapy it in fact is. In practice, the distinctions between individual clinical therapy, generalized therapy, and education become increasingly indeterminate. This is especially true since each individual call may become the occasion for focusing on a particular problem and strategies for its resolution and also for drawing broad generalizations about sexuality and human relations.

Throughout the program, confession on the part of participants—including those who call in or write, the actors in the staged therapy sessions, and the celebrity guests—enables Dr. Ruth repeatedly to rehearse two simultaneous and intertwined discourses. One of these has to do with the techniques of the body, while the other pertains to issues of morality and limits. Certainly, as a celebrity sexologist Dr. Ruth is best known for the former of these. She is the person who has brought frank talk about sex into the media in the name of education. Thus, discussion of masturbation (for men and women), use of words like "penis" and "vagina," and extensive talk about the importance of foreplay, to take only a few examples, are essential, persistent components of the program (and of all of her appearances on television). Yet this candid talk is accompanied by a strong and equally persistent moral framework. Both on the phone and with the celebrity guests, Dr. Ruth herself frequently giggles in the course of her discussions. In this way she suggests that there is something amiss in all this discussion of sex and pleasure, that it is indeed forbidden territory. There is at least the hint that she is being a "bad little girl" by discussing all of this so openly and frankly in public. (In this context, Dr. Ruth's diminutive size—which works hand-in-hand with the "bad little girl" impression—and pronounced German accent guarantee that she is not identified with conventional representations of feminine beauty in mass culture.)

More overtly, Dr. Ruth has certain perspectives and opinions about relationships that are stated over and over again in the course of her programs. These include the following: whatever two mature, healthy adults consent to constitutes acceptable behavior, as long as no one is coerced; you should have sex only in the context of a monogamous, loving relationship; and you should remain a virgin until you find someone you really love, and with whom you expect to sustain a monogamous, loving relationship. Moreover, Dr. Ruth does impose limits, believing, for example, that no activity that might cause physical pain or injury to one of the participants in a sexual relationship should be sanctioned. (Though sadomasochistic activities are rarely discussed, it is very clear that she does not approve of them.) Sexual pleasures and sexual techniques are thus circumscribed and confined in conventional social relationships.[25] For all Dr. Ruth's image as the popular maven of sexual openness, the values that she purveys are relatively conventional.

Dr. Ruth does not shy away from giving specific, directive advice to some individuals who call her, drawing precise conclusions about the nature of one case or another. In these instances her authority holds sway. For example, one woman called in and explained that she was living with her fiancé, who had recently been staying out all night, and she did not know what to do. Dr. Ruth assured her that she was "the luckiest woman in Texas" because this happened before she was married and had children. "You are very fortunate that this is happening now, and not later in your life." This diagnostic interpretation of the caller's situation accompanied instructions to the woman to break up with her fiancé immediately. The woman ended the conversation by thanking Dr. Ruth for her advice. "I knew you'd help me make up my mind." In other words, Dr. Ruth was able to confirm what the woman already suspected she should do.

Dr. Ruth is also eager to promote general understanding of issues related to human sexuality, contributing to a "sexually literate" society. Some questions from callers set the groundwork for this approach with their informational format: What is the definition of masturbation? What is the difference between a transvestite and a transsexual? Other questions provoke directive advice for the individual caller that can in turn be generalized to apply to viewers with similar problems. The program itself strives to confirm that on a routine basis the advice given to individual callers in fact extends to many viewers. This is, for example, the clear purpose behind Larry Angelo's reading at the start of one show a letter from a housewife with three children who writes that she has been able to correct a problem in her sex life thanks to the advice offered to another caller to the show. Still, information and advice, however directive, are not equivalent to therapy, although to the degree that they are diagnostic, they may initiate a therapeutic process.

While Dr. Ruth does not actually practice therapy on the air, the importance and variety of potential channels for therapy are centrally and repeatedly endorsed. Dr. Ruth suggests to many of the callers that they need to talk further, with a social worker, a therapist, a minister, a mother, an aunt, or *somebody*. What is frequently and forcefully underscored is the necessity of talking with someone else, of continuing to talk. A litany of sources to consult accompanies most of the recommendations made to callers. Sometimes the recommendation specifies a marriage counselor, sex therapist, or physician; other times a general list from which a sympathetic ear can be selected is offered instead. In these instances, therapeutic discourse is not the sole possession of Dr. Ruth, or of television. Yet as the originary instigator of confession, television therapy is rhetorically situated as a significant force in the initiation and dissemination of therapy in the everyday life of program viewers. And, crucially, while sex is important as the focus of all this talk, talk itself increasingly becomes its own end.

Dr. Ruth situates herself simultaneously as an authoritative source of information and as a conduit for articulating problems that can finally be dealt with only elsewhere. Indeed to the extent that she cannot officially "do therapy" over the telephone on television, what she does is serve as a relay for personal confessions in a public medium. In any case, her essential function as one who motivates confessions is crucial. And her ability to function in this way hinges on her celebrity personality. Callers frequently initiate (or end) their phone calls by telling her how much they enjoy her program, a statement of fan club adoration that usually elicits appreciative giggles from the program's star.[26] One episode opened with a discussion between Larry Angelo and Dr. Ruth about her forthcoming role in a feature film starring Sigourney Weaver and Gerard Depardieu, foregrounding her additional talents as a dramatic actor/star.

This combination of authorizing roles for Dr. Ruth—as therapist, celebrity, and actor—is reinforced in the dramatized therapy sessions that occur in the course of the show. These staged sessions have Dr. Ruth playing herself (Dr. Ruth Westheimer, sex therapist), while actors assume the roles of patients. The sessions are extremely brief relative to the traditional fifty-minute session, but they are significantly longer than most of the phone calls or letters to which she responds (though overall more time is devoted to phone calls than to the dramatic sequences in each show). In these sessions Dr. Ruth and her patient review the latter's primary problems, and

Dr. Ruth recommends a course of action for the patient. This provides an opportunity for the most extreme intervention and directive advice on the part of Dr. Ruth, since she is dealing with the patient face to face. Yet this occurs in a context where the simulational apparatus is fully apparent, as dramatic reenactment.

These scenes provide the most familiar and conventional form of therapy offered on the show, which is immediately identified with fictional simulation and with the dramas we see elsewhere on television. In the process, any clear-cut distinction between professional therapeutic competence and dramatic acting talent is also called into question. The sessions are introduced with a voice-over narration by Larry Angelo that briefly explains the situation in each: a couple's sex life has suffered since they had a baby; a woman's fiancé called off the wedding and announced that he is gay; a nineteen-year-old male is unhappy and embarrassed because he is a virgin, though he now has a girlfriend with whom he would like to have sex. Although the show presents only short scenes of therapy, it implies a larger context extending beyond the confines of the program that sanctions aggressive intervention. Dr. Ruth will give patients specific advice or instructions for the next week, clearly indicating that she expects to see the patient again even though the program does not usually show follow-up sessions. In this way the representational practices of the program demonstrate therapy in one-time sessions, even though an ongoing therapeutic relation is implied via narrative conventions (e.g., verbal references to prior or future sessions). This in turn raises questions about whether the program resembles or replicates therapy in the conventional sense or whether we can even continue to conceive of conventional therapy as a reference point of normative understanding.

The associations between Dr. Ruth as a therapist and the domain of dramatic fiction extend beyond the dramatized therapy sessions on the show. As mentioned earlier, Dr. Ruth's status as the star of the show is repeatedly underscored by the people who phone in. Over and over again, calls begin with people enthusiastically extolling her talents and talking about how much they like her. In part this is a device to get the conversation going, a way to forestall talking about "the problem." But it also prolongs the call, extending the time the caller and Dr. Ruth-as-celebrity are in contact. During the calls, Dr. Ruth herself may raise models from fiction in her responses to questions posed by callers.

In one instance, a woman explains that she has a lot of one-night stands. She enjoys the sex a lot but realizes this is not good for her self-respect. She concludes by asking Dr. Ruth how normal this behavior is. The doctor responds by saying that this is not normal at all; the caller will never find a relationship this way. Dr. Ruth is glad to hear the woman has moved back home as a way of curbing her activity and is lucky she is getting out "without getting hurt like Mr. Goodbar." The reference here is to Judith Rossner's *Looking for Mr. Goodbar,* a best-selling novel (made into a film by the same name), based on a real event but fictionalized for popular consumption, in which a woman who routinely engages in one-night stands is murdered by a man she picks up for casual sex. Finally, Dr. Ruth has published articles in *TV Guide* addressed to characters on prime-time fiction series, giving advice on sex and relationships.[27] All of these combine to suggest that therapy is intimately linked to popular fictional scenarios. In this case, television and therapy find their common ground in dramatic fiction.

Throughout the show Dr. Ruth maintains a posture of moral authority, a position supported by her dual identification as doctor and celebrity. While she insists that any activity engaged in by two consenting adults is acceptable, she even more frequently harps on the importance of a loving, monogamous relationship as the necessary foundation for good sex. All of this must be displayed and diagnosed through confession. Here the talk is more important than the act, since it is through confessional discourse and therapeutic relations that the program exists to begin with. The sexual act (itself multiple: good sex, orgasm, masturbation, what-have-you) is not necessarily the goal, but rather the alibi, of the discourse that maintains the program on television. And it is the maintenance of the program that sustains Dr. Ruth's own visibility as a star/celebrity and sustains viewers as participants within the therapeutic problematic.

## *The Program = The Relation*

The nature of therapy that emerges through the course of *Good Sex!,* and through the course of *TV Guide* articles that address questions of television and therapeutic discourse, results in an impression of therapy in which the positions of therapist and patient are increasingly relativized. On her show Dr. Ruth is the expert authority as well as the star. As such, she controls the terms of discourse most of the time. And yet her position is decentered in a number of ways. In the first place, and most obviously, her position is affected by the structural and institutional demands of commercial television programming. Whatever she accomplishes is done according to perceived demands for entertainment, and within the circumscribed time limits afforded by the medium. Phone conversations are cut short if they go on too long or if an ad break is coming up. The dramatized therapy session fits into the time between two commercial breaks. In all of these contexts, Dr. Ruth's interventions are controlled and delimited by the medium that carries her. Moreover, in the course of the show there are a variety of indications that she is not the authority, or certainly not the only authority, in place.

Co-host Larry Angelo plays a prominent role as facilitator throughout the program, even though he does not speak to callers directly. He is the one who signals transitions, announces impending commercial breaks, and highlights upcoming segments. In this capacity, he speaks with the authority of the medium itself, rather than with the medical/therapeutic authority of Dr. Ruth. He thereby helps to contain Dr. Ruth herself; and it is hardly accidental that the person who fills this role is male. He also helps promote

Dr. Ruth's star persona, a position that is from certain perspectives at odds with her identity and authority as a preeminent sexologist with advanced degrees. By the same token, according to the program's own logic, celebrity, medical authority, and dramatic talent are in fact mutually supporting.

The status of Dr. Ruth as the show's authority is further limited by the fact that she frequently appeals to her celebrity guests as experts in the domain of sex and relationships. She will ask people such as Willard Scott, Robert Klein, and Isaac Asimov questions about not only their careers but also marriage, parenthood, dating, romance, and so forth. Usually, one round of phone calls takes place while the celebrity guests are in the studio, and callers can pose questions to either Dr. Ruth or her guests. With this arrangement, Dr. Ruth initially assumes a position as the preeminent expert in affairs of the body, and simultaneously asserts her association with more generalized media celebrity, outside the domain of medical and psychological authority. The media celebrity guests are in turn situated—however provisionally—as experts in the domain of sex and relationships, especially since people who call in during these segments may address questions to them. When callers fail to include the celebrities in their questions, Dr. Ruth will often appeal directly to her guests to participate in the discussion, inviting them, in effect, to join her as a peer. In this way, the very strategies of the program blur the terms of distinction between psychologist-sexologist on the one hand and people whose reputation is based in the world of popular media culture (film, television, and literature) on the other.

The celebrities who participate may not even concur with Dr. Ruth in the course of handling a phone call. Such was the case on one show when her guests were the comedy team of Al Franken and Tom Davis (best known for their writing and appearances on *Saturday Night Live*). First they discussed their appearances on college campuses and the "Freudian" routines they incorporate in their act, such as theorizing the nasal stage of development (involving nasal expulsive/retentive behaviors) and exploring the problem of male penis shame, which causes men to hide their penises over and over again in dark places. During the subsequent call-in segment of the program, a woman called with the following problem. She was a nineteen-year-old virgin, engaged to a twenty-six-year-old man, with plans to marry in two years. They shared the same birthday, and he wanted to take her to Hawaii to celebrate. She came from a strict Italian family that opposed the trip. Dr. Ruth immediately said, "Don't go." Meanwhile, Franken and Davis together said, "Go. Go. Go Nina. . . . What are you, crazy?" Dr. Ruth repeated, "Absolutely don't go," while Franken and Davis continued, addressing both the caller and Dr. Ruth, "Go on, nineteen. Lie to your parents. She loves the guy. They're *going* to get married." Dr. Ruth concluded by asking the caller why she was waiting two years to get married, and proposed, "Tell your fiancé you want to get married in six months and go to Hawaii then."

There is never an attempt to reconcile one line of advice with the other. Since Franken and Davis are comedians the audience does not necessarily have to take them seriously; and their on-camera manner is less sober than that of Dr. Ruth. Nonetheless, as guests on her program, they are granted a degree of authority. This is supported by the opening of the sequence, in which Dr. Ruth appeals to them as experts, who can draw on their personal experiences on tour to comment on contemporary values, sex, and relationships among the college student population at large. At the same time, as men giving advice on sex—however apparently tongue-in-cheek—to a young single woman, their narrational position is not necessarily "disinterested."

To call in to Dr. Ruth is not just to seek help, but to become an educator and an authority for others.[28] Viewers who participate by calling in may be credited with the power to provoke educational and therapeutic effects. This recognition can come from both Dr. Ruth and other callers. Dr. Ruth herself frequently comments on the questions posed by her callers. She congratulates them on asking good questions or indicates that the problem they experience is shared by countless others. And in the course of a program, viewers who call in may refer to earlier calls and the value of the questions and comments that have been voiced by others. The callers are also associated with the celebrity status that comes from having appeared on television and from consorting with other celebrities.

This is tempered by the anonymity that callers can maintain, though most callers provide a first name and the city from which they are calling.[29] The information is sufficient to single out the caller as an individual with momentary star/authority status and yet protect the caller from receiving crank phone calls after speaking with Dr. Ruth and her guests. In all my viewing of *Good Sex!,* I encountered only a few occasions when callers failed to provide at least their first name. Thus, the confessional and therapeutic value of the program—and its value in generating star image and celebrity value—involves a circulation of discourse among a group of people rather than an exchange between two people fixed in the positions of therapist and patient.

Occasionally Dr. Ruth is unable to address the problems of a caller, failing as both educator and therapist; and at times she explicitly indicates her inability to help. When the problem described seems to require medical intervention, Dr. Ruth will instruct the caller to see a physician and cut the call short. (Such advice is given, for example, if the caller mentions experiencing pain during intercourse, or describes the common symptoms of a sexually transmitted disease.) There are also cases where the severity and specificity of the problem warrant a degree of intervention (including diagnosis and therapy) that Dr. Ruth is unable to deal with in the brief time allowed by the program. For example, one woman caller was extremely upset because her daughter had just told her that she was gay and that the daughter's best friend and roommate was in fact her lover. Dr. Ruth tried to tell the woman not to cut off the relationship with her daughter. "We do not know the aetiology of homosexuality. Some live the lifestyle all their life. For some—maybe not with your daughter—it's a phase." When the distraught mother interjected, "But she needs a *man*," Dr. Ruth immediately noted, "I can't do justice to this. There's not enough time. I would have to have you here, and hold your hand. Go and talk to a therapist. Not a friend. You do not need sympathy, you need help." Despite this insistence, the call went on for quite a while, although the mother, obviously rattled, was not reassured by any of Dr. Ruth's advice. Here, although Dr. Ruth immediately indicated that she could not help, the call ended up being the longest one she took all day. For the mother the phone call provided an important context for expressing her distress: "It's just such a *terrible* shock."

In some cases viewers may recognize lapses or mistakes in the exchanges between Dr. Ruth and her callers and feel that the therapist's responses are inadequate. In these instances, viewers who recognize the lapses are in a position of superiority over Dr. Ruth, since they are able to evaluate the caller-patient in different, and presumably more appropriate, terms. At times this takes the form of a breakdown in communication in the conversational exchange.[30] In these cases, the flow of information between caller and Dr. Ruth is out of phase. During a call from Barbara in Allentown, Dr. Ruth provides a series of recommendations to the woman, only to learn that the caller has already done what she suggests. The call begins with Barbara explaining that she and her husband have had sexual problems for over a year. They have two children. When her husband gets home from work, he wants sex but performs without any emotion. Dr. Ruth immediately suggests they see a sex therapist. Barbara explains that they went to see one that very morning, provoking a big fight in front of the children. After further questioning, Dr. Ruth recommends they see a man rather than a woman sex therapist. "But he chose a woman," Barbara explains. Dr. Ruth is forced to change her tactics at this point, since the directive advice has already been followed or has been considered and led to alternative results, tacitly rejected before the fact. In either case, her directive interventions here are not "working." Instead, Dr. Ruth says Barbara is lucky that her husband was willing to go to a sex therapist to begin with, and that she *must* give it another try and not let the problem slide.

Sometimes Dr. Ruth will just get things wrong. She may not hear the caller clearly or might misunderstand responses to her questions.[31] Even in the course of the dramatized therapy sessions, she can make a mistake. In the session with the nineteen-year-old male virgin, she at one point referred to his girlfriend by the wrong name. At other times, Dr. Ruth may try to rush through or brush off a call when the tone of voice and comments of the caller make it clear that she is not providing the degree of assistance and advice the caller thinks necessary. In this sense her support and advice are contingent on callers appropriately signaling their gratitude and satisfaction. One male caller was concerned that his wife was "preoccupied" with sex toys. Dr. Ruth suggests that he talk with a counselor. She says that if the wife uses the sex toys and experiences good lovemaking with him as a result, that is fine. "But I hear that there's something that is really bothersome to you," she adds. "Talk to somebody before it gets worse." The man tries to continue the conversation, but she cuts him off, explaining that she has to go, and thanks him for calling. In this case, the really bothersome something is never actually defined; nor is the caller given the opportunity to explain the nature of the problem.

In some cases, Dr. Ruth herself *is* able to probe and get callers to define the gist of their concerns more precisely. In other cases, callers themselves gradually reveal dimensions of the problem that redefine Dr. Ruth's (and the viewer's) understanding of what is at stake. One male caller explained that he had recently moved to the United States. Back in his native Britain, he had sex with his wife almost daily, but here in the States he finds that he is nervous and feels he "can't satisfy her." Dr. Ruth provides a reasonable response to this problem, suggesting that the move might have made him nervous, but cautioning him that if things do not improve in a few more weeks, he should consult a sex therapist. The caller then explains that what is really bothering him is that he found his supervisor's handkerchief in the closet and thinks his wife is having an affair. He goes on to express his anger toward his supervisor, declaring "I feel like shooting him in the groin or something." The sudden shift in tone, and unexpected redefinition of the nature of the problem, throws Dr. Ruth. Rather than pursuing this new perspective on the man's situation, she fairly quickly gets him off the phone: "No, no, no. Just do go and see a professional and see what the story is. Thanks a lot for calling."

All of these various interactions between Dr. Ruth and callers combine to relativize the ways in which audiences might understand the therapeutic transaction. It becomes a process where individual viewers can find many different terms of appeal and places of access. This understanding of therapy as a narrational strategy is aided and abetted by articles such as those found in *TV Guide* that encourage viewers to recognize the generality of therapeutic discourses in television in the first place and also suggest that the value and structure of these discourses are not fixed once and for all. One can participate in a counseling show like *Good Sex!* in literal terms—by calling in for example—or by finding a place in the confessional interchange. This place can be one of symbolic identification with someone posing questions or with the response; it can involve a recognition of expert authority or of the failures thereof. The apparatus itself provides the terms for the therapeutic relationship that becomes the appeal of the medium, the appeal to watch, in the first place. It is not even necessary to grant television and its counselors authority in order to participate. If someone watches long enough to judge them in the first place, that person is already a participant in the therapeutic transaction—listening to others' confessions and the therapists' responses in order to evaluate whether or not to take them seriously.

Identification per se is not necessarily required. Counseling programs obviously cater to voyeuristic pleasures—hearing and watching others speak about problems and confusions in the domain of sex and interpersonal relationships. The therapeutic relation structuring the discourse authorizes this voyeurism, for the therapeutic relation requires the presence of an interlocutor in the first place. And while this "first place" is initially given over to the therapist-star—in this case study, Dr. Ruth—the program goes on to relativize that position and mobilize the television viewer as an essential participant. The apparatus itself thereby claims the therapeutic relation as its own.

And it is therapy as a strategic discourse, as a relational transaction, rather than as a fixed hierarchical relation, that counts. For it provides a structure within which one can always appeal for help from elsewhere even while one listens to others' confessions. At the same time, it provides the comfort of authority, however relativized or provisional. Indeed part of the comfort is that as long as one recognizes the therapeutic as the discourse in dominance, one can choose to believe oneself as an authority among all the competing confessional voices. It is within this discursive context that the public is willing—and is invited—to participate. The therapeutic transaction provides the grounds for understanding the appeal that counseling programs hold for viewers and also the readiness with which individuals are willing to confess personal problems to the mass television audience. . . .

## Notes

1. The following *TV Guide* articles, listed here in chronological order, were consulted in writing this [essay]: Joanmarie Kalter, "No Problem Is Too Intimate for Your TV Therapist," June 4, 1983; Candy Justice, "Now TV Offers a Parenting Manual," July 28, 1984; John P. Docherty, M.D., "A Psychiatrist Looks at Prime Time," August 4, 1984; James E. Gardner, Ph.D., "Does Your Teen-ager *Need* a $95.00 Shirt—or Just Want One?" May 4, 1985; Teresa Kochmar Crout, "Your Therapy Could Be Watching *Dallas* or *Dynasty*," December 14, 1985; Willard Gaylin, M.D., "Prime Time on the Couch: A Psychiatrist Wonders What's Happening to Romantic Love," October 4, 1986; Louise Bernikow, "Is TV a Pal—or a Danger—for Lonely People?" October 25, 1986; David Hellerstein, M.D., "Can TV Cause Divorce?" September 26, 1987; Dr. Ruth Westheimer, "Dr. Ruth Advises: David—Quit Your Job! Maddie—Stop Playing Games!" October 24, 1987; Michael Leahy, "TV's Psychology Shows: More Hype Than Help?" November 21, 1987; Dr. Ruth Westheimer, "Dr. Ruth to The Golden Girls: How to Keep That Spice and Sparkle in Your Lives," April 30, 1988; David Hellerstein, M.D., "Now the Psychiatrist Can Do More Than Just Listen," October 8, 1988; Gerald Goodman, Ph.D., "Successful Intimacy?: Watch *L.A. Law*'s Stuart and Ann," October 29, 1988; Dr. Joyce Brothers, "If You Want to Be a Better Parent . . . ," March 4, 1989; Robert Coles, M.D., "How Television's Stories Help Us," June 3, 1989; Dr. Susan Amsterdam, "Midlife Crises in Men: Are Women to Blame?" July 22, 1989; Dr. Joyce Brothers, "The Shows That'll Make You Feel Better," July 29, 1989; Dr. Joyce Brothers, "Why We Need to Laugh," November 11, 1989; Dr. Joyce Brothers, "How TV Adds Spice to Your Life," February 10, 1990.

2. In the late 1980s, in particular, this genre was highly visible in the broadcasting industry. Ads for such programs in the pages of *Broadcasting* include ones for *Strictly Confidential*, starring Dr. Susan Forward, author of the best-seller *Men Who Hate Women and the Women Who Love Them; Getting in Touch*, with Dr. David Viscott; and *Mr. Romance*. "At one point there were as many as six half-hour shows—

either reenactments of therapy sessions or live sessions with an audience and a therapist—in the planning stages" (*Broadcasting,* January 19, 1987, p. 110).

3. This is not the full extent of Dr. Ruth's media career. She started on radio and became famous for her program *Sexually Speaking* before moving to television. In addition to these two programs, she has had other shows, both specials and regular series, on the Lifetime Channel.

4. Torres, "Melodrama, Masculinity, and the Family," cites two articles that discuss the use of taped episodes of *thirtysomething* in therapy: Patricia Hersch, "*thirtysomething*therapy," *Psychology Today,* October 1988, pp. 62–64, and Phoebe Hoban, "All in the Family: TV's *thirtysomething* Hits Home," *New York Magazine,* February 29, 1988, pp. 48–52. The specific context of these citations suggests that Torres sees this as a singular and symptomatic aspect of *thirtysomething* rather than as a tendency to be associated with television more broadly.

5. Crout, "Your Therapy," pp. 14–15.

6. Ibid., p. 15.

7. Ibid., p. 18.

8. Ibid.

9. Kalter, "No Problem Is Too Intimate," p. 6.

10. Ibid.

11. Brothers, "How TV Adds Spice to Your Life," and Gaylin, "Prime Time on the Couch."

12. Goodman, "Successful Intimacy?" and Hellerstein, "Can TV Cause Divorce?"

13. Hellerstein, "Can TV Cause Divorce?" p. 6.

14. Ibid., p. 7.

15. Bernikow, "Is TV a Pal?" p. 5.

16. Ibid., p. 6.

17. Ibid.

18. Leahy, "TV's Psychology Shows," p. 50.

19. Bernikow, "Is TV a Pal?" p. 5.

20. Justice, "Now TV Offers a Parenting Manual," p. 30.

21. Quoted in Leahy, "TV's Psychology Shows," p. 51.

22. For example, after the networks canceled *The Days and Nights of Molly Dodd,* Lifetime started showing reruns of the series and then began producing original episodes.

23. At least for the time being, sports programming, news, and public affairs are not centrally implicated in the therapeutic problematic. However, the airing of a number of documentaries germane to psychiatry and therapy was covered in *TV Guide* articles, becoming part of the discourse on television and therapy; see Hellerstein, "Now the Psychiatrist Can Do More," and Amsterdam, "Midlife Crises in Men." Also, in relation to sports, the emphasis on previews and postmortems of games, speaking with players and coaches, could certainly be analyzed in terms of the therapeutic problematic and confessional discourse.

24. Larry Angelo's function on the program is defined in terms of communications scholarship by Crow in "Conversational Pragmatics." A more sociologically oriented analysis is offered by Banks, "Listening to Dr. Ruth."

25. Her public stand on homosexuality has changed somewhat. Early on, she was quite critical of homosexuality and was criticized by the gay press for her position, which led to protests to advertisers, to Lifetime, and at her personal appearances. As a result, she became less critical, although her "neutrality" in some contexts

could be taken in any number of ways. Homosexuality was not a frequent issue of discussion on *Good Sex!*, but was occasionally raised by callers. In these instances, Dr. Ruth would routinely—even formulaically—declare, "We do not know the aetiology of homosexuality." This statement could be interpreted as a refusal to take a firm stand about whether or not homosexuality is "natural." But it also leaves room for accepting homosexuality, no matter what its aetiology. By the time Dr. Ruth began to appear in syndication, AIDS had become a more visible problem; and she did assume a leading role in promoting AIDS education on television, including having Dr. Mathilde Krim, one of the leading specialists in this area, as one of her guests. The view of sex and relationships promoted on *Good Sex!* certainly presumed heterosexuality, the monogamous couple, and family relations as a norm.

26. For more detail on the structure of phone discussions on the program, see Crow, "Conversational Pragmatics."

27. Westheimer, "Dr. Ruth Advises," and Westheimer, "Dr. Ruth to The Golden Girls."

28. This is made explicit on another counseling show, *Couples,* . . .

29. Interestingly, this is the same information that callers provide when they talk on the Home Shopping Club, . . .

30. See Crow, "Conversational Pragmatics."

31. Ibid., pp. 475–78.

# Securing the Middle Ground: Reporter Formulas in *60 Minutes*

## RICHARD CAMPBELL

Since *60 Minutes* became a Nielsen ratings success during the mid-1970s, critics and journalists have tried to account for the phenomenon of a popular news program (see Arlen, 1977; Black, 1981; Funt, 1980; Kahn, 1982; Madsen, 1984, Moore, 1978; Stein, 1979; Weisman, 1983). One explanation points to the CBS decision to air the program on Sunday evenings in the fall following pro football and against children's programming on other networks. Another explanation suggests the importance of the arrival on the program of star reporter Dan Rather, fresh from the Watergate beat. Mike Wallace (1984, pp. 352–53) argues that viewers discovered the program when they were forced "to stay home on Sundays instead of visiting relatives or going for a late afternoon drive" during the 1973–1974 Arab oil embargo and subsequent fuel shortage. Don Hewitt, creator and executive producer of the program, credits *60 Minutes'* strong connection to narrative tradition as a major factor for the program's popularity:

> Documentaries were getting the same rating whether they were on ABC, CBS, or NBC . . . the same 15 to 20 percent share of the audience. I said to myself, "I'll bet if we made it multisubject and we made it personal journalism—instead of dealing with issues we told stories; if we packaged

From *Critical Studies in Mass Communication*, Campbell, Richard. "Securing the Middle Ground: Reporter Formulas in 60 *Minutes*." December 1987. Pages 325–350. Copyright by the Speech Communication Association. Reprinted by permission of the publisher.

reality as well as Hollywood packages fiction, I'll bet we could double the rating. ("Father of '60 Minutes,' " 1981, p. 15)

In this study, I argue that not only are the meanings of *60 Minutes* embedded in the general narrative tradition Hewitt discusses but that the narratives of the show extend that tradition to carry a mythology for Middle America.

I define myths not as apocryphal stories associated with earlier cultures but as vital contemporary creators of meaning and value. Myths allow "a society to use factual or fictional characters and events to make sense of its environment, both physical and social. . . . They endow the world with conceptual values which originate in their language" (Hartley, 1982, p. 30). *60 Minutes* is a myth maker, constructing and reconstructing modern myths that signify "conceptual, intellectual or 'objective' values" (p. 26).

Myths help us make sense of experience. Within the human condition, as Edmund Leach (1968, p. 547) argues, "there are certain fundamental contradictions . . . with which all human beings must come to terms," and "myth provides a way of dealing with these universal puzzles." Similarly, Claude Lévi-Strauss (1967, p. 226) argues that myth provides a logical model in the form of a narrative structure that resolves abstract conflicts such as life and death, good and evil, tradition and change, and nature and culture. Mythic narratives mediate their own constructed tensions and allow a culture to come to terms with contradiction and ambiguity.

The mythic narrative pattern provides the shape for many news reports, including those of *60 Minutes*. The pattern picks up people, issues, events, and other experiences and transforms them into narratives of two-dimensional conflicts. The reporters of *60 Minutes* perform as mediators who resolve tensions between nature and culture, good and evil, hero and villain, individual and institution, the abstract and concrete, presence and absence, and ultimately between their own reports and the reality those reports represent.

Here I offer an interpretive reading of *60 Minutes* as narrative and metaphor. With an emphasis on formula, myth, and text, my approach is linked closely to literary formalism (see Strine, 1985). I am, however, indebted to recent developments in cultural studies focusing on the interrelationships between text and audience (see Allen, 1987; Fiske, 1986; Hall, 1980; Jensen, 1987; Katz and Liebes, 1984; Morley, 1980; Newcomb, 1984; Radway, 1984). Although my analysis is textual, I do not mean to foreclose variant audience positions toward *60 Minutes* (e.g., note the "Letters" segment at the end of each program for selected oppositional readings). I do believe, however, that analysis of text and analysis of audience response still may be treated as separate questions. Whereas John Fiske (1986, p. 404) argues that "meanings occur only in the encounter between texts and subjects," I argue that producers of texts, the producers of *60 Minutes,* construct formulaic ways to read the text, maps for meaning, and I am concerned here with interpreting those maps. Although viewers may read *60*

*Minutes* in a variety of ways, they do so in relation to the program's formulas which offer a mediated middle ground (rather than a "dominant position") to engage. While variant negotiated and oppositional readings pose an intriguing research problem, I concentrate on interpreting the mythic narrative maps or formulas that feature *60 Minutes* reporters in their metaphorical performances as detectives, analysts, and tourists. I take this path because I believe with Horace Newcomb and Paul Hirsch (1983, p. 53) that "research and critical analysis . . . must somehow define and describe the inventory that makes possible the multiple meanings extracted by audiences, creators, and network decision makers."

Individuals and cultures construct identities and reality through the narrative process. As Joan Didion (1979, p. 11) suggests, we live "by the imposition of a narrative line upon disparate images, by 'ideas' with which we have learned to freeze the shifting phantasmagoria which is our actual experience." Narrative enables us to make sense of our own phantasmagoria because, in contrast to that experience, narrative is a familiar, concrete, and objectified structure. Narratives then *are* metaphors, shaping and containing the bodiless flow of experience within the familiar boundaries of plot, character, setting, problem, resolution, and synthesis.

Since its inception in 1968, *60 Minutes* has built its narratives around conceptual formulas that feature reporters chiefly in three roles: as detective, as analyst, and as tourist.[1] These metaphorical stances allow reporters to convert the abstract into the concrete, the unfamiliar into the familiar, and the contradictory into the clear, thereby resolving the narrative conflict each episode creates. These metaphors of the reporter present us with figures of modernity who offer mediated interpretations of experience, maps for negotiating our world.

### *Reporter as Detective*

It is not mere coincidence that the fictional detective and the commercial reporter first appeared in the mid-nineteenth century. Both were products of a gradual cultural paradigm shift from romanticism to realism, from religion to science. Both were products of a utilitarian culture that took shape around the notion of conforming to "what is" rather than imagining "what ought to be" (Schudson, 1978). Many nineteenth century reporters associated their practices with the ideal of science, and the classical detective story, as personified by cerebral, rational detectives such as Arthur Conan Doyle's Sherlock Holmes and Edgar Allan Poe's Auguste Dupin, can be read similarly as an expression of faith in the "scientific mind" to gather facts and solve problems (see Cawelti, 1976; Nevins, 1970). By the end of the nineteenth century the identities and methods of both reporters and detectives were bound to a view of a world teeming with data waiting to be discovered, rationalized, organized, and transformed into frames of reference for the public to use in the explanation of immediate experience. Both news reports and detective stories in the nineteenth century then can be interpreted as vehicles for the new rationality of a utilitarian culture "in which the normative order moved from a set of commandments to do what is right to a set of prudential warnings to adapt realistically to what is" (Schudson, 1978, p. 121).

*60 Minutes* not only extends this tradition of faith in rationality, it has merged reporting practices with the literary traditions of the fictional detective. The detective formula describes a central narrative pattern in *60 Minutes,* featuring a variety of narrative characteristics that on the surface, at least, closely resemble those of the classical detective story (see Cawelti, 1976). As in classical detective stories, the *60 Minutes* detective episodes include the identification of a criminal situation followed by a series of actions to make sense of it: identifying victims, villains, and bystanders (who provide evidence and obstacles), reconstructing the factors contributing to the transgressions, revealing the perpetrators, and explaining the crime. The *60 Minutes* reporter, posed in front of the traditional *60 Minutes* storybook, introduces himself or herself and the "crime," which may range from political intrigue to deviation from Middle American values to murder. Next the reporter identifies major characters and settings, reconstructs the crime, and confronts a villain. Generally, there are characters who refuse to talk to the reporter and characters who try to hinder the search for evidence. But in the end the reporter fits together the puzzle and solves the crime. Instead of Sherlock Holmes revealing the crime's patterns for a befuddled Dr. Watson, however, the *60 Minutes* reporter ends the narrative posed again in front of the storybook frame, explaining in a direct address to the viewer the missing evidence, the fate of the villains, and any apparent contradictions.

The *60 Minutes* reporters carry no weapon but rely on rational analysis and their ability to outwit the other characters. They wear trench coats frequently when reconstructing crimes, a detail Don Hewitt notes with pride: "Those scenes of Mike Wallace on a stakeout in a trench coat are great if they produce anything . . . " (Henry, 1986, p. 29). And like their fictional counterpart, the *60 Minutes* detectives often succeed where traditional investigative agencies, victims of mediocrity or inferior intelligence, fail. Finally, following the tradition of the classical detective, the audience is privy to very little personal information about the reporters; we learn little about their private lives, their relationship to the stories they tell, or their own moral judgments of the stories. While unlike classical detectives, the *60 Minutes* reporters are real, they create and enact personae guided by audience expectations that detectives reveal the truth.

## The Detective Formula

• *60 Minutes* detective episodes reveal the following characteristics: (1) introduction of the detective and the crime, (2) reconstruction of the crime scene and the search for clues, (3) confrontation of villains and witnesses, and (4) explanation of the solution and/or the denouement. Three general conflicts serve to organize these narratives: individual versus institution, honesty versus deception, and safety versus danger.

*Introducing the Crime.* Detective episodes often explicitly frame the narrative *as* a detective story. For example, in "Land Fraud and a Murder" (3/2/75), a story about organized crime activity in Arizona, Morley Safer introduces "a new and frightening development," a tale of "dubious land sales involving shady characters," and "a story almost too hot to handle." [2] In "Warning: May Be Fatal" (12/14/75), a story about "potential lethal pollution" of chemical plant workers, Dan Rather suggests, "What we have here in no small way is a whodunit." Occasionally, too, reporters make direct allusions to the similarities between their narratives and the classical detective fiction of Doyle or Poe. Rather, for instance, begins "Equal Justice?" (8/24/80), a story about a black New Jersey political candidate who is allegedly framed for kidnapping, this way: "Tonight the strange case of Mims Hackett." And Mike Wallace introduces "The Stolen Cezannes" (10/14/79) similarly: "The case of the stolen Cezannes is not just the tangled tale of . . . three purloined paintings." In all of these episodes, the story begins with the reporter's introduction and the revelation of some crime that sets him off to unravel the mystery.

*Searching for Clues.* The reporter in these episodes sets off to unravel the mystery, stopping first at the place of the crime, where the crime's details and intrigue are recalled, then leading us to areas where clues reside and where the crime is resolved. In these scenes, *60 Minutes* visually mediates the conflict between safety and danger by displaying the reporter at the scene of the crime, a place once full of peril but now rendered safe by the passing of time and the reporter's presence. For example, Morley Safer in "Land Fraud and a Murder" takes us to a dark stairwell in a public parking garage to reconstruct a murder that has been connected to organized crime in Arizona. He points to blood-stained clues and shows us a newspaper photo with the victim lying in the very place that Safer now safely occupies. The episode also features safe, intimate interviews where characters recount details and clues. In "The Death of Edward Nevin" (2/17/80), Dan Rather poses on a rooftop in San Francisco to reconstruct the crime: how the government carried out a secret germ warfare experiment in 1950 that led to the death of an innocent man. In a trench coat, Rather demonstrates how villainous government agents collected dangerous bacteria samples. Again this public, once dangerous, place is now secure, filled by the reporter's presence. As in most *60 Minutes* episodes, the reporter, featured in an intimate mid-chest shot on the safe sets of a CBS studio, presents the final mediation and denouement.

*Confronting Villains.* The third detective element includes a confrontation segment where the reporter waylays a villain, an unwitting representative of an evil institution, or a befuddled witness or bystander. Because the reporter assumes certain risks in order to resolve a problem, the reporter's actions again embody and mediate the tension between safety and danger. For example, in "The Mystery in Building 213" (7/20/75), a story about spy satellites, the CIA, and national security, Mike Wallace approaches a building that is guarded by armed police and surrounded by a tall, barbed wire fence. This ominous setting reveals a public place full of potential intrigue and risk. Wallace, undaunted, confronts a security guard in the parking lot outside the building and suggests, "Everybody around here says it's a CIA building." The guard, appearing confused and guilty, refuses to link the building to the CIA, leaving this connection for Wallace to make later. Another example, "From Burgers to Bankruptcy" (12/3/78), tells a story about deception in the food franchising business. Wallace, in a trench coat, confronts an executive from a burger franchising company and tries to elicit a response. When the executive no longer will talk, Wallace stakes out his restaurant and in the parking lot confronts an employee (who nervously tells Wallace, "I'll watch what I'm saying") with information about his employer's past. These dramatic confrontations serve to display the reporter in some *danger* so that the tension later can be balanced against the *safety* provided by the reporter's revelation of and apparent solution to the crime.

Another aspect of confrontation scenes is the "probing, tough interview." In the detective episode, confrontation works to establish the oppositional structure and the values implicit in that structure. For instance, "Handcuffing the Cops?" (6/22/80) presents a prison confession scene between alleged murderer Barry Braeseke, whose constitutional rights were technically violated by police during an arrest, and Wallace:

> *Braeseke:* I was in my room and I had my rifle with me, and I came downstairs, and I walked into the family room and the family was watching the TV set with their back to me.
> *Wallace:* Hm-hmm.
> *Braeseke:* And then I started firing the rifle. I was standing behind my— my Dad and I shot him.
> *Wallace:* Through the head?
> *Braeseke:* Yes.
> *Wallace:* And then right away—?
> *Braeseke:* Almost instantly my mother. (CBS News, 1980, p. 629)

Here Wallace confronts the villain in his own dangerous space but elicits truth that renders the setting safe.

*Solving the Crime.* The explanation scene is one of the chief patterns of action in the detective formula, representing "the goal toward which the story has been moving" (Cawelti, 1976, p. 88). That goal is the resolution of narrative tension, "the pleasure of seeing a clear and meaningful order emerge out of what seemed to be random and chaotic events" (p. 89). As detective writer P. D. James (1984) notes regarding the formula, "No matter how difficult the problem, there is a solution. All of this is rather comforting in an age of pessimism and anxiety." In *60 Minutes,* interpreting events and identifying criminals serves to resolve major narrative conflicts. So, for example, in "Your Money or Your Life" (3/19/78), a world of normalcy is represented by Americans pitted against a world of deviance—foreign "terrorists" who threaten U.S. businessmen abroad and citizens at home. Dan Rather mediates this opposition by explaining how we can minimize threats from terrorists. In this narrative, Rather is located squarely between the worlds of normalcy and deviance. He is clearly neither a part of the deviant world of terrorism nor a member of an unsuspecting, victimized public. The reporter is the hero of this story, mediating the tensions between a normal world of middle class values and a deviant world of terrorism set upon destroying those values. As detached hero, Rather, however, does empathize with the normal, vulnerable world portrayed in the story. At one point he responds to a statement about terrorism in the United States: "Do you believe that? That scares the hell out of me." Such dramatic vulnerability actually heightens the reporter's credibility as a heroic figure who can live with contradictions and manage to interpret and resolve them.[3]

In a less dramatic example, "The Grapes of Wealth" (10/14/79), Morley Safer confronts wealthy southern California farmers who, through a bureaucratic loophole, qualify for special low interest loans designed to "save the family farm." *60 Minutes* juxtaposes expensive cars, homes, tennis courts, and horse stables owned by these farmers with narration about their acceptance of "low interest disaster loans" intended for poorer, struggling farmers. While the opposition between safety and danger is not displayed prominently in this episode, the reporter still must clarify tension between honesty and deception by portraying the practices of wealthy farmers set against honest, unsuspecting middle class farmers whom Safer represents in the narrative. In the final scene, the reporter, posed in front of the storybook frame, enumerates the deceptive practices. This ending or solution scene where the reporter wraps up the case is typical of most detective episodes, just as it is typical in much classical detective fiction where there is often a denouement featuring the "actual apprehension and confession of the criminal." But that denouement need not be an ingredient in the formula since "the classical story is more concerned with the isolation and specification of guilt than with the punishment of the criminal" (Cawelti, 1976, p. 90).

In some episodes, however, there is a clear denouement. "Another Elvis?" (8/12/79), for instance, presents a story about characters who pay money to shady record companies in return for recording careers. The disreputable president of a Nashville company (who, we learn, had a previous criminal record), is apprehended and subsequently confesses to the reporter, who has caught the villain in a lie. Mike Wallace reports at the end of the episode that this character has "quit the business" and now "thanks" *60 Minutes* for turning his life in an honest direction. In this episode, the reporter dissolves tension by placing himself between innocent victims and insensitive villains. Ultimately, he wraps up the case, reinstitutes safety, and reaffirms honesty in the denouement when the villain promises to reform.

### Detective as Mediator

A major function of the detective formula in the mythic framework of *60 Minutes* is to display the reporter amid conflicts between safety and danger. Reporters, of course, always emerge from these dramatic situations safely by either resolving the crime or at least presenting an interpretation. They thus straddle and mediate the conflict. Implicit in their mediation are moral values such as security, democracy, honesty, loyalty, and justice. Safety affirms these middle class values that gain definition only in relation to the danger of insecurity, communism, dishonesty, disloyalty, and injustice.

A second narrative conflict frequently used in the detective formula is between individual and institution. In "Titan" (11/8/81), where the crimes are government ineptitude and insensitivity, the narrative creates the tension through a victimized character, a former Air Force sergeant reprimanded by his superiors after he tried to investigate a toxic leak and fatal explosion in a U.S. missile silo. According to the ex-sergeant, "I went down there for God, my country, the flag, my job—everything—I didn't go down there for any other reason. I gave them my all. And what did I get from them [the Air Force]? A letter of reprimand. . . . What about the little guy?" *60 Minutes* affirms the ex-sergeant's status as little guy by juxtaposing him with the absent Air Force who, as Safer tells us, "absolutely refuses to answer any questions." The reporter mediates the tension between the individual and institution by supporting the side of the innocent, unsuspecting victim and by affirming values of honesty and loyalty against the Air Force, the institutional villain.

A third major tension, between honesty and deception, organizes detective narratives. In "Taking on the Teamsters" (12/3/78), Safer's narration creates a Robin Hood metaphor by opposing individuals, the "dissident leader" of "the people" and "his small band," who are running for labor union offices against the long-entrenched institutional union heads who are under surveillance for possible payoff schemes and organized crime connections. Here again the presence of the small band of dissidents at various meetings contrasts sharply with the absence of the current union leaders, who, Safer tells us, "refused absolutely" to be interviewed. In their place are aerial shots of their expensive homes and swimming pools and accompanying narration tells of the alleged criminal activity. The reporter mediates the conflict by implicitly supporting and affirming the honest, hard-work ethic of the dissident band of union individuals against corrupt, indulgent, institutional leaders. This ability to mediate derives from the reporter's values. The allegiance to honesty provides a basis for moral superiority over deceptive institutions.

An important subconflict in these detective episodes features the *presence* of victims, heroes, and the *60 Minutes* reporter and the *absence* of the villainous persons or institutions. Presence versus absence generally takes two forms on *60 Minutes:* (1) certain parties refuse to be filmed or interviewed, and (2) certain parties simply are not interviewed by *60 Minutes*

even though their absence is used as a major opposition in the narrative
(see Shaw, 1987). The absent, or faceless, villain generally represents some
form of business, government, or labor institution. For example, in "Distressed" (5/3/81), a story about a flourishing Florida county listed by HUD
(Department of Housing and Urban Development) as an economically distressed area, no top bureaucrat from HUD would grant Safer an interview.
He tells us that one institutional representative said, "No way will I sit
down for an interview." *60 Minutes* then introduces a General Accounting
Office report that criticizes HUD for using twenty-year-old statistics to determine economically distressed areas. Safer's narration, filmed outside the
HUD building, displays the agency as a big, insensitive, inefficient bureaucracy under no individual's control. In contrast, the reporter's presence affirms values of efficiency, justice, and common sense against the inept, un-
just, and illogical countervalues of the absent bureaucracy.

*60 Minutes* reinforces the narrative function of reporters as mediators
by a style that gives them more visual or frame space in which to operate.
With few exceptions, *60 Minutes* reporters are shot at a greater distance than
the characters they interview. Frequently, in reaction or question shots,
characters appear in extreme close-ups (usually with the top of the head cut
from the frame), while the reporter appears in medium close-up shots (normally shown from the middle of the chest up). The greater space afforded
the reporters in *60 Minutes* supports their function as mediators since it
endows them with a medium position, with the appearance of more control
over their place in the narrative. The greater distance granted the reporter
also reinforces the posture of detachment. Often the reporter's hands, shoulders, and head are free in the frame, whereas other characters—unsuspecting victims or devious villains who are shot in tighter close-up shots—have
less room to maneuver within the frame. As Meyrowitz (1985, p. 102)
notes, interviewees are shot so tightly "that any move of the head makes it
appear" that the subjects are "trying to escape scrutiny." With heads and
sometimes chins cut off, these characters are reduced to eyes, noses, lips,
foreheads, sweat, facial twitches, and tears. The greater space granted to the
reporters of *60 Minutes* places them in a position to see the larger picture,
control their environment, secure a middle ground, and thereby mediate
narrative tension.

What empowers the classical detective to find clues and resolve problems is superior intelligence and detachment, a "lack of moral or personal
involvement in the crime he is called upon to investigate" (Cawelti, 1976,
p. 83). The *60 Minutes* reporters display these traits whenever they assume
a more positive position than villains or inefficient institutions, the counterparts to the bungling and inefficient police of detective fiction. The greater
visual space afforded the reporters' portrayal as *individuals* (rather than as
institutional representatives of CBS News) grants them superior narrative
positions.

James Fernandez (1974, p. 124) argues that a cultural "mission of metaphor" is to move the subject of the metaphor, in this case the reporter,

"into optimum position in quality space," the space of the classical detective who is characterized by both superiority and detachment. In *60 Minutes,* the reporter occupies optimum position for mediating contradiction. The reporter receives metaphoric status as a heroic detective who champions Middle American individualism and integrity in the face of heartless bureaucracy. As Robert Bellah and his colleagues (1985, p. 149) note in *Habits of the Heart,* historically the detective first appeared as a popular cultural hero "when business corporations emerged as the focal institutions of American life. The fantasy of a lonely, but morally impeccable, hero corresponds to doubts about the integrity of self in the context of modern bureaucratic organization." Depicting *60 Minutes* reporters as individual loners, apart from a team of producers, researchers, and editors who construct the story, apart from the powerful CBS media corporation, allows the reporters' institutional identities to remain hidden. This portrayal enhances their preferred narrative position as mediators of tension, clarifiers of doubt, and affirmers of individualism.

### Reporter as Analyst

Like the detective, the analyst too has assumed a position in the quality space of our culture. The development of the analyst's social influence also can be traced to the late nineteenth century during a cultural shift from religion to science, a geographical shift from rural to urban living, and an economic shift from a producer to a consumer society. The analyst emerged from the changes as a prominent promoter of a "therapeutic ethos" that offered renewal of "a sense of selfhood that had grown fragmented, diffuse, and somehow 'unreal' " (Lears, 1983, p. 4). The analyst gained legitimacy by riding the coattails of prestige bestowed upon medical science and the ability of science in general to solve problems. Jackson Lears (pp. 6–11) identifies several crises and problems around the turn of the century including urbanization, which introduced the "anonymity of the city"; technological advance, which brought "prepackaged artificiality" along with "unprecedented comfort and convenience"; institutionalization, which created "an interdependent national market economy" and cut people off from ties to the land and "primary experience"; and secularization, which displaced religion and isolated people from traditions. The analyst stepped into these crises to try and restore a sense of individuality offering "harmony, vitality, and the hope of self-realization." The analyst offered, and continues to offer, cures for and mediations of cultural tension. As Bellah and his colleagues (1985, p. 47) point out, "the very term *therapeutic* suggests a life focused on the need for cure."

A second *60 Minutes* formula features the reporter as analyst who offers cures for contemporary tensions. In general, the reporter assumes two roles in these *60 Minutes* analyst episodes: that of the social or cultural commentator, and that of the psychoanalyst or therapist. And in any given analyst episode, the reporter may perform both roles. In each case, however, the reporter is, first of all, an interpreter who recasts, rearranges, and retells the stories of others. Such narrative interpretation, psychologist Roy Schafer (1981, pp. 31–32) suggests, is the essence of analysis:

> [People] tell the analyst about themselves and others in the past and present. In making interpretations, the analyst retells these stories. In the retelling, certain features are accentuated while others are placed in parentheses. . . .
>
> The analyst's retellings progressively influence the what and how of the stories told. . . . The analyst establishes new, though often contested or resisted, questions that amount to regulated narrative possibilities.

In many episodes of *60 Minutes,* reporters function precisely as such analysts when they construct narratives about the lives of others.

One result of the reporter's analytical interpretations is to endow narratives with closure. The desire for closure in narrative, as Hayden White (1981, p. 20) argues, is a desire for moral meaning:

> The demand for closure in the historical story is a demand . . . for moral meaning, a demand that sequences of real events be assessed as to their significance as elements of a *moral* drama. Has any historical narrative ever been written that was not informed not only by moral awareness but specifically by the moral authority of the narrator?

The affirmation of values or morality, both explicitly and implicitly, is a chief function of *60 Minutes* where, as in other forms of factual storytelling, "narrativity . . . is intimately related to, if not a function of, the impulse to moralize reality" (White, 1981, p. 14).

The *60 Minutes* analyst episodes reflect the impulse to moralize reality in mediating three key dramatic tensions: success versus failure, tradition versus change, and the personal versus the social. While the tension between the personal and social realms is a variant of individualism versus institution, the different terms call attention to the more general nature of this opposition as it is worked out in the analyst formula. In the detective episodes, the institution typically is a government or business organization. In the analyst episodes, bureaucracies give way to a broader notion of "society" as a set of demands and rules that "individuals" must confront. In the detective episodes, characters serve mainly as a function of plot and their individual behavior and attitudes are rarely explored in depth. In the analyst episodes, a single character generally is the focus of the fourteen- to fifteen-minute story. Whereas detective episodes include on the average eight to ten characters who are named and interviewed, in the analyst episodes only two main characters are usually featured, the subject of the story and the reporter.

## The Analyst Formula

Four characteristics of the analyst formula are: (1) the role of the reporter as social or psychological analyst, endowing the narrative with closure and moral meaning, (2) the personalized role of the analyst, (3) the treatment of characters as heroic or villainous, and (4) the style of the reporter as inquisitor, asking tough questions.

*Social and Psychological Analysts.* The role of the analyst on *60 Minutes* is that of the social commentator who analyzes and interprets political, economic, and cultural trends in relationship to the interview subjects, the major characters in these narratives. For example, in "Martin Luther King's Family at Christmas" (12/24/68), Mike Wallace tells a story about King's family and analyzes the effect of King's death on the civil rights movement which, in Wallace's words, was "leaderless" and "rudderless" without King. Wallace's role in this episode, the reporter's function in the social analyst episodes, is not so much to analyze historical events as it is to offer a narrative that explains the present and predicts future directions of social phenomena (in this case, the civil rights movement). In another example, "New York Yankee" (5/3/81), Harry Reasoner offers social commentary on Yankee owner George Steinbrenner, who is characterized by Reasoner as "almost impossible to work for." He "can't hit, run, or throw, but boy can he talk," Reasoner tells us. The reporter offers insights and forecasts trends regarding labor–management relations in baseball. Reasoner predicts ominously that Steinbrenner "should be heading toward another successful season—but don't bet on it." Reasoner's analysis implicitly rejects negative aspects of Steinbrenner's character (egocentrism, physical mediocrity, and self-indulgence), aspects that counter Middle American values.

A second analytic role features the reporter as psychoanalyst or therapist. Moving away from *public* events and issues, the reporter as psychoanalyst probes the *private,* emotional world of the characters in order to reaffirm values. Again in "Martin Luther King's Family at Christmas," Wallace probes the emotions of the King children by asking how they are adjusting to their father's assassination. The narrative then affirms the ability of the private family to overcome the chaos and conflict inherent in a public assassination. In another example,"The Shah of Iran" (10/14/76) episode allows Wallace to practice some confrontational psychology by reading to the Shah from his CIA psychological profile: "A brilliant but dangerous megalomaniac who is likely to pursue his own aims in disregard of U.S. interests." Wallace asks the Shah, "Should I go on?" and then concludes the scene with narration that analyzes the Shah's mental characteristics by juxtaposing them implicitly with positive values such as humility and democracy. As a final illustration, in "Anderson of Illinois" (2/17/80), Morley Safer overtly displays the therapist persona when he conducts a kind of verbal Rorschach test on 1980 presidential candidate John Anderson: "For an underdog, Anderson is a gentle soul who does not speak harshly of his rivals, even when you play the candidate game with him. I'll give you a name, and you give me an answer. Ronald Reagan?" Safer offers three more names to complete the test. In so doing, Safer as therapist constructs a story that implicitly supports Anderson's kindness and modesty as values for living. In each of these episodes, the reporter probes the emotional states of interviewees in order to offer a cure for the larger sociopolitical problem that these narratives are constructing as reality.

*The Intimate Reporter.* A second characteristic of the analyst formula features the reporter as a character with subjective likes and dislikes. One visual instance stands out, a scene from "Lena Horne" (12/27/81) where reporter Ed Bradley and interviewee Horne walk hand in hand across a busy New York street. Although not always as vividly, reporter intimacy also surfaces in other analyst episodes. In "Madame Minister" (9/19/82), actress and Greek cultural minister Melina Mercouri takes Harry Reasoner "on a date" to a Greek nightclub. In "Mister Right" (12/14/75), Mike Wallace, in a sweater and jacket, interviews presidential candidate Ronald Reagan during a jeep ride on Reagan's California ranch. In these examples, the analyst persona displays a more involved personal stance in the narrative.

A visual device that personalizes the reporter also comes into play in the analyst episodes. In contrast to the detective formula where the reporter is typically shot no closer than medium range, in the analyst formula the camera zooms in considerably closer in reaction or question shots with the interview subject. For example, in "More Than a Touch of Class" (4/7/74), an interview with actress Glenda Jackson in a London pub, Wallace appears three times in close-up shots and in one extreme close-up where part of his forehead is cut off; in none of the detective episodes featuring Wallace is he shot this tightly.

*Interviewee as Hero or Villain.* A third characteristic of the analyst episodes, the treatment of characters as heroes and villains, also involves the use of visual space. Interview subjects in these episodes are treated as either representatives of middle class values (heroic) or countervalues (villainous). In certain analyst episodes, the interview subject receives equal visual space with the reporter. These episodes generally portray the interviewee heroically, that is, the subject appears also as a mediator resolving oppositions, as an enforcer of or model for basic values. For example, in "Extremism in the Defense of Liberty" (7/20/75), Karl Hess, who has "progressed" from his job as Barry Goldwater's "arch-conservative" speech writer to inner-city social reformer, receives visual space similar to Morley Safer, who appears mostly in waist-up medium and long shots. Again in "What Became of Eldridge Cleaver?" (5/18/75), the camera helps us follow Cleaver's career from black revolutionary to reformed activist. In early shots in this episode, Mike Wallace asks Cleaver about his 1960s threat to "take off the heads" of established political figures; here Cleaver appears in extreme close-ups. Once Wallace establishes that "Cleaver is no longer a black symbol of resistance," no longer a villain, the camera pulls back and Cleaver too appears in waist-up medium shots. He now appears in similar visual space as Wallace. He is a mainstream, American figure who can resolve central contradictions in his own life and serve as a model for others.

When an interview subject appears as a deviant from basic values and a polarizer of tensions, the camera denies visual space (this is also true for victims, those who are caught in gray areas between conflicts they cannot resolve). For example, in "The Shah of Iran," Wallace battles the Shah in an interview that features tensions between American democracy and a foreign dictatorship. The villainous Shah appears almost totally in extreme close-ups that, at times, cut both his forehead and chin from the frame. Wallace, who is featured in nearly forty shots in this episode, receives more visual space. In most shots, we can see the reporter's hands at work in the frame, demonstrating that Wallace is in more control of the space around him. And it is Wallace who affirms American ideals in the face of the Shah who stands for foreign, tyrannical countervalues.

Another visual technique that separates heroic from villainous or victimized subjects involves the camera's depiction of private artifacts, a visual technique that personalizes interview subjects. In "The Empress" (5/18/75), for instance, a profile story of the Empress of Iran, shots create a contrast between public ceremonies and private, more intimate moments featuring the empress with her children at a birthday party. Other shots in this episode also help resolve a conflict between public and private through intimate glimpses of a living room featuring family artifacts that portray the empress as a model of simplicity, modesty, and motherhood in spite of her regal position and wealth. In contrast, the camera mostly ignores the private, personal lives of villains. For instance, in "The Shah of Iran," we do not see him with his children, at birthday parties, in an intimate living room, nor are we told stories from his youth. The questions and the visuals focus not on the private but on the sociopolitical nature of his public, anti-American, anti-democratic postures.

*Reporter as Inquisitor.* A final characteristic of the analyst formula features the tough confrontational questions that generally probe deviations from basic values. The reporters display this characteristic by confronting and questioning characters who are portrayed, at least during a part of the interview, as representing countervalues. In "General Ky and Big Minh (10/21/71), for instance, Mike Wallace asks General Ky if he plans to overthrow the South Vietnamese government. "No coups on your mind?" Wallace asks as he confronts Ky on questions of disloyalty, anti-democracy, and deception. In "Anderson of Illinois," Morley Safer challenges the presidential candidate on his plan to cut federal spending: "Well, so what are you saying? Abandon the cities?" At the end of this episode, Safer couches the tough question in negative comments made by other reporters:

> It's kind of interesting . . . the perception of John Anderson. For example, this *Washington Post* headline is "Dream Candidate Going Nowhere." Tom Wicker, a liberal columnist in *The New York Times,* says the man running on either side can't win. Conventional wisdom? Leo Durocher says nice guys finish last. What do you say to all that? (CBS News, 1980, p. 339)

A major function of the tough question scenes in the mythic framework of *60 Minutes* is to locate the reporter in the middle, between us and them, between private and public tensions. Indeed, as Ian Connell (1978, p. 83) has noted,

> The "hard," "tough" style of interviewing . . . was legitimized as an attempt "to get at the facts" on behalf of the public. This adoption of a "watchdog" role on behalf of the ordinary voters also led to the attempted identifications with "us," and the attempts to articulate the kinds of questions that "we" would ask of "our" powerful representatives if "we" only could.

In the analyst formula, posing tough questions includes the viewer in the reporter's point of view, clarifies narrative tension, and sets the stage for mediation.

### Analyst as Mediator

Within the mythic framework, the analyst episodes call for resolution of central narrative tensions. As in the detective episodes, a chief mission of the reporter here is to mediate fundamental, emotional, and conceptual oppositions. In "Martin Luther King's Family at Christmas," Mike Wallace mediates tension between life and death in the closing scene: "Martin Luther King left a legacy for all of us that we cannot fail to understand especially at this Christmas season. But he left us, too, Coretta King." Mrs. King then discusses "creative suffering" and argues that the tragedy "doesn't mean we will sit around and bathe in our grief." Wallace, as mediator, celebrates the presence of Coretta King and her family as a solution to contradiction and ambiguity. Other episodes also offer the family as a solution for social as well as personal conflicts and suffering. Resolution and moral meaning for the audience in these episodes are enabled through the probing, guiding questions of the reporter.

Conflicts between person and society, tradition and change, which also shape many analyst episodes, are often resolved through the presence of a character whose moral code once deviated from Middle American values. In "What Became of Eldridge Cleaver?" Cleaver emerges as a man who formerly wanted to decapitate U.S. leaders but who, five years later, wants to return from exile in Paris to attend his father's funeral and the U.S. bicentennial celebration. In this episode, Wallace switches discussion from Cleaver's personal loss to the social change and "success" of the entire Black Panther organization. He offers a symmetrical closure, the Panthers following Cleaver's repentant course. Instead of violent revolutionaries, analyst Wallace describes a "changed outfit," now "working," "laboring," "muted." They too have become "nonviolent" reformers hard at work "inside the system." The Panthers' adjustment from deviance to normalcy obscures Cleaver's personal loss and closes the narrative with a positive image, parallel to Cleaver's own repentance and return to normalcy. Within the structure of this analyst formula, narrative mediation occurs as Wallace sorts positive values from their negative counterparts, thereby endowing the narrative with moral meaning: Cleaver and the Panthers must pay a price when they overstep the bounds of a society's rules, but once that price is paid, society accepts them back into the mainstream.

The Cleaver episode offers a metaphorical framework for transforming the abstract and less familiar categories of deviance and normalcy, tradition and change, into the more concrete and familiar terms of a mythic narrative structure. Wallace introduces the concept and value of repentance in order to resolve opposition between the concrete entities of the story, a radical Cleaver and the Black Panthers versus established government leaders, and the abstract concepts both sets of characters represent: revolution and status quo, deviance and normalcy, change and tradition. The transformation of Cleaver's idiosyncratic experience into the myth of the prodigal son resuming his place in normal society requires that Wallace become a mediating term in the mythic narrative between opposed sets of moral values. It is not merely the introduction of the concept of repentance that resolves this particular mythic narrative but the performance of Wallace domesticating deviant attitudes and comforting the audience.

In the analyst episodes, as in the detective episodes, metaphor locates the reporter in preferred narrative space. That quality space is provided by the legitimacy that comes through associating the reporter with those who know about social processes and psychological states, social commentators and therapists. Within the structure of the analyst formula, mediation occurs as reporters rescue reality from apparent contradiction, thereby endowing and enriching the narrative referent with moral meaning and sense.

## —— *The Reporter as Tourist*

The same sets of nineteenth century historical conditions that prepared the way for the analyst also contributed to the importance of the tourist to twentieth century culture. Urbanization and secularization, technology and "prepackaged artificiality" severed connections to "primary experience" and older traditions (Lears, 1983). These changes initiated a twentieth century search for the past, for lost identity, for authentic experience. Dean MacCannell (1976, p. 3) describes this quest and defines modern tourists as "sightseers, mainly middle-class, who are at this moment deployed throughout the entire world in search of experience." What tourists implicitly seek, according to MacCannell (p. 3), is to resolve the contradictions between nature and civilization, tradition and modernization: "For moderns, reality and authenticity are thought to be elsewhere: in other historical periods and other cultures, in purer, simpler lifestyles." Dean O'Brien (1983, p. 9) finds MacCannell's quest for authentic experience at the center of reporting: "journalists, also in search of authenticity, travel from one source, event and sight to another," trying "to penetrate the markers and the 'staged authenticity' of front regions (news releases, government hand-outs, pseudo-events) to reach genuine back regions." Audiences, of course, vicariously accompany reporters in this contradictory quest to get beyond the artificial. Certain episodes of *60 Minutes* embody the quest for authentic experience, whether it be in a foreign land or in America's heartland.

## The Tourist Formula

The tourist stories portray the reporter (1) acting as surrogate in exploring and describing the new or unfamiliar, (2) searching for authenticity: trying to recover the past, the natural, and smashing through the facade that is modern civilization, and (3) confronting villains, usually portrayed as either bureaucracies or modernity itself (often in the guise of Americanization). The reporter in these episodes mediates three major narrative conflicts: tradition and modernization, nature and civilization (or rural and urban), and individual and institution.

The episodes that portray the tourist in Middle America often begin with explicit acknowledgment of their mythic narrative structure. For example, Dan Rather as our tourist surrogate, in jeans, travels the country in "Wild Cat Trucker" (2/22/76). He begins the episode: "There's a new brand of folk hero around these days—the wildcat trucker. Like the cowboy and the gold miner and the aerial barnstormer of an earlier era, he's taking his place in Americana." Again in "Charity Begins at Home" (2/4/79), a story about a town that "takes on the state of New Jersey over the issue of welfare," Morley Safer, in the manner of Thornton Wilder, leads us on a tour of small-town America. "This is one of those *Our Town* kind of stories, and if you were writing it as fiction, there are certain things you'd have to include" such as an ice cream parlor and a hardware store. "The tale we have to tell is one of those late twentieth century American dramas."

*Acting as Surrogate.* In describing a guiding myth of modernity, MacCannell (1976, p. 159) notes that "the position of the person who stays at home in the modern world is morally inferior to that of a person who 'gets out' often," because "authentic experiences are believed to be available only to those moderns who try to break the bonds of their everyday existence and begin to 'live.'" To a certain extent, television and television news mediate the bind experienced by those of us caught in the routine of everyday life who also seek authenticity outside our homes. The *60 Minutes* reporters function in part as surrogates, as tour guides, who take us to unfamiliar, foreign places and remind us of "reality and authenticity elsewhere" (MacCannell, 1976, p. 16). Thus in "Paris Was Yesterday" (4/22/73), when Mike Wallace dines at a Parisian cafe and narrates a sightseeing tour about old and new Paris, he becomes our surrogate, rendering this unfamiliar place accessible and familiar. The reporter, as our representative in this foreign land, helps us "break the bonds of everyday existence and begin to 'live.'" In "Oman" (8/24/80), a story about a U.S. military base in the Middle East, Wallace dons shorts and relaxes with a newspaper at poolside. In voice-over narration, he comments about his tourist role in the narrative: "Even Americans aren't especially welcome as tourists," a comment that implicitly includes the viewers in partnership with Wallace as he dares to tour exotic lands.

*Searching for Authenticity.* The episode "How To Live To Be 100" (7/5/81) illustrates a second feature of the tourist formula. Here Morley Safer comments that a particular community of Russians live long because of their *authentic* lifestyle. The reporter lists the community's rules and values: "food from earth, not from a can; hard physical labor, not the so-called leisure years, and above all an unbreakable belief in family life that makes age more important than youth or wealth." A visual counterpart to this narration features a 107-year-old man bathing in the natural, authentic setting of a cold mountain stream. Similarly, in "Yanks in Iran" (1/2/77), Mike Wallace talks to disgruntled American citizens who came to Iran in search of identity and authenticity, but failed to find them. One character, who is leaving Iran after eight years, tells Wallace about other Americans who "think they're coming to the promised land." He then discusses the expenses, drugs, "smog, noise, cars," and other symbols of modernization that have disrupted the Americans' quest. These explicit symbols of *artificiality* pose a set of countervalues to moderation, good health, and small-town pastoralism, which are affirmed in this narrative.

In the heartland episodes, part of what is accomplished is not only the authentication of the reporter's presence, but the verification that these places display genuine small-town markers and symbols: the ice cream parlor, the hardware store, the meeting hall, the church, the diner, and main street. The established authenticity and individuality of the small town then is contrasted with the artificiality and impersonality of the big city. For example, the opening of "Away From It All?" (8/1/76) establishes a quest for an authentic experience that can be found only in a small-town setting. The reporter here asks if we ever had the feeling that we "wanted to get away from it all—pollution, taxes, pressure." Shots of neon signs and congested urban streets, the symbols for the artificial, impersonal big city, are then contrasted visually with shots of a lake, a church, an inn, barns, horses, and a waterfall in Dairy, New Hampshire.

Whereas in the foreign-place episodes, the villain is typically modernization or foreign values, in these heartland episodes the villain is more often an institution (located in the artificial city rather than the natural country). Rarely represented visually, the invisible institution becomes more menacing. For example, in "Rural Justice" (2/22/76), the absent values of the legal profession are opposed to the law and order code of small-town South Carolina magistrates. Contrasted with the urban legal system that demands rural magistrates hold law degrees are these small-town judges, one of whom moonlights as a night watchman. In emphasizing the importance of humility and individualism, this rural judge tells Morley Safer that he refuses to wear a judicial robe "because it would scare people," and he wants to "make them feel at home. . . . I take this little magistrate's job to heart." While five rural judges visually are portrayed, no one from big-city legal institutions appears in this story. The concrete presence of individual magistrates, who connote familiarity and friendliness, contrast here with the law profession which, reinforced by its menacing *absence,* connotes unfamiliarity and unfriendliness.

*Confronting Modernity and Bureaucracy.* Aligned with the search for authenticity is a third aspect of the tourist formula, modernization or bureaucracy as the villain that conceals or destroys authentic experience. In "Rolls-Royce" (6/22/80), which offers us a tour of a British Rolls factory, Safer describes the effects of modernization on authentic experience: "The trouble really is that nothing these days is built to last. . . . We live most of our lives in a junk society. Our durables aren't very durable. But when something is built by hand out of materials given by nature, old-fashioned pride is maintained." Later in the episode, Safer, who describes the Rolls auto factory as a "cottage industry" in contrast to the sprawling technology of an American auto plant, queries a British automobile "craftsman": "Well, what's the difference between this and a stamped-out car?" The worker responds, "Well, a stamped-out car is just a stamped-out car, isn't it? I mean, anybody can build them." In continuing the indictment of the U.S. car industry, Safer asks, "How would you like to work in one of the big auto plants and run a machine that simply punched out one of those doors every ten seconds?" The man responds, "Well, I think it would bore me within two or three hours. . . . I'd sooner use my hands and make it myself." The American auto industry symbolizes bureaucracy and the negative dimension of modernization. A modern villain, it holds values counter to tradition, to craftsmanship, and to quality that Safer finds in another modern corporation but one that represents pre-modern values.

The tension between individual and institution also plays a significant role in the heartland episodes, helping structure our readings of these stories. In "Charity Begins at Home," Safer's opening narration sets up a conflict between a small New Jersey town attempting to transform welfare into "workfare" and ready to confront the state. In the episode, a character explicitly sets up the conflict between individual and institution for Safer: "I think the bureaucracy worships paper gods as opposed to seeing the needs of people." A recurrent variation of individual versus institution (or small town versus big city) pits *place* against *space* rather than person against the larger social system. O'Brien (1983) argues that *space* is more abstract, more hostile, less familiar and tranquil, more open and dangerous than *place,* which connotes security, stability, familiarity, serenity, closeness, and friendliness. And these connotations accompany the verbal and visual significations of small-town versus big-city conflict that underpin this set of *60 Minutes* stories. For example, in "Away From It All?" which looks at a small New Hampshire town facing the encroachment of urbanization, the episode's final narration explicitly acknowledges the opposition: "The scent of apple blossoms and the hum of honeybees are being replaced by the smell of asphalt and the noise of traffic. . . . Pastures are being turned into shopping centers." The visual accompaniment to the reporter's voice-over reveals the clutter of neon signs and fast food restaurants. In this episode, the pastoral serenity of an apple blossom setting represents the stability and authenticity of *place,* while the clutter of the "neon, asphalt" urban environment stands for the instability and artificiality of *space.*

## Tourist as Mediator

The tourist formula establishes central contradictions that organize many *60 Minutes* narratives. The conflict between tradition and modernization is displayed in "Paris Was Yesterday," where Mike Wallace and Janet Flanner, who wrote from Paris for *The New Yorker* for fifty years, recall the authentic Paris of the 1930s. Flanner comments, "Paris was more gracious in its pleasure—more customary. You knew who you were." This authentic, personal vision opposes shots of modern Paris overrun by litter, junkyards, crowds, and quick meal signs; modern buildings are "bogus new towers. . . . hatched by vipers" that lack the majesty of the authentic Eiffel Tower. In "Yanks in Iran," shots of traditionally dressed Iranians and plain white homes oppose shots of modern billboards and a sound track of radio ads for Pepsi, 7-Up, Caterpillar tractors, and baseball. Tradition versus modernization also frames "Rolls-Royce," where Safer refers to the car as "a British institution with an almost mythical tradition." After he tours the Rolls plant, he visits a woman in Ireland who has been driving the same Rolls since 1927. Her complaints about the modernization of newer Rolls models ("not half as dignified," "too complicated") gain support from the reporter: "They said it [the new Rolls] looked like nothing more than something out of Detroit in the fifties. . . . [The] Rolls had finally fallen victim to the fickle hands of modernity." Safer here identifies modernity and the U.S. auto industry as villains whose presence has undermined traditional values represented by the older Rolls.

Through opposition between tradition and modernization, *60 Minutes* constructs a map for meaning and sense. The reporter, in a metaphorical role as tourist, stalks authenticity, chastises modernity, and transforms the unfamiliarity of foreign space into a now familiar place. The reporter resolves the conflict between tradition and modernization by affirming places that suggest authenticity, simplicity, family, honesty, security, common sense, and other virtues of tradition.

While this tension between tradition and modernization generally features events, issues, and ideas in opposition, nature versus civilization is a tension of locale where natural settings oppose artificial cities. In "The Oil Kingdom" (6/9/74), which presents a tour of Saudi Arabia, shots of nomadic tribes herding sheep in the desert compete with crowded city street scenes and giant modern oil rigs springing from that same desert. Nature, visually portrayed by the desert, animals, and nomadic tribes, opposes civilization, visually portrayed by a modern city and oil rig technology. Similarly, in "Seward's Folly" (9/1/74), an Alaskan travelogue, an artificial enclosure houses and shields oil company workers from the stark, cold Alaskan tundra. American workers inside who are protected by that modern, artificial environment appear in contrast to the natural, native Eskimos who endure the tundra outside. This episode also juxtaposes spectacular mountains and waterfalls with scenes of prostitutes soliciting on the streets of Anchorage; the scenes display modern, urban values as undesirable, indeed villainous. The reporter, as mediator, once again resolves the meaning of the opposing images by framing elements associated with nature as real experience and elements associated with civilization as artificial. Symbols for nature emerge as affirmations of simplicity, purity, pastoralism, and unadorned beauty.

This romanticized vision of pure nature and false civilization occasionally gives way to a conflict where brutal natural elements become the villain in a test of the ingenuity and endurance of courageous individuals . Nature assumes this role, however, only in relation to individuals, not institutions. Again in "Seward's Folly," workers from oil companies tell their stories about surviving as a family in the brutal Alaskan frontier. Tour guide Morley Safer tells us that the growth of the oil industry, which these workers represent, could eventually "bring a dentist to town." Here civilization and modernization receive a personal face in the presence of an individual, a dentist. In this section of the narrative, nature opposes individuals who by courage, ingenuity, endurance, hard work, true grit, and a sense of family rise up to defeat this brutal villain. It is only by replacing institutions with workers and civilization with the individual dentist, however, that nature's villainous aspects emerge to create narrative tension.

In the heartland episodes, the mediation of individualism versus bureaucracy occurs when the reporter makes a connection between the story's characters and the promise of a better social system created by individuals. In "Wild Cat Trucker," Dan Rather mediates the conflict by placing himself at its center, traveling with the trucker, facing confrontations with the law, and surviving to summarize the spirit of individualism manifested by independent truckers. The visual frame that accompanies this concluding narration features a shot of a trucker, who Rather calls a cowboy, riding into the sunset with the popular mid-1970s song "Convoy" fading out in the audio background. Less dramatic mediation takes place in "Charity Begins At Home," where Morley Safer mediates between the state and a small town. Here he interviews individuals in town and a representative from the state over the issue of welfare. Even though the courts have ordered the town to comply with the state law in this particular case, Safer presents enough evidence (more than ten local characters and lots of filmed small-town footage) supporting individualism to offset the position of the state (represented by one character in an impersonal office). Safer mediates the tension by siding with small-town values and by offering *hope* in his concluding narration: "The battle is not over." In most concluding scenes, the reporter mediates narrative tension by suggesting that individuals hold answers to institutional and social problems.

Finally, a minor opposition that is significant in these tourist episodes features conflicts between near and far or us (the United States) and them. When these tensions structure a story, generally the villain is some "backward," foreign, or alien set of values. In "Yugoslavia" (2/17/80), for example, Dan Rather interviews a foreign couple who affirm and mirror middle class American values in contrast to the apparently more repressive, centralized values of communism. This narrative portrays a view of communist Yugoslavia as a U.S. ally modeled on our own culture. Rather tells the Yugoslav couple, "It strikes me, as you talk, that your life here is very much like middle class life in the United States or England." *60 Minutes* then reveals a scene in a Yugoslav nightclub with the couple out for an evening of dinner and dancing. With "When the Saints Go Marching In" playing in the background, Rather offers this voice-over narration: "The scene is Korcula. It could be Saturday night in Kalamazoo. Dancing to American jazz music mixed with Yugoslav rock. In many other important ways, this is happening in Yugoslav society." With this episode, *60 Minutes* merges the themes of its foreign land and American heartland formulas. Rather mediates tensions between capitalism and communism, us and them, as he presents this communist country as small-town, middle class America. Its people support consumerism and speak proudly of newly purchased middle class conveniences. These individuals from a foreign country are heroic through association with values of democracy, individualism, and capitalism. This portrayal ultimately helps Rather resolve narrative opposition. The Soviet Union, as the agent for villainous communism, emerges here as a distant institutional force (visually unrepresented in this story) that has lost the support of individuals or at least the support of this couple who, in spite of their life in a communist society, seem to prefer Middle American values.

As in the detective and analyst episodes, the dichotomies posed in the tourist formula require resolution, not merely termination, of the narrative. One function of journalism is that it "stimulates by bringing opposites together in a single context, thus suggesting the possibility of synthesis" (O'Brien, 1983, p. 18). Reporters perform that synthesis in *60 Minutes*, acting as mediators of tension by affirming a set of basic values and thereby bring closure to the story. "Rolls-Royce" offers the best example. Here contradictions between humanity and technology, nature and civilization, are resolved as the Rolls acquires human and animal characteristics. In the narration, the Rolls receives a personality and pedigree. A lab technician dressed like a veterinarian or doctor probes a Rolls with a stethoscope. The plant's director refers to the car's "gestation period," and Morley Safer compares the plant's testing facilities to an "incubation ward." As the Rolls—a product of civilization and technology—takes on personal and natural traits of humans and animals, technology becomes natural. Machine becomes person.

The central conflict in the tourist episodes is between place and space. The mythic narrative structure transforms abstract, unknown space into concrete, known place, resolving conflict between the two and mediating contradiction. The reporter, as our surrogate, locates the self in the new or unknown and through the narrative transforms that space into a familiar place. *60 Minutes* draws a map that helps its audience to distinguish local from alien, authenticity from artifice, villain from heroes, and basic values from ideologies.

## Conclusion

The mythic narratives of *60 Minutes* offer weekly mediation between the personal and social spheres. In its handling of changing social attitudes toward race, men and women, war, the economy, foreign powers, among other issues, *60 Minutes* mediates the conflict between tradition and change by applying familiar formulas to a wide variety of experiences. The detective formula, for example, can accommodate raw experience as diverse as waterfront crime, a Vietnam colonel, spy satellites, racial protests, child pornography, valium, diamond scams, the Teamsters, horse doping, art theft, cocaine, and the Nazis, all within repetitive and familiar structures. Other *60 Minutes* formulas provide similar flexibility to assimilate a variety of unfamiliar realities into a handful of familiar representations. At the center of these formulas is the device of metaphor, transforming complex, abstract experience into more manageable, concrete formulas.

These metaphors also serve *60 Minutes* by rescuing the reporters from a tide of criticism leveled at the profession since Watergate. The power of *60 Minutes* resides in replacing reporters who work for bureaucratic news organizations with lone detectives, analysts, and tourists. It is important that the institutional identities of the *60 Minutes* reporters remain concealed; they appear to be operating as individuals, independent of their bureaucratic organizations. This narrative strategy is crucial in order to transform the reporters into heroes and elevate them to a position once-removed from the routine realm of institutional reporting. While the profession more and more draws public criticism as an invader of privacy and a threat to other individual rights, the metaphorical roles performed by the *60 Minutes* reporters, located somewhere outside Middle America, situate them more comfortably on a symbolic map located *within* the space of a middle class mythology.

The mythology is not without contradictions. Celebrated and admired for the performance of independent roles, the reporters also dramatize an institutional function as an arm of social justice that acts on behalf of individuals to right wrongs committed by villains and bureaucracies. Mike Wallace (1984, p. 450) argues that viewers regard the program as a dramatic "unofficial ombudsman":

[B]y the late 1970s . . . I kept bumping into people who jumped at the chance to alert me to some scandal or outrage that was ripe for exposure on *60 Minutes*. They would give me vivid accounts of foul deeds and the culprits perpetrating them, and urge me to take appropriate action: "You really should look into this, Mike. . . ."

Certainly a large number of episodes conclude with *60 Minutes* serving explicitly as ombudsman. In "The Selling of Col. Herbert" (2/4/73), for example, Wallace calls for a public hearing concerning an alleged Army cover-up during the Vietnam War. In "Savak" (3/6/77), Wallace asks for a Senate investigation into the activities of Iranian secret police operating in the United States. And in "Equal Justice?" *60 Minutes* tracks down a witness (who police could not find) and, as a result, a federal judge reconsiders an earlier conviction. Leah Ekdom (1981, p. 149) classifies these kinds of stories as "heroic press narratives," that portray the press as an ideal, "the fourth estate, which monitors and checks government and other abuses of power, and calls our attention to instances in which the country fails to live up to its values." This heroic role of *60 Minutes* unites individual concerns and perceptions with a sense of social mission and order.

Like much of the news media, *60 Minutes* narrowly conceives its role as responsible ombudsman, as heroic press. The techniques of *60 Minutes* do reveal injustices, but they conceal as well. Some of what they conceal are the elements involved in constructing stories: the values of reporters and the economic self-interests of news organizations. While these elements mediate embedded conflicts between product and process, journalists and their audiences routinely assume that there is little distance between the narrative *product* and the raw experience that the product represents and accept one for the other without questioning the *process* of transformation. The taken for grantedness of formulaic narrative referents is so embedded in our consciousness that Hall (1980, p. 129) argues that "the event [reality] must become a 'story' before it can become a *communicative event*."

I nonetheless disagree with research that fails to recognize the social value of news formulas and accuses news institutions, particularly television news, of ideological control. In a society as pluralistic as our own, institutions create centers and provide frameworks for making sense of experience. As Newcomb and Hirsch (1983, p. 49) note, "one of the primary functions of the popular culture forum, the television forum, is to monitor the limits and effectiveness of this pluralism. . . ." Peter Dalgren (1981, p. 111) is correct when he says that television journalism is often caught in a double bind "between its attempt to appear as an independent, critical agent, and its commitment to the prevailing social arrangements," but he is wrong when he argues that "TV news mystifies rather than clarifies." Dalgren and others disregard the formulaic and metaphoric power of news to provide a *center*, a cultural forum for discussion, that transforms complexity and contradiction through the narrative process. Television news helps to create a common reality and to clarify issues and events within that reality.

In *Habits of the Heart*, Bellah and his colleagues (1985, p. 279–81) also miss the mark when they categorize television as part of the "culture of separation" because of its "disconnectedness" to reality, its "extraordinary discontinuity," and its failure to support "any clear set of beliefs or values." The mythic narrative impulse in television's "fictional" and "factual" programs seeks to resolve fundamental, archetypal conflicts between good and evil, tradition and change, individual and institution, and nature and culture. Not only does television offer a coherent vision that generally affirms individualism in the face of bureaucratic oppression, but it does so through the *coherence, continuity* and *assurance* of its formulas, through its romances, its mysteries, its melodramas, and its news.

Other studies also overlook or discredit diversity among audiences, arguing that news is merely a hegemonic device, a way for existing power structures to dominate. Fiske (1986, p. 392) points to the "failure of ideological criticism" to address both "the polysemy of the television text" and "the diverse subcultures in a society." While *60 Minutes* structures centered readings through its formulas, it cannot control the multiplicity of audience responses to that centering. And the program itself, in part, celebrates viewer diversity through the weekly "Letters" segment that offers viewpoints and values that oppose the program (*60 Minutes,* however, does frequently try to recenter this opposition by providing a roughly equal smattering of letters that support the program from week to week). I conceive of viewer response on a horizontal model (or as a map metaphor) rather than the vertical model of dominant, negotiated, and oppositional readings proposed by Hall (1980).[4] While *60 minutes* attempts to position us at the center of its map, viewers still may find their own location on or off that map. As Newcomb (1986, p. 223) notes: "Whatever the messages and meanings of television, we, the viewers may read them in our own ways, receive them only as raw material for our own uses, bend them to our purposes, subvert, parody, distort, and otherwise appropriate them at will."

The mythic narrative formulas of *60 Minutes* provide a map or mythology for the middle class and affirm that individuals through adherence to Middle American values can triumph over institutions that deviate from central social norms. The term "middle" in middle class signifies a mediation between a variety of ambiguous oppositions: up and down, high and low, here and there, liberal and conservative, tradition and change, nature and culture, individual and institution. The electronic media, by taking us to once inaccessible places, "strip away layers of social behavior" to create a new *middle region* reality that merges the ordinary and extraordinary, the usual and unusual, the public and private (Meyrowtiz, 1985, p. 311). (*60 Minutes'* preferred shot is the *middle* or *medium* shot that symbolically locates reporters advantageously in a region between close-ups and long shots where narrative conflict is mediated.)

In *60 Minutes,* stories affirm a set of central values: allegiances to family, education, religion, capitalism, health, democracy, competition, work, honesty, loyalty, duty, fidelity, moderation, fairness, team play, efficiency, simplicity, authenticity, discipline, common sense, modesty, humility, security, cooperation, and ingenuity. These stories celebrate the integrity of the individual and middle class norms. Don Hewitt, creator of *60 Minutes,* has argued that part of the reason for the program's success is his own ties to Middle America:

> My strength is that I have the common touch. I don't know why this is, because most of the people I hang around with are pretty elite. But Kiwanians, Rotarians, I understand them. . . . Maybe it's because I grew up in New Rochelle, the small town that George M. Cohan wrote "45 Minutes From Broadway" about. It was very Middle American. My father was in the advertising business and worked for Hearst. My mother was a housewife. We were middle class. (Henry, 1986, p. 28)

Hewitt's small-town, middle class history taps into a fundamental mythic impulse in American culture, a nostalgic yearning to retreat from the "large scale organizational and institutional structures" that rob our lives of "meaning and coherence" (Bellah et al., 1985, p. 204).

The formulas and metaphors of *60 Minutes* penetrate deeply into American consciousnous. The detective, for instance, taps into our desires for truth, honesty, and intrigue. The analyst helps us come to terms with our inner self, with order, and with knowledge about experience. The tourist cherishes tradition, nature, and authenticity. These metaphorical transformations of the reporter offer us figures of and for modernity, carriers of ways of knowing and interpreting complexity. In this way they offer the possibility of enriching rather than merely simplifying experience. Walker Percy (1975, p. 70) argues that while one function of metaphor is to mediate, "to diminish tension," more importantly metaphor is "a discoverer of being." Through metaphor we discover who we are. What *60 Minutes* ultimately offers its large audience through the detective, analyst, and tourist metaphors is the comfort, concreteness, and familiarity of a middle ground, a center to go back to (or start out from) each week. The power of its formula and metaphor is to reveal and conceal, transform and deform, enrich and simplify *experience* and secure a sense of place, a *middle ground,* where we map out meanings and discover once again *who we are.*

## Notes

1. This analysis is based on a study of 154 *60 Minutes* episodes from 55 programs broadcast between 1968 and 1983, held by the Library of Congress (Motion Picture, Broadcasting, and Recorded Sound Division) in Washington. In addition to the three major formulas, I also identified a minor referee formula and four combination episodes in which the reporter assumes a dual role (e.g., detective/analyst or tourist/analyst). Of the 154 episodes analyzed, 5 (mostly from the earlier years) did not fit into these formulas; I labeled these nonformulaic narratives as experimental.

At the time of my research, the *60 Minutes* collection at the Library of Congress was complete to 1983, except for twelve shows from 1968 and 1969 that were not copyrighted at the Library by CBS. It should be noted that *60 Minutes* was scheduled erratically by CBS at this time; the program was shown every other Tuesday evening and ran from late September to early or mid-June. The library's collection includes more than 500 *60 Minutes* programs available for viewing on 16mm film (1968–75) or ¾-inch videotape (since 1975). The library is about one to two years behind in cataloging more recent episodes. CBS News, which has a complete collection, said it did not have the facilities to accommodate my viewing and research requests. Viewing of more recent episodes suggests that the formulas remain intact, although more combination formulas featuring the reporters in dual metaphoric roles (rather than in the pure formulas described here) now dominate the program's structure.

In viewing each episode from the fifty-five sample programs, I utilized a viewing sheet on which I described the narrative sequence and the role of reporters as mediators of conflict. I identified each episode by (1) number in sample, (2) air date, (3) episode title, (4) episode position in program, (5) producer, and (6) reporter. I also summarized the chronological structure of each episode and identified the major characters in each narrative. I made notes on the reporter's location in the story both in terms of what he said, when he said it, and his relationships to the camera, other characters, and the setting. In addition, I broke down each episode into the following narrative elements: subject matter, plot, setting/locations, resolution/summary, conflicts, and values.

2. The numbers in parentheses refer to the broadcast dates of the *60 Minutes* programs I studied. CBS News published the transcripts of *60 Minutes* programs from the 1979–80 television season as a book, *60 Minutes Verbatim,* and quotations from those programs refer the reader to that text (see CBS News under references). All other quotations from the programs come from my own viewing notes.

3. In this episode, Rather briefly resembles the hard-boiled American private detective typified by the works of Dashiell Hammett and Raymond Chandler in the 1920s and 1930s (see Cawelti, 1976). In this model, the detectives are vulnerable, reflective, and far less detached; they are morally and emotionally involved in their investigations, and they are as interested in "what ought to be" as "what is." In hard-boiled fiction, detectives emerge as characters defined and determined by an explicit value system rather than as mere functions of plot and dramatic tension. The world of clues becomes a means for revealing aspects of a more personal, more complex world. *60 Minutes,* however, rarely looks to this model which would violate journalistic codes of moral neutrality.

4. I am grateful to Robert Allen for suggesting this conceptualization.

## *References*

Allen, R. (1987). Reader-oriented criticism and television. In R. Allen (ed.), *Channels of discourse: Television and contemporary criticism* (pp. 74–112). Chapel Hill: University of North Carolina Press.

Arlen, M. (1977, November 23). The prosecutor. *The New Yorker,* pp. 166–73.

Bellah, R. N., Madsen, R., Sullivan, W. M., Swindler, A., and Tipton, S. T. (1985). *Habits of the heart: Individualism and commitment in American life.* Berkeley: University of California Press.

Black, J. (1981, April/May). The stung. *Channels,* pp. 43–46.

Cawelti, J. (1976). *Adventure, mystery, and romance: Formula stories as art and popular culture.* Chicago: University of Chicago Press.

CBS News. (1980). *60 Minutes verbatim.* New York: Arno Press.

Connell, I. (1978). Monoply capitalism and the media. In S. Hibbin (Ed.), *Politics, ideology and the state* (pp. 69–98). London: Lawrence & Wishart.

Dalgren, P. (1981). TV news and the suppression of reflexivity. In E. Katz and T. Szecsko (eds.), *Mass media and social change* (pp. 101–14). Beverly Hills: Sage.

Didion, J. (1979). *The white album.* New York: Simon and Schuster.

Ekdom, L. (1981). "An interpretive study of the news: An analysis of news forms." Unpublished doctoral dissertation, University of Iowa.

Father of "60 Minutes": Taking the heat as no. 1.(1981, April 3). *Chicago Tribune,* sec 2, p. 15.

Fernandez, J. (1974). The mission of metaphor in expressive culture. *Current Anthropology,*15: 119–45.

Fiske, J. (1986). Television: Polysemy and popularity. *Critical Studies in Mass Communication,* 3: 391–408.

Funt, P. (1980, November). Television news: Seeing isn't believing. *Saturday Review,* pp. 30–32.

Hall, S. (1980). Encoding/decoding. In S. Hall, P. Willis, D. Hobson, and A. Lowe (eds.), *Culture, media, language* (pp. 128–38). London: Hutchinson.

Hartley, J. (1982). *Understanding news.* London: Methuen.

Henry, W. A., III. (1986, May). Don Hewitt: Man of the hour. *Washington Journalism Review,* pp. 25–29.

Himmelstein, H. (1984). *Television myth and the American mind.* New York: Praeger.

James, P. D. (1984). Interview on PBS's *Mystery* series, *Cover her face,* episode four [television interview]. WGBH, Boston.

Jensen, K. B. (1987). Qualitative audience research: Toward an integrative approach to reception. *Critical Studies in Mass Communication,* 4: 21–36.

Kahn, E. J., Jr. (1982a, July 19). Profiles: The candy factory—Part I. *The New Yorker,* pp. 40–41, 47–49, 54–61.

––––– (1982b, July 26). Profiles: The candy factory—Part II. *The New Yorker,* pp. 38–42, 45–46, 50–55.

Katz, E., and Liebes, T. (1984). Once upon a time, in Dallas. *Intermedia,* 12: 28–32.

Leach, E. (1968). Claude Lévi-Strauss: Anthropologist and philosopher. In R. Manners and D. Kaplan (eds.), *Theory in anthropology: A source book* (pp. 541–51). Chicago: Aldine.

Lears, T. J. J. (1983). From salvation to self-realization: Advertising and the therapeutic roots of consumer culture, 1880–1930. In R. W. Wightman and T. J. J. Lears (eds.), *The culture of consumption: Critical essays in American history, 1880–1980* (pp. 1–38). New York: Pantheon.

Lévi-Strauss, C. (1967). The structural study of myth (C. Jacobson and B. Grundfest Schoef, trans.). In *Structural anthropology* (pp. 202–28). Garden City., N.Y.; Anchor-Doubleday.

MacCannell, D. (1976). *The tourist: A new theory of the leisure class.* New York: Schocken Books.

Madsen, A. (1984). *60 Minutes: The power and the politics of America's most popular TV news show.* New York: Dodd, Mead, & Company.

Meyrowitz, J. (1985). *No sense of place: The impact of electronic media on social behavior.* New York: Oxford University Press.

Moore, D. (1978, January 12). 60 Minutes. *Rolling Stone,* pp. 43–46.

Morley, D. (1980). *The "nationwide" audience*. London: British Film Institute.

Nevins, F. M., Jr (Ed). (1970). *The mystery writer's art*. Bowling Green, Ohio: Bowling Green University Popular Press.

Newcomb, H. (1984). On the dialogic aspects of mass communication. *Critical Studies in Mass Communication,* 1: 34–50.

Newcomb, H. (1986). American television criticism, 1970–1985. *Critical Studies in Mass Communication,* 3: 217–28.

Newcomb, H., and Hirsch, P. (1983). Television as a cultural forum: Implications for research. *Quarterly Review of Film Studies* (3): 45–56.

O'Brien, D. (1983, September). The news as environment. *Journalism Monographs,* 85.

Percy, W. (1975). Metaphor as mistake. In *The message in the bottle* (pp. 64–82). New York: Farrar, Straus & Giroux.

Radway, J. A. (1984). *Reading the romance: Women, patriarchy, and popular literature*. Chapel Hill: University of North Carolina Press.

Schafer, R. (1981). Narration in the psychoanalytic dialogue. In W. J. T. Mitchell (ed.), *On narrative* (pp. 25–49). Chicago: University of Chicago Press.

Schudson, M. (1978). *Discovering the news*. New York: Basic Books.

Shaw, D. (1987, March 28). Grading "60 Minutes": It's still going strong—But needs help in one key area. *TV Guide*, pp. 4–10.

Stein, H. (1979, May 6). How "60 Minutes" makes news. *The New York Times Magazine*, pp. 28–30, 74–90.

Strine, M. S. (1985). The impact of literary criticism. *Critical Studies in Mass Communication,* 2: 167–75.

Wallace, M., and Gates, G. P. (1984). *Close encounters: Mike Wallace's own story*. New York: Berkeley Books.

Weisman, J. (1983, April 16). "60 Minutes"—How good is it now? *TV Guide*, pp. 5–14.

White, H. (1981). The value of narrativity in the representation of reality. In W. J. T. Mitchell (ed.), *On narrative* (pp. 1–23). Chicago: University of Chicago Press.

# Defining Media Events: High Holidays of Mass Communication

## DANIEL DAYAN AND ELIHU KATZ

This [study] is about the festive viewing of television. It is about those historic occasions—mostly occasions of state—that are televised as they take place and transfix a nation or the world. They include epic contests of politics and sports, charismatic missions, and the rites of passage of the great— what we call Contests, Conquests, and Coronations. In spite of the differences among them, events such as the Olympic Games, Anwar el-Sadat's journey to Jerusalem, and the funeral of John F. Kennedy have given shape to a new narrative genre that employs the unique potential of the electronic media to command attention universally and simultaneously in order to tell a primordial story about current affairs. These are events that hang a halo over the television set and transform the viewing experience.

We call them collectively "media events," a term we wish to redeem from its pejorative connotations. Alternatively, we might have "television ceremonies," or "festive television," or even "cultural performances" (Singer, 1984). These telecasts share a large number of common attributes which we shall attempt to identify. Audiences recognize them as an invitation— even a command—to stop their daily routines and join in a holiday experience. If festive viewing is to ordinary viewing what holidays are to the everyday, these events are the high holidays of mass communication. Conceptually speaking, this [study] is an attempt to bring the anthropology of ceremony (Durkheim, 1915; Handelman, 1990; Lévi-Strauss, 1963; Turner, 1985) to bear on the process of mass communication.

## *Television Genres*

Until very recently, television was thought to be saying nothing worthy of humanistic analysis. To propose that television—like the other media—deals in "texts" and "genres" seemed to be conferring too much dignity. Viewers were thought to be watching not programs but television. They were assumed to be passive and unselective, satisfied with stories intended for an undifferentiated audience with a short attention span. Social scientists studied television the way they had studied radio; they searched for mass response to persuasion attempts, or to the images of race, sex, or occupation, or to acts of violence. Some concentrated on long-run effects, taking note of the substitute environment with which TV envelops heavy viewers. Others focused on the effect of television on social institutions, such as politics.

Yet producers and audiences alike routinely assume the existence of television genres. The broadcasters themselves, and the TV listings in newspapers and magazines, regularly classify programs by type: news, documentary, sports, action, adventure, Western, situation comedy, soap opera, variety show, game show, talk show, children's cartoon, and the like. Researchers in mass communications employ these categories too, almost as uncritically. With the exception of soap opera that dates to radio (Herzog, 1941; Arnheim, 1944; Warner, 1962; Katzman, 1972; Modleski, 1982; Cassata, 1983; Cantor and Pingree, 1983; and Allen, 1985), very little serious work has been done on the characteristics of these forms, how they differ from one another, how they relate to corresponding forms in other media, what their messages are, and how these messages are communicated.

Systematic study of the news as a genre of broadcasting has recently begun to rival interest in the soap opera (Epstein, 1973; Tuchman, 1978; Fiske and Hartley, 1978; Schlesinger, 1978; Gans, 1979; Graber, 1984; Morse, 1985). Certain political forms—national conventions, presidential debates, political advertising—have also gained attention, and the situation comedy is having its day (Marc, 1989; Taylor, 1989). Still, until recently, and with only occasional exceptions, social studies of television have treated the medium as a whole or in terms of discrete stimuli, without paying serious attention to its component forms. The publications of Horace Newcomb (1974) in the United States and Raymond Williams (1975) in England represent major turning points in the mapping of television territory.

It is striking how different is the study of film. Cinema studies approach film with a literary perspective, as texts to be classified and decoded, sociologically, politically, and psychoanalytically.[1] The same kind of classificatory effort has been applied—although not always uncondescendingly—to the other genres of popular culture. In *Adventure, Mystery, Romance* (1976) John Cawelti elaborates on the dominant genres (he prefers to speak of formulas) of popular fiction. In adventure stories—chivalric tales, war novels, mysteries—Cawelti finds the message of triumph over death, injustice, and the dangerous enemy. The classic detective story stands out in his category of mystery and leads the reader to a desirable and rational restoration of order and of pacification of the unknown. Romance teaches the all-sufficiency of love, celebrating monogamy and domesticity.

Following Cawelti, Newcomb (1974) attempted to delve into the formulas of television. This was the first time, to our knowledge, that a scholar classified television programs systematically: he analyzed the programs in each category and generalized about what they had in common. In the process he proposed a much broader generalization: that television, as a medium, imposes an element of "familism" on each of the genres which it has inherited from the other media of popular culture. In other words, says Newcomb, the Western, the action adventure, and the detective story, not just the soap opera or the situation comedy, is domesticated by television as if to attune the medium as a whole to the nuclear family, television's original viewing group.

## Television with a Halo

Even those like Williams and Newcomb, who pioneered in the classification of television genres, approach the viewing experience not in terms of discrete programs but in terms of the patterned sequences of stimuli (images, issues, messages, stories) that constitute an evening's viewing. They prefer to speak of "strips" (Newcomb and Hirsch, 1983), "flow" (Williams, 1975), compounded interruption (Houston, 1984), relentless messages (Gerbner et al., 1979), moving wallpaper, and mindless chewing gum (Hood, 1967; Csikszentmihalyi and Kubey, 1981).

Even if it is true that most of television melds into some such seamless "supertext" (Browne, 1984), there are certain types of programs that demand and receive focused attention (Liebes and Katz, 1990). Media events are one such genre. Unique to television, they differ markedly from the genres of the everynight.

Readers will have no trouble identifying the kinds of broadcasts we have in mind.[2] Every nation has them. Our sample of a dozen of these events, internationally, includes the funerals of President Kennedy and Lord Louis Mountbatten, the royal wedding of Charles and Diana, the journeys of Pope John Paul II and Anwar el-Sadat, the debates of 1960 between John Kennedy and Richard Nixon, the Watergate hearings, the revolutionary changes of 1989 in Eastern Europe, the Olympics, and others. We have studied accounts and video recordings of these events, and have ourselves conducted empirical research into five of them.[3]

The most obvious difference between media events and other formulas or genres of broadcasting is that they are, by definition, not routine. In fact, they are *interruptions* of routine; they intervene in the normal flow of broadcasting and our lives. Like the holidays that halt everyday routines, television events propose exceptional things to think about, to witness, and to do. Regular broadcasting is suspended and preempted as we are guided by a series of special announcements and preludes that transform daily life into something special and, upon the conclusion of the event, are guided back again. In the most characteristic events, the interruption is *monopolistic,* in that all channels switch away from their regularly scheduled programming in order to turn to the great event, perhaps leaving a handful of independent stations outside the consensus. Broadcasting can hardly make a more dramatic announcement of the importance of what is about to happen.

Moreover, the happening is *live.* The events are transmitted as they occur, in real time; the French call this *en direct.* They are therefore unpredictable, at least in the sense that something can go wrong. Even the live broadcast of a symphony orchestra contains this element of tension. Typically, these events are *organized outside the media,* and the media serve them in what Jakobson (1960) would call a phatic role in that, at least theoretically, the media only provide a channel for their transmission. By "outside" we mean both that the events take place outside the studio in what broadcasters call "remote locations" and that the event is not usually initiated by the broadcasting organizations. This kind of connection, in real time, to a remote place—one having major importance to some central value of society, as we shall see—is credited with an exceptional value, by both broadcasters and their audiences (Vianello, 1983). Indeed, the complexity of mounting these broadcasts is such, or is thought to be such, that they are hailed as "miracles" by the broadcasters, as much for their technological as for their ceremonial triumphs (Sorohan, 1979; Russo, 1983).[4]

The organizers, typically, are public bodies with whom the media cooperate, such as governments, parliaments (congressional committees, for example), political parties (national conventions), international bodies (the Olympics committee), and the like. These organizers are well within the establishment. They are part of what Shils (1975) calls the center. They stand for consensual values and they have the authority to command our attention. It is no surprise that the Woodstock festival—the landmark celebration of protesting youth in the sixties—was distributed as a film rather than as a live television event.

Thus, the League of Women Voters and the two major political parties organized the presidential debates in 1976 and 1980; the palace and the Church of England planned and "produced" the royal wedding; the Olympics are staged by the International Olympics Committee. There may be certain exceptions to this rule: the European Broadcasting Union organizes the annual Eurovision Song Contest, for example, and the Super Bowl—the American football championship—involves a direct organizational input on the part of American broadcasters. But on the whole, these events are not organized by the broadcasters even if they are planned with television "in mind." The media are asked, or ask, to join.

Of course, there may well be collusion between broadcasters and organizers, as was evident in the Gerald Ford–Jimmy Carter debate in Philadelphia, for example, when the TV sound failed and the ostensibly local meeting in a hired hall was suspended until the national broadcast could be resumed. And a state-operated broadcasting system (Poland, for example; *not* England or Israel) may be indistinguishable from the organizers. But the exceptions only serve to prove the rule.

These events are *preplanned*, announced and advertised in advance. Viewers—and, indeed, broadcasters—had only a few days notice of the exact time of Sadat's arrival in Jerusalem (Cohen, 1978); Irish television advertised the Pope's visit to Ireland a few weeks in advance (Sorohan, 1979); the 1984 Los Angeles Olympics were heralded for more than four years. Important for our purpose is that advance notice gives time for anticipation and preparation on the part of both broadcasters and audiences. There is an active period of looking forward, abetted by the promotional activity of the broadcasters.

The conjunction of *live* and *remote*, on the one hand, and *interrupted* but *preplanned*, on the other, takes us a considerable distance toward our definition of the genre. Note that live-and-remote excludes routine studio broadcasts that may, originate live, as well as feature programs such as *Roots* or *Holocaust*. The addition of interruption excludes the evening news, while preplanned excludes major news events—such as the attempted assassination of a pope or a president, the nuclear accident at Three Mile Island, and, at first glance (but we shall reconsider this), the so-called television revolutions in Romania and Czechoslovakia. In other words, our corpus is limited to ceremonial occasions.

Returning to the elements of definition, we find that these broadcast events are presented with *reverence* and *ceremony*. The journalists who preside over them suspend their normally critical stance and treat their subject with respect, even awe. Garry Wills (1980) called media coverage of the Pope, including that of the written press, "falling in love with love" and "The Greatest Story Ever Told." He was referring to the almost priestly role played by journalists on the occasion, and we find a reverential attitude characteristic of the genre as a whole. We have already noted that the broadcast transports us to some aspect of the sacred center of the society (Shils, 1975).

Of course, the very flow of ceremonial events is courtly and invites awe. There is the playing of the national anthem, the funereal beat of the drum corps, the diplomatic ceremony of being escorted from the plane, the rules of decorum in church and at Senate hearings. The point is that in media events television rarely intrudes: it interrupts only to identify the music being played or the name of the chief of protocol. It upholds the definition of the event by its organizers, explains the meaning of the symbols of the occasion, only rarely intervenes with analysis and almost never with criticism. Often advertising is suspended. There are variations: the live broadcast of Sadat's arrival in Jerusalem was treated differently by Israeli television than by the American networks, which had more explaining to do (Zelizer, 1981). While we shall have occasion to point out these differences, they are outweighed by the similarities.

Even when these programs address conflict—as they do—they celebrate not conflict but *reconciliation.* This is where they differ from the daily news events, where conflict is the inevitable subject. Often they are ceremonial efforts to redress conflict or to restore order or, more rarely, to institute change. They call for a cessation of hostilities, at least for a moment, as when the royal wedding halted the street fighting in Brixton and the terror in Northern Ireland. A more permanent truce followed the journeys of Sadat to Jerusalem and the Pope to Argentina. These events applaud the *voluntary* actions of great personalities. They celebrate what, on the whole, are establishment initiatives that are therefore unquestionably *hegemonic.* They are proclaimed *historic.*

These ceremonials *electrify very large audiences*—a nation, several nations, or the world. They are gripping, enthralling. They are characterized by a *norm of viewing* in which people tell each other that it is mandatory to view, that they must put all else aside. The unanimity of the networks in presenting the same event underlines the worth, even the obligation, of viewing. They cause viewers to *celebrate* the event by gathering before the television set in groups, rather than alone. Often the audience is given an active role in the celebration. Figuratively, at least, these events induce people to dress up, rather than dress down, to view television. These broadcasts *integrate* societies in a collective heartbeat and evoke a *renewal of loyalty* to the society and its legitimate authority.

## A More Parsimonious Approach to Definition

Despite its heaviness, we shall argue that the elements in our definition are "necessary," and that no subset of them is "sufficient" without the others.[5] This hypothesis does not mean that the elements cannot exist without one another, but they are not then what we call media events; they are something else.

Consider, for example, the *live* broadcasting of an event which is not *preplanned*—say, the live reporting of the leaking atomic energy plant at Three Mile Island (Veron, 1981). The leakage is a great *news* event, but not one of the great *ceremonial* events that interest us. Thus, we are interested here in the Kennedy funeral—a great ceremonial event—and not the Kennedy assassination—a great news event. The messages of these two broadcasts are different, their effects are different, they are presented in quite a different tone. Great news events speak of accidents, of disruption; great ceremonial events celebrate order and its restoration. In short, great news events are another genre of broadcasting, neighbor to our own, that will help to set the boundaries of media events.[6]

Consider an event that fails to *excite* the public or one that is not presented with *reverence* by the broadcasters. Such events do not qualify according to the definition, but they are particularly interesting because they suggest a pathology of media events, of which the former is an event "manqué" and the latter an event "denied" by the broadcasters. . . .

Thus, by converting the elements of the definition into a typology—where elements are variously present or absent, or present in varying degree—we can identify alternative genres of broadcasting that differ from one another by virtue of a particular element. Examination of these alternative forms and the conditions of their occurrence will help define our own events by providing boundary markers.

One additional operation, methodologically speaking, can be performed on the definition. By transforming the elements into variables, one can note which elements correlate with which others. Doing so, one might ask whether, say, the degree of *reverence* invoked by the presenter correlates with the degree of viewer *enthrallment*.

Presentation of the genre can be formulated more elegantly, by grouping the elements of the definition into broader categories. The linguistic categories of syntactics, semantics, and pragmatics are useful for this purpose.

Syntactically, media events may be characterized, first, by our elements of interruption, monopoly, being broadcast live, and being remote. These are components of the "grammar" of broadcasting. The cancellation of regularly scheduled programs and the convergence of channels are the most dramatic kinds of punctuation available to broadcasters. They put a full stop to everything else on the air; they combine the cacophony of many simultaneous channels into one monophonic line. Of course, these elements also carry semantic meaning: they speak of the greatness of the event. And they have a pragmatic aspect as well: the interruption of the sequence of television puts a stop to the normal flow of life.

The live and remote broadcast takes us back and forth between the studio and some faraway place. Such broadcasts employ special rhetorical forms and the technology required to connect the event and the studio. The language is the language of transportation—"We take you now to . . ." Both pictures and words are slowed to a ceremonial pace, and aesthetic considerations are unusually important. The pictures of media events, relative to their words, carry much more weight than the balance to which we are accustomed in the nightly news, where words are far more important than pictures (Altman, 1986; Katz, Adoni, and Parness, 1977). The centrality of these various elements of syntax is immediately apparent when one compares the event itself to each subsequent representation of the event—the wrap-up, the news, and the eventual anniversaries: length is drastically cut; pace is speeded up; words reassert their importance; references to the heroic logistics of the broadcast disappear. Syntactic unpredictability (that matches the semantic uncertainty) is smoothed over.

The fact that the event is situated *outside* broadcasting organizations, both physically and organizationally, implies a network of connections that differ from the everyday. The specialists of outside broadcasting deploy their

OB units—as the British call them—and the studio now serves as intermediary between its people in the field and the audience, allowing some of the dialogue of stage directions between studio and field to become part of the spectacle. Their cadence is *reverential* and *ceremonial.*

The interruption, when it comes, has been elaborately advertised and *rehearsed.* It entails a major commitment of manpower, technology, and resources on the part of organizers and broadcasters. It comes not as a complete surprise—as in major newsbreaks—but as something long anticipated and looked forward to, like a holiday. In order to make certain that the point of this ritual framing will not be lost on the audience, the broadcasters spend hours, sometimes days, rehearsing the audience in the event's itinerary, timetable, and symbolics. Even one-time events can be ritualized in this way.

The meaning of the event—its semantic dimension—is typically proposed by its organizers and shared by the broadcasters—although this point requires elaboration. . . . Of course, each event is specific in this regard. For example, the royal wedding was proposed as a Cinderella story, the moon landings as the new American frontier, and the papal diplomacy as a pilgrimage. Regardless of the specifics of each event, the genre as a whole contains a set of core meanings, often loudly proclaimed. Thus, all such events are hailed as *historic;* they strive to mark a new record, to change an old way of doing or thinking, or to mark the passing of an era. Whether it is the Olympics or Watergate, Sadat or the Pope, the turning-point character of the event is central.

The event features the performance of symbolic acts that have relevance for one or more of the core values of society (Lukes, 1975). By dint of the cooperation between organizers and broadcasters, the event is presented with *ceremonial reverence,* in tones that express sacrality and awe.

The message is one of *reconciliation,* in which participants and audiences are invited to unite in the overcoming of conflict or at least in its postponement or miniaturization. Almost all of these events have heroic figures around whose *initiatives* the reintegration of society is proposed.

Pragmatically, the event *enthralls very large audiences.* A nation or several nations, sometimes the entire world, may be stirred while watching the superhuman achievement of an Olympic star or an astronaut. Sadat electrified the people of Israel, and the Pope revived the spirit of the Polish people. These are thrilling events, reaching the largest audiences in the history of the world. They are shared experiences, uniting viewers with one another and with their societies. A *norm of viewing* accompanies the airing of these events. As the day approaches, people tell one another that viewing is obligatory, that no other activity is acceptable during the broadcast. Viewers actively *celebrate,* preferring to view in the company of others and to make special preparations—unusual food, for example—in order to partake more fully in the event.

The genre is best defined, then, at the intersection of the syntactic, the semantic, and the pragmatic. And, as was argued above, we shall contend that all three elements are "necessary." If we chose to apply the pragmatic criterion alone, the events so defined would include television programs that enthralled very large audiences, such as the early miniseries or perhaps even key episodes of programs such as *Dallas*. They might also include films that attracted large, sometimes cultish audiences such as the *Rocky Horror Picture Show* or *Woodstock;* these were indeed compulsory viewing for certain segments of the population and invited widespread participation. If syntactics were the sole criterion, major news events would demand to be included. By the same token, if the genre were defined in terms of the semantic alone, we should number among media events all those films and programs that claim to be historic, preach reconciliation, celebrate initiative, and are produced and presented with reverence. Films of the Olympics by Leni Riefenstahl, Kon Ichikawa, or Claude Lelouch, for example, might therefore qualify.

Hence our insistence on defining the corpus of events in terms of all three linguistic categories, an insistence further justified by the fact that we are dealing with ceremonial performances and that no such performance can be described in terms of its text alone. A ceremony interrupts the flow of daily life (syntactics); it deals reverently with sacred matters (semantics); and it involves the response (pragmatics) of a committed audience.

## *Why Study Media Events?*

Implicit in this definition of the genre are answers to the question, Why study media events? The student of modern society—not just of television—will find a dozen or more powerful reasons for doing so. Let us spell them out.

1. The live broadcasting of these television events attracts the *largest audiences in the history of the world*. Lest we be misunderstood, we are talking about audiences as large as 500 million people attending to the same stimulus at the same time, at the moment of its emission. It is conceivable that there were cumulative audiences of this size prior to the electronic age—for the Bible, for example. Perhaps one might have been able to say that there were several hundred million people alive on earth who had read, or heard tell of, the same Book. But it was not until radio broadcasting—and home radio receivers—that simultaneity of exposure became possible. The enormity of this audience, together with the awareness by all of its enormity, is awesome. It is all the more awesome when one realizes that the subject of these broadcasts is ceremony, the sort which anthropologists would find familiar if it were not for the scale. Some of these ceremonies are so all-encompassing that there is nobody left to serve as out-group. "We Are the World" is certainly the appropriate theme song for media events. To enthrall such a multitude is no mean feat; to enlist their assent defies all of the caveats of media-effects research.[7]

2. The power of these events lies, first of all, in the rare *realization of the full potential of electronic media technology*. Students of media effects know

that at most times and places this potential of radio and television are capable of reaching everybody simultaneously and directly; their message, in other words, can be total, immediate, and unmediated. But this condition hardly ever obtains. Messages are multiple; audiences are selective; social networks intervene; diffusion takes time. On the occasion of media events, however, these intervening mechanisms are suspended. Interpersonal networks and diffusion processes are active before and after the event, mobilizing attention to the event and fostering intense hermeneutic activity over its interpretation. But during the liminal moments, totality, and simultaneity are unbound; organizers and broadcasters resonate together; competing channels merge into one; viewers present themselves at the same time and in every place. All eyes are fixed on the ceremonial center, through which each nuclear cell is connected to all the rest. Social integration of the highest order is thus achieved via mass communication. During these rare moments of intermission, society is both as atomized and as integrated as a mass-society theorist might ever imagine (Kornhauser, 1959).

3. Thus, the media have power not only to insert messages into social networks but to create the networks themselves—to atomize, to integrate, or otherwise to design social structure—at least momentarily. We have seen that *media events may create their own constituencies.* Egypt and Israel were united for Sadat's visit not only by images of the arrival of the leader of a theretofore hostile Arab nation, but by means of an ad hoc microwave link between the broadcasting systems of the two countries.[8] Similarly, the royal wedding reunited the British Empire, and Third World nations joined the first two worlds for the Olympics. That media events can talk over and around conventional political geography reminds us that media technology is too often overlooked by students of media effects in their distrust of hypotheses of technological determinism. Papyrus and ancient empire, print and the Protestant Reformation, the newspaper and European nationalism, the telegraph and the economic integration of American markets are links between attributes of communication technologies and social structures. They connect portability, reproducibility, linearity, simultaneity, on the one hand, to empire, church, nation, market, on the other.

By extension, it can be seen that the "center" of these media-engendered social structures is not bound by geography either. In the case of media events, the center—on which all eyes are focused—is the place where the organizer of a "historic" ceremony joins with a skilled broadcaster to produce an event. In this sense, Britain is often the center of the world; one has only to compare the broadcast funeral of the assassinated Mountbatten with the broadcast funeral of the assassinated Sadat or India's Indira Gandhi to understand why.

4. Conquering not only space but time, media events have the power to declare a holiday, thus to play a part in the *civil religion.* Like religious holidays, major media events mean an interruption of routine, days off from work, norms of participation in ceremony and ritual, concentration on some central value, the experience of communitas and equality in one's immediate

environment and of integration with a cultural center. The reverent tones
of the ceremony, the dress and demeanor of those gathered in front of the
set, the sense of communion with the mass of viewers, are all reminiscent
of holy days. The ceremonial roles assumed by viewers—mourner, citizen,
juror, sports fan—differentiate holiday viewing from everyday viewing and
transform the nature of involvement with the medium. The secret of the
effectiveness of these televised events, we believe, is in the roles which viewers bring with them from other institutions, and by means of which passive
spectatorship gives way to ceremonial participation. The depth of this involvement, in turn, has relevance for the formation of public opinion and
for institutions such as politics, religion, and leisure. In a further step, they
enter the collective memory.

5. *Reality is uprooted* by media events. If an event originates in a particular location, that location is turned into a Hollywood set. The "original"
is only a studio. Thus conquering space in an even more fundamental way,
television causes events to move off the ground and "into the air." The era
of television events, therefore, may be not only one in which the reproduction is as important as the original, as Benjamin (1968) proposed, but also
one in which the reproduction is more important than the original.

Sometimes the original is inaccessible to live audiences because it is
taking place in London, or because it is taking place on the moon, for
example. Even more fundamental are those events that have no original
anywhere because the broadcast is a montage originating in several different
locations simultaneously. The "reality" of Kennedy's debating Nixon when
one was in New York and the other in California is not diminished for its
being in the air, and in the living room. Prince Charles, at the church, is
waiting for Lady Diana as her carriage is drawn through the streets of London. This is reality. But it is an invisible reality that cannot be apprehended
as such because it is happening simultaneously at different places. No one
person can see all of it, that is, except the television director and hundreds
of millions in their homes.

6. The process of producing these events and telling their story relates
to the arts of television, journalism, and narration. Study of the rhetorical
devices for communicating festivity, enlisting participation, and mobilizing
consensus demands answers to the questions of how television manages to
project ritual and ceremony in the two-dimensional space of spectacle. Essential to an understanding of these events—in addition to the readiness of
the audience to assume ceremonial roles—is an analysis of how the story is
framed, how interest is sustained, how the event aggregates endorsements,
how the broadcasting staff is deployed to give depth to the event, how
viewers interact with the screen, what tasks are assigned to the viewers.
Media events give insight into the *aesthetics of television production,* together
with an awareness of the nature of the contract that obtains between organizers and broadcasters.

The audience is aware of the genre of media events. We (and certain
fellow researchers) recognize the constituent features of this rare but recur-

rent narrative form, and so do producers and viewers. The professional networks of producers buzz with information on the extraordinary mobilization of manpower, technology, aesthetics, and security arrangements required
to mount a media event.[9] At the same time, the networks of viewers carry
word of the attitudes, rehearsals, and roles appropriate to their celebration.
The expectation that certain events in the real world will be given media-
events treatment is proof of public awareness of the genre. Israelis appealed
to the High Court of Justice demanding that the war-crimes trial of John
Demjanjuk be broadcast live.[10]

   7. Shades of *political spectacle*. Are media events, then, electronic incarnations of the staged events of revolutionary regimes and latter-day versions
of the mass rallies of fascism? We think not, even if they might seem to be.
It is true that media events find society in a vulnerable state as far as indoctrination is concerned: divided into nuclear cells of family and friends, disconnected from the institutions of work and voluntary association, eyes and
ears focused on the monopolistic message of the center, hearts prepared
with room. This is reminiscent, *mutatis mutandis,* of the social structure of
a disaster that strikes at night, or of a brainwashing regimen. The threshold
of suggestibility is at its lowest the more isolated the individual is from
others, the more accessible he or she is to the media, the more dependent
the person is, the more the power to reward conformity or punish deviation
is in the hands of the communicator.

   Nevertheless, media events are not simply political manipulations.
Broadcasters—in Western societies—are independent of, or at least legally
differentiated from, government. They can, and sometimes do, say no to an
establishment proposal to mount an event. Journalists need convincing before suspending professional disbelief, and even commercial interest sometimes acts as a buffer. Second, public approval is required for an event to
succeed; official events cannot be imposed on the unwilling or unbelieving.
Third, individuals are not alone, not even alone with family, but in the
company of others whom they invite to join in the thrill of an event and
then to sit in judgment of it. Some societies provide public space for such
discussion and interpretation; others provide only living rooms and tele-
phones. Family friends, home, and living-room furniture are not a likely
context for translating aroused emotion into collective political action. Fourth,
the audience, too, has veto power. Oppositional readings are possible and
hegemonic messages may be read upside down by some. These checks and
balances filter the manipulative potential of media events and limit the vulnerability of mass audiences.[11]

   Still, the question of hegemonic abuse must be asked continually. Almost all of these events are establishment initiated, and only rarely, one
suspects, do the broadcasters say no. Instead, journalists—sometimes reluctantly—put critical distance aside in favor of the reverent tones of presen-
ters. Broadcasters thus share the consensual occasion with the organizers
and satisfy the public—so we have hypothesized—that they are patriots after
all.

8. When media events are seen as a *response to prior events* or to social crisis, the link to public opinion is evident. Thus, certain media events have a commemorative function, reminding us—as on anniversaries—of what deserves to be remembered. Others have a restorative function following social trauma. The most memorable of them have a transformative function inasmuch as they illustrate or enact possible solutions to social problems, sometimes engendering yet further events which actually "change the world." In the restorative domain, media events address social conflict—through emphasizing the rules (as in Contests), through praising the deeds of the great in whom charisma is invested (Conquests), and through celebrating consensual values (as do Coronations).

9. At the same time, certain events have an *intrinsically liberating* function, ideologically speaking; they serve a transformative function. However hegemonically sponsored, and however affirmatively read, they invite reexamination of the status quo and are a reminder that reality falls short of society's norm. Taking place in a liminal context, evoking that climate of intense reflexivity which Victor Turner characterized as the "subjunctive mode of culture," their publics exit the everyday world and experience a shattering of perceptions and certainties. Even if the situations in which they are immersed are shortlived and do not institutionalize new norms, at least they provoke critical awareness of the taken-for-granted and mental appraisal of alternative possibilities. They possess a normative dimension in the sense of displaying desirable alternatives, situations which "ought to" exist but do not. These are previews, foretastes of the perhaps possible, fragments of a future in which the members of society are invited to spend a few hours or a few days. Activating latent aspirations, they offer a peek into utopia.

10. One wonders whether the media-events genre is not an expression of a *neo-romantic desire for heroic action* by great men followed by the spontaneity of mass action. In this sense, media events go beyond journalism in highlighting charisma and collective action, in defiance of established authority. The dissatisfaction with official inaction and bureaucratic ritualism, the belief in the power of the people to do it themselves, the yearning for leadership of stature—all characterize media events. We can join Sadat or the Pope and change the world; the people can unite to save Africans from starvation by supporting "Live Aid." The celebration of voluntarism—the willful resolve to take direct, simple, spontaneous, ostensibly nonideological action—underlies media events, and may indeed constitute part of their attraction. The desire for spontaneous action, of course, recalls the erratic rhythm of arousal and repose predicted by the theory of mass society (Kornhauser, 1959). In the telling of media events, establishment heroes are made to appear more defiant than they actually were. But media events and collective action may be more than a dream. The escalation of interaction among public opinion, new or old leadership, and the mass media fanned the revolutions of Eastern Europe in the fall of 1989.[12]

11. The *rhetoric of media events* is instructive, too, for what it reveals not only about the difference between democratic and totalitarian ceremo-

nies, but also about the difference between journalism and social science, and between popular and academic history. The media events of democracies—the kind we consider here—are persuasive occasions, attempting to enlist mass support; they take the form of political contests or of the live broadcasting of heroic missions—those that invite the public to embrace heroes who have put their lives and reputations on the line in the cause of a proposed change.[13] The ceremonies of totalitarian societies (Lane, 1981) are more commemorative. They also seek to enlist support, but for present and past; the First of May parade was a more characteristic media event in postwar Eastern Europe (Lendvay, Tolgyesi, and Tomka, 1982) than a space shot.[14] Terrorist events contrast with both of these in their display not of persuasion but of force, not of majesty but of disruption and provocation.

The rhetoric of media events contrasts—as does journalism, generally—with academic rhetoric in its emphasis on great individuals and apocalyptic events. Where social science sees long-run deterministic processes, journalism prefers heroes or villains who get up one morning resolved to change the world. Where academic historians see events as projective as underlying trends, journalists prefer a stroboscopic history which flashes dramatic events on and off the screen.

12. Media events *privilege the home.* This is where the "historic" version of the event is on view, the one that will be entered into collective memory. Normally the home represents a retreat from the space of public deliberation, and television is blamed, perhaps rightly, for celebrating family and keeping people home (Newcomb, 1974). When it is argued that television presents society with the issues it has to face, the retort, "narcotizing dysfunction"—that is, the false consciousness of involvement and participation—is quick to follow (Lazarsfeld and Merton, 1948). Yet the home may become a public space on the occasion of media events, a place where friends and family meet to share in both the ceremony and the deliberation that follows. Observational research needs to be done on the workings of these political "salons." Ironically, critical theorists, newly alert to the feminist movement, now see in the soap opera and other family programs an important "site of gender struggle," and their derision of the apolitical home is undergoing revision.

But there is more to politics than feminism, and we need empirical answers to the question of whether the home is transformed into a political space during and after a media event. In fact, we need basic research on who is home and when (in light of the growing number of one- and two-person households), who views with whom, who talks with whom, how opinion is formed, and how it is fed back to decision-makers. These everyday occasions of opinion formation should then be compared with media events. It is hard to believe, but nevertheless true, that the study of public opinion has become disconnected from the study of mass communication.

13. Media events preview the *future of television*. When radio became a medium of segmentation—subdividing audiences by age and education—television replaced it as the medium of national integration. As the new

media technology multiplies the number of channels, television will also become a medium of segmentation, and television-as-we-know-it will disappear. The function of national integration may devolve upon television ceremonies of the sort we are discussing here. By that time, however, the nation-state itself may be on the way out, its boundaries out of sync with the new media technology. Media events may then create and integrate communities larger than nations. Indeed, the genre of media events may itself be seen as a response to the integrative needs of national and, increasingly, international communities and organizations.

Certain multinational interests have already spotted the potential of international events and may sink the genre in the process. Some combination of the televised Olympics and televised philanthropic marathons inspired the effort to enlist worldwide aid to combat famine in Africa. Satellite broadcasters already transmit live sports events multinationally (Uplinger, 1990; but see Mytton, 1991). Aroused collective feeling must be a great lure to advertisers, and one wonders whether the entry of the commercial impresario into the arena of these events does not augur ill for their survival as necessarily occasional, and heavily value-laden, "high holidays."

## Notes

1. The typologies of relationship among storyteller, camera, actors, and audience have all been carefully dissected. And the ideological implications of changes within and among film genres—for continuity and change in values and social structure—have been the subject of continuous speculation. Somehow cinema studies have fallen to the humanist and broadcasting to the social scientist. The rich theorizing of the humanist and the careful empiricism of the social scientist are only now being combined in the analysis of television. The humanists have found interest (and at last legitimacy) in the study of television, and certain social scientists have responded by building conceptual bridges and methods to facilitate joint work (for example, Schudson, 1978). It was in the aftermath of the antiestablishmentarianism of the 1960s that this began to happen, with the politicization of media studies. It led to a renewal of the dialogue between humanists and social scientists that characterized the early days of the study of mass communication, and to renewed interest in critical schools of media studies such as the German-American Frankfurt School, American "cultural studies," and the work of the semiologists in France.

2. The earliest postwar examples include the American presidential conventions of 1948 and 1952, the return of General Douglas MacArthur from the Pacific in 1951, Senator Estes Kefauver's hearings on crime the same year, the inaugurations of Harry Truman and Dwight Eisenhower, Truman's guided tour of the White House in 1952, and, most memorable of all, the coronation of Elizabeth II on June 22, 1953. For details on the broadcasts of these events, see Russo (1983). The analysis by Lang and Lang (1953) of the MacArthur Day procession is the pioneering classic of media-events research.

3. We observed and collected relevant material on the occasion of the first presidential debates (Katz and Feldman, 1962), Sadat's journey to Jerusalem (Katz, Dayan, and Motyl, 1983), and the royal wedding (Dayan and Katz, 1982). We went to Poland to examine materials on the Pope's first visit (Dayan, Katz, and Kerns, 1984).

One of our colleagues made an empirical study of the television audience of the 1984 Olympics (Rothenbuhler, 1985, 1988, 1989).

4. Russo's (1983) dissertation sketches the organizational and technological solutions to the problems of the live broadcasting of special events from remote locations. He shows, for example, how the series of space shots stimulated important developments in video technology that were subsequently employed in quite different contexts, such as coverage of the Kennedy assassination and funeral. We shall repeatedly draw on Russo's work.

5. Huizinga (1950) and Caillois (1961) face the same sort of problem with their multifaceted definition of play. Their definition is an overview of the different types of games. Thus, play is an activity (1) entered voluntarily (except when ceremonial roles or social pressures force participation); (2) situated "outside" ordinary life and accompanied by an awareness of its "unreality"; (3) intensely absorbing to the participants; (4) bounded in time and space; (5) governed by rules of order; (6) uncertain of outcome; and (7) promoting the formation of social groupings. Some of these elements also define media events; others are characteristics of ordinary television viewing (Stephenson, 1967). The problems of specifying the social conditions that give rise to play, the types of play, and the function of play for individual and society are closely related to the problems that confront the present project.

6. Russo (1983, p. 42) distinguishes three genres of television news—regular news, documentaries, and special events—but does not make the distinction between major news events and major ceremonial events that is central to our argument. "Special events broadcasting," Russo's subject, "refers to a genre or type of news coverage which deals with live origination of a major news story."

7. The appendix to [*Media Events,* Dayan and Katz, 1992] attempts to compare the effects of media events with those highlighted in the various traditions of research on media effects.

8. President Sadat's visit to Jerusalem was covered by an unprecedented number of newspapers and broadcasting organizations, mobilizing about 1,500 newsmen—580 from the United States alone. The live broadcast began at Ben Gurion Airport on Saturday night, November 19, 1977. Television followed the visitor and his hosts almost continually for most of the forty-four hours of his visit, to sites such as the Al Aqsa mosque, the Yad Vashem memorial, the Church of the Holy Sepulchre, and, most important, to his address before the Knesset. The live pictures provided by Israeli television were relayed to Egypt and to Western broadcasting organizations, which supplied their own commentaries and experts including Abba Eban (ABC) and John Kissinger (NBC). American television journalists played catalytic roles, beginning with Walter Cronkite's parallel interviews with Menachem Begin and Anwar el-Sadat on the eve of the visit. While the visit was in progress, exclusive joint interviews of the two protagonists were broadcast by CBS (Walter Cronkite), NBC (John Chancellor), and ABC (Barbara Walters). Egyptian television produced a live broadcast of Sadat's triumphant return to Cairo, which was relayed to Israel over the microwave link, with a commentary in Hebrew by an Egyptian broadcaster. Audiences for the broadcasts included almost all Israelis; only 3 percent recalled that they did not view the arrival ceremony (Israel Institute of Applied Social Research [Peled], 1979). It is estimated that perhaps 3 million Arab viewers saw Israeli television from across the border, and Israeli radio claimed 50 million Arab listeners. In the United States, the Knesset speech attracted an estimated 30 million viewers (Nasser, 1978), and in France, 58 percent of a 1978 viewing panel reported having seen at least part of the trip (Centre d'Etudes d'Opinion,

1981). Two years later, 87 percent of Israelis thought that it was "important" for them to have seen the live broadcast, and 94 percent of them thought it a "historic moment" (Peled, 1979).

9. In anticipation of the Pope's visit in 1979, Irish television (Radio Telefis Eirann) consulted the prior experience of Poland and Mexico. Equipment valued at 12 million pounds and including forty-four cameras was assembled for the occasion, in the biggest collective effort RTE had ever made (Sorohan, 1979; Gleeson, 1979). Russo (1983, p. 320) estimated that the Kennedy assassination and funeral involved virtually all of the 2,000 employees of the three U.S. network news organizations, at a cost of perhaps $32 million. The BBC's royal wedding required some sixty television cameras in the largest outside-broadcasting operation in its history (Griffin-Beale, 1981). Media events involve a huge amount of sharing and pooling of equipment among broadcasters. A great sense of pride in the "miracle" of accomplishment typically follows these events (Sorohan, 1979; Todorovic, 1980).

10. The trial of Adolf Eichmann, recorded on film by Alan Rosenthal, was later transferred to videotape but was not broadcast during the period of the trial.

11. Compare Martin's (1969, p. 93) discussion of the checks on divinely validated temporal power when the church—like the media, in the case of media events—has "the capacity both to compound and challenge . . . temporal authority." Martin points to "institutional checks," "symbolic checks," and, rather in despair, to the conclusion that restraint "is realized less in those who perform the acts of temporal and spiritual union than by those suffering and excluded groups who recognize that an ideal has been violated. It is the poor and meek of the earth who realize that crucial limiting concepts are available whereby they can assert the crown rights of the Redeemer against the rights of the crown." For a conclusion opposite to the one drawn here, see Zelizer (in press).

12. The Live Aid event was broadcast on July 13, 1985. It originated from multiple locations, including Wembley in the United Kingdom, Philadelphia, and Moscow. The event lasted sixteen hours and was diffused via thirteen satellites to 150 countries, including India and China. The international audience was estimated at 650 million (presumably based on some percentage of the 2 billion potential in the countries reached). The U.S. figure was 180 million. The event cost 3.7 million dollars and 4 million pounds, which was obtained through sale of TV rights and corporate donations. Estimates of money raised vary from $70 million to $147 million (Hickey, 1987).

13. Only in the fall of 1989 did Czech television openly defy the already crippled communist government by presenting live broadcasts of the mass rallies in Wenceslas Square. "If he [Jiri Hrabovsky, new anchor of Czech television] and a few hundred colleagues hadn't stuck their necks out earlier and broadcast the people's protests, Czechoslovakia's rout of communism might not have gone so far. People made this revolution. Television has spread it" (Newman, 1989). . . .

14. Yet there was a discussion over whether to risk publicizing failure in the live broadcasting of American space shots (Russo, 1983). Note also that, subsequent to the Challenger disaster, the Soviets offered a live broadcast of a space shot. . . .

### References

Allen, R. C. 1985. *Speaking of soap operas.* Chapel Hill: University of North Carolina Press.

Altman, R. 1986. Television/sound. In T. Modleski, ed. *Studies in entertainment: Critical approaches to mass culture.* Bloomington: Indiana University Press.

Arnheim, R. 1944. World of the daytime serial. In P. F. Lazarsfeld and F. N. Stanton, eds. *Radio research: 1942–1943.* New York: Duell, Sloan & Pearce.

Benjamin, W. 1968a. The work of art in the age of mechanical reproduction. In Hannah Arendt, ed. *Illuminations.* Trans. Harry Zohn. New York: Harcourt Brace Jovanovich.

———. 1968b. What is epic theater? In Hannah Arendt, ed. *Illuminations.* Trans. Harry Zohn. New York: Harcourt Brace Jovanovich.

Browne, N. 1984. The political economy of television's supertext. *Quarterly Review of Film Studies* 9(3): 174–83.

Caillois, R. 1961. *Man, play, and games.* New York: Free Press of Glencoe.

Cantor, M. G., and S. Pingree. 1983. *The soap opera.* Beverly Hills: Sage.

Cassata, M. B. 1983. *Life on daytime television: Tuning-in American serial drama.* Norwood, N.J.: Ablex.

Cawelti, J. 1976. *Adventure, mystery, romance: Formula stories as art and popular culture.* Chicago: University of Chicago Press.

Centre d'Etudes d'Opinion. 1981. Les grands événements historiques à la télévision. *Cahiers de la Communication* (1)1: 51–61.

Cohen, N. 1978. President Sadat's visit to Jerusalem: Broadcasting aspects. *EBU Review* 29: 8–12.

Csikszentmihalyi, M., and R. W. Kubey. 1981. Television and the rest of life: A systematic comparison to subjective experience. *Public Opinion Quarterly* 45: 317–28.

Dayan, D., and E. Katz. 1982. Rituel publics à usage privé: métamorphose télévisée d'un mariage royal. *Les Annales: Economie, Société, Civilisation.* Abridged and revised in English as Electronic ceremonies: Television performs a royal wedding. In M. Blonsky, ed. *On signs.* Baltimore: Johns Hopkins University Press, 1985.

———. 1992. *Media events: The live broadcasting of history.* Cambridge: Harvard University Press.

Dayan, D., E. Katz, and P. Kerns. 1984. Armchair pilgrimages: the trips of Pope John Paul II and their television public. *On Film* 13: 25–34. Reprinted in M. Gurevitch and M. Levy, eds. *Mass communication review yearbook.* Vol. 5. Beverly Hills: Sage. 1985.

Durkheim, E. 1915. *The elementary forms of the religious life: A study in religious sociology.* Trans. J. W. Swain. London: Allen & Unwin.

Epstein, E. J. 1973. *News from nowhere: television and the news.* New York: Vintage Books.

Fiske, J., and J. Hartley. 1978. *Reading television.* London: Methuen.

Gans, H. J. 1979. *Deciding what's news: a study of CBS Evening News, NBC Nightly News, Newsweek, and Time.* New York: Pantheon Books.

Gerbner, G., L. Gross, N. Signorielli, M. Morgan, and M. Jackson-Beeck. 1979. The demonstration of power: Violence profile no. 10. *Journal of Communication* 29(3): 177–96.

Gleeson, P. 1979. The chieftains, Bernadette, and a choir of 6,000. *Irish Broadcasting Review,* pp. 42–43.

Graber, D. A. 1984. *Processing the news: how people tame the information tide.* New York: Longman.

Griffin-Beale, C. 1981. The royal wedding day on ITV. *Broadcast* (UK), July 27, no. 1118: 15–17.

Handelman, D. 1990. *Models and mirrors: Towards an anthropology of public events.* New York: Cambridge University Press.

Herzog, H. 1941. On borrowed experience: An analysis of listening to daytime sketches. *Studies in Philosophy and Social Science* 9: 45–65.

Hickey, N. 1987. The age of global TV. *TV Guide,* October 3, pp. 5–11.

Hood, S. C. 1967. *A survey of television.* London: Heinemann.

Houston, B. 1984. Viewing television: The metapsychology of endless consumption. *Quarterly Review of Film Studies* 9(3): 183–95.

Huizinga, J. 1950. *Homo ludens: A study of the play element in culture.* New York: Roy.

Jakobson, R. 1960. Linguistics and poetics. In T. Sebeok, ed. *Style in language.* New York: Wiley.

Katz, E., and S. Feldman. 1962. The Kennedy–Nixon debates: A survey of surveys. In S. Kraus, ed. *The great debates: Background, perspectives, effects.* Bloomington: Indiana University Press.

Katz, E., H. Adoni, and P. Parness. 1977. Remembering the news: What the picture adds to recall. *Journalism Quarterly* 54: 231–39.

Katz, E., D. Dayan, and P. Motyl. 1983. Television diplomacy: Sadat in Jerusalem. In G. Gerbner and M. Seifert, eds. *World communications.* New York: Longman.

Katzman, N. 1972. Television soap operas: What's been going on anyway? *Public Opinion Quarterly* 36: 200–211.

Kornhauser, W. 1959. *The politics of mass society.* New York: Free Press of Glencoe.

Lane, C. 1981. *The rites of ruler: Ritual in industrial society—the Soviet case.* New York: Cambridge University Press.

Lang, K., and G. E. Lang. 1953. The unique perspective of television. *American Sociological Review* 18: 3–12.

Lazarsfeld, P. F., and R. K. Merton. 1948. Mass communication, popular taste, and organized social action. In L. Bryson, ed. *Communication of ideas,* pp. 95–118. New York: Harper & Row.

Lendvay, J., J. Tolgyesi, and M. Tomka. 1982. First of May: A Hungarian media event. Paper presented at the World Congress of Sociology, Mexico City.

Lévi-Strauss, C. 1963. The effectiveness of symbols. In C. Lévi-Strauss, *Structural anthropology.* Vol. 1. New York: Basic Books.

Liebes, T., and E. Katz. 1990. *The export of meaning: Cross-cultural readings of Dallas.* New York: Oxford University Press.

Lukes, S. 1975. Political ritual and social integration. *Sociology* 9(2): 289–308.

Marc, D. 1989. *Comic visions: Television comedy and American culture.* Boston: Unwin Hyman.

Martin, D. A. 1969. *The religious and the secular: Studies in secularization.* London: Routledge and Kegan Paul.

Modleski, T. 1982. *Loving with a vengeance.* Hamden, Conn.: Action Books.

Morse, M. 1985. Talk, talk, talk: The space of discourse on television. *Screen* (26)2: 2–17.

Mytton, G. 1991. A billion viewers can't be right. *InterMedia* 19/8: 10–12.

Nasser, M. 1979. Sadat's television manipulation. Unpublished. Annenberg School for Communication, University of Southern California, Los Angeles.

Newcomb, H. 1974. *TV: The most popular art.* New York: Anchor.

Newcomb, H., and P. Hirsch. 1983. Television as a cultural forum: Implications for research. *Quarterly Review of Film Studies* 8: 48–55.

Newman, B. 1989. Switching channels: Czechoslovakia's TV in a flash became free as it covered uprising, and other articles. *Wall Street Journal,* November 27, December 5.

Peled, T. 1979. Dynamics of public opinion from Sadat's visit to Jerusalem through President Carter's announcement of the Israeli-Egyptian agreement. Jerusalem: Israel Institute of Applied Social Research (research report for ABC News).

Rothenbuhler, E. 1985. Media events, civil religion, and social solidarity: the living room celebration of the Olympic Games. Ph.D. dissertation, Annenberg School for Communication, University of Southern California, Los Angeles.

———. 1988. The living room celebration of the Olympic Games. *Journal of Communication* 38: 61–81.

———. 1989. Values and symbolism: public orientations to the Olympic media event. *Critical studies in mass communication.* Vol. 6, pp. 138–57.

Russo, M. A. 1983. CBS and the American political experience: A history of the CBS News special events and election units, 1952–1968. Ph.D. dissertation, New York University; Ann Arbor: University microfilm.

Schlesinger, P. 1978. *Putting reality together: BBC News.* London: Constable.

Schudson, M. 1978. *Discovering the news: A social history of American newspapers.* New York: Basic Books.

Shils, E. 1975. *Center and periphery: Essays in macrosociology.* Chicago: University of Chicago Press.

Singer, M. 1984. *Man's glassy essence: Explorations in semiotic anthropology.* Bloomington: Indiana University Press.

Sorohan, J. 1979. Pulling off a broadcasting miracle with nine weeks' notice. *Irish Broadcasting Review:* 46–47.

Stephenson, W. 1967. *The play theory of mass communication.* Chicago: University of Chicago Press.

Taylor, E. 1989. *Prime-time families: Television culture in postwar America.* Berkeley: University of California Press.

Todorovic, A. 1988. The funeral of Marshal Tito. *European Broadcasting Union Review* 31: 25–28.

Turner, V. 1985. Liminality, Kabbala, and the media. *Religion* 15: 205–17.

Uplinger, H. 1989. Global TV: What follows live aid? *InterMedia* 17(6): 17.

Veron, E. 1981. *Construire l'événement: Les médias et l'accident de Three Mile Island.* Paris: Minuit. (With the collaboration of J. Dana and A. F. de Ferrière.)

Vianello, R. 1986. The power politics of live television. *Journal of Film and Video* 37(3): 26–40.

Warner, W. L. 1962. *American life: Dream and reality.* Chicago: University of Chicago Press.

Williams, R. 1975. *Television: Technology and cultural form.* New York: Schocken Books.

Zelizer, B. 1981. The parameters of broadcast of Sadat's arrival in Jerusalem. Master's thesis, Communications Institute, Hebrew University of Jerusalem.

———. In press. From home to public forum: media events and the public sphere. *Journal of Film and Video.*

# Alternative Television in the United States

## WILLIAM BODDY

As economic and political changes redefine the institutions of contemporary broadcasting in the United States and elsewhere, a sharpening debate has emerged around notions of alternative structures and modes of production for television, a debate which might usefully reflect on the history of independent television work in the United States. The defining terms commonly attached to video work produced outside of the American broadcast institutions—"independent video" and "artist's video"—beg distinctions less often argued than asserted within their critical communities. The defining absent term, the overbearing institution of commercial television, has remained a fugitive and implicit figure in much of the writing on American video since the 1960s. Despite a sometimes wilful hermeticism in the discourse around such alternative work, a brief account of its telescoped history in the United States can illuminate relations between independent video and the dominant television industry, as well as the changing profile of American commercial television as the moving target for oppositional practice.

The stylistically disparate works of independent video in its first two decades suggest the specific historical conditions of their making and reception. Nam June Paik's "prepared" TV sets, sculptures and performances of the 1960s and early 1970s incarnated a Neo-Dada aesthetic which asserted the personal gesture against the implacable flow of traditional commercial television. The "prepared" TV sets and the video sculptures and installations

From William Boddy, "Alternative Television in the United States," Screen 31:1, 1990, pp. 91–101. Reprinted with permission.

of many video artists of the time re-contextualized the familiar domestic appliance as contingent, vulnerable, and incongruous. Several early gallery works used live, delayed, and taped materials in monitor installations, challenging viewers' relation to the familiar TV screen through direct interaction with the video image. While animated by a revolt against the by-then hegemonic place of commercial broadcasting in defining the television apparatus, in retrospect much of the Neo-Dada video work also seems to replay wider contemporary anxieties accompanying the installation of the TV set as a domestic object in the American home.[1]

Despite the long, if limited, use of video in gallery installations, it was the introduction of half-inch portable video recording equipment in the United States in 1968 which largely defined subsequent independent video production. Video makers from within and outside the art world created a growing production community which remained almost entirely marginalized by the institutions of broadcast television. This gap, and the mutual antagonism between the two groups, was greater than that between the contemporary Hollywood cinema and independent filmmakers in the United States. Broadcast television rejected independent work on aesthetic, political, and even technological grounds (broadcasters decreed small-format video technically unfit for broadcast and warned that it could endanger their broadcast licenses by violating technical standards for transmission), and the exclusion united diverse independent producers in common marginality, creating a surprisingly close-knit community which took up the tasks not only of production, but also of distribution, exhibition, critical exegesis and publicity of the new work. An early video art show at a Fifty-seventh Street gallery in New York City in 1969, for example, seems to have led directly to the formation of Raindance Corporation (a small-format video research and production group), Global Village (a still-extant media center in New York City), and at least two other production groups. Raindance in 1971 published *Guerrilla Television,* a production handbook *cum* philosophical manifesto, as well as four years of a periodical entitled *Radical Software.* Other video production groups included Ant Farm (founded in 1968), Videofreex (1969), People's Video Theatre (1969), TVTV (1972), the Media Access Center at the Portola Institute (publishers of the *Whole Earth Catalogue*), and similar groups across the country.[2] The activities and discourses of the proclaimed "video guerrillas" were initially visionary and at times grandiose, promoting storefront theatres, traveling video troupes and schemes for electronic democracy via interactive cable, although some of the groups later became producers of video-verité documentaries for public television and of dramatic pilots for the commercial networks.

A pronounced strain of technological euphoria and utopianism animated the early guerrilla television movement, and its rhetoric was at times more ecological than political. The anti-political scientism of its manifestoes represents a clear break with the earlier Soviet and European political and artistic avant gardes of the 1920s as well as with the contemporary practice of Godard and other European filmmakers. Raindance, in its *Guerrilla*

*Television* handbook, discussed the name of their magazine, *Radical Software:*

> Most people think of something "radical" as being political, but we are
> not. We do, however, believe in post-political solutions to cultural prob-
> lems which are *radical* in their discontinuity with the past. Thus, our use
> of the adjective acts to lure people from an old context (the political) into
> our own.[3]

The post-political attitude produced an ironic echo of the agit-prop film-
trains of the Russian civil war in *Guerrilla Television*'s description of the
plans of the video collective Media Bus (formed when Videofreex abandoned New York City for a communal home in the country) to bring small-
format video to the countryside:

> They plan to extend their life style to Media Buses . . . , touring from
> town to town giving video shows and making tape. This is more or less
> analogous to rock road tours and will probably become self-sustaining when
> cable-TV opens up; a travelling troupe of video technicians and enter-
> tainers being paid to do local gigs at local TV studios or shoot tape in
> local communities.[4]

From the political education of the agit-prop film-trains for Soviet artists
like Eisenstein and Dziga Vertov to the vision of the video guerrilla as
touring rock star of the 1970s, an enormous contextual shift has obviously
occurred. At times the technological determinism of *Guerrilla Television*
obliterates history and politics altogether in an inversion of social causality:
"Broadcast television is structurally unsound. The way it is used is the result
of its inherent characteristics. Those attributes create the political and economic environment which determines the nature of programming, not visa-
versa."[5] The technological determinism of the work also seems at times to
find a technocratic fix on the political inequality of U.S. society: "The original American political system is as elegant as any ever designed. . . . The
state-of-the-art of government is way behind that of information technology. We could be voting by cable-TV and legislators could use computer-
retrieval systems to research government programs."[6]

  While the video guerrillas and other independent producers in the early
years of video in the United States were never large in numbers, they had
an important effect on the changing popular conception of American television. Network television, its commercial structures undisturbed and its
profits soaring since the 1950s, seemed at the height of its power at the end
of the 1960s. But by the late 1960s network hegemony began to be challenged not only by the manifestoes of the video guerrillas, but by political
attacks from the Nixon administration over purported liberal bias in network news, and by the economic prospect of the new delivery technologies
of cable, low power television, and direct broadcast satellites. The threat of
the new television delivery services took a decade to challenge the networks'
hold on American audiences, however, and while the networks were de-

nounced as dinosaurs by the video guerrillas in the early 1970s, it wasn't until 1982 that the trade journal *Channels* used the same term in a legitimate if apocalyptic question, "Are the Networks Dinosaurs?" a question which has often been restated in the subsequent broadcast trade press.[7]

It was possible to predict too quickly the extinction of the commercial networks; while the networks' grasp on the prime-time audience did decline approximately fifteen percent in the 1970s, they still captured about three-quarters of American viewers. Moreover, most of the audience loss was not at the hands of exotic new technologies, but to independent stations largely offering reruns of previous network programming. Nevertheless, the new delivery technologies were widely hailed in the popular press of the 1970s as harbingers of a "media revolution," bringing diversity and new viewer sovereignty to television in the home. In the popular press, cable television was the chief public benefactor, promising not only more channels of conventional programming, but specialized cable networks for diverse tastes.

Above all, however, it was the idea of interactivity which fueled the extraordinary attention the new television technologies received in the 1970s. Echoing the language of the same video artists, cable entrepreneurs contrasted the traditional audience-subject of network television with one prophesized for interactive cable, where not only would viewers actively seek out and select from a myriad of program choices, but would also accomplish information retrieval, shopping, and voting from their homes. Richard DeCordova argues that cable's promised new subjectivity offered a transcendence and unity new to the television spectator, though similar to that effected in a different manner via the subject positioning of the classical fiction film.[8] That the transcendent subject traded electronic omniscience for physical immobility and more prefect commodification was suggested in a 1981 trade handbook for interactive television:

> In the coming age of interactive television . . . the tremendous range of TV viewers' options in the home will contrast ironically to their shrinking options outside the home. . . . Unlike the expanded range of video services, the scope of our lives will narrow.[9]

Notwithstanding the visions of utopian subjectivity promised by interactive cable, the economic forces behind interactive television included the appeal to Hollywood program producers of a new pay-per-view box office and the appeal to advertisers of the more refined viewer demographic data possible with interactive cable's continuous surveillance of viewer activity. The triviality of the cable industry's wider notions of interactivity is suggested in the much-publicized Warner-Amex CUBE system of the late 1970s which occasionally offered instant viewer polls on issues inspired by its programming, enacting Jean Baudrillard's model of television's simulation of communication, "a speech that answers itself via the simulated detour of a response."[10]

The rhetorical similarities between the technological visions of some video guerrillas and the entrepreneurs of the booming cable industry of the

1970s seem disquieting in retrospect. The wishful thinking about the autonomy of technology and the refusal of history and politics among independent video makers may have inadvertently enlisted them as the avant garde for an (un)reconstructed communications industry only too happy to lead a "media revolution" which would leave existing power relations untouched.

Of all the alternative distribution vehicles envisioned by American independent video makers in the 1960s and 1970s, the clear survivor is public access cable. That it has so survived within a vastly altered and hostile cable industry and in the face of successful political attacks on other public interest values of broadcasting is due in part to the peculiar circumstances of cable television in the United States. Although cable had been around since the late 1940s, it was the success of satellite-delivered pay-cable networks like Home Box Office in the mid-1970s that attracted popular interest and major capital in the service. Because cable operators require access to public rights of way, local governments became cable's primary regulators in the form of local franchise agreements, which set rates and terms of service, frequently including the provision of public access: non-commercial channels reserved on a first-come, first-served, non-discriminatory basis, with editorial control ceded to the program producer. As the cable industry underwent a rapid consolidation from small owners to highly-capitalized multiple system operators in the late 1970s, the competition for large urban franchises became fierce. With channel capacity exceeding available programming services, cable operators looked at public access as an inexpensive bargaining chip with franchise officials, who themselves saw political advantage in other forms of access programming controlled by local governmental and educational institutions. The heady, speculative cable boom of the late 1970s touched off a "franchise war" among firms bidding for big-city markets; in the words of one company president:

> The theory was to get the franchise, particularly in the large cities. Promise anything—large city fees, low subscriber rates, major sophisticated systems, two-way communications capability—anything to get the franchise. Change it years later when the system was built and running. Use the franchise to raise new funds. It was standard procedure.[11]

One result of the financial and political conditions of the growth of the American cable industry in the past fifteen years has been the opening up of a public space unique within all the dominant commercial media, print and electronic: subsidized free speech in the form of public access to a community's cable channels. The result of local franchise agreements has been television programming unconstrained by program executives, cable management, or advertisers. Despite the opportunity such access obviously provides political activists, the Left in the United States has been slow to engage with public access, and public access programmers and producers remain largely isolated from progressive independent producers, media analysts, and grassroots organizations. An examination of the experience of one production collective in progressive public access cable may illuminate some of the political and public interest stakes in public access.

Paper Tiger Television is a video production collective which has created over 200 weekly thirty-minute programs for public access cable in Manhattan since 1981. While each program typically examines a single media text or issue, the entire body of Paper Tiger's work represents a broad analysis of the structures and products of the American mass culture industries. Though the shows, most of them produced live in a modest studio, are aimed primarily at the local cable television audience, tapes circulate to schools and to other public access programmers.

Paper Tiger Television's goal has been not only to analyze the American communications industry but also provide a model and network for other progressive public access programmers. The underfunded and politically precarious support of independent programming on public broadcasting has made public access television an important battleground in the 1980s. The continuing consolidation of the cable industry into a handful of giant multiple system operators, and their moves into ownership positions in cable programming firms, has created a growing hostility toward public access from vertically-integrated cable operators dizzy at the prospect of increasingly lucrative commercial cable program services. The federal Cable Franchise Policy and Communications Act of 1984 stripped much of the powers of local authorities in cable regulation and allows cable operators to usurp "underused" public access channels, and the Reagan and Bush administrations have given the communications industry unprecedented opportunities to roll back anti-trust and public interest standards in broadcasting.[12] The President of the National Association of Broadcasters announced to his members shortly after Reagan's inauguration:

> The country has rediscovered that "the business of America is business"—something lost sight of since Calvin Coolidge offered that definition some fifty years ago.
>
> There is no question that the current climate favors us—we must take advantage of it by taking charge. . . . I can say to you today that the opportunity is present to set the agenda for our own industry, create a strategic plan that will guarantee our growth, and gain control of our own destiny. Our timing is right—and there's no time like the present.[13]

Against this background of economic and political change within the American communications industry, Paper Tiger's manifesto outlines the following project:

> The power of mass culture rests on the trust of the public. This legitimacy is a paper tiger. Investigation into the corporate structures of the media and a critical analysis of their content is one way to demystify the information industry. Developing a critical consciousness about the commu-

nications industry is a necessary first step towards democratic control of information resources.[14]

The program of Paper Tiger typically examine issues of ownership (including concentration within media industries, interlocking boards of directors, and the background of individual executives); composition and strategies of advertisers; and the demographics of readership or audiences. Many of the early shows examined the business press and other upscale publications, and at times the economic facts behind the publication seemed to be left to speak for themselves about the role of economic structures in determining meaning in media texts. The danger of such economism, however, needs to be understood against the prejudices of mainstream American media studies and the persistence of a number of popular myths about the media, including the heroic figure of the lone crusader-journalist and autonomous creative artist, independent of institutional and ideological contexts; the belief in journalistic objectivity and ideological "balance," and a general hostility to economic analysis. Given this context, the simple insistence upon the relevance of economic, political, and ideological determinations of media texts is a first step in challenging conventional assumptions about the media.

Paper Tiger has subsequently widened scope beyond the print press to address general themes across different media and take on more specifically populist media forms, like the weekly tabloid, *The National Enquirer,* romance novels, and daytime television serials. In addition, the series has occasionally celebrated alternative periodicals like *Covert Action Newsletter,* as well as independent documentary filmmakers, early black cinema in the United States, and popular theatre in Nicaragua.

The political profile of Paper Tiger Television is somewhat eclectic: the group's name recalls Mao's guerrilla stance against superpower hegemony, and the manifesto's assertion of the importance of the reproduction of ideology is compatible with the ideas of Gramsci, Althusser, economistic marxism, and anarchism. Paper Tiger's eclecticism is also due to its loose collective structure and work process; ideas for programs are developed by the group months in advance, but generally each show is coordinated by one or two members who work with the program's host on research and preparation of non-studio segments, props, and dramatic vignettes. On the evening of a live show, members arrive at the studio thirty or sixty minutes before transmission, sort out specific production tasks, prepare the set and props, cue tapes, and try to set up some of the camera shots and transitions.

While the subject, tone, and political line varies according to the specific host and program, Paper Tiger Television does have a consistent visual style and attitude. These derive from pragmatic considerations and production constraints as well as from some of the lessons learnt from alternative video in the last twenty years. The eccentric visual style of the show is designed in part to mark the rejection of the standard fare not only of commercial television but also of most educational and public access programming. Rather than create pinched, low-budget imitations of a standard interview show or newscast, Paper Tiger employs a television style that is strikingly unconventional. Traditional television standards of casting, decorum, language, and political discourse are violated. The first six Paper Tiger programs featured media scholar Herbert Schiller "reading" the *New York Times,* and something of the illustrated lecture style remains in the live shows. The critic speaks directly to camera, but the conventional props of the television lecturer are absent from the show. The backgrounds are non-illusionistic, cartoon-like painted flats. In many cases loosely-related dramatic action is staged next to the presenter: a female analysand lies on an analyst's couch; an Asian-American woman makes herself up in white-face; two players battle over control of the world in a parlor game; a viewer watches TV and falls asleep. The clash of presentational styles extends from the studio sequences to include superimposed character-generated text, staged and edited sequences inserted as breaks into the shows, on-the-street interviews and other documentary material, and frequent appropriation, via the soundtrack and visual inserts, of mass media images and texts.

The anti-realism of the dramatic sequences and the manipulation and juxtaposition of disparate texts and voices at its best creates a didactic television both Brechtian and carnivalesque. Some of the determinants of Paper Tiger's style are practical: relying on donated labor and small budgets precludes substantial rehearsal and post-production time and foregrounds the theatricality of live production. The legal status of public access allows unconstrained use of copyrighted material on Paper Tiger, from defaced corporate reports and re-edited daytime serials and, in *Joan Does Dynasty,* a performer electronically keyed into the space of the prime-time series' scenes. While hand-lettered titles generally open and close the show, several programs have used electronic special effects, especially keys. The occasionally elaborate use of keys on several Paper Tiger shows nevertheless remains consistent with the program's style: the keys and switching are done live, generally in ways which both draw attention to themselves as devices and demystify their construction. The overall effect of Paper Tiger's style is engaging and casual; most live shows include wide shots of the studio crew, and little attempt is made to provide the seamlessness of traditional television. Programs typically end with an outline of the show's costs, generally under $300. Some of the attributes of the program's irreverent style echo a reflexive, anti-commercial genre of earlier independent video work, from Nam June Paik's 1960s Neo-Dada gestures to Shirley Clarke's real-time video improvisations, Dara Birnbaum's off-air collage tapes, and Richard Serra's deadpan *Television Delivers People* of the 1970s.

A logical outgrowth of the work of Paper Tiger was an ongoing project begun in 1986, Deep Dish TV, which put together public access programming from across the United States into ten sixty-minute shows around the topics of labor, housing, women's issues, the farm crisis, racism, U.S. policy in Central America, disarmament, children's tapes, and popular culture. The series enlisted over 150 tapes by producers from thirty states, and was designed to mitigate the isolation of local public access programmers by link-

ing them both with grassroots political groups organized around the programs' issues, and with other independent video makers. A second goal was
to offer to access programmers stylistic alternatives to the standard talk-
show public access format, and to demonstrate to independent video producers the viability of satellite-delivered public access television as an outlet
for their work. While part of the series' success was the demonstration that
small-format video, including third-generation half-inch tapes, could be successfully distributed by satellite, the design of Deep Dish TV was quite
different from the single-event satellite link-ups of video artists like Nam
June Paik's 1984 *Good Morning Mr. Orwell.* As opposed to Paik's unique
satellite performance, Deep Dish TV enlisted the traditional television concept of regularly-scheduled series programming for new works and voices
in order to encourage ongoing work.

The strand of optimism about the progressive possibilities of the tools
of mass media from Bertolt Brecht and Walter Benjamin to Hans Magnus
Enzensberger has informed many U.S. independent video makers since the
1960s. While Brecht's brief 1932 essay, "Radio as an Apparatus of Communication," and Benjamin's "The Work of Art in the Age of Mechanical
Reproduction" were often cited aphoristically by many video artists, the
lesson taken was most often the impropriety of the imbalance in the electronic media of few transmitters and many receivers. While many early video
artists and video guerrillas called for an inversion of the model, their asser-
tions often seemed based more on a leap of technological faith than on
extending what the earlier media theorists diagnosed as the tension between
the social formation and communications technology. Paper Tiger Television is committed to a form of television which is both decentralized and
personal, but it is also precisely concerned with the historical and political
determinations of the dominant media in U.S. society. Through a commitment to collective and grassroots production of small format video and to
public access television, which has remained consistent from the video guerrillas of the 1960s through Paper Tiger Television in the 1980s, the collective lessons and pitfalls of twenty years of independent video in the United
States can still illuminate and inspire.

## Notes

This is a revised version of an article which originally appeared in *Communications,*
no. 48.

1. For a useful discussion of Nam June Paik's career, see Martha Gever, "Pomp
and Circumstances: The Coronation of Nam June Paik," *Afterimage,* no. 10 (1982),
pp. 12–16; for an account of the anxieties which accompanied the positioning of
American commercial television in domestic space, see Lynn Spigel, "Installing the
Television Set: Popular Discourses on Television and Domestic Space, 1948–1955,"
*Camera Obscura,* no. 16 (January 1988), pp. 11–48.

2. Michael Shamberg and Raindance Corporation, *Guerrilla Television* (New
York: Holt, Reinhart and Winston, 1971), pp. 10–20; Deirdre Boyle, "Guerrilla

Television," in *Transmission,* ed. Peter D'Agostino (New York: Tanam Press, 1985), pp. 203–13.

3. Shamberg, np.
4. Ibid., p. 52.
5. Ibid., p. 32.
6. Ibid., p. 30.
7. Les Brown, "Are the Networks Dinosaurs?" *Channels,* June–July 1982, pp. 26–29, 57.
8. Richard DeCordova, "Cable Technology and the Utopian Subject," *On Film,* no. 12 (Spring 1984), pp. 23–27.
9. John Witek, *Response Television: Combat Television of the 1980s* (Chicago: Crain Press, 1981), p. v.
10. Jean Baudrillard, "Requiem for the Media," in *Video Culture: A Critical Investigation* (Rochester: Visual Studies Workshop Press, 1986), p. 129.
11. Burt I. Harris, president of Harris Cable Corporation, in 1975, quoted in Edward V. Dolan, *TV or CATV: A Struggle for Power* (Port Washington,N.Y.: National University Publications/Associated Faculty Press, Inc., 1984), p. 84.
12. For a discussion of the impact of the Reagan administration on the television and film industries in the United States, see Douglas Gomery, "The Reagan Record," *Screen,* vol. 30, nos. 1/2 (Winter/Spring 1989), pp. 92–99.
13. Vincent T. Wasilewsky, speech at the National Association of Broadcasters convention, April 12, 1981.
14. Paper Tiger Television, *Cat*a*log* (New York: Paper Tiger Television, 339 Lafayette Street, New York, NY 10012, 1987), back cover.

# III

# THE RECEPTION
# CONTEXTS OF
# TELEVISION

Issues, ideas, and arguments appearing in this section surround some of the most widely and hotly debated topics in the study of television today. Even the choice of a title for this grouping presents a problem. Is "reception" an adequate term to describe the interaction of viewers with the various forms of television discussed in the previous division? Is "reception" too skewed a term? Does it imply, too strongly perhaps, that those who sit before the set, or who walk around it, glance at it, listen to it, all those who in some way "watch" are actually "constructing the text" as they view? Or is "reception" merely synonymous with "viewing"? Is the very notion of "audience" a construction that favors the television industry? Would we be better off thinking in terms of "viewing publics" or "interpretive communities?"

Critical analysis addressing these questions—and many more in many other formulations—has become central to Television Studies. After decades of worrying over what it might mean to actually study viewer interactions with television, and decades of decrying traditional behavioral or social psychological research attempting to answer these questions, scholars from various backgrounds in the humanities have now attempted to deal with the issues themselves.

This is why Ien Ang's overview is helpful in beginning the explorations of the problems presented here. She not only demonstrates the ways in which industry researchers and marketing agencies actually create notions of "audience" that benefit their purposes, she also points out that academic researchers and critics engage in the same sorts of constructions. She goes on to show how the very history of this topic can be discerned from the variations and fundamental differences among the varying points of view and varying definitions. Her essay provides, then, something of a map with which to examine the essays that follow.

Ellen Seiter presents her analysis of some of these particular problems in terms of difficulties faced in interpreting information gathered in an actual interview session. In a powerfully self-conscious exploration of her own analytical strategies she demonstrates what is by now a commonplace—that the "data" we gather from audiences become another "text" to be interpreted. It is precisely the "problematic" nature of this process that has led to increasingly complex discussions of how scholars and critics in this area are doing their work, and "working on" viewers and users of television, even as they struggle to develop their own models of research and analysis.

Self-consciousness is at the center of John Fiske's essay as well, taking as he does his own "viewing practices" as part of the "data" for his analysis. Fiske's work has been at the heart of the most recent debates over viewer activity. He has been defined as the strongest proponent for a position suggesting that viewers are "free" to make personal meanings from materials offered them by television. While some arguments have overstated his positions, it is well that polarities can be established in such debates, for ultimately, as positions shift and merge, we usually discover that our "boundary positions" were more helpful as guideposts than as definitions. Here Fiske explores the various interpretive procedures required of critics who attempt to combine several types and "levels" of evidence in making claims about the interpretation and use of television by audiences.

His general position is roundly questioned by Celeste Condit, who is more concerned with establishing the constraints and limitations on viewers' interpretive freedom than with establishing that some freedom does exist. Even though she does not directly address the Fiske essay included here, her questions do put us at the heart of the wide-ranging debates over audience activity. For there are few researchers who would question the existence of active mental processes among television viewers, few who question even the possibility of viewer rejection of meanings or the ability to "rewrite" those meanings in personal contexts. The issue, as refined, has to do with the degrees of viewer freedom. Most importantly, we must remember that this issue is not debated merely in order to explore the process, to discover how individuals and groups use television. The issue is fundamental because of its ideological implications. This debate is part of the much larger one regarding how television does its ideological work in helping to shape consciousness—individual, group, social, cultural—with how television works to maintain or ameliorate existing social inequities.

This why the section closes with two essays that offer more detailed studies of actual viewing practices. Henry Jenkins examines these practices in the context of "fan behavior." He tracks and cites the work of *Star Trek* fans who literally rewrite the television program, the implied fictional "world" of that program, and the extensions of that world. These fans have appropriated and plundered the television offering in order to create imaginative constructs that far exceed the "source material." Significantly for our understanding of the essays in this section, he shows that even among these television users (perhaps a better term perhaps than viewers or audiences) there are debates mirroring the more academic discussions. Questions and arguments among fans concern "how far one should go" in "using" *Star Trek* for individual or group purposes. These discussions, carried out in fanzines and newsletters are not unlike those suggested by an exploration of the "limits of polysemy."

Similarly, debates—visible or submerged—within households also pose the academic questions in a different, more grounded manner. David Morley's discussion of the role of gender as a determining factor in the uses of television is a summary chapter from a much longer study of how families use TV. We might argue that this study is somehow more telling of the "normal" experience with television than that presented by Jenkins. But what these two essays make clear, and what all the essays in this section confront in different ways, is that the very definition of "normal"—common, typical, ordinary, choose your term—is an exceptionally difficult process when it comes to the uses of television by those who live and interact with it. Morley's essay presents significant variations in television use, variations through which the analyst must move very carefully in order to construct generalizations.

As with the essays in Part I, these offer models of variable use to students. Studying audiences, fans, and other viewing groups may be, in some instances, more easily arranged than can be the resources for studying production. But the conceptual problems are formidable. These essays offer a partial introduction to those problems and even more partial answers to the questions rising from them. Only by continuing this discussion will we come closer to more precise understanding of what television actually means and does in personal and social experience.

# Understanding Television Audiencehood

## IEN ANG

*We must make allowance for the complex and unstable process whereby discourse can be both an instrument and an effect of power, but also a hindrance, a stumbling-block, a point of resistance and a starting point for an opposing strategy.*

Michel Foucault (1980a: 101)

### Beyond the Institutional Point of View

First, a short recapitulation of what [my book, *Desperately Seeking the Audience,*] argued. . . . Television audiencehood is a pervasive social and cultural reality in the late twentieth century. In a multitude of ways, sometimes routine sometimes exceptional, television plays an intimate role in shaping our day-to-day practices and experiences—at home but also outside it; at work; at school; in our conversations with friends, family, and colleagues; in our engagements with society, politics, and culture. However, our understanding of what all these practices and experiences mean, what they imply and implicate, has remained scant. We do not have a sophisticated vocabulary with which we can usefully speak about the nuanced predicaments of our television-saturated world, either in a general sense or in particular instances. The "we" I am incorporating here refers to the rather loose community of scholars, critics, and other intellectuals, whose general task in a free and democratic society it is to interpret and comment upon important developments and events that affect our common conditions of existence. Given television's conspicuousness in contemporary culture and society, this poverty of discourse, this lack of understanding is rather embarrassing indeed, if not downright scandalous.[1]

In this book, I . . . tried to relate this lack of understanding with the preponderance of the institutional point of view in existing knowledge about

From *Desperately Seeking the Audience,* Routledge, 1991. Reprinted with permission.

the television audience. This point of view is primarily advanced and materialized in the knowledge produced within the television institutions—knowledge that is explicitly aimed at facilitating the institutions' ambition to "get" the audience. Institutional knowledge is not interested in the social world of actual audiences; it is in "television audience," which it constructs as an objectified category of others to be controlled. This construction has both political and epistemological underpinnings. Politically, it enables television institutions to develop strategies to conquer the audience so as to reproduce their own mechanisms of survival; epistemologically, it manages to perform this function through its conceptualization of "television audience" as a distinct taxonomic collective, consisting of audience members with neatly describable and categorizable attributes.

What I hope[d] to [show], however, is that even in its own terms institutional knowledge is lacking. Institutional knowledge does not only offer us limited insight into the concrete practices and experiences of television audiencehood; it is also ultimately unable to supply the institutions with the definitive guarantee of control they so eagerly seek.

[I also] traced the tendency toward increasing rationalization of institutional knowledge about the audience within European public service broadcasting organizations. It is a tendency characterized by a change in emphasis from ideological, normative, and philosophical knowledge, in which "television audience" is defined in terms of "what it needs," to empiricist, factual, and informational modalities of knowledge, pre-eminently demonstrated by the mounting prominence of audience research, and audience measurement in particular. This change signifies an eclipse of the classic idea of "serving the public" in favor of a more market-oriented approach, in which "television audience" is defined in terms of "what it wants." Public service institutions no longer address the television audience as "citizens," but as "consumers"—at least at a general, organizational level.

. . . [H]owever, . . . the taxonomic construction of "television audience" (or segments of it) along the purely objective axis of size in order to come to a streamlined empirical map of it is less unproblematic than it seems: as the pre-eminent form of institutional knowledge in commercial television institutions, ratings discourse is too replete with ambiguities and contradictions to function as the perfect mechanism to regulate the unstable institution–audience relationship. Epistemologically, the whole controversy around the people meter suggests one thing: namely, that in the end the boundaries of "television audience," even in the most simple, one-dimensional terms, are impossible to determine. Those boundaries are blurred rather than sharply demarcated, precarious rather than absolute.

This dissolution of "television audience" as a solid entity became historically urgent when "anarchic" viewer practices such as zapping and zipping became visible, when viewing contexts and preferences began to multiply, in short when the industry, because of the diversification of its economic interests, had to come to terms with the irrevocably changeable and capricious nature of "watching television" as an activity. However, from the institutional point of view this proliferation and dispersal of forms of television audiencehood can only be seen as a problem, because it only makes it more difficult for the television institutions to bring their relationship to the audience under control. Therefore, television institutions can only be reluctant to give up their calculated ignorance of the dynamic complexity of the social world of actual audiences. Instead, they are likely to continue to quest for encompassing, objectified constructions of "television audience"— as the continued search for the perfect audience measurement technology suggests.

If we abandon the institutional point of view, however, the current disruption brought about by the changing television landscape becomes the historical backdrop that provides us with an excellent opportunity finally to take seriously the challenge of developing understandings that can do justice to the differentiated subtleties of television audiencehood. In order to do this, we must resist the temptation to speak about the television audience as if it were an ontologically stable universe that can be known as such; instead, our starting point must be the acknowledgement that the social world of actual audiences consists of an infinite and ever expanding myriad of dispersed practices and experiences that can never be, and should not be, contained in any one total system of knowledge. To round off this book, then, I shall explore the epistemological and political consequences of such an acknowledgement.

## The Academic Connection

Academic mass communication researchers . . . have often all too easily complied to the institutional point of view in their attempts to know the television audience, not necessarily in a political sense, but all the more in an epistemological sense. This is a bold statement that needs to be substantiated. Perhaps I can best do this by pointing at the cognitive authority inhabited by ratings discourse, not only within television institutions, but also within the academic communication research community.

If ratings discourse derives its effectivity from its assumption that "television audience" is a taxonomic collective consisting of the sum of audience members defined exclusively in terms of their measurable "viewing behavior," this very assumption has also predominated in the search for knowledge about the television audience in academic discourse. For instance, in their prestigious overview of social scientific television research, *Television and Human Behavior*, Comstock et al., to come to a "depiction of the audience," have decided to draw heavily on data from the A.C. Nielson company "because of their freshness and comprehensiveness" (1978: 86). The chapter concerned goes on to extrapolate "trends and patterns" from the data, which are taken to "reflect the social phenomena of time use and taste." Through an array of impressive-looking charts, figures, tables, and graphics, representing things like average hours of viewing per week or by time of day, for different demographic groups, for different types of programs, and so on, a sense of total overview of "television audience" is created—a comprehensive map on which all important "facts" are systematically identified and classified.

In a book simply and revealingly entitled *The Television Audience: Patterns of Viewing,* Goodhart et al. have carried out an even more sophisticated discursive streamlining of "television audience." It is based upon (mainly British) audience measurement data and is, again, presented as a systematic study of "how the viewer actually *behaves*" (1975: vii, italics in original). Applying advanced statistical techniques, the authors have managed to construct a dazzling range of curious forms of aggregated audience behavior, such as "audience flow" (the extent to which the same audience watches subsequent television programs), "repeat-viewing" (the extent to which the same people view different episodes of the same program), and "channel loyalty" (the extent to which viewers show a consistent preference for one channel over another). Their mathematical *tour de force* leads the authors to conclude that "instead of being complex . . . viewing behaviour and audience appreciation appear to follow a few general and simple patterns operating right across the board" (Goodhardt et al. 1975: 127). As a result, Goodhardt et al. claim to be capable of mapping audience behavior with all but law-like precision. For example, what they call the "duplication of viewing law" signifies that for any two programs the level of "duplication" or overlap in their audiences can be predicted on the basis of the ratings of the programs, and not on their content: "people who watch one particular western are no more likely to watch other westerns than are other viewers" (ibid.: 129).[2] A statistically constructed, objective "fact" about viewing behavior is thus established without any reference to the subjectivity of viewers. In this instance the streamlining process has become all but complete: "television audience" is reduced almost entirely to a set of objective regularities, and seems to be more or less completely purged from any subjective peculiarities in people's engagements with television.[3]

The point here is not necessarily to refute the scientific quality of these conscientiously conducted research projects or to reject their value out of hand, but more positively to examine exactly what is known through such totalizing inquiries into the television audience, to query the discursive horizon they construct, as well as what vanishes beyond that horizon.

Thus Goodhardt et al.'s duplication of viewing law has led them to conclude that watching television does not have much to do with people's preferences and is a quite aselective practice: "there are discrepancies between what viewers say or feel they would like to watch and what they watch in practice" (ibid.: 127). But to infer from here that what people say or feel they would like to watch is of little interest or significance—because it is not reflected in manifest, measurable "viewing behavior"—would be a truncated kind of knowledge, which all too easily reveals its complicity with the institutional point of view. Several authors (e.g., Gunter and Svennevig 1987) have pointed to the "mediating influence of others" as an explanation for why what people watch does not correlate with what people say they would like to watch: especially when people watch in groups, they often end up watching programs not of their own choosing, because the choice has been imposed upon them by the dominant individual in the group— usually the father when the group is the family (Lull 1982; Morley 1986). Such discovery of "intervening variables" between "viewing behavior" and "program choice" already begins to subvert the decontextualized, one-dimensional definition of "watching television" that is implied in knowledge that takes for granted the institutional point of view. As David Morley (1986: 19) has remarked, "to expect that we could treat the individual viewer making programme choices as if he or she were the rational consumer in a free and perfect market is surely the height of absurdity when we are talking of people who live in families." More generally, what begins to become visible here is the uneven and variable everyday context in which the practices and experiences of television audiencehood are shaped and take on meaning for actual audiences.

Ironically George Comstock, one of the leading figures in mainstream communication research and principal author of *Television and Human Behavior,* has himself inadvertently recognized the limited vision of the knowledge produced by academic researchers. In an article revealingly titled, "Television and Its Viewers: What Social Science Sees," he notes:

> It is sometimes said that very little is known about television and people beyond the popularity of the former and the fickleness of taste of the latter. This is not really true, if one is willing to accept a scientific definition of "knowing." That is, there is a great deal "known" if one is willing to define the concept of knowing as a state in which there is verifiable evidence that disposes an observer toward one or another set of possible facts or explanations without establishing that such is the case with absolute certainty. (Comstock 1981: 491)

He then goes on to review the findings which fit in his definition of scientific knowledge, only to come to the conclusion that "There is no general statement that summarizes the scientific literature on television and human behavior, but if it is necessary to make one, perhaps it should be that television's effects are many, typically minimal in magnitude, but sometimes major in social importance" (ibid.: 504). Even the most fervent supporter of positivist social science must admit that such a conclusion, with its abstracting emphasis on quantified generalizations ("many," "minimal," "major"), can at best be called disappointingly trivial. Is this really all that can be said about "television and its viewers"?

One might object that I have given a rather unfair picture of the accomplishments of communication research here. Indeed, I do not want to risk the danger of slighting all the more focused research efforts that have been made by generations of communication scholars into the television audience, for example, in the contexts of the uses and gratifications approach(which roughly tries to explain "viewing behavior" in terms of people's needs or motives) (e.g., Blumler and Katz 1974; Rosengren et al.

1985), and of the cultivation analysis perspective (which roughly tries to examine the effects of "viewing behavior" on people's conceptions of social reality) (e.g., Gerbner 1969; Gerbner et al. 1986). However, even though these research traditions originated in a genuine interest in what watching television implies for the audience not the institutions, I would argue that they unwittingly tend to deepen, rather than challenge the institutional point of view, because they overwhelmingly hold on to the conceptual assumption that "television audience" is a given taxonomic grouping of serialized individuals who can be described and categorized in terms of measurable variables: not only the conventional variables of ratings discourse but also a host of other ones (depending on the research project concerned): sociodemographic variables, personality variables, television use variables, function variables, gratification variables, effect variables, and so on.[4] The kind of knowledge about the television audience generated from such research strategies, mostly using techniques of multivariate analysis, is generally directed toward condensing measured repertoires of individual responses into aggregated types of audience activity or experience, ultimately resulting in the isolation of distinct viewer types.

One of the most famous viewer types constructed by communication researchers is the "heavy viewer," on whom all sorts of concerns are projected. Thus, Comstock et al. (1978: 309) take some pains in singling out four demographic categories who "typically are heavier viewers of television": these categories are "females, blacks, those of lower socioeconomic status, and the elderly," about whom the authors speculate that "because of psychological and social isolation [they are] particularly susceptible to influence by television." Indeed, in a substantive review of the international research conducted on "heavy viewing" from 1945 to the present, Frissen (1988) has shown that communication scholars have been relentlessly preoccupied with describing and explaining "heavy viewing" as a problematic behavioral phenomenon, related to invariably negative and disturbing psychosocial characteristics such as depression, anxiety, lack of ambition, fatalism, alienation, and so on, resulting in a so-called "heavy viewer syndrome" (Gerbner and Gross 1976). Combined with repeated attempts to set viewers who presumably suffer from this syndrome apart from other groups, "heavy viewers" tend to be objectified as a category of stereotyped others. In one recent representative article, which sets out to construct a typology of European television viewers using a combination of the uses and gratifications approach and cultivation analysis, the researchers permit themselves to typify the category as follows:

> For those viewers who are interested in all types of programme [i.e., "nonselective" viewers], watching television is probably a habit, a ritualized way of occupying free time; little time would then be left over for other activities and for the use of other media. An ideal type in this category would be an older manual worker with a low level of education and low income. The image that emerges is of an unresourceful, uncritical, passive person who apparently prefers the world of television to his own world. (Espe and Seiwert 1986: 320)

Such a characterization, which is by no means untypical (see Frissen 1988), can only be made from a distant, exterior perspective on this trumped-up audience category.[5] So, what started as a genuine interest in viewers apparently unrelated to the institutional concern for audience control, ends up foregrounding a discourse that is just as objectifying and othering as institutional knowledge! By concentrating so heavily on differentiating between groups of viewers, academic communication researchers are driven toward drawing "fictions" of rigid, reified audience categories, a kind of knowledge that forecloses understanding of the concrete practices and experiences of people because those "fictions" are regarded as reflecting essential viewer identities that are taken to sufficiently explain certain patterns of "viewing behavior." Consequently, in a paradoxical leap of argument it is (some categories of) the viewers that are implicitly put on trial, not the institutions that provide the programming—as the case of the "heavy viewer" suggests.[6]

The epistemological limitations of the pull toward generalized categorization implied in the search for viewer types can be illustrated, in an anecdotal but telling fashion, by . . . the couch potatoes. . . . They are self-proclaimed heavy viewers, who cannot be understood by referring to the academically constructed fiction of this type of viewer. Faced with the idiosyncratic, self-reflective, witty, utterly recalcitrant "behavior" of the couch potatoes—and there is no reason to dismiss them as "atypical" in advance—communication researchers are ultimately left with empty hands, or better, want of words. This suggests that the pull toward categorization should at least be complemented by the opposite one of *particularization* (Billig 1987): rather than reducing a certain manifestation of "viewing behavior" to an instance of a general category, we might consider it in its particularity, treat it in its concrete specificity, differentiate it from the other instances of the general category. Only then can we begin to understand the multiple practices and experiences of actual audiences, rather than get stuck with abstracted, simplified fictions of categories of "television audience." Only then can we go beyond (statistical) "significance without much signification," as James Anderson (1987: 371) has put it.

More fundamentally, the very notion of "viewing behavior" which undergirds any taxonomic demarcation of "television audience" and its partitioning into fixed categories needs to be questioned. In her review, Frissen (1988: 149) has concluded that after decades of researching heavy viewing, "communication researchers apparently have not yet reached the point where they describe and explain heavy viewing from an explicit theoretical standpoint."[7] What's more, she found that even the very operationalization of "heavy viewing" itself has been done rather sloppily, in purely arbitrary and pragmatic ways, in some cases in terms of number of hours watching (e.g., two hours, three hours, four hours, ten quarters of an hour per day, sixteen and a half hours per week), in other cases just by slicing a certain percentage of the total population that spends the most time watching television (e.g., 25 percent, 30 percent) (ibid.: 142). This raises the question whether "heavy viewing," as a type of "viewing behavior," is not an artifact designed to simplify the researcher's task rather than an actually existing "syndrome" of a definite category of people. As a generalized concept, it is devoid of meaning.

To avoid the unwarranted construction of such artifacts which is one of the liabilities of taxonomic thinking, we should seriously recognize that "watching television" is always in excess of the sum of the isolatable, measurable "viewing behavior" variables in which it is operationalized. It should be seen as a complex and dynamic cultural process, fully integrated in the messiness of everyday life, and always specific in its meanings and impacts.

### From the Point of View of Actual Audiences

My critique of mainstream communication research, then, is directed at "the overly condensed character of the variables" (Fielding 1989: 9) used in too many research projects which claim to try to examine the television audience, resulting in quite simplistic, empiricist assumptions about what "watching television" implies as an activity. As Pierre Bourdieu has remarked,

> the absence of . . . preliminary analysis of the social significance of the indicators can make the most rigorous-seeming surveys quite unsuitable for a sociological reading. Because they forget that the apparent constancy of the products conceal the diversity of the social uses they are put to, many surveys on consumption impose on them taxonomies which are sprung straight from the statisticians' social unconscious. (Bourdieu 1984: 21)

The solution is not simple:

> the only way of completely escaping from the intuitionism which inevitably accompanies positivistic faith in the nominal identity of the indicators would be to carry out a—strictly interminable—analysis of the social value of each of the properties or practices considered—a Louis XV commode or a Brahms symphony, reading *Historia* or *Le Figaro,* playing rugby or the accordion and so on. (Bordieu: 20–21)

What Bourdieu calls for, in other words, is the evocation of the irreducible dynamic complexity of cultural practices and experiences, and "watching television" is no exception.[8]

Consider, for example, a woman's account of the TV viewing habits of her family, mentioned by Hermann Bausinger in his article, "Media, Technology and Daily Life" (1984: 344): "Early in the evening we watch very little TV. Only when my husband is in a real rage. He comes home, hardly says anything and switches on the TV." Here, comments Bausinger, "watching television" has a very particular meaning, profoundly immersed in "the specific semantic of the everyday": "pushing the button doesn't signify, 'I would like to watch this,' but rather, 'I would like to hear and see nothing.' "

This example suggests clearly that "viewing behavior" can only be adequately accounted for when it is grounded in the concrete situation in which it takes place. "Watching television" is always behavior-in-context, a generic term for heterogeneous kinds of activities whose multifarious and shifting meanings can only be understood in conjunction with their contexts. Of course, "context" itself cannot be reduced to a fixed number of "background" variables, because contexts are indefinite, and indefinitely extending in time and space.

The practical consequences of this fundamental undecidability of "watching television" and its "contexts," can already be traced in the increasing difficulties encountered by the ratings services to measure "television audience". . . . . Developments within the television industry and the changing television landscape have forced the ratings services to expand the scope of their operationalization of "viewing behavior" so as to include an ever increasing number of individual and situational variables. To date, it is unclear when and where, if ever, this stubborn search for more encompassing objectification of "television audience" will stop. Theoretically, however, it is clear that the loss of control in the audience measurement endeavor signifies a fundamental epistemological crisis. Karin Knorr-Cetina has given a fine characterization of this crisis:

> As the fineness of the grid and the number of relevant attributes increases, we are less likely to guess what the outcome of each arrangement of attributes that marks a social situation will be. This is one way in which we can make sense of the definitive role and the unpredictable dynamics of the situation. (Knorr-Cetina 1989: 28)

For all practical purposes, the consequences of this crisis cannot be acknowledged as such within the television institutions because their very existence depends upon clearcut measures of "television audience." But they can be taken up by academic scholars as a starting point for charting new avenues for understanding television audiencehood. Rather than despair over the insolubility of the crisis, I suggest, we should gladly embrace it, and develop another kind of knowledge on its ruins. Such alternative knowledge— knowledge that is constructed "from the point of view of actual audiences"—would differ from established knowledge, not only in substance, but also in its political uses. It goes without saying that I can only sketch the outlines of this alternative, for it is only in the process of developing it and articulating it that we will be able to refine and sharpen its focus.

So far as epistemology is concerned, we can follow Knorr-Cetina's (1989) proposal to adopt the principle of "methodological situationalism" in developing new research strategies, in replacement of the methodological individualism that underlies most research on the television audience (see also Lindlof and Meyer 1987). Giving analytic primacy to concrete situations of television audiencehood rather than to decontextualized forms of "viewing behavior" implies a recognition that the social world of actual audiences only takes shape through the thoroughly situated, context-bound ways in which people encounter, use, interpret, enjoy, think, and talk about television.

The analysis of micro-situations of television audiencehood should take precedence over either individual "viewing behavior" or totalized taxonomic collectives such as "television audience" because micro-situations cannot be reduced to the individual attributes of those participating in the situation. Thus, the viewer as such does not exist as the stable and unproblematic source of "viewing behavior." As John Fiske (1989: 57) has put it, "any one viewer . . . may at different times be a different viewing subject, as constituted by his or her social determinants, as different social alliances may be mobilized for different moments of viewing." In other words, rather than conceiving viewers as having a unified individuality that is consistent across circumstances, they should be seen as inhabiting multiple and mobile identities that fluctuate from situation to situation.[9] Furthermore, situations not psychological dispositions (needs, preferences, attitudes, and so on) tend to determine the kind of "viewing behavior" that people actualize. For example, Bausinger (1984: 349) refers to the apparently contradictory situation of "the same man who swears because the sports programme has been delayed by ten minutes because of the Pope's visit, then spends the sports programme working on the flower stand he is making, and hardly notices the programme." This suggests that in everyday contexts the distinction between viewing and non-viewing is radically blurred. In day-to-day reality audience membership is a fundamentally vague subject position;[10] people constantly move in and out of "television audience" as they integrate "viewing behavior" proper with a multitude of other concerns and activities in radically contingent ways.

Pushed to the extreme, the principle of methodological situationalism holds that "we cannot ever leave [the] micro-situations" (Knorr-Cetina 1989: 32) in which "watching television" is practiced and experienced in an indefinite number of spaces, at indefinite times. However, this does not mean that micro-situations are completely self-contained, merely following their own, unique principles of organization. On the contrary, micro-situations as interrelated in many different ways and things happening in such situations often transcend the immediate situation. Occurrences in a situation always have references to, and implications for, other situations (Knorr-Cetina: 36).

One obvious situation-transcending factor is presented by the institutionally-defined constraints placed upon the structural conditions in which watching television can be practiced in the first place.[11] The very framework of broadcast television implies the imposition of when and what people can watch. Prevalent scheduling and programming practices impose a temporal arrangement based upon predictability, regularity, and repetition (Scannell 1988); the very composition of the menu of programs being served is determined by the institutions. People cannot, in whatever situation they watch television, outdo these constraints; they can only negotiate with their terms and develop fragmentary tactics to subvert those constraints without ever escaping them (Silverstone 1990). As a result, all micro-situations of watching television are virtually connected to one another in so far as they have to realize themselves in relation to given institutional constraints. This, of course, is the ultimate power of the television institutions, the ultimate basis of their control over the audience.

However, the fact that instances of watching television are "controlled" in this way "does not catapult them out of micro-situations," as Knorr-Cetina (1989: 36) has put it. Thus, recognition of the situational dependency of actual audience practices and experiences can shed a new light on the fundamental unpredictability of "viewing behavior"—that irritating "fickleness" of the audience that television industry managers often complain about. It is only in concrete situations that people do or do not comply to the rules for "watching television" that the institutions implicitly lay down through their scheduling and programming strategies in their attempts to conquer the audience as effectively as possible. Statistically constructed differentiations between categories of viewers notwithstanding, concrete viewers sometimes zap or zip, sometimes don't. This time they watch the commercials, another time they don't. Sometimes, when the situation is right, they decide to watch an educational program attentively in order to learn from it, at other times they wouldn't bother.

Seen from this perspective, then, variability rather than consistency of "viewing behavior" is the order of the day. From this perspective, what are called "viewing habits" do not represent a more or less static set of characteristics moored in an individual or a group; they are no more than the temporary and superficial snapshots of a never-ending, dynamic, and complex process in which "the fine-grained interrelationships between meaning, pleasure, use and choice" (Hall, in Morley 1986: 10) are shaped in millions of situations.[12] From this perspective, "television audience" is a nonsensical category, for there is only the dispersed, indefinitely proliferating chain of situations in which television audiencehood is practiced and experienced— together making up the diffuse and fragmentary social world of actual audiences.

This brings me to the broader implications for the kind of knowledge emerging from the thoroughly ethnographic thrust of this perspective.[13] Obviously, emphasis on the situational embeddedness of audience practices and experiences inevitably undercuts the search for generalizations that is often seen as the ultimate goal of scientific knowledge. In a sense, generalizations are necessarily violations to the concrete specificity of all unique micro-situations; therefore, it is knowledge about particulars not the general that this perspective tends to highlight. As Stephen Tyler (1986: 131–32) has put forward in a suggestive metaphor, "It is not just that we cannot see the forest for the trees, but that we have come to feel that there are no forests where the trees are too far apart, just as patches make quilts only if the spaces between them are small enough." This is not to imply that as researchers we can say something only about one singular micro-situation— one tree or patch—at a time. We can, through some procedure of comparative analysis, look for what situations have in common and in what ways they differ (cf. Marcus and Fischer 1986). But it is unwarranted to add up the results into an ever more generalized, comprehensive system of knowledge that comprises the forest or quilt, that is, the whole social world of actual audiences, because the very fluid nature of that world resists full representation. The epistemology this implies cannot be reconciled with received notions of cumulative scientific progress, and the partiality at stake is stronger than the normal scientific dictates that we study problems piecemeal, that we must not overgeneralize, that the best picture is built up by an accretion of rigorous evidence (Clifford 1986). There is no whole picture that we can strive to gradually "fill in," because actual audiences are temporally and spatially dispersed and continuously changing formations that can never be pinned down as such.

In other words, if the alternative knowledge we have in mind here can be said to ensue from the standpoint of actual audiences, it should be stressed that by taking it up we cannot presume to be speaking with the authentic voice of the "real" audience, because there is no such thing. Rather, "the standpoint of actual audiences" is a discursively constructed, virtual position from which we can elaborate always partial and provisional understandings that evoke the dynamic complexity of television audiencehood rather than imprisoning it in static grids of information. Which brings me to a last point: what relevance can such understandings have, not only within the academic world of communication scholarship but, crucially, outside it?

## *The Politics of Television Audiencehood*

What I have tried to uncover here is the profundity of the gap between the institutional point of view on the one hand, and the virtual standpoint of actual audiences on the other. From the institutional point of view, watching television is the decontextualized, measurable viewing behavior that is taken to be the indicator for the existence of a clear-cut "television audience" out there; from the virtual standpoint of actual audiences watching television is the ill-defined shorthand term for the multiplicity of situated practices and experiences in which television audiencehood is embedded. It is a gap that can be understood in terms of what has been referred to, in a variety of theoretical contexts, as the opposition between macro and micro, the formal and the informal, control and creativity, structure and agency, strategy and tactics, communication as transmission and communication as ritual, the view from the top and the view from the bottom. It is also a gap that gives rise to opposing types of knowledge: one that strives toward prediction and control, and another that aims at reaching what could be called ethnographic understanding, a form of interpretive knowing that purports to increase our sensitivity to the particular details of the ways in which actual people deal with television in their everyday lives.

Meanwhile, television institutions and actual audiences remain locked into one another insofar as the former still to a large degree determine and constrain what the latter can see on their TV sets. These institutional constraints are being thoroughly reshuffled by recent changes in the television landscape. The European public service institutions, especially, are facing a severe crisis, not only in practical (economic, organizational) terms, but also in terms of their normative founding philosophies. The institutions themselves, as we have seen, have responded to this crisis by adopting the discourse of the marketplace in their approach of the audience: defining "television audience" as a collection of consumers rather than citizens, thinking in terms of "what the audience wants" rather than "what it needs." The residual markers of difference are formulated within, not beyond the boundaries of this overall consumerist framework: "diversity" and "quality." Whither public service broadcasting? This is a complicated and multifaceted political issue which cannot be fully addressed in this context. But I would like to end this [essay] with a few notes that shed light on this issue from the perspective I have tried to develop in the previous pages.

What contribution can ethnographic understandings of the social world of actual audiences make to assess the dilemma? To put it bluntly: little, in a direct sense at least. It cannot—and should not—give rise to prescriptive and legislative solutions to established policy problems, precisely because the ironic thrust of ethnography fundamentally goes against the fixities of the institutional point of view. What it can do, however, is encourage public debate over the problems concerned, by informing critical discourses on television—as a cultural form, as a medium ever more firmly implanted in the everyday texture of modern society—that are independent from established institutional interests. Seriously taking up the virtual standpoint of actual audiences is likely to highlight the limitations of any particular institutional arrangement of television, and can thus serve as a vital intellectual resource for the democratization of television culture.

Let us assume that there is something truly worthwhile to be lost if the seemingly unstoppable process of commercialization would wipe out all institutional undertakings of television provision that are not based upon the overall motive of profit making. Let us assume that some sense of "public service" should indeed be upheld against the risk for all cultural and social values to be subsumed to purely economic ones. But such political judgements, which presumably must eventually lead to decision making at the level of macro-institutional policy, need not necessarily concur with a defence of the existing institutional embodiments of the public service idea. Nicholas Garnham has usefully remarked that a much more profound cultural politics is at stake here:

> In the battle for the hearts and minds of the public over the future of public service broadcasting it is important to stress that the historical practices of supposedly public service institutions, such as the BBC, do not necessarily correspond to the full potential of public service and may indeed . . . be actively in opposition to the development of those potentials. (Garnham 1983: 24)

What we should discuss, then, is what that "full potential of public service" could be in a time so engrossed with "free enterprise" that the very idea of public service broadcasting seems hopelessly old-fashioned, at least when we persist to conceive it in its conventional, historically-rooted institutional form. As Graham Murdock (1990: 81) has asked, "can we arrive at an alternative definition of public broadcasting which is capable of defending and extending the cultural resources required for citizenship?"

It is in this respect that ethnographic understanding of the social world of actual audiences may feed the imagination needed to come to such alternatives. We have seen how public service institutions have generally originated in some idealized, mostly rather patronizing concept of what "serving the public" means, but that it proved to be impossible to uphold such utopian, philosophical definitions of the "full potential of public service" in the dirty reality of broadcasting practice. I would suggest that we should take this dirty reality seriously if we want to come to new visions of public service television, that do not prematurely comply with the limitations imposed by existing institutional arrangements. This dirty reality, of course, is ultimately nothing other than the intransigence of the social world of actual audiences.

Take the issue of "quality," one of the spearheads of modern-day public service institutions such as the BBC and VARA. "Quality" as formally defined and operationalized by the institutions in their programming decisions may well not at all correspond with what in practical terms counts as "quality" in the social world of actual audiences. We should realize, as Charlotte Brunsdon (1990) has remarked, that people constantly make their own judgments of quality when they watch television, judgments which can vary from situation to situation, depending on the type of satisfaction they look for at any particular time. From this perspective, "quality" is not a fixed standard of value on which the professional broadcaster holds a patent, but is a radically contingent criterion of judgment to be made by actual audiences in actual situations, "something that we all do whenever we channel hop in search of an image or sound which we can identify as likely, or most likely, to satisfy" (Brundson: 76).

This is to point out that there is much more to "quality" than the assurance from the broadcasters that they will try to provide us with what they define as quality programs: apart from professional quality, presumably a formal characteristic of programs, there is "lived" quality as it were, related to the concrete ways in which television is inserted in people's everyday lives. To put it differently, rather than being seen as a predetermined yardstick, "quality" should be posed as a problem, a problem of value whose terms should be explicated and debated, contested and agreed upon in an ongoing public and democratic conversation about what we, as publics, expect from our television institutions. Ethnographic understanding of the social world of actual audiences can help enrich that conversation because it foregrounds a discourse on quality that takes into account the situational practices and experiences of those who must make do with the television provision served them by the institutions—an open-ended discourse that conceives quality as something relative rather than absolute, plural rather than singular, context-specific rather than universal, a repertoire of aesthetic, moral, and cultural values that arises in the social process of watching television rather than through criteria imposed upon from above.

A similar case can be made about "diversity," the second ideal which contemporary public service broadcasting claims to represent. Several problems pertain to how this concept is generally treated. First of all, defending diversity is often conflated with the expansion of consumer choice, that is, with quantitative rather than with qualitative diversity. It is in this respect that public service discourse becomes almost interchangeable with commercial discourse. Of course, "diversity" is often defined in more formal terms, that is, in terms of the broad range of program genres that public service institutions are obliged to transmit, corresponding to the multiplicity of functions that broadcasting is supposed to fulfil, that is, "information," "entertainment" "drama," "education," and so on. However, such formal diversity easily overlooks the fact that from the standpoint of actual audiences, these functions often overlap: a popular drama series for example can in some situations, for some people, be more "informative" or "educative" than a news or current affairs program. A formal conceptualization of diversity, in other words, can easily be out of touch with the concrete experiences of those who watch the programs. Such formalism can be softened by a second, more sociological definition of the ideal of diversity, namely in terms of the responsibility of providing programs aimed at a variety of "target groups," including those minority groups that are not of interest for advertisers and are thus not well served in a commercial system. This is a laudable idea, but precisely by equating the concept of diversity with a more or less fixed range of sociologically observable categories within the population (e.g., ethnic minorities, women, the elderly, and so on) one risks objectifying those categories and their presumed needs and preferences, and seeing "diversity" as a static prescript rather than a dynamic and flexible cultural principle which, as Murdock (1990: 81–82) has put forward, aspires to "engage with the greatest possible range of contemporary experience" and "offer the broadest possible range of viewpoints on these experiences and the greatest possible array of arguments and contexts within which they can be interpreted and evaluated." In short, in a programming policy that takes seriously the dynamic complexity of television audiencehood the principle of "diversity," like that of "quality," cannot be institutionally predetermined, but should imply a constant and ongoing responsiveness towards and engagement with what is going on at all levels of the larger culture.

From such a perspective, we can only suspend judgment about the most desirable institutional arrangement of television provision in the 1990s and beyond; there is no guarantee that more commercial offerings would necessarily lead to lower quality and less diversity, and that a defense of public service broadcasting based upon established footings would necessarily be the best way to promote these values. Such relativist pragmatism may sound unsatisfactory to those who want unambiguous, once-and-for-all pros and cons. Against this I would argue that ethnographic understanding can be useful precisely because it can be potentially disturbing for the existing institutions, by keeping them from being too arrogant and self-assured about themselves, too self-contained in their cultural policies. More positively, I would suggest that the stance of relativist pragmatism endorsed by ethnography is the only way to create a democratic element in the organization of our television culture, in the sense of enlarging people's opportunity to deliberate and choose, in endlessly varied ways, for what they consider the "best" television.[14] Against this background, institutional solutions for the regulation of the changing television landscape, especially in Europe, should be sought not in establishing fixed, formalist definitions of quality and diversity, but in securing more flexible conditions in which a plurality of qualities can find their expression.[15] The further political task would then be the construction of institutional arrangements that can meet these conditions.

But this is not the end of the story. What ethnographic understanding of the social world of actual audiences also enables is a critique of facile nominalist notions of "the consumer," "the market," and "what the audience wants" that seem to have been so pervasively embraced by television institutions of all kinds. The ascendancy of these notions in public service institutional settings has been accompanied by a waning of normative discourse on what "serving the public" should be about, and the adoption of purely empiricist forms of "feedback." However, the streamlined information delivering this "feedback" ignores and obscures the fact that actual audiences are never merely a collection of consumers who happily choose to watch "what they want." Indeed, ethnographic knowledge can provide us with much more profound "feedback" because it can uncover the plural and potentially contradictory meanings hidden behind the catchall measure of "what the audience wants." It can help us to resist succumbing to all too triumphant allegations that commercial success means the victory of the sovereign consumer, for what is discursively equated with "what the audience wants" through ratings discourse is nothing more than an indication of what actual audiences have come to accept in the various, everyday situations in which they watch television. It says nothing about the heterogeneous and contradictory interminglings of pleasures *and frustrations* that television audiencehood brings with it.

The importance of ethnographic discourse, in short, lies in its capacity to go beyond the impression of "false necessity" (Unger 1987) as prompted by the abstracted empiricism of taxonomized audience information. It promises to offer us vocabularies that can rob television audiencehood of its static muteness, as it were.

We are living in turbulent times: the television industries and the governments that support them are taking aggressive worldwide initiatives to turn people into ever more comprehensive members of "television audience." At the same time, television audiencehood is becoming an ever more multifaceted, fragmented, and diversified repertoire of practices and experiences. In short, within the global structural frameworks of television provisions that the institutions are in the business to impose upon us, actual audiences are constantly negotiating to appropriate those provisions in ways amenable to their concrete social worlds and historical situations. It is both the dynamic complexity and the complex dynamics in the interface of this dialectic that ethnographic understanding can put into discourse—a never-ending discourse that can enhance a truly public and democratic conversation about the predicaments of our television culture.

## *Notes*

1. For an historical and sociological analysis of the task of intellectuals as "interpreters" in (post)modern culture and society, see Bauman (1987).

2. For a methodological critique of this "finding," see Wober and Gunter (1986).

3. A similarly objectivist and totalizing picture is offered in Barwise and Ehrenberg (1988). It should be noted that these authors have a business school background; it is therefore not surprising that they tend to formulate problems in terms relevant to the business world (advertisers, the marketers).

4. See, for other critiques of these research traditions, Elliott (1974); Bybee (1987); Newcomb (1978); Hirsch (1980; 1981).

5. It is interesting to note that Espe and Seiwert (1986) did not find it necessary to paint an equally detailed image of what they call "information viewers." In this case, they limit themselves to stating that for this viewer type "the medium represents a source of political and cultural information, further education, intellectual stimulation and debate" (1986: 321). Why not add to this, for example, the image of the snobbish, hypocritical, or elitist television-hater? Is it perhaps because the researchers themselves identify with this "high quality" viewer type?

6. Thus, although a lot of concern over "heavy viewing" was related to the presumed harmful effects of watching too much violence on television and so overtly instigated by anti-institutional motives, generally the researchers, certainly in the American context "were . . . unwittingly co-opted into various economic and political agendas" (Gans 1980: 79). According to Gans, the researchers were not only being used to support the politicians and interest groups seeking to alter network programming, but were also indirectly involved in the networks' attempts to maintain the status quo. See also Rowland (1983).

7. According to Frissen (1988), this lack of explicit theorization is due to the failure of the researchers concerned to take a critical and conscious stance toward the common sense discourse of public concern about the harmful effects of watching "too much" television. They have therefore unwarrantedly adopted the terms of that discourse without reflecting on their conceptual validity. Gans (1980: 78) also criticizes the lack of reflection on what he calls "the metaphysical assumptions of effects research," with its automatic emphasis on "bad" effects of the media.

8. This is not the place to discuss extensively the methodological implications of this change of perspective. In general, my critique implies a profound questioning of the large-scale, quantitative survey research methods that have been preferred in mainstream communication research, based as it is precisely on the aggregation of data and condensed operationalizations of variables. However, "quantitative" and "qualitative" research procedures can—and will—for the time being exist parallel to each other in the social sciences. It is important to develop more careful and theoretically sound conceptions of the relative applicability and limitations of both types of research methods and the kind of data they engender. Bourdieu's (1984) work is an excellent example of the sophisticated use of a combination of survey and eth-

nographic data. His work also points to the importance of extensive, non-empiricist, theoretically-informed interpretation—and thus, story telling—if one wants to make sense of any kind of data, something which tends to be denied by supporters of positivism. In general, "quantitative" methods can be seen as relevant for charting general, structural patterns (although, as we have seen, in insensitive hands the patterns found risk to be reified), while "qualitative" methods are indispensible for understanding forms of cultural meaning and practical consciousness that are hidden behind large-scale patterns. See, for comprehensive treatments of this subject, grounded in explicit theoretical or epistemological considerations, Giddens (1984: 281–347); Sayer (1984); Anderson (1987).

9. See, for a related, poststructuralist critique of psychology's conception of the individual as an integrated subject with a unified identity, Henriques et al. (1984).

10. See, for an explication of the importance of "vagueness" in daily patterns of social interaction, Lindlof and Meyer (1987: 25).

11. Other situation-transcending factors include the traces of cultural positionings and identifications that people "bring into" and actualize within concrete situations, such as those along the lines of gender, class, ethnicity, generation, and so on, as well as cultural ideologies as to the meaning of television as a social and aesthetic phenomenon. See, for example, my discussion of "the ideology of mass culture" in Ang (1985a).

12. The limited predictability that sometimes does occur in measured viewing behavior, allowing for the statistical construction of "viewing habits," can be explained by the fact that many people, due to the routines in which they have to organize their everyday lives, tend to reproduce similar situations for their recurring television watching practices. Two comments can be made on this. First, these (daily, weekly, and so on) practices are of course never completely similar, (thick) descriptions of each individual situation will certainly reveal subtle differences (the telephone rings, the mood is different, and so on); second, it is fair to expect that when people's living conditions change, they are likely to change their viewing situations as well, thereby altering their "viewing habits."

13. [T]his change of perspective has already begun to be explored in some recent developments in audience research, developments that are characterized by a distinctively ethnographic interest in the social world of actual audiences, in the contradictory practices and experiences of people living with television. Ethnography has recently been conceptualized as more than just a research method, but as a practice of inquiry and writing particularly suitable to do justice to the complex and dynamic character of contemporary cultural life. See, for example, Clifford and Marcus (1986); Marcus and Fischer (1986); Van Maanen (1988).

14. I base my endorsement of relativist pragmatism on the work of Richard Rorty (1989).

15. In this respect, both the commercial notion of consumer choice and the public service idea of representational diversity are indispensable values. Richard Collins (1989a) has usefully noted that while the old public service broadcasting order was based upon the idea of "internal" diversity (that is, the provision of a range of programs within a limited number of channels run by institutions given the mandate to do so), the new television landscape opens up the possibility of extending "external" diversity (in which a plurality of channels provides strongly "branded," single type programming). While there is no definitive answer to the question of the respective benefits and losses of both types of regulating of diversity, the new situation

does offer more opportunity of choice and this improvement should not be underestimated, certainly not in the age of postmodernity, with its orientation toward increasing individual freedom and cultural pluralism (D. Harvey 1989). As Collins (1989a: 13) has pointed out in discussing these issues, quoting Brecht, "the good old things are not always preferable to the bad new ones."

## *References*

Anderson, J. A. (1987) *Communication Research: Issues and Methods*. New York: McGraw-Hill.

Bausinger, H. (1984) "Media, Technology and Daily Life."*Media, Culture and Society*. 6, 4: 343–51.

Billig, M. (1987) *Arguing and Thinking*. Cambridge: Cambridge University Press.

Blumler, J. G., and Katz, E. (eds.) (1974) *The Uses of Mass Communications*. Beverly Hills, Calif.: Sage.

Bourdieu, P. (1984) *Distinction*. Trans. Richard Nice. Cambridge, Mass.: Harvard University Press.

Brunsdon, C. (1990) "Problems with Quality." *Screen* 31, 1: 67–90.

Clifford, J. (1986) "Introduction: Partial Truths." In J. Clifford and G. E. Marcus (eds.), *Writing Culture: The Poetics and Politics of Ethnography*. Berkeley: University of California Press.

Comstock, G. (1981) "Television and Its Viewers: What Social Science Sees." In G. C. Wilhoit and H. de Bock (eds.), *Mass Communication Yearbook*. Vol. 1. Beverly Hills, Calif.: Sage.

Comstock, G., Chaffee, S., Katzman, N., McCombs, M., and Roberts, D. (1978) *Television and Human Behavior*. New York: Columbia University Press.

Espe, H., and Seiwert, M. (1986) "European Television-Viewer Types: A Six Nation Classification by Program Interests." *European Journal of Communication* 1, 3: 301–25.

Fielding, N. G. (1989) "Between Micro and Macro." In N. G. Fielding (ed.), *Actions and Structure: Research Methods and Social Theory*. London: Sage.

Fiske, J. (1989) "Moments of Television: Neither the Text Nor the Audience." In E. Seiter, H. Borchers, G. Kreutzner, and E. M. Warth (eds.), *Remote Control: Television, Audiences, and Cultural Power*. London and New York: Routledge.

Foucault, M. (1980) *The History of Sexuality*. Vol. 1:*An Introduction*. Trans. Robert Hurley. New York: Vintage/Random House.

Frissen, V. (1988) "Towards a Conceptualization of Heavy Viewing." In K. Renckstorf and F. Olderaan (eds.), *Communicatiewetenschappelijke Bijdragen 1987–1988*. Nijmegen: Katholoieke U. Nijmegen.

Garnham, N. (1983) "Public Service versus the Market." *Screen* 24, 1: 70–80.

Gerbner, G. (1969) "Towards 'Cultural Indicators': The Analysis of Mass Mediated Public Message Systems." *AV Communication Review* 17, 2: 137–48.

———, and Gross, L. (1976) "The Scary World of TV's Heavy Viewer." *Psychology Today*. April.

———, Gross, L., Morgan, M., and Signorielli, N. (1986) "Living With Television: The Dynamics of the Cultivation Process." In J. Bryant and D. Zillman (eds.), *Perspectives on Media Effects*, Hillsdale, N.J., Erlbaum.

Goodhardt, G. J., Ehrenberg, A. S. C., and Collins, M. A. (1975) *The Television Audience: Patterns of Viewing*. Westmead: Saxon House.

Gunter, B., and Svennevig, M. (1987) *Behind and in Front of the Screen: Television's Involvement with Family Life.* London and Paris: John Libbey.

Knorr-Cetina, K. (1989) "The Micro-Social Order." In N. F. Fielding (ed.) *Actions and Structure: Research Methods and Social Theory.* London: Sage.

Lindlof, T. R., and Meyer, J. P. (1987) "Mediated Communication as Ways of Seeing, Acting, and Constructing Culture: The Tools and Foundations of Qualitative Research." In T. R. Lindlof (ed.), *Natural Audiences: Qualitative Research of Media Uses and Effects.* Norwood, N.J.: Ablex.

Lull, J. (1982) "How Families Select TV Programs: A Mass Observational Study." *Journal of Broadcasting* 26, 2: 801–11.

Marcus, G. E., and Fischer, M. M. J. (1986) *Anthropology as Cultural Critique.* Chicago and London: University of Chicago Press.

Morley, D. (1986) *Family Television: Cultural Power and Domestic Leisure.* London: Comedia.

Murdock, G. (1990) "Television and Citizenship: In Defense of Public Broadcasting." In A. Tomlinson (ed.), *Consumption, Identity, Style.* London and New York: Comedia/Routledge.

Rosengren, K. E., Wenner, L. A. and Palmgreen, P. (eds.) (1985) *Media Gratifications Research.* Beverly Hills, Calif.: Sage.

Scannell, P. (1988) "Radio Times: The Temporal Arrangement of Broadcasting in the Modern World." In P. Drummond and R. Paterson (eds.), *Television and Its Audience.* London: BFI.

Silverstone, R. (1990) "Television and Everyday Life: Towards an Anthropology of the Television Audience." In M. Ferguson (ed.), *Public Communication: The New Imperatives.* London: Sage.

Tyler, S. A. (1986) "Post-Modern Ethnography: From Document of the Occult to Occult Document." In J. Clifford and G. E. Marcus (eds.), *Writing Culture,* Berkeley: University of California Press.

Unger, R. M. (1987) *False Necessity.* Cambridge: Cambridge University Press.

# Making Distinctions in TV Audience Research: Case Study of a Troubling Interview

## ELLEN SEITER

Discussions about the television audience have proliferated recently. After a period of preoccupation with textual analyses of television, audience studies have been an attempt, in part, to verify empirically the kinds of ideological readings constructed by (white and middle-class) critics.[1] The new critical interest in television audiences can be traced to 1980 when David Morley published his study of *Nationwide,* but it was in 1986 that the debate on television audiences emerged as the focus of scholarly attention at gatherings such as the International Television Studies Conference. Recently, the debate about audience studies has taken place at a high level of abstraction, as witness Martin Allor's useful essay "Relocating the site of the audience" published with four responses in a recent issue of *Critical Studies in Mass Communication.*[2] In this [essay] I wish to discuss the political issues of audience studies in the terms laid out in Pierre Bourdieu's *Distinction,* and to discuss the problem of the "self-reflexive" researcher in the context of one case study. By doing so, I hope to encourage a change within the current academic discussion of audiences. I feel we need to take up more concretely the problems of research as a practice.

I believe cultural studies must focus on the differences in class and cultural capital which typify the relationship between the academic and the subject of audience studies. Nowhere is this more vivid than in the study of television. The problem emerges most clearly when we discuss empirical

From *Cultural Studies,* vol. 4, no. 1. Reprinted with permission of the author, *Cultural Studies,* and Routledge.

research as a practice. Here I describe an interview with two white men that illustrates some of the political problems of interpretation in audience studies. But in doing so I also make a gesture that scholars in television and film studies endlessly repeat in our work. I completely ignore racial difference, and thereby contribute to the racism which permeates academic discourse.

This is the story of a ninety-minute interview that I conducted with co-researcher Hans Borchers as part of a larger study on television soap operas carried out in Oregon during the summer of 1986.[3] Out of twenty-six such interviews we conducted, I have chosen this interview because of my interest in gender and class, and in the discomfort many people feel when they identify themselves as members of the television audience. This interview took place in the home of Jim Dubois, sixty-two years old, and his housemate Larry Howe, about fifty-five years old. (I have changed their names here.) Both men are retired graphic artists, who moved to Eugene from California two years ago. The interview was conducted by myself (an American film and television professor in my early thirties) and Hans Borchers (a German American studies professor in his early forties). All four of us are white. I wish to describe how our subjects reacted to us as interviewers, how the inferior television audience was persistently identified as female, and how the power differential between us as academic interviewers and Jim and Larry as subjects related to the playing out of class difference during the interview.

Throughout the interview, it was uppermost in these men's minds that we were academics. For them, it was an honor to talk to us and an opportunity to be heard by persons of authority and standing. They made a concerted effort to appear cosmopolitan and sophisticated. For them, our visit offered a chance to reveal their own personal knowledge, and their opinions about society and the media. They had no interest whatsoever in offering us interpretative, textual readings of television programs, as we wanted them to do. In fact, they exhibited a kind of "incompetence" as viewers in this regard: they were unable to reproduce critical categories common to *TV Guide* and academic television criticism.[4] All fiction shows could be labeled soap opera: situation comedies and medical shows, alike. Yet many television shows were seen by Mr. Howe to conform to a more personal, master narrative about the painful relations between generations, which stemmed from his own bitter experience as a father.

This interview made me personally uncomfortable, because of my age and my gender, and because of my status as an academic. When we talk about examining our own subjectivities as researchers, we also need to ask what it means to ask someone else about television viewing. Television watching can be a touchy subject, precisely because of its association with a lack of education, with idleness and unemployment, and its identification as an "addiction" of women and children. This interview exemplifies the defensiveness that men and women unprotected by academic credentials may feel in admitting to television viewing in part because of its connotations of feminine passivity, laziness, and vulgarity.

For me, the interview raised profound methodological questions about "unstructured" interviewing. Our goal, which we discussed as a research team at great length, was to hear whatever soap opera viewers wanted to tell us. We wanted to follow digressions, to be receptive to unanticipated areas of discussion. (The psychoanalytic scenario was clearly in the back of our minds.) But our subjects wanted to present themselves in a good light. Though we were strangers, they knew we were academics and to a large extent that dictated the kinds of things that were said to us. That is especially prominent in this interview where our "subjects" were doubly defensive as men and as members of an older generation.

I am going to describe the dynamics and the sequence of conversational events that took place when we visited Mr. Howe and Mr. Dubois's home at some length. When we extract quotes from this context, much of their significance is lost. Very often, the meaning of statements as part of an exchange (and an unequal one) between researcher and subject is obscured.

While this interview is atypical in many respects of our experiences with soap opera viewers, especially women, it vividly demonstrates something that happened in all the interviews. People often compare their own television viewing to that of the imagined mass audience, one that is more interested, more duped, more entertained, more gullible than they are. Academics as television viewers are no exception to this rule. The imagination of that other television viewer is deeply implicated in the class/gender system. For Mr. Howe and Mr. Dubois, television's contaminating effects were directly and persistently related to the feminine (a tendency they share with academic writing about the mass media, from the Frankfurt School on).[5] This interview also exemplifies the extent to which television as a "mass" form is viewed in a very different way by those without access to college educations and more authentic bourgeois culture. For working-class and petit-bourgeois viewers, television is alternately relied on as a source of education and condemned for its failure to confirm and to replicate experience.

### The Interview

> Torn by all the contradictions between an objectively dominated condition and would-be participation in the dominant values, the petit bourgeois is haunted by the appearance he offers to others and the judgment they make of it. He constantly overshoots the mark for fear of falling short, betraying his uncertainty and anxiety about belonging in his anxiety to show or give the impression that he belongs. . . . He is bound to be seen as the man of appearances, haunted by the look of others and endlessly occupied with being seen in a good light.[6]

Mr. Howe and Mr. Dubois began the interview with a disclaimer about the amount of time spent viewing and the insistence that they only watch soap operas occasionally because they are usually out in the afternoons. This was an unusual start for the interview because Mr. Howe had answered a newspaper advertisement asking to interview soap opera viewers. Mr. Dubois explained that he only watched occasionally when seized by the distant hope that "maybe today we'll see something good happen rather than something tragic happening. . . . I guess soap operas make you glad that you don't have more problems than you do." Mr. Dubois then offered an excuse for why they do watch: to see the houses and locations on the shows: "That home is beautiful. We'd like to own it." The vineyard on *Falcon Crest* is shot on location in Santa Barbara, where Larry used to live.

This being said in defense of their interest in that show, Mr. Howe moved on to a statement of his preferences, based in part on a critique of soap opera in general:

> I think the soap operas that I can tolerate the most are the ones that don't take place in one room for the whole episode—where they stand there and talk each other to death. And never move outside. I like to see some automobile travel, some outdoor scenery, not just two people or three people at a table sitting there yakking each other to death.

Despite our resolutions to be non-directive, we reverted to the tactic of trying to elicit a comparative critical scheme (and ended up sounding like network focus group supervisors). We asked: "which are the worst that way?" (Here we desperately try for some comparative rating, an attempt to be systematic.) The answer is an equation of this problem with women and with tragedy. Mr. Dubois explained: "I think the afternoon ones have a tendency to be too tragic . . . it is more tragic in the afternoon, because afternoons are geared to woman, and woman likes to cry and likes to see somebody else's tragedy. . . . 'Oh, I'm not doing so bad, Look at her.' And they relate to that, whereas I, as a man, can't relate to that."

Mr. Howe then broke in with a kind of mini-lecture typical of the interview, in which he described to us the origins of television before the First World War, and the Berlin Olympic telecasts. Mr. H was trying to display his knowledge about a topic he (erroneously) assumed we would be interested in and could confirm his expertise at (a mistaken assumption given our lack of knowledge or interest in the *technological* origins of television). We politely listened but did not reinforce this kind of talk. We were more receptive to Mr. D when he returned to the topic of soap opera. Still attributing the tragic sensibility to women, but now generalizing about the problems of illness, he told us this story: "I'd never been in the hospital in my life—so it's always an adventure—until recently. I had a minor heart attack. I was rushed to the hospital in emergency. So then all the *General Hospital* things that went on, that I had seen, sort of became real to me. I couldn't believe that I was in the hospital playing the part of a very, very sick man." As an older viewer, illness is the most authentic fiction television presents to Mr. D.

Mr. H broke in for five minutes with a discussion of medical students inviting Robert Young to be graduation speaker because they had confused Marcus Welby with a real doctor. We were feeling more and more uncomfortable with Mr. Howe. He sounded eccentric to us, and his story mixed up. I believed *he* must be wrong, not the educated medical students. But this story allowed him to formulate his strong objections to TV: "it just goes to show you how they can trivialize, that's the word, they can trivialize anything, and make it seem insignificant." He applied this tendency to trivialize to priests, nuns, doctors, and then—wanting to spark our interest—professors.

Mr. H proceeded to draw an opposition between professors and scriptwriters, who "misinform the public," "when they set themselves up as something to be admired or respected. I read about what high ratings they have, well I get resentful because of the fact . . . well what are they measuring themselves against . . . what is the mentality of the people that are watching?" Along with the imagined other viewers, many people we interviewed felt this enormous antagonism towards the television industry's creative personnel. Writers were constantly coming under attack for boring the audience, or patronizing them.

At this point, Mr. Dubois made a revelation: Mr. H is a self-taught man. Mr. D proudly described how much research his friend had done, and how television material instigated his research activities. "When he sees something that is definitely a mistake and wrong, he'll check on it." Mr. H explained: "I can't get into secret government archives. I can't get past security clearances on military bases. But there are ways of finding things out. Societies that I have knowledge of, the people who are members of societies. Through letter writing. Library textbooks." Mr. D continued to praise him: "He's an avid reader, so he knows his stuff. Also he has, he really has a photographic memory . . . which I rely on and he relies on."

This story about research unsanctioned by the academy, convinced me that Mr. H was a crackpot. Returning to his anger at the box, Mr. Howe told a story about Elvis Presley once firing a "45 caliber six-shooter" at his television set.

In his comprehensive empirical study of French cultural differences, Bourdieu discusses the difference between legitimate and illegitimate self-education, between the kind of autodidacticism practiced by academics, and the kind that Mr. Howe does:

> There is nothing paradoxical in the fact that in its ends and means the educational system defines the enterprise of "legitimate autodidacticism" that is ever more strongly demanded as one rises in the educational hierarchy (between sections, disciplines, and specialties and so forth, or between levels). The essentially contradictory phrase "legitimate autodidacticism" is intended to indicate the difference in kind between the highly valued "extra-curricular" culture of the holder of academic qualifications and the illegitimate extra-curricular culture of the autodidact. Illegitimate extra-curricular culture, whether it be the knowledge accumulated by the self-taught or the "experience" acquired in and through practice, outside the control of the institution specifically mandated to inculcate it and officially sanction its acquisition, like the art of cooking or herbal medicine,

craftsmen's skills or the stand-in's irreplaceable knowledge, is only valorized to the strict extent of its technical efficiency, without any social added-
value, and is exposed to legal sanctions whenever it emerges from the domestic universe to compete with the authorized competences. (Bordieu,
*Distinction,* p. 25)

Information is routinely given as one of the "uses and gratifications" of
media use. But I think it is difficult for academics involved in television
studies to imagine the frustration and anger provoked by a dependency on
television for education and a lifelong exclusion from elite forms of higher
education. Some of the flashes of rage in this interview on Mr. Howe's part,
and many of the stunned silences on mine and Hans's, stemmed from our
face-to-face confrontation of this difference in access to cultural capital. I
emphasize this point because it is an arena where the distinction between
legitimate and illegitmate autodidacticism is one we as academics have an
interest in maintaining.

I had already placed Mr. Dubois and Mr. Howe in class terms by their
home furnishings—carefully maintained, early American matching pieces, of
which they were proud. Taste in things like clothes, food and furniture, and
patterns of consumerism is one of the earliest acquired and most striking
forms of class distinction, according to Bourdieu. At this point in the interview, Mr. Dubois revealed that he had a career in the fashion industry and
used the television, in a specialized professional way, to learn about style:

they show the home, they show the furniture, well this is a little bit, it's
not all a loss, it's a little bit of an education for me. I know what early
American furniture looks like, or Renaissance furniture looks like. I know
what pseudo-modern furniture looks like . . . so its sort of educational for
me . . . even though the plot may be absolutely ridiculous.

Then the conversation turned to a brief discussion of the various things
that bother them about television: too much sex; too many commercial
interruptions. On guard against a perception on our part that they were too
familiar with television, Mr. Howe then drew a distinction between themselves and "others who have TVs in every room . . . if we have other
things to do we just go and do them . . . otherwise we'd just sit around
and watch TV all day." Mr. D: "if it's raining and we can't get out to do
the things that we want to do out of doors, that's when we'll watch. As I
told you, most afternoon programs are geared to ladies. That's when they
show all the gushy. . . ." Mr. H: "They can stand there and iron or they
can stand there and prepare food or whatever and watch TV. . . ." Despite
the fact that these men have lived in households without women for twenty
years, domestic work is seen as the domain of women, the domain of women
watching television

Hans asked if they talk with other viewers about television, since this
aspect has been conspicuously absent in this, compared to our other inter-

views. Mr. D talked about a friend who is a real fan, who arranges her lunch hour around them. This led to the following exchange:

> *Mr. H:* Well, you know, where the word came from . . . . it's actually true . . . it's a shortening of the word fanatic.
> *Mr. D:* She was brought up with soaps, because her mother, Mrs. Applebaum, a lovely lady, I really love her . . . she was an avid watcher . . . she was a housewife and this was her passion.
> *Mr. H:* Is that the lady that every time she coughs or sneezes bubbles come out of her mouth? He said she was brought up on soaps. . . .

Hans, cutting short the joke which we obviously did not find funny, broke in to ask whether soaps have changed, commenting that most of our interviews had been with younger people. Mr. H speculated that "the young people see through soap operas and consider them to be so absurd that they don't bother with them." But Hans contradicted him: "They watch them alright." Mr. D then confirmed this, displaying his knowledge about the young and about fashion trends: "They talk today's language . . . they use the slang . . . they'll also use even in the commercials, as well as the story itself . . . hard rock sounds and stuff like that. The clothes that they wear is the updated clothes." Mr. D then mentioned that he called up his son, a lawyer, to verify the implausibility of one of the legal plots on the show. Again, the theme of "research" appears, and Mr. D lets us know that he has access to real knowledge, and is not being taken in by television.

Mr. H then asked us about the popularity of American TV in Europe, something he has read about and which horrifies him. This allowed him to share with us more of his knowledge. He has traveled in Germany, he has a keen interest in the Soviet Union and subscribes to *Soviet Life*. He told us the history of the Carl May society. We listened politely but uninterestedly to these things. But at this moment Mr. H was telling us what he wanted to talk about. For him, talking to academics and to a European professor was a rare opportunity. For our part, we had no real interest in his encyclopedic knowledge or in his generalizations about twentieth-century culture, which repeatedly took the form of the statement "instead of art imitating life, life imitates art," woven throughout his comments.

To force him back into the position of the everyday television fan, we asked what other shows they watch. Mr. H continued in his educated tone: "private detectives type adventure shows, police shows, travelogue shows, historical and archeological productions, such as Jacques Cousteau . . . the discovery that dolphins have an intelligence now thought to be as high or perhaps even higher than human beings." This list is not only entirely respectable in terms of bourgeois cultural norms, it is also (not coincidentally) composed entirely of "gender genres." And his preoccupation with education (read class distinctions), voiced throughout the interview, surfaces here again in the fascination with *intelligence,* albeit in animals. Mr. D contributed that they like to watch documentaries about animals and about castles. "They take you through every old castle . . . we're interested in all sorts of old buildings . . . and one of our goals and dreams is that we want to someday buy a Victorian old home and try to restore it."

This returned Mr. H to a recounting of his travels in Germany. But Mr. D continued to describe the show about castles: "And that's what amazes me . . . the art, the paintings, the old paintings, and where they came from."

Mr. H interrupted here to mention his favorite show, *Murder She Wrote*. He explained that they have surmised that the show is shot in the actress's own home on the East Coast, in "a beautiful area and a beautiful home." We asked why he said it was like a soap opera, still trying, even after an hour, to draw out these critical categories. He explained that there is "a lot more talking than any action of any kind . . . that program is just very talkative." But Mr. D interrupted, "Larry, it's logical talk though. They are trying to solve a crime. At least a crime is committed." This led to a debate about whether *Hill Street Blues* and *Barney Miller* are soap operas, with Jim explaining to Larry the differences. "They label it a comedy. They don't label it a serious drama or a series."

Now Mr. H returned to his theme which sounded like something he planned to say before we arrived, "it's got to be negative and down in order to be . . ." (soap opera). Mr. D explained again "I think it was originally started for ladies that were doing their laundry and their wash and they were using this soap."

Sensing the end of the interview. Mr. H now brought up the most personal and the most passionate judgment of television. As in many interviews we conducted, feelings of real anger or despair surfaced at the end of the conversation, as more personal details from the past were brought in.

> One point I will say. If you want to call it resentment you could call it that . . . it is always the parents' fault. Never the kids that are old enough to make their own decisions. They are always made to look good. Or victims of the way their parents raised them. . . . Just as though they had no control over their own lives as all brainless boobs. All the father's fault. Father never knows anything. He's a boob, an idiot. Only mother knows how to make decisions. Or the kids are the heroes always. . . .

He continued:

> *Father Knows Best* was a satire on father knowing best. He was made to look like a ridiculous buffoon. That was another . . . you could call it a situation comedy, if you want. I think it borderline soap opera. But always the kids come out looking perfect. . . . Looking back now, I never suspected what my parents had to go through. I was one of those that was sympathizing with the kids until I became a parent and had children of my own. Now all of a sudden I'd just as soon slap the little monsters around. . . .

Mr. Dubois got in the last words before we broke up the interview: "That's my theme song. I know what it is like to be young but you don't know what it is to be old."

## *Conclusion*

In writing up this interview, I have done extensive editing. I have attributed intentions and feelings to others. I have bolstered some generalizations I wanted to make with the authority of the real empirical subject. I have emphasized a couple of points, the role of gender and the role of cultural capital, which I would like to pursue here. Before doing so, let me also argue that audience studies might be helped at this point in time by the publication of unedited transcripts along with analyses. In a partial fashion, I hope to have demonstrated here how certain statements, if taken out of context in a discussion of something like "genres" or sitcoms might create quite a different impression than when the interview is taken in sequence. Clearly, more concern should be paid to language in audience studies, and to the social context which produces it.

What I have described here involves many "errors" in terms of the goals of unstructured, "ethnographic" interviewing. These problems were accentuated by a lack of rapport in the interview, but I do not believe that these errors are avoidable. We must pay more attention to methodology in empirical audience studies, and at the very least describe these methods more fully. But the problem will not disappear by adopting a more "correct" method of interviewing. The social identities of academic researchers and the social identities of our TV viewing subjects are not only different, they are differently valued. We cannot lose sight of the differences that exist between us and our subjects outside of our discussions about television. This interview underscores the fact that the differences that may be played out in conversation between interviewer and subject (I use the term advisedly) are antagonistic differences, based on hierarchically arranged cultural differences. Recognizing such distinctions will be difficult for academics, Marxists or not, because of our highly homogeneous work environment, and our intensive professional socialization.

Television may be particularly fascinating to us because it seems to provide access to the "other," the working-class, the female audience, the fantasized agents of revolutionary change. But it will not suffice to imagine ourselves to *be* this other audience, or to adopt the position of the enthusiastic fan, as John Fiske has recommended. "Slumming it" on our part merely obscures the fact that we are in a dominant relation in terms of access to cultural capital. Sitting down face to face, doing interviews with people different from us, *does* help raise consciousness about these issues. Empirical study *is* necessary to understand television viewing. But we must not pretend that the differences we find will be sympathetic or ideologically correct or even comprehensible from our own class and race and gender positions.

As a feminist, I did not find Mr. H to be a particularly sympathetic subject (my irritation at his sexist jokes being, of course, a marker of my own class background). The currents of hostility and aggression which ran through Mr. H's conversation, culminating in the discussion of his children, were disturbing to me. In my other work I have tended to focus on interviews in all-female groups where a certain level of rapport was established, and where the subjects were much more malleable to my direction of the interview, more forthcoming, more interesting because they related the media to their personal histories. Feminist linguists have found some evidence that women interrupt less often and rarely speak in lengthy monologues: we certainly found this true in our interviews.

But Mr. H and Mr. D remind us about an important lesson in terms of the "gendered spectator." They offer a vivid example of the denigration of women as an ego-defense. I am reminded of Bourdieu's point that

> explicit aesthetic choices are in fact often constituted in opposition to the choices of the groups closest in social space, with whom the competition is most direct and most immediate, and more precisely, no doubt, in relation to those choices most clearly marked by the intention of marking distinction vis-à-vis lower groups. (p. 60)

The two men offer two different positions from which to do so (there are many more). Mr. H occupies the more machismo tradition (and appears the most ignorant about television). Television appeals to women because it is downbeat, emotional, talky. He is for history, for research, for facts, for action. He despises or resists character identification. He would rather "prove himself" by talking about his educational exploits. He dominates the conversation repeatedly. Mr. D sets up an equally rigid dichotomy between himself and the woman as spectator. But he is more involved with television and approaches it from the position of an aesthete: he is interested in [topics] such as clothes, fashion, home design, architecture, art, forms which may challenge heterosexual male norms. What is curious here is the similarity in relationship to the construction of the audience "other as female" despite these differences. It suggests the importance of mapping out the interplay of class and gender differences on the field of cultural consumption.

The interview was full of miscommunications, and these miscommunications were often based on class differences, on the unequal possession of cultural capital. When interviewing those less educated, or with less ambitions to appear educated, than Mr. H, this has been less of a problem. It is precisely because he aspired to bridge the gap which separated him from us that he made more mistakes. His system of references—Soviet life, German castles, secret societies, the military—was very different from our own, but we could recognize it as undistinguished. Throughout the interview, our status of academics compelled Mr. H to range widely over topics, to boast, to wander, to interrogate us about our opinions and knowledge. We did not want to surrender control of this verbal "exchange," and this increased the undercurrent of hostility in the interview. The gap in class between Mr. Howe and us, with its different priorities—material artifacts over ideas, history over sociology, encyclopedic knowledge over theory—is a relationship based not just in difference but in the antagonism of competing class values.

> The struggle between the dominant fractions and the dominated fractions tends, in its ideological retranslation, to be organized by oppositions that are almost superimposible on those which the dominant vision sets up between the dominant class and the dominated classes: on the one hand, freedom, disinterestedness, the "purity" of sublimated tastes, salvation in the hereafter; on the other, necessity, self-interest, base material satisfactions, salvation in this world. (Bourdieu, *Distinction,* p. 254)

Bourdieu argues that nothing is better able to express social differences than the field of cultural goods—and here I would substitute television—because "the relationship of distinction is objectively inscribed within it, and is reactivated, intentionally or not, in each act of consumption, through the instruments of economic and cultural appropriation which it requires" (*Distinction,* p. 226). Obviously, Bourdieu's ideas would have to be reworked to suit the United States. But his work attests to the importance of relating television to other cultural fields, food, art, clothes, furniture, films, newspapers, and so on, the better to see class distinctions at work. In this broader context, it may be possible to avoid portraying television as a paragon of cultural pluralism (something for everyone, separate but equal, everybody happy). Though it is farthest from our intentions, critical scholars engaged in empirical cultural studies have borne a slight resemblance to market researchers. The challenge is to investigate popular tastes *and* explain how these tastes are distributed in relations of domination. To do so will also necessitate recognizing our own dominance and our own class interests within the system of cultural distinctions.

### Notes

Thanks to Roy Metcalf for his extensive help on the project and his excellent transcription; and to Gabriele Kreutzner, Eva-Maria Warth, and Hans Borchers, my co-workers on the Soap Opera Project.

1. Charlotte Brunsdon, "Text and audience," in E. Seiter, H. Borchers, G. Kreutzner, and E. Warth (eds.) *Remote Control: Television, Audiences and Cultural Power* (London: Routledge, 1989), pp. 116–29.

2. Martin Allor, "Relocating the site of the audience," *Critical Studies in Mass Communication* 5: 3 (1988), 217–33.

3. This study has been written up in the essay "Don't treat us like we're so stupid and naive: Towards an ethnography of soap opera viewers," in Seiter, Borchers, Kreutzner, and Warth, eds., *Remote Control,* pp. 223–47.

4. In this sense they were the opposite of Umberto Eco's "model readers." See my "Eco's TV guide: The soaps," *Tabloid* 5 (1982).

5. See Andreas Huyssens, "Mass culture as woman: Modernism's other," in *After the Great Divide* (Bloomington: Indiana University Press, 1986).

6. Pierre Bourdieu, *Distinction: A Social Critique of the Judgement of Taste,* trans. Richard Nice (Cambridge, Mass.: Harvard University Press, 1984), p. 253.

## The Transcript

Because so little ethnographic data is published in full, the transcript of the interview upon which this article is based is included in the hope that others will find it useful to their research.

> *HB:* As I told you on the phone, we're a team of four people and the whole thing is funded by a German research foundation. The reason we are doing these interviews is to find out how people watch soap opera, what people's watching experiences are?
>
> *ES:* We don't have a set plan of questions that we want to ask you. We are interested in what *your* opinions are about soap opera.
>
> *Mr. H:* We're being recorded right now?
>
> *HB:* Yes. So which are your favorite soap operas? Which are the soap operas you watch?
>
> *Mr. D:* Well since we spend most of our afternoons out we prefer to watch any of the soaps that are on in prime time. The reason why, I think both Larry and I, have a tendency to watch these things occasionally, is because we are both semi-retired and some of the programs, other than soaps, are very very bad . . . that we figured, well, maybe today we'll see something good happen rather than something tragic happening, which the soaps have a tendency to do . . . they're always fighting with each other, they're always against each other. As a matter of fact, I think that the characters are written in that way. Instead of trying to make you feel better they make you feel worse. And we enjoy some of them. *Dynasty,* we watch *Dynasty.* And some of the new ones that are coming up . . . the new one with Charlton Heston . . . we both happen to like Charlton Heston's performances, and that's why we watch. *The Colby's* is the same. Of course, it's a spin-off of all the others. And after a while, after watching a few years, you begin to realize that it's repetitive. They repeat. If they don't have a certain problem in one, they'll have it in another. And . . . another reason why I think that I usually watch serials . . . I like to see how they work out their problems. I like to see how they work it out.
>
> *Mr. H:* Well, I guess, soap operas, in one way, make you glad that you don't have any more problems in life than you do. To make a comparison between your own life's ups and downs and whatever they have manufactured into each individual soap opera. Everything seems to be so terrible, you know, it's a tragedy about to happen all the time. Somebody's ill and dying or a child dies stillborn or a marriage is about to break up because of another woman or another man. It's just that, I guess, the soap operas would die on the vine if they were geared to happy events. They have to gear to bad events, because evidently the human animal is only interested in hearing and seeing the bad. You couldn't have a happy ending; you couldn't have an enjoyable program

with people just normally enjoying themselves or children behaving normally—they have to behave abnormally. Adults have to act crazy and be on drugs or liquor in order for people to enjoy it. And that gives the people, not necessarily me, but most people, I think, have a vicarious sense of pleasure that. . . . Thank God that's not the way I lead my life . . . or. . . . There but for the grace of God go I. . . . That's what it's geared to—the negative, the black side, the dark side of life.

*HB:* May I ask which soap operas are you referring to? The nighttime soap operas?

*Mr. H:* Some daytime early hours and other night time. *ATWT* [As the World Turns] is one. *Dynasty* is another, like he mentioned.

*Mr. D:* *Falcon's Crest* is one of my favorites. It just so happens that both of us like certain scenes . . . other than old houses, like in *FC*. That home is beautiful. We'd like to own it.

*Mr. H:* *Dallas* is a favorite. Not necessarily, but I mean it's the least bad of most of the rest. It's the most . . . the one you can tolerate the most.

*Mr. D:* It doesn't, you see, *Dallas* doesn't put women in a good light. They are all lousy. They are all full of hate, envy, and jealousy. They are all out for . . . *Dallas* tends to make you feel that these women are out to power hunt. And they want to take it away from men. But *FC* has its good points and its bad points. One of the good points is that it's a vineyard . . . and Larry comes from Santa Barbara where there are an awful lot of vineyards. So those scenes are familiar to him and I've been there and I like Santa Barbara—at least I used to. Anyway, so the scenes that they show in the vineyards, and the story involved around the family struggle on the show interests me, and interests him.

*ES:* You mentioned the house on *Falcon Crest*. Which house is that?

*Mr. D:* Her home. Jane Wyman's home. The big house. The big estate.

*Mr. H:* Used to be a Roman Catholic . . . like a monastery, many years ago. And it was sold and then it was . . . the area around it was converted into an area of a large amount of growing of grapes for wine purposes. They had their own separate wine cellars where the . . . oh, I remember now . . . did you ever hear of a Roman Catholic order called the Christian Brothers? That's the name of the order. In fact, they even have that label on their wines. Christian Brothers. But, what I was going to say is that, I think the soap operas that I can tolerate the most are the ones that don't take place in one room for the whole episode— where they stand there and talk each other to death. And never move outside. I like to see some automobile travel, some outdoor scenery, not just two people or three people at a table sitting there yakking each other to death.

*ES:* Which ones are the worst that way?

*Mr. D:* I think the afternoon ones have a tendency to be too tragic. They pull you down, whereas the nighttime ones have some element of surprise or adventure or sometimes, occasionally, things turn out well. The girl gets the guy that she's after and she finds that divorcing her husband is not as bad as she thought it would be. That she's not going to commit suicide. Those things, I seem to think that they make it more tragic in the afternoon, because afternoons are geared to woman, and woman likes to cry and they like to see somebody else's tragedy. It makes, as

Larry said, it makes their own lives more tolerable. "Oh, I'm not doing so bad. Look at her." And they relate to that, whereas I, as a man, can't relate to that. But I can relate to the . . . evening ones which show these powerful guys who are greedy and out for power and out to destroy and full of hate and everything else. It's man's inhumanity to man— and I can understand that. Not that I relate to it . . . I think it's disgusting . . . I think it's terrible. JR is hateful. I realize that most of these writers, that write soaps, try to gear it to make it more exciting that way. So they exaggerate everything. They exaggerate the want of power. And although I shouldn't say exaggerate, maybe I've led a very sheltered life, and I haven't seen it that exaggerated in my life time.

*HB:* When you say exaggerate do you mean exaggerate compared to actual life, business life, for example?

*Mr. D:* I think they make it more exciting that way. Now I've been in business for a long time and I know that there's thievery and I know that there's competition and I know that people are really pitted against each other, but I don't think they tend to think violently, as I say this is one of the exaggerations on soaps. They think violently. If a guy gets mad at this woman because she's trying to take his business away—he slaps her around. I mean, to me, I've never seen it—maybe it does happen—maybe it does happen. That's what I say, maybe I've been sheltered. Maybe I feel, because I'm nice and I wouldn't do that they should be nice. I realize that it is fiction and so I accept it as fiction. I can't really say that I would feel that this is true to life to me.

*HB:* What do you think of the women on these shows who don't let themselves be slapped around, who fight back?

*Mr. D:* They can be pretty vicious too. Joan Collins is the most vicious of them all. She don't let anything get in her way. She's a strong woman. Maybe there are women like that. Maybe there are. I haven't ever met any. So that's why I feel its sort of vicious.

*Mr. H:* I was just going to remark on a phrase, that I've heard a lot of times, about how art imitates life. I've seen an increasing tendency, at least in this country, cause I haven't traveled to other countries for a number of years, whereby there is a tendency to turn things completely around backwards. To the point where life is imitating art. That's not the way it should be, that's upside-down. Life should not imitate art— art should imitate life. More and more people are being . . . I call it a subtle form of brainwashing . . . where people are concluding that that is the way to behave because they saw it on television, that's accepted behavior. They're are not thinking for themselves. They're allowing that thing over there to do their thinking for them. I don't do that—I refuse to do that. That's where I say that more and more life is imitating art and I don't like that trend at all. And specially is that bad with kids— impressionable little kids who grow up. This is the first completely, 100 percent, television generation, as it has been called now. Completely, from the time of infancy on up to adulthood, complete television generation, first time. More and more, kids do some terrible things and then they tell the police later on. . . . Well, I. . . . How did you get such a bizarre idea. . . . Well, I saw it on TV. Life is imitating art, supposedly art. And another thing I don't like about these soap operas; they commercial you to death. You know what I mean. Every ten seconds they break into the story, they chop it up, and you don't know what's happening anyhow, because while the commercial is running so is the soap opera running. So that when they stop with the commercial and put you back on the soap opera you've lost track of where you were. Instead of stopping the episode of soap opera right where they cut in for the commercial . . . they don't . . . because they're in a timeslot, see. And they can only run so many minutes. So what they are doing is making a sandwich out of it. They are putting the commercial on top of the soap opera and running both of them together at the same time. And that's chopping off the soap opera, and what have you got left; when you've got a dozen commercials in the space of one soap opera.

*HB:* I'd like to ask you when did you first encounter TV or soap opera, for that matter?

*Mr. H:* Well, I guess, the first one about thirty years ago. They had 'em.

*Mr. D:* They also had them on radio.

*Mr. H:* Television has been around a number of years. Did you know television was around before World War II. In three countries—the United States, England, and in Germany all had viable working television networks. Did you know that the first Olympics . . . not the first . . . but the 1936 Berlin Olympics were televised back to Berlin. 1936! I would say that my television watching started thirty years ago, but my soap opera . . . I never looked . . . I never . . . what I'm trying to say is I never searched them out. They just happened to come along. That goes back to about 1960, I'd say.

*HB:* Which show was that, do you remember?

*Mr. H:* No, I don't. That's how important it was.

*Mr. D:* At that time, I believe, Larry, if you remember correctly, they came out with all these hospital and doctor. . . .

*Mr. H:* That's right. You're right. You broke the log jam there. *Ben Casey* was one. *Dr. Kildare* was another. Let's see, what else? *General Hospital,* that was another. That was a biggy. Now you start jogging the. . . .

*Mr. D:* And everybody was taking their pulses and. . . .

*Mr. H:* They did that to death on those hospital shows.

*Mr. D:* After a while, I guess, both the writers and directors and everybody else, and even the actors, were sick of them, because they had too many. Some of them are still on. *General Hospital* is still on.

*HB:* Do you still watch it?

*Mr. D:* Occasionally.

*HB:* Does it make sense to watch occasionally?

*Mr. D:* Occasionally. Not really. Sometimes . . . in *General Hospital,* I notice, what they do is they fill you in. They give you, at least, a synopsis, of part of the story that you haven't seen or may have missed. So, but occasionally also you'll all of a sudden you'll see a new character and you'll wonder where they came from. But . . . in those hospital series . . . it seems they get involved with one or two cases . . . that are new . . . and they complete them. Either the guy dies or he lives or the lady has her baby or whatever. Or she gets her disease and it is cured or it's terminal. So when they tell you it's going to be terminal already and you sob a little bit . . . you cry a little bit . . . and you feel sorry for the poor dear and these things are terrible. See now those things are more or less true to life. Those hospital things, I mean, we all have had family that have departed from chronic illnesses and stuff. I've never been in the hospital in my life—so it's always an adventure—until recently. I had a minor heart attack. I was rushed to the hospital in emergency. So then all the *GH* things that went on, that I had seen, sort of became real to me. I couldn't believe that I was in the hospital playing the part of a very, very sick man.

*Mr. H:* Want to hear something incredible that happened a few years ago regarding one soap opera? So incredible that you won't believe it. It happened in regard to this, uh, soap opera that starred Robert Young as a guy named Dr. Marcus Welby. Now the graduating class of Columbia University in New York . . . of the medical school department of Columbia University . . . now this has to do with life imitating art. That guy has been an actor all his adult life, he has no medical training at all, except reading cues and reading the script. The graduating class, and this was not a spoof either, they were quite serious about it . . . they had been indoctrinated, or at least several of them in positions of power in the graduating class . . . to bring whoever they chose as a speaker . . . a valedictorian speaker at their graduating class. So they actually extended an invitation to Robert Young, in Hollywood, to speak . . . to come and make a commencement address at their graduating class, in the late 1970s . . . and he was on his way to doing it until the authorities at the school, at Columbia U., found out about it. They went right through the roof. The only person authorized to speak at any graduating class at a medical school is an MD. Not a professor. Not a faculty member. Not an outsider. And certainly not an actor. An MD. A fully qualified doctor. But these kids . . . or I say kids, they're in their late twenties . . . were so indoctrinated with this idea that he was a real qualified doctor. And when they were asked later . . . they were called in on the carpet and asked if this was their idea of some kind of a practical joke or something. They looked wide-eyed. No it wasn't a practical joke, they were deadly serious. They actually called for a movie actor to come in and give a commencement address on medicine . . . on what we could expect in the future in the medical world. The guy knows nothing about that—except his cue cards and his script. This is what I mean by . . . not art imitating life but life imitating art. That's how deeply these soap operas among others affect people's thinking. And I still think that that is a classic case of how incredible these things are in influencing people's lives.

*HB:* Are you saying . . . would this have wider application on television or just on soap opera?

*Mr. H:* Wider. Not just soap operas. But in this case it was Marcus Welby, MD. Now they put MD on the tail end of that program. . . . I don't know how they got that by the medical association or medical society or AMA. Because the guy was in no way an MD. But it just goes to show you how they can trivialize, that's the word, they can trivialize anything, and make it seem insignificant. They do it with priests. They do it with nuns. They do it with doctors. With scientific people . . . professors . . . who spend all their lives studying intently and dedicating themselves to the pursuit of that single goal . . . and then they bring in some jerk-off actor, who nobody has seen before, and they put a white smock on him and call him Professor This or Professor That or Professor The Other, you know. Something so ridiculous that several times in these soap operas. . . . I'm not saying I'm so smart and they're so dumb, but I am saying that several times there have been some statements made, that were no doubt part of the script, that were so stupid that, unless you were a complete dunderhead, you would catch it. And realize that the scriptwriters don't really know what they are talking about.

*HB:* The way you speak about soap opera and soap opera actors you sound pretty resentful, I must say.

*Mr. H:* Only resentful when they misinform the public. When they set themselves up as something to be admired or respected. I read about what high ratings they have, well, I get resentful because of the fact. . . . Well what are they measuring themselves against to get those high ratings? What is the mentality of the people that are watching, if that's what they constitute high ratings. Like that Marcus Welby episode, what does it take to get high ratings like that? I mean, there are those medical students, that already invited a man on his way there. He had to cancel out at the last minute. Wouldn't you get resentful if your medical school was used for that?

*HB:* Do you ever have discussions among each other after having watched a soap opera?

*Mr. D:* You see, Larry likes to research an awful lot. He's into that. And when he sees something that is definitely a mistake and wrong, he'll check on it.

*Mr. H:* Whatever is necessary. What I can do. I can't get into secret government archives. I can't get past security clearances on military bases. But there are ways of finding things out. Societies that I have knowledge of, the people who are members of societies. Through letter writing. Library textbooks.

*Mr. D:* He's an avid reader, so he knows his stuff. Also he has, he really has a photographic memory . . . which I rely on and he relies on.

*Mr. H:* I'm not really resentful as much as I am . . . direct and maybe a little blunt. Because when it comes to something that is just a bunch of poppycock I don't believe . . . it's like a friend told me . . . a good simile is. . . . Some people, he said, insist on washing and perfuming the garbage before they throw it out. You know what I mean. In this blunt way, what he was saying is, that some things that are so useless and silly are made to seem important . . . and valuable. And in other ways, important things are trivialized 'til they become nothing . . . and that's what I . . . I won't go along with. You might think that that would practically say that I never watch soap operas, 'cause a lot of that is trivial. But I watch them because like I first said, a lot of times it makes you glad that you don't have any more problems in your life than you do.

*HB:* Do you ever get mad at a show to the point where you talk to the screen?

*Mr. H:* No. Nor do I get as mad as Elvis Presley once got at his television. He was watching with a couple of friends one time, in Nashville, Tenn., one time, and this is a documented fact . . . and he saw something that he didn't like . . . whether it was a commercial that broke in on his particular program . . . he, normally, went around frequently with a 45 caliber six-shooter in his waist band . . . he pulled that out and he shot his television set out . . . right in the dining room of his big mansion in Nashville, Tenn. That was a well-known story. Then he told his friends. . . . Take it out and throw it on the dump. But his two friends, then and there, decided they didn't want to work for him any more . . . they didn't know but what he might take a shot at them sometime if he got mad at them.

*Mr. D:* There are other times where, as I mentioned before, some of these new series, that are being on television today, are shot on location and I, in my small way, have been involved in fashion vision, which was commercialized by going to department stores and making short films, commercial films, of the fashions or the sales that they were having or whatever they were having. And it was started by a friend of mine and I went in with him, but that lasted a very, very short while . . . 'til we lost a lot of money. We lost a lot of time. So I've been behind the scene, these things. And that's when I started to notice things . . . when you're on location, and you go to a different location, you find things that you feel are for you. And on these programs that I watch, occasionally, they show the area, they show the home . . . to fit in with the character. . . . they show the furniture. So little by little I felt . . . well, this is a little bit . . . it's not all a loss. It's a little bit of an education for me. I know what early American furniture looks like, or Renaissance furniture looks like. I know what pseudo-modern furniture looks like. So it's sort of educational for me . . . even though the plot may be absolutely ridiculous.

*ES:* What is there about the furniture and the homes on *Dynasty,* or *Colbys?*

*Mr. D:* I like them. Of course, when they go into the sexual scenes and give you this business of a round bed . . . I've never had a desire for a round bed. Also, I feel that, as I said before, that evidently the writers have to do that. They over-exaggerate . . . even in the sex scenes . . . such wild passion. And then the next part of the series they're fighting and you know. Course that happens too—I guess love can turn to hate. For many reasons. One thing I do object to when it comes to families . . . when there is such violent hatred between a father and a son or a father and a daughter that she actually comes out . . . "I want to kill my own father." That bothers me. That bothers me, but I also realize that this is fiction, and maybe the writers have to do that. Another objection I have, and Larry has mentioned it already, is that when you finally get interested and you want to relate to it . . . you want to get involved with the story . . . a commercial comes on. So you go wash a dish or . . . go to the bathroom.

*HB:* Another topic we're interested in is how *do* you watch soap operas? do you do other things or do you concentrate . . . ?

*Mr. D:* It depends. It depends on how much time or where our minds are. If we have other pressing things to do and we can watch as well as accomplish something else without just sitting and watching . . . we'll do that.

*Mr. H:* Normally we don't do anything in here and that's where the TV set is . . . we only have one . . . some people have a TV set in every room so they can do this and that and they can stand there and watch a small set. But we would have to do everything in here in order to be able to do that. So we . . . if we have other things to do we just go and do them. If we expect to get anything done, otherwise we'd just sit around and watch TV and nothing else gets done.

*Mr. D:* Occasionally we do have the time and we're a little bit tired of running around all afternoon, especially in warm weather, so we'll . . . all of a sudden we'll decide we'll relax. Tonight we'll watch television and forget about all this other stuff. And then we'll do it the next day. I was advised to do that by my doctor. He said don't pile it up.

*HB:* So when you watch a daytime soap opera it's accidental?

*Mr. D:* It all depends. If it's raining and we can't get out to do the things that we want to do out of doors . . . that's when we'll watch in the afternoon. And as I told you, most afternoon programs are geared to ladies. That's when they show all the gushy. . . .

*Mr. H:* They can stand there and iron or they can stand there and prepare food or whatever and watch TV.

*HB:* Do you ever get a chance to talk to other viewers about soap opera?

*Mr. D:* I have a very, very good friend . . . been a friend for about forty-five years I've known her . . . she is now working in a hospital in LA. That lady has been watching soaps most of her life, and she's such an avid watcher that even at work she goes into her boss and tells him. . . . My program is on. And that's when she takes her break. They have a little TV set in the lunch room. Her boss doesn't mind. Shirley . . . she works in the Veteran's Hospital . . . her husband . . . that's why they have a TV . . . her husband sells them . . . he works for a Japanese company. He's one of their top salesman . . . so that's why they have a TV set.

*Mr. H:* Well, you know where the word fan came from . . . it's actually true . . . it's a shortening of the word fanatic.

*Mr. D:* She was brought up with soaps, because her mother, Mrs. Applebaum, a lovely lady, I really love her, . . . she was an avid watcher . . . she was a housewife and this was her passion.

*Mr. H:* Is that the lady that every time she coughs or sneezes bubbles come out of her mouth? . . . he said she was brought up with soaps.

*Mr. D:* One thing I want to tell you, while we're on the subject of soaps. They used to have a satire called *Soap.* That was on, at least when it started, it was on late at night because they used all kinds of foul language and very sexy talk and all kinds of really . . . almost porno situations . . . so they didn't want young people watching. Now they are showing it again, and it's about three o'clock in the afternoon. So, occasionally, when I see this is on, I turn it on and I wonder how they really get away with it. At three o'clock in the afternoon there are children home.

*Mr. H:* It shows you the passage of a few years. . . .

*Mr. D:* It's very funny and I know it's satirical and sometimes they go to an extreme to hook you, to grab you.

*HB:* You just reminded me of a question I wanted to ask before. Most of the people we have interviewed are younger, younger people of twenty, twenty-five, thirty years of age and they don't have the soap opera experience you have. Would you say that soap opera has changed?

*Mr. H:* I think the young people of today, whether this is good or bad I'm not necessarily prepared to say, but I think the young people see through soap operas and consider them to be so absurd that they don't bother with them. The plotting is so thin and so threadbare . . . it's like tissue paper . . . and they . . . as they put it . . . it doesn't turn them on. But if a rock and roll music show came on that would turn them on. All that banging and beating and screaming and wailing and yowling and howling and growling that would turn them on. But a soap opera . . . two people talking themselves to death . . . they'd sit there. . . .

*HB:* They watch them alright.

*Mr. D:* I'll tell you why. Because they have changed a great deal. Each . . . every ten years they gear it to a certain type of viewer, so they do have . . . you'll see . . . and even in the soaps they don't talk the same way. They talk today's language . . . they use the slang . . . they'll use also, even in the commercials, as well as the story itself, they'll use occasionally hard rock sounds and stuff like that. The clothes that they wear is the updated clothes. That's the same way it's been even many, many years ago. The clothing of each generation, or each ten years, of course fashion changes. The whole outlook seems to change. They go . . . I have two sons . . . I went through the hippie period with them . . . the era . . . and then I went through other . . . now one of my sons is a lawyer. So occasionally, when I see a soap with some legal background in it, I discuss it with my son. Because he doesn't watch them too much—he's too busy being a lawyer. But I tell him . . . is this true . . . and he'll tell me if there is a case that they tried to tell you about on your soap . . . and I ask him . . . Ron is this factual? Could this really happen—could a guy really get off? . . . so he says. . . . In these days, he probably could get off. No matter what crime he's committed and as he says. . . . Because the prisons are overcrowded, the system . . . the criminal system is . . . we don't have enough policemen around and things like that. So things change and I think that soaps change with them . . . they have to. They have to in order to make them even a tiny bit credible.

*HB:* Speaking of clothes. What do you think of the clothes on *Dynasty*, Krystle's clothes, for example?

*Mr. D:* Nothing but the best. Those women are so gorgeous. I wish I was young again.

*HB:* I'm probably not telling you anything new when I tell you that *Dynasty* is also very popular in Germany.

*Mr. H:* Is it? That brought to mind a question I was going to ask you about how much I have noticed since I was over in that part of the world . . . I don't know what is going on now, I only know what I read. It seems like Europe has been so heavily influenced by American television, but a subtle change has come about in this country . . . many of the things that in this country are almost ridiculed or put down as being silly or stupid are taken with far greater credibility over in Europe than they are here. And that's peculiar, there is another example of instead of art following life, life following art. I find that reversing of reality . . . is what it is. A reversal of reality. What should be often isn't and what should not be is. I think to myself. . . . My God . . . I mean this may sound very unflattering, but I think to myself. . . . My God, what is going on over there? Are people losing their minds or something? Can't they discriminate between the real and the fantasy? Because some of it is so palpably absurd, not just the soaps . . . I'm talking about TV in general. That . . . you think, how could such and such a show be popular? 'Cause you talk to people, like here or where I used to live . . . a group of people . . . and all of them are against a certain show . . . well you can say. . . . Well tastes change between certain people. What you and your friends don't like somebody else would like. Yet some things are so really bad that you can't imagine how they could be popular.

*HB:* Not just *Dynasty. Dallas, Falcon Crest, Bonanza.*

*Mr. D:* Well *Bonanza* is . . . used to be my favorite, because I like westerns.

*Mr. H:* You also have the Karl May Society. Have you ever heard of that? [Tape ends]

*Mr. H:* In Russia, no less, they have the equivalent of a Karl May Society. They actually get out and have summer encampments every year . . . they dress up in cowboy . . . in western costumes, authentic down to the last detail, also in various Indian costumes, authentic down to the last detail. They even have connections with various Indian tribes in this country to send them eagle feathers, of a certain type, not from eagles that have been killed but from eagles that have been salvaged out of the woods or whatever—because they want the authentic thing. And in Leningrad and Kiev and Odessa and all over Russia they have this network of enthusiasts. . . . My deep interest, as he said, when I get into something, I've got to get to the bottom of it. Find out about it . . . I won't rest until I do find out about it . . . and then I got a copy of a magazine called *Soviet Life.* From a friend. . . . Have you seen this magazine? It's a color . . . a full-color publication and it looks like what used to be *Life* magazine in this country. And in that particular issue was confirmation of what I had heard. About seven pages were devoted to text and photographs of these Russian citizens, male and female, youngsters, adults, young people in their twenties, dressed up as Indians on the one hand (Kiowas, Paiutes, Apaches, Commanchees) and on the other hand as cowboys and original federal soldiers dating back to our American Civil War. Now I can't believe any of this. You know what started it, believe it or not, there's a thread of continuity between your interview, with me and Jack, and what got this society going in Russia, about twenty years ago. It was an American soap opera devoted to the wild west. It wasn't *Dynasty* or *Falcon Crest* or *Dallas* and I'm not sure it was *Big Valley,* but it was . . . it might have been *Bonanza.* Because that came out at a very crucial moment, shortly after WWII and when Russia was recovering from the effects of WWII. I don't know what triggered it off but all of a sudden . . . it's just like after a rain, all of a sudden the grass springs up. . . .

*HB:* What's the connection between *Bonanza* and *Karl May*?

*Mr. H:* Karl May is another thing entirely. Karl May pre-dates television. Karl May is a horse of a different color. But without television there is that same preoccupation with the American wild west . . . and did you know that the favorite viewing of both Adolph Hitler and Joseph Stalin, in their private movie theaters in the Reichschancellory in Berlin or the Kremlin in Russia, were American cowboy movies. And this, I think, was a take off into later versions of soap operas that dealt with the wild west. Did you know there are, and have been, a number of soap operas. But not Carl May, but I mean this Russian version of it almost sprang completely from television and a television soap opera. He may be right, I can't put my finger on it, but it can all be traced back to one soap opera. After the Second World War. It started in about 1955. About ten years after the war. And it had Russian in the subtitles printed on the screen. . . . It was still in English . . . but Russian subtitles. And all of a sudden enough people got interested, so intently in this . . . again, another example of instead of art imitating life, life imitating art. So it does have a great effect, in some cases, soap operas have a great effect on people's thinking.

*HB:* To return for the last few minutes of our interview to American television. What other shows do you appreciate Mr. Howe? Are there other shows you *do* appreciate opposed to these others?

*Mr. H:* Numerous. For instance, private detectives type adventure shows, police shows, travelogue shows, historical and archeological productions, such as Jacques Cousteau. There are a couple of others that are involved, having to do with the sea, with dolphins. The discovery that dolphins have an intelligence now thought to be as high or perhaps even higher than human beings. So they can learn a language. They have an experimental school in Florida where they're teaching dolphins to speak the English language. Really fantastic things.

*Mr. D:* We get a great deal of pleasure out of watching certain documentaries, specially about animals. We are both animal lovers. Not only about animals—they have a new thing that has been on television recently—about the old English castles. Which is fascinating. It's usually a series—they usually have about three or four of them. They take you through every old castle . . . we're interested in all sorts of old buildings . . . and one of our goals and dreams is that we want to some day buy a Victorian old home and try to restore it.

*Mr. H:* I would like to get back some day and see my favorite castle in all of Germany. The number one. By myself. But I have one favorite castle that I want to see again. One that was built by King Ludwig of Bavaria.

*Mr. D:* So they take you through the home and they tell you all about the artifacts. And that's what amazes me . . . the art, the paintings, the old paintings and where they came from. And I'm very surprised that in a lot of these old mansions there are a lot of the Dutch and German and French paintings from way back are on those walls. Besides the old English paintings and it's fascinating.

*Mr. H:* Don't forget to tell them about a program that's a favorite with you, well it is with me too, and it's almost like a soap opera and it's shot in a movie actress's . . . what Jack believes and I tend to agree with him . . . in her own home on the East Coast. It's called *Murder, She Wrote.* It's very much like a soap opera. What's her name . . . Angela Lansbury. She's the detective lady. It's taken strictly from Agatha Christie's books. And the interior of the home, the kitchen and everything . . . after seeing several episodes . . . Jack became convinced that it was shot on location . . . right in her own home. And when he's talking about paintings and all these artifacts and everything . . . it's a very beautiful area and a very beautiful home.

*HB:* Why do you say it's like a soap opera?

*Mr. H:* Because of the way the script seems to go. There is . . . it is a trademark of soap operas that there is a lot more talking than any action of any kind. Chase scenes or searches out in the woods or . . . so much of it is talking in a room between two or three people. That program is just very talkative. That's why it resembles a soap opera. Maybe it's not meant to be, but. . . .

*Mr. D:* Larry, it's logical talk though. They are trying to solve a crime. At least a crime is committed.

*Mr. H:* I'm not knocking it—I'm just saying it reminds me of a soap opera.

*HB:* What do you think of *Hill Street Blues?* That's a detective show I watch.

*Mr. H:* I saw it last night. Sometimes they have good episodes and other times they're too silly for words. I don't know what happens to the script writers . . . they must have a bad day or something.

*Mr. D: Hill Street Blues* with me has a tendency to be a little too violent. A little bit too much violence for me, so I wouldn't consider it among my favorite shows.

*Mr. H: Barney Miller,* that's almost a soap opera type.

*Mr. D:* But that's comedy.

*HB:* I was going to say, isn't that a sitcom?

*Mr. H:* Almost. Would you say that's a separate category? Can't soap operas also be funny?

*Mr. D:* They label it a comedy. They don't label it a serious drama or a series. Like we're used to watching within old movies . . . we used to go and watch serials.

*Mr. H:* In other words, it's got to be negative and down in order to be. . . . Which leads you to wonder why it was labeled soap opera in the first place. Why that name? Unless it was being put down as a form of entertainment.

*Mr. D:* I think it was originally started for ladies that were doing their laundry and their wash and they were using this soap.

*Mr. H:* Here's another category . . . this was films now not television . . . they applied it to TV . . . but first it became known in films, cowboys movies were, at one point, back about the late 1940s or 50s all of a sudden became known as horse operas. There again is the use of the word opera, like as to ridicule the type of show. On one hand you ridicule it by putting the word soap in front of the word opera and on the other hand you ridicule the cowboy movie by putting the word horse in front of it. Because opera is serious drama as well as music. So to sort of trivialize it, I think, I suspect, I'm not positive, but I suspect, as a means of trivializing it or putting it down you call it a soap opera on the one hand and a horse opera on the other. I've long wondered about that.

*HB:* You think *Hill Street Blues* falls into the category of soap operas?

*Mr. H:* Borderline. Borderline.

*HB:* It does have this little love thing going on between . . . what are the names. Furillo and. . . .

*Mr. H:* One point I will say, if you want to call it resentment you could call it that, I do resent the point that they seem to be so intent on . . . in most soap operas dealing with juveniles and juvenile problems of any kind and that is that always, always, always the parents come out looking bad. It is always the parents' fault. Never the kids that are old enough to make their own decisions. They are always made to look good. Or victims of the way their parents raised them. Just as though they had no control over their own lives at all. Like they were brainless boobs. It's all the parents' fault. All the father's fault. Father never knows anything. He's a boob—an idiot. Only mother knows how to make decisions. Or on the other hand mother is a drunk, she runs around with other men and father has to try to hold the family together. The kids are the heroes, always.

*HB:* Hasn't this always been the case? In television shows of the sixties?

*Mr. H:* To a large degree. *Father Knows Best* was a satire on father knowing best. He was made to look like a ridiculous buffoon. *My Three Sons* was another. That was another . . . you could call it a situation comedy, if you want. I think it borderline soap opera. But always the kids come out looking perfect. No fault can be found with them. Only the parents. It's just like one guy told me . . . He was a product of that generation. . . . He said. . . . Looking back now, I never suspected what my parents had to go through. I was one of those that was sympathizing with the kids until I became a parent and had children of my own. Now all of a sudden I'd just as soon slap the little monsters around. God, what a change in him.

*Mr. D:* That's my theme song. I know what it's like to be young but you don't know what it is to be old.

*Mr. H:* Orson Welles made a very good . . . well-known song . . . a 45 r.p.m. with that title.

*Mr. D:* I'd like to get a tape made and send it to both my kids.

*HB:* Well, this has been very interesting and helpful.

# Ethnosemiotics:
# Some Personal and
# Theoretical Reflections

## JOHN FISKE

An anecdote to start with: in 1988 I conducted a small-scale ethnographic study of viewer readings of *The NewlyWed Game* (Fiske, 1989). On the show four couples were asked about the wives' sexual compliance to their husbands' "romantic needs." The two wives who were least compliant were non-white. The two compliant wives were white. The female responses from which the contestants had to choose the most appropriate all had a racial accent—they were "Yes, master," "No way, José," and "Get serious, man." "Master" bore the accent of slavery, "man" of blacks and "José" of hispanics. Gender politics and racial politics were mapped onto each other.

I studied three audiences. Myself as fan; graduate and undergraduate students, some of whom were regular viewers of the program, some not; and self-identified fans contacted by advertising. These audiences were over-whelmingly white middle class, but of both genders—although the third "audience" was overwhelmingly female. The readings revealed by my eth-nographic methodologies were largely concerned with the gender politics, and almost totally ignored the racial.

### *Autoethnography*

So what were these methodologies? First I attempted through theoretically structured introspection to study my own responses. I knew that I watched

From *Cultural Studies,* vol. 4, no. 1. Reprinted with permission of the author, *Cultural Studies,* and Routledge.

the program regularly with considerable enjoyment. My theory of popular culture did not situate me as a cultural dupe, but as an active selector and user of the resources provided by the cultural industries out of which to produce my popular culture and my pleasure. My theory also told me that this pleasurable culture was produced at the interface between the program and my everyday life, and that the meanings involved in it would be produced by the interdiscourse between the social discourses in the text and those through which I made sense of my "self," my social relations, and my social experience. These discourses worked not only to circulate meanings but also to constitute "me" as both a social agent in the reproduction and regeneration of those meanings, and also as the social agency through which they circulated.

I think I could discern in myself three main discursive practices by which I produced sense and pleasure out of this interface between the mass-produced text and my everyday life. There was a professional discourse which blurred the distinction between the domestic and academic—as a professional theorist of the media and popular culture I was interested in this as an instance in which I could trace larger cultural and political forces at work, much as a linguist can trace macro-linguistic systems in a single utterance. It is, obviously, this discourse which is dominating my account of my investigation of myself as fan. This professional discourse was in some ways contradicted, in others completed, by a more popular one. I have vulgar tastes; the garish, the sensational, the obvious give me great pleasure, not least because they contradict the tastes and positioning of the class to which, objectively, I "belong," that of the white, middle-class male with acquired rather than inherited cultural capital. Taste is as political as knowledge. As a fan and an academic I moved between my professional and popular discursive strategies in both my viewing of the program and my investigation of my viewing.

My third discursive strategy was much harder to describe—it involved those more semantic discourses by which I made sense of the topics that both infused my daily life and were called up by the program. Central to these was the discourse of gender, though any discourse necessarily overlaps into those situated close to it in that discursive matrix around which we, as social agents, move as we struggle to make sense of our experiences within the socio-political system. So, intersecting with the discourse of gender were, in my case, those of age, class, and, to a lesser extent, race, as I used my viewing of the program as a pleasurable part of the constant social process of making sense of myself as a social agent, a sense that had to encompass the position of being a middle-aged, middle-class white male, once married, now in an important heterosexual, non-marital, mixed-age relationship, living successfully in patriarchy and capitalism while being deeply dissatisfied with both, yet being concerned to produce areas of personal pleasure and happiness within which to make functional, usable meanings of my gender, class, age, and sexuality. This meaning process is cultural even though I trace it in the realm of the personal because the sense I make of myself is not only reproduced and regenerated from social resources, both discursive and textual, but is then inevitably put into social circulation in myriad ways as I move through my daily histories and encounters. My first investigation, then, was of myself, not as an individual, but as a site and as an instance of reading, as an agent of culture in process—not because the reading I produced was in any way socially representative of, or extrapolable to, others, but because the process by which I produced it was a structured instance of culture in practice; it was an instance of enunciation of a systemic social resource; it was *langue* manifesting itself in *parole*. What makes it theoretically valid is its systematicity and its amenability to theorization, not its representativeness in an empiricist sense. Neither I nor my readings are typical, but the process by which I produced them is evidence of a cultural system.

Its systematicity is evidenced further by the physical environment in which I viewed the program. The discursive mix that structures my social self, and my readings of television, not surprisingly also structures my living room. My academic discourse has dictated not just the prominence of books in the room, but also the sort of books—academic, paperback, to be read and used, rather than coffee table books to be looked at and to give pleasure as objects. These books are functional books. But displaying them at all is the taste of a fraction of the middle classes—what Bourdieu (1984) calls the dominated dominant, those with high cultural, but low economic capital, and it signals a particular professional blurring of the distinction between work and leisure.

This same class position is evidenced by the antique furniture, much of it bought at auctions rather than through dealers. This, again, as Bourdieu has shown, is a typical cultural practice of those whose position in the social space is similar to mine. The antiques, however, are inflected by my left-wing inclinations in a number of ways. They are what is known as "country" antiques. Thus the table is not by Sheraton nor by a craftsman working directly from his book of designs but by an anonymous local carpenter: it is evidence of how Sheraton's style was modified and absorbed into popular taste—simplified, made more conventional, the signs of its individual artistic signature erased—the movement is from the bourgeois to the popular, from the artist to the anonymous craftsman.

More important perhaps, for me at least, antique furniture, unlike modern manufactured furniture, can be extracted from the commodity system of industrial capitalism. It resists the notion of individual possession. That table is 200 years old; it has passed through many households and will pass through many more: rather than possessing it I feel that I hold it in trust, I have a social responsibility towards it; it is situated within the communal and the historical rather than the individual. A similar reluctance to enter uncritically into the capitalist commodity economy can be seen in the home-made TV stand, and also in the TV, the VCR, and the audio cassette player, all of which are the cheapest I could find that would serve the function I wanted them to. They are (relatively) low tech, low consumerism. On the wall are poster reproductions of Aboriginal paintings. These clearly carry specifics of my personal history—I spent some years in Australia, and they are both a souvenir and a public display of that Australianness. They also exist with my class discourse—as Leal shows, "peasant" or native art in urban, middle-class homes bears class-specific meanings which are absolutely different from those it bears in the society in which it was produced. But in conjunction with my antiques they enter not only a middle-class discourse, but also a left-wing inflection of it. They, too, carry meanings of the communal, of the unpossessability of the anti-commodity. But these meanings, like all which oppose the dominant, can only exist in conditions of struggle. The posters are cheap (democratic) reproductions of the "originals" (the word exists only in the high bourgeois aesthetic, not at all in the Aboriginal, nor in the popular) which *are* possessed by a white millionaire: each reproduction is captioned: "Aboriginal paintings from Central Australia: from the collection of Robert Holmes à Court." The commodified originals are now part of the economo-cultural investment of a major capitalist: their poster reproductions enter the mass-market system (though their profit went to the Aboriginal peoples, not to Holmes à Court). Despite this powerful incorporating process, however, they can still carry enough meaning of a completely contradictory culture: erasure is never total, recovery always possible—it depends on the process and context of reading.

The contradiction between my professional, class tastes and my more vulgar, popular ones must be what led me to place cheap plastic toy TVs on top of the real one, as though to provide myself with a reminder of the contradiction between my vulgar tastes and my respectable ones with their "message" that TV is essentially trashy. Like all signs these toys are multidiscursive—they mean differently in different discourses: in my populist discourse they bear meanings of garishness and fun; they are also functional—as souvenirs they remind me of the trips on which I bought them. But in my middle-class discourse they comment on the trashiness of TV; in my academic discourse their irony enables me to distance myself from both the middle-class put-down of TV and the populist reluctance to theorize its own pleasures. Besides the TV are rubber caricatures of Reagan and Thatcher which squeak discordantly when squeezed, but their "silent" speech is equally eloquent. They are from a British satirical TV puppet show, *Spitting Images,* and are not only a sign of my left-wing politics, but also that those politics can use television as part of their discourse, as part of their social circulation. They also indicate the fun and offensiveness that the process can produce. They are akin to my reading of the *NewlyWed Game* in their pleasurable mix of progressive politics, fun, offensiveness, and a populist vulgarity.

This excursion into autoethnography has had a number of aims. I wanted to explore, in a specific instance, the idea of a symbolic environment that is constructed by a social agent out of the socially available resources, and that equally constructs that agent as a social member and marks his (in this case) position in the social space. The practices of dwelling in a house by which we make the landlord's or builder's structure our own for the time of our dwelling become, for de Certeau (1984) emblematic of how popular culture works in capitalist societies. We make our own spaces within the place of the other. An environment, therefore, is a contradictory mix of ours and theirs, it is what we make out of their resources (it is therefore a homologue, on a smaller scale, of its use in geographic discourse, where it refers to what a society makes of its natural resources). Television viewing produces a symbolic environment which occurs within a physical environment that is, in its turn, equally symbolic. The meanings and pleasures of television are accented by the semiotics of its place of reception. The two intersect and inform each other at all levels, and simultaneously and continuously merge into the social agent, in this case me, who both produces and is produced by them. The contradictory structure of discourse that organizes me as a social agent also organizes my living room and the meanings and pleasures of my television viewing. An environment, then, cannot be separated from the process of inhabiting it, for it exists only in that process. As I am the inhabitant of my environment, it is only I that can experience it from the inside. Using this "native" experience in an academic discourse of course means that I must externalize it and distance myself from it in order both to describe and theorize it. But to produce an autoethnography, I have to be able to move in and out of my domestic environment, I have to be able to bring different, distancing discourses to bear upon my experience, to make that experience both private and public, to account for it as both a specific cultural practice and a systemic instance. Environments can be observed and interpreted up to a point from the outside, but they can only be experienced from the inside, and an autoethnography may be able to offer both perspectives.

But any ethnography must include data that lies outside the ethnographer's: it is necessarily empirical, though not empiricist. It must deal with data that has a material existence in the social world—but the problem is the nature of that existence, particularly if that existence has been generated by and for the process of ethnography. Much ethnographic data has been produced specifically for the investigator, which does not invalidate it, but it does urge caution and self-awareness in its interpretation (in which Seiter's article [1990] is exemplary). Such data usually consist of words (in interviews, questionnaires, letters, phone calls, or conversations). These are recorded and turned into texts which can be analyzed discursively and related intertextually both to the original text (such as a TV program) that occasioned them and to the everyday life of their producers (as Neumann and Eason do [1990]).

Such data differ from those "found," but not produced, by the ethnographer. These may be letters to TV producers (D'Acci, 1989), to fanzines or the press, or the domestic environment of everyday life (Leal, [1990]). These sometimes blur into the previous category as when "real" conversations are recalled for the ethnographer (e.g., Hobson in Brown, 1989; Radway, 1984; Morley, 1986), and sometimes the ethnographer herself attempts to make the "produced" into the "found" by initiating a "real" conversation and taking part in it as participant as well as ethnographer (Brown, 1987). This often becomes a form of participant observation (Hobson, 1982; Palmer, 1986) when the ethnographer is present during the production of the data but does not generate it directly, though her or his presence must affect it to a greater or lesser extent. In a useful extension of this method, the subjects may be asked to comment on the data noted by the investigator.

Autoethnography produces yet another kind of data, ones in which the ethnographer is both producer and product, and which extend the data available from that which exists in material circulation in the social world to include that in the interior, personal world. Critiques of cultural ethnography from high theory (particularly its ideological and psychoanalytic forms) claim that its reliance on observable data exclude it from taking account of some of the most important forces in human and social experience, forces which can only be explained by the theorization of macro-experience, that which transcends and shapes the specificities of both the everyday and the individual. A properly theoretical autoethnography may deflect some of this critique, or at least the more valid points in it. It may be able to investigate ways in which socio-cultural forces or the play of meanings, pleasures, and power are experienced and actualized both externally (in objects, symbols, practices) and internally in the social- and self-consciousness of the agent, in the organizations of discourse that structure both the interior and the exterior, both the social and the personal and that set the two in a relationship of analogic continuity rather than one of homologic similarity and difference. What autoethnography may be able to do, then, is to open up the realm of the interior and the personal, and to articulate that which, in the practices of everyday life, lies below any conscious articulation. (I had never made myself conscious of the meanings of my living room until I wrote this piece, not was I aware of their analogic relationship with my TV viewing.)

It may also avoid the privileging of the theory and the theorist. Both psychoanalytic and ideological theories finally allow the theorist a privileged insight into the experiences of their subjects that is not available to the subjects themselves. They both position the theorist-investigator in a way that is disconcertingly similar to that of the old-fashioned imperialist ethnographer who descended as a white man [sic] into the jungle and bore away, back to the white man's world, "meanings" of native life that were unavailable to those who actually lived it. Autoethnography may offer a way of coping with the theoretical, ethical, and political problematics shared, in very different ways, by both high theory and empiricist (or imperialist) ethnography. I find traces of autoethnography in that way that some cultural ethnographers (e.g., Hobson, 1982; Radway, 1984; Brown, 1987; Ang, 1985; Lee and Cho, [1990]) share the cultural experiences, politics, and pleasures of their subjects. The ethnographer in these cases becomes part of the community of viewers or readers, participates in some of their cultural experiences and thus begins to include her own experience as part of that which is to be studied and thus some of the insights that an autoethnography may be able to make more explicit. They all show, in various ways, the ability to move into and out of their own (auto)ethnography. It is probably not coincidental that all of these workers are women working within a patriarchal academy—for their ethnography requires them to step down (even if momentarily) from the position of privilege and power which patriarchal academia insists is the proper position for the "objective" scholarly investigator and to align themselves with their disempowered subjects. It is also no coincidence that my call for an extension of this methodology (with its politics, ethics, and theory) comes from a left-wing, progressive academic (albeit a male, though hopefully not too masculine a one).

## *Ethnosemiotics*

My second method was to show groups of students a recording of the program and to play as little part as possible in the discussion that followed. But, of course, my influence pervaded it. All the students knew me and my theories, all were in a professional relationship to me. But this does not mean that they were totally dominated by me—some rejected my theories; many negotiated toughly with my point of view; many were fans of the program long before they met me, and, if I was able to move between the discursive practices of academic and fan, then so too were they. To the extent that a training in semiotics and discourse analysis gives me some purchase on the structures of a text, it ought to give me some purchase, too, on the new text produced by my students in discussion. In particular, it ought to enable me to identify within it the play of a popular discourse, the play of an academic discourse, and the interplay between them (see Fiske, 1989).

My third method was, through advertising, to call up yet more texts, both by telephone and letter. The telephoned ones were like the classroom ones, influenced by my presence, albeit in a different and more distant form. The ones that came in the form of letters, however, were influenced by the writer's knowledge of academia rather than by my particular presence within it. Shortly I will analyze one of them in some detail, but before that I wish to distinguish them from, but relate them to, the original text of the program. Let me refer to the program as an industrial text, one produced and distributed for profit by the cultural industries, and the words of the viewers as popular texts produced out of the resources of the industrial as the cultural competencies of everyday life were brought to bear upon it. The industrial text is a commodity that exists as electronic patterns on a tape that can be sold and resold, stored, distributed, used and reused. It is a product composed of signifiers. Popular texts, however, have no such physical presence—they exist only in their moments of reading, which are *their* moments of reproduction and circulation. They are elusive, they disappear as fast as they are produced, they are ephemeral and live only in their moments and contexts of production. They are thus both difficult to study and to account for in the socio-politics of culture. To study them we almost inevitably have to record them—on tape or on paper or even in our memory, to freeze their state of being constantly-in-process, and thus to give them a material presence like that of the industrial text. This necessarily distorts one of their centrally definitive features, but it does allow them, when frozen, to be taken out of their context and into the study. This points to the second, apparently inevitable, distortion—decontextualization. It is not that the popular text is produced in a different context from that in which it is studied, but that the distinction of text and context is itself invalid. The context of the reception of the industrial text is not just an environment within which the popular text is produced—it is itself part of that text, just as the industrial text is part of that environment, for the popular text can only exist contextually. Popular culture exists only in its process, and that process is inherently contextual for it is social not symbolic or linguistic. Insofar as semiotics and discourse analysis deal with texts that are, however falsely, separable from their contexts, the academic studying them must always engage in a hermeneutic reconstruction which can reinsert into the analysis, however hypothetically and tentatively, some of these elements of the context that cannot be turned into a text for analysis. If this is not possible, and it can never be completely so, the analyst must leave signposted gaps in the analysis.

A letter from a fan of the *NewlyWed Game* will serve as an example. It reads:

> Dear Mr. Fiske,
>
> Although I'm not sure exactly what you want to know, I can tell you, with some embarrassment, that I have become an avid viewer of *The NewlyWed Game*. I don't know if I can really explain why, but I know I started watching it over three months ago, mostly because the time it comes on fits my schedule well. After I come home from work, I make dinner while watching the evening news, and then at 7:00 it's ready, and I can sit down and eat dinner while watching *The NewlyWed Game*.
>
> Probably the thing I like about it is that I can relate a lot of that trivial domestic stuff to my relationship with my own husband, and it's a little uncomfortable realizing how much we are like those couples on the show. In fact, I usually will answer the questions, and then guess how my husband would respond. If he's home at that time and watches the show with me, I urge him to answer the questions, so we can compare our answers. Actually, we do learn something about one another, not always what we want to know. So, for that reason I enjoy the show. Another reason might be that I probably like to see people making fools of themselves. Well, maybe not fools, but just being themselves with all their peculiar idiosyncracies. It makes for good comedy in my opinion.
>
> The last reason has got to be because I just can't stand game shows like *Wheel of Fortune* or *Jeopardy*. I don't get excited about games of chance or gambling, or that type of thing.
>
> I just thought of something else: it has to do with the MC of the show. He definitely goads the contestants to get them going. Stirring up

conflict between the husband and wife does seem like part of the show's appeal.

That about covers it, Mr. Fiske. For what its worth, I also watch *The New Dating Game* which follows *The NewlyWed Game,* but I don't enjoy that as much. If I were single, I probably would prefer that one more.

There are a number of points to make here. I do not wish to focus on the pleasure she produces for herself as she urges her husband to watch with her, so that they can learn things about each other that they do not always want to know. Rather I wish to focus on some of the methodological issues.

How do we interpret her desire to write to me, which led her to keep my ad until she had time to? In my ad I identified myself as a fan, I tried to signal my popular discourse as well as my academic one: following Ang (1985) I identified myself as a fan as well as a researcher.

I intend to "read" her letter through my academic discourse to attempt to demonstrate a critical ethnosemiotics, a semiotic reading of ethnographic data. First she *wants* to write to me: not knowing what I want to know, she focuses on what interests her—her embarrassment at her "avid" viewing. I trace here the ideology of popular culture (Ang, 1985) at work. She "knows" that this show is low in the socio-cultural hierarchy of taste, but yet, despite her college education, she finds pleasure in it, and she may find that my interest offers a form of legitimation. Her discourse shows evidence of this ideology of popular culture which leads us to denigrate ourselves for enjoying what is socially derided. This is typically explained by a metaphor of addiction: so the phrase "I have become" hints at an addictive process beyond the writer's control—a loss of control that is firmly contradicted later in the letter when she frees herself from this particular value system.

Then she begins to contextualize the program—first by making it part of her domestic routine and in particular her nightly transition from the public to the private. At work she plays a publicly determined role, then, in cooking dinner, plays an equally externally determined role in domestic, rather than waged, labor. Watching the news, with its emphasis on the public sphere and the roles people play in it, seems appropriate as she labors in her social role of wife. *The NewlyWed Game* completes her transition to her personal sphere within the domestic—I can hear her relief in the words "it's ready, and I can sit down"—the word "can" suggests she is at last doing something she chooses to—its absence from the phrase "I make dinner" is significant. It signals the change from labor to leisure, a change her husband made as he stepped through the door. The program is part of her entry into that sphere of her life where she has some control: where popular culture is produced.

The discourse in the next paragraph is fascinating. First it reproduces the embarrassment expressed in the first paragraph—"Probably," "trivial," "domestic," "stuff." It passes through this quickly, signaling that the values that cause it have to be rejected if the writer's pleasures are to be investigated. Her discourse is that of patriarchy here with its implicit assumption that "trivial domestic stuff" constitutes the world of the feminine in opposition to the important public world of the masculine reported on by the news. Once she has rejected this discourse and its values, her confidence in her own grows visibly. The verbs and qualifiers become stronger "In *fact* I usually will," "I urge him," "Actually, we *do* learn," and she concludes emphatically "So, for that reason I enjoy the show." The gradual self-empowerment that we can trace in the tone of her discourse here parallels that produced by her watching of the show. She "urges" her husband to answer the questions, presumably against his reluctance. She associates this reluctance with the revelation of parts of their relationship that are repressed and though, as a good wife, she claims that what they learn is not always what they want to know, her confident next sentence shows that only in her role as *wife* does she not want to know; in the sphere of her life where she makes *her* popular culture—which is politically opposed to her wifely role—*she* does. The repression of strains in the relationship is to the advantage of the masculine not the feminine, partly because it is a mechanism of making the relationship conform better to the patriarchal ideology of the couple, and partly because the blame for such strains is ideologically placed as part of the feminine role. She returns to this point later, to re-emphasize the stirring up of conflict between husband and wife as part of the show's appeal—and "stirring up" implies bringing up into visibility what is "normally" kept invisible and undisturbed.

My respondent's admission of her pleasure in the revelation of the repressed in her own marriage leads her immediately to its equivalent in the couples on the show. Again she admits her embarrassment and uses discourse that bears values that differ from hers—so she uses it hesitantly: "might be that I probably . . . ." The switch to her values comes in the discursive association of "making fools of themselves" with "just being themselves." This is a small moment of recognition that masculinity and femininity are roles that people play in a constant masquerade, and that when the ideological roles slip people simultaneously "make fools of themselves" and "are themselves": the pleasure comes when the masks slip and their ideology is revealed. The phrase "just being themselves" indicates her sense that there is a social "reality" that is repressed by the roles: the laughter caused by the eruption of this "reality" into the role is subversive. In almost every paragraph of the letter there is evidence of a (partial) adoption of patriarchal discourse followed by a pleasurable shift away from it that is the discursive equivalent of the conditions of her everyday life: she lives in patriarchy but finds spaces of her own within which to evade or oppose it.

## The Dual Problem of Ethnography

The problems of an ethnosemiotics, then, can be organized into two main categories: problems of generating the popular (con)texts and converting them into a form that, while making them available for analysis, distorts two of their centrally defining features—their processuality and their contextuality. This leads to the second group of problems, those concerned with the interpretation of these "new" texts and the analysis of the process by which they were produced. The interpretation of any text is inherently fraught with problems, but at least a recorded text of a TV program is, in the realm of the signifiers, the same as that produced and distributed by the industry. A recording of a discussion, a photograph of the room in which the program was viewed, or a letter, are all representations or reproductions of a popular text that were called up for a particular function that differs from that of their original production; they were generated for academic study, not for the purposes of everyday life. They were produced in a different context, for contexts are as imbricated in function as they are in texts. This does not mean that the new functionalizing context has no connection with the original one, but that the differences and connections need to be taken into the analysis and theoretically reconstructed in it. I attempted in my autoethnography to go some way towards contextualizing a moment of reading of an industrial text. I hope I showed some of the ways in which the context is a text itself and in its intertextual relationship with the original text inflected it towards certain readings and away from others.

But this text or context was not called up by the ethnographer. The letter was. How do I take account in my analysis that it was a response to my advertisement, and otherwise would never have existed? I suppose I can take some comfort in its tone—friendly, almost as a fan to a fellow fan. The phrase "Well, that's about it, Mr. Fiske," coupled with her inability to know what I wanted, might be taken as evidence that she was minimizing the differences between the context of writing a letter to an unknown academic and that of watching the program. The analysis has revealed some discursive evidence that writing it was pleasurable in a way that paralleled and reproduced the pleasure of watching the program. So I may be able to argue that my analysis of the letter can be transposed relatively directly into an understanding of how she watches the program and how her watching is part of the condition of her daily life. Seiter [1990] gives the best example I have yet seen of an analysis of ethnographic data in which she takes full account of the presence of the interviewers, of their gender politics, and of their status as professors. She is acute at identifying possible differences between what the viewers tell academics about their TV viewing and the way they watch it in their everyday lives. Of course, she has no way of describing this difference, for her only data is the interview, but she can and does identify those parts of the interview which appear to have been called up by the ethnographers' academic status. This does not mean that the interviewers gave fake accounts of their TV viewing, but it does point to their foregrounding of certain practices and pleasures over others.

Both Tulloch and Moran (1984) and Hodge and Tripp (1986) in their work with schoolchildren's viewing are aware that when the interviews and screenings take place in school the discourse about TV is necessarily influenced by its context and by the way that the interviewer is inevitably posi tioned, to a greater or lesser extent, as a teacher. Krisman (1987) in her study of working-class girls in a London school found a qualitative difference in the data obtained from discussions when she was present, and one which she was unable to attend, so asked one of the girls to operate the tape recorder in her absence.

Articulating such methodological and hermeneutic problems in ethnography is an important first step, even though articulating them is a long way from solving them. None the less, there seems little point in taking the extreme position of writing off all ethnographic data as invalid because of the problems of collecting and interpreting it. It is important that the process of collection and interpretation should be critically commented upon in the study. It is also important that various forms of ethnographic data should be correlated and their analysis combined with a textual analysis. This does not necessarily mean that we should give greater weight to the similarities among various forms of data, for ethnography is often concerned with the investigation of differences and specificities, but it does mean that we should be able to explain such specificities as instances of culture in process in which similar cultural and social forces are negotiated with in specific ways. We should, in other words, see them not as representative, but as systematic.

But the ethnographer has to work not only with the data obtained, but also with absences. Semiotics tells us that what is absent from a text is as significant as what is present. The problem faced by the ethnographer is whether or not to intervene in the production of data in order to fill a perceived absence, or whether to leave the absence as significant and face the problems of interpreting it.

This bring me back to the problem of the discourse of race in my study. The industrial text contains it: the images of people on the screen and the words they use are conventionally part of that discourse. All signs are, of course, multidiscursive for they are part of a variety of discourses—the words "master," "José," and "man" circulate in the discourses of gender and class as much as that of race, and it was clearly the one of gender that my respondents, and I, selected as the most useful/meaningful to us. This may be because in the experiences of our white, middle-class, everyday lives, which is where our popular culture circulates, we had little need to make sense of race (or class) relationships, but much call on our ability to make sense of gender ones. It is also true, of course, that the explicit focus of the program is on gender.

Ethnosemiotics must consider not only which resources of the industrial text are used, but also those which are not (*contra* those who ignore the industrial text, e.g., Morley, 1986). The absence of the discourse of race is a hermeneutic problem. Maybe it was, in fact, present in some readings but was repressed from their articulated, re-presented form. If so, this was a systematic repression that may be traced to a racist ideology.

This, then, poses a methodological problem. Should I have drawn my subjects' attention to it and thus "called up" a use of the discourse that they, for whatever reasons, had chosen to ignore or repress? It is difficult to re-member back to my own first viewing of the tape, but I think the discourse of race was part of my enjoyment of the black couple's discomfiture—their *style* of interaction and reaction was entertaining and amusing: one white use of the discourse was to make sense of blacks as spectacles of entertain-ment for whites. I am not sure that I gave much weight to the hispanicness of the other non-white couple. I think, too, that I may have found it mar-ginally amusing and therefore relieving that it was the black man who was least masterful of all. There may have been some relief in the contradiction of the white myth of the physical superiority of black sexuality; but I don't think that I experienced any pleasure in the placing of the challenge to patriarchal power in the non-white women, though, clearly, this is a pos-sible popular use of this juncture of the discourses of gender and race. I do recall distinctly, however, because it reproduced my dissatisfaction with pa-triarchy, my pleasure in the resistances of the two (non-white) wives and my displeasure at the two compliant white ones and my dislike of their husbands. What I do not recall at all is whether my pleasure was produced, in part, by the racial dimension. I doubt it, despite the fact that my racial politics are, I like to think, progressive. But these politics pertain more to my public and professional life; unlike gender politics they impinge less on the mundanity of my day-to-day existence.

The process by which I turned this industrial text into my popular cul-ture involved a selective use of its discursive resources and, as far as I could discern, the primary criteria for this selection were ones of relevance to my everyday life and the social terrains I had to negotiate most frequently and most problematically within it. Everyday life is, as Willis shows so clearly [1990], highly politicized not just because it occurs within a social structure whose only stable characteristic is its unequal distribution of power and privilege, but also because any personal negotiation of our immediate social relations is a necessary part of our larger politics—the micro-political is where the macro-politics of the social structure are made concrete in the practices of everyday life.

The popular text is where the potential meanings, pleasures, and poli-tics of the industrial text are actualized, and it is only in their actualizations that we can identify them. This involves an ethnography of social agents which runs the risk, which we must guard against at all costs, of allowing itself to be incorporated into the ideology of individualism. The social agent is not the individual of individualism but a socially constructed site crossed by a number of intersecting and sometimes contradictory discourses that has been produced by his or her particular trajectory through social space over his or her lifetime. But though social agents must, because of the con-tradictory nature of social forces, negotiate actively their own trajectory and their meanings of social experience, they cannot produce the discursive re-sources nor the social structure within which experience occurs, and mean-ings are made: they cannot choose the socio-historical terrain which their trajectory traverses. They are agents, not subjects, because their activity is not confined to making do with the determinations provided (active, creative, and underestimated though this may be), but what they *do* with these determinations feeds back in however small a way into the structure that produces them. How we use our language modifies, eventually, the linguistic system itself; so, too, our use of the social system is not just a product of that system but a producer of it as well. Any system is modified by each and every one of its uses.

Macro-social systems have forces of reaction built into them, whether they be systems of representation or more directly social or political ones: they tend to maintain the current distribution of power (economic, gender, moral, aesthetic, racial, or whatever). Social progress or change can come, then, either from the overthrow of the system at the macro-structural level or from the progressive uses of the system at the micro-politics of everyday life. Ethnography is concerned to trace the specifics of the uses of a system, the ways that the various formations of the people have evolved of making do with the resources it provides. Ethnosemiotics is concerned with interpreting these uses and their politics and of tracing in them instances of that larger system through which culture (meanings) and politics (action) intersect. It is concerned, then, not just with meanings in process but with meanings in action. Refusing the distinction between text and context requires the parallel refusal of the distinction between meanings and behaviors.

In the letter from the fan of the *NewlyWed Game,* the discursive progression from "Probably the thing I like about it" through "I urge him to answer the questions" to "So, for that reason I enjoy the show" evidences a trajectory of meanings that continues from watching the show into social action within the micro-politics of her marriage. So, too, the continuities that connect my construction and activity as a social agent with my pleasures in the program and the symbolic environment of my living room will not stop in the realm of the semiotic or cultural, but will extend themselves into the socio-behavioral. An ethnosemiotics that focuses on cultural processes will necessarily dissolve boundaries between the interior, the personal and the social, for these boundaries are unhelpful constructions: they are constantly crossed by the play of meanings in process which occurs simultaneously within and across all of their territories. The interior is the personal is the social is the political is the interior. Similarly an ethnosemiotics is a textual analysis is an autoethnography is an ethnosemiotics. And through it all we catch glimpses of the play of culture which is ultimately our quarry.

### References

Ang, I. (1985) *Watching Dallas.* London: Methuen.

Bourdieu, P. (1984) *Distinction: A Social Critique of the Judgement of Taste.* Cambridge, Mass.: Harvard University Press.

Brown, M. E. (1987) "The politics of soaps: pleasure and feminine empowerment," *Australian Journal of Cultural Studies,* 4, 2: 1–26.

D'Acci, J. (1989) "Women, 'woman' and television: the case of *Cagney and Lacey*." Dissertation, University of Wisconsin-Madison.

de Certeau, M. (1984) *The Practice of Everyday Life*. Berkeley: University of California Press.

Fiske, J. (1989) *Understanding Popular Culture*. Boston: Unwin Hyman.

Hobson, D. (1982) *Crossroads: The Drama of a Soap Opera*. London: Methuen.

Hobson, D. (in press) in M. E. Brown (ed.) *Television and Women's Culture*, London: Sage.

Hodge, R., and Tripp, D. (1986) *Children and Television*. Cambridge: Polity.

Krisman, A. (1987) "Radiator girls: the opinions and experiences of working-class girls in a London comprehensive." *Cultural Studies*, 1, 2: 219–29.

Leal, O. F. (1990) "Popular Taste and Erudite Repertoire: The Place and Space of Television in Brazil." *Cultural Studies*, 4, 1: 19–29.

Lee, M., and Cho, C. H. (1990) "Women Watching Together: An Ethnographic Study of Korean Soap Opera Fans in the United States." *Cultural Studies*, 4, 1: 30–44.

Morley, D. (1986) *Family Television*. London: Routledge/Comedia.

Newman, M. and Eason, D. (1990) "Casino World: Bringing It All Back Home." *Cultural Studies*, 4, 1: 45–60.

Palmer, P. (1986) *The Lively Audience: A Study of Children Around the Television Set*. Sydney: Allen & Unwin.

Radway, J. (1984) *Reading the Romance: Feminism and the Representation of Women in Popular Culture*. Chapel Hill: University of North Carolina Press.

Seiter, E. (1990) "Making Distinctions in TV Audience Research: Case Study of a Troubling Interview," *Cultural Studies*, 4, 1: 61–84.

Tulloch, J., and Moran, A. (1984) "*A Country Practice:* Approaching the Audience." Paper delivered at the Australian Communication Association Conference, Perth, 1984.

# The Rhetorical Limits of Polysemy

## CELESTE MICHELLE CONDIT

The recent, energetic critical program focused on the receivers of mass communication emphasizes the autonomy and power of audiences to exert substantial control of the mass communication process and hence to exercise significant social influence. The polysemic character of texts, these studies argue, allows receivers to construct a wide variety of decodings and thereby prevents simple domination of people by the messages they receive (Fiske, 1986; Hall, 1980; Morley, 1980; Radway, 1986).

These theoretical claims are supported by substantial evidence demonstrating the active character of audience viewing. The theoretical conclusions, however, overstate the evidence because they oversimplify the pleasures experienced by audience members. As many of the preeminent scholars in critical audience studies themselves admit, audiences are not free to make meanings at will from mass mediated texts (Fiske, 1987c, pp. 16, 20, 44). Consequently, the pleasures audiences experience in receiving texts are necessarily complicated. In this essay, I employ a multidimensional rhetorical critique of a single television text to suggest that the ability of audiences to shape their own readings, and hence their social life, is constrained by a variety of factors in any given rhetorical situation. These factors include audience members' access to oppositional codes, the ratio between the work required and pleasure produced in decoding a text, the repertoire of available texts, and the historical occasion, especially with regard to the text's positioning of the pleasures of dominant and marginal audiences. I conclude that mass media research should replace totalized theories of polysemy and audience power with interactive theories that assess audience reactions as part of the full communication process occurring in particular rhetorical configurations.

## Critical Studies of the Audience

Audience-centered critical research argues that viewers and readers construct their own meanings from texts. Audiences do not simply receive messages; they decode texts. Members of mass audiences are therefore not mere "cultural dupes" of message producers. As John Fiske (1987c, p. 65) describes the process, viewers have the "ability to make their own socially pertinent meanings out of the semiotic resources provided by television." As a consequence, "viewers have considerable control, not only over its meanings, but over the role that it plays in their lives" (p. 74). Janice Radway (1984, p. 17) makes a similar argument about mass-produced fiction: "Because reading is an active process that is at least partially controlled by the readers themselves, opportunities exist within the mass-communication process for individuals to resist, alter, and reappropriate the materials designed elsewhere for their purchase."

Critical audience analysts position their work as a radical break with the history of critical media studies, which they depict as having emphasized the power of the media to impose a dominant ideology or to control beliefs and behaviors (Fiske, 1986; Morley, 1980; Radway, 1986). The new studies indicate that disparate audiences do not decode messages in uniform ways (Katz and Liebes, 1984; Morley, 1980; Palmer, 1986), in the precise directions critics have suggested they might (Radway, 1984), or even as the messages authors seemed to have intended (Hobson, 1982; Steiner, 1988).[1] These studies conclude that the texts which link producers' intended messages with actual audiences are not univocal. Reworking structuralist insights, they emphasize that all texts are polysemic (Fiske, 1986; Newcomb, 1984), that is, capable of bearing multiple meanings because of the varying intertextual relationships they carry (especially Bennett and Woollacott, 1987) and because of the varying constructions (or interests) of receivers.

The study of the polysemic character of texts has thus included two research schools, often not clearly distinguished. Works in the American school (Kellner, 1982; Newcomb, 1984) emphasize the variety of ideological positions contained within the mass media. In contrast, the British approach highlights the variety of decodings possible from a single text or message (e.g., Burke, Wilson, and Agardy, 1983; Morley, 1980).

Whether based in the variety of available texts or in the flexibility of decoding processes, polysemy has been taken to be a widespread or even dominant phenomenon, bearing significance for theories of social change. Rather than portraying the mass media as the channel of oppression generated through the top-down imposition of meanings, such a perspective allows for the suggestion that the pleasures of the popular media might in fact be liberating. Radway (1984, p. 184), for instance, claims that because of the pleasure women derive from romance reading, "they at least partially reclaim the patriarchal form of the romance for their own use." Fiske (1987c, p. 239) finds similar pleasures and effects operative in television: "The pleasure and the power of making meanings, of participating in the mode of representation, of playing with the semiotic process—these are some of the most significant and empowering pleasures that television has to offer." Fiske argues that, even without the additional step of circulating one's own representations, these pleasures may offer a real resistance to the dominant ideology. Escape, he indicates (p. 318), may itself be liberating, because to escape from dominant meanings is to construct one's own subjectivity, and that is an important step in more collective moves toward social change. Fiske (p. 230) concludes:

> While there is clearly a pleasure in exerting social power, the popular pleasures of the subordinate are necessarily found in resisting, evading, or offending this power. Popular pleasures are those that empower the subordinate, and they thus offer political resistance, even if only momentarily and even if only in a limited terrain.

Recent critical audience studies thus repudiate prior portrayals of television as a sinister social force in favor of a celebration of the ability of audiences, enabled by the broad referential potentiality of texts, to reconstruct television messages. Television, because it is popular, therefore becomes a force for popular resistance to dominant interests.

These audience studies and the theories they are generating offer a useful counterbalance to the flat assertion that messages produced by elites necessarily dominate social meaning-making processes. Nonetheless, the scope and character of audience power have not yet been delimited, and I believe they are as yet overstated.[2] It is clear that there are substantial limits to the polysemic potential of texts and of decodings. If television offered a true "semiotic democracy" (Fiske, 1987c, p. 236), we would have to assume either that television—with all the distortions described by the last fifty years of quantitative and critical analysts—is in fact an accurate producer of the popular interest or that it will soon reform itself to be such. This seems either too dark a description or too optimistic a forecast. The underlying agonistic theory common to British cultural studies, postmodern theory, and American rhetorical studies offers a more appropriate line of approach. We need to begin to describe the precise range of textual polysemy and the power held by the audience in its struggle with texts and message producers.

These limits ought to be found both in production conditions (Meehan, 1986) and in texts. As a rhetorical critic, I focus my attention on the latter, exploring, in a variety of ways, the communication event occasioned by the broadcast on November 11, 1985, of an episode of *Cagney & Lacey* concerning the topic of abortion. Because rhetorical criticism focuses on language usage as a means of distributing power among a particular group of agents who are uniquely situated in a communication process (e.g., McGee, 1982), this critique examines two particular audience members for the program, then the specific political codes made present in the message, and, finally, the historical occasion of the broadcast. While this case study leads to a focus on television, the implications extend to other national mass media as well.

## *The Polysemous* Cagney & Lacey

My own viewing of the abortion episode leads me to describe the central plot as follows: police detectives Cagney and Lacey help a pregnant woman (Mrs. Herrera) to enter an abortion clinic where pickets (led by Arlene Crenshaw) are blocking access. Lacey, married and pregnant, eagerly helps Mrs. Herrera, while Cagney, feeling conflicted, resists any assistance beyond that necessitated by her job. When the abortion clinic is bombed and a vagrant dies as a result, the detectives investigate and locate the bomber, who, in a climactic scene, threatens to blow up herself and the detectives. She gives up when confronted with the inconsistency of killing Lacey's "preborn" child for a Pro-life cause.[3]

Two viewers, selected from a larger project I am conducting, offer particularly interesting responses to the episode. The two were college students recruited through local-scale organizations active in the abortion controversy. They were asked to view the program and to respond, during commercial breaks and after the program ended, to my open-ended and nonjudgmental questions. These two college students and their responses are not presented because of their "typicality." I do not claim their responses are representative but rather that they are suggestive of new questions that must be asked in order to gain an accurate picture of the relative power of encoding and decoding as social processes. The first respondent, whom I call "Jack," was the leader of the student Pro-life group. A twenty-one-year-old male, first exposed to the abortion issue through a required essay in a Catholic all-boys high school, Jack described himself as not being a particularly successful student and as having a life goal of becoming a major league baseball umpire. "Jill," a first year student, was the daughter of a feminist mother. Active in the student Pro-choice organization, her goal was to complete a doctorate. Neither of these two leaders of politically active groups had seen the episode previously, but both reported having heard about it and having talked about it in their organizations when the episode originally was broadcast. While Jill displayed more familiarity with the series, Jack showed more knowledge about the political preferences and activities of the actresses and producers and reported having read about the episode in newspapers and magazines.

At one important level, the eighteen single-spaced pages of transcripts provided by these two opposed activists confirm the polysemy thesis. Their replies to my questions agreed less than 10 percent of the time. For example, when asked about the fairness of the presentation, Jill replied, "Yeah, I think it is fair," whereas Jack said, "I think it's really grossly unfair." Jill responded to Arlene Crenshaw by saying, "I don't like her. I don't respect her," whereas Jack listed her as his favorite character, noting that she was the "lone good-guy type of figure in the show." Similarly, Jill claimed that the value of "family" was "definitely portrayed as positive," whereas Jack concluded, "I don't think they take a very pro-family type response." Throughout their interviews, Jill and Jack provided virtually diametrically opposed opinions of the episode.

There were, nonetheless, important elements in their responses which lead me to suggest that the term "polyvalence" characterizes these differences better than does the term "polysemy." Polyvalence occurs when audience members share understandings of the denotations of a text but disagree about the valuation of those denotations to such a degree that they produce notably different interpretations. In this case, it is not a multiplicity or instability of textual meanings but rather a difference in audience evaluations of shared denotations that best accounts for the two viewers' discrepant interpretations. Careful listening and examination of the transcripts make it clear that neither Jill nor Jack misunderstood the program, and they did not decode the images and words as holding different denotations. Their plot summaries, although extremely rough, were not inconsistent. More important, perhaps, each advocate was able to predict what the other's response to the program would be. If we accept the premise that understanding is effectively assessed by the ability to predict another's interpretation, this is an important test that both pass. After claiming that the episode "presents both sides of the story," for example, Jill admitted that "I'm sure that a lot of Pro-life people would hate it because it ends up that they are criminals at the end." Jack shared the ability to reflect on how the text might be read by others with different values: "A lot of people . . . would say, 'oh, it's great, it's a fair portrayal, it presents our side very well and does a good job of the other one too,' whereas the Pro-lifers would say 'it's a terrible portrayal, it's absolutely biased against our side.'" On another occasion, in talking about his preference for Arlene Crenshaw as a character, he noted, "You know, obviously, coming from my point of view, I can see if I was *[sic]* pro-abortion, she'd be like the 'bad guy.'"

On a number of specific counts, it further becomes clear that both viewers shared a basic construction of the denotations of the text. Both described Cagney as the character "in the middle." Both cited the transformation of the lieutenant's attitudes. Both noted the poverty and minority status of Mrs. Herrera. Ultimately, in spite of their different attitudes toward the episode, there was nothing in their responses to suggest that they did not share a basic understanding of the story line or even of what the program was trying to convey.

This finding is consistent with other major audience studies. In David Morley's partial transcripts of interviews surrounding the program *Nation-*

*wide* (1980), I detect little fundamental inconsistency in the denotations processed by the viewers; instead it is the valuation of those denotations, and the attached connotations that viewers draw upon, which become important (see also Eco, 1979, pp. 54–56). The response of Morley's group members to an interview with Ralph Nader seems to be typical. Even groups which are opposed in their attitudes toward Nader, the program, and in their life conditions share a basic understanding of what the interview denoted. Likewise, in Radway's contrast of the professional critics and the Smithton readers (1984), it is not that the two sets of readings are inconsistent but simply that the critics devalue any patriarchal codings, whereas the Smithton women accept some of those codings as consistent with their values. The only instance in which true shifts in denotations are recorded, to my knowledge, is Elihu Katz and Tamar Liebes's study of Middle Eastern readers of *Dallas* (1984), and in this case it requires massive cross-cultural differences and language shifts to produce such discrepant interpretations.

The emphasis on the polysemous quality of texts thus may be overdrawn. The claim perhaps needs to be scaled back to indicate that responses and interpretations are generally polyvalent, and texts themselves are occasionally or partially polysemic. It is not that texts routinely feature unstable denotation but that instability of connotation requires viewers to judge texts from their own value systems. Different respondents may similarly understand the messages that a text seeks to convey. They may, however, see the text as rhetorical—as urging positions upon them—and make their own selections among and evaluations of those persuasive messages. As I note in the conclusion, this will have profound implications for the practice of academic critical reading. For clarity, then, we might reserve the term "intertextual polysemy" to refer to the existence of variety in messages on mass communication channels, the terms "internally polysemous" or "open texts" for those discourses which truly offer unstable or internally contradictory meanings, and the term "polyvalence" to describe the fact that audiences routinely evaluate texts differently, assigning different value to different portions of a text and hence to the text itself. Such revisions imply the need to generate a more careful account of the actual social force of popular or mass communication. Such an endeavor begins with a more detailed exploration of audience interpretations.

## Audiences: Groups of Individuals

The claim that audiences have the ability to create their own empowering responses to mass mediated texts loses little of its force when it is acknowledged that the polysemic freeplay of discourse has been overestimated. Whether deriving from decoding processes related to denotation or connotation, critical audience studies have indicated fairly clearly that viewers can construct a variety of responses to any given mass mediated text. The central issue remains, however, to what extent do these responses constitute liberating pleasure and social empowerment? The situation of audiences as members of groups in a social process constructs some fundamental limits to these pleasures and powers which can now be explored.

The proposition that decoding a message always requires work is a fundamental postulate supporting the claim that audiences have control of the mass communication process. As Morley (1980, p. 10) puts it, "The production of a meaningful message in the TV discourse is always problematic 'work.' " The work receivers must do inserts them into a position of influence in relationship to the text. Such accounts, however, fail to note that decoding requires *differential* amounts of work for different audience groups. Jack's responses to *Cagney & Lacey* consumed more than twice the space and time of Jill's replies. Jill was positioned to give a reading of the text that was dominant or only slightly negotiated (e.g., she objected to the tokenizing of minorities in the program and the lack of women in the more powerful job hierarchies). Jack was required to provide a largely oppositional reading.

Not only did Jack's interpretation require more time and space, and visibly more effort (his nonverbal behavior was frequently tense and strained), it showed itself to be more incomplete and problematic in other ways. Frequently, Jack's responses departed from the program altogether to provide the background of a fairly extended Pro-life argument. In reacting to the abortion clinic's male physician, Jack cited the doctor's story about a twelve-year-old girl who came in to get an abortion, arguing:

> . . . little does he tell them now, however, that it is easier for the younger, anywhere from a twelve- to eighteen-year-old, statistically and medically, to bear children than it is for women who are over twenty-five or thirty, only because it's like, their bodies are ripe and just developing, as opposed to either at the peak or really past that. See, they don't want to get into that; he just talks about how terrible it allegedly or supposedly is for the young women to have children. So it's the best thing to do, get them in there, you know, do the abortion, and get them out, no worries. Do they ever talk about post-abortion counseling that that doctor might do? . . . Are they willing to go so far as to say that he just does the abortion and have [*sic*] nothing more to do with her?

Jack thus worked very hard to oppose his own ideology to the program. At times this entailed distortions of the truth which were probably unintentional. For example, Jack's statistics are skewed. More important for Jack, at times he was simply unsuccessful at producing a consistent response. At several key points he was reduced to a position of virtual incoherence, and he indicated his frustration in nonverbal ways. For example, at one point he became trapped between his denial that normal Pro-life people are violent and his attempt to project how the network should portray abortion clinic bombers. He concluded:

> If I was [*sic*] nuts enough to bomb, I'd go about it real calmly, talk to them, and wait until they dig up some more information before I went, got overly nervous. I think they did a good job of portraying her as, well,

see, she was involved in the sixties and seventies and all these demonstrations, the typical type. Why couldn't they portray, if they are going to, a
bomber who is just an average everyday American? They did a good job
of portraying her as an extreme fanatic. That is to say that, see, they're all
like this. They're the type who did that and they'll do this again. It's rather
illogical.

Jack was unable to come up with a consistent characterization of clinic
bombers. He described them as "nuts" yet asked that they be portrayed as
"an average everyday American," displaying his difficulty in putting together a response to the text that was persuasive (either to himself or to
me). Jill did not show such strain in her interpretations.

Finally, Jill and Jack differed with regard to the chief tests they put to
the text. For Jill, the recurrent test was "Is this realistic?" Accusing the text
of committing errors, she argued that the portrayal of the Pro-life leaders
and the bomber as women was inaccurate, but that Pro-lifers in fact generated violence, and so on. For Jack, the reality criterion emphasized motives
rather than facts. His most frequent strategy was to talk about what the text
omitted: the character of the fetus, the "ripeness" of young women, the
poor quality of counseling the women received. For Jill, therefore, the negotiation process was simply one of relatively minor factual corrections. For
Jack, the process was a matter of filling in major motivational absences in
the text (see Sholle, 1988; Wander, 1984).

For Jack, in short, the work of interpreting the text and resisting its
persuasive message was much more difficult than the accommodative response was for Jill. Although these differences may have been caused by
factors other than their political positions (e.g., differential academic ability
or familiarity with the series), they provide grounds for considering the
important possibility that oppositional and negotiated readings require more
work of viewers than do dominant readings. This possibility is reinforced
as well by work with public speakers (Lucaites and Condit, 1986). Three
factors give impact to the difference in audience work load: its silencing
effects, its reduction of pleasure, and its code dependence.

The first consequence of the greater work load imposed on oppositionally situated audience groups is the tendency of such burdens to silence
viewers. In its most stark form, this leads to turning off the television, a
widespread phenomenon, especially among minority groups (Fiske, 1987c,
p. 312; Morley, 1980, p. 135). If the particular range of television's textual
polysemy excludes marginal group messages, and if oppositional reading
requires comparatively oppressive quantities of work, then minority groups
are indeed silenced, even as audiences, and therefore discriminated against
in important ways.

Another consequence of this work load is disproportional pleasure for
oppositional and dominant readers. As Fiske (1987c, p. 239) points out, it
is clearly the case that viewers can take great pleasure in constructing oppositional readings, simply because of the human joy in constructing representations. Nonetheless, this does not mean that the pleasures of the text

are fairly distributed. Jill indicated that she enjoyed the episode of *Cagney & Lacey* very much and that she found it "powerful," and her nonverbal response indicated a restful, enjoyable experience. Jack, on the other hand, clearly took some pleasure in his ability to argue against the text, but he also displayed clear signs of pain and struggle in that decoding. Jack's relative displeasure may be widely shared, given that even a popular program enlists only 20 million viewers out of a population of over 240 million and that most of those viewers are simultaneously engaged in other activities (Meyrowitz, 1985, p. 348). The disparity is made pernicious given that the most highly sought audiences have the characteristics of more elite groups: more money and hence more attractiveness to advertisers (Feuer, 1984a, p. 26; Kerr, 1984, p. 68). Programs are tailored for the greater pleasure of a relative elite.

A similar disparity of pleasures in the mass publishing industry is suggested by Radway (1984, pp. 104, 165–67) when she reveals that the repressed pornography that producers of romance believe to be attractive to women may not actually be their primary interest and that an extremely different genre of stories might bring greater pleasures to these audiences. As Fiske (1987c, p. 66) notes, to be popular enough to gain economic rewards, mass media must attract a fairly large audience. That popularity, however, is only relative to other programming the producers are willing to construct. Hence, the trade-off among what marginal audience groups want, what other audience groups want, and what the producers are willing to give them as a compromise may still retain a great deal of control for producers and dominant groups.

Mass mediated texts might be viewed, therefore, not as giving the populace what they want but as compromises that give the relatively well-to-do more of what they want, bringing along as many economically marginal viewers as they comfortably can, within the limitations of the production teams' visions and values. If so, the differential availability of textual pleasures and the costs in pain become as important as any absolute statements about viewer abilities. It is not enough to argue that audiences can do the work to decode oppressive texts with some pleasure. We need to investigate how much more this costs them and how much more silencing of oppositional groups this engenders. In addition, we need to understand better the various conditions that best enable oppositional decoding.

A third consequence of the differential work load required of viewing groups provides further clues to the variability of audience experiences. Among oppositional readers of the *Cagney & Lacey* text, Jack was in a particularly empowered position. As a leader of a Pro-life group, he was experienced in producing Pro-life representations and had access to a large network of oppositional codes. This experience and access were evident throughout his interview (as in the instance where he used Pro-life rhetoric to point out gaps in the doctor's story). The utility of such experience and skill in helping viewers to produce self-satisfying decodings is echoed throughout the audience literature. Morley (1980, p. 141) especially notes the enhanced ability of shop stewards to produce oppositional codings more successfully than do rank and file union members. Importantly, most of the content-based audience research thus far taps into audiences where group leadership exists and where audience members have access to counter-rhetorics. Radway's study (1984) relies on a group centered on Dorothy Evans, who encodes negotiated readings, giving access to a resistive code to her group members. Linda Steiner's study (1988) relies similarly on a site, *Ms.* magazine, where oppositional rhetorics are provided.

In sum, the strongest evidence about the actualization of audiences' abilities to decode messages to their own advantage comes from studies that select audiences or conditions in which we would expect the receivers to be relatively advantaged as opponents to the message producers. Moreover, in cases with the weakest access to group organization, it also seems that oppositional interpretations are weaker. In his study of adolescent female responses to Madonna, for example, Fiske (1987a, p. 274; 1987c, p. 125) suggests that the young girls are only "struggling" to find counter-rhetorics. They experience, therefore, only limited success at resistance.

The commonalities in these studies suggest two conclusions. First, there is a need for research to assess the typicality of oppositional readings. The tendency to notice successful oppositional decodings may have led scholars to overplay the degree to which this denotes typical behavior. Correctives could come from comparing audiences with different access to oppositional codes on a particular topic and from studies of the relative degrees of oppositionality in typical decodings. Only if a strong and pervasive response to dominant messages can be demonstrated can we assert that the limited repertoire of mass mediated messages really coexists with a semiotic democracy.

Second, these commonalities also reestablish the importance of leadership and organized group interaction. Leadership always has been largely a matter of the ability to produce rhetorics that work for a group. While being human may mean having the ability to encode and decode texts (Burke, 1966), it is not the case that all human beings are equally skilled in responding to persuasive messages with counter-messages. The masses may not be cultural dupes, but they are not necessarily skilled rhetors. Here, another fragment from the abortion communication event is instructive.

In interviews of abortion activists in California, Kristin Luker (1984, p. 111) noticed an interesting phenomenon. The women who became abortion activists reported one factor that led to replacing their guilt and negative feelings about abortions with active campaigning for a right to choose. It was not the experience of abortion per se; many of them had had abortions long before the change in their attitude. It was, they said, the ability of a few articulate rhetors that had been instrumental in helping them to resist the prior, dominant views. The presentation of different codings had helped them resist the dominant rhetorics. If popular media are read oppositionally only to the extent that countermedia exist to help audiences decode dominant messages, the mass media's role in social change processes may be extremely limited. In this case, Fiske (1987c, p. 326) is not wrong when he concludes that "resistive reading practices that assert the power of the subordinate in the process of representation and its subsequent pleasure pose a direct challenge to the power of capitalism to produce its subjects-in-ideology." It is simply that we do not yet know how widespread such resistive interpretive practices are or can be, given the more substantial obstacles outlined here. In contrast, we should weigh the power that these texts give to dominant audiences.

## Codes and the Public

The disproportional viewing pleasures experienced by elite groups might present only a minor social problem if turning off the television set sufficiently closed down the influence of its texts. However, even in such relative silence, the television texts continue to go about constructing hegemony in important ways. This becomes evident if we shift our perspective so that the important audience for television is no longer individual viewers (even grouped by social interest) but "the public."

The term "public" is highly contested (Bitzer, 1987; Goodnight, 1987; Hauser, 1987; McGee, 1987). By "public" I mean those members of a nation-state who have had their interests articulated to a large enough mass of people to allow their preferred vocabulary legitimacy as a component in the formation of law and behavior. I suggest that television's political functions are not confined to its address to the pleasures of individuals. In addition, television "makes present" particular codings in the public space (Perelman and Olbrechts-Tyteca, 1971). Once such codings gain legitimacy they can be employed in forming public law, policy, and behavior. Even if they are not universally accepted, their presence gives them presumption (the right to participate in formulations, and even the need for others to take account of them in their policy formulations). Crucially, the upscale audience courted by television advertisers is also the group most likely to constitute the politically active public (e.g., "Young Blacks Have," 1987). Hence, television, or any mass medium, can do oppressive work solely by addressing the dominant audience that also constitutes the public.

It is because television "makes present in public" a vocabulary that prefers the dominant audience's interests that the dominant audience gets the most pleasure from television and that television actively promotes its interests. The fact that other groups can counter-read this discourse, and enjoy doing so, does not disrupt the direct functions of governance that television serves for dominant groups. A return to the case of the broadcast of abortion practices will explain this point more thoroughly.

Prime time television addressed the practice of abortion in clearly patterned ways. The very few, highly controversial programs concerning abortion in the sixties and early seventies occasioned sponsor withdrawal, boycotts of sponsors who did not withdraw, and extended editorial comment by opponents of legalized abortion (Condit, 1987). Probably as a consequence of this extra-popular control mechanism, a second round of abortion programs did not appear until the mid-eighties, more than a decade after abortion had been legalized through the actions of state legislators and the Supreme Court. For many years, television producers were dissuaded from making present the practice of abortion. When abortion reappeared, it did so with a dominant-preferring code firmly, if cleverly, in place.

The evolution of prime time television's treatment of abortion between the years 1984 and 1988 was such that it began to include more problematic cases of abortion, and it featured distinctive types. Nonetheless, the main clump of programs between 1984 and 1986 constructed a limited repertoire of meanings.

Different viewers, with different viewing habits, may have found themselves introduced to abortion in the mass culture in one of three ways. For viewers who enjoyed "family" programs, *Call to Glory, Webster, Family, Dallas,* or *Magruder and Loud* provided episodes in which prominent female characters found themselves unintentionally pregnant, decided against having an abortion, and then were relieved of the consequences of that decision through miscarriage or the discovery that they were not pregnant after all. Fans of MTM productions, and their liberal values (Feuer, 1984b), would have been introduced to abortion in a different manner. On *Cagney & Lacey, Hill Street Blues,* and *St. Elsewhere,* professionals supported the choices of transitory female characters to have abortions, and confronted the violence of the Pro-life movement. Finally, viewers might have first encountered televised abortion in a more sharply conflicted manner through *Spenser for Hire, L.A. Law,* or the second episode of *St. Elsewhere.* In these programs, central women characters made highly contested choices to undergo abortions.

Prime time television thus introduced the public to the practice of abortion with a polysemic voice. The mass mediated *message* itself appeared to bring different textual resources to different audiences. As I have previously argued, however, this textual polysemy had very clear limits (Condit, 1987). Regardless of whether the program was primarily "pro" or "anti," abortion was portrayed as a morally problematic act that was, nonetheless, the woman's choice. Although female characters decided in favor of and against abortions in a wide variety of problem situations, the abortions presented in prime time were never those of women in optimal reproductive situations. Women in caring, financially secure marriages did not abort healthy fetuses. Moreover, the practicalities of abortion were absent. There was no direct mention of the problem of payment, the pain of the operation, or the real but difficult alternatives of adoption or contraception.

As a consequence, dominant group vocabularies and practices were normalized (Condit, in press). Career women could get abortions and feel more comfortable with the practice, even though their role or obligation as mothers was not erased. This was both an attractive enactment of career women's own reproductive practices and a discursive instantiation of their "choices" in the public vocabulary. The power distributed through such reinforcement is immense; it is virtually the social glue that allows dominant groups to coordinate their efforts in a democracy and thereby maintain power. Moreover, the reinforcement shields dominant groups from understanding the ways that different conditions might make different practices necessary or right for others. In the face of such a public culture, it was relatively easy for the Reagan administration, in its second term, to withdraw virtually all indirect financial support of abortion *and* of family planning in both the national and international arenas.

In contrast, prime time television neither informed the poor about how to finance abortions nor told the young how to avoid needing them and why they might want to avoid such a need. No constructive efforts on their behalf provided useful information or created pleasurable self-validation for these other groups of women. Hence, even if other groups were active interpreters of these programs, in order to seek legitimacy or cultural sympathy for their own practices they would have had to do double work—deconstructing the dominant code and reconstructing their own. In addition, to effect favorable policies, marginal groups also would have had to make a public argument in some other, *less pleasurable* arena, counterposing their interests and vocabulary to this now-dominant vocabulary. Finally, even if they were able to present an equally attractive argument, they would still, at best, be able to win a compromise with this already legitimated dominant code (a position they might not have faced absent its broadcast).

In sum, television disseminates and legitimates, in a pleasurable fashion, a political vocabulary that favors certain interests and groups over others, even if by no other means than consolidating the dominant audience by giving presence to their codes. Given the interest of advertisers in dominant economic groups, the ability of marginal groups to break this grip seems particularly unlikely. Fiske's conclusion (1987c, p. 319) that homogenization will lead to the inclusion of these other groups presumes much about the demographics of television audiences that is yet to be established. It also rests on imprecise definitions of "the popular" which do not seem to distinguish who the dominant elites are (the rich or the middle and well-to-do working class [e.g., in automotive unions]?) and who the "resisting populace" might be (secretaries or the unemployed?). Further examination of that relationship will require more careful studies of the economic side of this question. If, however, maximal economic return can be purchased through appeal to dominant audiences, then the fact that programs also attract oppositional readers around the globe may be only of minimal importance. In short, the jury is still out on the "popularity" of the mass media.

A second political consequence of television's coding of abortion practices has to do with the dissemination of new information to individuals. It can be explored through a turn to the third component of rhetorical events.

## *Occasions: Historical Agents*

Historical agents are embedded in particular occasions with specific power relationships, communicated through ideologies. Recent interpretations of ideology have begun to explore its character as information (Foucault, 1972; Lyotard, 1984; Sholle, 1988). In place of the old "ideology versus science" equation, some analyses suggest that one of the primary ways that ideology functions is by making present or dominant certain pieces of information to certain audiences. On this account, one important function of the broadcast of the *Cagney & Lacey* abortion episode was the degree to which it gave access to new and useful information about the practice of abortion.

To make such an evaluation of the *Cagney & Lacey* episode requires an accounting of the historical situation and self-consciousness about criteria. In the mid-eighties, it was clear that the legality of abortion was widely shared knowledge. Less widely shared was information of many kinds: about the types of women who have abortion and their reasons, about the experience of the operation, about women's control over their sexuality and fertility, and perhaps about the character of the fetus. *Cagney & Lacey* distributed some of this information (especially about the wide variety of "good" women who had abortions) but not others (the character of the operation and of the fetus).

The social impact of the program was in part a matter of the particular information it disseminated to different groups, even to groups able to decode the program through their own value structures. Jack, for example, was forced by the program's presence to confront the fact of abortion's "so-called social acceptability by too many people." Television programs distribute varying sorts of information about abortion, even to viewers who wish to change that practice and who actively and negatively decode the program.

Evaluating the impact of *Cagney & Lacey* on this learning dimension might seem to imply survey research, but that approach is unlikely to be cost effective. Research in the "direct effects" tradition of rhetorical studies indicated the virtual impossibility of quantitatively tracking learning and persuasion impacts on large audiences (e.g., Baran and Davis, 1975). Most important, in historical studies, scholars can never go back and get the kind of data that would meet the tests of quantitative-style knowledge claims. Further, academics are rarely prescient enough to know what programming is important with enough lead time to prepare for such surveys. Knowledge claims thus must be critically based.

Historically based evaluations need to take into account a more sensitive gauge than has been applied previously (and this might well be the most important moral of Radway's work on romance novels). Rather than describing a text and its readings simply as good or bad, critics need to develop judgments of better and worse. From this perspective, *Cagney & Lacey* should be evaluated on comparative grounds. First, it should be placed as the earliest of the second wave of the televisualization of abortion. Second, it should be compared to other programs and entertainment media. On this scale, the episode was far more conservative in the amount of information it provided than *St. Elsewhere,* with its far greater detail about the experience and emotions of having an abortion and inclusion of the issue of contraception. However, it was far more informative than episodes such as *Webster* or *Call to Glory,* neither of which ever directly even named abortion as "the option" nor dealt with the consequences.

Such an evaluation process will lead not to a condemnation or simple praise of a program but to a calibrated understanding of the particular role it played in introducing certain limited pieces of information to different ranges of audiences at different times. Critical analysis should therefore, at least at times, be rhetorical; it should be tied to the particularity of occasions: specific audiences, with specific codes or knowledges, addressed by specific programs and episodes (McGee, 1982; Wichelns, 1972). Such an approach does not deny the wisdom of also exploring the intertextuality of programs, the stripped character of the viewing experience (Newcomb, 1984, p. 44), and the disengaged character of much viewing. It merely adds one additional vector to our understanding.

### Evaluation

After considering the historical moment, the public code constructed, and the range of audience readings, we might be in a position to provide an evaluation of *Cagney & Lacey.* I wish to turn that evaluation to the key criterion on which I see Fiske, Hall, Morley, Radway, and others (but probably not Newcomb) converging, that is, the judgment of a mass communication event based on its "resistance" to the dominant ideology. This judgmental criterion rests on the assumption that academics have a duty to the society that pays their salaries to try to produce a better world. This is a duty widely accepted for the ever more technically oriented scientists, although with admittedly different procedures. In the humanities and social sciences, however, the execution of that criterion is eternally and politically controversial, and that deters us from encouraging scholars in communication studies to undertake endeavors of a sort we virtually demand from scholars in natural studies. I nonetheless support such efforts.

For many years, critics interested in bringing about positive social changes assumed that the deconstruction of the dominant ideologies contained in popular and political texts was the best contribution toward human progress. This kind of criticism gradually became too predictable to suit the tastes of an academic machine that voraciously devours "new ideas" in preference to the good execution of old ones. Furthermore, at its worst, and too frequently, such criticism merely imposed the ideology/methodology of a particular political preference upon dominant texts, threatening to produce nothing but a blanket condemnation of the status quo rather than insight into how to improve society.

Today, with the rise of attention to audiences, such a textual approach has come under further attack. Fiske (1987c, p. 64; see also 1987b) writes, for example:

> Textual studies of television now have to stop treating it as a closed text, that is, as one where the dominant ideology exerts considerable, if not

total, influence over its ideological structure and therefore over its reader. Analysis has to pay less attention to the textual strategies of preference or closure and more to the gaps and spaces that open up to meanings not preferred by the textual structure.

In placing enormous faith in the capacity of audiences to resist, however, a similar blindness may be on its way to being produced on the other side. We can endlessly generate studies that demonstrate that clever readers can take pleasure in reconstructing texts, but this does not certify that mass communication in general functions as a force for positive social change.

The assumption that pleasure liberates is too simplistic on a myriad of counts. To begin with, Fiske's argument (1987c, p. 19) is based on the premise that "Pleasure results from a particular relationship between meanings and power. Pleasure for the subordinate is produced by the assertion of one's social identity in resistance to, in independence of, or in negotiation with, the structure of domination." This is a flat assertion with no support. It is based on the claim that "escape" is always escape from the dominant ideologies' subjective positioning of the marginal person (p. 317). While Fiske documents that this kind of escape can and does sometimes occur, he does not demonstrate that it is the only or primary kind of pleasure to be gained from a text by a subordinate. There are a wide variety of pleasures; some of them are merely temporary escape from truly painful thoughts and activities, and these do not challenge the subjective identity television programs present. The most important of these pleasures is what Kenneth Burke (1969, p. 19) has called "identification." One can fully identify with the rich patrons of *Dynasty*, enjoying the vicarious experience of opulence, without building any oppositional identity. I have revelled in such play, the pleasure coming from a temporary "giving in" rather than from resistance. My female career-oriented students generally admit relishing the Cinderella myth offered by *An Officer and a Gentleman*. Such pleasurable identification does not require that we naively confuse reality and our own position (a different thesis which Fiske [1987c, pp. 44–47, 63–72] argues against forcefully and accurately). We know that we are not as rich as Krystal and will never be. Nonetheless, we can enjoy playing as if we were. This kind of pleasure offers only temporary escape.

I would not willingly deny any of us such pleasures. Human life is hard, under capitalism or any other system human beings have yet devised. Radway's Smithton readers need a pleasurable escape from their oppressive husbands and demanding children. However, we should be very cautious about our portrayal of such escape as liberating. Attention to the discrepancies between critical readings of television's embedding of subjects in patriarchy and those subjects' own readings (the opening of Radway's book) should not obscure the realization that both personal pleasure and collective domination can go on at the same time (the conclusion of Radway's book). We need to make a clear distinction between the personal or "private" experience of pleasure which temporarily liberates us from the painful conditions of our lives and the collectivized pleasures which, in the right historical conditions, may move us toward changing those conditions. Because of the character of the mass media, both are social pleasures, but *collectivized* (grouped, internally organized through communication production) action and pleasure are essential to social change. Alterations in subjectivity may indeed provide a first step in that latter process, but it is an extremely limited step, and it is not the case that all pleasurable readings produce such resisting subjectivities (Sholle, 1988, p. 33). Moreover, if the cost of mildly altered subjectivities is complacence, the potential for change may be offset. Television does not, therefore, simply offer "a set of forces for social change" (Fiske, 1987c, p. 326); television is engaged in a set of social forces within which actors may or may not promote social change.

To assess the social consequences of a mass communication event requires, consequently, that we dispense with the totalized concept of "resistance." It is not enough to describe a program or an interpretation of a program as oppositional. It is essential to describe what particular things are resisted and how that resistance occurs. In part, this requires taking more seriously the melding of liberal interest group theory and Marxism evident in Fiske's work (1987c, p. 16). Fiske's explicit political theory dismantles views of politics that portray it either as an evenhanded barter between various interest groups (the classic liberal account) or as the dominance of a unified, all-powerful elite. Instead, he argues (1987c, p. 16), as do I (in press), that politics is a battle and barter among a wide range of groups, each of which is differently and unevenly empowered. Unfortunately, like most other audience theorists, Fiske does not carry this theory through into his analysis. Instead, he reduces the multiplicity of differently empowered groups to "the dominant" and "the resistant." Such a totalized concept of resistance from a system is at odds with a theory that posits a wide range of groups with a wide range of investments in the system they share. Given that perspective, for my interpretation of *Cagney & Lacey*, I offer the following evaluations.

From the perspective of women like Jill, the decisions by the production team headed by Barney Rosenzweig, which resulted in this particular treatment of abortion on prime time television, were mildly progressive. Jill's interpretation needs to be supplemented by that of other women, but for her the program portrayed powerful characterizations of "good" women having abortions and reaffirmed the evaluation that abortion was not a repudiation of familial love. Most important, it affirmed that even though abortion is the morally problematic termination of the potential of a growing creature's life, it is always the woman who must weigh the principles and factors involved to make the decision. This is perhaps surprisingly mild progressive ideological work for a production team that dealt in outstanding detail with the experience of rape and that treated the fallout of the AIDS crisis on single adults with gingerly directness. The program, however, was the leader of the second wave of telecasts and took a great deal of public criticism even for these steps. For Jill's group, it accomplished some important ideological work.

For women in poverty and women of color the program is more mixed. It explicitly affirmed the choices of a particular minority woman, but it did not deal with the ways in which poor women might fund abortion or contraception. It did not deal with the options provided by extended families or with the importance of motherhood in different cultures. It offered a sugary and unrealistic moral, "have an abortion so you can go to school and get off welfare," that may have appealed to latent racism in white audiences more than assisting poor women with real options. In the face of such silences, the Republican administration could continue its largely hidden work in pro-natalism by dismantling funding for family planning. From the perspective of these groups of viewers, this restricted presentation of abortion represents a serious political shortcoming of this episode.

The situation is much grimmer from Jack's perspective as a clearly marginal viewer of this text, and in many ways a person whom I sensed to be involved in popular culture (especially sports), but disempowered by the dominant political economy. I find it difficult to argue that Jack found his reading of this text, resistant and skilled as it was, to be either a predominantly pleasurable or liberating experience. Jack expressed the following general response to the program's significance: "I think it's a [sic] pretty much a devastating blow, not that it's totally going to stop the movement, but it set us back." For Jack, as for other relatively unempowered males, especially of Pro-life positions, *Cagney & Lacey* did not promote the social changes they preferred. Even their resistant readings left them with the feeling of oppression by the media.

*Cagney & Lacey*'s broadcast about abortion broke new ideological ground, inserting new political codes into the public culture. It was thus a progressive but not radical text that tended to oppose the interests of marginally positioned traditional males. It favored the interests of career women but only marginally supported other groups of women.

I have, of course, stacked the deck here by probing readings that scramble the left's general presumption that marginal readers of texts are the potential source of liberation, the groups with whom we, as academics, ought to identify and praise. I have done so to heighten my point that "resistance" and the metaphor of a "dominant system" is a bad way to phrase what it is those interested in social change should praise. History creates "hegemonies," but hegemonies are not equivalent to dominant ideologies. A hegemony is a negotiation among elite and nonelite groups and therefore always contains interests of nonelite groups, though to a lesser degree. To resist the power of dominant groups may be safe, but to resist the hegemony that is constructed in negotiation with those groups is always also to resist what is partially of one's own interests. The totalizing concept of resistance should give way to the recognition and analysis of historically particular acts in order to bring about specific social changes. This shift will require academics to affirm particular goals rather than simply to critique that which is.

## Conclusions

Recent reemphasis on the audience as an important component of what happens in the process of mass communicating is a useful redress of an old imbalance. We should avoid, however, totalizing the audience's abilities. The receiver's political power in mass mediated societies is dependent upon a complex balance of historically particular forces which include the relative abilities of popular groups and their access to oppositional codes, the work/ pleasure ratio of the available range of the media's intertextual polysemy, the modifications programs make in the dominant code, and the degree of empowerment provided to dominant audiences.

To scholars, this balance of forces presents a series of challenges. There is a need to explore more precisely the relative decoding abilities of audiences and their access to counter-rhetorics. There is also a need to continue to explore what texts "make present," even without regard to their "seams," through careful historically grounded studies of the particular issue contents of television programming. There is, finally, a need to explore the "occasion" of a discourse in terms other than the family viewing context (contexts emphasized in Fiske [1987c, p. 239] and Morley [1986, p. 14]). Different families and different members of families are always embedded in larger political occasions that create collective experiences across family walls. Unless we ask about the particular contents of particular sets of programs, the relationship of those contents to the stasis of the issue for viewers and for the larger society at the time of broadcast, we will not be able to assess fully television's roles in the process of social change for its various constituencies.

There are additional implications for scholars as teachers. One of the primary ways through which we can bring about positive social change is through our teaching of undergraduates for whom our arcane battles about research protocols are rightfully boring and meaningless. For our students, decoding alternatives, through painful effort, can become pleasurable resources they can use throughout life. A perspective that emphasizes the receiver's placement within a complexly balanced process suggests the need to continue to use classrooms to teach students a range of decodings for possible texts, a project that may include increasing their ideological range (the ability to see *An Officer and a Gentleman* as *Cinderella*, Sonny Crocket as a 1980s John Wayne, *Dallas* as the costs inherent to capitalism). It might also include familiarizing students with the history of the various issue contents of the mass media. Studies of the participation of news and entertainment programming, in particular social movements and issues, might be added to genre studies and analyses of private audiences (e.g., Hallin, 1986; Rushing, 1986a, 1986b).

As a whole, the effort to gain a more variegated picture of audiences is an important one. However, the tendency to isolate the audience from the communication process and then pronounce the social effects of mass communication based on the ability of some receivers to experience pleasure in producing oppositional decoding is undesirable. It simply repeats the error of message-dominated research which attempted to describe the mass media's influence solely by investigating texts (or, in other research strands, presumed intents of sources). Audience members are neither simply resistive nor dupes. They neither find television simply pleasurable, simply an escape, nor simply obnoxious and oppressive. The audience's variability is a consequence of the fact that humans, in their inherent character as audiences, are inevitably situated in a communication *system,* of which they are a part, and hence have some influence within, but by which they are also influenced. To study the role of that communication system in the processes that change our humanity and the system itself therefore requires a multiplicity of approaches to the critical analysis of the massive media.

## Notes

1. I am aware that to locate intent in television programs is a difficult matter because of the multiplicity of inputs into such productions. However, this multiplicity does not negate the fact that messages have sources and therefore some collection of intended meanings. To abrogate the use of the term simply because intent is complex would be to ignore an important component of the communication process.

2. Radway (1984; 1986) begins such a delimitation with regard to her case study of romance readers.

3. I choose the terms "Pro-life" and "Pro-choice" because they are the names employed by the members of the respective movements to define themselves.

## References

Baran, S. J., and Davis, D. K. (1975). The audience of public television: Did Watergate make a difference? *Central States Speech Journal,* 26: 93–98.

Bennett, T., and Woollacott, J. (1987). *Bond and beyond: The political career of a popular hero.* New York: Methuen.

Bitzer, L. F. (1987). Rhetorical public communication. *Critical Studies in Mass Communication,* 4: 425–28.

Burke, J., Wilson, H., & Agardy, S. (1983). *A Country Practice and the child audience—A case study.* Melbourne: Australian Broadcasting Tribunal.

Burke, K. (1966). Definition of man. In *Language as symbolic action* (pp. 3–24). Berkeley: University of California Press.

——— (1969). Identification. In *A rhetoric of motives* (pp. 55–59). Berkeley: University of California Press.

Condit, C. (1987). Abortion on television: The "system" and ideological production. *Journal of Communication Inquiry,* 11: 47–60.

——— (in press). *Decoding abortion rhetoric: Communicating social change.* Urbana: University of Illinois Press.

Eco, U. (1979). Denotation and connotation. In *A theory of semiotics* (pp. 54–57). Bloomington: Indiana University Press.

Feuer, J. (1984a). MTM enterprises. An overview. In J. Feuer, P. Kerr, and

T. Vahimagi (eds.), *MTM: "Quality television"* (pp. 1–31). London: British Film Institute.

——— (1984b). The MTM style. In J. Feuer, P. Kerr, and T. Vahimagi (eds.), *MTM: "Quality Television"* (pp. 32–60). London: British Film Institute.

Fiske, J. (1986). Television: Polysemy and popularity. *Critical Studies in Mass Communication,* 3: 391–408.

——— (1987a). British cultural studies and television. In R. C. Allen (ed.), *Channels of discourse* (pp. 254–89). Chapel Hill: University of North Carolina Press.

——— (1987b). *Cagney and Lacey:* Reading character structurally and politically. *Communication,* 9: 399–426.

——— (1987c). *Television culture.* New York: Methuen.

Foucault, M. (1972). *The archaeology of knowledge.* New York: Pantheon.

Goodnight, G. T. (1987). Public discourse. *Critical Studies in Mass Communication,* 4: 428–32.

Hall, S. (1980) Encoding/decoding. In S. Hall, D. Hobson, A. Lowe, and P. Willis (eds.), *Culture, media, language* (pp. 128–38). London: Hutchinson.

Hallin, D. C. (1986). *The "uncensored war": The media and Vietnam.* New York: Oxford University Press.

Hauser, G. A. (1987). Features of the public sphere. *Critical Studies in Mass Communication,* 4: 437–41.

Hobson, D. (1982). *Crossroads: The drama of a soap opera.* London: Methuen.

Katz, E., and Liebes, T. (1984). Once upon a time in *Dallas. Intermedia,* 12: 28–32.

Kellner, D. (1982). TV, ideology, and emancipatory popular culture. In H. Newcomb (ed.), *Television: The critical view* (3d ed., pp. 386–421). New York: Oxford University Press.

Kerr, P. (1984). The making of (the) MTM (show). In J. Feuer, P. Kerr, and T. Vahimagi (eds.), *MTM: "Quality television"* (pp. 61–98). London: British Film Institute.

Lucaites, J., and Condit, C. (1986, November). *Equality in the martyrd black vision.* Paper presented at the meeting of the Speech Communication Association, Chicago.

Luker, K. (1984). *Abortion and the politics of motherhood.* Berkeley: University of California Press.

Lyotard, J. F. (1984). *The postmodern condition: A report on knowledge.* Minneapolis: University of Minnesota Press.

McGee, M. (1982). A materialist's conception of rhetoric. In R. E. McKerrow (ed.), *Explorations in rhetoric* (pp. 23–48). Scott, Foresman and Company.

McGee, M. C. (1987). Power to "the people." *Critical Studies in Mass Communication,* 4: 432–37.

Meehan, E. R. (1986). Conceptualizing culture as commodity: The problem of television. *Critical Studies in Mass Communication,* 3: 448–57.

Meyrowitz, J. (1985). *No sense of place.* New York: Oxford University Press.

Morley, D. (1980). *The "Nationwide" audience: Structure and decoding.* London: British Film Institute.

——— (1986). *Family television: Cultural power and domestic leisure.* London: Comedia.

Newcomb, H. (1984). On the dialogic aspects of mass communication. *Critical Studies in Mass Communication,* 1: 34–50.

Palmer, P. (1986). *The lively audience: A study of children around the TV set*. Sydney: Allen & Unwin.

Perelman, P., and Olbrechts-Tyteca, L. (1971). *The new rhetoric: A treatise on argumentation*. Notre Dame: University of Notre Dame Press.

Radway, J. (1984). *Reading the romance: Woman, patriarchy, and popular literature*. Chapel Hill: University of North Carolina Press.

———— (1986). Identifying ideological seams: Mass culture, analytical method, and political practice. *Communication*, 9: 93–123.

Rushing, J. (1986a). Mythic evolution of "The new frontier" in mass mediated rhetoric. *Critical Studies in Mass Communication*, 3: 265–96.

———— (1986b). Ronald Reagan's "Star Wars" address: Mythic containment of technical reasoning. *Quarterly Journal of Speech*, 72: 415–33.

Sholle, D. J. (1988). Critical studies: From the theory of ideology to power/knowledge. *Critical Studies in Mass Communication*, 5: 16–41.

Steiner, L. (1988). Oppositional decoding as an act of resistance. *Critical Studies in Mass Communication*, 5: 1–15.

Wander, P. (1984). The third persona: An ideological turn in rhetorical theory. *Central States Speech Journal*, 35: 197–216.

Wichelns, H. (1972). The literary criticism of oratory. In R. L. Scott and B. Brock (eds.), *Methods of rhetorical criticism: A twentieth century perspective* (pp. 27–60). New York: Harper & Row.

Young blacks have higher voting rate than 18–24 whites. (1987, October 7). *The Champaign-Urbana News-Gazette*, p. A13.

# *Star Trek* Rerun, Reread, Rewritten: Fan Writing as Textual Poaching

## HENRY JENKINS III

In late December 1986, *Newsweek* (Leerhsen, 1986, p. 66) marked the twentieth anniversary of *Star Trek* with a cover story on the program's fans, "the Trekkies, who love nothing more than to watch the same seventy-nine episodes over and over." The *Newsweek* article, with its relentless focus on conscpicuous consumption and "infantile" behavior and its patronizing language and smug superiority to all fan activity, is a textbook example of the stereotyped representation of fans found in both popular writing and academic criticism, "Hang on: You are being beamed to one of those *Star Trek* conventions, where grown-ups greet each other with the Vulcan salute and offer in reverent tones to pay $100 for the autobiography of Leonard Nimoy" (p. 66). Fans are characterized as "kooks" obsessed with trivia, celebrities, and collectibles; as misfits and crazies; as "a lot of overweight women, a lot of divorced and single women" (p. 68). Borrowing heavily from pop Freud, ersatz Adorno, and pulp sociology, *Newsweek* explains the "Trekkie phenomenon" in terms of repetition compulsion, infantile regression, commodity fetishism, nostalgic complacency, and future shock. Perhaps most telling, *Newsweek* consistently treats *Trek* fans as a problem to be solved, a mystery to be understood, rather than as a type of cultural activity that many find satisfying and pleasurable.[1]

From *Critical Studies in Mass Communication,* Jenkins, Henry III. "*Star Trek* Rerun, Reread, Rewritten: Fan writing as Textual Poaching." June 1988. Pages 85–107. Copyright by the Speech Communication Association. Reprinted with permission of the publisher.

Academic writers depict fans in many of the same terms. For Robin Wood (1986, p. 164), the fantasy film fan is "reconstructed as a child, surrendering to the reactivation of a set of values and structures [the] adult self has long since repudiated." The fan is trapped within a repetition compulsion similar to that which an infant experiences through the *fort/da* game. A return to such "banal" texts could not possibly be warranted by their intellectual content but can only be motivated by a return to "the lost breast" (p. 169), by the need for reassurance provided by the passive reexperience of familiar pleasures. "The pleasure offered by the *Star Wars* films corresponds very closely to our basic conditioning; it is extremely reactionary, as all mindless and automatic pleasure tends to be. The finer pleasures are those we have to work for" (p. 164). Wood valorizes academically respectable texts and reading practices at the expense of popular works and their fans. Academic rereading produces new insights; fan rereading rehashes old experiences.[2]

As these two articles illustrate, the fan constitutes a scandalous category in contemporary American culture, one that provokes an excessive response from those committed to the interests of textual producers and institutionalized interpreters and calls into question the logic by which others order their aesthetic experiences. Fans appear to be frighteningly out of control, undisciplined and unrepentant, rogue readers. Rejecting aesthetic distance, fans passionately embrace favored texts and attempt to integrate media representations within their own social experience. Like cultural scavengers, fans reclaim works that others regard as worthless and trash, finding them a rewarding source of popular capital. Like rebellious children, fans refuse to read by the rules imposed upon them by the schoolmasters. For fans, reading becomes a type of play, responsive only to its own loosely structured rules and generating its own types of pleasure.

Michel de Certeau (1984) has characterized this type of reading as "poaching," an impertinent raid on the literary preserve that takes away only those things that seem useful or pleasurable to the reader. "Far from being writers . . . readers are travellers; they move across lands belonging to someone else, like nomads poaching their way across fields they did not write, despoiling the wealth of Egypt to enjoy it themselves" (p. 174). De Certeau perceives popular reading as a series of "advances and retreats, tactics and games played with the text" (p. 175), as a type of cultural bricolage through which readers fragment texts and reassemble the broken shards according to their own blueprint, salvaging bits and pieces of found material in making sense of their own social experience. Far from viewing consumption as imposing meanings upon the public, de Certeau suggests, consumption involves reclaiming textual material, "making it one's own, appropriating or reappropriating it" (p. 166).

Yet, such wanton conduct cannot be sanctioned; it must be contained, through ridicule if necessary, since it challenges the very notion of literature as a type of private property to be controlled by textual producers and their academic interpreters. Public attacks on media fans keep other viewers in line, making it uncomfortable for readers to adopt such inappropriate strategies. One woman recalled the negative impact popular representations of the fan had on her early cultural life:

> Journalists and photographers always went for the people furthest out of mainstream humanity . . . showing the reader the handicapped, the very obese, the strange and the childish in order to "entertain" the "average reader." Of course, a teenager very unsure of herself and already labeled "weird" would run in panic. (Ludlow, 1987, p. 17)

Such representations isolate potential fans from others who share common interests and reading practices and marginalize fan-related activities as outside the mainstream and beneath dignity. These same stereotypes reassure academic writers of the validity of their own interpretations of the program content, readings made in conformity with established critical protocols, and free them from any need to come into direct contact with the program's crazed followers.[3]

In this essay, I propose an alternative approach to fan experience, one that perceives "Trekkers" (as they prefer to be called) not as cultural dupes, social misfits, or mindless consumers but rather as, in de Certeau's term, "poachers" of textual meanings. Behind the exotic stereotypes fostered by the media lies a largely unexplored terrain of cultural activity, a subterranean network of readers and writers who remake programs in their own image. "Fandom" is a vehicle for marginalized subcultural groups (women, the young, gays, etc.) to pry open space for their cultural concerns within dominant representations; it is a way of appropriating media texts and rereading them in a fashion that serves different interests, a way of transforming mass culture into a popular culture.

I do not believe this essay represents the last word on *Star Trek* fans, a cultural community that is far too multivocal to be open to easy description. Rather, I explore some aspects of current fan activity that seem particularly relevant to cultural studies. My primary concern is with what happens when these fans produce their own texts, texts that inflect program content with their own social experience and displace commercially produced commodities for a kind of popular economy. For these fans, *Star Trek* is not simply something that can be reread; it is something that can and must be rewritten in order to make it more responsive to their needs, in order to make it a better producer of personal meanings and pleasures.

No legalistic notion of literary property can adequately constrain the rapid proliferation of meanings surrounding a popular text. Yet, there are other constraints, ethical constraints and self-imposed rules, that are enacted by the fans, either individually or as part of a larger community, in response to their felt need to legitimate their unorthodox appropriation of mass media texts. E. P. Thompson (1971) suggests that eighteenth and nineteenth century peasant leaders, the historical poachers behind de Certeau's apt metaphor, responded to a kind of "moral economy," an informal set of consensual norms that justified their uprisings against the landowners and tax collectors in order to restore a preexisting order being corrupted by its avowed protectors. Similarly, the fans often cast themselves not as poachers but as loyalists, rescuing essential elements of the primary text misused by those who maintain copyright control over the program materials. Respecting literary property even as they seek to appropriate it for their own uses, these fans become reluctant poachers, hesitant about their relationship to the program text, uneasy about the degree of manipulation they can legitimately perform on its materials, and policing each other for abuses of their interpretive license. They wander across a terrain pockmarked with confusions and contradictions. These ambiguities become transparent when fan writing is examined as a particular type of reader-text interaction. My discussion consequently has a double focus: first, I discuss how the fans force the primary text to accommodate their own interests, and then I reconsider the issue of literary property rights in light of the moral economy of the fan community.

## Fans: From Reading to Writing

The popularity of *Star Trek* has motivated a wide range of cultural productions and creative reworkings of program materials: from children's backyard play to adult interaction games, from needlework to elaborate costumes, from private fantasies to computer programming. This ability to transform personal reaction into social interaction, spectator culture into participatory culture, is one of the central characteristics of fandom. One becomes a fan not by being a regular viewer of a particular program but by translating that viewing into some type of cultural activity, by sharing feelings and thoughts about the program content with friends, by joining a community of other fans who share common interests. For fans, consumption sparks production, reading generates writing, until the terms seem logically inseparable. In fan writer Jean Lorrah's words (1984, p. 1):

> Trekfandom . . . is friends and letters and crafts and fanzines and trivia and costumes and artwork and filksongs [fan parodies] and buttons and film clips and conventions—something for everybody who has in common the inspiration of a television show which grew far beyond its TV and film incarnations to become a living part of world culture.

Lorrah's description blurs all boundaries between producers and consumers, spectators and participants, the commercial and the home crafted, to construct an image of fandom as a cultural and social network that spans the globe.

Many fans characterize their entry into fandom in terms of a movement from social and cultural isolation, doubly imposed upon them as women within a patriarchal society and as seekers after alternative pleasures within dominant media representations, toward more and more active participation in a community receptive to their cultural productions, a community where they may feel a sense of belonging. One fan recalls:

> I met one girl who liked some of the TV shows I liked . . . but I was otherwise a bookworm, no friends, working in the school library. Then my friend and I met some other girls a grade ahead of us but ga-ga over *ST*. From the beginning, we met each Friday night at one of the two homes that had a color TV to watch *Star Trek* together. . . . . Silence was mandatory except during commercials, and, afterwards, we "discussed" each episode. We re-wrote each story and corrected the wrongs done to "Our Guys" by the writers. We memorized bits of dialog. We even started to write our own adventures. (Caruthers-Montgomery, 1987, p. 8)

Some fans are drawn gradually from intimate interactions with others who live near them toward participation in a broader network of fans who attend regional, national, and even international science fiction conventions. One fan writes of her first convention: "I have been to so many conventions since those days, but this one was the ultimate experience. I walked into that Lunacon and felt like I had come home without ever realizing I had been lost" (Deneroff, 1987, p. 3). Another remarks simply, "I met folks who were just as nuts as I was, I had a wonderful time" (Lay, 1987, p. 15).

For some women, trapped within low paying jobs or within the socially isolated sphere of the homemaker, participation within a national, or international, network of fans grants a degree of dignity and respect otherwise lacking. For others, fandom offers a training ground for the development of professional skills and an outlet for creative impulses constrained by their workday lives. Fan slang draws a sharp contrast between the mundane, the realm of everyday experience and those who dwell exclusively within that space, and fandom, an alternative sphere of cultural experience that restores the excitement and freedom that must be repressed to function in ordinary life. One fan writes, "Not only does 'mundane' mean 'everyday life,' it is also a term used to describe narrow-minded, pettiness, judgmental, conformity, and a shallow and silly nature. It is used by people who feel very alienated from society" (Osborne, 1987, p. 4). To enter fandom is to escape from the mundane into the marvelous.

The need to maintain contact with these new friends, often scattered over a broad geographic area, can require that speculations and fantasies about the program content take written form, first as personal letters and later as more public newsletters, "letterzines," or fan fiction magazines. Fan viewers become fan writers.

Over the twenty years since *Star Trek* was first aired, fan writing has achieved a semi-institutional status. Fan magazines, sometimes hand typed, photocopied, and stapled, other times offset printed and commercially bound, are distributed through the mails and sold at conventions, frequently reaching an international readership. *Writer's Digest* (Cooper, 1987) recently estimated that there were more than 300 amateur press publications that regularly allowed fans to explore aspects of their favorite films and television programs. Although a wide variety of different media texts have sparked some fan writing, including *Star Wars, Blake's Seven, Battlestar Galactica, Doctor Who, Miami Vice, Road Warrior, Remington Steele, The Man From*

*U.N.C.L.E., Simon and Simon, The A-Team,* and *Hill Street Blues, Star Trek* continues to play the central role within fan writing. *Datazine,* one of several magazines that serve as central clearing houses for information about fanzines, lists some 120 different *Star Trek* centered publications in distribution. Although fanzines may take a variety of forms, fans generally divide them into two major categories: "letterzines" that publish short articles and letters from fans on issues surrounding their favorite shows and "fiction-zines" that publish short stories, poems,and novels concerning the program characters and concepts.[4] Some fan-produced novels, notably the works of Jean Lorrah (1976a, 1978) and Jacqueline Lichtenberg (1976), have achieved a canonized status in the fan community, remaining more or less in constant demand for more than a decade.[5]

It is important to be careful in distinguishing between these fan-generated materials and commercially produced works, such as the series of *Star Trek* novels released by Pocket Books under the official supervision of Paramount, the studio that owns the rights to the *Star Trek* characters. Fanzines are totally unauthorized by the program producers and face the constant threat of legal action for their open violation of the producer's copyright authority over the show's characters and concepts. Paramount has tended to treat fan magazines with benign neglect as long as they are handled on an exclusively nonprofit basis. Producer Gene Roddenberry and many of the cast members have contributed to such magazines. Bantam Books even released several anthologies showcasing the work of *Star Trek* fan writers (Marshak and Culbreath, 1978).

Other producers have not been as kind. Lucasfilm initially sought to control *Star Wars* fan publications, seeing them as a rival to its officially sponsored fan organization, and later threatened to prosecute editors who published works that violated the "family values" associated with the original films. Such a scheme has met considerable resistance from the fan community that generally regards Lucas's actions as unwarranted interference in its own creative activity. Several fanzine editors have continued to distribute adult-oriented *Star Wars* stories through an underground network of special friends, even though such works are no longer publicly advertised through *Datazine* or sold openly at conventions. A heated editorial in *Slaysu,* a fanzine that routinely published feminist-inflected erotica set in various media universes, reflects these writers' opinions:

> Lucasfilm is saying, "you must enjoy the characters of the *Star Wars* universe for male reasons. Your sexuality must be correct and proper by my (male) definition." I am not male. I do not want to be. I refuse to be a poor imitation, or worse, someone's idiotic ideal of femininity. Lucasfilm has said, in essence, "this is what we see in the *Star Wars* films and we are telling you that this is what you will see." (Siebert, 1982, p. 44)

C. A. Siebert's editorial asserts the rights of fanzine writers to consciously revise the character of the original texts, to draw elements from dominant culture in order to produce underground art that explicitly challenges patriarchal assumptions. Siebert and the other editors deny the traditional property rights of textual producers in favor of a right of free play with the program materials, a right of readers to use media texts in their own ways and of writers to reconstruct characters in their own terms. Once characters are inserted into popular discourse, regardless of their source of origin, they become the property of the fans who fantasize about them, not the copyright holders who merchandise them. Yet the relationship between fan texts and primary texts is often more complex than Siebert's defiant stance might suggest, and some fans do feel bound by a degree of fidelity to the original series' conceptions of those characters and their interactions.

## Gender and Writing

Fan writing is an almost exclusively feminine response to mass media texts. Men actively participate in a wide range of fan-related activities, notably interactive games and conference planning committees, roles consistent with patriarchal norms that typically relegate combat—even combat fantasies—and organizational authority to the masculine sphere. Fan writers and fanzine readers, however, are almost always female. Camille Bacon-Smith (1986) has estimated that more than 90 percent of all fan writers are female. The greatest percentage of male participation is found in the "letterzines," like *Comlink* and *Treklink,* and in "nonfiction" magazines, like *Trek* that publish speculative essays on aspects of the program universe. Men may feel comfortable joining discussions of future technologies or military lifestyle but not in pondering Vulcan sexuality, McCoy's childhood, or Kirk's love life.

Why this predominance of women within the fan writing community? Research suggests that men and women have been socialized to read for different purposes and in different ways. David Bleich (1986) asked a mixed group of college students to comment, in a free association fashion, on a body of canonized literary works. His analysis of their responses suggests that men focused primarily on narrative organization and authorial intent while women devoted more energy to reconstructing the textual world and understanding the characters. He writes, "Women enter the world of the novel, take it as something 'there' for that purpose; men see the novel as a result of someone's action and construe its meaning or logic in those terms" (p. 239). In a related study, Bleich asked some 120 University of Indiana freshmen to "retell as fully and as accurately as you can [William] Faulkner's 'Barn Burning' " (p. 255) and, again, notes substantial differences between men and women:

> The men retold the story as if the purpose was to deliver a clear simple structure or chain of information: these are the main characters, this is the main action, this is how it turned out. . . . The women present the narrative as if it were an atmosphere or an experience. (p. 256)

Bleich finds that women were more willing to enjoy free play with the story content, making inferences about character relationships that took them well beyond the information explicitly contained within the text. Such data strongly suggest that the practice of fan writing, the compulsion to expand speculations about characters and story events beyond textual boundaries, draws heavily upon the types of interpretive strategies more common to the feminine than to the masculine.

Bleich's observations provide only a partial explanation, since they do not fully account for why many women find it necessary to go beyond the narrative information while most men do not. As Teresa de Lauretis (1982, p. 106) points out, female characters often exist only in the margins of male-centered narratives:

> Medusa and the Sphinx, like the other ancient monsters, have survived inscribed in hero narratives, in someone else's story, not their own; so they are figures or markers of positions—places and topoi—through which the hero and his story move to their destination and to accomplish meaning.

Texts written by and for men yield easy pleasures to their male readers, yet may resist feminine pleasure. To fully enjoy the text, women are often forced to perform a type of intellectual transvesticism, identifying with male characters in opposition to their own cultural experiences or to construct unwritten countertexts through their daydreams or through their oral interaction with other women that allow them to explore their own narrative concerns. This need to reclaim feminine interests from the margins of masculine texts produces endless speculation, speculation that draws the reader well beyond textual boundaries into the domain of the intertextual. Mary Ellen Brown and Linda Barwick (1987) show how women's gossip about soap opera inserts program content into an existing feminine oral culture. Fan writing represents the logical next step in this cultural process: the transformation of oral countertexts into a more tangible form, the translation of verbal speculations into written works that can be shared with a broader circle of women. In order to do so, the women's status must change; no longer simply spectators, these women become textual producers.

Just as women's gossip about soap operas assumes a place within a preexisting feminine oral culture, fan writing adopts forms and functions traditional to women's literary culture. Cheris Kramarae (1981, pp. 3–4) traces the history of women's efforts to "find ways to express themselves outside the dominant modes of expression used by men," to circumvent the ideologically constructed interpretive strategies of masculine literary genres. Kramarae concludes that women have found the greatest room to explore their feelings and ideas within privately circulated letters and diaries and through collective writing projects. Similarly, Carroll Smith-Rosenberg (1985) discusses the ways that the exchange of letters allowed nineteenth century women to maintain close ties with other women, even when separated by great geographic distances and isolated within the narrow confines of Victorian marriage. Such letters provided a covert vehicle for women to explore common concerns and even ridicule the men in their lives. Smith-Rosenberg (p. 45) concludes:

Nineteenth-century women were, as Nathaniel Hawthorne reminds us, "damned scribblers." They spoke endlessly to one another in private letters and journals . . . about religion, gender roles, their sexuality and men's, about prostitution, seduction, and intemperance, about unwanted pregnancies and desired education, about their relation to the family and the family's to the world.

Fan writing, with its circulation conducted largely through the mails, with its marketing mostly a matter of word of mouth, with the often collective construction of fantasy universes, and with its highly confessional tone, clearly follows within that same tradition and serves some of the same functions. The ready-made characters of popular culture provide these women with a set of common references for discussing their similar experiences and feelings with others with whom they may never have enjoyed face-to-face contact. They draw upon these shared points of reference to confront many of the same issues that concerned nineteenth century women: religion, gender roles, sexuality, family, and professional ambition.

## *Why* Star Trek?

While most texts within a male-dominated culture presumably have the capacity to spark some sort of feminine countertext, only certain programs have generated the type of extended written responses characteristic of fandom. Why, then, has the bulk of fan writing centered around science fiction, a genre that Judith Spector (1986, p. 163) argues until recently has been hostile toward women, a genre "by, for and about men of action"? Why has it also engaged other genres like science fiction (the cop show, the detective drama, or the western) that have represented the traditional domain of male readers? Why do these women struggle to reclaim such seemingly unfertile soil when there are so many other texts that more traditionally reflect feminine interests and that feminist media critics are now trying to reclaim for their cause? In short, why *Star Trek*?

Obviously, no single factor can adequately account for all fanzines, a literary form that necessarily involves the translation of homogeneous media texts into a plurality of personal and subcultural responses. One partial explanation, however, might be that traditionally feminine texts (the soap opera, the popular romance, the "women's picture," etc.) do not need as much reworking as science fiction and westerns in order to accommodate the social experience of women. The resistance of such texts to feminist reconstruction may require a greater expenditure of creative effort and therefore may push women toward a more thorough reworking of program materials than so-called feminine texts that can be more easily assimilated or negated.

Another explanation might be that these so-called feminine texts satisfy, at least partially, the desires of traditional women yet fail to meet the needs of more professionally oriented women. A particular fascination of *Star Trek* for these women appears to be rooted in the way that the program seems to hold out a suggestion of nontraditional feminine pleasures, of greater and more active involvement for women within the adventure of professional space travel, while finally reneging on those promises. Sexual equality was an essential component of producer Roddenberry's optimistic vision of the future; a woman, Number One (Majel Barrett), was originally slated to be the Enterprise's second in command. Network executives, however, consistently fought efforts to break with traditional feminine stereotypes, fearing the alienation of more conservative audience members (Whitfield and Roddenberry, 1968). Number One was scratched after the program pilot, but throughout the run of the series women were often cast in nontraditional jobs, everything from Romulan commanders to weapon specialists. The networks, however reluctantly, were offering women a future, a "final frontier" that included them.

Fan writers, though, frequently express dissatisfaction with these women's characterizations within the episodes. In the words of fan writer Pamela Rose (1977, p. 48), "When a woman is a guest star on *Star Trek*, nine out of ten times there is something wrong with her." Rose notes that these female characters have been granted positions of power within the program, only to demonstrate through their erratic emotion-driven conduct that women are unfit to fill such roles. Another fan writer, Toni Lay (1986, p. 15), expresses mixed feelings about *Star Trek*'s social vision:

> It was ahead of its time in some ways, like showing that a Caucasian, all-American, all-male crew was not the only possibility for space travel. Still, the show was sadly deficient in other ways, in particular, its treatment of women. Most of the time, women were referred to as "girls." And women were never shown in a position of authority unless they were aliens, i.e., Deela, T'Pau, Natira, Sylvia, etc. It was like the show was saying "equal opportunity is OK for their women but not for our girls."

Lay states that she felt "devastated" over the repeated failure of the series and the later feature films to give Lieutenant Penda Uhura command duties commensurate with her rank: "When the going gets tough, the tough leave the womenfolk behind" (p. 15). She contends that Uhura and the other women characters should have been given a chance to demonstrate what they could do when confronted by the same types of problems that their male counterparts so heroically overcome. The constant availability of the original episodes through reruns and shifts in the status of women within American society throughout the past two decades have only made these unfulfilled promises more difficult to accept, requiring progressively greater efforts to restructure the program in order to allow it to produce pleasures appropriate to the current reception context.

Indeed, many fan writers characterize themselves as "repairing the damage" caused by the program's inconsistent and often demeaning treatment of its female characters. Jane Land (1986, p. 1), for instance, characterizes her fan novel, *Kista,* as "an attempt to rescue one of *Star Trek*'s female characters [Christine Chapel] from an artificially imposed case of foolishness." Promising to show "the way the future never was," *The Woman's List,* a recently established fanzine with an explicitly feminist orientation, has called for "material dealing with all range of possibilities for women, including: women of color, lesbians, women of alien cultures, and women of all ages and backgrounds." Its editors acknowledge that their publication's project necessarily involves telling the types of stories that network policy blocked from airing when the series was originally produced. A recent flier for that publication explains:

> We hope to raise and explore those questions which the network censors, the television genre, and the prevailing norms of the time made it difficult to address. We believe that both the nature of human interaction and sexual mores and the structure of both families and relationships will have changed by the 23rd century and we are interested in exploring those changes.

Telling such stories requires the stripping away of stereotypically feminine traits. The series characters must be reconceptualized in ways that suggest hidden motivations and interests heretofore unsuspected. They must be reshaped into full-blooded feminist role models. While, in the series, Chapel is defined almost exclusively in terms of her unrequited passion for Spock and her professional subservience to Dr. McCoy, Land represents her as a fiercely independent woman, capable of accepting love only on her own terms, ready to pursue her own ambitions wherever they take her, and outspoken in response to the patronizing attitudes of the command crew. Siebert (1980, p. 33) has performed a similar operation on the character of Lieutenant Uhura, as this passage from one of her stories suggests:

> There were too few men like Spock who saw her as a person. Even Captain Kirk, she smiled, especially Captain Kirk, saw her as a woman first. He let her do certain things but only because military discipline required it. Whenever there was any danger, he tried to protect her. . . . Uhura smiled sadly, she would go on as she had been, outwardly a feminine toy, inwardly a woman who was capable and human.

Here, Siebert attempts to resolve the apparent contradiction created within the series text by Uhura's official status as a command officer and her constant displays of "feminine frailty." Uhura's situation, Siebert suggests, is characteristic of the way that women must mask their actual competency behind traditionally feminine mannerisms within a world dominated by patriarchal assumptions and masculine authority. By rehabilitating Uhura's character in this fashion, Siebert has constructed a vehicle through which she can document the overt and subtle forms of sexual discrimination that an ambitious and determined woman faces as she struggles for a command post in Star Fleet (or for that matter, within a twentieth century corporate board room).

Fan writers like Siebert, Land, and Karen Bates (1982; 1983; 1984), whose novels explore the progression of a Chapel–Spock marriage through

many of the problems encountered by contemporary couples trying to jug-
gle the conflicting demands of career and family, speak directly to the concerns of professional women in a way that more traditionally feminine works
fail to do.[6] These writers create situations where Chapel and Uhura must
heroically overcome the same types of obstacles that challenge their male
counterparts within the primary texts and often discuss directly the types of
personal and professional problems particular to working women. Land's
recent fan novel, *Demeter* (1987), is exemplary in its treatment of the professional life of its central character, Nurse Chapel. Land deftly melds action
sequences with debates about gender relations and professional discrimination, images of command decisions with intimate glimpses of a Spock–
Chapel marriage. An all-woman crew, headed by Uhura and Chapel, are
dispatched on a mission to a feminist separatist space colony under siege
from a pack of intergalactic drug smugglers who regard rape as a manly
sport. In helping the colonists to overpower their would-be assailants, the
women are at last given a chance to demonstrate their professional competence under fire and force Captain Kirk to reevaluate some of his command
policies. *Demeter* raises significant questions about the possibilities of male–
female interaction outside of patriarchal dominance. The meeting of a variety of different planetary cultures that represent alternative social philosophies and organizations, alternative ways of coping with the same essential
debates surrounding sexual difference, allows for a far-reaching exploration
of contemporary gender relations.

## From Space Opera to Soap Opera

If works like *Demeter* constitute intriguing prototypes for a new breed of
feminist popular literature, they frequently do so within conventions borrowed as much from more traditionally feminine forms of mass culture as
from *Star Trek* itself. For one thing, the female fans perceive the individual
episodes as contributing to one great program text. As a result, fan stories
often follow the format of a continuous serial rather than operating as a
series of selfenclosed works. Tania Modleski (1982) demonstrates the ways
that the serial format of much women's fiction, particularly of soap operas,
responds to the rhythms of women's social experience. The shaky financing
characteristic of the fanzine mode of production, the writers' predilections
to engage in endless speculations about the program content and to continually revise their understanding of the textual world, amplifies the tendency
of women's fiction to postpone resolution, transforming *Star Trek* into a
never ending story. Fan fiction marches forward through a series of digressions as new speculations cause the writers to halt the advance of their
chronicles, to introduce events that must have occurred prior to the start of
their stories, or to introduce secondary plot lines that pull them from the
main movement of the event chain. This type of writing activity has been
labeled a "story tree." Bacon-Smith (1986, p. 26) explains:

> The most characteristic feature of the story tree is that the stories do not fall in a linear sequence. A root story may offer unresolved situations, secondary characters whose actions during the main events are not described or a resolution is unsatisfactory to some readers. Writers then branch out from that story, completing dropped subplots, exploring the reactions of minor characters to major events.

This approach, characteristic of women's writing in a number of cultures, stems from a sense of life as continuous rather than fragmented into a series of discrete events, from an outlook that is experience centered and not goal oriented: "Closure doesn't make sense to them. At the end of the story, characters go on living in the nebulous world of the not yet written. They develop, modify their relationships over time, age, raise families" (p. 28).

Moreover, as Bacon-Smith's comments suggest, this type of reading and writing strategy focuses greater attention on ongoing character relationships than on more temporally concentrated plot elements. Long-time fan writer Lichtenberg (personal communication, August 1987) summarizes the difference: "Men want a physical problem with physical action leading to a physical resolution. Women want a psychological problem with psychological action leading to a psychological resolution." These women express a desire for narratives that concentrate on the character relationships and explore them in a "realistic" or "mature" fashion rather than in purely formulaic terms, stories that are "true" and "believable" and not "syrupy" or "sweet." Fan writers seek to satisfy these demands through their own *Star Trek* fiction, to write the type of stories that they and other fans desire to read.

The result is a type of genre switching, the rereading and rewriting of "space opera" as an exotic type of romance (and, often, the reconceptualization of romance itself as feminist fiction). Fanzines rarely publish exclusively action-oriented stories glorifying the Enterprise's victories over the Klingon-Romulan Alliance, its conquest of alien creatures, its restructuring of planetary governments, or its repair of potential flaws in new technologies, despite the prevalence of such plots in the original episodes. When such elements do appear, they are usually evoked as a background against which the more typical romance or relationship-centered stories are played or as a test through which female protagonists can demonstrate their professional skills. In doing so, these fan writers draw inspiration from feminist science fiction writers, including Johanna Russ, Marion Zimmer Bradley, Zenna Henderson, Marge Piercy, Andre Norton, and Ursula Le Guin. These writers' entry into the genre in the late 1960s and early 70s helped to redefine reader expectations about what constituted science fiction, pushing the genre toward greater and greater interest in soft science and sociological concerns and increased attention on interpersonal relationships and gender roles.[7] *Star Trek,* produced in a period when masculine concerns still dominated science fiction, is reconsidered in light of the newer, more feminist orientation of the genre, becoming less a program about the Enterprise's struggles against the Klingon-Romulan Alliance and more an examination of a character's efforts to come to grips with conflicting emotional needs and professional responsibilities.

Women, confronting a traditionally masculine space opera, choose to read it instead as a type of women's fiction. In constructing their own stories about the series characters, they turn frequently to the more familiar and comfortable formulas of the soap, the romance, and the feminist coming-of-age novel for models of storytelling technique. While the fans themselves often dismiss such genres as too focused upon mundane concerns to be of great interest, the influence of such materials may be harder to escape. As Elizabeth Segel (1986) suggests, our initial introduction to reading, the gender-based designation of certain books as suitable for young girls and others for young boys, can be a powerful determinant of our later reading and writing strategies, determining, in part, the relative accessibility of basic genre models for use in making sense of ready-made texts and for constructing personal fantasies. As fans attempt to reconstruct the feminine counter-texts that exist on the margins of the original series episodes, they, in the process, refocus the series around traditional feminine and contemporary feminist concerns, around sexuality and gender politics, around religion, family, marriage, and romance.

Many fans' first stories take the form of romantic fantasies about the series characters and frequently involve inserting glorified versions of themselves into the world of Star Fleet. The Bethann (1976, p. 54) story, "The Measure of Love," for instance, deals with a young woman, recently transferred to the Enterprise, who has a love affair with Kirk:

> We went to dinner that evening. Till that time, I was sure he'd never really noticed me. Sitting across the table from him, I realized just what a vital alive person this man was. I had dreamed of him, but never imagined my hopes might become a reality. But, this was real—not a dream. His eyes were intense, yet they twinkled in an amused sort of way.
> "Captain . . ."
> "Call me Jim."

Her romance with Kirk comes to an abrupt end when the young woman transfers to another ship without telling the captain that she carries his child because she does not want her love to interfere with his career.

Fans are often harshly critical of these so-called "Lieutenant Mary Sue" stories, which one writer labels "groupie fantasies" (Hunter, 1977, p. 78), because of their self-indulgence, their often hackneyed writing styles, their formulaic plots, and their violations of the established characterizations. In reconstituting *Star Trek* as a popular romance, these young women reshape the series characters into traditional romantic heroes, into "someone who is intensely and exclusively interested in her and in her needs" (Radway, 1984, p. 149). Yet, many fan writers are more interested in what happens when this romantic ideal confronts a world that places professional duty over personal needs, when men and women must somehow reconcile careers and marriage in a confusing period of shifting gender relationships. Veteran fan writer Kendra Hunter (1977, p. 78) writes, "Kirk is not going to go off into the sunset with anyone because he is owned body and soul by the Enterprise." *Treklink* editor Joan Verba (1986, p. 2) comments, "No believable character is gushed over by so many normally level-headed characters such as Kirk and Spock as a typical Mary Sue." Nor are the women of tomorrow apt to place any man, even Jim Kirk, totally above all other concerns.

Some, though by no means all, of the most sophisticated fan fiction also takes the form of the romance. Both Radway (1984) and Modleski (1982) note popular romances' obsession with a semiotics of masculinity, with the need to read men's often repressed emotional states from the subtle signs of outward gesture and expression. The cold logic of Vulcan, the desire to suppress all signs of emotion, make Spock and Sarek especially rich for such interpretations as in the following passage from Lorrah's *Full Moon Rising* (1976b, pp. 9–10):

> The intense sensuality she saw in him [Sarek] in other ways suggested a hidden sexuality. She had noticed everything from the way he appreciated the beauty of a moonlit night or a finely-cut sapphire to the way his strongly-molded hands caressed the mellowed leather binding of the book she had given him. . . . That incredible control which she could not penetrate. Sometimes he deliberately let her see beyond it, as he had done earlier this evening, but if she succeeded in making him lose control he would never be able to forgive her.

In Lorrah's writings, the alienness of Vulcan culture becomes a metaphor for the many things that separate men and women, for the factors that prevent intimacy within marriage. She describes her fiction as the story of "two people who are different physically, mentally, and emotionally, but who nonetheless manage to make a pretty good marriage" (p. 2). While Vulcan restraint suggests the emotional sterility of traditional masculinity, their alien sexuality allows Lorrah to propose alternatives. Her Vulcans find sexual inequality to be illogical and allow for very little difference in the treatment of men and women. (This is an assumption shared by many fan writers.) Moreover, the Vulcan mindmeld grants a degree of sexual and emotional intimacy unknown on earth; Vulcan men even employ this power to relieve women of labor pains and to share the experience of childbirth. Her lengthy writings on the decades-long romance between Amanda and Sarek represent a painstaking effort to construct a feminist utopia, to propose how traditional marriage might be reworked to allow it to satisfy the personal and professional needs of both men and women.

Frequently, the fictional formulas of popular romance are tempered by women's common social experiences as lovers, wives, and mothers under patriarchy. In Bates's novels, Nurse Chapel must confront and overcome her feelings of abandonment and jealousy during those long periods of time when her husband, Spock, is deeply absorbed in his work. *Starweaver Two* (1982, p. 10) describes this pattern:

The pattern had been repeated so often, it was ingrained. . . . Days would pass without a word between them because of the hours he labored and pored over his computers. Their shifts rarely matched and the few hours they could be together disappeared for one reason or another.

Far from an idyllic romance, Bates's characters struggle to make their marriage work in a world where professionalism is everything and the personal counts for relatively little. Land's version of a Chapel–Spock marriage is complicated by the existence of children who must remain at home under the care of Sarek and Amanda while their parents pursue their space adventures. In one scene, Chapel confesses her confused feelings about this situation to a young Andorian friend: "I spend my life weighing the children's needs against my needs against Spock's needs, and at any given time I know I'm shortchanging someone" (1987, p. 27).

While some male fans denigrate these types of fan fiction as "soap operas with Kirk and Spock" (Blaes, 1986a, p. 6), these women see themselves as constructing soap operas with a difference, soap operas that reflect a feminist vision. In Siebert's words (1982, pp. 44–45), "I write erotic stories for myself and for other women who will not settle for being less than human." Siebert suggests that her stories about Uhura and her struggle for recognition and romance in a male-dominated Star Fleet have helped her to resolve her own conflicting feelings within a world of changing gender relations and to explore hidden aspects of her own sexuality. Through her erotica, she hopes to increase other women's awareness of the need to struggle against entrenched patriarchal norms. Unlike their counterparts in Harlequin romances, these women refuse to accept marriage and the love of a man as their primary goal. Their stories push toward resolutions that allow Chapel or Uhura to achieve both professional advancement and personal satisfaction. Unlike almost every other form of popular fiction, fanzine stories frequently explore the maturing of relationships beyond the nuptial vows, seeing marriage as continually open to new adventures, new conflicts, and new discoveries.

The point of contact between feminism and the popular romance is largely a product of these writers' particular brand of feminism, one that, for the most part, is closer to the views of Betty Friedan than to those of Andrea Dworkin. It is a feminism that urges a sharing of feelings and lifestyles between men and women rather than radical separation or unresolvable differences. It is a literature of reform, not of revolt. The women still acknowledge their need for the companionship of men, for men who care for them and make them feel special, even as they are asking for those relationships to be conducted in different terms. Land's Nurse Chapel, who in *Demeter* is both fascinated and repelled by the feminist separatist colony, reflects these women's ambiguous and sometimes contradictory responses toward more radical forms of feminism. In the end, Chapel recognizes the potential need for such a place, for a "room of one's own," yet sees greater potential in achieving a more liberated relationship between men and women.

She learns to develop self-sufficiency, yet chooses to share her life with her husband, Spock, and to achieve a deeper understanding of their differing expectations about their relationship. Each writer grapples with these concerns in her own terms, yet most achieve some compromise between the needs of women for independence and self-sufficiency on the one hand and their needs for romance and companionship on the other. If this does not constitute a radical break with the romance formula, it does represent a progressive reformulation of that formula which pushes toward a gradual redefinition of existing gender roles within marriage and the work place.

## *The Moral Economy of Fan Fiction*

Their underground status allows fan writers the creative freedom to promote a range of different interpretations of the basic program material and a variety of reconstructions of marginalized characters and interests, to explore a diversity of different solutions to the dilemma of contemporary gender relations. Fandom's IDIC philosophy (Infinite Diversity in Infinite Combinations, a cornerstone of Vulcan thought) actively encourages its participants to explore and find pleasure within their different and often contradictory responses to the program text. It should not be forgotten, however, that fan writing involves a translation of personal response into a social expression and that fans, like any other interpretive community, generate their own norms that work to insure a reasonable degree of conformity between readings of the primary text. The economic risk of fanzine publishing and the desire for personal popularity insures some responsiveness to audience demand, discouraging totally idiosyncratic versions of the program content. Fans try to write stories to please other fans; lines of development that do not find popular support usually cannot achieve financial viability.

Moreover, the strange mixture of fascination and frustration characteristic of fan response means that fans continue to respect the creators of the original series, even as they wish to rework some program materials to better satisfy their personal interests. Their desire to revise the program material is often counterbalanced by their desire to remain faithful to those aspects of the show that first captured their interests. E. P. Thompson (1971, p. 78) has employed the term "moral economy" to describe the way that eighteenth century peasant leaders and street rioters legitimized their revolts through an appeal to "traditional rights and customs" and "the wider consensus of the community," asserting that their actions worked to protect existing property rights against those who sought to abuse them for their own gain. The peasants' conception of a moral economy allowed them to claim for themselves the right to judge the legitimacy both of their own actions and those of the landowners and property holders: "Consensus was so strong that it overrode motives of fear or deference" (pp. 78–79).

An analogous situation exists in fandom: the fans respect the original texts, yet fear that their conceptions of the characters and concepts may be jeopardized by those who wish to exploit them for easy profits, a category that typically includes Paramount and the network but excludes Rodden-berry and many of the show's writers. The ideology of fandom involves both a commitment to some degree of conformity to the original program materials as well as a perceived right to evaluate the legitimacy of any use of those materials, either by textual producers or by textual consumers. The fans perceive themselves as rescuing the show from its producers who have manhandled its characters and then allowed it to die. In one fan's words, "I think we have made *ST* uniquely our own, so we do have all the right in the world (universe) to try to change it for the better when the gang at Paramount starts worshipping the almighty dollar, as they are wont to do" (Schnuelle, 1987, p. 9). Rather than rewriting the series content, the fans claim to be keeping *Star Trek* alive in the face of network indifference and studio incompetence, of remaining true to the text that first captured their interest some twenty years before: "This relationship came into being be-cause the fan writers loved the characters and cared about the ideas that are *Star Trek* and they refused to let it fade away into oblivion" (Hunter, 1977, p. 77).

Such a relationship obligates fans to preserve a certain degree of fidelity to program materials, even as they seek to rework them toward their own ends. *Trek* magazine contributor Kendra Hunter (1977, p. 83) writes, "*Trek* is a format for expressing rights, opinions, and ideals. Most every imagin-able idea can be expressed through *Trek*. . . . But there is a right way." Gross infidelity to the series concepts constitutes what fans call "character rape" and falls outside of the community's norms. In Hunter's words (p. 75):

> A writer, either professional or amateur, must realize that she . . . is not omnipotent. She cannot force her characters to do as she pleases. . . . The writer must have respect for her characters or those created by others that she is using, and have a full working knowledge of each before committing her words to paper.

Hunter's conception of character rape, one widely shared within the fan community, rejects abuses by the original series writers as well as by the most novice fan. It implies that the fans themselves, not the program pro-ducers, are best qualified to arbitrate conflicting claims about character psy-chology because they care about the characters in a way that more commer-cially motivated parties frequently do not. In practice, the concept of character rape frees fans to reject large chunks of the aired material, including entire episodes, and even to radically restructure the concerns of the show in the name of defending the purity of the original series concept. What deter-mines the range of permissible fan narratives is finally not fidelity to the original texts but consensus within the fan community itself. The text that they so lovingly preserve is the *Star Trek* that they created through their own speculations, not the one that Roddenberry produced for network air play.

Consequently, the fan community continually debates what constitutes a legitimate reworking of program materials and what represents a violation of the special reader-text relationship that the fans hope to foster. The earliest *Star Trek* fan writers were careful to work within the framework of the information explicitly included within the broadcast episodes and to minimize their breaks with series conventions. In fan writer Jean Lorrah's words (1976a, p. 1), "Anyone creating a *Star Trek* universe is bound by what was seen in the aired episodes; however, he is free to extrapolate from those episodes to explain what was seen in them." Leslie Thompson (1974, p. 208) explains, "If the reasoning [of fan speculations] doesn't fit into the framework of the events as given [on the program], then it cannot apply no matter how logical or detailed it may be." As *Star Trek* fan writing has come to assume an institutional status in its own right and therefore to require less legitimization through appeals to textual fidelity, a new conception of fan fiction has emerged, one that perceives the stories not as a necessary expansion of the original series text but rather as chronicles of alternate universes, similar to the program world in some ways and different in others:

> The "alternate universe" is a handy concept wherein you take the basic *Star Trek* concept and spin it off into all kinds of ideas that could never be aired. One reason Paramount may be so liberal about fanzines is that by their very nature most fanzine stories could never be sold professionally. (L. Slusher, personal communication, August 1987)

Such an approach frees the writers to engage in much broader play with the program concepts and characterizations, to produce stories that reflect more diverse visions of human interrelationships and future worlds, to rewrite elements within the primary texts that hinder fan interests. Yet, even alternate universe stories struggle to maintain some consistency with the original broadcast material and to establish some point of contact with existing fan interests, just as more faithful fan writers feel compelled to rewrite and revise the program material in order to keep it alive in a new cultural context.

### Borrowed Terms: Kirk/Spock Stories

The debate in fan circles surrounding Kirk/Spock (K/S) fiction, stories that posit a homoerotic relationship between the show's two primary characters and frequently offer detailed accounts of their sexual couplings, illustrates these differing conceptions of the relationship between fan fiction and the primary series text.[8] Over the past decade, K/S stories have emerged from the margins of fandom toward numerical dominance over *Star Trek* fan fiction, a movement that has been met with considerable opposition from more traditional fans. For many, such stories constitute the worst form of character rape, a total violation of the established characterizations. Kendra Hunter (1977, p. 81) argues that "it is out of character for both men, and as such comes across in the stories as bad writing. . . . A relationship as complex and deep as Kirk/Spock does not climax with a sexual relationship." Other fans agree but for other reasons. "I do not accept the K/S homosexual precept as plausible," writes one fan. "The notion that two men that are as close as Kirk and Spock are cannot be 'just friends' is indefensible to me" (Landers, 1986, p. 10). Others struggle to reconcile the information provided on the show with their own assumptions about the nature of human sexuality: "It is just as possible for their friendship to progress into a love-affair, for that is what it is, than to remain status quo. . . . Most of us see Kirk and Spock simply as two people who love each other and just happen to be of the same gender" (Snaider, 1987, p. 10).

Some K/S fans frankly acknowledge the gap between the series characterizations and their own representations yet refuse to allow their fantasy life to be governed by the limitations of what was actually aired. One fan writes, "While I read K/S and enjoy it, when you stop to review the two main characters of *Star Trek* as extrapolated from the TV series, a sexual relationship between them is absurd" (Chandler, 1987, p. 10). Another argues somewhat differently:

> We actually saw a very small portion of the lives of the Enterprise crew through 79 episodes and some six hours of movies. . . . How can we possibly define the entire personalities of Kirk, Spock, etc., if we only go by what we've seen on screen? Surely there is more to them than that! . . . Since I doubt any two of us would agree on a definition of what is "in character," I leave it to the skill of the writer to make the reader believe in the story she is trying to tell. There isn't any limit to what could be depicted as accurate behavior for our heroes. (Moore, 1986, p. 7)

Many fans find this bold rejection of program limitations on creative activity, this open appropriation of characters, to be unacceptable since it violates the moral economy of fan writing and threatens fan fiction's privileged relationship to the primary text:

> [If] "there isn't any limit to what could be depicted as accurate behavior of our heroes," we might well have been treated to the sight of Spock shooting up heroin or Kirk raping a yeoman on the bridge (or vice-versa). . . . The writer whose characters don't have clearly defined personalities, [through] limits and idiosyncrasies and definite characteristics, is the writer who is either very inexperienced or who doesn't have any respect for his characters, not to mention his audience. (Slusher, 1986, p. 11)

Yet, I have shown, all fan writing necessarily involves an appropriation of series characters and a reworking of program concepts as the text is forced to respond to the fan's own social agenda and interpretive strategies. What K/S does openly, all fans do covertly. In constructing the feminine counter-text that lurks in the margins of the primary text, these readers necessarily redefine the text in the process of rereading and rewriting it. As one fan acknowledges, "If K/S has 'created new characters and called them by old names,' then all of fandom is guilty of the same" (Moore, 1986, p. 7). Jane

Land (1987, p. ii) agrees: "All writers alter and transform the basic *Trek* universe to some extent, choosing some things to emphasize and others to play down, filtering the characters and the concepts through their own perceptions."

If these fans have rewritten *Star Trek* in their own terms, however, many of them are reluctant to break all ties to the primary text that sparked their creative activity and, hence, feel the necessity to legitimate their activity through appeals to textual fidelity. The fans are uncertain how far they can push against the limitations of the original material without violating and finally destroying a relationship that has given them great pleasure. Some feel stifled by those constraints; others find comfort within them. Some claim the program as their personal property, "treating the series episodes like silly putty," as one fan put it (Blaes, 1987, p. 6). Others seek compromises with the textual producers, treating the original program as something shared between them.

What should be remembered is that whether they cast themselves as rebels or loyalists, it is the fans themselves who are determining what aspects of the original series concept are binding on their play with the program material and to what degree. The fans have embraced *Star Trek* because they found its vision somehow compatible with their own, and they have assimilated only those textual materials that feel comfortable to them. Whenever a choice must be made between fidelity to their program and fidelity to their own social norms, it is almost inevitably made in favor of lived experience. The women's conception of the *Star Trek* realm as inhabited by psychologically rounded and realistic characters insures that no characterization that violated their own social perceptions could be satisfactory. The reason some fans reject K/S fiction has, in the end, less to do with the stated reason that it violates established characterization than with unstated beliefs about the nature of human sexuality that determine what types of character conduct can be viewed as plausible. When push comes to shove, as Hodge and Tripp (1986, p. 144) recently suggested, "Nontelevisual meanings can swamp televisual meanings" and usually do.

## Conclusion

The fans are reluctant poachers who steal only those things that they truly love, who seize televisual property only to protect it against abuse by those who created it and who have claimed ownership over it. In embracing popular texts, the fans claim those works as their own, remaking them in their own image, forcing them to respond to their needs and to gratify their desires. Female fans transform *Star Trek* into women's culture, shifting it from space opera into feminist romance, bringing to the surface the unwritten feminine countertext that hides in the margins of the written masculine text. Kirk's story becomes Uhura's story and Chapel's and Amanda's as well as the story of the women who weave their own personal experiences into the lives of the characters. Consumption becomes production; reading becomes writing; spectator culture becomes participatory culture.

Neither the popular stereotype of the crazed Trekkie nor academic notions of commodity fetishism or repetition compulsion are adequate to explain the complexity of fan culture. Rather, fan writers suggest the need to redefine the politics of reading, to view textual property not as the exclusive domain of textual producers but as open to repossession by textual consumers. Fans continuously debate the etiquette of this relationship, yet all take for granted the fact that they are finally free to do with the text as they please. The world of *Star Trek* is what they choose to make it: "If there were no fandom, the aired episodes would stand as they are, and yet they would be just old reruns of some old series with no more meaning than old reruns of *I Love Lucy*" (Hunter, 1977, p. 77). The one text shatters and becomes many texts as it is fit into the lives of the people who use it, each in her or his own way, each for her or his own purposes.

Modleski (1986) recently, and I believe mistakenly, criticized what she understands to be the thrust of the cultural studies tradition: the claim that somehow mass culture texts empower readers. Fans are not empowered *by* mass culture; fans are empowered *over* mass culture. Like de Certeau's poachers, the fans harvest fields that they did not cultivate and draw upon materials not of their making, materials already at hand in their cultural environment; yet, they make those raw materials work for them. They employ images and concepts drawn from mass culture texts to explore their subordinate status, to envision alternatives, to voice their frustrations and anger, and to share their new understandings with others. Resistance comes from the uses they make of these popular texts, from what they add to them and what they do with them, not from subversive meanings that are somehow embedded within them.

Ethnographic research has uncovered numerous instances where this occurs. Australian schoolchildren turn to *Prisoner* in search of insight into their own institutional experience, even translating schoolyard play into an act of open subordination against the teachers' authority (Hodge and Tripp, 1986; Palmer, 1986). American kindergartners find in the otherness of Pee-Wee Herman a clue to their own insecure status as semisocialized beings (Jenkins, in press). British gay clubs host *Dynasty* and *Dallas* drag balls, relishing the bitchiness and trashiness of nighttime soap operas as a negation of traditional middle class taste and decorum (Finch, 1986). European leftists express their hostility to Western capitalism through their love–hate relationship with *Dallas* (Ang, 1986). Nobody regards these fan activities as a magical cure for the social ills of post-industrial capitalism. They are no substitution for meaningful change, but they can be used effectively to build popular support for such change, to challenge the power of the culture industry to construct the common sense of a mass society, and to restore a much-needed excitement to the struggle against subordination.

Alert to the challenge such uses pose to their cultural hegemony, textual producers openly protest this uncontrollable proliferation of meanings from their texts, this popular rewriting of their stories, this trespass upon their literary properties. Actor William Shatner (Kirk), for instance, has said of *Star Trek* fan fiction: "People read into it things that were not intended. In *Star Trek*'s case, in many instances, things were done just for entertainment purposes" (Spelling, Lofficier, and Lofficier, 1987, p. 40). Producers insist upon their right to regulate what their texts may mean and what types of pleasure they can produce. Yet, such remarks carry little weight. Undaunted by the barking dogs, the "no trespassing" signs, and the threats of prosecution, the fans already have poached those texts from under the proprietors' noses.

## Notes

1. An earlier draft of this essay was presented at the 1985 Iowa Symposium and Conference on Television Criticism: Public and Academic Responsibility. I am indebted to Cathy Schwichtenberg, John Fiske, David Bordwell, and Janice Radway for their helpful suggestions as I was rewriting it for *CSMC*. I am particularly indebted to Signe Hovde and Cynthia Benson Jenkins for introducing me to the world of fan writing; without them my research could not have been completed. I have tried to contact all of the fans quoted in this text and to gain their permission to discuss their work. I appreciate their cooperation and helpful suggestions.

2. For representative examples of other scholarly treatments of *Star Trek* and its fans, see Blair (1983), Greenberg (1984), Jewett and Lawrence (1977), and Tyre (1977). Attitudes range from the generally sympathetic Blair to the openly hostile Jewett and Lawrence.

3. No scholarly treatment of *Star Trek* fan culture can avoid these pitfalls, if only because making such a work accessible to an academic audience requires a translation of fan discourse into other terms, terms that may never be fully adequate to the original. I come to both *Star Trek* and fan fiction as a fan first and a scholar second. My participation as a fan long precedes my academic interest in it. I have sought, where possible, to employ fan terms and to quote fans directly in discussing their goals and orientations toward the program and their own writing. I have shared drafts of this essay with fans and have incorporated their comments into the revision process. I have allowed them the dignity of being quoted from their carefully crafted, well-considered published works rather than from a spontaneous interview that would be more controlled by the researcher than by the informant. I leave it to my readers to determine whether this approach allows for a less mediated reflection of fan culture than previous academic treatments of this subject.

4. The terms "letterzine" and "fictionzine" are derived from fan discourse. The two types of fanzines relate to each other in complex ways. Although there are undoubtedly some fans who read only one type of publication, many read both. Some letterzines, *Treklink* for instance, function as consumer guides and sounding boards for debates about the fictionzines.

5. Both Lorrah and Lichtenberg have achieved some success as professional science fiction writers. For an interesting discussion of the relationship between fan writing and professional science fiction writing, see Randall (1985).

6. Although a wide range of fanzines were considered in researching this essay, I have decided, for the purposes of clarity, to draw my examples largely from the

work of a limited number of fan writers. While no selection could accurately reflect the full range of fan writing, I felt that Bates, Land, Lorrah, and Siebert had all achieved some success within the fan community, suggesting that they exemplified, at least to some fans, the types of writing that were desirable and reflected basic tendencies within the form. Further, these writers have produced a large enough body of work to allow some commentary about their overall project rather than localized discussions of individual stories. I have also, wherever possible, focused my discussion around works still currently in circulation and therefore available to other researchers interested in exploring this topic. No slight is intended to the large number of other fan writers who also met these criteria and who, in some cases, are even better known within the fan community.

7. I am indebted to K. C. D'alessandro and Mary Carbine for probing questions that refined my thoughts on this particular issue.

8. The area of Kirk/Spock fiction falls beyond the project of this particular paper. My reason for discussing it here is because of the light its controversial reception sheds on the norms of fan fiction and the various ways fan writers situate themselves toward the primary text. For a more detailed discussion of this particular type of fan writing, see Lamb and Veith (1986), who argue that K/S stories, far from representing a cultural expression of the gay community, constitute another way of feminizing the concerns of the original series text and of addressing feminist concern within the domain of a popular culture that offers little space for heroic action by women.

## References

Ang, I. (1986). *Watching Dallas*. London: Methuen.
Bacon-Smith, C. (1986, November 16). Spock among the women. *The New York Times Book Review*, pp. 1, 26, 28.
Bates, K. A. (1982). *Starweaver two*. Missouri Valley, Iowa: Ankar Press.
——— (1983). *Nuages one*. Tucson, Ariz.: Checkmate Press.
——— (1984). *Nuages two*. Tucson, Ariz.: Checkmate Press.
Bethann. (1976). The measure of love. *Grup*, 5: 53–62.
Blaes, T. (1986a). Letter. *Treklink*, 5: 6.
——— (1987). Letter. *Treklink*, 9: 6–7.
Blair, K. (1983). Sex and *Star Trek*. *Science Fiction Studies*, 10: 292–97.
Bleich, D. (1986). Gender interests in reading and language. In E. A. Flynn and P. P. Schweickart (eds.), *Gender and reading: Essays on readers, texts and contexts* (pp. 234–66). Baltimore: Johns Hopkins University Press.
Brown, M. E., and Barwick, L. (1987, May). *Fables and endless generations: Soap opera and women's culture*. Paper presented at a meeting of the Society of Cinema Studies, Montreal.
Caruthers-Montgomery, P. L. (1987). Letter. *Comlink*, 28: 8.
Chandler, M. (1987). Letter. *Treklink*, 8: 10.
Cooper, C. (1987, February). Opportunities in the "media fanzine" market. *Writer's Digest*, p. 45.
de Certeau, M. (1984). *The practice of everyday life*. Berkeley: University of California Press.
de Lauretis, T. (1982). *Alice doesn't: Feminism, semiotics, cinema*. Bloomington: Indiana University Press.

Deneroff, L. (1987). A reflection on the early days of *Star Trek* fandom. *Comlink,* 28: 3–4.

Finch, M. (1986). Sex and address in *Dynasty. Screen,* 27: 24–42.

Greenberg, H. (1984). In search of Spock: A psychoanalytic inquiry. *Journal of Popular Film and Television,* 12: 53–65.

Hodge, R., and Tripp, D. (1986). *Children and television: A semiotic approach.* Cambridge: Polity Press.

Hunter, K. (1977). Characterization rape. In W. Irwin and G. B. Love (eds.), *The best of Trek 2* (pp. 74–85). New York: New American Library.

Jenkins, H. (in press). "Going bonkers!" Children, play and Pee-Wee. *Camera Obscura,* 18.

Jewett, R., and Lawrence, J. S. (1977). *The American monomyth.* Garden City, N.Y.: Anchor Press.

Kramarae, C. (1981). *Women and men speaking.* Rowley, Mass.: Newburry House.

Lamb, P. F., and Veith, D. L. (1986). Romantic myth, transcendence, and *Star Trek* zines. In D. Palumbo (ed.), *Erotic universe: Sexuality and fantastic literature* (pp. 235–56). New York: Greenwood Press.

Land, J. (1986). *Kista.* Larchmont, N.Y.: Author.

——— (1987). *Demeter.* Larchmont, N.Y.: Author.

Landers, R. (1986). Letter. *Treklink,* 7: 10.

Lay, T. (1986). Letter. *Comlink,* 28: 14–16.

Leerhsen, C. (1986, December 22). *Star Trek*'s nine lives. *Newsweek,* pp. 66–73.

Lichtenberg, J. (1976). *Kraith collected.* Grosse Point Park, Mich.: Ceiling Press.

Lorrah, J. (1976a). *The night of twin moons.* Murray, Ky.: Author.

——— (1976b). *Full moon rising.* Bronx, N.Y.: Author.

——— (1978). The Vulcan character in the NTM universe. In J. Lorrah (ed.), *NTM collected* (Vol. 1, pp. 1–3). Murray, Ky.: Author.

——— (1984). *The Vulcan academy murders.* New York: Pocket Books.

Ludlow, J. (1987). Letter. *Comlink,* 28: 17–18.

Marshak, S., and Culbreath, M. (1978). *Star Trek: The new voyages.* New York: Bantam Books.

Modleski, T. (1982). *Loving with a vengeance: Mass-produced fantasies for women.* Hamden, Conn.: Archon Books.

——— (1986). *Studies in entertainment: Critical approaches to mass culture.* Bloomington: Indiana University Press.

Moore, R. (1986). Letter. *Treklink,* 4: 7–8.

Osborne, E. (1987). Letter, *Treklink,* 9: 3–4.

Palmer, P. (1986). *The lively audience.* Sidney, Australia: Unwyn & Allen.

Radway, J. (1984). *Reading the romance: Women, patriarchy and popular literature.* Chapel Hill: University of North Carolina Press.

Randall, M. (1985). Conquering the galaxy for fun and profit. In C. West (Ed.), *Words in our pockets* (pp. 233–41). Paradise, Calif.: Dustbooks.

Rose, P. (1977). Women in the federation. In W. Irwin and G. B. Love (Eds.), *The best of Trek 2* (pp. 46–52). New York: New American Library.

Schnuelle, S. (1987). Letter. *Sociotrek,* 4: 8–9.

Segel, E. (1986). Gender and childhood reading. In E. A. Flynn and P. P. Schweickart (eds.), *Gender and reading: Essays on readers, texts and contexts* (pp. 164–85). Baltimore: Johns Hopkins University Press.

Siebert, C. A. (1980). Journey's end at lover's meeting. *Slaysu,* 1: 28–34.

——— (1982). By any other name. *Slaysu,* 4: 44–45.

Smith-Rosenberg, C. (1985). *Disorderly conduct: Gender in Victorian America.* New York: Knopf.

Snaider, T. (1987). Letter. *Treklink,* 8: 10.

Spector, J. (1986). Science fiction and the sex war: A womb of one's own. In J. Spector (ed.), *Gender studies: New directions in feminist criticism* (pp. 161–83). Bowling Green, Ohio: Bowling Green State University Press.

Spelling, I., Lofficier, R., and Lofficier, J-M. (1987, May). William Shatner, captain's log: *Star Trek V. Starlog,* pp. 37–41.

Thompson, E. P. (1971). The moral economy of the English crowd in the 18th century. *Past and present,* 50: 76–136.

Thompson, L. (1974). *Star Trek* mysteries—Solved! In W. Irwin and G. B. Love (eds.), *The best of Trek* (pp. 207–14). New York: New American Library.

Tyre, W. B. (1977). *Star Trek* as myth and television as myth maker. *Journal of Popular Culture,* 10: 711–19.

Verba, J. (1986). Editor's corner, *Treklink,* 6: 1–4.

Whitfield, S. E., and Roddenberry, G. (1968). *The making of Star Trek.* New York: Ballantine Books.

Wood, R. (1986). *Hollywood from Vietnam to Reagan.* New York: Columbia University Press.

# Television and Gender

## DAVID MORLEY

[My] interviews [with families of television viewers] identified the following major themes, which recur across the interviews with [those] different families, where I can point to a reasonable degree of consistency of response. Clearly, the one structural principle working across all the families interviewed is that of gender. These interviews raise important questions about the effects of gender in terms of:

- power and control over program choice
- viewing style
- planned and unplanned viewing
- amounts of viewing
- television-related talk
- use of video
- "solo" viewing and guilty pleasures
- program type preference
- channel preference
- national versus local news programming
- comedy preferences

Before going on to detail my findings under these particular headings I would first like to make some general points about the significance of the empirical differences which my research revealed between the viewing hab-

From *Family Television*, Comedia (Routledge) 1986, pp. 146–172. Reprinted with permission.

its of the men and women in the sample. As will be seen below, the men and women offer clearly contrasting accounts of their viewing habits—in terms of their differential power to choose what they view, how much they view, their viewing styles, and their choice of particular viewing material. However, I am not suggesting that these empirical differences are attributes of their essential biological characteristics as men and women. Rather, I am trying to argue that these differences are the effects of the particular social roles that these men and women occupy within the home. Moreover, as I have indicated, this sample primarily consists of lower middle-class and working-class nuclear families (all of whom are white) and I am not suggesting that the particular pattern of gender relations within the home found here (with all the consequences which that pattern has for viewing behavior) would necessarily be replicated either in nuclear families from a different class or ethnic background, or in households of different types with the same class and ethnic backgrounds. Rather, it is always a case of how gender relations interact with, and are formed differently within, these different contexts.

However, aside from these qualifications, there is one fundamental point which needs to be made concerning the basically different positioning of men and women within the domestic sphere. It should be noted that in the earlier chapters of this [study] there was much emphasis on the fact that this research project was concerned with television viewing in its domestic context. The essential point here is that the dominant model of gender relations within this society (and certainly within that sub-section of it represented in my sample) is one in which the home is primarily defined for men as a site of leisure—in distinction to the "industrial time" of their employment outside the home—while the home is primarily defined for women as a sphere of work (whether or not they also work outside the home). This simply means that in investigating television viewing in the home one is by definition investigating something which men are better placed to do wholeheartedly, and which women seem only to be able to do distractedly and guiltily, because of their continuing sense of their domestic responsibilities. Moreover, this differential positioning is given a greater significance as the home becomes increasingly defined as the "proper" sphere of leisure, with the decline of public forms of entertainment and the growth of home-based leisure technologies such as video, and so forth.

These points are well illustrated in research by Ann Gray into women's viewing and the use of video in the home. Gray argues that many women do not really consider themselves as having any specific leisure time at all in the home and would feel too uncomfortably guilty to "just" sit and watch television when there always are domestic tasks to be attended to.[1]

When considering the empirical findings summarized below, care must be taken to hold in view this structuring of the domestic environment by gender relations as the backdrop against which these particular patterns of viewing behavior have developed. Otherwise we risk seeing this pattern as somehow the direct result of "essential" or biological characteristics of men and women per se. As Charlotte Brunsdon has put it, commenting on research in this area we could

> mistakenly . . . differentiate a male—fixed, controlling, uninterruptible—gaze, and a female—distracted, obscured, already busy—manner of watching television. There is some empirical truth in these characterisations, but to take this empirical truth for explanation leads to a theoretical short circuit. . . . Television is a domestic medium—and indeed the male/female differentiation above is very close to the way in which cinema and television have themselves been differentiated. Cinema, the audiovisual medium of the public sphere [demands] the masculine gaze, while the domestic (feminine) medium is much less demanding, needing only an intermittent glance. This, given the empirical evidence . . . offers us an image of male viewers trying to masculinise the domestic sphere. This way of watching television, however, seems not so much a masculine mode, but a mode of power. Current arrangements between men and women make it likely that it is men who will occupy this position in the home.[2]

From this perspective we can then see the empirical differences between the accounts of their viewing behavior offered by the men and women in this sample as generated within this structure of domestic power relations.

## Power and Control over Program Choice

Masculine power is evident in a number of the families as the ultimate determinant on occasions of conflict over viewing choices ("we discuss what we all want to watch and the biggest wins. That's me. I'm the biggest," Man, Family 4). More crudely, it is even more apparent in the case of those families who have an automatic control device. None of the women in any of the families use the automatic control regularly. A number of them complain that their husbands use the channel control device obsessively, channel flicking across programs when their wives are trying to watch something else. Characteristically, the control device is the symbolic possession of the father (or of the son, in the father's absence) which sits "on the arm of Daddy's chair" and is used almost exclusively by him. It is a highly visible symbol of condensed power relations (the descendant of the medieval mace perhaps?). The research done by Peter Collett and Roger Lamb in which they videotaped a number of families watching television over an extended period shows this to comic effect on at least one occasion where the husband carries the control device about the house with him as he moves from the living room to the kitchen and then engages in a prolonged wrestling match with his wife and son simultaneously so as to prevent them from getting their hands on it.[3]

> *F2 Daughter:* Dad keeps both of the automatic controls—one on each side
> of his chair.
> *F3 Woman:* Well, I don't get much chance, because he sits there with the
> automatic control beside him and that's it . . . I get annoyed because I

can be watching a program and he's flicking channels to see if a program on the other side is finished, so he can record something. So the television's flickering all the time, while he's flicking the timer. I just say, "For goodness' sake, leave it alone." I don't get the chance to use the control. I don't get near it.

*F15 Woman:* No, not really. I don't get the chance to use the automatic control. I leave that down to him. It is aggravating, because I can be watching something and all of a sudden he turns it over to get the football result.

*F9 Daughter:* The control's always next to dad's chair. It doesn't come away when Dad's here. It stays right there.

*F9 Woman:* And that's what you do [her husband], isn't it? Flick, flick, flick—when they're in the middle of a sentence on the telly. He's always flicking it over.

*F9 Man:* The remote control, oh yes, I use it all the time.

*F9 Daughter:* Well, if you're in the middle of watching something, Dad's got a habit of flicking over the other side to see the result of the boxing.

*F8 Woman (to Son):* You're the keeper of the control aren't you?

*F8 Son:* Either me or Dad has it. I have it mostly.

*F16 Woman:* Yes, he uses it a lot . . . the remote control.

*F16 Man:* Oh yes, quite a bit. I think, Oh I'll just see what's on the other side.

In most of these families, the power relations are fairly clear. The man in F8 helpfully explains their family's way of resolving conflicts over viewing preferences:

*F8 Man:* We normally tape one side and watch what I want to watch.

Interestingly, the main exceptions to this overall pattern concern those families in which the man is unemployed while his wife is working. In these cases it is slightly more common for the man to be expected to be prepared to let other family members watch what they want to when it is broadcast, while videotaping what he would like to see, in order to watch that later at night or the following day—given that his timetable of commitments is more flexible than those of the working members of the family. Here we begin to see the way in which the position of power held by most of the men in the sample (and which their wives concede) is based not simply on the biological fact of being men but rather on a social definition of a masculinity of which employment (that is, the "breadwinner" role) is a necessary and constituent part. When that condition is not met, the pattern of power relations within the home can change noticeably.[4]

One further point needs to be made in this connection. It has to be remembered that this research is based on people's accounts of their behavior, not on any form of direct observation of behavior outside of the interview context itself. It is noteworthy that a number of the men show some anxiety to demonstrate that they are "the boss of the household" and their very anxiety around this issue perhaps betokens a sense that their domestic power is ultimately a fragile and somewhat insecure thing, rather than a fixed and permanent "possession" which they can always guarantee to hold with confidence. Hence perhaps the symbolic importance to them of physical possession of the channel control device.

## Styles of Viewing

One major finding is the consistency of the distinction between the characteristic ways in which men and women describe their viewing activity. Essentially the men state a clear preference for viewing attentively, in silence, without interruption "in order not to miss anything." Moreover, they display puzzlement at the way their wives and daughters watch television. This the women themselves describe as a fundamentally social activity, involving ongoing conversation, and usually the performance of at least one other domestic activity (ironing, etc.) at the same time. Indeed, many of the women feel that to just watch television without doing anything else at the same time would be an indefensible waste of time, given their sense of their domestic obligations. To watch in this way is something they rarely do, except occasionally, when alone or with other women friends, when they have managed to construct an "occasion" on which to watch their favorite program, video, or film. The women note that their husbands are always "on at them" to shut up. The men can't really understand how their wives can follow the programs if they are doing something else at the same time.

> *F2 Man:* We don't talk. They talk a bit.
>
> *F2 Woman:* You keep saying sshh.
>
> *F2 Man:* I can't concentrate if there's anyone talking while I'm watching. But they can, they can watch and just talk at the same time. We just watch it—take it all in. If you talk, you've missed the bit that's really worth watching. We listen to every bit of it and if you talk you miss something that's important. My attitude is sort of go in the other room if you want to talk.
>
> *F5 Man:* It really amazes me that this lot [his wife and daughters] can talk and do things and still pick up what's going on. To my mind it's not very good if you can do that.
>
> *F5 Woman:* Because we have it on all the time it's like second nature. We watch, and chat at the same time.
>
> *F18 Woman:* I knit because I think I am wasting my time just watching. I know what's going on, so I only have to glance up. I always knit when I watch.
>
> *F15 Woman:* I can generally sit and read a book and watch a film at the same time and keep the gist of it. If it's a good film it doesn't bother me. I'm generally sewing or something like that.
>
> *F9 Man:* I like to watch it without aggravation. I'd rather watch on my own. If it's just something I want to watch, I like to watch everything with no talking at all.
>
> *F9 Woman:* Every now and again he says, "Ssshhh shut up." It's terrible. He comes in . . . from a pool match and he'll say, "Shut up, please shut up!"

*F9 Man:* You can't watch anything in peace unless they're all out. Half the time they start an argument and then you've missed easily twenty minutes of it . . . usually the catchphrase which you've got to listen to find out what's going to happen in the program. Sometimes I just go upstairs. It's not worth watching.

*F11 Woman:* I can't think of anything I'll totally watch. I don't just sit and watch. I'll probably sew—maybe knit. I very rarely just sit—that's just not me.

*F11 Man:* She tends to watch and do another activity at the same time.

*F11 Woman:* He is totally absorbed by it.

*F11 Man:* If I like it, I am in there in the action, feeling every blow, running every mile—especially something like live football.

*F8 Woman:* I suppose we do chat a little bit. We tend to chat. We don't sit in complete silence. We chat through everything, don't we?

*F8 Man:* I try to listen . . . and I'll try and listen . . . then I'll miss half of it.

*F12 Woman:* There is always something else, like ironing. I can watch anything while I'm doing the ironing. I've always done the ironing and knitting and that. . . . You just sit down and watch it, whereas you've got things to do, you know, and you can't keep watching television. You think, Oh my God, I should have done this or that.

*F4 Man:* I like to watch in peace and quiet, but there's not much chance. I do like to watch in silence.

*F4 Daughter:* All we get is "Shut up and don't fidget." That's what he says when he's watching.

*F17 Man:* I like to sit and concentrate. You lose the atmosphere if the children are mucking about.

*F17 Woman:* I do a bit of knitting, and crocheting. He gives a running commentary while I'm doing the dishes after dinner, or when I'm in the kitchen. I know what's going on on the television, when he's with the boys. I know what's going on by his running commentary.

Charlotte Brunsdon, commenting on this and other research in this area, provides a useful way of understanding the behavior reported here. As she argues, it is not that the women have no desire ever to watch television attentively, but rather that their domestic position makes it almost impossible for them to do this unless all the other members of the household are "out of the way":

> The social relations between men and women appear to work in such a way that although the men feel OK about imposing their choice of viewing on the whole of the family, the women do not. The women have developed all sorts of strategies to cope with television viewing that they don't particularly like. . . . However, the women in general seem to find it almost impossible to switch into the silent communion with the television set that characterises so much male viewing. Revealingly, they often speak rather longingly of doing this, but it always turns out to require the physical absence of the rest of the family.[5]

Again we see that these distinctive viewing styles are not simply characteristics of men and women as such but, rather, characteristics of the domestic roles of masculinity and femininity. The comments about the physical conditions under which women feel able to view attentively are explored further in the section below on solo viewing.

## Planned and Unplanned Viewing

It is the men, on the whole, who speak of checking through the paper (or the teletext) to plan their evening's viewing. Very few of the women seem to do this at all, except in terms of already knowing which evenings and times their favorite series are on and thus not needing to check the schedule. This is also an indication of a different attitude to viewing as a whole. Many of the women have a much more take-it-or-leave-it attitude, not caring much if they miss things (except for their favorite serials).

> *F7 Man:* Normally I look through the paper because you [his wife] tend to just put on ITV, but sometimes there is something good on the other channels, so I make a note—things like films and sport.
>
> *F14 Man:* I just read the paper to see—and we might say there's something good on about 7:30 or 8PM—and we might turn it on. Otherwise it stays off.
>
> *F14 Woman:* I don't read newspapers. If I know what's going to be on, I'll watch it. He tends to look in the paper. I don't actually look in the paper to see what's on.

One extreme example of this greater tendency for the men to plan their viewing in advance in this way is provided by the man in F3, who at points sounds almost like a classical utilitarian aiming to maximize his pleasure quotient (in terms both of viewing choices and calculations of program time in relation to video tape availability, etc.):

> *F3 Man:* I've got it on tonight on BBC, because it's *Dallas* tonight and I do like *Dallas,* so we started to watch *EastEnders* . . . and then they put on *Emmerdale Farm* because I like that, and we record *EastEnders*—so we don't have to miss out. I normally see it on a Sunday anyway . . . I got it all worked out to tape. I don't mark it in the paper, but I register what's in there. Like tonight it's *Dallas* then at 9PM it's *Widows,* and then we've got *Brubaker* on till the news. So the tape's ready to play straight through . . . what's on at 7:30PM? Oh, *This Is Your Life* and *Coronation Street.* I think BBC is better to record because it doesn't have the adverts. *This Is Your Life* we'll record because it's only on for half an hour, whereas *Dallas* is on for an hour, so you only use half an hour of tape. . . . Yeah, Tuesday if you're watching the other program means you're going to have to cut it off halfway through. I don't bother, so I watch the news at 9PM . . . yes, because there's a film at 9PM on a Tuesday, so what I do, I record the film so I can watch *Miami Vice,* so I can watch the film later.

Or, as he puts it elsewhere, "Evening times, I go through the paper, and I've got all my programs sorted out."

Again, the exceptions to this tendency are again themselves systematic—it is the women in F13 and F16 (who both in fact occupy the traditionally "masculine" position in the family) who do take responsibility for planning their (and their families') viewing.

> *F13 Woman:* No, well, I jealously guard the newspaper because people read the programs out to me and I can't get it into sequence, you know, how many of them are running, so I won't be separated from the paper. I am the program controller. . . . He doesn't know what the programs are all about and I say, "I think you might like this," so we give it a go, and see if he likes it or not . . . Yes, I tell him what I've heard about it and whether he'd enjoy it.
>
> *F16 Man:* Oh yes, I say to her, "What's on the television?" of a night. If there is something worth watching, then I will make a point of watching it.

## Amounts of Viewing

In a number of these families it is acknowledged by both partners that the husband watches television far more than his wife.

> *F12 Woman:* I always say he is a TV addict. He'd have it on all day long.
>
> *F17 Man:* I watch more than she does.
>
> *F9 Woman:* I don't like television. It's very rare . . . it's got to be something very good for me to want to see it. It bores me. He's the one that likes the TV. When he goes out it won't be on.
>
> *F9 Man:* I watch telly—quite a bit of the time. If I'm in I'll always have the telly on. The nights I stay in I do watch the television. If I come in from work I'll turn on the TV. It's more like a habit.
>
> *F9 Woman:* Whereas I'll put records on. It's got to be something really good for me to put the telly on . . . I'm hard to please, I think, over anything on the television.

The women, on the whole, display far less interest in television in general except for the particular soap operas, which they are following.

> *F8 Woman:* I don't think about television an awful lot. If I miss something, I miss it. I don't . . . if I was doing something else, well, you know—I miss it.
>
> *F3 Woman:* I can do a crossword and forget it . . . I am happy with what I see. He watches them films after we've gone to bed.
>
> *F11 Woman:* It really is a question of if there's something he wants to watch, I don't mind. I'm just not really that interested.

It might be objected that my findings in this respect conflict with (and are therefore perhaps "invalidated" by) the common survey finding that women report more viewing hours than men. However, I would argue that this is to do with the fact that in most families women are simply at home more than men and are therefore "available" as viewers more than men. My point is that while women are there, in front of (or rather, to the side, or in earshot of) the set, their dominant viewing practice is much more "bitty" and much less attentive than that of men. This is partly because there are so few programs on, apart from soap operas, which they really like, and partly because their sense of guilt about watching television while surrounded by their domestic obligations makes it hard for them to view attentively. Thus, while more women may be "available" to view television more of the time and their potential viewing hours, considered as a mere matter of quantity, may be greater than those for men, when we consider attentive viewing (the key issue for this research project) their reported viewing is lower than that of men's.

## Television Related Talk

Women seem to show much less reluctance to "admit" that they talk about television to their friends and workmates. Very few men (see below for the exceptions) admit to doing this. It is as if they feel that to admit that they watch too much television (especially with the degree of involvement that would be implied by finding it important enough to talk about) would be to put their very masculinity in question (see the section on program type preference below). The only standard exception is where the men are willing to admit that they talk about sport on television. All this is clearly related to the theme of gender and program choice and the "masculinity/ femininity" syllogism identified there. Some part of this is simply to do with the fact that femininity is a more expressive cultural mode than is masculinity. Thus even if women watch less, with less intent viewing styles, none the less they are inclined to talk about television *more* than men, despite the fact that the men watch more of it, more attentively.

> *F1 Woman:* Actually my mum and my sister don't watch *Dynasty* and I often tell them bits about it. If my sister watches it, she likes it. And I say to her, "Did you watch it?" and she says no. But if there's something especially good on one night—you know, you might see your friends and say, "Did you see so and so last night?" I occasionally miss *Dynasty*. I said to a friend, "What happened?" and she's caught me up, but I tend to see most of the series. Marion used to keep me going, didn't she? Tell me what was happening and that.
>
> *F2 Man:* I might mention something on telly occasionally, but I really don't talk about it to anyone.
>
> *F5 Woman:* At work we constantly talk about *Dallas* and *Dynasty*. We run them down, pick out who we like and who we don't like. What we think should happen next. General chit-chat. I work with quite a few girls, so we have a good old chat. . . . We do have some really interesting discussions about television [at work]. We haven't got much else in common, so we talk a lot about television.
>
> *F6 Woman:* I go round my mate's and she'll say, 'Did you watch *Coronation Street* last night? What about so and so?' And we'll sit there discussing it. I think most women and most young girls do. We always sit down and it's "Do you think she's right last night, what she's done?"

Or, "I wouldn't have done that," or "Wasn't she a cow to him? Do you reckon he'll get . . . I wonder what he's going to do?" Then we sort of fantasize between us, then when I see her the next day she'll say, "You were right," or "See, I told you so."

*F16 Woman:* Mums at school will say, "Have you seen any good videos?" And when *Jewel in the Crown* was on, yes, we'd talk about that. When I'm watching the big epics, the big serials, I would talk about those.

*F8 Daughter:* I like to watch *Brookside*, it's my favorite program . . . 'cause down the stables everyone else watches it—it's something to chat about when we go down there . . .

*F17 Man:* If we do talk, it'll be about something like a news program—something we didn't know anything about—something that's come up that's interesting.

*F18 Woman:* I'll talk about things on telly to my friends. I do. I think it is women who talk about television more so than men. I work with an Indian girl and when *Jewel in the Crown* was on we used to talk about that, because she used to tell me what was different in India. *Gandhi* we had on video. She told me what it was like and why that was interesting. Other than that it's anything. She went to see *Passage to India* and she said it was good, but it was a bit like *Jewel in the Crown*.

*F18 Man:* I won't talk about television at work unless there'd been something like boxing on. I wouldn't talk about *Coronation Street* or a joke on Benny Hill, so other than that, no.

There is one exception to this general pattern—in F10. In this case it is not so much that the woman is any less willing than most of the others in the sample to talk about television but simply that her program tastes (BBC2 drama, etc.) are at odds with those of most of the women on the estate where she lives. However, in describing her own dilemma, and the way in which this disjunction of program tastes functions to isolate her socially, she provides a very acute account of why most of the mothers on her estate do spend so much time talking about television.

*F10 Woman:* Ninety-nine percent of the women I know stay at home to look after their kids, so the only other thing you have to talk about is your housework, or the telly—because you don't go anywhere, you don't do anything. They are talking about what the child did the night before or they are talking about the telly—simply because they don't do anything else.

In the main, the only television material that the men will admit to talking about is sport. The only man who readily admits to talking to anyone about other types of television material is the man in F11, who is a Civil Service manager. Primarily he talks about television at work, quite self-consciously, as a managerial device, simply as a way of "opening up" conversations with his staff, so he can find out how they're getting on, "using it as a topic, rather than you actually wanting to discuss what was on TV," or "as the first sort of gambit for establishing rapport . . . it's always a very good common denominator—"What did you see on the telly last night?"

Interestingly, beyond this conscious use of television as a conversational device in his role at work this middle-class man, exceptionally in my sample, also admits to having the kind of conversations with his men friends about fictional television programs which, on the whole, only the women in my sample are prepared to admit to doing. Thus, this man is a keen fan of *Hill Street Blues* and will readily discuss with his friends issues such as "Should Renko marry? Should Furillo go back on the drink?" Even if there is a conscious tone of self-mocking irony in his account of their discussions, most of the men will not admit to having conversations of this type with the friends (especially about fictional television) at all.

The issue of the differential tendency for women and men to talk about their television viewing is of considerable interest. It could be objected that, as my research is based only on respondents' accounts of their behavior, the findings are unreliable insofar as respondents may have misrepresented their actual behavior—especially when the accounts offered by my respondents seem to conflict with established survey findings. Thus in principle it could be argued that the claims many of the male respondents make about only watching "factual" television are a misrepresentation of their actual behavior, based on their anxiety about admitting to watching fictional programs. However, even if this were the case, it would remain a social fact of some interest that the male respondents felt this compulsion to misrepresent their actual behavior in this particular way. Moreover, this very reluctance to talk about some of the programs they may watch itself has important consequences. Even if it were the case that men and women in fact watched the same range of programs (contrary to the accounts they gave me), the fact that the men are reluctant to talk about watching anything other than factual programs or sport means that their viewing experience is profoundly different from that of the women in the sample. Given that meanings are made not simply in the moment of individual viewing, but also in the subsequent social processes of discussion and "digestion" of material viewed, the men's much greater reluctance to talk about (part of) their viewing will mean that their consumption of television material is of a quite different kind from that of their wives.

### Technology—The Use of the Video Machine

None of the women operate the video recorder themselves to any great extent, relying on husband or children to work it for them. This is simply an effect of their cultural formation as "ignorant" and "disinterested" in relation to machinery in general, and is therefore an obvious point, but one with profound effects none the less. Videos, like automatic control panels, are the possessions of fathers and sons (and occasionally of teenage daughters whose education has made them more confident with machinery than their mothers).

*F2 Woman:* There's been things I've wanted to watch and I didn't under-stand the video enough. She [the daughter] used to understand it more than us.

*F3 Woman:* I'm happy with what I see, so I don't use the video much. I mean lots of the films he records I don't even watch. He watches them after we've gone to bed.

*F6 Man:* I use it most—me and the boys more than anything—mostly to tape the racing, pool, programs we can't watch when they [the women] are watching.

*F6 Woman:* Usually as my son goes out he'll leave me a little list with the girls—not with me, because I wouldn't do it 'cause I don't understand it. Well, I haven't got the patience, and I'll say to the girls, "Tape that for me." Otherwise I don't, very rarely, tape it, no, I'll leave them to tape it, because well, I'm all fingers and thumbs. I'd probably touch the wrong key. That's why I won't touch it.

*F8 Woman:* I don't think about it an awful lot, do I? If I miss something, I miss it. I don't, if I was doing something else, well, you know, I just miss it.

*F9 Woman:* I can't use the video. I tried to tape *Widows* for him and I done it wrong. He went barmy. I don't know what went wrong . . . I always ask him to do it for me because I can't. I always do it wrong. I've never bothered with it.

It is worth noting that these findings have also received provisional confir-mation in the research that Ann Gray has conducted. Given the primary fact of their tangential relation to the video machine, a number of conse-quences seem to follow—for instance, that it is common for the woman to make little contribution to (and have little power over) decisions over hiring video tapes; that it is rare for the woman actually to go into a video tape shop to hire tapes; that when the various members of the family all have their "own" blank tape on which to tape time-shifted material it is common for the woman to be the one to let the others "tape over" something on her tape when theirs is full, and so forth.

As Gray puts it:

> The relationship between the viewer and the television [or video machine] . . . is . . . a relationship which has to be struggled for, won or lost in the dynamic and often chaotic processes of family life . . . The VCR is . . . purchased or rented for use within these already existing structures of power and authority relations between household members, with gender being one of the most significant variations . . . women and men have differential access to technology in general and to domestic technology in particular . . . when a new piece of technology is purchased . . . [for example, the video] it is often already inscribed with gender expectations.[6]

Given that many women routinely operate sophisticated pieces of domestic technology, it is clearly these gender expectations, operating alongside, and framing, any particular difficulties the woman may experience with the spe-cific technology of video that have to be understood as accounting for the alienation which most of the women in the sample express towards the video recorder.

Clearly there are other dimensions to the problem—from the possibility that the expressions of incompetence in relation to the video fall within the classic mode of dependent femininity which therefore "needs" masculine help, to the recognition, as Gray points out, that some women may have developed what she calls a "calculated ignorance" in relation to video, lest operating the video should become yet another of the domestic tasks expected of them.[7]

## *"Solo" Viewing and Guilty Pleasures*

A number of the women in the sample explain that their greatest pleasure is to be able to watch "a nice weepie," or their favorite serial, when the rest of the family aren't there. Only then do they feel free enough of their domestic responsibilities to "indulge" themselves in the kind of attentive viewing which their husbands engage in routinely. Here we enter the territory identified by Brodie and Stoneman . . . , who found that mothers tended to maintain their role as "domestic manager" across program types, as opposed to their husbands' tendency to abandon their manager/parent role when viewing material of particular interest to them. The point is expressed most clearly by the woman in F7 who explains that she particularly enjoys watching early morning television at the weekends—because, as these are the only occasions on which her husband and sons "sleep in," these are, by the same token, the only occasions when she can watch television attentively, without keeping half an eye on the needs of others.

Several of these women will arrange to view a video or film with other women friends during the afternoon. It is the classically feminine way of dealing with conflict—in this case over program choice—by avoiding it, and "rescheduling" the program (often with someone's help in relation to the video) to a point where it can be watched more pleasurably.

> *F5 Woman:* That's one thing we don't have on when he's here, we don't have the games programs on because he hates them. If we women are here on our own—I love it. I think they're lovely . . . If I'm here alone, I try to get something a bit mushy and then I sit here and have a cry, if I'm here on my own. It's not often, but I enjoy that.
>
> *F6 Woman:* If I get a good film on now, I'll tape it and keep it, especially if it's a weepie. I'll sit there and keep it for ages—especially in the afternoon, if there's no one here at all. If I'm tired, I'll put that on—especially in the winter—and it's nice then, 'cause you sit there and there's no one around . . . We get those *Bestsellers* and put them together so you get the whole series together, especially if it's late at night. You're so tired—it's nice to watch the whole film together. We try and keep them, so of an afternoon, if you haven't got a lot to do, you can sit and watch them.
>
> *F7 Woman:* If he's taped something for me I either watch it early in the morning about 6AM . . . I'm always up early, so I come down and

watch it very early about 6 or 6:30 Sunday morning. Now I've sat for an hour this afternoon and watched *Widows*. I like to catch up when no one's here—so I can catch up on what I've lost . . . I love Saturday morning breakfast television. I'm on my own, because no one gets up till late. I come down and really enjoy that program.

*F15 Woman:* I get one of those love stories if he's not in.

*F15 Man:* Yes, I don't want to sit through all that.

*F15 Woman:* Yes, it's on his nights out. It doesn't happen very often.

*F18 Woman:* I don't work Mondays and quite often my friends will get a film and watch it up here. I've done that about three times. A lot of my friends haven't got videos. They can't afford them. So it's something special to them.

My findings in this respect are very clearly supported by Ann Gray's research. Gray argues that her women respondents do have definite preconceptions as to what constitutes a "film for men" as against a "film for women," and on this basis she also develops a typology of viewing contexts, for masculine and feminine viewing (jointly and separately) along with a typology of types of films and programs "appropriate" to these different viewing contexts. Her point is that quite different types of viewing material are felt to be appropriate to the different viewing contexts of the whole family together, male and female partners together, male alone, and female alone. Moreover, she argues that among her respondents, women will only usually watch the kinds of material which they particularly like when their partner is out of the house (at work or leisure), whereas the men will often watch the material which they alone like while their partner is there—she simply would busy herself around the house, or sit without really watching.[8]

As Gray notes, women who are at home all day in fact have obvious opportunities then to view alone, but for many of them daytime television viewing is seen as a kind of "drug" to which they feel, guiltily, that they could easily become addicted.

These comments bring us back to the issue already considered concerning the sense in which the home simply is not a sphere of leisure for women, and thus the ways in which their viewing is constrained by guilt and obligation. However, beyond these considerations there is another dimension which is perhaps even more fundamental. As Ann Gray expresses it, summarizing her research in this area, "It is the most powerful member of the household who defines this hierarchy of serious and silly, important and trivial, which leaves women and their pleasures downgraded, objects and subjects of fun and derision, having to consume [the films and programs they like] almost in secret."[9]

What is at issue here is the guilt that most of these women feel about their own pleasures. They are, on the whole, prepared to concede that the drama and soap opera they like is "silly" or "badly acted" or inconsequential—that is, they accept the terms of a masculine hegemony which defines their preferences as having a low status. Having accepted these terms, they then find it hard to argue for their preferences in a conflict (because, by definition, what their husbands want to watch is more prestigious). They then deal with this by watching their programs, where possible, on their own, or only with their women friends, and will fit such arrangements into the crevices of their domestic timetables.

> *F3 Woman:* What I really like is typical American trash I suppose, but I love it . . . All the American rubbish, really. And I love those Australian films. I think they're really good, those.
> *F17 Woman:* When the children go to bed he has the ultimate choice. I feel guilty if I push for what I want to see because he and the boys want to see the same thing, rather than what a mere woman would want to watch . . . if there was a love film on, I'd be happy to see it and they wouldn't. It's like when you go to pick up a video, instead of getting a nice sloppy love story, I think I can't get that because of the others. I'd feel guilty watching it—because I think I'm getting my pleasure whilst the others aren't getting any pleasure, because they're not interested.
> *F10 Woman:* I would want to watch the sloppy films. He hates them. We just watched a film recently, in the afternoon . . . The school bus is hit by a train and the woman loses her legs. I mean, I don't mind watching that, I know they're going to end up happy.
> *F14 Woman:* I read a lot of crap and all the scandal stuff—Harold Robbins and all that. The last one I really enjoyed, which I shouldn't have enjoyed, was *Lace*. I read it from cover to cover.
> *F9 Woman:* I like tear-jerkers, things that are really sad—more so than anything funny, because I don't like things funny. I like, really, tear-jerkers—something like *Thorn Birds*. Yes, I loved that one. It was really great.

## Program Type Preference

My respondents displayed a notable consistency in this area, whereby masculinity was primarily identified with a strong preference for "factual" programs (news, current affairs, documentaries) and femininity identified with a preference for fictional programs. The observation may be banal, but the strength of the consistency displayed here was remarkable, whenever respondents were asked about program preferences, and especially when asked which programs they would make a point of being in for, and viewing attentively.

> *F6 Man:* I like all documentaries . . . I like watching stuff like that . . . I can watch fiction but I am not a great lover of it.
> *F6 Woman:* He don't like a lot of serials.
> *F6 Man:* It's not my type of stuff. I do like the news, current affairs, all that type of stuff.
> *F6 Woman:* Me and the girls love our serials.
> *F6 Man:* I watch the news all the time, I like the news, current affairs, and all that.
> *F6 Woman:* I don't like it so much.

*F6 Man:* I watch the news every time, 5:40PM, 6PM, 9PM, 10PM, I try to watch.

*F6 Woman:* I just watch the main news so I know what's going on. Once is enough. Then I'm not interested in it.

*F4 Man:* I watch the news almost every night. I like that *Question Time.* That's the sort of thing. The news programs, I tend to watch the news every time it comes on, on the different sides.

*F7 Man:* The news—I always watch the 10PM news. I like documentaries.

*F17 Woman:* Things I like least are things like *World in Action,* when it's more political. The *Money Program*—there's too much talking. On the whole I don't bother too much with those kind of programs. I don't like documentaries. I like something with a story, entertainment, variety.

*F10 Man:* I must admit I prefer more factual television. I enjoy some of the *TV Eye* series. We have just watched the *Trojan War*—that was brilliant. I enjoy series like that—like *Life On Earth*—wildlife programs, and *World in Action . . .* I like to know about things because basically I am a working-class man and I like to know what is happening. I like to know what is happening to me personally . . . I do enjoy watching factual programs. I think I would much rather watch a factual program.

*F11 Man:* I like debate, research programs such as *QED, TV Eye, Panorama, World in Action,* and nature programs. It has to be a subject I am interested in—*Time Watch, Survival.*

Moreover the exceptions to this rule (where the wife prefers "factual programs," etc.), are themselves systematic. This occurs only where the wife, by virtue of educational background, is in the dominant position in terms of cultural capital. One clear instance of this occurs in F13, where the man is a relatively uneducated council flat caretaker, but his wife is a highly literate woman who has recently started to attend college as a mature student. In this family the usual pattern of responses in my sample is reversed:

*F13 Woman:* I like *Newsnight* very much. And *Question Time.* I might say to him, "Look, this is about council estates in Wandsworth," in which case . . . (that is, she thinks he ought to be interested).

*F13 Man:* Yeah, then I'll watch it but if I'm sitting by myself, I'll watch *World in Action* if it's on, but I'd rather come in and watch a game of snooker than come in and watch *World in Action.*

*F13 Woman:* It's the same with *Horizon.* I mean that was very interesting the other night. That was very interesting to me, but he would never have sat through the first five minutes.

As will be seen later, for the same reasons, this woman's responses are "out of line" with those of most women in the sample in relation to a number of the key themes identified. In this particular connection, it is not simply that she is more interested in factual programming than most of the women in the sample, but also that she is less interested in the kind of fictional programming that most of the women respondents are very keen on.

> *F13 Woman: Crossroads* doesn't appeal to me—it's terrible acting, and *EastEnders,* they make me want to scream, they seem so false. *Dallas* and *Dynasty* don't appeal to me, because he's too ridiculous that bloke J.R.

Interestingly, this very same refrain is taken up by the only other woman in the sample who is also now a mature student.

> *F4 Woman:* There's a thing about Australian doctors—it's like a bad *Crossroads,* but this is even worse.

One of the only two other women in the sample who had this negative view of soap opera is not a mature student, but did stay on at school to do A-levels (unlike most of the women interviewed). She is strongly influenced by her husband's political interests and activities and describes herself as "not a feminist, but . . ."

> *F10 Woman:* Before we were married, if *Dallas* was on I never went out . . . but now it has changed. I don't watch any of the soap operas at all. I'm not interested in them now.

What is interesting is that living on a council estate where most of the other women do not share her views of life, not watching this type of programming makes her social life very different, as it means she has one less "thing in common" to talk about to the other mothers on the estate.

> *F10 Woman:* The other mothers watch *Dallas* and *Dynasty* and all that. They can't understand why I don't watch *Crossroads.* . . . Even though I'm a mum, I feel out of it because they don't watch what I do. They watch *Crossroads* and *EastEnders, Gems, The Practice.*

The only other woman to take this negative view of soap opera is the woman in F16. In this family, the wife is again more highly educated than her husband. In this woman's particular case it is prestigious television serials (*Jewel in the Crown,* etc.) which she prefers, "not these 'soapy' ones."

The argument also extends further. First, there is the refrain among the men that watching fiction in the way that their wives do is an improper and almost "irresponsible" activity, an indulgence in fantasy of which they disapprove (compare nineteenth-century views of novel-reading as a "feminizing" activity). This is perhaps best expressed in the words of the couples in F1 and F6, where in both cases the husbands clearly disapprove of their wives' enjoyment of "fantasy" programs.

> *F1 Woman:* That's what's nice about it [*Dynasty*]. It's a dream world, isn't it?
> *F1 Man:* It's a fantasy world that everybody wants to live in, but that— no, I can't get on with that.

The husband in F6 takes the view that watching television in a way is an abrogation of civil responsibility.

> *F6 Man:* People get lost in TV. They fantasize in TV. It's taken over their lives. . . . People today are coming into their front rooms, they shut

their front door and that's it. They identify with that little world on the box.

*F6 Woman:* To me, I think telly's real life.

*F6 Man:* That's what I'm saying. Telly's taken over your life.

*F6 Woman:* Well, I don't mind it taking over my life. It keeps me happy.

The depth of this man's feelings on this point are confirmed later in the interview when he is discussing his general leisure pursuits. He explains that he now regularly goes to the library in the afternoons, and comments that he "didn't realize the library was so good—I thought it was all just fiction." Clearly, for him, "good" and "fiction" are simply incompatible categories.

Secondly, the men's program genre preference for factual programs is also framed by a sense of guilt about the fact that watching television is a "second-best" choice in itself—in relation to a strong belief (not shared in the same way by the majority of the women) that watching television at all is "second-best" to "real" leisure activity:

> *F4 Man:* I'm not usually here. I watch it if there's nothing else to do, but I'd rather not. . . . In the summer I'd rather go out. I can't bear to watch TV if it's still light.
>
> *F16 Man:* I like fishing. I don't care what's on if I'm going fishing. I'm not worried what's on the telly then.
>
> *F11 Man:* If it's good weather we're out in the garden or visiting people . . . I've got a book and a crossword lined up for when she goes out, rather than just watch television.

It is of note that these last quotes are all from families where the men have a particularly strong feeling that "just watching television" is not, on the whole, an acceptable activity. All of these three quotes come from men [whose] particular concern about not "wasting time" just "watching television" seems also to be related to their class position.

Moreover, when the interviews move to a discussion of the fictional programs that the men do watch, consistency is maintained by their preference for a "realistic" situation comedy (a realism of social life) and a rejection of all forms of romance.

These responses seem to fit fairly readily into a kind of "syllogism" of masculine/feminine relationships to television:

| *Masculine* | *Feminine* |
| --- | --- |
| Activity | Watching television |
| Fact programs | Fiction programs |
| Realistic fiction | Romance |

Again, it may be objected that my findings in this respect exaggerate the "real" differences between men's and women's viewing and underestimate the extent of "overlap" viewing as between men and women. Certainly my respondents offer a more sharply differentiated picture of men's and women's viewing than is ordinarily reported by survey work, which seems to show substantial numbers of men watching "fictional" programs and equally substantial numbers of women watching "factual" programs.

However, this apparent contradiction largely rests on the conflation of "viewing" with "viewing attentively and with enjoyment." If we use the first definition, then we can expect considerable degrees of overlap as between men's and women's "viewing." Once we use the second definition, the distinctions as between men's and women's preferred forms of viewing become much more marked. Moreover, even if this were not the case, and it could be demonstrated that my respondents had misrepresented their behavior to me (offering classical masculine and feminine stereotypes which belied the complexity of their actual behavior), it would remain as a social fact of considerable interest that these were the particular forms of misrepresentation which respondents felt constrained to offer of themselves—and these tendencies (for the men to be unable to admit to watching fiction) themselves have real effects in their social lives.

Further it could be objected that the fact that the respondents were interviewed *en famille* may have predisposed them to adopt stereotyped familial roles in the interviews which, if interviewed separately, they would not adhere to—thus again leading to a tendency towards misleading forms of classical gender stereotyping. However, this was precisely the point of interviewing respondents *en famille*—as it was their viewing *en famille* which was at issue, specifically in respect of the ways in which their familial roles interact with their roles as viewers. Accounts which respondents might give of their behavior individually would precisely lack this dimension of family dynamics and role-playing.

More fundamentally, if one poses the issue as one in which "real" behavior (as monitored by survey techniques) is counterposed to "unreliable" accounts offered by respondents, one runs the risk of remaining perpetually stuck at the level of external measurements of behavior which offer no insight or understanding into what the observed behavior means to the people concerned. Thus, monitoring techniques may seem to show that many women are "watching" factual television (as measured in terms of physical presence in front of the set) when, as far as they are concerned, they are in fact paying little or no attention to what is on the screen (not least because it is often a program which they did not themselves choose to watch), as revealed by their comments when asked to give their own accounts of their viewing behavior. Moreover, it is only through viewers' own accounts of why they are interested (or disinterested) in particular types of programs that we can begin to get any sense of the criteria they employ in making the particular viewing choices they do.

## *Channel Preferences*

There is a tendency for men to claim to prefer BBC (and in some cases BBC2) rather than ITV, and for women to do the opposite. This has got some connection with the images of BBC as "educational" and ITV as "entertaining"—thus their preference is simply a homology with the program genre preference explained above. These statements about "channel loyalty" are not to be interpreted as simple empirical statements—not least because the men, having stated a clear preference for BBC, will often then go on to enthuse about programs which are in fact on ITV. However, these statements are still very significant, not only as indicators of the primary connotations associated with each channel. The point is that when the men then refer (unwittingly) to ITV programs, they tend to be speaking of programs which are in fact on ITV but which they would have expected to be on BBC in terms of their understanding of what type of programs each channel primarily offers. The converse is true for many of the women, who state a clear preference for ITV and then often go on to enthuse about BBC programs. Again, the BBC programs they enthuse about tend to be the type of entertainment programs which are primarily associated in their minds with ITV. Whilst this tendency is not consistent across all the families (for instance, in some, neither husband nor wife like BBC) the only cases of the converse pattern (where the wife prefers BBC and the husband ITV) are those where the wife is, because of her educational background, in the "masculine" position.

> *F11 Woman:* Oh yes, I put ITV on. It's habit. Old habits die hard. You tend to watch ITV, so you think that'll be more likely to be what you want to watch. It certainly is for me anyway. I would probably look at ITV first—I'm sure I do. It's not conscious, but on reflection I'm sure I do.
>
> *F12 Woman:* ITV it's got to be for me. BBC will only go on for *Wogan* or something like that.
>
> *F12 Man:* I watch BBC2 quite a lot.
>
> *F5 Woman:* We're inclined to watch ITV more. He always puts on BBC2, no matter what's on.
>
> *F2 Man:* They all tend to watch ITV, but I think BBC2's got a lot of good things.
>
> *F15 Woman:* I think *Dallas* is the only regular program we watch on BBC.
>
> *F8 Woman:* We don't watch BBC, apart from if there is a good film on. It's mainly ITV. We don't watch a lot of BBC programs.
>
> *F16 Man:* Perhaps I watch BBC a bit more, though Channel Four is beginning to get a better sort of program.
>
> *F1 Woman:* We tend to watch ITV more—well, I do.
>
> *F1 Man:* Sometimes I like the documentaries on BBC.

Once again, it is the interview in F13 which provides the clearest exception to this general pattern.

> *F13 Woman:* You notice the rubbish creeping on to the BBC now as well. *Cover Up*, it really belongs on ITV. *Minder*'s on ITV. That's quite amusing but I don't watch it very much.
>
> *F13 Man:* I don't go straight to Channel Four, but she does.
>
> *F13 Woman:* I don't think Channel Four talk down to you so much. I like things like *The Young Ones* and stuff like that, and the other channels

are stuffy. I mean ITV is just soap, soap, and more soap generally. And Channel Four seems to be more adult. Even the films they show are much better. That's the sort of thing you would see on Channel Four you wouldn't see anywhere else. If there's anything new on and I don't know what it is, then I'll give it a go if it's on Channel Four, but if it's on ITV I might well not try so hard to watch it.

## National and Local News Programming

As has been noted, it is men and not women who tend to claim an interest in news programming. Interestingly, this pattern varies when we consider local news programs, which a number of the women claim to like. In several cases they give very cogent reasons for this—as they don't understand what the "pound going up or down" is about, and as it has no experential bearing on their lives they're not interested in it. However, if there has been a crime (for instance, a rape) in their local area, they feel they need to know about it, both for their own sake and their children's sakes. This connects directly to their other expressed interest—in programs like *Police Five,* or programs warning of domestic dangers. In both these kinds of cases the program material has a practical value to them in terms of their domestic responsibilities, and thus they will make a point of watching it. Conversely, they frequently see themselves as having no practical relation to the area of national and international politics presented in the main news and therefore don't watch it.

The clearest expression of this perspective is that offered by the woman in F9. As she explains, local news is of considerable interest to her: "Sometimes I like to watch the news if it's something that's gone on—like where that little boy's gone and what's happened to him. Otherwise I don't, not unless it's local only when there's something that's happened local." Whereas national news just "gets on her nerves." "I can't stand *World in Action* and *Panorama* and all that. It's wars all the time. You know, it gets on your nerves."

Her explanation of precisely why she doesn't view the national news is worth considering in some detail, as she explains her perfectly cogent reasons for not watching it. "What I read in the papers and listen to the news is enough for me. I don't want to know about the Chancellor somebody in Germany and all that. When I've seen it once I don't want to see it again. I hate seeing it again—because it's on at breakfast time, dinner time, and tea time, you know, the same news all day long. It bores me. What's going on in the world? I don't understand it all, so I don't like to listen to that. I watch—like those little kids—that gets to me, I want to know about it. Or if there's actually some crime in Wandsworth, like rapes and all the rest of it I want to read up on that, if they've been caught and locked away. As for like when the guy says the pound's gone up and the pound's gone down, I don't want to know about all that, 'cause I don't understand it. It's complete ignorance really. If I was to understand it all, I would probably get interested in it."

However, her response is merely the clearest expression of what is a very common pattern among most of the women in the sample.

> *F15 Woman: Reporting London*—more than anything in the news sort of line—we watch that. That's what's happening where we're living, in this city.
>
> *F2 Woman:* I like the one that's just after the main news for about five minutes—the *Thames News*. I've got something about that—I have to see it. Yes, it's only on for about five minutes. They tell you all the news, not as much as the other one, though they seem to tell you more, you know.
>
> *F3 Woman:* I like *Thames News*. I watch the six o'clock news and *Thames News* again on Friday with Michael Aspel.
>
> *F5 Woman:* ITV's best because that's *London Weekend*. It tells you what's going on in your area, so you're more interested.
>
> *F6 Woman:* I just watch the main news. So I know what's going on. Once is enough—then I'm not interested in it.
>
> *F14 Woman:* I'm not interested in news and all that stuff.
>
> *F8 Woman:* I find the news really depressing. We'll watch the 6PM news maybe. That's about the only time we have it on.
>
> *F3 Woman:* The only sort of film I would say to them to watch is an educational film, like if the police was to put one on warning children about strangers, then I'd make sure they'd watch it. I mean when Jimmy Saville used to have them on a Sunday—*Dangers in the Home,* I used to say to them, "Now watch this because it's interesting. You can learn from them." And that crime program—*Crimewatch*—that's a good program to watch because it gives you some idea of what to look out for . . . what the kids should look out for as well.

## Comedy Preferences

Quite simply, a significant number of the women interviewed display a strong dislike of "zany" comedy as a genre, and of *The Young Ones* in particular as an instance of this genre. On the other hand, their husbands, sons, and teenage daughters all tend to like this type of comedy very much.

The main point here seems to be that for women for whom maintaining domestic order is their primary responsibility and concern, comedy of this kind is seen as something of an insult, in so far as it is premised on the notion that domestic disorder is funny. That is why, for instance, the woman in F9 dislikes it to the extent that she does. As the person in the household who is responsible when she comes in from work for getting everybody's socks clean and their food cooked, she does not find domestic disorder particularly amusing.

> *F1 Man:* We've got virtually the same taste apart from *The Young Ones.* I really like that. She hates it.

*F3 Woman: The Young Ones*—I think that's a revolting programme. The
things they come out with in it. But you [her daughter] think it's funny,
you think it's really funny, don't you?

*F9 Woman: Kenny Everett* can't make me laugh. *Lenny Henry* doesn't make
me laugh. You see, I don't find all that funny really . . . That's sickly
[*The Young Ones*]. I think they're real sick. If that's on I do the washing,
anything.

Again, the women who prove exceptions to this basic pattern are themselves systematic. It is the women who have moved out of their traditional,
feminine position who do not conform to this basic pattern of response.

*F14 Woman:* I like *The Young Ones* of course—though I don't actually
turn the TV on to watch it.

In her case her recent work experience at a drama college and her present involvement in setting up her own business go some part of the way
towards explaining why she does not conform to the general pattern of
feminine responses in this particular respect. Indeed her own prioritization
of her career over her domestic responsibilities is evident in the very layout
of her own home: she is far less "houseproud" than most of the other women
in the sample.

The other two women who are at odds with the standard pattern are
the women from F10 and F13. As has already been mentioned, the woman
in F10 has very different tastes in television programming from most of the
other mothers that she knows. Their failure to grasp what it is that she likes
about *The Young Ones* is simply the sharpest edge of this fundamental divide.

*F10 Woman:* We were watching *The Young Ones* the other night. Oh, we
did laugh. You see, none of the mums watch that. It's too intellectual
for them. Now that has me in stitches, where they are locking him in
the fridge and he falls out. And to them it's lost—it does not mean a
thing. It must be on Monday night, because I went out on Tuesday and
I was still laughing about it, and I asked if they'd seen it and they all
asked what's *The Young Ones?* You can't explain it. Even though I'm a
mum I feel out of it because they don't watch what I do.

The third exception is the woman student in F13 who also positively
likes zany comedy of one kind or another.

*F13 Woman:* I don't think Channel Four talk down to you so much. And
I like things like *The Young Ones* and stuff like that. The other channels
are so stuffy.

It is worth noting that all of these three women who constitute the
exceptions are in identifiably different cultural positions than most of the
women in the sample. These are precisely the kind of systematic exceptions
that I would claim prove the rule.

## Notes

1. See Ann Gray, "Women and Video" in Helen Baehr and Gillian Dyer (eds.) *Boxed In: Women On and In Television,* Routledge Kegan Paul, 1987.

2. Charlotte Brunsdon, *Women Watching Television,* paper to Women and The Electronic Mass Media Conference. Copenhagen, 1986, unpublished.

3. Peter Collett and Roger Lamb, *Watching People Watching Television,* Report to the Independent Broadcasting Authority, 1986.

4. See D. Marsden and E. Duff, *Workless: Some Unemployed Men and Their Families,* Penguin, 1975, for more on these issues.

5. Brunsdon, p. 5.

6. Gray, "Women and Video."

7. Ibid.

8. Ibid.

9. Ibid.

# IV

## OVERVIEWS

Again, headings and designations are difficult. Philosophies of television might be more appropriate to describe this section, but that weighty term seems to overwhelm the medium. Theory is clearly a potential descriptor, and certainly "theory" has been the major offering in Television Studies as in many other fields in recent years. As the essays in the previous sections have demonstrated in their own foregrounding of theoretical issues, even the most applied analyses of TV require that writers position themselves fairly precisely within ongoing debates about how we define our questions, establish our categories, choose and apply our evidence.

Still, theory, like philosophy implies either understanding or predictive power that is missing from most academic, scholarly discourse, even when practiced by theorists and philosophers. In my view it is preferable, partic-ularly in the humanities, to consider our attempts at understanding as parts of a grand, ongoing discussion, a mental conversation—made visible in writing—through which we attempt to improve the contributions we and our students can make as citizens.

It should not be surprising, then, that the essays in this section often deal first with the ways in which television, as a large, complex, varied, and variable entity enters social life. My essay, written with Paul Hirsch, remains my best understanding of the medium. As often pointed out, it does rely too heavily on too simple a notion of social and cultural pluralism. It does not fully address the issue of power. That issue is clearly implied, however, and perhaps by shifting the term "forum" to "arena," with images of con-flict made more central, the model is made even more useful. The central problem faced by this essay, a problem illustrated throughout this collection with other models, is how we should construct an analytical approach to the contradictions and variations within television. In our view, these contradictions locate, identify, construct, and contextualize conflicts and contradictions in society at large.

Todd Gitlin disagrees strongly, suggesting that almost all the narrative strategies and economic organization of television works to mask or defuse those contradictions and conflicts. Where they are recognized, his model suggests they are rapidly and successfully contained. These two essays remain clear statements of fundamental issues underlying and directing much of what has gone on in Television Studies in the last decade.

Similarly, David Thorburn's essay, "Television Melodrama," outlines many of the fundamental issues regarding both the aesthetics and the cultural uses of television, issues addressed by many other writers since his first definition of them. His central question—how is television a distinctive medium, with its own distinctive aesthetic properties—continues to drive many, if not most close textual analyses of television content. Moreover, the historical and production oriented studies offered in Part I often focus on the same questions from a different point of view. How did, and how does television emerge as a distinctive form? Thorburn's essay outlines many of the topics that are approached in production studies.

Formal distinction is also surely a central problem in Jane Feuer's work as represented in her own essay on melodrama and serial form. But she places equal emphasis on social and ideological questions. Not content to let the these topics remain at the general level they sometimes take in Thorburn's essay, she seeks to relate formal and ideological issues more specifically and directly. Taken together, these two essays provide insightful overviews to guide us in understanding television's expressive functions. But they also suggest the need for still more precise application of formal questions to more and more program types, and an equal need to relate these analyses to a concern with the real ideological power of television. Television Studies has entered, and become central to, the humanities' ongoing discussion of expressivity and power at a fairly late date. As a field of study, it can hardly afford easily compartmentalized approaches to its problems. While individuals may emphasize one aspect or another, most must recognize the interaction of all these questions in any given study.

This is why Eileen Meehan argues that we must find strategies for discussing television as both cultural and economic product. Her outline of categories, matching terms from one vocabulary with those from another shows once again the "focusing" aspect of television. This medium is the locus of conjunctions, drawing the social-ideological, and the aesthetic-cultural into powerful relationships. Either without the other, as Meehan argues, is deficient.

But we must now also recognize that even the most perceptive of these overviews are grounded in a model of television that calls them into question. Indeed, they are included here to serve, in part, a historical function.

They are reminders, more than specifically historical studies, of a television that "was." The passing of the network era, the coming of the new technologies and the new economic organization, with their attendant new aesthetic possibilities and new forms of narrative, present us with the difficulty of beginning again.

These earlier overview models of television are by no means useless. Nor will they disappear. Like reruns on a targeted cable channel, they are layered over with new meaning, given new contexts. What does a "cultural forum" look like in a 50-or 500-channel television world? How does the hegemonic process continue to operate in this new context? How can we apply aesthetic and formal analysis when forms shift from week to week? How do the cultural and the economic intersect and interact under the new technological regimes of individualized media devices? These are the questions that Television Studies must now confront.

The final four essays, then, are attempts to describe, define, understand, indeed predict some answers, some ways of analyzing and understanding the newer forms of television.

For Michael Saenz this means understanding how the actual practice of constructing cultural categories can be described in the new context. His essay should be read in conjunction with those in Part III, for it has strong implications about the processes of reception. But it is also historical in tone, chronicling significant shifts in the organization of television that can lead to shifts in experience.

John Caughie sets such changes not so much in historical context, though that element is present, as in the context of geography and social difference. He asks whether or not—and how—our experiences of television are rooted in our own social and personal experience as members of specific societies and cultures. Travel is at the heart of his analysis, and he argues that we can "see" differently when we compare not only the organization of television systems, but the ways in which those varying organizations variously organize our experience with the medium. He reminds us that there are indeed still such differences at work. But he also suggests that television, and travel, allow us to blend these differences into new mixtures.

James Hay also recognizes those differences, but goes farther, suggesting that their grounding in actual geography and social organization might be disappearing. Multichannel, multinational television is creating, in his view, a new geography of the imagination. How shall we "travel" in this new landscape? What are its meanings and markers? Do cities now take their identity more from their location in the television schedule than on the land? Do nations know themselves by forms of citizenship, history, and language, or by programming strategy and satellite connections?

Clearly, the questions and issues posed by Saenz, Caughie, and Hay bring us again to considerations of postmodernism as a descriptive category, and raise the question of how we might best use it for analytical and explanatory purposes. Once again we are faced with the conundrum of how television serves to define both causes and effects within postmodernism.

It is into this context that Eric Michaels projects his study of Aboriginal television. And with his essay, we come, in some ways, full circle. As Lynn Spigel reminded us that we "learned" to live with and use television, Michaels shows us that this process continues. As suggested in the Introduction, his work condenses many of the issues addressed throughout this book and points directions for how our work on and with those issues might be refined. As is often the case with the best anthropologically grounded work, it is the comparison that teaches us how little we know, in this case about television—and perhaps about postmodernism. Perhaps we are neither so modern, nor so post, as we might imagine.

Or perhaps criticism, finally, is, at its best, an exercise of the imagination that sorts out paths of understanding. Michaels's account of the coming of television to Yuendumu offers paths that are not truly optional. They must be followed. The best purpose of the essays in this collection, old and new, is to point us toward the future.

# Television as a Cultural Forum

## HORACE NEWCOMB
## AND PAUL M. HIRSCH

A cultural basis for the analysis and criticism of television is, for us, the bridge between a concern for television as a communications medium, central to contemporary society, and television as aesthetic object, the expressive medium that, through its storytelling functions, unites and examines a culture. The shortcomings of each of these approaches taken alone are manifold.

The first is based primarily in a concern for understanding specific messages that may have specific effects, and grounds its analysis in "communication" narrowly defined. Complexities of image, style, resonance, narrativity, history, metaphor, and so on are reduced in favor of that content that can be more precisely, some say more objectively, described. The content categories are not allowed to emerge from the text, as is the case in naturalistic observation and in textual analysis. Rather they are predefined in order to be measured more easily. The incidence of certain content categories may be cited as significant, or their "effects" more clearly correlated with some behavior. This concern for measuring is, of course, the result of conceiving television in one way rather than another, as "communication" rather than as "art."

The narrowest versions of this form of analysis need not concern us here. It is to the best versions that we must look, to those that do admit to a range of aesthetic expression and something of a variety of reception. Even when we examine these closely, however, we see that they often as-

Reprinted from *Quarterly Review of Film Studies,* Summer 1983, with permission of the publisher and authors. Copyright © 1983.

sume a monolithic "meaning" in television content. The concern is for "dominant" messages embedded in the pleasant disguise of fictional entertainment, and the concern of the researcher is often that the control of these messages is, more than anything else, a complex sort of political control. The critique that emerges, then, is consciously or unconsciously a critique of the society that is transmitting and maintaining the dominant ideology with the assistance, again conscious or unconscious, of those who control communications technologies and businesses. (Ironically, this perspective does not depend on political perspective or persuasion. It is held by groups on the "right" who see American values being subverted, as well as by those on the "left" who see American values being imposed.)

Such a position assumes that the audience shares or "gets" the same messages and their meanings as the researcher finds. At times, like the literary critic, the researcher assumes this on the basis of superior insight, technique, or sensibility. In a more "scientific" manner the researcher may seek to establish a correlation between the discovered messages and the understanding of the audience. Rarely, however, does the message analyst allow for the possibility that the audience, while sharing this one meaning, may create many others that have not been examined, asked about, or controlled for.

The television "critic" on the other hand, often basing his work on the analysis of literature or film, succeeds in calling attention to the distinctive qualities of the medium, to the special nature of television fiction. But this approach all too often ignores important questions of production and reception. Intent on correcting what it takes to be a skewed interest in such matters, it often avoids the "business" of television and its "technology." These critics, much like their counterparts in the social sciences, usually assume that viewers should understand programs in the way the critic does, or that the audience is incapable of properly evaluating the entertaining work and should accept the critic's superior judgment.

The differences between the two views of what television is and does rest, in part, on the now familiar distinction between transportation and ritual views of communication processes. The social scientific, or communication theory model outlined above (and we do not claim that it is an exhaustive description) rests most thoroughly on the transportation view. As articulated by James Carey, this model holds that communication is a "process of transmitting messages at a distance for the purpose of control. The archetypal case of communication then is persuasion, attitude change, behavior modification, socialization through the transmission of information, influence, or conditioning."[1]

The more "literary" or "aesthetically based" approach leans toward, but hardly comes to terms with, ritual models of communication. As put by Carey, the ritual view sees communication "not directed toward the extension of messages in space but the maintenance of society in time; not the act of imparting information but the representation of shared beliefs."[2]

Carey also cuts through the middle of these definitions with a more succinct one of his own: "Communication is a symbolic process whereby reality is produced, maintained, repaired, and transformed."[3] It is in the attempt to amplify this basic observation that we present a cultural basis for the analysis of television. We hardly suggest that such an approach is entirely new, or that others are unaware of or do not share many of our assumptions. On the contrary, we find a growing awareness in many disciplines of the nature of symbolic thought, communication, and action, and we see attempts to understand television emerging rapidly from this body of shared concerns.[4]

Our own model for television is grounded in an examination of the cultural role of entertainment and parallels this with a close analysis of television program content in all its various textual levels and forms. We focus on the collective, cultural view of the social construction and negotiation of reality, on the creation of what Carey refers to as "public thought."[5] It is not difficult to see television as central to this process of public thinking. As Hirsch has pointed out,[6] it is now our national medium, replacing those media— film, radio, picture magazines, newspapers—that once served a similar function. Those who create for such media are, in the words of anthropologist Marshall Sahlins, "hucksters of the symbol."[7] They are cultural *bricoleurs,* seeking and creating new meaning in the combination of cultural elements with embedded significance. They respond to real events, changes in social structure and organization, and to shifts in attitude and value. They also respond to technological shift, the coming of cable or the use of videotape recorders. We think it is clear that the television producer should be added to Sahlins's list of "hucksters." They work in precisely the manner he describes, as do television writers and, to a lesser extent, directors and actors. So too do programmers and network executives who must make decisions about the programs they purchase, develop, and air. At each step of this complicated process they function as cultural interpreters.

Similar notions have often been outlined by scholars of popular culture focusing on the formal characteristics of popular entertainment.[8] To those insights cultural theory adds the possibility of matching formal analysis with cultural and social practice. The best theoretical explanation for this link is suggested to us in the continuing work of anthropologist Victor Turner. This work focuses on cultural ritual and reminds us that ritual must be seen as process rather than as product, a notion not often applied to the study of television, yet crucial to an adequate understanding of the medium.

Specifically we make use of one aspect of Turner's analysis, his view of the *liminal* stage of the ritual process. This is the "inbetween" stage, when one is neither totally in nor out of society. It is a stage of license, when rules may be broken or bent, when roles may be reversed, when categories may be overturned. Its essence, suggests Turner, is to be found in its release from normal constraints, making possible the deconstruction of the "uninteresting" constructions of common sense, the "meaningfulness of ordinary life," . . . into cultural units which may then be reconstructed in novel ways, some of them bizarre to the point of monstrosity. . . . Liminality is the domain of the "interesting" or of "uncommon sense."[9]

Turner does not limit this observation to traditional societies engaged in the *practice* of ritual. He also applies his views to postindustrial, complex societies. In doing so he finds the liminal domain in the arts—all of them.[10] "The dismemberment of ritual has . . . provided the opportunity of theatre in the high culture and carnival at the folk level. A multiplicity of desacralized performative genres have assumed, prismatically, the task of plural cultural reflexivity."[11] In short, contemporary cultures examine themselves through their arts, much as traditional societies do via the experience of ritual. Ritual and the arts offer a metalanguage, a way of understanding who and what we are, how values and attitudes are adjusted, how meaning shifts.

In contributing to this process, particularly in American society, where its role is central, television fulfills what Fiske and Hartley refer to as the "bardic function" of contemporary societies.[12] In its role as central cultural medium it presents a multiplicity of meanings rather than a monolithic dominant point of view. It often focuses on our most prevalent concerns, our deepest dilemmas. Our most traditional views, those that are repressive and reactionary, as well as those that are subversive and emancipatory, are upheld, examined, maintained, and transformed. The emphasis is on process rather than product, on discussion rather than indoctrination, on contradiction and confusion rather than coherence. It is with this view that we turn to an analysis of the texts of television that demonstrates and supports the conception of television as a cultural forum.

This new perspective requires that we revise some of our notions regarding television analysis, criticism, and research. The function of the creator as *bricoleur,* taken from Sahlins, is again indicated and clarified. The focus on "uncommon sense," on the freedom afforded by the idea of television as a liminal realm helps us to understand the reliance on and interest in forms, plots, and character types that are not at all familiar in our lived experience. The skewed demography of the world of television is not quite so bizarre and repressive once we admit that it is the realm in which we allow our monsters to come out and play, our dreams to be wrought into pictures, our fantasies transformed into plot structures. Cowboys, detectives, bionic men, and great green hulks; fatherly physicians, glamorous female detectives, and tightly knit families living out the pain of the Great Depression; all these become part of the dramatic logic of public thought.

Shows such as *Fantasy Island* and *Love Boat,* difficult to account for within traditional critical systems except as examples of trivia and romance, are easily understood. Islands and boats are among the most fitting liminal metaphors, as Homer, Bacon, Shakespeare, and Melville, among others, have recognized. So, too, are the worlds of the Western and the detective story. With this view we can see the "bizarre" world of situation comedy as a means of deconstructing the world of "common sense" in which all, or most, of us live and work. It also enables us to explain such strange phenomena as game shows and late night talk fests. In short, almost any version of the television text functions as a forum in which important cultural topics may be considered. We illustrate this not with a contemporary program where problems almost always appear on the surface of the show, but with an episode of *Father Knows Best* from the early 1960s. We begin by noting that *FKB* is often cited as an innocuous series, constructed around unstinting paeans to American middle-class virtues and blissfully ignorant of social conflict. In short, it is precisely the sort of television program that reproduces dominant ideology by lulling its audience into a dream world where the status quo is the only status.

In the episode in question Betty Anderson, the older daughter in the family, breaks a great many rules by deciding that she will become an engineer. Over great protest, she is given an internship with a surveying crew as part of a high school "career education" program. But the head of the surveying crew, a young college student, drives her away with taunts and insensitivity. She walks off the job on the first day. Later in the week the young man comes to the Anderson home where Jim Anderson chides him with fatherly anger. The young man apologizes and Betty, overhearing him from the other room, runs upstairs, changes clothes, and comes down. The show ends with their flirtation underway.

Traditional ideological criticism, conducted from the communications or the textual analysis perspective, would remark on the way in which social conflict is ultimately subordinated in this dramatic structure to the personal, the emotional. Commentary would focus on the way in which the questioning of the role structure is shifted away from the world of work to the domestic arena. The emphasis would be on the conclusion of the episode in which Betty's real problem of identity and sex-role, and society's problem of sex-role discrimination, is bound by a more traditional conflict and thereby defused, contained, and redirected. Such a reading is possible, indeed accurate.

We would point out, however, that our emotional sympathy is with Betty throughout this episode. Nowhere does the text instruct the viewer that her concerns are unnatural, no matter how unnaturally they may be framed by other members of the cast. Every argument that can be made for a strong feminist perspective is condensed into the brief, half-hour presentation. The concept of the cultural forum, then, offers a different interpretation. We suggest that in popular culture generally, in television specifically, the raising of questions is as important as the answering of them. That is, it is equally important that an audience be introduced to the problems surrounding sex-role discrimination as it is to conclude the episode in a traditional manner. Indeed, it would be startling to think that mainstream texts in mass society would overtly challenge dominant ideas. But this hardly prevents the oppositional ideas from appearing. Put another way, we argue that television does not present firm ideological conclusions—despite its *formal* conclusions—so much as it *comments on* ideological problems. The conflicts we see in television drama, embedded in familiar and nonthreatening frames, are conflicts ongoing in American social experience and cultural history. In a few cases we might see strong perspectives that argue for the absolute correctness of one point of view or another. But for the most part the rhetoric of television drama is a rhetoric of discussion. Shows such as *All in the Family,* or *The Defenders,* or *Gunsmoke,* which raise the forum/ discussion to an intense and obvious level, often make best use of the medium and become highly successful. We see statements *about* the issues and it should be clear that ideological positions can be balanced within the forum by others from a different perspective.

We recognize, of course, that this variety works for the most part within the limits of American monopoly-capitalism and within the range of American pluralism. It is an effective pluralistic forum only insofar as American political pluralism is or can be.[13] We also note, however, that one of the primary functions of the popular culture forum, the television forum, is to monitor the limits and the effectiveness of this pluralism, perhaps the only "public" forum in which this role is performed. As content shifts and attracts the attention of groups and individuals, criticism and reform can be initiated. We will have more to say on this topic shortly.

Our intention here is hardly to argue for the richness of *Father Knows Best* as a television text or as social commentary. Indeed, in our view, any emphasis on individual episodes, series, or even genres, misses the central point of the forum concept. While each of these units can and does present its audiences with incredibly mixed ideas, it is television as a whole system that presents a mass audience with the range and variety of ideas and ideologies inherent in American culture. In order to fully understand the role of television in that culture, we must examine a variety of analytical foci and, finally, see them as parts of a greater whole.

We can, for instance, concentrate on a single episode of television content, as we have done in our example. In our view most television shows offer something of this range of complexity. Not every one of them treats social problems of such immediacy, but submerged in any episode are assumptions about who and what we are. Conflicting viewpoints of social issues are, in fact, the elements that structure most television programs.

At the series level this complexity is heightened. In spite of notions to the contrary, most television shows do change over time. Stanley Cavell has recently suggested that this serial nature of television is perhaps its defining characteristic.[14] By contrast we see that feature only as a primary aspect of the rhetoric of television, one that shifts meaning and shades ideology as series develop. Even a series such as *The Brady Bunch* dealt with ever more complex issues merely because the children, on whom the show focused, grew older. In other cases, shows such as *The Waltons* shifted in content and meaning because they represented shifts in historical time. As the series moved out of the period of the Great Depression, through World War II, and into the postwar period, its tone and emphasis shifted too. In some cases, of course, this sort of change is structured into the show from the beginning, even when the appearance is that of static, undeveloping nature. In *All in the Family* the possibility of change and Archie's resistance to it form the central dramatic problem and offer the central opportunity for dramatic richness, a richness that has developed over many years until the character we now see bears little resemblance to the one we met in the beginning. This is also true of *M\*A\*S\*H*, although there the structured conflicts have more to do with framing than with character development. In *M\*A\*S\*H* we are caught in an anti-war rhetoric that cannot end a war. A truly radical alternative, a desertion or an insurrection, would end the series. But it would also end the "discussion" of this issue. We remain trapped, like American culture in its historical reality, with a dream and the rhetoric of peace and with a bitter experience that denies them.

The model of the forum extends beyond the use of the series with attention to genre. One tendency of genre studies has been to focus on similarities within forms, to indicate the ways in which all Westerns, situation comedies, detective shows, and so on are alike. Clearly, however, it is in the economic interests of producers to build on audience familiarity with generic patterns and instill novelty into those generically based presentations. Truly innovative forms that use the generic base as a foundation are likely to be among the more successful shows. This also means that the shows, despite generic similarity, will carry individual rhetorical slants. As a result, while shows like *M\*A\*S\*H, The Mary Tyler Moore Show,* and *All in the Family* may all treat similar issues, those issues will have different meanings because of the variations in character, tone, history, style, and so on, despite a general "liberal" tone. Other shows, minus that tone, will clash in varying degrees. The notion that they are all, in some sense, "situation comedies" does not adequately explain the treatment of ideas within them.

This hardly diminishes the strength of generic variation as yet another version of differences within the forum. The rhetoric of the soap opera *pattern* is different from that of the situation comedy and that of the detective show. Thus, when similar topics are treated within different generic frames another level of "discussion" is at work.

It is for this reason that we find it important to examine strips of television programming, "flow" as Raymond Williams refers to it.[15] Within these flow strips we may find opposing ideas abutting one another. We may find opposing treatments of the same ideas. And we will certainly find a viewing behavior that is more akin to actual experience than that found when concentrating on the individual show, the series, or the genre. The forum model, then, has led us into a new exploration of the definition of the television text. We are now examining the "viewing strip" as a potential text and are discovering that in the range of options offered by any given evening's television, the forum is indeed a more accurate model of what goes on *within* television than any other that we know of. By taping entire weeks of television content, and tracing various potential strips in the body of that week, we can construct a huge range of potential "texts" that may have been seen by individual viewers.

Each level of text—the strip as text, the television week, the television day—is compounded yet again by the history of the medium. Our hypothesis is that we might track the history of America's social discussions of the past three decades by examining the multiple rhetorics of television during that period. Given the problematic state of television archiving, a careful study of that hypothesis presents an enormous difficulty. It is, nevertheless, an exciting prospect.

Clearly, our emphasis is on the treatment of issues, on rhetoric. We recognize the validity of analytical structures that emphasize television's skewed demographic patterns, its particular social aberrations, or other "unrealistic distortions" of the world of experience. But we also recognize that in order to make sense of those structures and patterns researchers return again and again to the "meaning" of that television world, to the processes and problems of interpretation. In our view this practice is hardly limited to those of us who study television. It is also open to audiences who view it each evening and to professionals who create for the medium.

The goal of every producer is to create the difference that makes a difference, to maintain an audience with sufficient reference to the known and recognized, but to move ahead into something that distinguishes his show for the program buyer, the scheduler, and most importantly, for the mass audience. As recent work by Newcomb and Alley shows,[16] the goal of many producers, the most successful and powerful ones, is also to include personal ideas in their work, to use television as all artists use their media, as means of personal expression. Given this goal it is possible to examine the work of individual producers as other units of analysis and to compare the work of different producers as expressions within the forum. We need only think of the work of Quinn Martin and Jack Webb, or to contrast their work with that of Norman Lear or Gary Marshall, to recognize the individuality at work within television making. Choices by producers to work in certain generic forms, to express certain political, moral, and ethical attitudes, to explore certain sociocultural topics, all affect the nature of the ultimate "flow text" of television seen by viewers and assure a range of variations within that text.

The existence of this variation is borne out by varying responses among those who view television. A degree of this variance occurs among professional television critics who like and dislike shows for different reasons. But because television critics, certainly in American journalistic situations, are more alike than different in many ways, a more important indicator of the range of responses is that found among "ordinary" viewers, or the disagreements implied by audience acceptance and enthusiasm for program material soundly disavowed by professional critics. Work by Himmleweit in England[17] and Neuman in America[18] indicates that individual viewers do function as "critics," do make important distinctions, and are able, under certain circumstances, to articulate the bases for their judgments. While this work is just beginning, it is still possible to suggest from anecdotal evidence that people agree and disagree with television for a variety of reasons. They find in television texts representations of and challenges to their own ideas, and must somehow come to terms with what is there.

If disagreements cut too deeply into the value structure of the individual, if television threatens the sense of cultural security, the individual may take steps to engage the medium at the level of personal action. Most often this occurs in the form of letters to the network or to local stations, and again, the pattern is not new to television. It has occurred with every other mass medium in modern industrial society.

Nor is it merely the formation of groups or the expression of personal points of view that indicates the working of a forum. It is the *range* of response, the directly contradictory readings of the medium, that cue us to its multiple meanings. Groups may object to the same programs, for example, for entirely opposing reasons. In *Charlie's Angels* feminists may find yet another example of sexist repression, while fundamentalist religious groups may find examples of moral decay expressed in the sexual freedom, the personal appearance, or the "unfeminine" behavior of the protagonists. Other viewers doubtless find the expression of meaningful liberation of women. At this level, the point is hardly that one group is "right" and another "wrong," much less that one is "right" while the other is "left." Individuals and groups are, for many reasons, involved in making their own meanings from the television text.

This variation in interpretive strategies can be related to suggestions made by Stuart Hall in his influential essay, "Encoding and Decoding in the Television Discourse."[19] There he suggests three basic modes of interpretation, corresponding to the interpreter's political stance within the social structure. The interpretation may be "dominant," accepting the prevailing ideological structure. It may be "oppositional," rejecting the basic aspects of the structure. Or it may be "negotiated," creating a sort of personal synthesis. As later work by some of Hall's colleagues suggests, however, it quickly becomes necessary to expand the range of possible interpretations.[20] Following these suggestions to a radical extreme it might be possible to argue that every individual interpretation of television content could, in some way, be "different." Clearly, however, communication is dependent on a greater degree of shared meanings, and expressions of popular entertainment are perhaps even more dependent on the shared level than many other forms of discourse. Our concern then is for the ways in which interpretation is negotiated in society. Special interest groups that focus, at times, on television provide us with readily available resources for the study of interpretive practices.

We see these groups as representative of metaphoric "fault lines" in American society. Television is the terrain in which the faults are expressed and worked out. In studying the groups, their rhetoric, the issues on which they focus, their tactics, their forms of organization, we hope to demonstrate that the idea of the "forum" is more than a metaphor in its own right. In forming special interest groups, or in using such groups to speak about television, citizens actually enter the forum. Television shoves them toward action, toward expression of ideas and values. At this level the model of "television as a cultural forum" enables us to examine "the sociology of interpretation."

Here much attention needs to be given to the historical aspects of this form of activity. How has the definition of issues changed over time? How has that change correlated with change in the television texts? These are important questions which, while difficult to study, are crucial to a full understanding of the role of television in culture. It is primarily through this sort of study that we will be able to define much more precisely the limits of the forum, for groups form monitoring devices that alert us to shortcomings not only in the world of television representation, but to the world of political experience as well. We know, for example, that because of heightened concern on the part of special interest groups, and responses from the creative and institutional communities of television industries, the "fictional" population of black citizens now roughly equals that of the actual population. Regardless of whether such a match is "good" or "necessary," regardless of the nature of the depiction of blacks on television, this indicates that the forum extends beyond the screen. The issue of violence, also deserving close study, is more mixed, varying from year to year. The influence of groups, of individuals, of studies, of the terrible consequences of murder and assassination, however, cannot be denied. Television does not exist in a realm of its own, cut off from the influence of citizens. Our aim is to discover, as precisely as possible, the ways in which the varied worlds interact.

Throughout this kind of analysis, then, it is necessary to cite a range of varied responses to the texts of television. Using the viewing "strip" as the appropriate text of television, and recognizing that it is filled with varied topics and approaches to those topics, we begin to think of the television viewer as a *bricoleur* who matches the creator in the making of meanings. Bringing values and attitudes, a universe of personal experiences and concerns, to the texts, the viewer selects, examines, acknowledges, and makes texts of his or her own.[21] If we conceive of special interest groups as representatives of *patterns* of cultural attitude and response, we have a potent source of study.

On the production end of this process, in addition to the work of individual producers, we must examine the role of network executives who must purchase and program television content. They, too, are cultural interpreters, intent on "reading" the culture through its relation to the "market." Executives who head and staff the internal censor agencies of each network, the offices of Broadcast Standards or Standards and Practices, are in a similar position. Perhaps as much as any individual or group they present us with a source of rich material for analysis. They are actively engaged in gauging cultural values. Their own research, the assumptions and the findings, needs to be re-analyzed for cultural implications, as does the work of the programmers. In determining who is doing what, with whom, at what times, they are interpreting social behavior in America and assigning it meaning. They are using television as a cultural litmus that can be applied in defining such problematic concepts as "childhood," "family," "maturity," and "appropriate." With the Standards and Practices offices, they interpret *and* define the permissible and the "normal." But their interpretations of behavior open to us as many questions as answers, and an appropriate over-view, a new model of television is necessary in order to best understand their work and ours.

This new model of "television as a cultural forum" fits the experience of television more accurately than others we have seen applied. Our assump-tion is that it opens a range of new questions and calls for re-analysis of older findings from both the textual-critical approach and the mass com-munications research perspective. Ultimately the new model is a simple one. It recognizes the range of interpretation of television content that is now admitted even by those analysts most concerned with television's presenta-tion and maintenance of dominant ideological messages and meanings. But it differs from those perspectives because it does not see this as surprising or unusual. For the most part, that is what central storytelling systems do in all societies. We are far more concerned with the ways in which television contributes to change than with mapping the obvious ways in which it maintains dominant viewpoints. Most research on television, most textual analysis, has assumed that the medium is thin, repetitive, similar, nearly identical in textual formation, easily defined, described, and explained. The variety of response on the part of audiences has been received, as a result of this view, as extraordinary, an astonishing "discovery."

We begin with the observation, based on careful textual analysis, that television is dense, rich, and complex rather than impoverished. Any selec-tion, any cut, any set of questions that is extracted from that text must somehow account for that density, must account for what is *not* studied or measured, for the opposing meanings, for the answering images and sym-bols. Audiences appear to make meaning by selecting that which touches experience and personal history. The range of responses then should be taken as commonplace rather than as unexpected. But research and critical analysis cannot afford so personal a view. Rather, they must somehow de-fine and describe the inventory that makes possible the multiple meanings extracted by audiences, creators, and network decision makers.

Our model is based on the assumption and observation that only so rich a text could attract a mass audience in a complex culture. The forum offers a perspective that is as complex, as contradictory and confused, as much in process as American culture is in experience. Its texture matches that of our daily experiences. If we can understand it better, then perhaps we will better understand the world we live in, the actions that we must take in order to live there.

## Notes

The authors would like to express their appreciation to the John and Mary R. Markle Foundation for support in the preparation of this [essay] and their ongoing study of the role of television as a cultural forum in American society. The ideas in this [essay] were first presented, in different form, at the seminar on "The Mass Production of Mythology," New York Institute for the Humanities, New York University, February 1981. Mary Douglas, Seminar Director.

1. James Carey, "A Cultural Approach to Communications," *Communications* 2 (December 1975).

2. Ibid.

3. James Carey, "Culture and Communications," *Communications Research* (April 1975).

4. See Roger Silverstone, *The Message of Television: Myth and Narrative in Contemporary Culture* (London: Heinemann, 1981), on structural and narrative analysis; John Fiske and John Hartley, *Reading Television* (London: Methuen, 1978), on the semiotic and cultural bases for the analysis of television; David Thorburn, *The Story Machine* (Oxford University Press: forthcoming), on the aesthetics of television; Himmleweit, Hilda et al., "The Audience as Critic: An Approach to the Study of Entertainment," in *The Entertainment Functions of Television,* ed. Percy Tannenbaum (New York: Lawrence Erlbaum Associates, 1980) and W. Russel Neuman, "Television and American Culture: The Mass Medium and the Pluralist Audience," *Public Opinion Quarterly,* 46: 4 (Winter 1982), pp. 471–87, on the role of the audience as critic; Todd Gitlin, "Prime Time Ideology: The Hegemonic Process in Television Entertainment," *Social Problems* 26: 3 (1979), and Douglas Kellner, "TV, Ideology, and Emancipatory Popular Culture," *Socialist Review* 45 (May–June, 1979), on hegemony and new applications of critical theory; James T. Lull, "The Social Uses of Television," *Human Communications Research* 7: 3 (1980), and "Family Communication Patterns and the Social Uses of Television," *Communications Research* 7: 3 (1979), and Tim Meyer, Paul Traudt, and James Anderson, "Non-Traditional Mass Communication Research Methods: Observational Case Studies of Media Use in Natural Settings, *Communication Yearbook IV,* ed. Dan Nimmo (New Brunswick, N.J.: Transaction Books), on audience ethnography and symbolic interactionism; and, most importantly, the ongoing work of The Center for Contemporary Cultural Studies at Birmingham University, England, most recently published in *Culture, Media, Language,* ed. Stuart Hall et al. (London: Hutchinson, in association with The Center for Contemporary Cultural Studies, 1980), on the interaction of culture and textual analysis from a thoughtful political perspective.

5. Carey, 1976.

6. Paul Hirsch, "The Role of Popular Culture and Television in Contemporary Society," *Television: The Critical View,* ed. Horace Newcomb (New York: Oxford University Press, 1979, 1982).

7. Marshall Sahlins, *Culture and Practical Reason* (Chicago: University of Chicago Press, 1976), p. 217.

8. John Cawelti, *Adventure, Mystery, and Romance* (Chicago: University of Chi-

cago Press, 1976), and David Thorburn, "Television Melodrama," *Television: The Critical View* (New York: Oxford University Press, 1979, 1982).

9. Victor Turner, "Process, System, and Symbol: A New Anthropological Synthesis," *Daedalus* (Summer 1977), p. 68.

10. In various works Turner uses both the terms "liminal" and "liminoid" to refer to works of imagination and entertainment in contemporary culture. The latter term is used to clearly mark the distinction between events that have distinct behavioral consequences and those that do not. As Turner suggests, the consequences of entertainment in contemporary culture are hardly as profound as those of the liminal stage of ritual in traditional culture. We are aware of this basic distinction but use the former term in order to avoid a fuller explanation of the neologism. See Turner, "Afterword," to *The Reversible World,* Barbara Babcock, ed. (Ithaca: Cornell University Press, 1979), and "Liminal to Liminoid, in Play, Flow, and Ritual: An Essay in Comparative Symbology," *Rice University Studies,* 60: 3 (1974).

11. Turner, 1977, p. 73.

12. Fiske and Hartley, 1978, p. 85.

13. We are indebted to Prof. Mary Douglas for encouraging this observation. At the presentation of these ideas at the New York Institute for the Humanities seminar on "The Mass Production of Mythology," she checked our enthusiasm for a pluralistic model of television by stating accurately and succinctly, "there are pluralisms and pluralisms." This comment led us to consider more thoroughly the means by which the forum and responses to it function as a tool with which to monitor the quality of pluralism in American social life, including its entertainments. The observation added a much needed component to our planned historical analysis.

14. Stanley Cavell, "The Fact of Television," *Daedalus* 3: 4 (Fall 1982).

15. Raymond Williams, *Television, Technology and Cultural Form* (New York: Schocken, 1971), p. 86 ff.

16. Horace Newcomb and Robert Alley, *The Television Producer as Artist in American Commercial Television* (New York: Oxford University Press, 1983).

17. Ibid.

18. Ibid.

19. Stuart Hall, "Encoding and Decoding in the Television Discourse," *Culture, Media, Language* (London: Hutchinson, in association with The Center for Contemporary Cultural Studies, 1980).

20. See Dave Morley and Charlotte Brunsdon, *Everyday Television: "Nationwide"* (London: British Film Institute, 1978), and Morley, "Subjects, Readers, Texts," in *Culture, Media, Language.*

21. We are indebted to Louis Black and Eric Michaels of the Radio-TV-Film department of the University of Texas-Austin for calling this aspect of televiewing to Newcomb's attention. It creates a much desired balance to Sahlin's view of the creator as *bricoleur* and indicates yet another matter in which the forum model enhances our ability to account for more aspects of the television experience. See, especially, Eric Michaels, *TV Tribes,* unpublished Ph.D. dissertation, University of Texas-Austin, 1982.

# Prime Time Ideology:
# The Hegemonic Process
# in Television
# Entertainment

## TODD GITLIN

Every society works to reproduce itself—and its internal conflicts—within its cultural order, the structure of practices and meanings around which the society takes shape. So much is tautology. In this [essay] I look at contemporary mass media in the United States as one cultural system promoting that reproduction. I try to show how ideology is relayed through various features of American television, and how television programs register larger ideological structures and changes. The question here is not, What is the impact of these programs? but rather a prior one, What do these programs mean? For only after thinking through their possible meanings as cultural objects and as signs of cultural interactions among producers and audiences may we begin intelligibly to ask about their "effects."

The attempt to understand the sources and transformations of ideology in American society has been leading social theorists not only to social-psychological investigations, but to a long overdue interest in Antonio Gramsci's (1971) notion of ideological hegemony. It was Gramsci who, in the late twenties and thirties, with the rise of fascism and the failure of the Western European working-class movements, began to consider why the working class was not necessarily revolutionary; why it could, in fact, yield to fascism. Condemned to a fascist prison precisely because the insurrectionary workers' movement in Northern Italy just after World War I failed, Gramsci spent years trying to account for the defeat, resorting in large measure to the concept of hegemony: bourgeois domination of the thought, the common sense, the life-ways and everyday assumptions of the working class. Gramsci counterposed "hegemony" to "coercion"; these were two analytically distinct processes through which ruling classes secure the consent of the dominated. Gramsci did not always make plain where to draw the line between hegemony and coercion; or rather, as Perry Anderson shows convincingly (1976),[1] he drew the line differently at different times. Nonetheless, ambiguities aside, Gramsci's distinction was a great advance for radical thought, for it called attention to the routine structures of everyday thought—down to "common sense" itself—which worked to sustain class domination and tyranny. That is to say, paradoxically, it took the working class seriously enough as a potential agent of revolution to hold it accountable for its failures.

Because Leninism failed abysmally throughout the West, Western Marxists and non-Marxist radicals have both been drawn back to Gramsci, hoping to address the evident fact that the Western working classes are not predestined toward socialist revolution.[2] In Europe this fact could be taken as strategic rather than normative wisdom on the part of the working class; but in America the working class is not only hostile to revolutionary *strategy,* it seems to disdain the socialist *goal* as well. At the very least, although a recent Peter Hart opinion poll showed that Americans abstractly "favor" workers' control, Americans do not seem to care enough about it to organize very widely in its behalf. While there are abundant "contradictions" throughout American society, they are played out substantially in the realm of "culture" or "ideology," which orthodox Marxism had consigned to the secondary category of "superstructure." Meanwhile, critical theory—especially in the work of T. W. Adorno and Max Horkheimer—had argued with great force that the dominant forms of commercial ("mass") culture were crystallizations of authoritarian ideology; yet despite the ingenuity and brilliance of particular feats of critical exegesis (Adorno, 1954, 1974; Adorno and Horkheimer, 1972), they seemed to be arguing that the "culture industry" was not only meretricious but wholly and statically complete. In the seventies, some of their approaches along with Gramsci's have been elaborated and furthered by Alvin W. Gouldner (1976; see also Kellner, 1978) and Raymond Williams (1973), in distinctly provocative ways.

In this [essay] I wish to contribute to the process of bringing the discussion of cultural hegemony down to earth. For much of the discussion so far remains abstract, almost as if cultural hegemony were a substance with a life of its own, a sort of immutable fog that has settled over the whole public life of capitalist societies to confound the truth of the proletarian telos. Thus to the questions, "Why are radical ideas suppressed in the schools?", "Why do workers oppose socialism?" and so on, comes the single Delphic answer: hegemony. "Hegemony" becomes the magical explanation of last resort. And as such it is useful neither as explanation nor as guide to action. If "hegemony" explains everything in the sphere of culture, it explains nothing.

Concurrent with the theoretical discussion, but on a different plane, looms an entire sub-industry criticizing and explicating specific mass-cultural products and straining to find "emancipatory" if not "revolutionary" meanings in them. Thus in 1977 there was cacophony about the TV version of *Roots;* this year the trend-setter seems to be TV's handling of violence. Mass media criticism becomes mass-mediated, an auxiliary sideshow serving cultural producers as well as the wider public of the cultural spectacle. Piece by piece we see fast and furious analysis of this movie, that TV show, that book, that spectator sport. Many of these pieces have merit one by one, but as a whole they do not accumulate toward a more general theory of how the cultural forms are managed and reproduced—and how they change. Without analytic point, item-by-item analyses of the standard fare of mass culture run the risk of degenerating into high-toned gossip, even a kind of critical groupie-ism. Unaware of the ambiguity of their own motives and strategies, the partial critics may be yielding to a displaced envy, where criticism covertly asks to be taken into the spotlight along with the celebrity culture ostensibly under criticism. Yet another trouble is that partial critiques in the mass-culture tradition don't help us understand the *hold* and the *limits* of cultural products, the degree to which people do and do not incorporate mass-cultural forms, sing the jingles, wear the corporate T-shirts, and most important, permit their life-worlds to be demarcated by them.

My task in what follows is to propose some features of a lexicon for discussing the forms of hegemony in the concrete. Elsewhere I have described some of the operations of cultural hegemony in the sphere of television news, especially in the news's framing procedures for opposition movements (Gitlin, 1977a,b).[3] Here I wish to speak of the realm of entertainment: about television entertainment in particular—as the pervasive and (in the living room sense) *familiar* of our cultural sites—and about movies secondarily. How do the *formal* devices of TV prime time programs encourage viewers to experience themselves as anti-political, privately accumulating individuals (also see Gitlin, 1977c)? And how do these forms express social conflict, containing and diverting the images of contrary social possibilities? I want to isolate a few of the routine devices, though of course in reality they do not operate in isolation; rather, they work in combination, where their force is often enough magnified (though they can also work in contradictory ways). And, crucially, it must be borne in mind throughout this discussion that the forms of mass-cultural production do not either spring up or operate independently of the rest of social life. Commercial culture does not *manufacture* ideology; it *relays* and *reproduces* and *processes* and *packages* and *focuses* ideology that is constantly arising both from social elites and from active social groups and movements throughout the society (as well as within media organizations and practices).

A more complete analysis of ideological process in a commercial society would look both above and below, to elites and to audiences. Above, it would take a long look at the economics and politics of broadcasting, at its relation to the FCC, the Congress, the President, the courts; in case studies and with a developing theory of ideology it would study media's peculiar combination and refraction of corporate, political, bureaucratic, and professional interests, giving the media a sort of limited independence—or what Marxists are calling "relative autonomy"—in the upper reaches of the political-economic system. Below, as Raymond Williams has insisted, cultural hegemony operates within a whole social life-pattern; the people who consume mass-mediated products are also the people who work, reside, compete, go to school, live in families. And there are a good many traditional and material interests at stake for audiences: the political inertia of the American population now, for example, certainly has something to do with the continuing productivity of the goods-producing and -distributing industries, not simply with the force of mass culture. Let me try to avoid misunderstanding at the outset by insisting that *I will not be arguing that the forms of hegemonic entertainment superimpose themselves automatically and finally onto the consciousness or behavior of all audiences at all times:* it remains for sociologists to generate what Dave Morley (1974)[4] has called "an ethnography of audiences," and to study what Ronald Abramson (1978) calls "the phenomenology of audiences" if we are to have anything like a satisfactory account of how audiences consciously and unconsciously process, transform, and are transformed by the contents of television. For many years the subject of media effects was severely narrowed by a behaviorist definition of the problem (see Gitlin, 1978a); more recently, the "agenda-setting function" of mass media has been usefully studied in news media, but not in entertainment. (On the other hand, the very pervasiveness of TV entertainment makes laboratory study of its "effects" almost inconceivable.) It remains to incorporate occasional sociological insights into the actual behavior of TV audiences[5] into a more general theory of the interaction—a theory which avoids both the mechanical assumptions of behaviorism and the trivialities of the "uses and gratifications" approach.

But alas, that more general theory of the interaction is not on the horizon. My more modest attempt in this extremely preliminary essay is to sketch an approach to the hegemonic thrust of some TV forms, not to address the deflection, resistance, and reinterpretation achieved by audiences. I will show that hegemonic ideology is systematically preferred by certain features of TV programs, and that at the same time alternative and oppositional values are brought into the cultural system, and domesticated into hegemonic forms at times, by the routine workings of the market. Hegemony is reasserted in different ways at different times, even by different logics; if this variety is analytically messy, the messiness corresponds to a disordered ideological order, a contradictory society. This said, I proceed to some of the forms in which ideological hegemony is embedded: *format and formula; genre; setting and character type; slant;* and *solution.* Then these particulars will suggest a somewhat more fully developed theory of hegemony.

## Format and Formula

Until recently at least, the TV schedule has been dominated by standard lengths and cadences, standardized packages of TV entertainment appearing, as the announcers used to say, "same time, same station." This week-to-weekness—or, in the case of soap operas, day-to-dayness—obstructed the development of characters; at least the primary characters had to be preserved intact for next week's show. Perry Mason was Perry Mason, once and for all; if you watched the reruns, you couldn't know from character or set whether you were watching the first or the last in the series. For commercial and production reasons which are in practice inseparable—and this is why ideological hegemony is not reducible to the economic interests of elites—the regular schedule prefers the repeatable formula: it is far easier for production companies to hire writers to write for standardized, static characters than for characters who develop. Assembly-line production works through regularity of time slot, of duration, and of character to convey images of social steadiness: come what may, *Gunsmoke* or *Kojak* will check in to your mind at a certain time on a certain evening. Should they lose ratings (at least at the "upscale" reaches of the "demographics," where ratings translate into disposable dollars),[6] their replacements would be—for a time, at least!—equally reliable. Moreover, the standard curve of narrative action—stock characters encounter new version of stock situation; the plot thickens, allowing stock characters to show their standard stuff; the plot resolves—over twenty-two or fifty minutes is itself a source of rigidity and forced regularity.

In these ways, the usual programs are performances that rehearse social fixity: they express and cement the obduracy of a social world impervious to substantial change. Yet at the same time there are signs of routine obsolescence, as hunks of last year's regular schedule drop from sight only to be supplanted by this season's attractions. Standardization and the threat of evanescence are curiously linked: they match the intertwined processes of commodity production, predictability, and obsolescence in a high-consumption society. I speculate that they help instruct audiences in the rightness and naturalness of a world that, in only apparent paradox, regularly requires an irregularity, an unreliability which it calls progress. In this way, the regular changes in TV programs, like the regular elections of public officials, seem to affirm the sovereignty of the audience while keeping deep alternatives off the agenda. Elite authority and consumer choice are affirmed at once—this is one of the central operations of the hegemonic liberal capitalist ideology.

Then, too, by organizing the "free time" of persons into end-to-end interchangeable units, broadcasting extends, and harmonizes with, the industrialization of time. Media time and school time, with their equivalent units and curves of action, mirror the time of clocked labor and reinforce the seeming naturalness of clock time. Anyone who reads Harry Braverman's *Labor and Monopoly Capital* can trace the steady degradation of the work process, both white and blue collar, through the twentieth century, even if Braverman has exaggerated the extent of the process by focusing on managerial *strategies* more than on actual work *processes*. Something similar has happened in other life-sectors.[7] Leisure is industrialized, duration is homogenized, even excitement is routinized, and the standard repeated TV format is an important component of the processs. And typically, too, capitalism provides relief from these confines for its more favored citizens, those who can afford to buy their way out of the standardized social reality which capitalism produces. Thus Sony and RCA now sell home video recorders, enabling customers to tape programs they'd otherwise miss. The widely felt need to overcome assembly-line "leisure" time becomes the source of a new market—to sell the means for private, commoditized solutions to the time-jam.

Commercials, of course, are also major features of the regular TV format. There can be no question but that commercials have a good deal to do with shaping and maintaining markets—no advertiser dreams of cutting advertising costs as long as the competition is still on the air. But commercials also have important *indirect* consequences on the contours of consciousness overall: they get us accustomed to thinking of ourselves and behaving as a *market* rather than a *public,* as consumers rather than citizens. Public problems (like air pollution) are propounded as susceptible to private commodity solutions (like eyedrops). In the process, commercials acculturate us to interruption through the rest of our lives. Time and attention are not one's own; the established social powers have the capacity to colonize consciousness, and unconsciousness, as they see fit. By watching, the audience one by one consents. Regardless of the commercial's "effect" on our behavior, we are consenting to its domination of the public space. Yet we should note that this colonizing process does not actually require commercials, as long as it can form discrete packages of ideological content that call forth discontinuous responses in the audience. Even public broadcasting's children's shows take over the commercial forms to their own educational ends—and supplant narrative forms by herky-jerky bustle. The producers of *Sesame Street,* in likening knowledge to commercial products ("and now a message from the letter B"), may well be legitimizing the commercial form in its discontinuity and in its invasiveness. Again, regularity and discontinuity, superficially discrepant, may be linked at a deep level of meaning. And perhaps the deepest privatizing function of television, its most powerful impact on public life, may lie in the most obvious thing about it: we receive the images in the privacy of our living rooms, making public discourse and response difficult. At the same time, the paradox is that at any given time many viewers are receiving images discrepant with many of their beliefs, challenging their received opinions.

TV routines have been built into the broadcast schedule since its inception. But arguably their regularity has been waning since Norman Lear's first comedy, *All in the Family,* made its network debut in 1971. Lear's contribution to TV content was obvious: where previous shows might have made passing reference to social conflict, Lear brought wrenching social issues into the very mainspring of his series, uniting his characters, as Michael Arlen once pointed out, in a harshly funny *ressentiment* peculiarly appealing to audiences of the Nixon era and its cynical, disabused sequel.[8] As I'll argue below, the hegemonic ideology is maintained in the seventies by *domesticating* divisive issues where in the fifties it would have simply *ignored* them.

Lear also let his characters develop. Edith Bunker grew less sappy and more feminist and commonsensical; Gloria and Mike moved next door, and finally to California. On the threshold of this generational rupture, Mike broke through his stereotype by expressing affection for Archie, and Archie, oh-so-reluctantly but definitely for all that, hugged back and broke through his own. And of course other Lear characters, the Jeffersons and Maude, had earlier been spun off into their own shows, as *The Mary Tyler Moore Show* had spawned *Rhoda* and *Phyllis*. These changes resulted from commercial decisions; they were built on intelligent business perceptions that an audience existed for situation comedies directly addressing racism, sexism, and the decomposition of conventional families. But there is no such thing as a strictly economic "explanation" for production choice, since the success of a show—despite market research—is not foreordained. In the context of my argument, the importance of such developments lies in their partial break with the established, static formulae of prime time television.

Evidently daytime soap operas have also been sliding into character development and a direct exploitation of divisive social issues, rather than going on constructing a race-free, class-free, feminism-free world. And more conspicuously, the "mini-series" has now disrupted the taken-for-granted repetitiveness of the prime time format. Both content and form mattered to the commercial success of *Roots;* certainly the industry, speaking through trade journals, was convinced that the phenomenon was rooted in the series' break with the week-to-week format. When the programming wizards at ABC decided to put the show on for eight straight nights, they were also, inadvertently, making it possible for characters to *develop* within the bounds of a single show. And of course they were rendering the whole sequence immensely more powerful than if it had been diffused over eight weeks. The very format was testimony to the fact that history takes place as a continuing process in which people grow up, have children, die; that people experience their lives within the domain of social institutions. This is no small achievement in a country that routinely denies the rich texture of history.

In any event, the first thing that industry seems to have learned from its success with *Roots* is that they had a new hot formula, the night-after-night series with some claim to historical verisimilitude. So, according to *Broadcasting*, they began preparing a number of "docu-drama" series, of which 1977's products included NBC's three-part series *Loose Change* and *King*, and its four-part *Holocaust*, this latter evidently planned before the *Roots* broadcast. How many of those first announced as in progress will actually be broadcast is something else again—one awaits the networks' domestication and trivializing of the radicalism of *All God's Children: The Life of Nate Shaw,* announced in early 1977. *Roots'* financial success—ABC sold its commercial minutes for $120,000, compared to that season's usual $85,000 to $90,000—might not be repeatable. Perhaps the network could not expect audiences to tune in more than once every few years to a series that began one night at eight o'clock, the next night at nine, and the next at eight again. In summary it is hard to say to what extent these format changes signify an acceleration of the networks' competition for advertising dollars, and to what extent they reveal the networks' responses to the restiveness and boredom of the mass audience, or the emergence of new potential audiences. But in any case the shifts are there, and constitute a fruitful territory for any thinking about the course of popular culture.

## Genre[9]

The networks try to finance and choose programs that will likely attract the largest conceivable audiences of spenders; this imperative requires that the broadcasting elites have in mind some notion of popular taste from moment to moment. Genre, in other words, is necessarily somewhat sensitive; in its rough outlines, if not in detail, it tells us something about popular moods. Indeed, since there are only three networks, there is something of an oversensitivity to a given success; the pendulum tends to swing hard to replicate a winner. Thus *Charlie's Angels* engenders *Flying High* and *American Girls,* about stewardesses and female reporters respectively, each on a long leash under male authority.

Here I suggest only a few signs of this sensitivity to shifting moods and group identities in the audience. The adult western of the middle and late fifties, with its drama of solitary righteousness and suppressed libidinousness, for example, can be seen in retrospect to have played on the quiet malaise under the surface of the complacency of the Eisenhower years, even in contradictory ways. Some lone heroes were identified with traditionally frontier-American informal and individualistic relations to authority (Paladin in *Have Gun, Will Travel,* Bart Maverick in *Maverick*), standing for sturdy individualism struggling for hedonistic values and taking law-and-order wryly. Meanwhile, other heroes were decent officials like *Gunsmoke's* Matt Dillon, affirming the decency of paternalistic law and order against the temptations of worldly pleasure. With the rise of the Camelot mystique, and the vigorous "long twilight struggle" that John F. Kennedy personified, spy stories like *Mission: Impossible* and *The Man From U.N.C.L.E.* were well suited to capitalize on the macho CIA aura. More recently, police stories, with cops surmounting humanist illusions to draw thin blue lines against anarcho-criminal barbarism, afford a variety of official ways of coping with "the social issue," ranging from *Starsky and Hutch's* muted homoeroticism to *Barney Miller's* team pluralism. The single-women shows following from *Mary Tyler Moore* acknowledge in their privatized ways that some sort of feminism is here to stay, and work to contain it with hilarious versions of "new life styles" for single career women. Such shows probably appeal to the market of "upscale" singles with relatively large disposable incomes, women who are disaffected from the traditional imagery of housewife and helpmeet. In the current wave of "jiggle" or "T&A" shows patterned on *Charlie's Angels* (the terms are widely used in the industry), the attempt is to appeal to the prurience of the male audience by keeping the "girls" free of romance, thus catering to male (and female?) backlash against feminism. The black sitcoms probably reflect the rise of a black middle class with the purchasing power to bring forth advertisers, while also appealing *as comedies*—for conflicting reasons, perhaps—to important parts of the white audience. (Serious black drama would be far more threatening to the majority audience.)

Whenever possible it is illuminating to trace the transformations in a genre over a longer period of time. For example, the shows of technological prowess have metamorphosed over four decades as hegemonic ideology has been contested by alternative cultural forms. In work not yet published, Tom Andrae of the Political Science Department at the University of California, Berkeley, shows how the Superman archetype began in 1933 as a menace to society; then became something of a New Dealing, anti-Establishmentarian individualist casting his lot with the oppressed and, at times, against the State; and only in the forties metamorphosed into the current incarnation who prosecutes criminals in the name of "the American way." Then the straight-arrow Superman of the forties and fifties was supplemented by the whimsical, self-satirical Batman and the Marvel Comics series of the sixties and seventies, symbols of power gone silly, no longer prepossessing. In playing against the conventions, their producers seem to have been exhibiting the self-consciousness of genre so popular among "high arts" too, as with Pop and minimal art. Thus shifts in genre presuppose the changing mentality of critical masses of writers and cultural producers; yet these changes would not take root commercially without corresponding changes in the dispositions (even the self-consciousness) of large audiences. In other words, changes in cultural ideals and in audience sensibilities must be harmonized to make for shifts in genre or formula.

Finally, the latest form of technological hero corresponds to an authoritarian turn in hegemonic ideology, as well as to a shift in popular (at least children's) mentality. The seventies generation of physically augmented, obedient, patriotic superheroes *(The Six Million Dollar Man* and *The Bionic Woman)* differ from the earlier waves in being organizational products through and through; these team players have no private lives from which they are recruited task by task, as in *Mission: Impossible,* but they are actually *invented* by the State, to whom they owe their lives.

Televised sports too is best understood as an entertainment genre, one of the most powerful.[10] What we know as professional sports today is inseparably intertwined with the networks' development of the sports market. TV sports is rather consistently framed to reproduce dominant American values. First, although TV is ostensibly a medium for the eyes, the sound is often decisive in taking the action off the field. The audience is not trusted to come to its own conclusions. The announcers are not simply describing events ("Reggie Jackson hits a ground ball to shortstop"), but interpreting them ("World Series 1978! It's great to be here"). One may see here a process equivalent to advertising's project of taking human qualities out of the consumer and removing them to the product: sexy perfume, zesty beer.

In televised sports, the hegemonic impositions have, if anything, probably become more intense over the last twenty years. One technique for interpreting the event is to regale the audience with bits of information in the form of "stats." "A lot of people forget they won eleven out of their last twelve games. . . ." "There was an extraordinary game in last year's World Series. . . ." "Rick Barry hasn't missed two free throws in a row for 72 games. . . ." "The last time the Warriors were in Milwaukee Clifford Ray *also* blocked two shots in the second quarter." How *about* that? The announcers can't shut up; they're constantly chattering. And the stat flashed on the screen further removes the action from the field. What is one to make of all this? Why would anyone want to know a player's free throw percentage not only during the regular season but during the playoffs?

But the trivialities have their reason: they amount to an interpretation that flatters and disdains the audience at the same time. It flatters in small ways, giving you the chance to be the one person on the block who already possessed this tidbit of fact. At the same time, symbolically, it treats you as someone who really knows what's going on in the world. Out of control of social reality, you may flatter yourself that the substitute world of sports is a corner of the world you can really grasp. Indeed, throughout modern society, the availability of statistics is often mistaken for the availability of knowledge and deep meaning. To know the number of megatons in the nuclear arsenal is not to grasp its horror; but we are tempted to bury our fear in the possession of comforting fact. To have made "body counts" in Vietnam was not to be in control of the countryside, but the U.S. Army flattered itself that the stats looked good. TV sports shows, encouraging the audience to value stats, harmonize with a stat-happy society. Not that TV operates independently of the sports event itself; in fact, the event is increasingly organized to fit the structure of the broadcast. There are extra time-outs to permit the network to sell more commercial time. Michael Real of San Diego State University used a stopwatch to calculate that during the 1974 Super Bowl, the football was actually moving for—seven minutes (Real, 1977). Meanwhile, electronic billboards transplant the stats into the stadium itself.

Another framing practice is the reduction of the sports experience to a sequence of individual achievements. In a fusion of populist and capitalist dogma, everyone is somehow the best. This one has "great hands," this one has "a great slam dunk," that one's "great on defense." This indiscriminate commendation raises the premium on personal competition, and at the same time undermines the meaning of personal achievement: everyone is excellent at something, as at a child's birthday party. I was most struck by the force of this sort of framing during the NBA basketball playoffs of 1975, when, after a season of hearing Bill King announce the games over local KTVU, I found myself watching and hearing the network version. King's Warriors were not CBS's. A fine irony: King with his weird mustache and San Francisco panache was talking about team relations and team strategy; CBS, with its organization-man team of announcers, could talk of little besides the personal records of the players. Again, at one point during the 1977 basketball playoffs, CBS's Brent Musburger gushed: "I've got one of the greatest players of all time [Rick Barry] and one of the greatest referees of all time [Mendy Rudolph] sitting next to me! . . . I'm surrounded by experts!" All in all, the network exalts statistics, personal competition, expertise. The message is: The way to understand things is by storing up statistics and tracing their trajectories. This is training in observation without comprehension.

Everything is technique and know-how; nothing is purpose. Likewise, the instant replay generates the thrill of recreating the play, even second-guessing the referee. The appeal is to the American tradition of exalting means over ends: this is the same spirit that animates popular science magazines and do-it-yourself. It's a complicated and contradictory spirit, one that lends itself to the preservation of craft values in a time of assembly-line production, and at the same time distracts interest from any desire to control the goals of the central work process.

The significance of this fetishism of means is hard to decipher. Though the network version appeals to technical thinking, the announcers are not only small-minded but incompetent to boot. No sooner have they dutifully complimented a new acquisition as "a fine addition to the club" than often enough he flubs a play. But still they function as cheerleaders, revving up the razzle-dazzle rhetoric and reminding us how uniquely favored we are by the spectacle. By staying tuned in, somehow we're "participating" in sports history—indeed, by proxy, in history itself. The pulsing theme music and electronic logo reinforce this sense of hot-shot glamor. The breathlessness never lets up, and it has its pecuniary motives: if we can be convinced that the game really is fascinating (even if it's a dog), we're more likely to stay tuned for the commercials for which Miller Lite and Goodyear have paid $100,000 a minute to rent our attention.

On the other hand, the network version does not inevitably succeed in forcing itself upon our consciousness and defining our reception of the event. TV audiences don't necessarily succumb to the announcers' hype. In semi-public situations like barrooms, audiences are more likely to see through the trivialization and ignorance and—in "para-social interaction"—to tell the announcers off. But in the privacy of living rooms, the announcers' framing probably penetrates farther into the collective definition of the event. It should not be surprising that one fairly common counter-hegemonic practice is to watch the broadcast picture without the network sound, listening to the local announcer on the radio.

## Setting and Character Type

Closely related to genre and its changes are setting and character type. And here again we see shifting market tolerances making for certain changes in content, while the core of hegemonic values remains virtually impervious.

In the fifties, when the TV forms were first devised, the standard TV series presented—in Herbert Gold's phrase—happy people with happy problems. In the seventies it is more complicated: there are unhappy people with happy ways of coping. But the set itself propounds a vision of consumer happiness. Living rooms and kitchens usually display the standard package of consumer goods. Even where the set is ratty, as in *Sanford and Son*, or working-class, as in *All in the Family*, the bright color of the TV tube almost always glamorizes the surroundings so that there will be no sharp break between the glorious color of the program and the glorious color of the commercial. In the more primitive fifties, by contrast, it was still possible for a series like *The Honeymooners* or *The Phil Silvers Show* (Sergeant Bilko) to get by with one or two simple sets per show: the life of a good skit was in its accomplished *acting*. But that series, in its sympathetic treatment of working-class mores, was exceptional. Color broadcasting accomplishes the glamorous ideal willy-nilly.

Permissible character types have evolved, partly because of changes in the structure of broadcasting power. In the fifties, before the quiz show scandal, advertising agencies contracted directly with production companies to produce TV series (Barnouw, 1970). They ordered up exactly what they wanted, as if by the yard; and with some important but occasional exceptions—I'll mention some in a moment—what they wanted was glamor and fun, a showcase for commercials. In 1954, for example, one agency wrote to the playwright Elmer Rice explaining why his *Street Scene*, with its "lower class social level," would be unsuitable for telecasting:

> We know of no advertiser or advertising agency of any importance in this country who would knowingly allow the products which he is trying to advertise to the public to become associated with the squalor . . . and general "down" character . . . of *Street Scene*. . . .
>
> On the contrary it is the general policy of advertisers to glamorize their products, the people who buy them, and the whole American social and economic scene. . . . The American consuming public as presented by the advertising industry today is middle class, not lower class; happy in general, not miserable and frustrated. . . . (Barnouw, 1970:33)

Later in the fifties, comedies were able to represent discrepant settings, permitting viewers both to identify and to indulge their sense of superiority through comic distance: *The Honeymooners* and *Bilko*, which capitalized on Jackie Gleason's and Phil Silvers's enormous personal popularity (a personality cult can always perform wonders and break rules), were able to extend dignity to working-class characters in anti-glamorous situations (see Czitrom, 1977).

Beginning in 1960, the networks took direct control of production away from advertisers. And since the networks are less provincial than particular advertisers, since they are more closely attuned to general tolerances in the population, and since they are firmly in charge of a buyer's market for advertising (as long as they produce shows that *some* corporation will sponsor), it now became possible—if by no means easy—for independent production companies to get somewhat distinct cultural forms, like Norman Lear's comedies, on the air. The near-universality of television set ownership, at the same time, creates the possibility of a wider range of audiences, including minority-group, working-class and age-segmented audiences, than existed in the fifties, and thus makes possible a wider range of fictional characters. Thus changes in the organization of TV production, as well as new market pressures, have helped to change the prevalent settings and character types on television.

But the power of corporate ideology over character types remains very strong, and sets limits on the permissible; the changes from the fifties through the sixties and seventies should be understood in the context of essential cultural features that have *not* changed. To show the quality of deliberate choice that is often operating, consider a book called *The Youth Market,* by two admen, published in 1970, counseling companies on ways to pick "the right character for your product":

> But in our opinion, if you want to create your own hardhitting spokesman to children, the most effective route is the superhero-miracle worker. He certainly can demonstrate food products, drug items, many kinds of toys, and innumerable household items. . . . The character should be adventurous. And he should be on the right side of the law. A child must be able to mimic his hero, whether he is James Bond, Superman or Dick Tracy; to be able to fight and shoot to kill without punishment or guilt feelings. (Helitzer and Heyel, 1970)

If this sort of thinking is resisted within the industry itself, it's not so much because of commitments to artistry in television as such, but more because there are other markets that are not "penetrated" by these hard-hitting heroes. The industry is noticing, for example, that *Roots* brought to the tube an audience who don't normally watch TV. The homes-using-television levels during the week of *Roots* were up between 6 and 12 percent over the programs of the previous year (*Broadcasting,* January 31, 1977). Untapped markets—often composed of people who have, or wish to have, somewhat alternative views of the world—can only be brought in by unusual sorts of programming. There is room in the schedule for rebellious human slaves just as there is room for hard-hitting patriotic-technological heroes. In other words—and contrary to a simplistic argument against television manipulation by network elites—the receptivity of enormous parts of the population is an important limiting factor affecting what gets on television. On the other hand, network elites do not risk investing in *regular* heroes who will challenge the core values of corporate capitalist society: who are, say, explicit socialists, or union organizers, or for that matter born-again evangelists. But like the dramatic series *Playhouse 90* in the fifties, TV movies permit a somewhat wider range of choice than weekly series. It is apparently easier for producers to sell exceptional material for one-shot showings—whether sympathetic to lesbian mothers, critical of the 1950s blacklist or of Senator Joseph McCarthy. Most likely these important exceptions have prestige value for the networks.

### Slant

Within the formula of a program, a specific slant often pushes through, registering a certain position on a particular public issue. When issues are politically charged, when there is overt social conflict, programs capitalize on the currency. ("Capitalize" is an interesting word, referring both to use and to profit.) In the program's brief compass, only the most stereotyped characters are deemed to "register" on the audience, and therefore slant, embedded in character, is almost always simplistic and thin. The specific slant is sometimes mistaken for the whole of ideological tilt or "bias," as if the bias dissolves when no position is taken on a topical issue. But the week-after-week angle of the show is more basic, a hardened definition of a routine situation *within which* the specific topical slant emerges. The occasional topical slant then seems to anchor the program's general meanings. For instance, a 1977 show of *The Six Million Dollar Man* told the story of a Russian-East German plot to stop the testing of the new B-1 bomber; by implication, it linked the domestic movement against the B-1 to the foreign Red menace. Likewise, in the late sixties and seventies, police and spy dramas have commonly clucked over violent terrorists and heavily armed "anarchist" maniacs, labeled as "radicals" or "revolutionaries," giving the cops a chance to justify their heavy armament and crude machismo. But the other common variety of slant is sympathetic to forms of deviance which are either private (the lesbian mother shown to be a good mother to her children) or quietly reformist (the brief vogue for *Storefront Lawyers* and the like in the early seventies). The usual slants, then, fall into two categories: either (a) a legitimation of depoliticized forms of deviance, usually ethnic or sexual; or (b) a delegitimation of the dangerous, the violent, the out-of-bounds.

The slants that find their way into network programs, in short, are not uniform. Can we say anything systematic about them? Whereas in the fifties family dramas and sit-coms usually ignored—or indirectly sublimated—the existence of deep social problems in the world outside the set, programs of the seventies much more often domesticate them. From *Ozzie and Harriet* or *Father Knows Best* to *All in the Family* or *The Jeffersons* marks a distinct shift for formula, character, and slant: a shift, among other things, in the image of how a family copes with the world outside. Again, changes in content have in large part to be referred back to changes in social values and sensibilities, particularly the values of writers, actors, and other practitioners: there is a large audience now that prefers acknowledging and domesticating social problems directly rather than ignoring them or treating them only indirectly and in a sublimated way; there are also media practitioners who have some roots in the rebellions of the sixties. Whether hegemonic style will operate more by exclusion (fifties) than by domestication (seventies) will depend on the level of public dissensus as well as on internal factors of media organization (the fifties blacklist of TV writers probably exercised a chilling effect on subject matter and slant; so did the fact that sponsors directly developed their own shows).

## Solution

Finally, cultural hegemony operates through the solutions proposed to difficult problems. However grave the problems posed, however rich the imbroglio, the episodes regularly end with the click of a solution: an arrest, a defiant smile, an I-told-you-so explanation. The characters we have been asked to care about are alive and well, ready for next week. Such a world is not so much fictional as fake. However deeply the problem is located within society, it will be solved among a few persons: the heroes must attain a solution that leaves the rest of the society untouched. The self-enclosed world of the TV drama justifies itself, and its exclusions, by "wrapping it all up." Occasional exceptions are either short-lived, like *East Side, West Side,* or independently syndicated outside the networks, like Lear's *Mary Hartman, Mary Hartman.* On the networks, *All in the Family* has been unusual in sometimes ending obliquely, softly, or ironically, refusing to pretend to solve a social problem that cannot, in fact, be solved by the actions of the Bunkers alone. The Lou Grant show is also partial to downbeat, alienating endings.

Likewise, in mid-seventies mass-market films like *Chinatown, Rollerball, Network,* and *King Kong,* we see an interesting form of closure: as befits the common cynicism and helplessness, society owns the victory. Reluctant heroes go up against vast impersonal forces, often multinational corporations like the same Gulf & Western (sent up as "Engulf & Devour" in Mel Brooks's *Silent Movie*) that, through its Paramount subsidiary, produces some of these films. Driven to anger or bitterness by the evident corruption, the rebels break loose—only to bring the whole structure crashing down on them. (In the case of *King Kong,* the great ape falls of his own weight—from the World Trade Center roof, no less—after the helicopter gunships "zap" him.) These popular films appeal to a kind of populism and rebelliousness, usually of a routine and vapid sort, but then close off the possibilities of effective opposition. The rich get richer and the incoherent rebels get bought and killed.

Often the sense of frustration funneled through these films is diffuse and ambiguous enough to encourage a variety of political responses. While many left-wing cultural critics raved about *Network,* for example, right-wing politicians in Southern California campaigned for Proposition 13 using the film's slogan, "I'm mad as hell and I'm not going to take it any more." Indeed, *the fact that the same film is subject to a variety of conflicting yet plau-*

*sible interpretations may suggest a crisis in hegemonic ideology.* The economic system is demonstrably troubled, but the traditional liberal recourse, the State, is no longer widely enough trusted to provide reassurance. Articulate social groups do not know whom to blame; public opinion is fluid and volatile, and people at all levels in the society withdraw from public participation.[11] In this situation, commercial culture succeeds with diverse interest groups, as well as with the baffled and ambivalent, precisely by propounding ambiguous or even self-contradictory situations and solutions.

### The Hegemonic Process in Liberal Capitalism

Again it bears emphasizing that, for all these tricks of the entertainment trade, the mass-cultural system is not one-dimensional. High-consumption corporate capitalism implies a certain sensitivity to audience taste, taste which is never wholly manufactured. Shows are made by guessing at audience desires and tolerances, and finding ways to speak to them that perpetuate the going system.[12] (Addressing one set of needs entails scanting and distorting others, ordinarily the less mean, less invidious, less aggressive, less reducible to commodity forms.) The cultural hegemony system that results is not a closed system. It leaks. Its very structure leaks, at the least because it remains to some extent competitive. Networks sell the audience's attention to advertisers who want what they think will be a suitably big, suitably rich audience for their products; since the show is bait, advertisers will put up with—or rather buy into—a great many possible baits, as long as they seem likely to attract a buying audience. In the news, there are also traditions of real though limited journalistic independence, traditions whose modern extension causes businessmen, indeed, to loathe the press. In their 1976 book *Ethics and Profits,* Leonard Silk and David Vogel quote a number of big businessmen complaining about the raw deal they get from the press. A typical comment: "Even though the press is a business, it doesn't reflect business values." That is, it has a certain real interest in truth—partial, superficial, occasion- and celebrity-centered truth, but truth nevertheless.

Outside the news, the networks have no particular interest in truth as such, but they remain sensitive to currents of interest in the population, including the yank and haul and insistence of popular movements. With few ethical or strategic reasons not to absorb trends, they are adept at perpetuating them with new formats, new styles, tie-in commodities (dolls, posters, T-shirts, fan magazines) that fans love. In any case, it is in no small measure because of the economic drives themselves that *the hegemonic system itself amplifies legitimated forms of opposition.* In liberal capitalism, hegemonic ideology develops by domesticating opposition, absorbing it into forms compatible with the core ideological structure. Consent is managed by absorption as well as by exclusion. The hegemonic ideology changes in order to remain hegemonic; that is the peculiar nature of the dominant ideology of liberal capitalism.

Raymond Williams (1977) has insisted rightly on the difference between two types of non-hegemonic ideology: *alternative* forms, presenting
a distinct but supplementary and containable view of the world, and *oppositional* forms, rarer and more tenuous within commercial culture, intimating
an authentically different social order. Williams makes the useful distinction
between *residual* forms, descending from declining social formations, and
*emergent* forms, reflecting formations on the rise. Although it is easier to
speak of these possibilities in the abstract than in the concrete, and although
it is not clear what the emergent formations are (this is one of the major
questions for social analysis now), these concepts may help organize an agenda
for thought and research on popular culture. I would add to Williams's own
carefully modulated remarks on the subject only that there is no reason a
priori to expect that emergent forms will be expressed as the ideologies of
rising *classes,* or as "proletarian ideology" in particular; currently in the United
States the emergent forms have to do with racial minorities and other ethnic
groups, with women, with singles, with homosexuals, with old-age subcultures, as well as with technocrats and with political interest groups (loosely
but not inflexibly linked to corporate interests) with particular strategic goals
(like the new militarists of the Committee on the Present Danger). Analysis
of the hegemonic ideology and its rivals should not be allowed to lapse into
some form of what C. Wright Mills (1948) called the "labor metaphysic."

One point should be clear: the hegemonic system is not cut-and-dried,
not definitive. It has continually to be reproduced, continually superimposed, continually to be negotiated and managed, in order to override the
alternative and, occasionally, the oppositional forms. To put it another way:
major social conflicts are transported *into* the cultural system, where the
hegemonic process frames them, form and content both, into compatibility
with dominant systems of meaning. Alternative material is routinely *incorporated:* brought into the body of cultural production. Occasionally oppositional material may succeed in being indigestible; that material is excluded
from the media discourse and returned to the cultural margins from which
it came, while *elements* of it are incorporated into the dominant forms.

In these terms, *Roots* was an alternative form, representing slaves as un-
blinkable facts of American history, blacks as victimized humans and humans nonetheless. In the end, perhaps, the story is dominated by the chance
for upward mobility; the upshot of travail is freedom. Where Alex Haley's
book is subtitled "The Saga of an American Family," ABC's version carries
the label—and the self-congratulation—"The *Triumph* of an American Family." It is hard to say categorically which story prevails; in any case there is
a tension, a struggle, between the collective agony and the triumph of a
single family. That struggle is the friction in the works of the hegemonic
system.

And all the evident friction within television entertainment—as well as
within the schools, the family, religion, sexuality, and the State—points back
to a deeper truth about bourgeois culture. In the United States, at least,
hegemonic ideology is extremely complex and absorptive; it is only by ab-

sorbing and domesticating conflicting definitions of reality and demands on it, in fact, that it remains hegemonic. In this way, the hegemonic ideology of liberal capitalism is dramatically different from the ideologies of pre-capitalist societies, and from the dominant ideology of authoritarian social-ist or fascist regimes. What permits it to absorb and domesticate critique is not something accidental to capitalist ideology, but rather its core. *The he-gemonic ideology of liberal capitalist society is deeply and essentially conflicted in a number of ways.* As Daniel Bell (1976) has argued, it urges people to work hard, but proposes that real satisfaction is to be found in leisure, which ostensibly embodies values opposed to work.[13] More profoundly, at the center of liberal capitalist ideology there is a tension between the affirmation of patriarchal authority—currently enshrined in the national security state—and the affirmation of individual worth and self-determination. Bourgeois ideology in all its incarnations has been from the first a contradiction in terms, affirming "life, liberty and the pursuit of happiness," or "liberty, equality, fraternity," as if these ideals are compatible, even mutually depen-dent, at all times in all places, as they were for one revolutionary group at one time in one place. But all anti-bourgeois movements wage their battles precisely in terms of liberty, equality, or fraternity (or, recently, sorority); they press on liberal capitalist ideology *in its own name.*

Thus we can understand something of the vulnerability of bourgeois ideology, as well as its persistence. In the twentieth century, the dominant ideology has shifted toward sanctifying consumer satisfaction as the pre-mium definition of "the pursuit of happiness," in this way justifying cor-porate domination of the economy. What is hegemonic in consumer capi-talist ideology is precisely the notion that happiness, or liberty, or equality, or fraternity can be affirmed through the existing private commodity forms, under the benign, protective eye of the national security state. This ideolog-ical core is what remains essentially unchanged and unchallenged in tele-vision entertainment, at the same time the inner tensions persist and are even magnified.

### Notes

An earlier version of this paper was delivered to the 73rd Annual Meeting of the American Sociological Association, San Francisco, September 1978. Thanks to Vic-toria Bonnell, Bruce Dancis, Wally Goldfrank, Karen Shapiro, and several anony-mous reviewers for stimulating comments on earlier drafts.

1. Anderson has read Gramsci closely to tease out this and other ambiguities in Gramsci's diffuse and at times Aesopian texts. (Gramsci was writing in a fascist prison, he was concerned about passing censorship, and he was at times gravely ill.)

2. In my reading, the most thoughtful specific approach to this question since Gramsci, using comparative structural categories to explain the emergence or ab-sence of socialist class consciousness, is Mann (1973). Mann's analysis takes us to structural features of American society that detract from revolutionary consciousness and organization. Although my [essay] does not discuss social-structural and histor-ical features, I do not wish their absence to be interpreted as a belief that culture is

all-determining. This [essay] discusses aspects of the hegemonic culture, and makes no claims to a more sweeping theory of American society.

3. In Part III of the latter, I discuss the theory of hegemony more extensively. Published in *The Whole World is Watching: Mass Media and the New Left, 1965–70,* Berkeley: University of California Press, 1980.

4. See also, Willis (n.d.) for an excellent discussion of the limits of both ideological analysis of cultural artifacts and the social meaning system of audiences, when each is taken by itself and isolated from the other.

5. Most strikingly, see Blum's (1964) findings on black viewers putting down TV shows while watching them. See also Willis's (n.d.) program for studying the substantive meanings of particular pop music records for distinct youth subcultures; but note that it is easier to study the active uses of music than TV, since music is more often heard publicly and because, there being so many choices, the preference for a particular set of songs or singers or beats expresses more about the mentality of the audience than is true for TV.

6. A few years ago, *Gunsmoke* was cancelled although it was still among the top ten shows in Nielsen ratings. The audience was primarily older and disproportionately rural, thus an audience less well sold to advertisers. So much for the networks' democratic rationale.

7. Borrowing "on time," over commensurable, arithmetically calculated lengths of time, is part of the same process: production, consumption, and acculturation made compatible.

8. The time of the show is important to its success or failure. Lear's *All in the Family* was rejected by ABC before CBS bought it. An earlier attempt to bring problems of class, race, and poverty into the heart of television series was *East Side, West Side* of 1964, in which George C. Scott played a caring social worker consistently unable to accomplish much for his clients. As time went on, the Scott character came to the conclusion that politics might accomplish what social work could not, and changed jobs, going to work as the assistant to a liberal Congressman. It was rumored about that the hero was going to discover there, too, the limits of reformism—but the show was cancelled, presumably for low ratings. Perhaps Lear's shows, by contrast, have lasted in part *because they are comedies:* audiences will let their defenses down for some good laughs, even on themselves, at least when the characters are, like Archie Bunker himself, ambiguous normative symbols. At the same time, the comedy form allows white racists to indulge themselves in Archie's rationalizations without seeing that the joke is on them.

9. I use the term *loosely* to refer to general categories of TV entertainment, like "adult western," "cops and robbers," "black shows." Genre is not an objective feature of the cultural universe, but a conventional name for a convention, and should not be reified—as both cultural analysis and practice often do—into a cultural essence.

10. This discussion of televised sports was published in similar form (Gitlin, 1978b).

11. In another essay I will be arguing that forms of pseudo-participation (including cult movies like *Rocky Horror Picture Show* and *Animal House,* along with religious sects) are developing simultaneously to fill the vacuum left by the declining of credible radical politics, and to provide ritual forms of expression that alienated groups cannot find within the political culture.

12. See the careful, important, and unfairly neglected discussion of the tricky needs issue in Leiss (1976). Leiss cuts through the Frankfurt premise that commod-

ity culture addresses false needs by arguing that audience needs for happiness, diversion, self-assertion, and so on are ontologically real; what commercial culture does is not to invent needs (how could it do that?) but to insist upon the possibility of meeting them through the purchase of commodities. For Leiss, all specifically human needs are social; they develop within one social form or another. From this argument—and, less rigorously but more daringly from Ewen (1976)—flow powerful political implications I cannot develop here. On the early popularity of entertainment forms which cannot possibly be laid at the door of a modern "culture industry" and media-produced needs, see Altick (1978).

13. There is considerable truth in Bell's thesis. Then why do I say "ostensibly"? Bell exaggerates his case against "adversary culture" by emphasizing changes in avant-garde culture above all (Pop Art, happenings, John Cage, etc.); if he looked at *popular* culture, he would more likely find ways in which aspects of the culture of consumption *support* key aspects of the culture of production. I offer my discussion of sports as one instance. Morris Dickstein's (1977) affirmation of the critical culture of the sixties commits the counterpart error of overemphasizing the importance of *other* selected domains of literary and avant-garde culture.

## References

Abramson, Ronald (1978) Unpublished manuscript, notes on critical theory distributed at the West Coast Critical Communications Conference, Stanford University.

Adorno, Theodor W. (1954) "How to look at television." *Hollywood Quarterly of Film, Radio and Television*. Spring. Reprinted 1975: 474–88 in Bernard Rosenberg and David Manning White (eds.), *Mass Culture*. New York: Free Press.

——— (1974) "The stars down to earth. The Los Angeles Times Astrology Column." *Telos* 19. Spring 1974 (1957): 13–90.

Adorno, Theodor W., and Max Horkheimer (1972) "The culture industry: Enlightenment as mass deception." Pp. 120–167 in Adorno and Horkheimer, *Dialectic of Enlightenment* (1944). New York: Seabury.

Altick, Richard (1978) *The Shows of London*. Cambridge: Harvard University Press.

Anderson, Perry (1976) "The antinomies of Antonio Gramsci." *New Left Review* 100 (November 1976–January 1977): 5–78.

Barnouw, Erik (1970) *The Image Empire*. New York: Oxford University Press.

Bell, Daniel (1976) *The Cultural Contradictions of Capitalism*. New York: Basic Books.

Blum, Alan F. (1964) "Lower-class Negro television spectators: The concept of pseudo-jovial scepticism." Pp. 429–435 in Arthur B. Shostak and William Gomberg (eds.), *Blue-Collar World*. Englewood Cliffs, N.J.: Prentice-Hall.

Braverman, Harry (1974) *Labor and Monopoly Capital: The Degradation of Work in the Twentieth Century*. New York: Monthly Review Press.

Czitrom, Danny (1977) "Bilko: A sitcom for all seasons." *Cultural Correspondence* 4: 16–19.

Dickstein, Morris (1977) *Gates of Eden*. New York: Basic Books.

Ewen, Stuart (1976) *Captains of Consciousness*. New York: McGraw-Hill.

Gitlin, Todd (1977a) "Spotlights and shadows: Television and the culture of politics." *College English* April: 789–801.

———. (1977b) " 'The whole world is watching': Mass media and the new left, 1965–70." Doctoral dissertation, University of California, Berkeley.

————. (1977c) "The televised professional." *Social Policy* (November/December): 94–99.

————. (1978a) "Media sociology: The dominant paradigm." *Theory and Society* 6:205–53.

———— (1978b) "Life as instant replay." *East Bay Voice* (November–December): 14.

Gouldner, Alvin W. (1976) *The Dialectic of Ideology and Technology*. New York: Seabury.

Gramsci, Antonio (1971) *Selections From the Prison Notebooks*. Quintin Hoare and Geoffrey Nowell Smith (eds.) New York: International Publishers.

Helitzer, Melvin, and Carl Heyel (1970) The Youth Market: Its Dimensions, Influence and Opportunities for You. Quoted pp. 62–63 in William Melody, *Children's Television* (1973). New Haven: Yale University Press.

Kellner, Douglas (1978) "Ideology, Marxism, and advanced capitalism." *Socialist Review* 42 (November–December): 37–66.

Leiss, William (1976) *The Limits to Satisfaction*. Toronto: University of Toronto Press.

Mann, Michael (1973) *Consciousness and Action Among the Western Working Class*. London: Macmillan.

Mills, C. Wright (1948) *The New Men of Power*. New York: Harcourt, Brace.

Morley, Dave (1974) "Reconceptualizing the media audience: Towards an ethnography of audiences." Mimeograph, Centre for Contemporary Cultural Studies, University of Birmingham.

Real, Michael R. (1977) *Mass-Mediated Culture*. Englewood Cliffs, N.J.: Prentice-Hall.

Silk, Leonard, and David Vogel (1976) *Ethics and Profits*. New York: Simon and Schuster.

Williams, Raymond (1973) "Base and superstructure in Marxist cultural theory." *New Left Review:* 82.

————. (1977) *Marxism and Literature*. New York: Oxford University Press.

Willis, Paul (n.d.) "Symbolism and practice: A theory for the social meaning of pop music." Mimeograph, Centre for Contemporary Cultural Studies, University of Birmingham.

# Television Melodrama

## DAVID THORBURN

*I remember with what a smile of saying something daring and*
*inacceptable John Erskine told an undergraduate class that some day we*
*would understand that plot and melodrama were good things for a novel*
*to have and that* Bleak House *was a very good novel indeed.*

Lionel Trilling, *A Gathering of Fugitives*

Although much of what I say will touch significantly on the medium as a
whole, I want to focus here on a single broad category of television programming—what *TV Guide* and the newspaper listings, with greater insight
than they realize, designate as "melodrama." I believe that at its increasingly
frequent best, this fundamental television genre so richly exploits the conventions of its medium as to be clearly distinguishable from its ancestors in
the theater, in the novel, and in films. And I also believe, though this more
extravagant corollary judgment can only be implied in my present argument, that television melodrama has been our culture's most characteristic
aesthetic form, and one of its most complex and serious forms as well, for
at least the past decade and probably longer.

*Melo* is the Greek word for music. The term *melodrama* is said to have
originated as a neutral designation for a spoken dramatic text with a musical
accompaniment or background, an offshoot or spin-off of opera. The term
came into widespread use in England during the nineteenth century, when
it was appropriated by theatrical entrepreneurs as a legal device to circumvent statutes that restricted the performances of legitimate drama to certain
theaters. In current popular and (much) learned usage, *melodrama* is a res-
olutely pejorative term, also originating early in the last century, denoting a
sentimental, artificially plotted drama that sacrifices characterization to ex-

Reprinted from *Television as a Cultural Force,* ed. Douglass Cater and Richard Adler,
by permission of Praeger Publishers, Aspen Institute Program on Communications
and Society, and the author. Copyright © 1976 by David Thorburn.

travagant incident, makes sensational appeals to the emotions of its audience, and ends on a happy or at least a morally reassuring note.

Neither the older, neutral nor the current, disparaging definitions are remotely adequate, however. The best recent writings on melodrama, drawing sustenance from a larger body of work concerned with popular culture in general, have begun to articulate a far more complex definition, one that plausibly refuses to restrict melodrama to the theater, and vigorously challenges long-cherished distinctions between high and low culture—even going so far as to question some of our primary assumptions about the nature and possibilities of art itself. In this emerging conception, melodrama must be understood to include not only popular trash composed by hack novelists and filmmakers—Conrad's forgotten rival Stanley Weyman, for example; Jacqueline Susann; the director Richard Fleischer—but also such complex, though still widely accessible, artworks as the novels of Samuel Richardson and Dickens, or the films of Hitchcock and Kurosawa. What is crucial to this new definition, though, is not the actual attributes of melodrama itself, which remain essentially unchanged; nor the extension of melodrama's claims to prose fiction and film, which many readers and viewers have long accepted in any case. What is crucial is the way in which the old dispraised attributes of melodrama are understood, the contexts to which they are returned, the respectful scrutiny they are assumed to deserve.[1]

What does it signify, for example, to acknowledge that the structure of melodrama enacts a fantasy of reassurance, and that the happy or moralistic endings so characteristic of the form are reductive and arbitrary—a denial of our "real" world where events refuse to be coherent and where (as Nabokov austerely says) harm is the norm? The desperate or cunning or spirited stratagems by which this escape from reality is accomplished must still retain a fundamental interest. They must still instruct us, with whatever obliqueness, concerning the nature of that reality from which escape or respite has been sought. Consider the episode of the Cave of Montesinos in *Don Quixote*, in which the hero, no mean melodramatist himself, descends into a cavern to dream or conjure a pure vision of love and chivalry and returns with a tale in which a knight's heart is cut from his breast and salted to keep it fresh for his lady. This is an emblem, a crystallizing enactment, of the process whereby our freest, most necessary fantasies are anchored in the harsh, prosaic actualities of life. And Sancho's suspicious but also respectful and deeply attentive interrogation of Quixote's dream instructs us as to how we might profitably interrogate melodrama.

Again, consider the reassurance-structure of melodrama in relation to two other defining features of the form: its persistent and much-contemned habit of moral simplification and its lust for topicality, its hunger to engage or represent behavior and moral attitudes that belong to its particular day and time, especially behavior shocking or threatening to prevailing moral codes. When critics or viewers describe how television panders to its audience, these qualities of simplification and topicality are frequently cited in evidence. The audience wants to be titillated but also wants to be confirmed in its moral sloth, the argument goes, and so the melodramatist sells stories in which crime and criminals are absorbed into paradigms of moral conflict, into allegories of good and evil, in which the good almost always win. The trouble with such a view is not in what it describes, which is often accurate enough, but in its rush to judgment. Perhaps, as Roland Barthes proposes in his stunning essay on wrestling, we ought to learn to see such texts from the standpoint of the audience, whose pleasures in witnessing these specta-, cles of excess and grandiloquence may be deeper than we know, and whose intimate familiarity with such texts may lead them to perceive as complex aesthetic conventions what the traditional high culture sees only as simple stereotypes.[2]

Suppose that the reassuring conclusions and the moral allegorizing of melodrama are regarded in this way, as *conventions,* as "rules" of the genre in the same way that the iambic pentameter and the rimed couplet at the end of a sonnet are "rules" for that form. From this angle, these recurring features of melodrama can be perceived as the *enabling conditions* for an encounter with forbidden or deeply disturbing materials: not an escape into blindness or easy reassurance, but an instrument for seeing. And from this angle, melodrama becomes a peculiarly significant public forum, compli- cated and immensely enriched because its discourse is aesthetic and broadly popular: a forum or arena in which traditional ways of feeling and thinking are brought into continuous, strained relation with powerful intuitions of change and contingency.

This is the spirit in which I turn to television melodrama. In this cate- gory I include most made-for-television movies, the soap operas, and all the lawyers, cowboys, cops and docs, the fugitives and adventurers, the fraternal and filial comrades who have filled the prime hours of so many American nights for the last thirty years.[3] I have no wish to deny that these entertain- ments are market commodities first and last, imprisoned by rigid timetables and stereotyped formulas, compelled endlessly to imagine and reimagine as story and as performance the conventional wisdom, the lies and fantasies, and the muddled ambivalent values of our bourgeois industrial culture. These qualities are, in fact, the primary source of their interest for me, and of the complicated pleasures they uniquely offer.

Confined (but also nourished) by its own foreshortened history and by for- mal and thematic conventions whose origins are not so much aesthetic as economic, television melodrama is a derivative art, just now emerging from its infancy. It is effective more often in parts of stories than in their wholes, and in thrall to censoring pressures that limit its range. But like all true art, television melodrama is cunning, having discovered (or, more often, stum- bled upon) strategies for using the constraints within which it must live.

Its essential artistic resource is the actor's performance, and one expla- nation—there are many others—for the disesteem in which television melo- drama is held is that we have yet to articulate an adequate aesthetics for the art of performance. Far more decisively than the movie-actor, the television- actor creates and controls the meaning of what we see on the screen. In order to understand television drama, and in order to find authentic standards for judging it as art, we must learn to recognize and to value the discipline, energy, and intelligence that must be expended by the actor who succeeds in creating what we too casually call a *truthful* or *believable* performance. What happens when an actor's performance arouses our latent faculties of imaginative sympathy and moral judgment, when he causes us to acknowledge that what he is doing is true to the tangled potency of real experience, not simply impressive or clever, but *true*—what happens then is art.

It is important to be clear about what acting, especially television-acting, is or can be: nothing less than a reverent attentiveness to the pain and beauty in the lives of others, an attentiveness made accessible to us in a wonderfully instructive process wherein the performer's own impulses to self-assertion realize themselves only by surrendering or yielding to the claims of the character he wishes to portray. Richard Poirier, our best theorist of performance, puts the case as follows: "performance . . . is an action which must go through passages that both impede the action and give it form, much as a sculptor not only is impelled to shape his material but is in turn shaped by it, his impulse to mastery always chastened, sometimes made tender and possibly witty by the recalcitrance of what he is working on."[4]

Television has always challenged the actor. The medium's reduced visual scale grants him a primacy unavailable in the theater or in the movies, where an amplitude of things and spaces offers competition for the eye's attention. Its elaborate, enforced obedience to various formulas for plot and characterization virtually require him to recover from within himself and from his broadly stereotyped assignment nuances of gesture, inflection, and movement that will at least hint at individual or idiosyncratic qualities. And despite our failure clearly to acknowledge this, the history of television as a dramatic medium is, at the very least, a history of exceptional artistic accomplishment by actors. The performances in television melodrama today are much richer than in the past, though there were many remarkable performances even in the early days. The greater freedom afforded to writers and actors is part of the reason for this, but (as I will try to indicate shortly) the far more decisive reason is the extraordinary sophistication the genre has achieved.

Lacking access to even the most elementary scholarly resources—bibliographies, systematic collections of films or tapes, even moderately reliable histories of the art—I can only appeal to our (hopefully) common memory of the highly professional and serious acting regularly displayed in series such as *Naked City, Twilight Zone, Route 66, Gunsmoke, The Defenders, Cade's County, Stoney Burke, East Side, West Side, The Name of the Game,* and others whose titles could be supplied by anyone who has watched American television over the past twenty or twenty-five years. Often the least promising dramatic formulas were transformed by vivid and highly intelligent performances. I remember with particular pleasure and respect, for example, Steve McQueen's arresting portrayal of the callow bounty hunter Josh Randall in the western series, *Wanted: Dead or Alive*—the jittery lean grace of his physical movements, the balked, dangerous tenderness registered by his voice and eyes in his encounters with women; the mingling of deference and menace that always enlivened his dealings with older men, outlaws and sheriffs mainly, between whom this memorable boy-hero seemed fixed or caught, but willingly so. McQueen's subsequent apotheosis in the movies was obviously deserved, but I have often felt his performances on the large screen were less tensely intelligent, more self-indulgent than his brilliant early work in television.

If we could free ourselves from our ingrained expectations concerning dramatic form and from our reluctance to acknowledge that art is always a commodity of some kind, constrained by the technology necessary to its production and by the needs of the audience for which it is intended, then we might begin to see how ingeniously television melodrama contrives to nourish its basic resource—the actor—even as it surrenders to those economic pressures that seem most imprisoning.

Consider, for example, the ubiquitous commercials. They are so widely deplored that even those who think themselves friendly to the medium cannot restrain their outrage over such unambiguous evidence of the huckster's contempt for art's claim to continuity. Thus, a writer in the official journal of the National Academy of Television Arts and Sciences, meditating sadly on "the total absence" of serious television drama, refers in passing to "the horrors of continuous, brutal interruption."[5]

That commercials have shaped television melodrama decisively is obvious, of course. But, as with most of the limitations to which the genre is subjected, these enforced pauses are merely formal conventions. They are no more intrinsically hostile to art than the unities observed by the French neoclassical theater or the serial installments in which so many Victorian novels had to be written. Their essential effect has been the refinement of a segmented dramatic structure peculiarly suited to a formula-story whose ending is predictable—the doctor will save the patient, the cop will catch the criminal—and whose capacity to surprise or otherwise engage its audience must therefore depend largely on the localized vividness and potency of the smaller units or episodes that comprise the whole.

Television melodrama achieves this episodic or segmented vividness in several ways, but its most dependable and recurring strategy is to require its actors to display themselves intensely and energetically from the very beginning. In its most characteristic and most interesting form, television melodrama will contrive its separate units such that they will have substantial independent weight and interest, usually enacting in miniature the larger patterns and emotional rhythms of the whole drama. Thus, each segment will show us a character, or several characters, confronting some difficulty or other; the character's behavior and (especially) his emotional responses will intensify, then achieve some sort of climactic or resolving pitch at the commercial break; and this pattern will be repeated incrementally in subsequent segments.

To describe this characteristic structure is to clarify what those who complain of the genre's improbability never acknowledge: that television melodrama is in some respects an *operatic* rather than a conventionally dramatic form—a fact openly indicated by the term *soap opera.* No one goes to Italian opera expecting a realistic plot, and since applause for the important arias is an inflexible convention, no one expects such works to proceed without interruption. The pleasures of this kind of opera are largely (though not exclusively) the pleasures of the brilliant individual performance, and good operas in this tradition are those in which the composer has contrived roles which test as fully as possible the vocal capacities of the performers.

Similarly, good television melodramas are those in which an intricately formulaic plot conspires perfectly with the commercial interruptions to encourage a rich articulation of the separate parts of the work, and thus to call forth from the realistic actor the full energies of his performer's gifts. What is implausible in such works is the continual necessity for emotional display by the characters. In real life we are rarely called upon to feel so intensely, and never in such neatly escalating sequences. But the emotions dramatized by these improbable plots are not in themselves unreal, or at least they need not be—and television melodrama often becomes more truthful as it becomes more implausible.

As an example of this recurring paradox—it will be entirely familiar to any serious reader of Dickens—consider the following generically typical episode from the weekly series, *Medical Center.* An active middle-aged man falls victim to an aneurysm judged certain to kill him within a few years. This affliction being strategically located for dramatic use, the operation that could save his life may also leave him impotent—a fate nasty enough for anyone, but psychologically debilitating for this unlucky fellow who has divorced his first wife and married a much younger woman. The early scenes establish his fear of aging and his intensely physical relationship with his young wife with fine lucid economy. Now the plot elaborates further complications and develops new, related central centers of interest. His doctor—the series regular who is (sometimes) an arresting derivation of his television ancestors, Doctors Kildare and Ben Casey—is discovered to be a close, longtime friend whose involvement in the case is deeply personal. Confident of his surgeon's skills and much younger than his patient, the doctor is angrily unsympathetic to the older man's reluctance to save his life at the expense of his sexuality. Next, the rejected wife, brilliantly played by Barbara Rush, is introduced. She works—by a marvelous arbitrary coincidence—in the very hospital in which her ex-husband is being treated. There follows a complex scene in the hospital room in which the former wife acts out her tangled, deep feelings toward the man who has rejected her and toward the woman who has replaced her. In their tensely guarded repartee, the husband and ex-wife are shown to be bound to one another in a vulnerable knowingness made in decades of uneasy intimacy that no divorce can erase and that the new girl-wife must observe as an outsider. Later scenes require emotional confrontations—some of them equally subtle—between the doctor and each wife, between doctor and patient, between old wife and new.

These nearly mathematic symmetries conspire with still further plot complications to create a story that is implausible in the extreme. Though aneurysms are dangerous, they rarely threaten impotence. Though impotence is a real problem, few men are free to choose a short happy life of potency, and fewer still are surrounded in such crises by characters whose relations to them so fully articulate such a wide spectrum of human needs and attitudes. The test of such an arbitrary contrivance is not the plausibility of the whole but the accuracy and truthfulness of its parts, the extent to which its various strategies of artificial heightening permit an open enactment of feelings and desires that are only latent or diffused in the muddled incoherence of the real world. And although my argument does not depend on the success or failure of one or of one dozen specific melodramas—the genre's manifest complexity and its enormous popularity being sufficient to justify intensive and respectful study—I should say that the program just described was for me a serious aesthetic experience. I was caught by the persuasiveness of the actors' performances, and my sympathies were tested by the meanings those fine performances released. The credibility of the young wife's reluctant, pained acknowledgement that a life without sex *would* be a crippled life; the authenticity of the husband's partly childish, partly admirable reverence for his carnal aliveness; and, especially, the complex genuineness of his ambivalent continuing bonds with his first wife—all this was there on the screen. Far from falsifying life, it quickened one's awareness of the burdens and costs of human relationships.

That the plots of nearly all current television melodramas tend, as in this episode of *Medical Center,* to be more artificially contrived than those of earlier years seems to me a measure not of the genre's unoriginality but of its maturity, its increasingly bold and self-conscious capacity to *use* formal requirements which it cannot in any case evade, and to exploit (rather than be exploited by) various formulas for characterization. Nearly all the better series melodramas of recent years, in fact, have resorted quite openly to what might be called a *multiplicity principle:* a principle of plotting or organization whereby a particular drama will draw not once or twice but many times upon the immense store of stories and situations created by the genre's brief but crowded history. The multiplicity principle allows not less but more reality to enter the genre. Where the old formulas had been developed exhaustively and singly through the whole of a story—that is how they became stereotypes—they are now treated elliptically in a plot that deploys many of them simultaneously. The familiar character-types and situations thus become more suggestive and less imprisoning. There is no pretense that a given character has been wholly "explained" by the plot, and the formula has the liberating effect of creating a premise or base on which the actor is free to build. By minimizing the need for long establishing or expository sequences, the multiplicity principle allows the story to leave aside the question of *how* these emotional entanglements were arrived at and to concentrate its energies on their credible and powerful present enactment.

These and other stratagems—which result in richer, more plausible characterizations and also permit elegant variations of tone—are possible because television melodrama can rely confidently on one resource that is always essential to the vitality of any artform: an audience impressive not simply in its numbers but also in its genuine sophistication, its deep familiarity with the history and conventions of the genre. For so literate an audience, the smallest departure from conventional expectations can become meaningful, and this creates endless chances for surprise and nuanced variation, even for thematic subtlety.

In his instructive book on American films of the forties and fifties, Michael Wood speaks nostalgically of his membership in "the universal movie audience" of that time. This audience of tens of millions was able to see the movies as a coherent world, "a country of familiar faces, . . . a system of assumptions and beliefs and preoccupations, a fund of often interchangeable plots, characters, patches of dialog, and sets." By relying on the audience's familiarity with other movies, Wood says, the films of that era constituted "a living tradition of the kind that literary critics always used to be mourning for."[6]

This description fits contemporary television even more closely than it does those earlier movies, since most members of the TV audience have lived through the whole history of the medium. They know its habits, its formulas, its stars, and its recurring character actors with a confident, easy intimacy that may well be unique in the history of popular art. Moreover, television's capacity to make its history and evolution continuously available (even to younger members in its universal audience) is surely without precedent, for the system of reruns has now reached the point of transforming television into a continuous, living museum which displays for daily or weekly consumption texts from every stage of the medium's past.

Outsiders from the high culture who visit TV melodrama occasionally in order to issue their tedious reports about our cultural malaise are simply not seeing what the TV audience sees. They are especially blind to the complex allusiveness with which television melodrama uses its actors. For example, in a recent episode of the elegant *Columbo* series, Peter Falk's adventures occurred onboard a luxury liner and brought him into partnership with the captain of the ship, played by Patrick Macnee, the smooth British actor who starred in the popular spy series, *The Avengers*. The scenes between Falk and Macnee were continuously enlivened not simply by the different acting styles of the two performers but also by the attitudes toward heroism, moral authority, and aesthetic taste represented in the kinds of programs with which each star has been associated. The uneasy, comic partnership between these characters—Falk's grungy, American-ethnic slyness contrasting with, and finally mocking, Macnee's British public school elegance and fastidiousness—was further complicated by the presence in the show of the guest villain, played by yet another star of a successful TV series of a few years ago—Robert Vaughn of *The Man From U.N.C.L.E.* Vaughn's character had something of the sartorial, upper-class *elan* of Macnee's ship's master but, drawing on qualities established in his earlier TV role, was tougher, wholly American, more calculating, and ruthless. Macnee, of course, proved no match for Vaughn's unmannerly cunning, but Falk-Columbo succeeded in exposing him in a climax that expressed not only the show's usual fantasy of working-class intelligence overcoming aristocratic guile, but also the victory of American versions of popular entertainment over their British counterparts.

The aesthetic and human claims of most television melodrama would surely be much weakened, if not completely obliterated, on any other medium, and I have come to believe that the species of melodrama to be found on television today is a unique dramatic form, offering an especially persuasive resolution of the contradiction or tension that has been inherent in melodrama since the time of Euripides. As Peter Brooks reminds us in his provocative essay on the centrality of the melodramatic mode in romantic and modern culture, stage melodrama represents "a popular form of the tragic, exploiting similar emotions within the context of the ordinary." Melodrama is a "popular" form, we may say, both because it is favored by audiences and because it insists (or tries to insist) on the dignity and importance of the ordinary, usually bourgeois world of the theatergoer himself. The difficulty with this enterprise, of course, is the same for Arthur Miller in our own day as it was for Thomas Middleton in Jacobean London: displacing the action and characters from a mythic or heroically stylized world to an ordinary world—from Thebes to Brooklyn—involves a commitment to a kind of realism that is innately resistant to exactly those intense passionate enactments that the melodramatist wishes to invoke. Melodrama is thus always in conflict with itself, gesturing simultaneously toward ordinary reality *and* toward a moral and emotional heightening that is rarely encountered in the "real" world.

Although it can never be made to disappear, this conflict is minimized, or is capable of being minimized, by television—and in a way that is simply impossible in the live theater and that is nearly always less effective on the enlarged movie-screen. The melodramatic mode is peculiarly congenial to television, its inherent contradictions are less glaring and damaging there, because the medium is uniquely hospitable to the spatial confinements of the theater and to the profound realistic intimacy of the film.

Few would dispute the cinema's advantages over the theater as realistic medium. As every serious film theorist begins by reminding us, the camera's ability to record the dense multiplicity of the external world and to reveal character in all its outer nuance and idiosyncrasy grants a visually authenticating power to the medium that has no equivalent in the theater. Though the stage owns advantages peculiar to its character as a live medium, it is clearly an artform more stylized, less visually realistic than the film, and it tests its performers in a somewhat different way. Perhaps the crucial difference is also the most obvious one: the distance between the audience and the actor in even the most intimate theatrical environment requires facial and vocal gestures as well as bodily movements "broader" and more excessive than those demanded by the camera, which can achieve a lover's closeness to the performer.

The cinema's photographic realism is not, of course, an unmixed blessing. But it is incalculably valuable to melodrama because, by encouraging understatement from its actors, it can help to ratify extravagant or intense emotions that would seem far less credible in the theater. And although television is the dwarf child of the film, constrained and scaled down in a great many ways, its very smallness can become an advantage to the melodramatic imagination. This is so because if the cinema's particularizing immediacy is friendly to melodrama, certain other characteristics of the medium are hostile to it. The extended duration of most film, the camera's freedom of movement, the more-than-life-sized dimensions of the cinematic image—all these create what has been called the film's mythopoeic tendency, its inevitable effect of magnification. Since the natural domain of melodrama is indoors, in those ordinary and enclosed spaces wherein most of us act out our deepest needs and feelings—bedrooms, offices, courtrooms, hospitals—the reduced visual field of television is, or can be, far more nourishing than the larger, naturally expansive movie-screen. And for the kind of psychologically nuanced performance elicited by good melodrama, the smaller television screen would seem even more appropriate: perfectly adapted, in fact, to record those intimately minute physical and vocal gestures on which the art of the realistic actor depends, yet happily free of the cinema's malicious (if often innocent) power to transform merely robust nostrils into Brobdingnagian caverns, minor facial irregularities into craterous deformities.

Television's matchless respect for the idiosyncratic expressiveness of the ordinary human face and its unique hospitality to the confining spaces of our ordinary world are virtues exploited repeatedly in all the better melodramas. But perhaps they are given special decisiveness in *Kojak,* a classy police series whose gifted leading player has been previously consigned almost entirely to gangster parts, primarily (one supposes) because of the cinema's blindness to the uncosmetic beauty of his large bald head and generously irregular face. In its first two years particularly, before Savalas's character stiffened into the macho stereotype currently staring out upon us from magazine advertisements for razor blades and men's toiletries, *Kojak* was a genuine work of art, intricately designed to exploit its star's distinctively urban flamboyance, his gift for registering a long, modulated range of sarcastic vocal inflections and facial maneuvers, his talent for persuasive ranting. The show earned its general excellence not only because of Savalas's energetic performance, but also because its writers contrived supporting roles that complemented the central character with rare, individuating clarity, because the boldly artificial plotting in most episodes pressed toward the revelation of character rather than shoot-em-up action, and because, finally, the whole enterprise was forced into artfulness by the economic and technolog-ical environment that determined its life.

This last is at once the most decisive and most instructive fact about *Kojak,* as it is about television melodrama generally. Because *Kojak* is filmed in Hollywood on a restricted budget, the show must invoke New York elliptically, in ingenious process shots and in stock footage taken from the full-length (and much less impressive) television-movie that served as a pilot for the series. The writers for the program are thus driven to devise stories that will allow the principle characters to appear in confined locations that can be created on or near studio sound-stages—offices, interrogation rooms, dingy bars, city apartments, nondescript alleys, highway underpasses, all the neutral and enclosed spaces common to urban life generally. As a result, *Kojak* often succeeds in projecting a sense of the city that is more compel-ling and intelligent than that which is offered in many films and television movies filmed on location: its menacing closeness, its capacity to harbor and even to generate certain kinds of crime, its watchful, unsettling accuracy as a custodian of the lists and records and documents that open a track to the very center of our lives. *Kojak*'s clear superiority to another, ostensibly more original and exotic police series, *Hawaii Five-O,* is good partial evidence for the liberating virtues of such confinement. This latter series is filmed on location at enormous expense and is often much concerned to give a flavor of Honolulu particularly. Yet it yields too easily to an obsession with scenic vistas and furious action sequences which threaten to transform the pro-gram into a mere travelogue and which always seem unnaturally confined by the reduced scale of the television screen.

That the characters in *Kojak* frequently press beyond the usual stereo-types is also partly a result of the show's inability to indulge in all the out-door muscle-flexing, chasing, and shooting made possible by location film-ing. Savalas's Kojak especially is a richly individuated creation, his policeman's cunning a natural expression of his lifelong, intimate involvement in the very ecology of the city. A flamboyant, aggressive man, Kojak is continually engaged in a kind of joyful contest for recognition and even for mastery with the environment that surrounds him. The studio sets on which most of the action occurs, and the many close-up shots in each episode, reinforce and nurture these traits perfectly, for they help Savalas to work with real subtlety—to project not simply his character's impulse to define himself against the city's enclosures but also a wary, half-loving respect for such imprison-ments, a sense indeed that they are the very instrument of his self-realization.

Kojak's expensive silk-lined suits and hats and the prancing vitality of his physical movements are merely the outer expressions of what is shown repeatedly to be an enterprise of personal fulfillment that depends mostly on force of intellect. His intelligence is not bookish—the son of a Greek immigrant, he never attended college—but it is genuine and powerfully self-defining because he must depend on his knowledge of the city in order to prevent a crime or catch a criminal. Proud of his superior mental quickness and urban knowingness, Kojak frequently behaves with the egotistical flair of a bold, demanding scholar, reveling in his ability to instruct subordinates in how many clues they have overlooked and even (in one episode) performing with histrionic brilliance as a teacher before a class of students at the police academy. Objecting to this series because it ratifies the stereotype of the super-cop is as silly as objecting to Sherlock Holmes on similar grounds. Like Holmes in many ways, Kojak is a man who realizes deeply private needs and inclinations in the doing of his work. Not law-and-order simplicities, but intelligence and self-realization are what *Kojak* celebrates. The genius of the series is to have conceived a character whose portrayal calls forth from Savalas exactly what his appearance and talents most suit him to do.

The distinction of *Kojak* in its first two seasons seems to me reasonably representative of the achievements of television melodrama in recent years. During the past season, I have seen dozens of programs—episodes of *Harry-O, Police Story, Baretta, Medical Center,* the now-defunct *Medical Story,* several made-for-TV movies, and portions at least of the new mini-series melodramas being developed by ABC—whose claims to attention were fully as strong as *Kojak*'s. Their partial but genuine excellence constitutes an especially salutary reminder of the fact that art always thrives on restraints and prohibitions, indeed that it requires them if it is to survive at all. Like the Renaissance sonnet or Racine's theater, television melodrama is always most successful when it most fully embraces that which confines it, when *all* the limitations imposed upon it—including such requirements as the sixty- or ninety-minute time slot, the commercial interruptions, the small dimensions of the screen, even the consequences of low-budget filming—become instruments of use, conventions whose combined workings create unpretentious and spirited dramatic entertainments, works of popular art that are engrossing, serious, and imaginative.

That such honorific adjectives are rarely applied to television melodrama, that we have effectively refused even to consider the genre in aesthetic terms is a cultural fact and, ultimately, a political fact almost as interesting as the artworks we have been ignoring. Perhaps because television melodrama is an authentically popular art—unlike rubber hamburgers, encountergroup theater or electric-kool-aid journalism—our understanding of it has been conditioned (if not thwarted entirely) by the enormous authority American high culture grants to avant-garde conceptions of the artist as an adversary figure in mortal conflict with his society. Our attitude toward the medium has been conditioned also by even more deeply ingrained assumptions about the separate high dignity of aesthetic experience—an activity we are schooled to imagine as uncontaminated by the marketplace, usually at enmity with the everyday world, and dignified by the very rituals of payment and dress and travel and isolation variously required for its enjoyment. It is hard, in an atmosphere which accords art a special if not an openly subversive status, to think of television as an aesthetic medium, for scarcely another institution in American life is at once so familiarly *un*special and so profoundly a creature of the economic and technological genius of advanced industrial capitalism.

Almost everything that is said or written about television, and especially about television drama, is tainted by such prejudices; more often it is in utter servitude to them. And although television itself would no doubt benefit significantly if its nature were perceived and described more objectively, it is the larger culture—whose signature is daily and hourly to be found there—that would benefit far more.

In the introduction to *The Idea of a Theater,* Francis Fergusson reminds us that genuinely popular dramatic art is always powerfully conservative in certain ways, offering stories that insist on "their continuity with the common sense of the community." Hamlet could enjoin the players to hold a mirror up to nature, "to show . . . the very age and body of the time his form and pressure" because, Fergusson tells us, "the Elizabethan theater was itself a mirror which had been formed at the center of the culture of its time, and at the center of the life and awareness of the community." That we have no television Shakespeare is obvious enough, I guess. But we do already have our Thomas Kyds and our Chapmans. A Marlowe, even a Ben Jonson, is not inconceivable. It is time we noticed them.[7]

### Notes

1. The bibliography of serious recent work on melodrama is not overly intimidating, but some exciting and important work has been done. I list here only pieces that have directly influenced my present argument, and I refer the reader to their notes and bibliographies for a fuller survey of the scholarship. Earl F. Bargainnier summarizes recent definitions of melodrama and offers a short history of the genre as practiced by dramatists of the eighteenth and nineteenth centuries in "Melodrama as Formula," *Journal of Popular Culture* 9 (Winter, 1975). John G. Cawelti's indispensable *Adventure, Mystery, and Romance* (Chicago, 1976) focuses closely and originally on melodrama at several points. Peter Brooks's "The Melodramatic Imagination," in *Romanticism: Vistas, Instances, Continuities,* ed. David Thorburn and Geoffrey Hartman (Cornell, 1973), boldly argues that melodrama is a primary literary and visionary mode in romantic and modern culture. Much recent Dickens criticism is helpful on melodrama, but see especially Robert Garis, *The Dickens Theatre* (Oxford, 1965), and essays by Barbara Hardy, George H. Ford, and W. J. Harvey in the Dickens volume of the Twentieth-Century Views series, ed. Martin Price (Prentice-Hall, 1967). Melodrama's complex, even symbiotic linkages with the economic and social institutions of capitalist democracy are a continuing (if implicit) theme of Ian Watt's classic *The Rise of the Novel* (University of California Press, 1957), and of Leo Braudy's remarkable essay on Richardson, "Penetration and Impenetrability in Clarissa," in *New Approaches to Eighteenth-Century Literature,* ed. Phillip Harth (Columbia University Press, 1974).

2. Roland Barthes, "The World of Wrestling," in *Mythologies,* trans. Annette Lavers (Hill and Wang, 1972). I am grateful to Jo Anne Lee of the University of California, Santa Barbara, for making me see the connection between Barthes's notions and television drama.

3. I will not discuss soap opera directly, partly because its serial nature differentiates it in certain respects from the prime-time shows, and also because this interesting subgenre of TV melodrama has received some preliminary attention from others. See, for instance, Frederick L. Kaplan, "Intimacy and Conformity in American Soap Opera," *Journal of Popular Culture* 9 (Winter, 1975); Renata Adler, "Afternoon Television: Unhappiness Enough and Time," *The New Yorker* 47 (February 12, 1972); Marjorie Perloff, "Soap Bubbles," *The New Republic* (May 10, 1975); and the useful chapter on the soaps in Horace Newcomb's pioneering (if tentative) *TV, The Most Popular Art* (Anchor, 1974). Newcomb's book also contains sections on the prime-time shows I am calling melodramas. For an intelligent fan's impressions of soap opera, see Dan Wakefield's *All Her Children* (Doubleday, 1976).

4. Richard Poirier, *The Performing Self* (Oxford, 1971), p. xiv. I am deeply indebted to this crucial book, and to Poirier's later elaborations on this theory of performance in two pieces on ballet and another on Bette Midler (*The New Republic*, January 5, 1974; March 15, 1975; August 2 and 9, 1975).

5. John Houseman, "TV Drama in the U.S.A.," *Television Quarterly* 10 (Summer, 1973), p. 12.

6. Michael Wood, *America in the Movies* (Basic Books, 1975), pp. 10–11.

7. Though they are not to be held accountable for the uses to which I have put their advice, the following friends have read earlier versions of this essay and have saved me from many errors: Sheridan Blau, Leo Braudy, John Cawelti, Peter Clecak, Howard Felperin, Richard Slotkin, Alan Stephens, and Eugene Waith.

# Melodrama, Serial Form, and Television Today

## JANE FEUER

"The indulgence of strong emotionalism; moral polarization and schematization; extreme states of being, situations, action; overt villainy, persecution
of the good, and final reward of virtue; inflated and extravagant expression;
dark plottings, suspense, breathtaking peripety" . . . What Peter Brooks
calls the "everyday connotations" of the term "melodrama"[1] describes almost perfectly the current form-in-dominance on American network television: the continuing serial or "soap opera." Although only a few years ago
there seemed to be no equivalent on prime-time television[2] to the film
melodramas of the 1950s recently rescued from obscurity by film theorists,
we now find the domestic melodrama encroaching upon the domain of the
sitcom and the cop show. At the same time, daytime soap operas are expanding, having risen to an astonishing peak of popularity.[3] Indeed awareness of their own significance seems to have reached the producers of daytime dramas. In a 1982 broadcast of the immensely popular daytime soap
*All My Children,* a young woman character, Silver Kane, begged her sister's
lover to get her a part on a soap, explaining what an honor that would be—
they even teach them in college, she tells him, as a form of "folk drama."

For the purposes of this article, I am choosing to stress the similarities
between daytime soaps and the prime-time continuing melodramatic serials
such as *Dallas, Dynasty, Falcon Crest, Knots Landing, Flamingo Road* and the
short-lived *Bare Essence.* I do this because I feel they have overriding simi-

From Jane Feuer, "Melodrama, Serial Form and Television Today," *Screen* 25:1,
1984, pp. 4–16. Reprinted with permission.

larities in terms of the theories I will discuss⌊Daytime and prime-time serials share a narrative form consisting of multiple plot lines and a continuing narrative (no closure). Both concentrate on the domestic sphere, although the prime-time serials also encompass the world of business and power (designed to appeal to the greater number of males in the evening viewing audience).⌉

However, there are significant differences between the two forms, some of which will come out in my discussion. In *The Soap Opera,* Muriel Cantor and Suzanne Pingree do not consider the prime-time programs or other related programs (e.g., limited serials such as the British *Forsyte Saga* or U.S. mini-series) to fit their definition of soap opera.[4] ⌈They believe the primary difference is one of "content." Prime-time serials, they state, have a less conservative morality, deal with power and big business, and contain more action. They believe the most significant difference is that daytime soap operas are manifestations of women's culture, and prime-time serials are not. Although these are significant points, much of what I will argue in this paper transcends the distinction between the two forms. I would also argue that due to the influence of prime-time serials, many daytime soaps have added amoral wealthy families and faster action. Moreover, by excluding so many distant relatives of the daytime soap (including the serial cop show *Hill Street Blues* and the serial medical show *St. Elsewhere*), Cantor and Pingree are unable to stress the pervasive influence of serial form and multiple plot structure upon *all* of American television. I will use the term "television melodrama" to encompass both and to exclude other program types which take the form of episodic series as opposed to continuing serials.

Fortunately, we need not start from scratch in studying the new form of the prime-time melodramatic serial. Melodrama has flourished before, and we can benefit from the body of ideas surrounding Hollywood melodramas of the 1950s. I would like to begin by surveying and offering a critique of that theory, and then go on to consider its possible applications to the prime-time continuing serials *Dallas* and *Dynasty*.

Initially, critical interest in the films of Sirk bore little relationship to interest in the still-despised genre in which he most often worked. Quite the contrary: interest in Sirk stemmed from an extreme formalist tendency in *auteur* criticism, an attempt to bypass the narrative level in order to capture pure expressivity through *mise-en-scène. Mise-en-scène* critics were drawn to Sirk (as to other melodramatists such as Minnelli and Ophuls) for the way in which his style seemed to transcend the narrative level. In a much more sophisticated way, interest in style seemingly for its own sake dominated expositions of Sirk's films in several articles in the 1972 Edinburgh booklet on Sirk,[5] extending positions taken in the Summer 1971 issue of *Screen.*[6] One of these essays, Paul Willemen's "Distanciation and Douglas Sirk," links the early interest in Sirk as a stylist to a new interest in a level of style which precludes audience identification in the usual sense.[7] Because this "Brechtian" position haunts Sirk criticism from this point on, it is worth summarizing in some detail.

Willemen explains Sirk's style as an "intensification" of generic practices, not as irony per se. Since he had to appeal to a mass audience, Sirk drew on Expressionist and Brechtian theatrical experience "not to break the rules . . . but to intensify them." According to Willemen, this was accomplished through the magnification of emotionality, use of pathos, choreography, and music, and through aspects of *mise-en-scène* such as "mirror-ridden walls." Such intensification puts a distance, though not necessarily one perceived by the audience, between "the film and its narrative pretext." Even if not perceived by the mass audience, Willemen argues, distanciation "may still exist within the film itself." According to this view, a discrepancy exists between the audience Sirk is aiming at and the audience he knows will come to his films. The "formalist" critics had also conceptualized a rupture between the narrative/dramatic and the filmic codes in Sirk. Willemen's hypothesis gives us an explanation for this rupture.

Following Willemen's logic, one must conceptualize a Sirk film as two films in one. The "primary" text, the one which the mass audience will read and which consists of the narrative level, is melodrama, pure and simple. Whereas the secondary text springs from the distance "intensification" opens up between Sirk's formal level and his narrative/dramatic generic level. For melodrama itself, according to this line of reasoning, lies fully within the "dominant ideology." Stylization, Willemen argues, "can also be used to parody the stylistic procedures which traditionally convey an extremely smug, self-righteous, and petit bourgeois world view paramount in the American melodrama."[8] Willemen proceeds to place Sirk's work in the category of films which—according to the well-known classification system of post-'68 *Cahiers*—turns out to be ambiguous in terms of the dominant ideology even though, at first glance, they may seem to rest fully within such an ideology.

This is the theoretical justification for Willemen's interest in Sirk and the way in which formalist readings of Sirk's films may be linked to a new interest in film as ideology. To put it in terms of the "two texts," the primary Sirk text is fully within the dominant ideology because the narrative/dramatic level consists of pure melodrama, indistinguishable from any run-of-the-mill Hollywood melodrama (indeed two of Sirk's most lauded films were remakes). It is the secondary text, which, through authorial intervention at the level of *mise-en-scène* is subversive of this dominant ideology. However, in actual practice, distinguishing between the two texts can be difficult.[9] If it is true that only an *auteur* such as Sirk is capable of bringing stylistic pressure to bear upon the purely ideological melodramatic material and thus causing it to "rupture" and reveal its own textual gaps in terms of the dominant ideology, then only an elite audience, indeed one already committed to subversive ideas, would be able to read the secondary text. Such a position does not explicate the spectator position melodrama allows for its intended audience.

Out of this impasse emerged a number of theoretical articles which, while retaining notions of distanciation and rupture, nevertheless shifted the emphasis from a specifically authorial and intentional subversive practice to the idea that melodrama *qua* melodrama contained the potential for exposing contradictions in the dominant ideology and for readings "against the grain." These new feminist and psychoanalytical readings open up the possibility of application to the distinctly nonauthorial texts of American network television.

A close relationship between melodrama as a form and the ideology of capitalism had already been stressed in Thomas Elsaesser's influential 1973 article "Tales of Sound and Fury: Observations on the Family Melodrama." Elsaesser traces the roots of the 1950s family melodrama to the eighteenth and nineteenth century sentimental novel in order to show that the form has always been embedded in a social context:

> an element of interiorisation and personalisation of what are primarily ideological conflicts, together with the metaphorical interpretation of class-conflict as sexual exploitation and rape, is important in all subsequent forms of melodrama.[10]

Elsaesser argues that melodrama functions as either subversive or escapist relative to the given historical and social context. It is also relative to where the emphasis lies—upon the ideological conflicts or upon the happy ending.[11] Several other critics took up this emphasis on the exposure of contradictions, although they disagree as to whether the form is ultimately subversive or not. Geoffrey Nowell-Smith gave the debate a psychoanalytical slant, arguing that the audience knew the happy endings in melodramas were often impossible: "a happy end which takes the form of an acceptance of castration is achieved only at the cost of repression."[12] According to Nowell-Smith's view,

> the importance of melodrama lies precisely in its ideological failure. Because it cannot accommodate its problems either in a real present or an ideal future, but lays them open in their contradictoriness, it opens a space which most Hollywood films have studiously closed off.[13]

Yet, other critics questioned the nature of the "space" thus opened. In a feminist reading of Sirk, Laura Mulvey suggested that melodrama as a form opens up contradictions in bourgeois ideology in the domestic sphere. However, she sees the purpose of opening ideological contradictions as providing a "safety valve" rather than as progressive. Mulvey believes this view of melodrama places it 'in the context of wider problems."[14] One of these wider problems would be the relationship between melodrama as a form and the capitalist social formation. According to Chuck Kleinhans, the raw material of any melodrama consists in exposing contradictions of capitalism in the personal sphere. Kleinhans believes that the main contradiction melodrama explores is the expectation that the family should fulfill all needs society can't fill. His conclusion is that melodramas offer artistic presentations of genuine problems but locate these problems in the family, the place where they can't be solved. He sees melodrama as serving an important function for women in capitalist society, but sees its form as ultimately self-defeating.[15]

A few currents run consistently through the shifting theoretical viewpoints just delineated. Melodrama seemed amenable to a variety of theoretical approaches because melodramas seemed to encourage different levels of reading to a greater extent than did other "classical narrative" films. Traditionally male-oriented genres such as the western or the gangster film did not problematize the reader in the same way as melodrama. Thus few articles appeared on "The Western and the Male Spectator." If one assumes, as early studies of male genres did, a nonproblematic and universalized male subject, then westerns and gangster films can be studied by means of the textually-based structuralism in vogue during the late 1960s and early 1970s. Melodrama, in problematizing questions of spectatorship and gender, demands reader-response based modes of analysis such as psychoanalysis.

Central to all the theoretical positions I have just enumerated is the concept of melodrama as creating an *excess*, whether that excess be defined as a split between the level of narrative and that of *mise-en-scène* or as a form of "hysteria," the visually articulated return of the ideologically repressed. Despite the changing theoretical stances, all see the excess not merely as aesthetic but as *ideological*, opening up a textual space which may be read against the seemingly hegemonic surface. The key text for the theorization of visual excess has tended to be Sirk's *Written on the Wind* with its intricately layered (and thus visually ruptured) mirror shots, phallic symbolism, and "hysterical" montage. More than any other film this oil dynasty saga seems to provide a prototype for *Dallas* and *Dynasty*. More than any other Sirk film *Written on the Wind* seems to occupy the same representational field as today's prime-time serial melodramas. Unlike Sirk's other melodramas and also unlike daytime soap operas, *Written on the Wind, Dallas*, and *Dynasty* focus on the capitalist ruling elite rather than the bourgeois family. The address is not so uniformly from one bourgeois to another as it is in other forms of melodrama.[16] (Although of course the representation of the upper classes is intended to be *read* by a bourgeois audience.) Despite the similarity of representational field, today's prime-time melodrama does not take the same visual and narrative form as *Written on the Wind*. Unlike the texts upon which much of the theory of film melodrama has been constructed, *Dallas, Dynasty,* and their imitators appear to lack visual excess as it has been described in the fifties family melodrama.[17] Moreover, they lack another element crucial to theories of textual deficiencies which run counter to the dominant ideology—that is to say, they lack closure.

Is there a potential for reading *Dallas* and *Dynasty* in terms of excesses and contradictions? Are these programs the conventional domestic melodramas of their time which now seek an *auteur* to subvert them? Or do they already contain the potential for subversive readings? In the analysis that follows I will focus on some of the conventions employed in episodes from the 1981–82 seasons of *Dallas* and *Dynasty*. (The dates used throughout are for U.S. seasons, rather than British transmissions.) My argument will be that excess needs to be defined not in terms of the norms for films of the fifties but rather in terms of those for television of the seventies.

(Seen in terms of their own medium, the seemingly simple *mise-en-scène* and editing style of the prime-time serials takes on a new signification.) Although *mise-en-scène* in *Dallas* and *Dynasty* does not take on the hysterical dimensions of a Sirk or Fassbinder film it does seem at the very least *opulent* compared to other prime-time programs and certainly compared to the daytime soaps. Budgetary considerations alone show the emphasis placed on *mise-en-scène*. According to one source, "*Dynasty* costs approximately one million dollars an hour because of the show's cavernous and opulent sets, not to mention the dazzling fashions worn by cast members."[18] While there is nothing inherently subversive about such splendor, it does serve to take the family dynasty serials outside the normal upper-middle-class milieu of most film and television melodrama. (The very rich portrayed in these narratives exceed the norms of their audience both economically and morally; luxurious *mise-en-scène* objectifies such excess. But in order to fulfil the theory that excess leads to a counter-current in the text, some authorial voice would need to use the visual excess against the narrative level. This does not appear to happen. Although the programs appear to be aware of their own splendid tackiness, they do not appear to set out explicitly to subvert any generic codes, as did the comic parody *Soap* or the ambiguously conventional version of daytime soaps *Mary Hartman, Mary Hartman*.

For *Dallas* and *Dynasty*, *mise-en-scène* would appear to function for the most part expressively, as in the so-called conventional film melodramas. For example, an unusually complex "layered" composition in *Dynasty* featured Alexis Carrington in the foreground of the frame arranging flowers in her ex-husband's drawing room as Krystle, the current Mrs. Carrington and Alexis's arch-rival, enters to the rear of the frame carrying an identical flower arrangement. The flowers externalize the emotions of the characters without in any way splitting the perception of the viewer. Another episode of *Dynasty* featured a classically Oedipal composition as Fallon, the father-fixated daughter, and her father Blake Carrington kiss over her baby's crib as Fallon's husband enters into the center of the composition. To be sure, character relationships of an "hysterical" nature are expressed, but the *mise-en-scène* represents this hysteria rather than being itself hysterical and thus calling into question that which is represented.

Excess in prime-time serials cannot easily center upon *mise-en-scène*, for television's limited visual scale places its representational emphasis elsewhere. Acting, editing, musical underscoring, and the use of the zoom lens frequently conspire to create scenes of high (melo)drama, even more so when these televisual conventions are overdetermined by heavily psychoanalytical representations. If, as David Thorburn has written, all television acting is operatic, then prime-time soap opera acting must be positively Wagnerian.[19] In fact it is the acting conventions of soap opera which are most often ridiculed for their excess, their seeming to transgress the norms for a "realistic" television acting style. Compared to Peter Brooks's description of melodramatic acting in the nineteenth century French theatre with its eye-rolling and teeth-gnashing, acting on TV serials approaches minimalism; nevertheless it appears excessive in comparison to the more naturalistic mode currently employed in other forms of television and in the cinema, just as the overblown "bad acting" in Sirk's films did for its time.[20] Yet both forms of melodramatic acting are in keeping with related conventions for distilling and intensifying emotion.

On *Dallas* and *Dynasty,* as on daytime soaps, the majority of scenes consist of intense emotional confrontations between individuals closely related either by blood or by marriage. Most scenes are filmed in medium close-up to give full reign to emotionality without obscuring the decor. The hyper-intensity of each confrontation is accentuated by a use of underscoring not found in any other TV genre, and by conventions of exchanged glances, shot duration, and the zoom lens. Although television does not often avail itself of the elaborate moving camera and mirror shots Sirk employed in the fifties (and Fassbinder in the seventies) to Brechtian effect, these televisual codes appear to serve many of the same functions in terms of exceeding the norms of their medium.

Following and exaggerating a convention of daytime soaps, *Dallas* and *Dynasty* typically hold a shot on the screen for at least a "beat" after the dialogue has ended, usually in combination with shot-reverse shot cuts between the actors' locked gazes. This conventional manner of closing a scene (usually accompanied by a dramatic burst of music) leaves a residue of emotional intensity just prior to a scene change or commercial break. It serves as a form of punctuation, signifying momentary closure, but it also carries meaning within the scene, a meaning connected to the intense interpersonal involvements each scene depicts. Another intensifying technique adapted from daytime drama is the use of zooms-in of varying speeds and durations, with the fast zoom-in to freeze frame being the most dramatic, as when it is used on a close-up of JR at the finale of most episodes of *Dallas.* For coding moments of "peak" hysteria, *Dallas* and *Dynasty* will employ repeated zooms-in to close-ups of all actors in a scene. Reserved for moments of climactic intensity, this technique was used to create the end-of-season cliffhanger for the 1981–82 season of *Dynasty*. In this case the climax was both narrative and sexual, with the zooms used on the injured Blake Carrington intercut with scenes of Alexis Carrington making love to Blake's enemy Cecil Colby. *Dallas* employed a similar device in a scene where JR finally accepts his father's death and we zoom repeatedly to a portrait of Jock on the wall at Southfork.

But it would be misleading to discuss clotural conventions as excessive without considering their relationship to the narrative/dramatic structure. For, as we have seen, moments of melodramatic excess relate to the serial structure of these dramas and occur as a form of temporary closure within and between episodes and even entire seasons. It is serial form, even more than visual conventions, which most distinguishes the contemporary television melodrama from its cinematic predecessors. And it is over the issue of serial form that arguments similar to the Brechtian and feminist positions on Sirk have been proposed in recent theories of daytime serials.

A concept of closure is crucial to an argument that the "happy endings" in Sirk's films fail to contain their narrative excess, allowing contradictions in the text to remain exposed. According to several articles on this subject, the contradictions seemingly burst through the weakly knit textual seams, rendering closure ineffective. In this view a successful closure of the narrative would be seen as ideologically complicit with a "smug, petit bourgeois" view of the world. However the Sirk melodramas question that world view by leaving contradictions unresolved.[21] But what becomes of this argument when the representational field of melodrama takes the form of a serial drama that has no real beginning or end but only (as one critic describes it) "an indefinitely expandable middle"[22]? Since serials offer only temporary resolutions, it could be argued that the teleological metaphysics of classical narrative structure have been subverted.[23] The moral universe of the primetime serials is one in which the good can never ultimately receive their just rewards, yet evil can never wholly triumph. Any ultimate resolution—for good or for ill—goes against the only moral imperative of the continuing serial form: the plot must go on. A moment of resolution in a serial drama is experienced in a very different way from the closure of a classical narrative film. Compare, for example, the ending of *Written on the Wind* to the remarriage of JR and Sue Ellen Ewing in the 1982–83 plotline of *Dallas*. When, at the end of the Sirk film, Rock Hudson and Lauren Bacall drive off together, the meaning is ambiguous because too much has been exposed to allow us to believe they will live simply and happily ever after. However, any speculation about the "afterlife" of the characters that a viewer might indulge in is just that, speculation. When on the other hand, Sue Ellen approaches the altar and JR, we feel a sense of impending doom (accentuated by having Cliff Barnes rise up in protest as a cliffhanger) that we *know* will be fulfilled in future plotlines.

Marriage—with its consequent integration into the social order—is never viewed as a symbol of narrative closure as it is in so many comic forms. Indeed to be happily married on a serial is to be on the periphery of the narrative. There are moments of equilibrium and even joy on TV serials, but in general we know that every happy marriage is eventually headed for divorce and that the very existence of the continuing serial rests upon the premise that "all my children" cannot be happy at once.[24] Thus the fate of various couples depends not upon any fixed and eternal character traits, for example, good/evil, happy/sad, but rather upon a curious fulcrum principle in relationship to other couples in the current plotline. Characters who represent the societal "good" of happy monogamy with a desire to procreate are just as miserable as the fornicators. During the 1982–83 season, the two marriages that seemed above the vagaries of intrigue—those of Pam and Bobby Ewing, and Blake and Krystle Carrington—were torn asunder by obviously contrived plot devices. Even the implicit moral goodness of a character such as Pamela was called into question. In the plutocracies of *Dallas* and *Dynasty,* as in the more bourgeois worlds of daytime soaps, happy marriage does not make for interesting plot complications.

From this it might be argued that prime-time family dynasty serials in particular offer a criticism of the institution of bourgeois marriage, since marital happiness is never shown as a final state. Wedded bliss is desirable but also unobtainable. Moreover, that cornerstone of bourgeois morality—marriage for love—also appears to be demystified. Both *Dallas* and *Dynasty* deal with the economics of multinational corporations but they do so in terms of the familial conflicts which control the destinies of these companies. This is typical of the domestic melodrama's oft-noted tendency to portray all ideological conflicts in terms of the family. However, *Dallas* and *Dynasty* also depict the family in economic terms, thus apparently demystifying the middle-class notion of marriage based upon romantic love (e.g., JR's re-marrying of Sue Ellen in order to regain control of his son and heir; the Byzantine interweavings of the Colby and Carrington empires in *Dynasty*). In one episode of *Dynasty,* Blake Carrington buys his wife, Krystle, a new Rolls Royce, telling her that he is giving her the Rolls because she is giving him a child. This would seem to reduce their love to a financial contract, thus exposing its material basis. Yet in a sense these characters are beyond bourgeois morality because they represent the ruling class. One critic has offered the interpretation that the transgressions of the nouveau riche decadents of prime-time ultimately serve to reinforce bourgeois norms:

> *Dallas, Dynasty* and *Falcon Crest* give us the satisfaction of feeling superior to them: We can look down on their skewed values and perverted family lives from the high ground of middle-class respectability. When Angela Channing *(Falcon Crest)* coolly threatens to disinherit her grandson if he won't wed a woman he despises (the marriage would tighten her hold on the valley's wine industry), our own superior respect for love and marriage is confirmed. The prime-time soaps also confirm the suspicion that great wealth and power are predicated on sin, and, even more satisfying, don't buy happiness anyway. [25]

How can the same programs yield up such diametrically opposed readings? According to two recent feminist studies, serial form and multiple plot structure appear to give TV melodrama a greater potential for multiple and aberrant readings than do other forms of popular narrative. [26] Since no action is irreversible, every ideological position may be countered by its opposite. Thus the family dynasty sagas may be read either as critical of the dominant ideology of capitalism or as belonging to it, depending upon the position from which the reader comes at it.

Of course most U.S. television programs are structured to appeal to a broad mass audience and to avoid offending any segment of that audience. The "openness" of TV texts does not in and of itself represent a salutory or

progressive stance. Nevertheless, I would argue that the continuing melodramatic serial seems to offer an especially active role for the spectator, even
in comparison to the previous decade's form-in-dominance, the socially-
conscious situation comedy of the early-mid seventies.[27] The popular press
bemoaned the transition from these "quality" sitcoms to "mindless" comedies and "escapist" serials later in the decade. The popular sitcoms of the
1970s—for example, Norman Lear's *All in the Family* and *Maude,* and MTM
Enterprises' *The Mary Tyler Moore Show* and *Rhoda*—were engaged with their
times, often to the point of encompassing overtly political themes with a
progressive bent. *Dallas* and *Dynasty* seem by contrast to be conservative
Republican programs. The article cited above goes on to argue that prime-
time soaps duplicate the imagery of Reaganism and reinforce its ideology.

> . . . both imply that the American dream of self-made success is alive and
> might be made well by releasing the frontier instincts of the wealthy from
> the twin shackles of taxes and regulation.[28]

Although the sitcoms contained overtly liberal "messages," their strong
drive toward narrative closure tended to mask contradictions and force a
false sense of social integration by the end of each episode. For example,
the problems raised by *All in the Family* had to have easy solutions within
the family so that a new "topical" issue could be introduced in the next
episode. TV critic Michael J. Arlen has described this phenomenon very
well in his essay, "The Media Dramas of Norman Lear."

> Modern, psychiatrically inspired or induced ambivalence may, indeed, be
> the key dramatic principle behind this new genre of popular entertainment.
> A step is taken, and then a step back. A gesture is made and then with-
> drawn—blurred into distracting laughter, or somehow forgotten. This seems
> especially true in the area of topicality . . . .[29]

(It is no accident, I believe, that Norman Lear's subsequent [1976] venture
into social satire took the form of the continuing serial *Mary Hartman,
Mary Hartman;* nor that both *Mary Hartman* and *Soap* [1977] blended situation comedy with elements of melodrama.)

Prime-time melodramas by contrast can never *resolve* contradictions by
containing them within the family, since the family is the very site of economic struggle and moral corruption. In these serials, the corruption of the
very rich much more often stands exposed and remains exposed. If, for
example, Blake Carrington reconciles with his homosexual son, it does not
represent an easy resolution to or liberal blurring of the challenge Stephen's
gayness poses to the disposal of the Carrington fortune. The temporary
reconciliation merely portends yet another breach between father and son
which does in fact ensue when Stephen takes his son and moves in with his
male lover.

To put it schematically, the 1970s sitcoms dealt with liberal "messages"
within a narrative form (the episodic series sitcom) limited by its own con-
servatism. The prime-time serials reverse this, bearing what appears to be a

right-wing ideology by means of a potentially progressive narrative form. This is not to imply that narrative forms *in themselves* structure the ideologies of an era. Quite the contrary. It would seem that the multiplication of social contradictions in the 1980s could not be expressed within the boundaries of the situation comedy. Narrative forms *do* have expressive limitations, and, in the case at hand, one can correlate a shift in the dominant narrative form of American network television with a shift in sensibilities outside the text. This is not to say, as many have argued, that the new serials represent a turning away from social concerns. The emergence of the melodramatic serial in the 1980s represents a *radical* response to and expression of cultural contradictions. Whether that response is interpreted to the Right or to the Left is not a question the texts themselves can answer.

## Notes

1. Peter Brooks, *The Melodramatic Imagination,* New Haven, Yale University Press, 1976, pp. 11–12.

2. American national network television is divided into two major time periods, each of which has its own corporate division and advertising policies, as well as specific program types. Daytime lasts from about 10 AM until 4 PM. Its main fare consists of soap operas, quiz programs, and talk shows. Prime-time, so called because it is the "prime" viewing period with the largest audiences and highest advertising rates, lasts from 8 to 11 PM. Most prime-time programs have traditionally been episodic series, with the major genres being the situation comedy and action-adventure drama. *Dallas* started a trend toward continuing serial dramas in prime-time.

3. U.S. daytime serials are broadcast in the late morning and early afternoon on all three networks, five days/week, fifty-two weeks/year. Each day about twenty-five million viewers, 80 percent of whom are women, watch them (*World Almanac,* 1982). In 1982 there were thirteen daytime soap operas on the air, most of an hour's duration.

4. Beverley Hills, Sage Publications, 1983.

5. Laura Mulvey and Jon Halliday (eds.), *Douglas Sirk,* Edinburgh Film Festival, 1972.

6. The formalist positions taken by David Grosz and Fred Camper are summarized in Jean-Loup Bourget, "Sirk and the Critics," *Bright Lights* 2, Winter 1977–78, pp. 6–11.

7. Paul Willemen, "Distanciation and Douglas Sirk," *Screen,* Summer 1971, vol. 12, no. 2, pp. 63–67. Republished in Mulvey and Halliday, *Douglas Sirk.*

8. Ibid., p. 28.

9. Especially since none of the authors I discuss offer as a "control group" a detailed comparison to other 1950s melodramas similar to those of Sirk, Ray, and Minnelli, for example, *Hilda Crane* (directed by Philip Dunne, 1956) or *Peyton Place* (directed by Mark Robson, 1957).

10. Thomas Elsaeeser, "Tales of Sound and Fury: Observations on the Family Melodrama," *Monogram,* no. 4, 1973, p. 3.

11. Ibid., p. 4.

12. Geoffrey Nowell-Smith, "Minnelli and Melodrama," *Screen,* Summer 1977, vol. 18, no. 2, p. 117.

13. Ibid., p. 118.

14. Laura Mulvey, "Douglas Sirk and Melodrama," *Movie,* no. 25, Winter 1977–78, p. 53.

15. Chuck Kleinhans, "Notes on Melodrama and the Family Under Capitalism," *Film Reader 3,* 1978, pp. 40–48.

16. As noted by Nowell-Smith, "Minelli and Melodrama."

17. I am indebted to the Melodrama Seminar at the 1981 British Film Institute Summer School for this point, and especially to Charlotte Brunsdon for the idea that "melodrama" consists of an "ideological problematic" *and* a "mode of address," so that it may manifest itself in different forms in different historical periods.

18. *Soap Opera Digest* 7, December 7, 1982, p. 141.

19. David Thorburn, "Television Melodrama," in Horace Newcomb (ed.), *Television: The Critical View,* 3d [and 5th] Ed., New York, Oxford University Press, 1982, p. 536.

20. Brooks, *Melodramatic Imagination,* p. 47.

21. See, for example, Christopher Orr, "Closure and Containment: Marylee Hadley in Written on the Wind," *Wide Angle* vol. 4, no. 2, pp. 28–35.

22. Dennis Porter, "Soap Time: Thoughts on a Commodity Art Form," *College English* 38, 1977, p. 783.

23. This issue is addressed in Tania Modleski, "The Search for Tomorrow in Today's Soap Operas," *Film Quarterly,* vol. 33, no. 1 (1979), pp. 17–18.

24. See Tania Modleski, *Loving With a Vengeance: Mass Produced Fantasies for Women,* Hamden, Conn., The Shoe String Press, 1982, p. 90.

25. Michael Pollan, "The Season of the Reagan Rich," *Channels of Communications* 2, November/December 1982, pp. 14–15.

26. Modleski, *Loving With a Vengeance,* and Ellen Seiter, "Eco's TV Guide—the Soaps," *Tabloid* 5, Winter 1982, pp. 35–43.

27. This is not to imply a quantitative conception of dominance. Rather, I'm referring to a hegemonic form, one which appears to be at the center of a decade's ideology.

28. Pollan, "Season of the Reagan Rich," p. 86.

29. Michael J. Arlen, "The Media Dreams of Norman Lear," *The View from Highway 1,* New York, Farrar, Straus & Giroux, 1974, p. 59.

# Conceptualizing Culture as Commodity: The Problem of Television

## EILEEN R. MEEHAN

Critical inquiries into television tend to fall into two broad categories. On the one hand, academics study corporate rivalries, production processes, technical innovation, and governmental pressures to reveal the material constraints that constitute the industrial environment of television. On the other, scholars explicate the interplay of audio, video, and narrative elements to uncover the modern mythos, symbolic representations, and ideologies that constitute American culture. On the one hand, companies struggle for profit, market control, and growth. On the other, members of the culture cast up and celebrate their Weltanschauung. On the one hand, political economy, on the other, cultural studies and—to add another handy cliche—never the twain shall meet.

Yet, the twain does meet. This is due, I believe, to the nature of television itself; television is not reducible solely to manufacture nor to artifact. Rather, television is a complex combination of industry and artistry. This is not an essentialist claim, but instead a recognition that the term television embraces a range of social practices bounded by material constraints. And this range of social practices places television within both the economic base and the ideological superstructure. As part of the base, television is characterized by relations of production that are typical of capitalism. Labor is

From *Critical Studies in Mass Communication*, Meehan, Eileen. "Conceptualizing Culture as Commodity: The Problem of Television." 1986. Pages 448–457. Copyright by the Speech Communication Association. Reprinted with permission of the publisher.

appropriated, surplus value is extracted, commodities are circulated, and profits are expropriated by capitalists. From this perspective, there is little difference between the manufacture of television and the manufacture of shoes. However, when television is treated as part of the superstructure, the differences between these two commodities become obvious. Drawing from the cultural fund and the conventions of realism, television presents selected images, worldviews, symbols, myths, truth claims, values, and visions. This representation of social life, especially with its seeming immediacy and intimacy, has great potential as a disseminator of dominant ideology and as a cultivator of hegemony. To capture this duality in mass culture production, analysts have used such terms as "dream factory" (Powdermaker, 1950), "consciousness industry" (Enzensberger, 1974), "industrialized folklore" (McLuhan, 1969). Drawing on Marxist theory, I use the terms "contradiction" and "contradictory institution" (Bottomore, Harris, Kiernan, and Miliband, 1983) to emphasize the fundamental nature of this rift between our experience of television as industry and as culture, which tends to be reflected in our analyses of television as base and as superstructure.

From this recognition, then, one must rethink television as constituted in contradiction as both culture industry and industrial culture. Television is always and simultaneously an artifact and a commodity that is both created and manufactured; television always and simultaneously presents a vision for interpretation and an ideology for consumption to a viewership that is always and simultaneously a public celebrating meaning and an audience produced for sale in the marketplace. Only by embracing these dualities can we explicate the contradictory fact of television.

As members of a scholarly community, our academic responsibility is clear. We must construct the most adequate understanding of empirical phenomena that our methods, theories, and creativities will allow. To do that, we must engage the phenomenon at its most basic level. For television, that requires an acceptance of the economic as determinate since, in our culture, television is first and foremost a business. Yet, while we recognize that economics set the parameters, we must also recognize that television is a very peculiar sort of industry—a culture industry that reprocesses the symbolic "stuff" from which dreams and ideologies are made. In the remainder of this [essay], I will argue for a reconstruction of television scholarship based on an integration of political economy and cultural studies. To do this, I will first analyze the economic constraints on the television industry with special attention to how these constraints serve as both an impetus for, and a limitation on, innovation in national television. Next, I will contextualize the question of representation within the intersection of culture and ideology as bounded by economics. From this synthesis, I will derive five conceptual categories for the study of television as both a culture industry and an industrial culture. But, I should add that these categories have already been presaged in these introductory remarks. Finally, I will argue that such a synthesis is crucial if we are to fulfill our larger public responsibility—a responsibility that requires a full, holistic account of television to guide progressive intervention and practice. So, let us start with the first analytic moment, that is, with television as a culture industry.

## *Television as Culture Industry*

In advanced capitalist countries, the creation of cultural artifacts is primarily an economic activity subject to the bounds of profitability, cost efficiency, oligopoly, and interpenetrating industries. Processes of production and distribution tend to be centralized, rationalized, and routinized, clearly placing such activities within the term "culture industry." As the major form of industrially produced and mass distributed culture, television attracts critical scrutiny as *the* exemplar of modern capitalism's industrialization of culture. However, not all televisual products are merely carbon copies; variation and innovation in form does occur (Ettema and Whitney, 1982). Similarly, not every televisual text simply and solely celebrates the dominant ideology that legitimates capitalism. The guidelines for alternative or even oppositional readings of the text may be found embedded in the televisual text itself (Fiske, 1985; Hall, 1982; Morley, 1980; Smith, 1985; Williams, 1977). The problem here is to account for creativity, innovation, and variation through an analysis of economic structures, structures which include such features as corporate rivalry within closed markets, rationalization of production, and so on. My analysis of industrial structure, then, will suggest that the very structure mitigates for bursts of innovation and creativity just as surely as it mandates duplication and imitation. To illuminate this dynamic, let us consider some of the relationships that directly surround the distribution and production of first-run, prime time commercial television series.

At the macroscopic level, television is a multilayered industry embedded in the information/entertainment sector of the economy. As such, television involves manufacturers of electronic equipment, companies that control distribution technologies (broadcast, cable, microwave, satellite, etc.), the national advertising industry comprised by agencies and manufacturers, the ratings industry, the national television industry proper, and a multiplicity of trade associations serving various configurations of these constituencies. While all of these elements have had varying degrees of influence over programming at different times, the most significant for current series production seems first to be the structure of the national industry and then its links to the advertising and ratings industries (Barnouw, 1966, 1968, 1970, 1978; Bergreen, 1980; Cantor, 1971; Ewen, 1976; Meehan, 1983; Mosco, 1979). For our purposes, the national television industry includes networks, stations, and producers, although here we will focus primarily on networks and producers.

The national industry has long been marked by centralization and, since 1972, has been constituted by a series of closed markets dominated by a handful of firms. In distribution, the three networks control the market, each vying with the other two for dominance in the ratings. Ratings measure the networks' productivity, that is, the networks' ability to attract the right sort of audience in cost efficient groups for sale to advertisers. As such, ratings are the tangible "proof" that the networks' intangible commodity— the audience—exists. Given the inherent conflict of interests between buyer/ seller over audience prices, the measurement of productivity is the purview of a separate, yet intertwined, ratings industry. Significantly, the production of national television ratings is generally dominated by a single firm. In economic terms, this monopoly is eminently rational since advertisers and networks require the same sorts of information in order to transact business. Thus, the commodity audience comes to be defined by the dominant rating firm's methodology. And that methodology is itself a function of economic pressures including cost efficiency and profitability of ratings production, corporate tactics used to gain and maintain monopoly status, manipulation of discontinuities in demand for ratings, and so on. This renders the sample audience a true commodity produced by the major ratings firm for sale to advertisers and networks, a commodity used by these purchasers to gauge the success of networks' schedules, to set prices for audiences, and to measure the success of individual programs (Cantor, 1980; Livant, 1979; Meehan, 1984; Murdock, 1978; Smythe, 1977). At this macroscopic level, then, is a closed market within which networks, national advertisers, and the dominant ratings firms pursue their own, particular interests. The intersection of these three industries sets up the networks' central problematics: how to exercise influence within this closed market and how to acquire programming that will attract the commodity audience. It is the latter problem that we will analyze here.

Since 1972, the production of television programming has been the purview of three types of companies (i.e., independent production companies, production companies that work with motion picture studios, and the studios themselves [Cantor, 1980; Guback and Dombkowski, 1976]). Prior to 1972, program production had been vertically integrated into the networks, which then discriminated against independent production in favor of programs owned by the networks themselves. However, State intervention at the level of anti-trust action forced divestiture of these internal production units, ostensibly leaving the marketplace open to free competition. Indeed, in this seeming intersection between the studios producing films, the independent filmmakers, and the independent companies producing television series, one might well expect the rough and tumble competition so dear to ideologists of the free market.

Precisely the contrary happened. Just as distribution has been oligopolized by the three networks, so too has production been oligopolized by a small number of program producers selected by the networks from the larger pool of available producers. As Cantor (1980) points out, the numbers of producers in both groups has remained quite stable over the years (four or five companies selected from twenty), despite the exit and entry of particular firms. Thus producers vie for inclusion in that inner circle. Inclusion, exclusion, and expulsion are matters of network fiat and these decisions are based on the probability/ability of a firm's product to deliver ratings (i.e., to produce the appropriate audience in the right numbers for the general advertisers that patronize network television). Here the producing firm is caught in a web of relationships where network decisions based on advertiser demand and the current market definition of the audience constrain the manufacture of programs.

The effect of these relationships for programming is significant indeed. In periods when the ratings sample is relatively stable, networks act rationally in the marketplace by selecting programs that imitate formats proven successful in attracting the commodity audience. At times of partial or total turnover in the ratings sample, networks act rationally when they either select programs that are innovative in ways which "ought" to appeal to the new commodity audience, or imitate innovations that have proven their appeal, or select programs that combine a "tried and true" format with some unusual twist. In fact, all of these tactics were used by the networks in their attempts to cope with the first, major overhaul of the A. C. Nielsen Company's sample completed in 1970—a year dubbed by industry pundits as the Year of Relevance—with varying degrees of success (Barnouw, 1975; Brown, 1971; Meehan, 1983).

Complicating this, however, is the networks' rivalry. Within the oligopoly, each network attempts not only to sell audiences, but also to differentiate itself from its rivals. Networks advertise themselves to both viewers and advertisers by promoting an image of the network to cultivate brand loyalty. Occasionally this means acting in a manner that in the short term appears economically irrational—for example, scheduling an innovative series or an expensive special or a serious documentary that includes controversial material. Both innovation and duplication can serve as tactics by which the networks negotiate their economic environment. In the structure of transindustrial relationships, then, lies the first structural impetus for creativity and constraint on that creativity in the manufacture of television programming.

The second set of structural factors becomes apparent by shifting the level of analysis to the relationships between networks and production companies. Like the networks, production companies operate in a series of economic relationships which include self-promotion as well as rivalry. But while networks' transindustrial relationships are between co-equals, production companies are directly subordinated in their industrial relationships with networks. Despite divestiture, networks exercise considerable control, albeit indirect, over program production. Most typically a network selects an idea and then selects a production company to develop that concept. Companies within the inner circle of producers are clearly favored by this practice, which also tends to stereotype certain companies as producers of particular types of shows. This encourages a division of labor within the oligopoly of producers with some companies manufacturing liberal sit-coms, others concentrating on drive-and-shoot shows, and so on. Thus, processes of product development are rationalized by the oligopoly and this rationalization encourages rote manufacture, imitation, and variation rather than bold creativity.

Significantly, producers may foster this practice themselves to both rationalize their own production processes and to carve out and control a niche within the oligopoly. Obviously this can be a mixed blessing as network demand for particular kinds of program product does fluctuate and companies too narrowly identified with a type of product may fall out of the oligopoly when network demand for that product fades. On the other hand, the division of labor does serve the producers' by differentiating their product from those of their rivals and by narrowing the already narrow competition within the inner circle by basing rivalry on generic specialization. Interestingly, this division of labor provides a structural impetus for auteurism in television production, thus encouraging a particular kind of creativity, which has been well documented in auteurist film studies and which has increasingly attracted the attention of television critics.

But as pointed out above, the business environment surrounding television production is by no means stable. Not only does the sample audience change, but intangibles like audience taste and willingness to endure overly familiar programs also change. Joining these factors is the networks' own need for self-promotion as well as the occasional need for more innovative programming. Finally, the struggle to achieve or maintain membership in the oligopoly should foster some innovative production by firms. These drives towards change are balanced, however, by drives that may be as strong or even stronger. For just as the network seeks out some innovation, so too it attempts to streamline the process of idea development by contracting with producers that specialize in a generic form. Similarly, just as producing firms attempt to capture trends, they also try to capture profitable genres. Thus innovation and imitation, variation and duplication are founded on an economic structure that facilitates alternating cycles of rote manufacture and bounded creativity.

### Innovation: Economic Context for Culture

But how does such bounded creativity occur? Within the constraints of oligopoly and intertwined industries, how do new visions, meanings, images, subjects, on so on get incorporated into televisual products? While structural analysis identifies the economic dynamics that support innovation, such analysis does not explain the possibility of innovation. For the orthodox political economist, this is where culture rears its ugly head. To account for innovation, we need a theory of culture. But, beyond this, we need a theory of culture to account both for congruence and slippage between the dominant ideology and televisual representations.

Now, obviously, limitations of space preclude the exposition of a full-blown theory of culture, particularly of one that is contextualized within the economic. Instead, I will reference the work of such cultural scholars as Williams (1980), Hall (1980), and the Birmingham School, materialist theorists such as Marx (1972, 1981), Gramsci (1973), and Althusser (1970), anthropologists such as Geertz (1973), Wallace (1970), Harris (1968), and Burridge (1969). From this general point of reference, I will then sketch the main features of a culture theory that are relevant to this project.

In this conception, culture is both relations of diversity and shared webs of meaning. Culture exists as a fund of meanings, images, understandings, and so on which human collectivities construct within the constraints of social structure, economic structure, concrete experience, socialization, overdetermination, and random error. These are the material bases upon which culture is reared. In most cultures, the first constraint is set by division of labor based on gender. In capitalist cultures, the second constraint is set by the hierarchical structure of class. In this capitalist culture, further constraints are introduced by an array of socioeconomic categories that discriminate among people according to region, dialect or linguistic group, race ethnicity, sexual preference, age grade, and so on. In the dynamic of division, we see structures that support both sharedness and diversity as each individual is socially constructed and simultaneously self-constructed across shared categories in ways that are both predictable and surprising, patterned and unique.

Now, given the hierarchical and exploitative economic structure of capitalism, relations among identified collectivities revolve around issues of legitimation, domination, and control. In the cultural arena, this means a struggle to legitimate collective systems of interpretation and understanding, that is, to legitimate ideologies. But just as relations between ideologies are in process, so too are the ideologies themselves as collectivities reinterpret, revise, re-present, and re-create the expression of their experiences. Hence, the ability of the capitalist class to legitimate its ideology as the rational, commonsensical view of the world is not so much a *fait accompli* as a continuing attempt to reinterpret the world and to gain privilege for that interpretation against the continuously changing array of alternative, oppositional, and emerging ideologies. And this constant shifting on the cultural scene, especially in tandem with economic supports for innovation, surfaces in television. Thus television becomes both a culture industry and an industrial culture.

## *Conceptualizing Television as Dualities in Contradiction*

This brings us full circle to the problem of conceptualizing culture as commodity, to the central problem presented by television. Precisely because of the dynamics that structure both the national television industry and the cultural fund, televisual programs do present meanings, characterizations, images, and so on that neither simply nor directly reflect the dominant ideology. Indeed, embedded within the televisual text may lie the guidelines for alternative or even oppositional readings of that program. For, while the selection of elements from the cultural fund is limited by the requirements of everyday business, the cultural fund remains broader than the dominant ideology. Also, economic drives for innovation, variation, and auteurism encourage producers to sift through the constantly shifting cultural fund for trends, gimmicks, and novelties. In some cases, producing companies simply abstract images, themes, and so on for reworking within terms of the dominant ideology. This rather instrumental use of culture is matched by a "humanistic" use where firms draw on the cultural and subcultural memberships of their employees to generate more innovative, "authentic," and ambiguous representations—in the hope that these innovations can be routinized in terms of production, that they can attract the commodity audience, that they can differentiate the firm and limit competition. As an industrial project, the production of culture is limited by the profitability and cost efficiency of bold innovation, of rote manufacture.

The important element here is the lived duality of this process. For just as producers are always and simultaneously making a commodity containing the most accessible content—that is, the dominant ideology—so too are producers constructing a cultural artifact whose ability to resonate with different collectivities depends on a combination of elements from the cultural fund. And these collectivities, these publics "read" the televisual text in a multiplicity of ways, sometimes consuming the text within the dominant ideology, sometimes re-coding the text entirely, sometimes mixing elements of dominant, alternative, oppositional, or emerging ideologies. But even as these publics grapple with the televisual text, some of us are abstracted for processing and sale as the commodity audience. Even as we find elements of oppositional ideologies within the artifact, producers may be planning to capitalize on those elements as a key to a new and profitable generic form of production. By counter-posing the dynamics between industrial culture and culture industry, we begin to unravel the complex layering of contradictory dualities that constitute television.

The problem of television, then, is how to capture these dualities and lay bare their interconnections. Solving this analytic difficulty requires a synthesis of both cultural and political economic studies of television precisely because television is both an industrial process and a cultural process. As a small step towards that necessary synthesis, I offer the five analytic categories that have guided this analysis for your consideration: manufacture/creation, commodity/artifact, ideology/culture, consumption/interpretation, audience/publics. In each dualism, I have priviledged the economic term in deference to the fact that television is primarily a business enterprise in this country. Yet, for a holistic understanding of television, a political economist can not afford to lose sight of the cultural dynamics of television—just as the culturalist cannot afford to overlook the economic base upon which televisual representations are constructed. Indeed, the time seems particularly apt for such a synthesis given the controversies raised by the Althusserian circle in both political economy and cultural studies. In the former, Althusserian structuralism has been reinterpreted to emphasize notions of indeterminancy and progressive intervention in an attempt to balance the tendency to depict capitalism in closed, functionalist terms (Mosco, 1982;

Schiller, 1977). Ironically, in cultural studies, Althusserian analyses of ideological apparati and the overdetermined subject have sparked heated denunciations over functionalist readings of the relationship between impersonal social structure and individual consciousness (Thompson, 1980).

This "ferment in the field" bodes well for a synthesis of political economy and cultural studies that would more fully integrate notions of human creativity, class struggle, ideology, and impersonal social structures. Such a synthesis might best be facilitated by organizing the process of synthesis around a concrete instance. I suggest the contradictory institution of television, particularly because an adequate theorization and study of television requires the integration of these critical approaches; this much we owe to our community of scholars and to the academy.

But there is more at stake here. Television is not simply an academic subject. The relative precision or imprecision of academic knowledge about television can have an impact on the publics, on public debate over television programming, and on policy. Until an adequate integration of economic and cultural critique are widely available, until each of the dualities and their interconnections are traced, the policy remains uninformed and—perhaps worse—stereotyped as the ignorant, slack-jawed mass of consumers who are themselves solely responsible for the proliferation of stultifying televisual products. While political economy correctly locates the boundaries of action within the economic structure of the industry, cultural studies reminds us that we—not some amorphous mob but we the folks—are the mass of consumers and that we live, create, and interpret culture. While political economy reminds us of the limitations of individuals and collectivities in the face of large scale, impersonal structures, cultural studies reminds us that we have created these structures through time and we can surely tear them down. Now is the time for academicians on both sides of critical communications to join together, to accept the challenge of our academic and our public responsibilities.

## References

Althusser, L. (1970). *For Marx* (B. Brewster, trans.). New York: Random House.
Barnouw, E. (1966). *A tower in Babel.* New York: Oxford University Press.
——— (1968). *The golden web.* New York: Oxford University Press.
——— (1970). *The image empire.* New York: Oxford University Press.
——— (1975). *Tube of plenty.* New York: Oxford University Press.
——— (1978). *The sponsor.* New York: Oxford University Press.
Bergreen, L. (1980). *Look now, pay later.* Garden City, N.Y.: Doubleday.
Bottomore, T., Harris, L., Kiernan, V. G., and Miliband, R. (eds.). (1983). *A dictionary of Marxist thought.* Cambridge, Mass.: Harvard University Press.
Brown, L. (1971). *Televi$ion.* New York: Harcourt Brace Jovanovich.
Burridge, K. (1969). *New heaven, new earth.* New York: Schocken Books.
Cantor, M. G. (1971). *The Hollywood TV producer.* New York: Basic Books.
Cantor, M. G. (1980). *Prime-time television.* Beverly Hills: Sage.
Enzensberger, H. M. (1974). *The consciousness industry.* New York: Seabury Press.

Ettema, J., and Whitney, D. C. (eds.). (1982). *Individuals in mass media organizations: Creativity and constraint.* Beverly Hills: Sage.

Ewen, S. (1976). *Captains of consciousness.* New York: McGraw-Hill.

Fiske, J. (1985, April). *Television and popular culture: Key trends in British and Australian experience.* Paper presented at the Iowa Symposium and Conference on Television Criticism, Iowa City, Iowa.

Geertz, C. (1973). *The interpretation of cultures.* New York: Basic Books.

Gramsci, A. (1973). *Selections from the prison notebooks* (Q. Hoare and G. N. Smith, trans.). New York: International Publishers.

Guback, T., and Dombkowski, D. (1976). Television and Hollywood: Economic relations in the 1970's. *Journal of Broadcasting,* 20: 511–27.

Hall, S. (1980). Cultural studies: Two paradigms. *Media, Culture and Society,* 2: 57–72.

Hall, S. (1982). The rediscovery of 'ideology': Return of the repressed in media studies. In M. Gurevitch, T. Bennett, J. Curran, and J. Woollacott (eds.), *Culture, society and the media* (pp. 56–90). New York: Methuen.

Harris, M. (1968). *The rise of anthropological theory.* New York: Crowell.

Livant, B. (1979). The audience commodity: On the blindspot debate. *Canadian Journal of Political & Social Theory,* 3: 91–106.

Marx, K. (1981). *A contribution to the critique of political economy.* New York: International Publishers.

Marx, K., and Engels, F. (1972). *The German ideology.* New York: International Publishers.

McLuhan, H. M. (1969). *The mechanical bride.* New York: Vanguard Press.

Meehan, E. R. (1983). *Neither heroes nor villains: Towards a political economy of the ratings industry.* Unpublished doctoral dissertation, University of Illinois, Urbana.

——— (1984). Ratings and the institutional approach: A third answer to the commodity question. *Critical Studies in Mass Communication,* 1: pp. 216–25.

Morley, D. (1980). *The "Nationwide" audience: Structure and decoding.* London: British Film Institute.

Mosco, V. (1979). *Broadcasting in the United States.* Norwood, N.J.: Ablex.

——— (1982). *Pushbutton fantasies.* Norwood, N.J.: Ablex.

Murdock, G. (1978). Blindspots about western Marxism: A reply to Dallas Smythe. *Canadian Journal of Political & Social Theory,* 2: 109–19.

Powdermaker, H. (1950). *Hollywood: The dream factory.* Boston: Little, Brown.

Schiller, H. (1977). *Mass communications and American empire.* New York: Augustus Kelly.

Smith, J. E. (1985, April). *Subtexts of resistance in ideological commodities: The case of Magnum, P.I.* Paper presented at the Iowa Symposium and Conference on Television Criticism, Iowa City, Iowa.

Smythe, D. (1977). Communications: Blindspot of western Marxism. *Canadian Journal of Political & Social Theory,* 1: 1–27.

Thompson, E. P. (1980). *The poverty of theory.* New York: Monthly Review Press.

Wallace, A. F. C. (1970). *Culture and personality.* New York: Random House.

Williams, R. (1977). *Marxism and literature.* New York: Oxford University Press.

——— (1980). *Problems in materialism and culture.* London: Verso.

# Television Viewing
# as a Cultural Practice

## MICHAEL K. SAENZ

This essay suggests that television viewing provides a prominent occasion
for viewers' construction of culture. That occasion—watching television do-
mestically, often with a family, during an evening prime time, for entertainment, primarily as an engagement of fiction, and as the offering of a national commercial network—is a highly particular kind of institutional
arrangement, one that has become, by social convention, strategically important in audiences' construction and accommodation of their culture in
general. Watching television, indeed, institutes a persistent social practice
through which audiences carry out considerable rhetorical, political, poetic,
cultural work. That work provides them with a continually problematized
store of "implicit social knowledge" (Taussig 1987, 303). People inscribe
portions of that knowledge into their lives partially and selectively, by their
subsequent actions. When their actions are played out as discursive strategies, television "induces the effects of power" (Foucault 1980b). When they
are played out tactically to nondiscursive (and potentially counterhege-
monic) ends, viewers end up "poaching" on television, which has thereby
served as a "proper place" for American culture (de Certeau 1984). In all
this, television does not act as a strict cause of social life, or reflection of it,
but as material used in making meaning and action, as a component of
"doxa" used in the production of social practices (Bourdieu 1986, 164). A
practice which produces other practices, television viewing suggests that doxic

From *Journal of Communication Inquiry,* 16:2 (Summer 1992), pp. 37–51. Reprinted with permission.

materials, far from being inherently static, can circulate, and that their pace and circulation is as fundamental a social quality as the pace and circulation of social exchange.

## Television and Social Relationality

Observers of American television seem taken by a persistent intuition that the medium is central to society and culture in the United States. The popular press commonly uses television programming to characterize American culture and its history, assuming that what appears on television gives unique (or at least handy) access to the "structure of feeling" (Williams 1977, 132) of a generation, a decade, or a cultural and political moment like the summer of Tienanmen Square or the dissolution of the USSR. Viewers too use television to construct peculiarly collective memories and associations. A prime time docudrama about AIDS or a show including a character with Down's Syndrome, is remarkable not only for the moral pathos of its topic, but also for the topic's depiction via an institution which constructs such a direct and widely shared sense of relation with current national life. An article about the same topics in the *New York Review of Books,* or *Newsweek,* or *The Austin American-Statesman,* does not reorganize the sociology of knowledge surrounding the issue in quite the same way—or, just as important, reorganize most readers' perception of the sociology of knowledge surrounding the issue. While the novel, journal, or film might use exclusive language, and prompt evaluation or appreciation in an idiom of private or intimate thoughts, television exists as an inclusive text whose appreciation also takes place in an idiom of cultural currency and controversy. Thus someone who discusses the latest poem she read is commonly regarded as disclosing something about herself—her tastes, her education, her priorities for personal time and effort, her literary experiences and interpretations. Someone discussing a television program, by contrast, usually engages a more casual discussion, centered less on the self, and far more likely to be about how "things" are (in politics, fashion, work, sex, entertainment, education, etc.). Evaluations of television are less likely to be conducted as demonstrations of the speaker's personal sensitivity or understanding, than topical commentary on perceived social facts—both the social facts existing as TV programming, and other social facts denominated by TV's content.[1] Colloquially, television is used and appreciated as a highly accessible metonym for American society.

The social currency viewers accord television is not attributable simply to the medium's representations of society, nor exclusively to its ability to propagandize, but also, fundamentally, to the opportunity it provides for exercising viewers' adeptness at cultural production. Television viewing, this argument suggests, prompts the invention of culture in those watching it (cf. Wagner 1981). Viewing serves as a generative symbolic practice, a conventional, socially marked occasion devoted to the production, by viewers, of peculiarly "cultural" apprehensions.

This generation of meanings is more than the formal production of a semiotic field; it incorporates trenchant historical relationships of power, wealth, and inequality. Anthropologist Michael Taussig develops the notion of "implicit social knowledge"—"an essentially inarticulable and imagistic, nondiscursive knowing of social relationality" (1987, 303)—on which people draw in articulating meanings. In Taussig's study of Columbian shamans, for example, he records that highlanders consider lowland shamans more powerful, lowlanders consider highland shamans more powerful, and Indians anywhere are considered more magical than non-Indians. Implicit social knowledge of the marginality of faraway people and of Indians, then, serves as the basis for imputing extraordinary and somewhat menacing power. Television similarly draws on Americans' implicit map of historical social relationships. Thomas Schatz remarks that prime time soap operas such as *Dynasty* and *Dallas,* for example, are invariably set in sunbelt cities where, presumably, their explorations of power within the business world and of family lineages in the domestic realm can take place in a setting of economic possibility and social contingency freed from Northeastern industrial decline and social hierarchy.[2] Just as the social difference created by geography and race serve to justify perceptions of shamanistic magic and mystery among Columbians, so social differences created by region and economic growth serve to premise certain stories of high stakes personal machination for American audiences. This is not to say that viewers believe that *Dallas* is Dallas, rather, that one's implicit social knowledge of Dallas provides a meaningful way of establishing a certain realm of fiction. Television viewing draws on and articulates viewers' store of implicit social knowledge.

Indeed, television generally provides viewers with an elaborately extended sense of "otherness," of social groups beyond the viewer's immediate ken, of stories outside the viewer's immediate world. Among the most powerful shapers of cultural order are the extent and strength of boundaries presumed to delimit one's cultural world (Douglas 1966; 1982); television invites a continual refiguration of such boundaries, a continual renewal of one's pragmatic, colloquial cosmology.

Saying this, it is possible to draw a parallel between television viewing and social processes such as nationality, ethnicity, and polity. Michael Fischer (1986), for instance, argues that, in the context of highly mobile, extended societies such as ours, ethnicity is as much a kind of remembrance as a form of regional affiliation or folkloric tradition. The ethnic seeks to reconstruct her identity, and establish a kind of ethical program for herself, by tying herself retrospectively to the world of her parents and the people surrounding her in youth. Similarly, the television viewer ties himself to memories—as well as to contemporary retrospectives—of previous generations and decades, developing constantly changing relationships with national, ethnic, or class culture (Lipsitz 1990; cf. Halbwachs 1980; Durkheim 1965; Gramsci 1971; Anderson 1985).

Suggestions of otherness are inherent, not only in television content, but in the practice of viewing television itself, especially on weeknights during prime time, when the nation of viewers tunes in to network programming. The medium's consequent cultural prominence is not simply a technological and demographic phenomenon. It resides too in the conventional expectation—among viewers as much as producers and advertisers—that prime time addresses a full cross section of American society. It is from this ability to implicate a nation of co-viewers that the medium's narratives gain much of their authority and become dominant sites for the display of symbols in American culture.

For television viewing, especially in prime time, entails attendance at a widely shared, collectively appreciated performance, an immediate delivery of textual material to a large and general audience.[3] The quality of performance emerges not only from simultaneous widespread attendance, but also from its conventions of presentation and use. In an analysis of network morning news shows, for example, Jane Feuer (1983) points to a significant kind of social experience intimately linked to broadcast television: the "liveness" of far-flung events brought together by television, a use of technology hardly inherent in the medium, but rather, one manifesting ideological presumptions about the medium's cultural institutional roles. Liveness, I suggest, is by no means limited to news, but is rather a feature of prime time fiction too, which, even when prerecorded, assumes a sense of collective immediacy and participation missing from more private, isolated, and individual enterprises like watching videotapes, attending a movie, or reading a novel. The sense of performance and cultural currency in television programming constitutes an important dimension of viewers' appreciation and evaluation of television drama as a prominent cultural event.

### *Viewers' Poetics and Television's Rhetorics*

Television, I have suggested, elaborates a fund of implicit knowledge which viewers turn to personal interpretive ends. It is useful to consider this elaboration and use of cultural knowledge as a kind of poetics. "Poetics" seems especially apt, because it connotes a reader actively involved in a sensible association of meanings; in an aesthetic appreciation of symbols entailing sentiment as well as cognition; and in a textual involvement that may range from ludic, to meditative and serious. It sidesteps the usual epistemological and ontological misgivings about television's truth (cf. Ewen 1976; Postman 1985) by de-emphasizing television programming's mimetic quality in favor of television viewing's generative properties. Too often "culture" implies a totality in which all imaginable meanings exist a priori, arranged into ideological categories or traditions. A focus on poetics ties the production of social relationality to viewers who are active cultural agents; it connotes a more contingent, ironic activity, informed not just by cognition and ideology, but by affect and aesthetics (cf. Bordwell 1989; Allen 1985, ch. 4; Fiske 1987).

Suggesting that television forms the basis for poetic activity among viewers does not imply that the medium presents its images freely, without interest or partiality. Widespread social conventions for evaluating television, in fact, include dispraise of a medium that is blatantly commercial and interested. Its "credibility"—its mimetic authority—is relatively low; it is "entertainment." Such characterization implies a critical, or at least parsimonious, use of television by its viewers, one which holds the manifest ideologies of television at bay, but which nevertheless engages television's symbolic production for ulterior (noncommercial, conceivably unhegemonic) reasons. Viewers, I suggest, develop an aesthetic enjoyment of television's rhetoric, without necessarily subscribing to the ideology supporting such rhetoric.

By rhetoric I do not only mean overt attempts at persuasion; I refer also to the general problem of a text's address of an audience (cf. Booth 1983). Regarding texts as rhetoric identifies them as social performances directed to audiences to some purpose, as social enactments serving certain pragmatic ends. Rhetoric, of course, is evident in advertisements, previews, cross-promotions, and station identifications. Some promotions—for network news, hospitals, churches, political candidates, non-profit councils, or corporations seeking to establish a distinctive image—operate through a rhetoric of direct address that tries to assert the authority of the institution that is speaking. Other commercials work less on an expository model than by dramatic transformation of their subjects in sound and image. Both often refer backward and forward in time from the viewing moment—to next week's game, to the last decade's great movie. Thus, in a typical night's viewing, the mode and temporality of narrational belief are interspliced with those of an elaborate (highly aestheticized) poetry of consumerism, and of rhetorics of authority. This does not suggest that television confuses viewers about their "reality"; the wonder, note Fiske and Hartley, is that "the audience apprehends the various levels and orders of discourse simultaneously and without confusion" (1978, 89), that each comprises a distinctive register of meaning, perception, and evaluation (cf. Bakhtin 1981; Newcomb 1984). The transition from one register to another is made automatically with practice, but it is felt, and invites continual reconsideration of the relations between narratives, rhetoric, and authority.

Television's informal "flow" (Williams 1974, 86–96) impels this reconsideration. Viewers' interpretations are constantly redirected by television. The allegorical and moral dimensions of a commercial may still be lingering when a newsbreak appears; the import of a newsbreak may be half-perceived before a program's narrative reappears in media res. Final evaluation of any given segment is delayed, attenuated, cut off, or redirected. Television, then, not only presents viewers a number of distinctive registers of meaning, but complicates the viewer's efforts to effect closure in a single register.

Such complication is well exemplified by the recent use of topical newsbreaks relating to fictional programming. During a well-publicized miniseries about a lower middle class homeless family, for instance, one Austin station devoted its newsbreak to a controversy about the local Salvation Army's efforts for the homeless in Austin, with more promised at ten. In the register of political economy, it was shameless opportunism. In the journalistic register, it was an opportune occasion to inform. In the register of narrative experience, it was touching: the docudrama's pathos had been effective, and the newsbreak underscored its truth.

Television's interpretive complication also extends to its mutual insinuation with the household. Altman (1987) discusses how television sound is enjoyed simultaneously with other activities, cueing visual attention to the set at specific moments. Such interaction between programming and household produces a hybrid, processual sense not only of the television story, but also of one's domestic pursuits. Occasional spontaneous conformity between events on television and around the household prompt any viewer, sooner or later, to ruminate over the social limits of coincidence. Television programming serves as an ever-unfolding event inviting comparison with one's own ongoing household, and establishing a readily accessible interpretive dialectic between the two.

Television, then, organizes cultural production in viewers as a dialogic engagement of distinctive registers of experience. It imbues the aesthetics of textual and narrational appreciation with an inherently social scope: aesthetic sensibility applied to television becomes an address of several registers of social meaning, and thus a critical kind of self-construction and orientation within simultaneous discourses.[4] Television remains a central institution in cultural formation because it offers socially prominent, narrative, and rhetorical touchstones which (much like religion) coordinate the specific historicity of its viewers, without determining their entire way of life. It is an ideological, hegemonic, narrational intervention—but a partial and ambiguous, hardly total one.

Different viewers, of course, are equipped quite differently for handling television's rhetorics. Condit (1989) properly emphasizes the "work" required to redirect and answer a rhetorical proposition, especially among viewers who have not been armed with an alternative vocabulary to combat the rhetoric. Acknowledging the work of watching television adjusts the impression given by "poetics" or "social relationality" that constructing meanings is always spontaneous, proliferative, and unproblematic. Though Condit does not press this reading, it connotes the Freudian notion of dream work, emphasizing the importance, not of the specific symbols involved in a dream (or moment of television), but of the dreamer's (the viewer's) particular preoccupations, of her own puzzles over certain condensed or displaced meanings which need sorting out. Watching television remains a practice which enacts a self-conscious working out of hegemonic and historical positions, within the gestures of narration and aesthetic rhetorical appreciation.

### Television's Circulation of Generalized Culture

Watching television, viewers gain not just particular textual pleasures, but an ongoing involvement with a more generalized experience: the feeling of being present at a busy, "live" cultural site. Along with other contemporary mass media, in fact, television serves to imbue Americans with a uniquely modern, generalized sense of cultural reality and experience.

Here I draw on Michel Foucault, who describes changes in other cultural institutions in the early modern period as just such a process of generalization (1979; 1980a). One example is the modern penal code, which replaced particular, retributive acts by monarchs with an extended system of moral impediment "general throughout the entire social body . . . [and] capable of coding all its behaviour " through carefully graded sentences (Foucault 1979, 94). The efficacy of the system rested, not on the divine imperative of the king, but on the "perfect certainty" that "to the idea of each crime is associated the idea of a particular punishment." Signs of punishment serving as obstacles to crime were aimed less at the criminal than at those who had not yet committed a crime. Crimes and their consequences became clearly articulated, widely circulated, and applied without exception to all. Foucault calls this new notion "generalized punishment," referring to its dispersion throughout the body politic, not only through a new judicial system, but in the consciences and calculations of individuals considering their own repertoires of action.

In the "generalized culture" promoted by television, I suggest, narrative events and meanings are likewise changed from particulars into signs—signs for moral speculation, signs of motivation, signs of how to represent. These signs are thought of as "capable of coding all behavior," hence the medium's apparent scope and inclusiveness. They are not as rationally articulated as penal codes, but they are, within the idiom of the several genres available, closely differentiated, offering a sense of position and modulation. Like signs of punishment aimed at the typical citizen, television's rhetorical signs are manifestly not just "about" their topics, but about what appeals to television's audience; they are strategies of address. In a "perfect certainty" of its own, everyone presumes that the medium is watched by all the nation, effecting an indisputably thorough display of its signs. Like the generalized sense of moral repercussion that replaced the monarch's ritual exercise of power in the realm of punishment, the generalized sense of culture proliferates a different kind of subject from the one constructed by appreciating art in museums, by attending church, or by reading philosophy. Television contributes (albeit in highly contingent ways) to the construction of certain kinds of subjectivity, to the (hardly unitary) construction of a self from the raw material of narratives, consumer choices, moral predispositions, and selected rituals of conduct. "Culture," under this view, is an artifice characteristic of a particular historical moment when previously separate symbolic practices become subsumed under a new category of generalized knowledge, aesthetic appreciation, and meaning. Television's incitement to produce "culture" forms part of a larger, historically specific mode of social reproduction.[5]

## *The Tactics of Watching TV*

Broadcast television narration, John Ellis notes, is distinct from cinema in being composed, not of scenes building causally, rhythmically and climactically into a free-standing thematic whole, but rather of segments circumspectly related, climaxing in internal "action-clinches," and never resolving the organizing problematic of the series, which, if all is as it should be, will continue next week. Substituting for the causality which joins scenes in cinematic narration is, on one hand, an effort to bridge segments with suspense.

> Our heroes perpetually encounter fresh incidents, and equally often find themselves suspended in a ambiguous position at the end of a segment (cue for commercial break). (Ellis 1982, 151)

Or, rather than fomenting suspense, a segment may produce what Peter Brooks calls (in a somewhat different context) melodrama's "mode of excess" (1984), its events "generating tidal waves of verbiage, of gossip, discussion, speculation, recrimination. Guilt, jealousy, worry and an immense curiosity about people is generated by this form" (151). Often "speculation abounds; the event is perfunctory; the mulling over of the repercussions is extended" (Ellis 1982, 151). The result is a "dispersed" and "extensive" narrative form whose "characteristic mode is not one of final closure or totalising vision" but rather "a continuous refiguration of events." Indeed, the ruminative narrative of the series remains, in marked distinction to film, "a constant basis for [more] events, rather than an economy of reuse directed towards a final totalisation" (145). "The TV series proposes a problematic [a premise] that is not resolved; narrative resolution takes place at a less fundamental level, at the level of the particular incidents . . . that are offered each week" (154). "Fundamentally," he concludes,

> the series implies the form of the dilemma rather than of resolution and closure. This perhaps is the central contribution that broadcast TV has made to the long history of narrative forms and narrativised perception of the world. (Ellis 1982, 154)

It is this open-ended, ruminative presentation of narratively posed problematics which I think especially entitles television fiction to be regarded as a site of culturally germane symbolic production. Fredric Jameson forwards the notion of the "ideologeme," an object able "to manifest itself either as a pseudoidea—[i.e., as] a conceptual or belief system, an abstract value, an opinion or prejudice—or as a protonarrative"; the ideologeme by definition is "susceptible to both a conceptual description and a narrative manifestation all at once" (1981, 87). This seems an apt description of the problematics—the "treatments"—which television producers must frame in creating their series:

> A black professor of history from the Northeast inherits a New Orleans restaurant upon the death of his long estranged father and decides to man-

age the establishment as a newcomer to the bar's black, lower-middle-class neighborhood, a community suffused with local mythology *(Frank's Place)*.

A company of eccentric doctors—including a Boston Brahmin prima donna, a Jewish patriarch, and a hopelessly mystic Irishman—presides over a formerly Catholic, now municipal (and in the final season, corporation-run) Boston hospital, whose black comedies of error take place only a stone's throw away from more effective (or at least soberer) medical centers *(St. Elsewhere)*.

A New Jersey Italian punk from a Mafia family grows up to be an undercover cop, indulging the playacting and confidence-man schemes of his sanctioned illegalities self-righteously, recklessly, in an elaborately excused form of male fantasy *(Wise Guy)*.

A Vietnam veteran leaves Naval intelligence to become a private eye, investigating crime in the quasi-Asian paradise of Hawaii, aided by his war buddies and living as hired help on an estate whose benign British major-domo helped build the bridge over the River Kwai as a World War II prisoner of war *(Magnum, P.I.)*.

These premises are almost ludicrously packed. But they are indeed bundles of ideologemes, supremely "susceptible to both a conceptual description and a narrative manifestation all at once." They are fashioned from the viewer's map of implicit social knowledge—about race, regionalism, kinship, age, ethnicity, gender, class, wealth, religion, education, health, lawfulness—and provide (as they are designed to) occasions for making such knowledge narratively explicit.

Television's "effects" then, lie in incrementally reshaping viewer's internalized poetics of implicit social knowledge. (This might be considered ideology, if "ideology" is understood as a bricolage of poeticized meanings, a strategic, common-sense accommodation to hegemonic philosophies, rather than a totalizing, transcendental, formally coherent system of action and belief.) That reshaping is frequently self-conscious because it emanates from a medium acknowledged as central to cultural production—but also because it is often displayed within ironical narrative frames. Of the shows cited above, the New Orleans bar show was comedy presented as an archly choreographed, nostalgic drama; the hospital program, a mordant satire that edged frequently into mythic surrealism (encounters with the dead were common). The third held its machismo at arms length by suspending "the action" for extended bouts of male soul-searching, articulately enunciated in penny-ante hood accents. The Hawaiian private eye show, like much detective fiction, was voiced in the first person, demanding the viewer to form a continually re-evaluated relationship with the diegetic narrator (often the protagonist was a lovable buffoon that couldn't see what the audience saw coming, other times he became an unexpected murderer).

I do not suggest that all television shows are ironic. But nothing inherent in the medium precludes such irony. Enjoying the shows enumerated above does not require the assumption of hegemonic views. Indeed, it is

more credible to suggest that programs are most ideological in their premises, in order to ensure a pretext comprehensible to a maximum number of
viewers; but that the ideology of the premise is increasingly subject to transformation in (to return to Ellis' term) each episode's "refiguring," until the
narrative's mode of refiguring supersedes the premise, replacing it with the
viewer's narrative experience of the story. This experience is itself subject to
scrutiny as a whole, when the viewer watches the same narrative later in the
institutionally arranged *déjà vu* of syndicated reruns or videotape. The viewers' relationship to this site of cultural construction therefore assumes an
increasingly intimate and biographical tone, enabling retrospective periodi-
zation or thematization of one's life according to collectively shared narrative performances.

In all these ways, television is a social institution set up to be, in de
Certeau's (1984) term, "poached." Television's perceived centrality to
American viewers emerges from its quality as a "proper place," an institution organized by empowered, capitalized fractions of society to their own
ends, but inhabited and appropriated to personal ends by the people who
attend to it (de Certeau 1984, xix–xx). Like architecture, television occupies
a definite and identifiable social space, suggesting certain confinements and
procedures, forwarding highly official allegories of what the space is like
and for. But like architecture's users, who always prefer to use the back
doors, block over the window, and eat lunch in the honorific rotunda, television's viewers do not always subscribe to television's priorities, setting up
instead their own tactics of use.

Indeed, de Certeau's comments on the "art of memory and circumstances" (1984, 82) suggest a necessary variation in television viewers' encounters with television. Every story, he asserts, begins as an establishment,
in a certain institutional, conventional, narrative "place"—the source of narrative "forces." These are applied to the listener's "fund of memories"; a
particular, "punctual" act of memory in the listener produces a founding
rupture or break, making possible "a transgression of the law of the place."
This rupture reconfigures the place itself, producing the narrative's effects.
De Certeau points to the economies played out among these actions. A
rather small force is all that is required to activate a larger fund of memories.
A larger fund of memories will require less time to discern its "punctual"
act. A quicker rupture will effect a larger transgression. To television's general refiguration of events suggested by Ellis above, each listener brings a
different fund of memories, slowing or accelerating the pace of television's
effects, altering the punctuality of its effects, and hence the nature of television (of a central site for cultural production) for that viewer's future
practices.

### Television as Dynamic Doxa

"Every established order," writes Bourdieu,

> tends to produce . . . the naturalization of its own arbitrariness. Of all the
> mechanisms tending to produce this effect, the most important and best

conceived is undoubtedly the dialectic of the objective chances and agents' aspirations, out which arises the sense of limits, commonly called the sense of reality. . . . This experience we will call doxa, so as to distinguish it from an orthodox or heterodox belief implying awareness and recognition of the possibility of different or antagonistic beliefs. (1986, 164)

Faith in inarticulable, self-evident, taken-for-granted doxic reality is what gives a culture's forms of magic, art, generosity, marriage, or education authority over the people subscribing to that culture. Because of such faith, Bourdieu suggests, social exchanges which actually serve to accrue capital are infallibly regarded as expressions of something else—mystical prowess, aesthetics, love, curiosity. These symbolic stores, Bourdieu argues, are capital nevertheless, supported and reproduced by the mode of production set in place by dominant class fractions. The particular kind of "misrecognition" ratifying any form of symbolic capital as something else establishes the value of exchanges and counter-exchanges, and consequently the value of people's strategies for social production—of the strategies they pursue in their everyday practices.

Bourdieu portrays doxa as static and entrenched, beyond discourse or opinion. But as an articulator of implicit social knowledge, a motivation to poetic invention, a narrative "force" reorganizing a central "place," television, I suggest, institutionalizes an ever-changing doxa. The faiths in social relationality produced by television are frequently changing. Television produces in the realm of symbolic capital what an article in the *The Wall Street Journal* about a displeased Federal Reserve produces in the realm of literal capital: a palpable set of preoccupations that alter the nature of exchange, and with it, potentially, the nature of the entire (financial or symbolic) economy, perhaps even to the point of outdating the Fed, or Wall Street, or an institution like the Berlin Wall, entirely. This potential to change social faith explains again the sense of centrality viewers may get from television. The changes are not necessarily dramatic, but are there. Almost all viewers have had television events—chaotic coverage of the San Francisco earthquake, shots of Jimmy Carter, Anwar Sadat, and Menachem Begin together at Camp David—dispel their staid cosmology. Even television's mode of cultural production is itself subject to change, as, for instance, technological changes (VCR, cable) alter the registers of its rhetoric. Few central cultural sites illustrate so well Marshall Sahlins's insistence that *"plus c'est la même chose, plus ça change"* (1981, 7). Television establishes—or at least offers a highly accessible figure for considering—an historicized doxa.

## The Social Practice of Watching TV

If television viewing is indeed an historically located institution for the production of culture, how are we to define the form it takes: as ritual, aesthetic, literary, poetic, rhetorical? I have suggested that television is a "cultural" hybrid entailing all these forms in the production of a new, more generalized sense of social relation. Most important, I have suggested that watching TV should be approached as a social practice, as a set of conventional habits and gestures subject to empirical observation, rather than as a monolithic agent automatically assumed to determine social reality or reflect social conditions. This vantage can alter the tenor of inquiry dramatically, and open new possibilities for study (Morley and Robins 1989).

Approaching television as a social practice, indeed, is in some ways like observing how people take meals. Meals are ubiquitous, routine, domestic events that embody subtle ideologies of providence and sociality, apply complex rhetorics of arrangement and display, and stage occasions for a great variety of special comportment. For all that, the boundaries and entailments of most meals are quite flexible and informal. Though they can be ritual, no one would argue that they are wholly alien impositions upon the culture of the household, even though they are almost exclusively supported by consumerism (indeed, meals are in a sense what consumerism is "about"). And though each of us has felt compelled to sit through meals uncomfortable or untenable for some combination of aesthetic and social reasons, no one, I think, would argue that the coercion (or complicity) in dining was covert; to the contrary, the meal provided a well marked (perhaps dreaded) occasion for mystification and intensity. Each of these points, I think, could be elaborated for the practice of viewing television.[6]

Conceiving of television viewing as an historically located symbolic practice allows us to sidestep a set of analytical dilemmas too commonly brought to bear on television. Writers addressing television as part of a more general critique of modernity (e.g., Schulte-Sasse 1988) frequently structure their arguments against mass produced narrations by presuming antinomies between the public and the private; between life that is "just lived" and life "consciously worked through" (1988, 196); between the psychological and the ideological (181); "between the realm of art and that of everyday interaction and communication" (196); between historical actuality and the ideological sophistries of "baroque" entertainment (182).

Realizing that viewing television is a social practice renders these antinomies inutil—as limiting for understanding television as asking, in another kind of analysis, whether eating meals is inherently public or private, psychological or ideological. The interest in either case should lie, not in settling such antinomies, but instead in the practices' constitutive nature, and their varieties. For watching television, if not exactly public, remains an irreducibly social and cultural process, one whose socially arranged agent happens to be the (not necessarily private) individual subject. As a colloquial narrative encounter of wholly implicit social knowledge, it is both "just lived" and "consciously worked through"; a simultaneous engagement of psychological and ideological work; a site of symbolic production that is everyday but which entails poetics as complex as that required for the aesthetic appreciation of art. It is a symbolically productive engagement in which baroque transformations are not mere gesture, but a meaningful kind of "evaluation"—or perhaps better said, a meaningful kind of valuation—carried out within a prominent site for the production of a generalized sense of culture. Watching television is an active social practice; and analyzing television a problem in analyzing a complex institution which continually establishes occasions for viewers' production of culture.

## Notes

1. As the expanded menu of television viewing brought by cable, VCRs, the Fox network, and independent stations divides the general television audience into more differentiated groups, however, it is likely that discussion and appreciation of television will become more individualized, and increasingly used to assert one's membership within certain social fractions (cf. Bourdieu 1984). A presage of this already occurs in many professionals' assertion that they primarily watch PBS.

2. These observations form part of Schatz's course on narrative strategies in film and television.

3. The sense of a wide general audience attending to the same text is being increasingly qualified, again, by the television audience's fragmentation among cable channels and other alternatives. Thus the choric dimension to television drama,and the sense of prime time as performance, is probably subject to subtle and profound change in the next years.

4. The novel is probably the best analog to the television text, in its lack of formal prescription, and its frequent locus in contemporary social problems.

5. In acting as a prominent medium for advertising, television especially serves to reproduce the patterns of cultural distinction discussed by Bourdieu (1984).

6. Anthropologists have long proposed that commensality is a fundamental gesture of sociality. It is also worth noting, then, the common practice of eating in front of the television (or, perhaps in competition with it, refusing to eat with the television on). Analogously, the public sites where televisions are commonly present are bars and restaurants. At home or in the bar, I suggest, television cues a choric sensibility fully commensurate with the conviviality of meal and drink.

## References

Allen, Robert. 1965. *Speaking of Soaps*. Chapel Hill: University of North Carolina Press.

Altman, Rick. 1987. Television sound. In *Television: The Critical View*, 4th ed., ed. Horace Newcomb, pp. 566–84. New York: Oxford University Press.

Anderson, Benedict. 1985. *Imagined Communities: Reflections on the Origin and Spread of Nationalism*. London: Verso.

Bakhtin, Mikhail. 1981. *The Dialogic Imagination: Four Essays by M. M. Bakhtin*, trans. Caryl Emerson and Michael Holquist. Austin: University of Texas Press.

Booth, Wayne. 1983. *The Rhetoric of Fiction*. Chicago: University of Chicago Press.

Bordwell, David. 1989. Historical Poetics of Cinema. In *The Cinematic Text: Methods and Approaches*, ed. R. Barton Palmer, pp. 369–98. New York: AMS Press.

Bourdieu, Pierre. 1984. *Distinction: A Social Critique of the Judgment of Taste*. Trans. Richard Nice. Cambridge, Mass. Harvard University Press.

———. 1986. *Outline of a theory of practice*. Trans. Richard Nice. Cambridge: Cambridge University Press.

Brooks, Peter. 1984. *The Melodramatic imagination.* New Haven: Yale University Press.

Condit, Celeste. 1989. The Rhetorical limits of polysemy. *Critical Studies in Mass Communication,* 6(2): 103–22.

de Certeau, Michel. 1984. *The Practice of everyday life.* Berkeley: University of California Press.

Durkheim, Emile. 1965. *The Elementary forms of the religious life.* Trans. Joseph Ward Swain. New York: Free Press.

Ellis, John. 1982. *Visible fictions: Cinema, television, video.* 2d ed. London: Routledge & Kegan Paul.

Ewen, Stuart. 1976. *Captains of consciousness: Advertising and the social roots of the consumer culture.* New York: McGraw-Hill.

Feuer, Jane. 1983. The concept of live TV: Ontology as ideology. In *Regarding television,* ed. E. Ann Kaplan, pp. 12–22. Fredrick, Md.: The American Film Institute and University Publications of America.

Fischer, Michael. 1986. Ethnicity and the post-modern arts of memory. In *Writing Culture: The Poetics and politics of ethnography,* ed. James Clifford and George Marcus, pp. 194–233. Chicago: University of Chicago Press.

Fiske, John. 1987. *Television Culture.* New York: Methuen.

Fiske, John, and Hartley, Jon. 1978. *Reading television.* London: Methuen.

Foucault, Michel. 1979. *Discipline and punish.* Trans. Alan Sheridan. New York: Viking Press.

———. [French, 1976]. 1980a. *The History of sexuality.* Vol. 1. Trans. Robert Hurley. New York: Viking Press.

———. 1980b. *Power/knowledge: Selected interviews and other writings, 1972–1977.* Ed. Colin Gordon. Trans. Colin Gordon, Leo Marshall, John Mepham, and Kate Soper. New York: Pantheon Books.

Gramsci, Antonio. 1971. *Selections from the prison notebooks.* Trans. Quintin Hoare and Geoffrey Nowell Smith. New York: International Publishers.

Halbwachs, Maurice. 1980. *The collective memory,* rev. ed. Ed. Mary Douglas. Trans. Francis J. Ditter, Jr., and Vida Yazdi Ditter. New York: Harper & Row.

Jameson, Fredric. 1981. *The Political unconscious.* Ithaca: Cornell University Press.

Lipsitz, George. 1990. *Time passages: Collective memory and American popular culture.* Minneapolis: University of Minnesota Press.

Morley, David, and Kevin Robins. 1987. Spaces of Identity: Communications technologies and the Reconfiguration of Europe. *Screen* 30(4): 10–33.

Newcomb, Horace. 1984. On the Dialogic Aspects of Mass Communication. *Critical Studies in Mass Communication* 1(1): 34–50.

Postman, Neil. 1985. *Amusing ourselves to death: Public discourse in the age of show business.* New York: Viking Press.

Sahlins, Marshall. 1981. *Historical metaphors and mythical realities: Structure in the early history of the Sandwich Islands.* Ann Arbor: University of Michigan Press.

Schulte-Sasse, Jochem. 1988. Can the disempowered read mass-produced narratives in their own voice? *Cultural Critique* (Fall): 171–99.

Taussig, Michael. 1987. *Shamanism, colonialism, and the wild man: A Study in terror and healing.* Chicago: University of Chicago Press.

Wagner, Roy. 1981. *The Invention of culture.* Rev. ed. Chicago: University of Chicago Press.

Williams, Raymond. 1974. *Television: Technology and cultural form.* New York: Schocken Books.

———. 1977. *Marxism and Literature.* Oxford: Oxford University Press.

# Playing at Being American: Games and Tactics

## JOHN CAUGHIE

Playing at being American. There's a kind of impudence here that pulls in two directions:

First: there's a conscious and deliberate echo of Venturi's *Learning from Las Vegas;* [1] the sense of an objectification, from the outside, of a mental landscape which has a special status in the cultural imaginary. But whereas Venturi may appeal to the democracy of the vernacular, my sense of this objectification is that what it in fact does is to confer on the observer a token of superiority, the "distant distinction" which Pierre Bourdieu sees as the distinguishing, aesthetically "ennobling" mark of the owner of cultural capital. "Nothing," says Bourdieu, "is more distinctive, more distinguished, than the capacity to confer aesthetic status on objects that are banal, or even 'common.'"[2] In the context of a "media imperialism" which has America as its center, this impudent objectification of an imaginary America which can be played at and with, claims such a distinction, seeking to reverse the current of imperialism. The empire strikes back as its center, the "colonized unconscious" *knows* its colonizer, the periphery creates the core as *its* other, the subaltern attempts speech. The rhetorical "tactic,"[3] then, is one of empowerment.

Second: if sarcasm is indeed the lowest form of wit, it may be because

---

From "Playing at Being American: Games and Tactics" by John Caughie in *Logics of Television,* ed. Patricia Mellencamp, Bloomington: Indiana University Press, 1992. This selection was originally commissioned by The Center for Twentieth Century Studies at the University of Wisconsin at Milwaukee. Reprinted with permission.

it's the last resort of the powerless. The cheekiness of the "Kynic," which Peter Sloterdijk[4] traces in the heirs of Diogenes as a defense against the cynical reason of a decaying Enlightenment, offers a certain satisfaction—rudeness rather than rationality as the rebuttal of idealism, farting in the face of the Platonic dialogue[5]—but the carnival of fools happens only under license, and it's business as usual next day. Impudence may confirm rather than subvert the "normal" relations of power. So the insubordination of playing at being American may, in fact, be nothing more than a licensed game, one of the permitted games of subordination.

Tactics of empowerment and games of subordination: a play of subjectivity and identification reminiscent of the oscillations of women's identification discussed by Laura Mulvey in her afterthoughts on "Visual Pleasure and Narrative Cinema"[6] or, perhaps more exactly, with the process of "double identification"—subject and object, seer and seen, both inextricably at once—argued by Teresa de Lauretis in *Alice Doesn't*.[7] I want to suggest that this play of subjectivity, oscillation or doubling, with all the sometimes pleasurable, sometimes painful contortions which either process involves, is a more adequate way of understanding the "colonized unconscious" than the simple and singular positioning which theories of media imperialism usually allow.

### Anecdote: Local Knowledge

Imagine, if you will: a remote village, high in the Spanish Pyrennees; on one side of the single unpaved street, a cluster of buildings, on the other side, fields; in the fields, women and men, scything, forking over the hay, raking with wooden rakes; in the very center of the village an open cow barn; immediately beside the cow barn, a cantina, bead-curtained and stone-floored; inside the cantina, a woman behind the bar, a girl (could she be barefoot?) sweeping the floor, and a television set. It is 11 A.M. on Wednesday, 23 July, 1987, and the television is on.

We (mountain-walking, non-Spanish-speaking "enlightened tourists") drink Coke and glance at the television. It is twenty minutes before either of us realizes that the person we are glancing at is indeed Joan Collins, and that this is indeed *Dynasty*, almost more realistic in Spanish, the melodramatic legitimized as foreign, other, operatic. The cultural anomaly—outside, there are cows in the main street, inside, the world of *Dynasty*—isn't enough to hold our interest for long, but our moves to leave are met with some kind of anxiety. By this time, half the population of the village seems to have congregated to watch television; there is anticipation of something big about to happen which we, particularly, should be witness to, and so we stay—politely. *Dynasty* finishes, and on comes, sure enough, to communal delight and recognition, the English Royal Wedding: Fergie and Andrew (Nancy Reagan, Princess Di as bit players), along with all the pageantry and the celebration of English nationhood and identity, which makes us, in our "distant distinction" (and our Scottishness), seek out otherness and difference in the first place.

The story is nothing special. Most of us probably have a version of it. It's the kind of story that we bring back in our luggage and drop into conversation to illustrate the thesis of cultural homogenization: there is no difference, the media have made it all the same out there. But if it's to be circulated as anything other than a cute epiphany, its meaning cannot be self-evident.

First, there's an appropriateness in these particular programs which gives some specificity to the kinds of marketable images which the United States and United Kingdom now sell, images which come to define them in the international imaginary, and which secure their places in the international image markets. *Dynasty* (with *Dallas*—the two, "Dallas n' Dynasty," as Richard Collins points out,[8] become shorthand, a codeword for the moral panic generated by the American penetration of the European psyche and the European markets) surely both celebrates and castigates advanced capitalist structures and relationships, offering desires and their punishments in the same movement, short-circuiting guilt, giving us things to wish for and rewarding us with confirmations of our own superiority because we don't have them: a playful mode, available to more than one reading. The Royal Wedding (an episode in a continuing series) marks out with wonderful precision what it is that Britain can still sell easily on the image market: the past, tradition, the spectacle of national pageantry, the costume dramas of a simpler class hierarchy, the nostalgia for a lost aristocracy: *Brideshead Revisited, The Jewel in the Crown, Upstairs, Downstairs,* "Masterpiece Theater"— even the soap operas—*Eastenders*—for all their qualities of recognition and familiarity in the home market, enter the U.S. market as a mild nostalgia for a lost community. So if Britain and America are indeed the exporters in the international dream market, they may be marketing dreams which are dreamed in quite diverse ways, a fantasy of the other as quaint, perhaps, rather than as compelling object of desire.

Second, and consequently, is *Dynasty,* dubbed into Spanish, watched by Spanish men and women in a Spanish village, still, only and simply, an American text (with all that textuality has come to mean)? Is the English Royal Wedding *celebrated* in Catalonia, the most fiercely independent region in mainland Europe outside the Basque territory? How can it mean the same thing in Spain, with that country's intensely contradictory history of monarchism and republicanism, as it does in the shires of Southern England, which may contain, albeit in isolated pockets, the only audience that is decoding it "correctly" in the way it was encoded?

The seduction of the thesis of homogenization and the eradication of cultural difference is hard to resist, and can continually be supported by empirical data on ownership and distribution. But at the level of theory (and experience), it seems continually to fall back on the belief in the passively receptive, "duped" consumer, the always obedient subject jumping into line at the call of interpellation. If we revise that view in favor of a more complex and various view of reading relations for the theorized domestic subject (female or male; audience, viewer, reader), we're clearly going to have to rethink it for the "colonized" subject as well.

A depressingly familiar scenario, then: the "enlightened tourist" seeking to penetrate the other without penetrating its otherness; in search of the virgin unknown, and constantly disappointed by its loss of innocence. This is the twist that makes travelers' tales, and it's founded on the same old paternalism: the owner of cultural capital laments the corruption of the other, who knows no better, despite our warnings, than to want to own the goods too.

The question of locality seems to me to have a particular urgency for television. It's increasingly accepted that theory and critique become most material when they are localized rather than universalized. For any film and television theory which has been attentive to feminism, the universal, neutered subject (the "it" that always concealed a "he") must give way to a subjectivity located within specific and "local" formations of gender and sexuality. I want to suggest that one of the localities from which theories can be materialized is the embarrassingly persistent category of the nation.

Now, clearly, nationalism has not had a particularly good press in the twentieth century; so it's important to say that the nation I'm talking about has less to do with the "nation state" of nineteenth-century nationalism, a legal and political "ego" presumed to function as a concrete subject, and more to do with the "imagined community" which Benedict Anderson describes,[9] this "imagining" qualifying in complex ways, rather than simply disqualifying, its subjectivity. (Parenthetically, the condensing image for me of this "imagining" is the road signs, encountered when driving through Arizona, that mark out tribal land—"Entering the Navajo Nation," "Leaving the Hopi Nation": a nationhood apparent less in its legal or economic status, which may be formal, and even derisory, than by the affective, subjective, and political aspiration which transforms as well as transcends its physical landscape.) The nationality in which I'm interested, then, and from which I speak (Scotland—in relation to the United States a periphery on the edge of a periphery, a community which has often suffered from too much imagining), is never given, never already has pride of place, but is uncertain, tentative, wary of its own double edges.

Clearly also, even this tentative, marginalized nationality as locality cannot smoothly be expropriated from feminism as locality. Although nationalism conceived in this way seems to resonate in important ways with feminist theory and experience (the pressure to smile complicitly at our own humiliation, the marketing of the images of our own subjection, landscape as body, excessive identification—national dress and the tourist trade—"double identification" and the masquerade—precisely "playing at being American"), the relation to power remains different, the identification more abstract, the options—to play or not to play—more open.

Nevertheless, despite the dangers, it's this subjectivity, with those resonances, that is the context here for a way of thinking the subjectivity of television from a local perspective. Much television writing, while it is sensitive (sometimes) to systems of gender, is strikingly insensitive to the specificities of national systems, an insensitivity which drags it toward universal theories and descriptions of television that ignore the extent to which viewing is formed within particular national histories and localized broadcasting systems. Paradoxically, this universality is most apparent in those rhetorical celebrations of television as the quintessence of postmodernity which, while proclaiming the end of grand narratives and universal theories, simultaneously universalize a local, national experience—the U.S. experience—as the essence of television, thus marginalizing all other experiences, and confusing the effect of a particular commercial arrangement with an inevitability of nature. For the record, I would argue that British television, and much European television, is still rooted in modernity, the concept and practice of public-service broadcasting part of an unbroken tradition of "good works" dating from the administration of capitalism in the latter part of the nineteenth century. While that tradition is clearly under threat from the readministration of capitalism and the redistribution of power in global markets, nevertheless the scenario of magical transformation—the marvelous vanishing act of deregulation: now you see "quality," now you don't—in both its optimistic and its pessimistic variants seems naive. More likely is a scenario in which transformation is uneven and diverse, continually modified by local conditions, local demands, expectations, resistances, and compromises, the future still bearing the residual traces of the way it always was. This diversity and local specificity seems important, not as a point of national pride, or nationalist pique, but as a challenge to notions of the indifference of an essentialized and universal television.

So: playing at being American. In general terms, the question of how U.S. television and the discourses which surround it figure in the empire of signs and representations and critical and theoretical practices that constitute television and our various relationships to it inside and outside America is clearly very important. I'm interested, generally, in the assumptions that ground the commonsense notion of the colonization of the unconscious or the imaginary, a notion which informs quite persistent national anxieties about the seductiveness of American popular entertainment, and about the dependency on U.S. production within the schedules of popular television. These anxieties surface at a number of levels from the genuinely political to the cynically populist. I'm interested in this from a Scottish or British or possibly European context, and have no real sense or experience of what it might mean in the context of the more specialized forms of economic and mental subjection involved in the exploitation of so-called Third World markets. But here, in this essay, at a more particular and fundamental level, I'm interested in the extent to which the desire to locate television within local perspectives complicates assumptions and theoretical formulations about reception and representation and interpellation and identification in quite significant ways.

## *Detour: Discovering America*

As a slight detour, in the context of an American conference and an American publication, it may be useful to continue the objectification of the American experience as other, and to identify, very loosely, some of the local variants of watching television in America which strike me with the enlightening force of anthropological surprise. To characterize the specificity, or, for me, the difference, of watching television in America (as opposed to watching American television—which one can do anywhere), I want to offer three fairly basic, obvious, banal observations. They have to do not with program style or content but with the structure of viewing. They are consciously naive snapshots which may crystallize my "tourist" sense of the radical otherness, from a habituated British perspective, not simply of American television conceived as programs but of the American "televisual" conceived as an experience of watching and viewing, and switching on and switching off and switching over, and spending time.

First, and most obvious, the technologies of plurality. Even without peripheral technologies, there is a plurality of choice which is still (and the temporality of the "still" has to be emphasized) unknown in Britain. This is so obvious that it barely needs comment, except to suggest that this plurality breaks down the pattern of channeled or even programmed viewing. While there is still the anticipation of favorite programs, or favorite clusters on particular evenings, for many of us and for much of the time the televisual in America seems to be a relatively unstructured activity, zapping more satisfying than planning. Much more characteristically, I think, than in Britain, the American televisual is contingent.

Second, the organization of time. In some sense cutting against the contingency, in another sense reinforcing it, I am struck by the regularity of the network schedules, by the clarity of temporal definition given to primetime, by the ease of cross-over in a temporal system in which each program begins and ends on the hour or half-hour, by the starkness of choice when like is scheduled against like—not just "Do I want to see a cop show or a sitcom?" but "Which cop show?" "Which sitcom?" There is no space here to consider the extent to which this conflation of time and genre and the resulting competition within generic boundaries may account for the continual inventiveness and self-reflexive fluidity of U.S. popular genres— an inventiveness which compares favorably with the repetition, stability, and even stasis of British generic conventions. Here, I am simply interested in the difference which American scheduling makes to the structuring of my time, particularly my primetime. It had never seemed odd before that a program such as *Dallas,* shown in Britain on BBC, should begin at 8:10 P.M. (because without commercials it lasts only fifty minutes), thus creating a ten-minute gap of "dead-time" between the end of a domestic soap opera on one channel and the beginning of *Dallas* on another. (Nor did it seem unusual that a film which lasted an hour and forty minutes in the cinema would also last an hour and forty minutes on television.) The specific nature of "flow" produced by the staggered scheduling of British television, with built-in resistance to clean channel cross-over—the risk of "dead-time"— seems to me to encourage a residual degree of channel loyalty (or inertia— it's easier to stay than to switch) quite uncharacteristic of U.S. television flow. And the relative absence of like-against-like scheduling organizes the movement from program to program in what seems like more structured, rational choices. The regularity of American television time, the opposition of like against like, dissolves my loyalties and draws me to the jumpy, nervy, mosaic gratifications of sampling. There's no need to insist on the generality of my secret practices and domestic rituals: the point is in the differences.

Third, and most important, the breaks and interruptions. For all the preparation that one receives from conventional wisdom about the annoy-ance of the commercial interruptions on American television, their effect still catches me a little unprepared. It is much less the regularity of the interruption, or the wonderful blatancy of their placement, more the specific way they interrupt, their effect on (or, more exactly, in) the text. In Britain, commercials interrupt programs, but in parentheses: a logo or a caption— *St. Elsewhere,* or "End of Part 1"—signals the suspension and resumption of the program, bracketing our disengagement, demarcating the adjustments in the form of our attention. On American television, Cagney looks out of frame, and the answering reverse-field is a commercial for Mack trucks. The space of the commercial is continuous from the space of the fiction: a split second of cognitive dissonance, but, for the unhabituated viewer, enough for a lurch in attention, a hiccup in the mental logic. Added to this, on the local stations, there are the breaks just before the final coda, or the station announcements over the credits, or station identification superimposed on the image. At one level, these are simply irritations of commercialism. At another level, though, they can be read symptomatically as little contests between commercial logic—the need to deliver audiences to advertisers— and narrative logic—the need to hold audiences in identification. What is so striking to someone raised in the protective shelter of public service is the visibility of the contest. I *experience* the effect which I had always known in theory: a quite radical destabilizing of the text as an autonomous and logical fictional space complete within its own boundaries. For a still "for-eign" viewer, the experience of watching American television is never sim-ply the experience of watching programs as texts in any classical sense, but is always also the experience of reading the specific forms of instability of an interrupted and interruptible space.

Part of the force for me of these banal reflections comes from the ex-perience of teaching a television course, using largely British material, for American undergraduate students, and finding my assumptions about what television is and what the experience of viewing is in tension, albeit produc-tive tension, with assumptions structured by a radically different economy of viewing and the televisual. There is a real risk in the theorizing and, particularly, in the teaching of television of opening up a gap between the television which is taught and theorized and the television which is experienced. Teaching seeks out the ordering regularities of theory. A television is constructed which is teachable, but may not be recognizable.

But another part of the force of this different watching is that of an estrangement effect, a seeing "as if for the first time," in which television viewing appears as a specific procedure, necessarily learned in response to specific textual practices. Each of my banal reflections crystallizes (for me) a specific uncertainty around the television text, the particular "strangeness" of American television giving that uncertainty a peculiar sharpness. Nothing theoretically new in the problem of textuality—except the force of experience, which opens the suspicion that a television theory which has still to find its own way of understanding textuality is not yet adequate to the difference of the television text, and the particular forms of its uncertainty and instability.

In theory and in criticism, what I'm suggesting as the instability of the television text, the blurring of its boundaries, the erosion of its integrity and its autonomy, are features completely familiar to poststructuralist and postmodernist criticism, not at all specific to television. The integrity of the autonomous text exists, at the theoretical level, only in formalist nostalgia (although in theoretical and critical practice the yearning for an isolation ward for critical cases still seems quite frequently to have material effects). All I want to suggest here is that the systematic qualifications and interruptions and erosions of the text-as-program, the text-as-unitary, build contingency into the structure of viewing in a way which faces the securities of the theoretical landscape with quite hard-edged problems. It may not be enough to qualify or even reverse the terms in order to accommodate television, sailing on a different part of the ocean of discourse but steering by the same stars; it may be that the metaphors which frame our theoretical discourse have to be checked for their continuing adequacy. I've suggested elsewhere,[10] in the context of theories of popular culture, that the economic metaphor, rooted in a nineteenth-century industrial economy, often brings with it a discourse of struggle which can be both romantic and virile, creating a romance of titanic forces. In its duality of production and consumption, it stabilizes the system, one term always already valued over the other, allowing a pendulum swing of critical values, but still swinging along the same axis. Here, in the context of thinking about television, I want to suggest that the spatial metaphor also—distance, positionality; a metaphor which can also be used to stabilize relationships in a diagrammatic geography—is, at the very least, up for review.

The spatial metaphor of position has been foundational for much film theory, either implicitly or explicitly. It allowed us to think not only positions but opposition, not only identification but distance, founding not only a textual economy but a politics of textuality: a kind of political geography whose orientation can be detected in the fact that while distance has been constantly recharted—detachment, distraction, passionate detachment, relaxed detachment, decentering—we seemed always to come back to the same old home port of identification, the wrong place to be. Politically, identification was passive, on the side of the consumer, where distance was active, on the side of production, and the simple pleasures of consumption were rejected in favor of the romance of radical resistance.

In his book *Horizons of Assent: Modernism, Postmodernism, and the Ironic Imagination,*[11] Alan Wilde distinguishes, very provisionally—"an ad hoc shorthand," he calls it, "deliberately 'inadequate'"[12]—three divisions of what he calls the "ironic imagination": "mediate irony," "disjunctive irony," and "suspensive irony." "Mediate irony" is the mode of satire, essentially, Wilde argues, a premodernist mode. It "imagines a world lapsed from a recoverable norm."[13] "Disjunctive irony" Wilde associates with the heroic agonism of the high modernists, fashioning meaning out of the fragments. "The ironist confronts a world that appears inherently disconnected and fragmented. . . . Disjunctive irony both recognizes the disconnections and seems to control them."[14] Finally, perhaps most appropriate for a consideration of television, is "suspensive irony," which Wilde associates with postmodernity:

> Suspensive irony . . . with its yet more radical vision of multiplicity, randomness, contingency, and even absurdity, abandons the quest for paradise altogether—the world in all its disorder is simply (or not so simply) accepted. . . . Ambiguity and paradox give way to quandary, to a low-keyed engagement with a world of perplexities and uncertainties, in which one can hope, at best, to achieve what Forster calls "the smaller pleasures of life," and Stanley Ilkin, its "small satisfactions."[15]

The interest of Wilde's work in the present context is not simply to discover three more terms we can happily instrumentalize for television studies, inaugurating debates on whether *St. Elsewhere* was disjunctive or suspensive, or discovering the mediate irony in *Family Ties*. Wilde himself is careful to insist that he is not constructing a teleology or even a rigid critical taxonomy, but that each of the ironic modes may be present in any text or any author in particular but shifting configurations and hierarchies. The classifications are intended to enjoy, he says, "a strictly performative function as discriminating and temporary instruments . . . , a 'truly empirical' sounding of the movements and sinuosities within the concrete appearances of single or grouped phenomena."[16]

In the context of the development of television studies, what I find attractive in Wilde's "sounding" of irony (while reserving judgment on the "assent" of some of his conclusions) is not simply the flexibility which he produces, but rather the extent to which his formulation of the "ironic imagination" echoes Peter Brooks in his formulation of an earlier "melodramatic imagination,"[17] offering a way of thinking *together* a historical consciousness in which subjects do their imagining *and* a set of determinate textual conditions and practices. Thus it's possible to argue that some kind of ironic imagination, an ironic "suspensiveness" perhaps, or what Sloterdijk calls "enlightened false consciousness,"[18] is something we increasingly bring to television, an ironic sensibility already formed outside the space of television, a function of our local histories of modernity and postmodernity, within or against which television always has to play; and, at the same time, to argue that the specific practices and procedures of television—the de-textualizing which it performs on its texts, the snaps in consciousness, the distractions, the physical rather than metaphorical spatial configurations of our rooms, the dipping in and out—provide very particular material and textual conditions for the production and play of an ironic imagination.

The irony which I'm talking about, then, has very little to do with authorial or institutional intention, and more with historical and textual condition. Such a concept of irony offers a way of thinking about dissocia-tion and engagement as simultaneous or, at least, temporally connected ac-tivities, a knowing play in the "movements and sinuosities" of television, outside the spatial metaphor in which texts assign us to a position. In the sense in which Michel de Certeau elaborates the terms, it allows us to think of viewing practices and relations as "procedures" rather than as places.[19] It allows us, that is, to think of the possibility of negotiating two (or more) "conditions" at once (in such a way that the metaphor of boundaried place becomes inadequate), of holding subjectivities in suspension or disjunction, of knowing but agreeing not to know, of a play of irony, of playing at being. . . .

This attention to irony, then, whether it be to ironic forms of attention as a principle of modern or postmodern readership or viewership, or to forms of television address which constitute the conditions of ironic suspen-siveness, is intended simply to open up within theory and criticism a more complicated, shifting, and sensitive way of thinking about how we might be relating to the appeals of television than is offered by the metaphors of place and position. It is emphatically not intended as a new methodology which can instrumentally be "applied" to television; nor am I suggesting that irony is a property of the television text, or that every television text is ironic, or that every television moment is a moment of play; nor, finally, am I suggesting that irony is an exclusive property of the televisual, al-though I am suggesting that the specific conditions of television produce the possibility of more ironic forms of attention than the conditions of, say, the cinematic. Less intensely fascinating in its hold than cinema, television seems to insist continually on an attention to viewing as mental activity and "knowingness" (almost a "street-wise" smartness), rather than to the obe-dience of interpellation or the affect of the "always already."

The implications of an attentiveness to this ironic imagination might be traced in distinct ways in analysis, and in theory and criticism. At the level of analysis, the question can be asked: If an ironic imagination is a charac-teristic mode for television, is there a way of identifying a characteristic figure within the routines of television rhetoric which supports it and puts it into play? I am thinking of the various ways in which the point-of-view shot has been used to figure out notions of cinematic identification. As a possible line of inquiry, if the point-of-view shot (which for various historical and practical reasons is relatively weak in television) is a foundational figure for cinematic identification, could it be that the reaction shot forms an equivalent figure for the ironic suspensiveness of television? As it has been argued, the point-of-view shot centers knowledge within the narrative space by identifying the look of the spectator with the look of a character. The reaction shot, as it is characteristically used in television, disperses knowledge, frequently registering it on the faces of a multiplicity of characters whose function may only be to intensify the event, to charge it with the emotional excess which Jane Feuer identifies in primetime melodrama,[20] but without the centered identification of the point-of-view shot: reaction without identification. Soap opera and melodrama may represent privileged instances here, but at the same time, a similar process can be detected in the cutaways to audience or panelists in studio discussions, game shows, or talk shows, or to spectators or managers in television sport. There isn't space here to develop the point, but in that gap between reaction and narrative identification may lie one of the ways in which irony is figured within the specific textual practices of television.

At the level of theory and criticism, it seems endemic to television writing that whereas film theory is marked by a sense of people trying to come to terms with their own, almost perverse, fascination—what Paul Willemen calls "cinephilia"[21]—television theory always seems to be written by people who can see the seduction but are not seduced. The critic's fascination with the audience in television writing may be due, at least in part, to a lack of fascination with the texts. An effect of this is that television viewers always risk being constructed as the determinate or indeterminate "other," reified as an object of investigation or special attention in roughly the same abstract, and even cynical, way in which election campaigns construct the housewife, the farmer, or the consumer. One of the things that attract me about the concept of an ironic play, of a kind of suspensiveness of knowing and not knowing, being and not being, is that it seems to identify the most characteristic way in which I respond to television, the "small pleasure" that I use television for. Even if this is simply the superior and privileged response of the intellectual, the owner of cultural capital, it is at least worth acknowledging. A television criticism which can identify with itself may avoid speaking for experiences it isn't having.

The argument, then, is that television produces the conditions of an ironic knowingness, at least as a possibility, which may escape the obedience of interpellation or cultural colonialism and may offer a way of thinking subjectivity free of subjection. It gives a way of thinking identities as plays of cognition and miscognition, which can account for the pleasures of playing at being, for example, American, without the paternalistic disapproval that goes with the assumption that it is bad for the natives. Most of all, it opens identity to diversity, and escapes the notion of cultural identity as a fixed volume for which, if something comes in from the outside, something from the inside must inevitably go out. But if it does all this, it does not do it in that utopia of guaranteed resistance which assumes the progressiveness of naturally oppositional readers who will get it right in the end. It does it, rather, within the terms hung in suspension at the beginning of this essay: tactics of empowerment, games of subordination, with neither term fixed in advance.

For this continual return to suspensiveness is, in every sense, double-edged. With one edge, it opens closed systems and dispenses with guarantees; with the other, it lays waste solid ground and exposes ideals, objectives, and aspirations to a sceptical or cynical paralysis. It echoes those familiar terms of postmodernity—ambivalence, indeterminacy, paradox—which challenge the accustomed systems of rationality that founded "enlightened" theory and politics. This challenge is common to much of contemporary cultural and social studies. But for a television studies which is still looking for a definition, however indefinite, and which, imbricated in every way with contemporaneity, cannot avoid the implications of the analysis of postmodernity, the "quandary" seems to represent a particularly numbing critical impasse.

The terms of the impasse can be indicated very schematically: value, politics, the text. What are the terms of decidability by which we can argue for or against values when indecidability is the rule of the game? How do we measure progressiveness without the certainty of the linear narrative of progress? And, underpinning all this, what is the television text anyway? Within the impasse may lie a certain discomfort in television studies on the part of the radical/popular/academic intellectual (for whom impasse and crisis are, of course, terms of self-dramatization). For many of us, formed historically in theory and ideology critique, in a political commitment to popular culture, and in a belief in our special (and "subversive") place at the cutting edge of the liberal academy, television, and particularly the television text, without the proper sanctification of a theory, may still be an unworthy object, the clarion call to "take television seriously" still a little desperate. The result is a series of displacements. The privileged objects of much television studies—the audience, the institutions, the market—are effective ways of displacing the theoretical problems of values, politics, and texts onto empirically testable bodies. Which is not at all to devalue such studies, but simply to remind ourselves that the question of television's textuality—untestable, uncertain, repressed—will keep returning.

The attempt to rediscover irony may be to contribute as much to the problem as to the solution. In its suspensiveness, or disjunction, its play of being and knowing, its ability to go either way, the ironic imagination may simply play out the amused quandary which dissolves held positions of value, principle, or politics. Attractive because it holds open the possibility of a "difference" which can play with an objectified dominance without being subjected to it, suspensiveness may also conceal the smile of an indifference to which nothing much matters anymore.

In trying to bring irony—as a term of analysis, theory, and criticism—into some kind of conjunction with nationality—as a term of culture and identity—I'm trying to ground the one in the other precisely as a way out of that quandary of indifference. Irony not as a universal theory but as a various and local "tactic"; nationality not as an always already but as a set of aspirations. If irony is a useful concept, even if only to hold at the back of the mind, it is not because it offers a guarantee of the resistance which John Fiske discovers everywhere in television viewing,[22] or even the romance of the guerrilla tactics of counterproduction and refunctioning which de Certeau finds in "consumer practices."[23] It is not a universal category, and its inflection depends on local conditions. The claims to usefulness would be much more modest. Simply, irony seems to me to offer a neglected aesthetic category which might bring together textual practices which could be identified by analysis, the peculiar disidentificatory *dispositif* of television as discursive practice, and a possible historically formed disposition of locally constituted audiences. It recognizes distance, but also the "incorrect" "small satisfactions" of being superior *and* being the dupe. As an experiential category, it seems to me to avoid the well-meaning cynicism of the audience as the intellectual's other, and implicates the critic in his/her own discourse as a material, rather than a transcendent, subject.

But what if playing at being American is the only game in town? Is there anything else we can play at being? The return, again, of the national question. The essay comes to seem somewhat perverse and awkward in its desire to hold on, till the very end, to that conjunction of irony and locality, an aesthetics and a cultural politics. Clearly, as an aesthetic category, ironic suspensiveness has very little to say about the economic and ideological power of the American film and television industries in the international television market. Almost to the contrary, the ironic imagination, as a mere playing at being, might seem to dissolve the importance of that dominance as a cultural issue. But the very difficulty of the conjunction seems to me to be important, setting up a number of resistances between the two terms. Irony, in the sense in which I would use it, approaches viewing as a procedure rather than a place, giving a name to the play of identity and distinction—sometimes pleasurable, sometimes contorted; sometimes conscious, sometimes unconscious; sometimes resistant, often compromised—of subject and object, seer and scene. The insistence on the national, or the local in any of its forms, prefers, if only as a possibility, difference and diversity to indifference and mere plurality. The continual return to locality, whether it be of nation, race, class, gender, or generation, resists the easy rationality of a general category or a universal theory. It is conscious of a less systematic specificity, to be determined by local readings of texts and conditions and histories and objectives, and it proposes a politics which is not guaranteed by textuality or by natural resistance but is open to historical conditions. For television, these conditions are, in their turn, enmeshed in the expectations, aspirations, and possibilities produced by particular histories of broadcasting and by particular legal, commercial, and political arrangements of regulation and deregulation.

One of the things which make television writing so difficult, particularly at an international level, is precisely the absence of an "international standard," of the sort that Hollywood's "classical cinema" has provided for writing about film. In that absence, general statements are always vulnerable to local and empirical knowledges and experiences. But since these are the conditions, it may be as well to find ways of coping with them. In her essay "Feminism and Critical Theory," Gayatri Spivak takes certain American feminists to task for their attitude to history: "As long as [they] understand 'history' as a positivistic empiricism that scorns 'theory' and therefore remains ignorant of its own, the 'Third World' as its object of study will remain constituted by those hegemonic First World intellectual practices."[24] It seems to me that there is something for television studies in that warning, in its insistence on the need for empirical and local understandings of histories and practices, dispositions and *dispositifs,* which poses the resistance to imperial theory.

## Notes

1. Robert Venturi et al., *Learning from Las Vegas* (Cambridge: MIT Press, 1972).

2. Pierre Bourdieu, *Distinction: A Social Critique of the Judgement of Taste* (Cambridge: Harvard University Press, 1984), p. 5.

3. For the distinction between "tactics" and "strategies," see Michel de Certeau, *The Practice of Everyday Life* (Berkeley: University of California Press, 1984), pp. 34–39.

4. Peter Sloterdijk, *Critique of Cynical Reason* (Minneapolis: University of Minnesota Press, 1987), particularly chapter 5, "In Search of Lost 'Cheekiness.' "

5. Ibid., p. 101.

6. Laura Mulvey, "Afterthoughts on 'Visual Pleasure and Narrative Cinema' inspired by *Duel in the Sun,*" *Framework* 15/16/17 (1981).

7. Teresa de Lauretis, *Alice Doesn't: Feminism, Semiotics, Cinema* (Bloomington: Indiana University Press, 1984), particularly chapter 5, "Desire in Narrative."

8. Richard Collins, "Wall-to-Wall *Dallas?* The US–UK Trade in Television," *Screen* 27: 3–4 (May–August 1986).

9. Benedict Anderson, *Imagined Communities: Reflections on the Origin and Spread of Nationalism* (London: Verso, 1983).

10. See John Caughie, "Popular Culture: Notes and Revisions," in Colin MacCabe, ed., *High Theory/Low Culture* (Manchester: Manchester University Press, 1986), pp. 161–62.

11. Alan Wilde, *Horizons of Assent: Modernism, Postmodernism, and the Ironic Imagination* (Baltimore: Johns Hopkins University Press, 1981).

12. Ibid., p. 9.

13. Ibid., pp. 9–10.

14. Ibid., p. 10.

15. Ibid.

16. Ibid., p. 9.

17. See Peter Brooks, *The Melodramatic Imagination: Balzac, Henry James, Melodrama, and the Mode of Excess* (New Haven, Conn.: Yale University Press, 1976).

18. Sloterdijk, pp. 5–6.

19. De Certeau, see particularly pp. 34 ff.

20. See Jane Feuer, "Melodrama, Serial Form, and Television Today," *Screen* 25: 1 (January–February 1984).

21. Paul Willemen, "The Desire for Cinema: An Edinburgh Retrospective," *Framework* 19 (1982).

22. See John Fiske, *Television Culture* (London: Methuen, 1987).

23. De Certeau, see particularly pp. xii ff.

24. Gayatri Chakrovorty Spivak, *In Other Worlds: Essays in Cultural Politics* (London: Methuen, 1987), pp. 81–82.

# Invisible Cities/ Visible Geographies: Toward a Cultural Geography of Italian Television in the 1990s

## JAMES HAY

There is in Italo Calvino's *Invisible Cities* at least one parable upon which I would like to reflect for just a moment as a way of shaping a discussion about the "cultural geography" of Italian TV in the 1990s and as a way of beginning to address some of the general issues involved in the study of nationhood and media borders. Calvino's text is constructed as a discourse between the late-Medieval Italian explorer and trader, Marco Polo, and the Mongolian emperor, Kublai Khan—two mythic figures who converse at the historic threshold of the European Renaissance and the emergence of global trade routes for Western commercial interests. Much of the text is given to Marco's accounts or narrative reconstructions of places—generally cities— through which he has passed before meeting the Great Khan. These accounts are, however, occassionally interrupted by a somewhat amused Khan who, having never visited these cities himself, wonders whether the cities in his empire described by Marco are really all that Marco says they are, or whether they are but exotic images produced by the vagabond eye of the Italian explorer.

The Kahn is particularly confused about the relations among the places Marco recounts. The cities seem too randomly selected and conjoined through Marco's wandering narrative (and narrative wandering). From his position of privilege, where he has formed an image of a "model," quintessential city, the Khan presses Marco to explain their commonality, their connectedness,

Reprinted with permission from the *Quarterly Review of Film and Video,* © Harwood Academic Publishers (Summer 1993).

their "code"—something that would guarantee his power, as emperor, to decipher them, to gauge and predict, to rule. At one point, the Khan even convinces himself that Marco's cities resemble one another, "as if the passage from one to another involved not a journey but a change of elements." The relation among these places particularly becomes an issue in their conversation when Marco carefully assembles before the Khan an array of natural objects and artifacts, each collected from a different city on his journey. Between Marco and the Khan, the objects (neither only metaphoric or metonymic any longer) *become* the cities; but across the black and white tiles of the floor where they have been arranged, they also can be moved and rearranged, in a game not unlike chess:

> Arranging the objects in a certain order on the black and white tiles, and occassionally shifting them with studied moves, the ambassador [Marco] tried to depict for the monarch's eyes the vicissitudes of his travels, the conditions of the empire, the prerogatives of the distant provincial seats.
>
> Kublai was a keen chess player; following Marco's movements, he observed that certain pieces implied or excluded the vicinity of other pieces and were shifted along certain lines. Ignoring the objects' variety of form, he could grasp the system of arranging one with respect to the others on the majolica floor. He thought: "If each city is like a game of chess, the day when I have learned the rules, I shall finally possess my empire, even if I shall never succeed in knowing all the cities it contains."

The Khan is repeatedly troubled because he has no way of refuting Marco's accounts; Marco's cities and his seemingly random, incomplete *passages* that link them are enigmas that can only be deciphered through the Khan's own *maps* of his vast empire. Thus Calvino's text offers a way of thinking about the power and politics, the enabling and constraining qualities, of Marco's narrative of places along his journey. Before hearing Marco's tales, the Khan had understood his empire and his place at the center of that empire through his own atlas—an official and totalizing dictionary of maps, each offering its own image of his cities and, perhaps more importantly for a Khan, their differences and interconnectedness. Marco's accounts of cities along his path constitute a matrix of emblems (a "logo-griph," to use Calvino's expression) through which the Khan must now imagine and decipher his empire, that is, the empire as *his*.

Because *Invisible Cities* offers a narrative about the interfacing of two symbolic systems or narrative practices *and* of two positions (two ways of thinking about positions) in a power system, it underscores how discursive relations result from, generate, and gradually transform relations of power and status, how discursive relations become sites for struggles to claim territory for constructing identity. The Khan's sense of discursive and political order, for instance, is predicated upon the affective power of an *essentialized* city-image that models the vast distance and circumference of an empire he can only imagine. Marco, however, explains that, as an explorer-merchant, he relies upon his own conception of a model city—one that is constituted *entirely* of "exceptions, exclusions, incongruities, contradictions" and that in this way includes all of the places he is recounting.

The Khan continually asks Marco if his cities are real or the result of Marco's narration and his own reading of them. Marco, on the other hand, repeatedly asks the Khan if his map has room for new emblems. Because the Khan's perspective and system (reinforced by his current collection of maps) is totalizing, he reproaches Marco because his descriptions of cities lack a sense of connectedness. Marco responds that he recognizes only the cities, not what lies outside them. While Marco's cities are texts (spaces that take form through the temporality of narration), the Khan's empire is an atlas-dictionary (the necessary embodiment of his desire for stability and fixity). Yet through their dialogue, one can't help but recognize that the empire, as map, is comprised only of those places for which there are words or that are imaginable through a narrative. One also comes to recognize that both Marco's travel narrative and the Khan's sense of domain are constrained and tested by each other's way of imaging and imagining spatial relations. In this sense, *Invisible Cities* calls attention to the provisional and contingent nature of maps and coordinates, while affirming the indispensability of spatial references. And while Marco's narrative does not quite produce a "map," it does offer an inventory/itinerary of places that, through an accumulation of meaning, through their position in his narratives and their relation to Venice (his starting point or "home") and his encounter with Kubla Khan, produce a narrative space within which he, the Khan, and the reader are caught.

I find this parable useful in a discussion about the current formation and transformation of "*media* borders" not necessarily because the postmodern world is prone to reimagining (as Umberto Eco might suggest) a "new" Middle Ages, but precisely because of the issues it raises about the discursive or narrative production of territory/domain, about the narrative or cultural politics of mapping and geography, about historic transformations in territoriality brought about by convergences or collisions between conflicting narrative systems and positions, about the transformation of material "places" into narrative or imagined "spaces" and about the production of space in particular places.

Many of these points have been developed more fully by other theorists of space. The discourse between Kublai Khan and Marco Polo plays out the distinction that Gilles Deleuze and Felix Guattari draw between the "rhizomatic" production of space through maps and the mimetic, permanent, and totalizing vision of place offered by "tracings," through Deleuze and Guattari emphasize what this distinction has to do with *politics* of spatial production.[1] The point above about the relation between "space" and "place" is elaborated by Michel de Certeau, who suggests that discourses/stories may traverse and circulate amidst places but they also produce and, through their circulation, performance, and practice, constitute a discursive or narrative space by establishing implicit contracts between interlocutors.[2] De Certeau thus wants to acknowledge not only the transformative capabilities of narrative, that is, that stories organize places and how one navigates a place, but also that the production of space is tied to particular places or locations. The French social theorist Henri Lefebvre is concerned with somewhat the same issue when he discusses how social relations are predicated upon complexly *coherent* yet *changing* relations among "spatial practices," "representations of space," and "representational spaces" (or among, what he also terms, "perceived," "conceived," and "lived" space)[3] Both Lefebvre and de Certeau, furthermore, share Calvino's interest in how understandings of the past and future are bound up with understandings of space. Whereas de Certeau is concerned with the dynamic relation among memories, stories, and spaces, Lefebvre directs his attention to how spatial production historically enables and constrains future spaces, ideologies, and social relations. While none of these critical treatments or dramatizations of spatial production or perception overtly concern the current media environment or current media practices, they introduce an important set of issues into a discussion of nationhood, cultural identity, and media borders that I want to address briefly here.[4]

Over the 1980s, the emergence of various media practices have significantly reorganized perceptions of place and territory through their role in the formation of *narrative networks*. Here I am thinking of how the proliferation of cable and satellite broadcasting facilitated the practice of narrowcasting and how the increasingly common integration of television broadcasting and telecommunication processes through fiber optics or the rapid expansion of domestic computer use have made way for a variety of "interactive" systems and networkings (from "1-900" shopping/dating networks to "computer-net" systems such as Bitnet and Prodigy). And while I want to recognize what these formations or media networkings have to do with the emergence of new industries and a new regime of consumption that have most frequently been discussed by theorists of neo- or post-Fordism,[5] I'm equally interested in how they have become the point of the formation of culture, identity, allegiances, and alliances in the 1990s across and within previously existing geographic and media borders. In other words, I want to consider strategies for understanding (in part through a new "media geography") how cultural identity is currently formed through the production of social space in this current media environment and, particularly, how current media networkings have produced territoriality, alliances, and allegiances that aren't quite confined to cities, nations, or geographic regions but that must still be imaged or imagined through these places (and through these places as signs). And because the production of allegiances through these narrative networks occurs within and around *already* recognized geographies, it becomes just as necessary to recognize how these formations produce understandings about relations to the past (i.e., how, in many cases, their role in the formation of social identity occurs through discourses about rupture, the "new," tradition, and "competing" identities or formations).

Considering some of the recent transformations of television broadcasting in the United States for instance, one could consider not only how

networks such as Black Entertainment Television and the Christian Broadcasting (a.k.a. the Family) Network have become the historic points for
imaging and producing Black and Christian culture/identity through cable
networking of narrative practices and styles, but also how WGN, TBS, or
TNN have transformed their broadcasting bases (Chicago, Atlanta, Nashville) into symbolic centers for the formation of television audiences and
into *coordinates* for a changing media geography of the nation. Atlanta and
Chicago, in particular, are cities whose "renaissance" over the 1980s had as
much to do with city replanning and the erection of tourist centers as with
these networks' role in conventionalizing ways of reading ("accessing") these
cities. A "superstation" such as WGN places "local Chicago news" into national circulation, organizes its programming and advertising through an
iconography of Chicago (recoding, for example, the forever-syndicated *Andy
Griffith Show* through the iconography of contemporary Chicago for a national audience) and, expanding the circuit of Cub baseball, sustains a network of "fans" whose bond is ritualistically acknowledged through Harry
Carey's greeting of visiting fans from locations outside Chicago throughout
the broadcast of each game (again, for a national audience).

Examples such as these, not to mention the circulation of metaphoric,
narrativized, and endlessly translated cities such as *Dallas,* make it necessary
that we begin to rethink issues such as nationhood and the formation of
cultural identity with greater regard to the ways that the current media
technologies or practices have significantly contributed to redefining the spatial
features of our environment. Most recently, the Clinton-Gore administration has suggested the possibility of a federally subsidized "electronic superhighway"—a venture they have likened to the transformation of national
life resulting from the federal highway systems since the New Deal programs of the 1930s.[6] Whether as policy, metaphor, or utopian vision, such
a project holds significant implications both for the broadcasting, telecommunications, and computer industries and for the flow and reception of
various traditional and emerging narrative forms. It also is a vivid example
of how the immediate future of national cultural politics is being played out
on and over a changing grid of spatial relations.

There may well be something of interest for media and cultural analysis
in the body of studies described as "postmodern geography" (work by Edward Soja, David Harvey, and Manuel Castells is perhaps the most well
known), particularly in their attempt to move issues about the social pro-
duction/perception of space closer to issues tied to postmodernity and in
their recognition that one needs new strategies for understanding the politics of mapping in the contemporary cultural environment.[7] But while their
interventions into debates about the relation between postmodernism,
Marxism, and geography rightly criticize media and cultural analysis for having
ignored for too long the significance of media and culture as the site where
social conceptions of space are produced, these same studies have seemed
particularly ill-equipped to discuss the complexity of these media and cultural processes, preferring instead to work within the generalizations of

postmodernist theory, as David Harvey does when, echoing familiar eulogizing by Jean Baudrillard, he sums up television by proposing that "the whole world's cuisine is now assembled in one place in almost exactly the same way that the world's geographical complexity is nightly reduced to a series of images on a static television screen."[8]

Where issues concerning postmodernism, media practices, and the reformulation of spatial relations have all been brought to bear more directly on the question of national identity, the analyses tend to over-emphasize the political economy of media and postmodernity—such that both are understood as broadly determined by the emergence of "late" (or in Lasch and Urry's terms, "disorganized") capitalism.[9] These studies rightly acknowledge the necessity of rethinking the formation of national identity and the concept of *community* within "postmodernism" (and, in so doing, considering postmodernism "as a question of geography" or spatial relations). Morley and Robins, in particular, have called attention to the role of new media technologies in producing a new kind of relationship between space and place: "through their capacity to transgress frontiers and subvert territories, they [new media technologies] are implicated in a complex interplay of deterritorialization and reterritorialization."[10] But while their attempt to understand nationhood through the formation of "electronic communities," which they discuss primarily within a global–local nexus, acknowledges the importance of discussing media as sites for the production of culture and identity, their wedding of postmodernism and post-Fordism explains spatial relations in only the broadest sense of culture and tends to ignore the discursive and textual features of spatial relations, culture, and identity.

Without refuting the general argument of these treatments of national identity "beyond Fordism" (and by returning to issues about the discursive production of space and place raised in *Invisible Cities*), I therefore want to reconsider the strategic implications of introducing this problematic of spatial production and perception into media and cultural studies of nationhood by outlining how some of the issues might be engaged in a discussion about the relation between spatial production and changing media practices in Italy. And I particularly want to consider how one might begin to see media practices as sites for the production of these relations. Let me restate the problematic. First, how have changing media practices, technologies, and networks produced a sense of territoriality and perceptions of space in everyday life? Second, how, on the one hand, has this process of territorialization involved the production and perception of space that may not be represented within an established or institutional geography of places? Third, how, on the other hand, does this process occur through particular places that can only be imaged or imagined through an existing lexicon, symbolic order, and narrative logic for coding space, even though this symbolic order may be reorganized (the traffic of symbols redirected) as a result of these new media practices? Fourth, how is this process of territorialization, reterritorialization, and deterritorialization a form of cultural politics in which media practices play a changing role?

In order to recognize these processes in Italy, one would have to consider the complex and changing role of film and television (as industrial, narrative, ideological, and cultural systems) in the formation of spatial perceptions, particularly with respect to historically competing, emerging, and residual ways of modeling—of imaging and imagining—spatial relations.

To begin, one might consider how the emergence of film in Italy occurred within Italian Futurism's discourses on the "modern" or modernistic nature of twentieth century Italian environment and life. Over the first four decades of the twentieth century, Italian Futurists attempted increasingly to articulate their "project" through a discourse on the relation between art and the Nation (and, by the 1920s, Fascism), while their verse, prose, theatrical set design and performance, public spectacle in the form of street parades, choreography, graphic art, sculpture, architectural visions, interior designs, photography, and even films, all agressively reformulated the neo-classical and Enlightenment perceptions of space that still dominated "institutional" Italian culture. They frequently described all of these forms as informed by and producing a new "geometric splendor"—an expression referring as much to their recasting or erasure of spatial relations through art as to their repositioning art in relation to traditional culture and contemporary Italy.

Their prolific production of manifestoes did as much to elaborate the historic significance of their artistic practice as to establish a circuit of recognition about Futurist sensibility in modern Italian life. Their celebration of Speed attempted to deflate the notion that time and space were static points and to concentrate on trajectories; they celebrated a "new" relation between linguistic and spatial simultaneity, in which elements move without relation to material geography and environmental conditions. These issues and concepts converge for Italian Futurists in their metaphor of the "Futurist city" and in neologisms such as *velo-citta'* (a play upon the Italian words for "velocity" and "city"); both were hyper-technological tropes for conceptualizing Futurist production of art as image-spaces in a Modern and Urban Italy (Enrico Prampolini's "Space Rhythms" or Giacomo Balla's Speed-series), for imagining a new urban architecture through Futurist design (Mario Chiatone, Antonio Sant'Elia, and Virgilio Marchi's City-series), for highlighting and celebrating the "modern" spatial relations and perspectives of Italian cities (Carlo Carra''s treatment of the Galleria in Milan), and for producing new ways of perceiving temporal and spatial relations in Italian cities whose architectonics and organization were still bound to spatial modeling in the Renaissance, Middle Ages, and even Classical period (Giacomo Balla's early paintings that refigured Rome, the Ancient City, through Futurist "abstract dynamism").

During the 1920s and 1930s, the emergence of a film industry, policy, and culture in Italy (i.e., the emergence of film as a national-popular cultural form) became the point of a crisis over signifying the relation between the City and the Country, between established and emerging myths of a rural and urban Italy.[11] The myth of the Nation, and of the nation as a Popular

Domain, was contingent upon cinema's increasing role in mediating these conflicting ways of imaging and imagining spatial relations. This mediation occurred as much through the production of film *narratives* as through the inauguration of traveling cinema caravans, which brought movies to rural Italy, and through the construction of Cinecitta' (Cinema City), which not only attempted to remap the territoriality of Hollywood—of Hollywood films in Italy—but which became the most public emblem of a city built upon the realization of a national-popular culture and the symbolic center of a "new order" and a new Empire. Clearly all of this contributed in those years to the meaning of public and everyday discourses about the *domain* of Fascism.

More than cinema, however, radio broadcasting during the 1920s and 1930s transformed Italy into a national "network" where the Futurist vision of spatial simultaneity became wedded, in part through Fascism, to the notion that the Nation was a unified place, or rather, a space without separate places. The differences among cities, between cities and towns, between the north and the south, between public and domestic life (between listening to a broadcast in a piazza versus receiving it in the home) could be imagined by listeners as having disappeared, as each place became a point of simultaneous reception.

If post-war film production rejected its pre-war legacy, it was perhaps most evident in its reimaging and imagining of the nation as cinematic space, particularly through the practice of location filming which not only "reinserted" localities onto the cinematic map of the Nation which had been produced over the 1920s and 1930s. Here I am thinking about films ranging from Rosselini's *Paisa'*, which literally remaps an Italy emerging from Allied occupation, to DeSica's narratives set in the "backstreets" of Rome and Milan, to films set in specific localities outside these famous, cinematically mythologized cities, to a "woman's film" such as *Girls of the Spanish Steps* in the mid-1950s that constructs a mythic *path* of everyday life in postwar Rome for a group of for working women.

It is also significant that Italian television emerges within, though quickly reorganizes, "neo-realist" film practices that were conventionalizing ways of perceiving the nation as a matrix of spatial relations. While the emergence of RAI television broadcasting (the first channel in the early 1950s and the second channel in 1961) may have initially operated within the practices and national networking of a national radio system, the number of Italians who owned television sets grew so slowly that the circuit of reception was much different than radio. The small number of television sets in Italy also encouraged group and public viewings, the spirit of which can be most vividly recognized in the popular late 1950s and early 1960s television gameshow, *Campanile sera*. Visually and narratively, this series articulated images of regional identity with national identity, weekly staging nationally-televised competition between two teams from different regions and cross-cutting between studio participants and on-location scenes of the competing towns' inhabitants gathered in the town piazza. In a sense, these new television broadcasting practices enabled the kind of image of national cohesion that filmmaking, and particularly neo-realist styles, had avoided.

Over the 1960s and 1970s, the increasing convergence, competition, and conflict in Italy between film and television, as cultural forms of spatial modeling, contributed to several significant reformulations of urban space and of the relation of regional culture to national identity. Some of the more acclaimed Italian films from the late 1950s and early 1960s, such as Antonioni's *La notte* (1961) and *L'eclisse* (1962) or Fellini's *La dolce vita* (1959), through narratives that chart the convolution, absurdity, and repetition of characters' paths in Rome and Milan, begin to work against the notion that urban space is organized around a center (or around a town piazza, as in *Campanile sera*). And in the wake of films such as Passolini's *Accatone* (1961) and *Mama Roma* (1962), and through later films such as *Brutti, sporchi, cattivi* (1976), the city becomes an invisible force in narratives set entirely in an urban periphery.

These trends, particularly this latter one, are historically significant in that they offer a cultural framework for the emergence of the third RAI channel during the 1970s—a channel that was virtually established by and identified with the Italian Communist Party in an effort to promote regional broadcasting and culture that had become marginalized by the traditional two RAI networks. While the older RAI channels' penchant for on-location productions (e.g., *Sabato sera*—weekend variety shows broadcast live from different locations; *Linea verde*—weekend live features about local rituals in agricultural areas; soccer broadcasting) had transformed various Italian localities into stages for public spectacle, they had always maintained their linkage with *national* broadcasting based in Rome and, through RAI personalities, performers, and commentators, reproduced "national" myths about the culture of that region. Thus by the 1970s, RAI television's *nationalization* or *centralization* of regional culture and its implementation of the PAL standard (as a kind of television border that prevented the reception of "foreign" broadcasting) became the bases for "national culture," even though the RAI's role/image as arbiter of national culture precipitated a series of spatial relations between Rome and other localities and between Italy and other national cultures that came to be increasingly challenged through RAI 3 and the emerging "private" stations and networks.

Local broadcasting, which became more common by the mid-1970s after the legalization of private cable broadcasting and as a result of initiatives by the Italian Left to "decentralize" broadcasting and encourage local political reform, undercut somewhat the RAI's historic effort to construct and claim (through production values and programming) the political and cultural "center" of Italian life. While the concept of "local" broadcasting suggested regional, "non-national" discursive formations and audiences, it also became the form of broadcasting for discourses and audiences (neo-Fascists, the ultra-Left, youth, feminists, homosexuals) constructed as *marginal* under the traditional RAI practices. In a sense, the ambiguity surrounding the meaning of "local" programming (and the great variety of local programming practices) made it incongruous with a broadcasting formation such as the RAI's, modeled in terms of a center–periphery nexus. And it is no coincidence that the reformulation of the "local"–"national" distinction during the late 1970s made way for the formation of multiple national cable networks, which were not necessarily given to narrowcasting practices but were certainly prone to constructing themselves as the mediators of the taste cultures of geographically and demographically defined audiences and consumers.

In the late 1970s, the emerging private networks, formed through the proliferating number of "local" networks, did not resist their image as regional programmers in order to avoid being perceived as at odds with the RAI's image as mediator of a national culture. But by the 1980s, with the enormous success of the *Dallas*-ethos in Italy, Berlusconi's private networks resituated national television and reimaged a national audience within a television practice that was at once regional and cosmopolitan. The private networks' initial flooding of Italian television with foreign television productions and films so redirected the traffic of symbols and texts for a national audience and contributed to the image of the private networks as cosmopolitan that in 1986 the RAI (in an affair whose financial ramifications resulted in both parlimentary and court battles) spent a reported $20 million to finance a two-week long series of live, evening broadcasts from New York featuring a cast of American celebrities famous in Italy.

The U.S.-Italian axis becomes particularly significant for Italian television broadcasting between 1982 and 1987, not only because U.S. films and telefilms account for an average of 80 percent of programming on the fledgling private networks but also because narrativized cities such as *Dallas* (on the private networks) or Miami (i.e., -*Vice* on the RAI) became coordinates for a new media and cultural geography—the mythoi that structured America in Italian popular culture, but also televisual cities whose "stories" were organized around *invisibile* (and quasi-legitimate) international markets. (A timely metaphor in Italy for the story and status of television itself in the age of cable and privatization!) Particularly interesting in this regard were the press releases and newspaper reviews of the premiere episode of *Miami Vice* in 1986, which directed much attention to the impact of Italian design and fashion styles (e.g., the series' use of designs by Ferrari, Versaci, and Memphis) and to the issue of reading "Italian" style (and the meaning of "made in Italy") through a popular American television series. As TV text in Italy, Miami became a complex of narrative spaces or scenes linked largely through the power of style; but in this case, style had as much to do with the sense of an affective alliance among a group of consumers (as was more the case in the United States) as with the cultural identity of a nation in the age of cable broadcasting.

The Madonna concert in Torino, sponsored by Coke and RAI 1 and broadcast by RAI 1 to a number of other European countries in 1987, is one of the most provocative examples of TV's role in resituating the nation within a new cultural geography. It is not simply another example of the

RAI's traditional role as broadcaster of such national-popular spectacles as the annual San Remo music festival, but more an effort to refashion its image by making it at once both cosmopolitan (particularly as that image is signified here by tapping into the construction of youth in Italian television culture through music video and its stars) *and* Italian. Particularly significant in this latter respect were the attempts just preceding Madonna's performance to underscore Madonna's "Italian-ness"—that is, Madonna's roots in Italy, the legacy of Italy in Madonna, and the RAI's mediation of both—by visiting her relatives' town in the Abruzzi (relatives, I might add, whom she had never met before) and then staging a reunion between Madonna and these relatives.

The distinction, common throughout the 1980s, between public and private television (and between Rome and Milan as television *centers*) has already become more complex. By 1991 there were six kinds of television networks in Italy: the three RAI networks, the national private networks that transmit from inside Italy, national private networks that broadcast expressly into Italy from other countries (e.g., Capo d'Istria, a Berlusconi sports network originating in Yugoslavia), quasi-national networks of loosely connected regional stations (e.g., Odeon), strictly regional stations, and a pay-TV network received by roughly 80 percent of Italy (Tele piu' uno, which broadcasts movies without commercial interruption).[12] Each of these systems is oriented around particular programming practices and genre preferences and each attempts to maintain its own "circuit of reception" that becomes an important context for understanding the cultural geography of Italy in the 1990s.

In the age of satellite broadcasting and amidst the economic, political, and cultural transformations that have produced and will be produced by the EEC, networking and circuits of reception increasingly occur across the borders of the Italian state. In Italy, television customers can purchase satellite dishes to receive broadcasts from throughout Europe. In several regions, cultural alliances are produced and maintained across state borders through TV broadcasting: Lombardy receives Swiss programming, while Piedmont and Liguria receive French programming. Since the late 1980s, Berlusconi has begun broadcasting ventures in France, Spain, and Germany and is a co-financier of a broadcasting company in Belgium; he has also attempted to co-produce, with other European producers, a mini-series set in Italy and other European countries that could be broadcast in different European countries. His Canale 5 has recently broadcast a quiz show, *Bellezza al bagno,* which, in the tradition of pan-European quiz shows such as *Europa Europa* and *Jeux sans fronte'res,* stages competition among contestants from different nations (though, significantly, all are nations/stations from his TV empire). Berlusconi has long sought to gain access to a satellite which would not only enable him to enter the trans-national traffic of sports culture in Europe but to link his stations across Europe into the first pan-European private network. And recently the RAI has instituted its own "RAI-Sat," a state-satellite network (there is some debate as to whether it constitutes a fourth state channel) that programs material from other European state-broadcasting while it broadcasts Italian productions, often in different languages and with subtitles, to other European countries.

There is a significant difference between the Nation realized simultaneously in various places (as was the case before the 1970s) and the Nation produced through multiple networks and amidst more localized and transnational programming and narrowcasting, each attempting to define culture and everyday life. Certainly it is tempting to claim, as Eco does, that "neotelevision" in Italy operates through a discursivity that is more postmodernist than modernist. But frequently such theorizing has more to say about how these television practices all produce a kind of grand cultural logic rather than how they become the terrain for a cultural politics—for the resistances and competition—through which identities, alliances, and allegiances are produced. Likewise, the problem with many current attempts to theorize media's role in the production of cultural identity "beyond Fordism" and in the age of a "global economy" is that they end up equivocating postmodernism (as a cultural logic) with post-Fordism (as the logic of late capitalism). Indeed it is very important to recognize that identities are formed through media and cultural practices that are now as much local as they are global. But this local–global model does not get at the complexity and transitoriness of media formations and networks or the cultural geographies produced through them; the examples in Italy that I cite above are not unequivocally local or trans-national. Theories of post-Fordist media and culture fail, furthermore, to acknowledge how the very concepts of "local" and "global" are always *constructed* through media and other cultural discourses and through emerging and residual ways of imagining spatial relations. And above all, attempts to theorize postmodernism and post-Fordism as the cultural and economic terrain for the formation of identities frequently closes off the issue of nationhood by over-emphasizing merely the formation of culture across borders and capitalism's formation of new global markets. While the broader transformations explained by theories of postmodernism and post-Fordism may have a great deal to do with understanding the media's role in the formation of cultural identity and geography, the issue of national identity and the struggle over marking or defining national borders are more crucial than ever. (In this regard, the Italian examples seem just as pertinent as the Clinton-Gore proposal mentioned above.) What seems necessary for current media and cultural studies, in this regard, is a critical geography that is just as much concerned with the highly localized and global production and perception of spatial relations as it is with how particular nations can be understood as political states, as economic and cultural spaces for multiple media formations, as signs that circulate along competing networks and through specific narrative practices, and as audiences whose perceptions of the nation, their locality, and the world are bound up with their implication in changing media practices. And this critical geography needs to come to terms with the media's role in mapping the distinctions and interconnectedness of places as much as it needs to develop strategies for recognizing its "invisible cities"—for recognizing, in short, that because media studies involves geography, it has a stake in re-mapping and reimagining alliances.

As the Calvino text reminds us, empires may be conceived through spa-tial models, but models are themselves assemblages of interchangeable co-ordinates. The activity of charting media and cultural formations, therefore, not only involves recognizing how and where model-building occurs but how this charting itself produces "logogriphs" for future empire-building, future cartography and future explorers of new passages:

> Kublai asked Marco: "You, who go about exploring and who see signs, can tell me toward which of these futures the favoring winds are driving us."
>
> [Marco responds:] "For these ports I could not draw a route on the map or set a date for the landing. At times all I need is a brief glimpse, an opening in the midst of an incongruous landscape, a glint of lights in the fog, the dialogue of two passersby meeting in the crowd, and I think that, setting out from there, I will put together, piece by piece, the perfect city, made of fragments mixed with the rest, of instants separated by intervals, of signals one sends out, not knowing who receives them. If I tell you that the city toward which my journey tends is discontinuous in space and time, now scattered, now more condensed, you must not believe the search for it can stop. Perhaps while we speak, it is rising, scattered, within the con-fines of your empire; you can hunt for it, *but only in the way I have said*" [italics mine].

Calvino never overtly elaborates the consequences of this historical con-vergence of two discursive positions and systems, though the text may be read as an outline for a cartography of "new times." And certainly the rela-tion between Kublai Kahn and Marco Polo has nothing and everything to do with the production of spatial relations in modern and postmodern Ital-ian culture. Therefore, just as *Invisible Cities* offers a way of thinking about the cultural geography of Italian television, this outline of the changing relation of nationhood and culture in Italy may offer a way of thinking about the consequences of the convergence that Calvino narrates.

### Notes

I want to express my appreciation to Franco Minganti and Waddick Doyle for their assistance in my research of Italian television.

1. See Gilles Deleuze and Felix Guattari, *On the Line*, trans. John Johnston (New York: Semiotext(e), 1983).

2. See Michel de Certeau, "Spatial Practices," *The Practice of Everyday Life*, trans. Steven Rendall (Los Angeles: University of California Press, 1984).

3. See Henri Lefebvre, *The Production of Space*, trans. Donald Nicholson-Smith (Oxford, Cambridge, Mass.: Blackwell, 1990). One could also consider Pierre Bour-dieu's conceptualization of "social space" as a "habitus"—a structured and structur-ing system for representing "place" or position. See particularly Bourdieu, "The

Economy of Practices," *Distinction: A Social Critique of the Judgement of Taste,* trans. Richard Nice (Cambridge, Mass.: Harvard University Press, 1984).

4. In the original version of this essay, I elaborate reasons why recent geo-political and geo-economic transformations, coupled with changing global, national, and local media practices that have resituated the place of film and television in everyday life, all have made it increasingly difficult to understand the complexity of spatial production surrounding media formations either through traditional Marxist and structuralist theories of film and television or through traditional political economies of mass media. I am nonetheless indebted to the recent work of two individuals: Eric Michaels's treatment of "aboriginality," Australia, and the production of culture through contemporary media and Meaghan Morris's provocative examples of how issues of spatial production can be brought into a discussion of nationhood, media, and postmodern culture. See, for instance, Eric Michaels's *For a Cultural Future* (Melbourne: Artspace, 1987) and Meaghan Morris's "At Henry Parkes Motel," *Cultural Studies* 2:1 (January 1988), pp. 1–47; "Tooth and Claw: Tales of Survival, and *Crocodile Dundee,*" *The Pirate's Fiancee: Feminism, Reading, and Post-modernism* (London, New York: Verso, 1988); "*Panorama:* The Live, the Dead and the Living," *Island in the Stream,* ed. Paul Foss (Leichhardt, New South Wales: Pluto Press, 1988).

5. Stuart Hall and Martin Jacques, eds., *New Times: The Changing Face of Politics in the 1990s* (London and New York: Verso, 1989).

6. See John Markoff, "Building the Electronic Superhighway," *New York Times,* January 24, 1993.

7. See Edward Soja, *Postmodern Geographies: The Reassertion of Space in Critical Social Theory* (London, New York: Verso, 1989); David Harvey, *The Condition of Postmodernity* (Oxford, Cambridge, Mass.: Basil Blackwell, 1989); Manuel Castells, *The Informational City* (Oxford, Cambridge, Mass.: Basil Blackwell, 1989).

8. Harvey, *Postmodernity,* p. 300. After devoting most of his book to Renaissance, Enlightenment, and Modernist conceptions of space and geography, Harvey's discussion of spatiality and postmodernism, in one of his last short chapters, is un-fortunately only one more reading of *Blade Runner*—a reading that focuses largely on the film's recombinant features, has relatively little to do with spatial relations, adds little to the numerous references to this film as a quintessential example of a postmodernist aesthetic, and generally smoothes out the film's ironies and contradictions.

9. Kevin Robins, "Reimagined Communities? European Image Spaces, Beyond Fordism," *Cultural Studies* 3 (May 1989), pp. 145–61; Kevin Robins and David Morley, "Spaces of Identity: Communications Technologies and the Reconfiguration of Europe," *Screen* 30 (Autumn 1989), pp. 10–34.

10. Robins and Morley, "Reimagined Communities?" p. 22.

11. See James Hay, *Popular Film Culture in Fascist Italy: The Passing of the Rex* (Bloomington: Indiana University Press, 1987).

12. There are currently plans for two additional pay-channels, Tele piu' 2 and 3—one entirely for sports programming and the other for "educational" programming.

# For a Cultural Future

## ERIC MICHAELS

On 1 April 1985, daily television transmissions began from the studios of the Warlpiri Media Association at the Yuendumu community on the edge of Central Australia's Tanami desert. Television signals, when broadcast as radio waves, assure a kind of mute immortality: they radiate endlessly beyond their site of creation, so this first program might be playing right now to the rings of Saturn. But it no longer exists at its point of origin in Australia. The message, the events behind it, their circumstances and meanings have mostly been ignored, and are likely to be forgotten. This is why I recall such events here: to reassert their significance and to establish in print their remarkable history.

All content the Warlpiri Media Association transmits is locally produced. Almost all of it is in the Warlpiri Aboriginal language. Some is live: schoolchildren reading their assignments, community announcements, old men telling stories, young blokes acting cheeky. The station also draws on a videotape library of several hundred hours of material that had been produced in the community since 1982 (a description of this material and the conditions of its production is one of the main purposes of this essay). Yuendumu's four-hour schedule was, by percentage, and perhaps absolute hours, in excess of the Australian content of any other Australian television station. The transmissions were unauthorized, unfunded, uncommercial, and

From *For a Cultural Future: Francis Jupurrla Makes TV at Yuendumu* by Eric Michaels, Melbourne: Artspace, 1987 (Art and Criticism Monograph Series, Vol. 3).

illegal. There were no provisions within the Australian Broadcasting Laws for this kind of service.

Yuendumu television was probably Australia's first public television service, although it might be misleading to make too much of this. Open Channel sponsored experimental community access TV transmissions in Melbourne for a few days during 1982. The Ernabella Aboriginal community had been experimenting along similar lines since 1984, and began their own daily service days after Yuendumu started theirs. Even before, throughout remote Australia, small transmitters had been pirated so that mining camps and cattle stations could watch Kung Fu and Action Adventure videos instead of the approved and licenced Australian Broadcasting Corporation (ABC) "high culture" satellite feed, which became available to outback communities in 1984–85. Issues of community access and local transmission had been on the agenda certainly since the Whitlam government, but they had lain dormant in Fraser's and then Hawke's official broadcasting agendas. In the 1980s, inexpensive home video systems proved subversive of the bureaucracy's tedious intents, while satellite penetration and sustained interest in production by independents had all kept issues of public television very much alive in the public arena.

Yuendumu's accomplishment must be seen in the context of these developments. Yuendumu's need and motivation to broadcast should also be considered in evaluating any claim to the accolade of pioneer. There was, in the early 1980s, a considerable creative interest among Aborigines in the new entertainment technology becoming available to remote communities. There was equally a motivated, articulate, and general concern about the possible unwanted consequences of television, especially among senior Aborigines and local indigenous educators. In particular, the absence of local Aboriginal languages from any proposed service was a major issue. Without traditional language, how could any media service be anything but culturally subversive? Native speakers of indigenous languages understood (a good deal better than anyone else) that this was something only they could correct.

There are more than twenty-two Aboriginal languages currently spoken in the Central Australian satellite footprint. The simple logistics of providing for all these languages on a single service indicate clearly a fundamental mismatch. The bias of mass broadcasting is concentration and unification; the bias of Aboriginal culture is diversity and autonomy. Electronic media are everywhere; Aboriginal culture is local and land-based. Only local communities can express and maintain linguistic autonomy. No one elsewhere can do this for the local community—not in Canberra, Sydney, or even Alice Springs. Indeed, Warlpiri speakers at Yuendumu make much of the distinctions between their dialect and the one spoken by the Lajamanu Warlpiri, 600 kilometres away on the other side of the Tanami desert. These differences are proper, for they articulate a characteristic cultural diversity. The problem of language signals a more general problem of social diversity that introduced media pose for indigenous peoples everywhere: how to respond to the insistent pressure towards standardization, the homogenizing tendencies of contemporary world culture?

Postmodernist critique provides very little guidance here. Indeed, the temptation to promote Warlpiri media by demonstrating a privileged authenticity—the appeal to traditionalism—which legitimates these forms, would be firmly resisted. How can we even employ such terms? "Authentic" and "inauthentic" are now merely labels assigned and reassigned as manipulable moves in a recombinatory game. Postmodernism may promote an appetite for primitive provenances, but it has proven to be an ultra-consumerist appetite, using up the object to the point of exhaustion, of "sophistication," so as to risk making it disappear entirely. It would be better to shift ontologies, by problematizing the very term "originality" and denying that any appeals to this category can be legitimated. Then, we refuse degrees of difference or value that might distinguish between the expressive acts (or even the persons) of Warlpiri videomakers, urban Koorie artists in Fitzroy, Aboriginal arts bureaucrats in Canberra, a Black commercial media industry in Alice Springs—none of these could be called more truly "Aboriginal" than any other. This redirects the mode of analysis to a different if somewhat more fashionable inquiry: who asks such a question, under what circumstances, and so forth? An analysis of the history of the official constitution of Aboriginality, as well as its *mise en discours,* its rhetorical and institutional deployment, might explain, for instance, current moves towards pan-aboriginalism. At least, it can provide a much needed caveat to the banal and profoundly racist war that is being waged in this country.

Despite the importance of subjecting Aboriginal expression to these critical debates, there is a danger that they lead away from the specific pleasures to be found in an encounter with the Warlpiri and their video. And it is these pleasures I want to describe, indeed promote as something more than the product of my own research interests—even if this risks reasserting authenticity, with an almost naive empiricist's faith, and employing a positivist's notebook to amass the particulars necessary to bring these "alien" texts into focus. What I am seeking is also a way to test critical theory's application to ethnographic subjects. It is possible that such an investigation will supersede the dilemma that tradition, ethnicity, and value have variously posed for empiricism.

Of course, to understand what happened with media at Yuendumu—and what didn't—one needs to describe more than just local circumstances. It will be necessary to reference the State and its interventions again and again. Warlpiri media, no less than its history, is the product of a struggle between official and unofficial discourses that seem always stacked in the State's favor. This might suggest a discouraging future for Yuendumu Television. Given the government's present policy of promoting media centralization and homogenization, we would expect that Yuendumu will soon be overwhelmed by national media services, including "approved" regional Aboriginal broadcasters who serve the State's objectives of ethnicization, standardization, even aboriginalization, at the expense of local language, representation, autonomy. If this scenario is realized, then Yuendumu's community station seems likely to join the detritus of other development projects which litter the contemporary Aboriginal landscape, and shocked Europeans will take this as one more example of Aboriginal intractability and failure of effort—if not genes. We won't know that the experience of television for remote Aborigines could have been any different: for example, a networked cooperative of autonomous community stations resisting hegemony and homogenization. Instead, we expect Warlpiri television to disappear as no more than a footnote to Australian media history, leaving unremarked its singular contribution to a public media, and its capacity to articulate alternative—unofficial—aboriginalities.

But something in Warlpiri reckoning confounds their institutionalisation and the grim prophecy this conveys. A similar logic predicted the disappearance of their people and culture generations ago, but proved false. A miraculous autonomy, almost fierce stubbornness, delivers the Warlpiri from these overwhelming odds and assures their survival, if not eventual victory.

### How Warlpiri People Make Television

Videomaker Francis Kelly does not like to be called by name. He prefers I call him Jupurrurla. This is a "skin" or subsection term which identifies him as a member of one of eight divisions of the Warlpiri people, sometimes called totemic groups by anthropologists. Better yet, I should call him *panji*, a Kriol word for brother-in-law. The term is a relative one, in both senses. It does not merely classify our identities, but describes our relatedness. It is through such identities and relationships that cultural expression arises for Warlpiri people, and the description of these must precede any discussion of Aboriginal creativity.

Warlpiri people are born to their skins, a matter determined by parentage, a result of marriage and birth. But I became a Japanangka—brother-in-law to Jupurrurla—by assignment of the Yuendumu Community Council. One result of my classification is that it enables Francis and me to use this term to position our relationship in respect to the broader Warlpiri community. It establishes that we are not merely two independent individuals, free to recreate ourselves and our obligations at each social occasion. Rather, we are persons whose individuality is created and defined by a preexisting order. In this case, being brothers-in-law means that we should maintain an amiable, cooperative reciprocity. We can joke, but only in certain ways and not others. We should give things to each other. My "sister" will become wives for Jupurrurla; I might take his "sisters" for wives in return.

These distinctions refer to a symbolic divisioning of the community into "two sides" engaged in reciprocal obligations. But it implies also a division of expressive (e.g., ceremonial) labor, and particular relations of ritual production reaching into all Warlpiri social life and action. For certain ceremonies this division is articulated as roles the Warlpiri name *Kirda* and

---

## WARLPIRI SUBSECTIONS

Understanding how Warlpiri people articulate their relationships, through marriage and descent is critical to explaining videomaking as a mode of cultural inscription and reproduction. A basic diagram of Warlpiri "skin names" is required to follow the argument.[1] In a simplistic (patrilineal) sense, the eight terms represent divisions of four father/son groups:

|  ONE SIDE | OTHER SIDE |
| --- | --- |
| *Japangardi* | *Jupurrurla* |
| *Japanangka* | *Jangala* |
| *Japaljarri* | *Jampijimpa* |
| *Jungarrayi* | *Jakamarra* |

(In the case of women, "N" is substituted for "J" in the initial position)

As these names designate marriage choices and imply descent lines, one can treat this diagram as a abstracted matrix from which an entire universe of social relations can be extrapolated. Both Aborigines and anthropologists do this, but the connection between these ideological models and social practice has been questioned in recent literature. The present treatment intends to remain mostly in the domain of social practice, and so avoids any explicit position on these more scholastic questions.

---

*Kurdungurlu,* two classes which share responsibility for ritual display: one to perform, the other to stage-manage and witness. The roles are situational and invertable, so identification as Kirda "Boss" and Kurdungurlu "Helper" may alter from event to event. As "brothers-in-law," Jupurrula and I will always find ourselves on different sides of this opposition, although the roles themselves may reverse from setting to setting. This see-saw balancing act that shifts roles back and forth over time affected all our collaborations.

Warlpiri brothers-in-law can also trace their relationships and their obligations more circuitously through mothers and grandparents. This describes a quite complex round of kin, which eventually encompasses the entire Warlpiri "nation" of over 5,000 people. It can even go beyond this, identifying marriages and ceremonial ties which relate to corresponding skin groups among Pitjanjatjara people to the south, Pintubi to the west, and Arrente, Anmatjarra, Kaitij, Warrumungu, and others to the north and east. Thus, a potentially vast social matrix is interpolated with every greeting.

I introduce the reader to Jupurrurla to promote a consideration of his art: videotaped works of Warlpiri life transmitted at the Yuendumu television station in this desert community 300 kilometres north-west of Alice Springs. But it is not quite correct to identify Jupurrurla as the author of these tapes, to assign him personal responsibility for beginning video production at Yuendumu, or for founding the Warlpiri Media Association, although these are the functions he symbolizes for us here.

In fact, the first videomaker at the Yuendumu community was Jupur-rurla's actual brother-in-law, a Japanangka. It was Japanangka who was already videotaping local sporting events when I first came to Central Australia early in 1983. It was Japanangka who responded to the prospect of satellite television by saying "We can fight fire with fire . . .", making a reference to the traditional Warlpiri ritual, the Fire Ceremony or *Warlukur-langu,* and assigning this name to their artist's association. But Japanangka, for all his talent and rage, found the mantle of "boss" for the video too onerous—for to be a boss is to be obligated. During 1983–84, a sensitive series of negotiations transferred the authority for Yuendumu video to Ju-purrurla. Part of Jupurrurla's success in this role resulted from his cleverness in distributing the resources associated with the video project. He trained and then shared the work with a Japaljarri, another Japanangka, a Japan-gardi, and a Jangala. Eventually, all the male subsections had video access through at least one of their members.

Obviously, the identification of an individual artist as the subject for critical attention is problematized where personhood is reckoned in this fashion. Rather than gloss over this issue, or treat it romantically within a fantasy of primitive collectivity, I want to assert more precisely its centrality for any discussion of Warlpiri expression. And I want to use this example to signal other differences between Aboriginal and European creative practices, differences which need to be admitted and understood if the distinctiveness and contribution of Warlpiri creativity is to be evaluated critically in a contemporary climate—the goal of this essay. These differences include what may be unfamilar to readers:

- ideological sources and access to inspiration;
- cultural constraints on invention and imagination;
- epistemological bases for representation and actuality;
- indistinctiveness of boundaries between authorship and oeuvre;
- restrictions on who makes or views expressive acts.

By describing these, I hope to avert some likely consequences of European enthusiasm for Aboriginal media which results in the appropriation of such forms to construct a generic "primitive" only to illustrate modern (and, more recently, postmodern) fantasies of evolutionary sequences. In a practical sense, I also want to subvert the bureaucratization of these forms, such as may be expressed in the training programs, funding guidelines, or development projects which claim to advance Aborigines, but always impose standards alien to the art (because these will be alien to the culture producing it). Wherever Australian officialdom appropriates a population, as it has attempted to do with the Aborigines, it quickly bureaucratizes such relationships in the name of social welfare. This assuredly defeats the emergence of these sovereign forms of expression, as it would defeat Jupurrurla's own avowed objective to create *Yapa*—that is, truly Warlpiri—media.

Those European Art practices since the Renaissance and Industrial Revolution which constitute the artist as an independent inventor/producer of original products for the consumption/use of public audiences, do not apply

to the Aboriginal tradition. This has been said before, in more or less accurate ways, and led to some questions (misplaced, I think) about the "authenticity" of "traditional" Aboriginal designs, such as those painted in acrylics
and now made available to the international art maker.[2] In the case of Jupurrurla's art, the implicit question of authenticity becomes explicit: Jupurrurla, in Bob Marley T-shirt and Adidas runners, armed with his video port-
apak, resists identification as a savage updating some archaic technology to
produce curiosities of primitive tradition for the jaded modern gaze. Jupurrurla is indisputably a sophisticated cultural broker who employs videotape
and electronic technology to express and resolve political, theological, and
aesthetic contradictions that arise in uniquely contemporary circumstances.
This will be demonstrated in the case of two videotapes which the Warlpiri
Media Association has produced: *Coniston Story* and *Warlukurlangu (Fire
Ceremony)*.

The choice of these two out of a corpus of hundreds of hours of tape
was a difficult one. Not even included is Jupurrurla's most cherished production, *Trip to Lapi-Lapi*, the record of a long trip into the Western Desert
to re-open country people had not seen for decades, a tape that often causes
its audiences to weep openly. Neither are the politically explicit tapes, those
documenting confrontations with government officials which became important elements in Warlpiri negotiating strategies: these tapes cause people
to shout, to address the screen, and then each other, sometimes provoking
direct action. But my purpose in this monograph is not to survey the whole
of the work, the various emergent genres, or the remarkable bush networks
which arose to carry these tapes when official channels such as the new
satellite were closed.[3] My choice of tapes was made because they illustrate
best some things about the Warlpiri mode of video production: how Jupurrurla and others discovered ways to fit the new technology to their particular information-based culture. The only way of beginning such an analysis is by locating the sources of Warlpiri expression in an oral tradition.

[At this point in his more extended essay Michaels offers brief discussions of "The Law" and "Restricted Expressions." For the Walpiri people
of Yuendumu video had to be inserted into these traditional categories,
understood as part of them. The definitions of these categories and their
application are, I believe, clear from the contexts of Michaels's analysis of
the Fire Ceremony.

He also preceeds his analysis of the Fire Ceremony events with a discussion of another complex tape and screening under the subheading "*Coniston Story:* Warlpiri Modes of Video Production." Inclusion of the full essay
was impossible for this volume, but the reader is urged to examine the full
text of Michaels's work.]

## The Fire Ceremony: For a Cultural Future

In 1972, anthropologist Nicolas Peterson and filmmaker Roger Sandall arranged with the old men to film a ritual of signal importance for the Warlpiri: Warlukurlangu, the Fire Ceremony. In a subsequent journal article,[4]

Peterson described these ceremonies in terms of the functions of Aboriginal social organization. He identified such Warlpiri ceremonies as a means of resolving conflict, or of negotiating disputes, a kind of pressure valve for the community as a whole. One pair of patrilines (or "side") of the community acts as Kurdungurlu, and arranges a spectacular dancing ground, delineated by great columns of brush and featuring highly decorated poles. The Kirda side paints up, and dances. Following several days and nights of dancing, they don elaborate costumes festooned with dry brush. At night they dance towards the fire, and are then beaten about with burning torches by the Kurdungurlu. Finally, the huge towers of brush are themselves ignited and the entire dance ground seems engulfed in flame. Following some period (it may be months or even years) the ceremony is repeated, but the personnel reverse their roles. The Kurdungurlu become Kirda, and receive their punishment in turn.

Visually and thematically, this ceremony satisfies the most extreme European appetite for savage theatre, a morality play of the sort Artaud describes for Balinese ritual dance—what could be more literally signaling through the flames than this? Yet I do not think the Peterson/Sandall film does this, partly due to the technical limitations of lighting for their black-and-white film stock, and partly because of the observational distance maintained throughout the filming. The effect is less dramatic, more properly "ethnographic" (and, perhaps wisely, politically less confrontative). It was approved by the community at the time it was edited in 1972 by Kim McKenzie, and joined other such films in the somewhat obscure archives of the Australian Institute of Aboriginal Studies, used mostly for research and occasional classroom illustration.

Remarkably, the ceremony lapsed shortly after this film was made. When I arrived at Yuendumu in 1983, the Fire Ceremony seemed little more than a memory. Various reasons were offered:

- one of the owners had died, and a prohibition applied to its performance;
- it had been traded with another community;
- the church had suppressed its performance.

These are not competing explanations, but may have in combination discouraged Warlukurlangu. The interdiction by the church (and the state, in some versions) was difficult to substantiate, though it was widely believed. Some of the more dramatic forms of punishment employed in the ceremony contradict Western manners, if not morals. There seemed to be some recognition among the Warlpiri that the Fire Ceremony was essentially incompatible with the expectations of settlement life, and the impotent fantasies of dependency and development they were required to promote. The Fire Ceremony was an explicit expression of Warlpiri autonomy, and for nearly a generation it was obscured. The question arises, as it does also in accounting for the ceremony's recent revival: what role did introduced media play in this history?

Yet Warlukurlangu persisted in certain covert ways. The very first videotape which the community itself directed in 1983 recorded an apparently casual afternoon of traditional dancing held at the women's museum. Such spontaneous public dance events are comparatively rare at Yuendumu. Dances occur in formal ceremonies, or during visits, in modern competitions and recitals, or in rehearsal for any of these. Yet this event appeared to meet none of these criteria. Equally curious was the insistence on the presence of the video camera. These were early days—Jupurrurla had not yet taken up the camera, and Japanangka and myself were having trouble arranging the shoot. A delegation of old men showed up at each of our camps and announced that we must hurry; the dancing wouldn't start till the video got there. What was taped was not only some quite spectacular dancing, but an emotional experience involving the whole community. When I afterwards asked some of the younger men the reason for all the weeping, they explained that people were so happy to see this dance again. I later discovered I had seen excerpts of the dances associated with the Fire Ceremony.

Some months later, I was invited to a meeting of the old men in the video studio. They had written to Peterson, asking for a copy of the film, and now were there to review it. I set up a camera and we videotaped this session. As it was clear that many of the on-film participants would now be dead, how the community negotiated this fact in terms of their review was very important. The question of the film's possible circulation was raised. Following a spirited discussion, the old men (as mentioned above) came to the decision that all the people who died were "in the background": the film could be shown in the camps. Outside, a group of women elders had assembled, and were occasionally peeking through the window. Some were crying. They did not agree that the deceased were sufficiently backgrounded, and it made them "too sorry to look." These women did not watch the film, but didn't dispute the right of the men to view or show it.

It became clear that the community was gearing up to perform the Fire Ceremony again for the first time in this generation. As preparations proceeded, video influenced the ritual in many ways. For example, the senior men announced that the Peterson film was "number one Law," and recommended that we shoot the videotape of exactly the same scenes in precisely the same order. (When this did not happen, no one in fact remarked on the difference.) Andrew Japaljarri Spencer, who acted as first cameraman, stood in Kurdungurlu relationship to the ceremony. This meant that he produced an intimate record of the ceremony from his "on the side" perspective.

We are at close-up range for some of the most dramatic moments, alongside the men actually administering the fiery punishments. Jupurrurla absented himself from this production. Although he was willing to do certain preproduction work, and subsequent editing and technical services, he would not act as cameraman because he would be a Kirda for this event. Quite sensibly, he pointed out that if Kirda were cameramen, the camera might catch on fire. Jupurrurla was not unaware, like many of the younger men, that he too might catch on fire, so at the climax of the ceremony they were nowhere to be found.

The tape of this major ceremony was copied the very next day and presented to a delegation from the nearby Willowra community, who were in fact in the midst of learning and acquiring the ceremony for performances themselves. This is a traditional aspect of certain classes of ceremonies. In oral societies where information is more valued than material resources, ceremonies can be commodities in which ritual information is a medium of exchange. This exchange may take years, and repeated performances, to accomplish. For instance, there was a dramatic (if not unexpected) moment when a more careful review of the tape revealed that one of the painted ceremonial poles had been rather too slowly panned, rendering its sacred design too explicit. This design had not yet been exchanged, and so the Willowra people might learn it—and reproduce it—from the tape. Runners went out to intercept the Willowra mob, and to replace their copy with one that had the offending section blanked out.

These new tapes of the Fire Ceremony circulated around the Yuendumu community, and in their raw state were highly popular. In fact, it became difficult to keep track of the copies. This was one of the motivations to proceed with broadcasting—more to assure the security of the video originals and provide adequate local circulation of tapes than to achieve any explicitly political intent. But perhaps there was a broader public statement to be made with the record of these events. I recommended, and was authorized to propose to the Australian Institute of Aboriginal Studies, that we edit together the tapes to produce an account which would describe both the ceremony and its reproductions. We had the Peterson film, the community dance, the review of the film, and the extraordinary footage of the 1984 performance. This seemed to me an excellent and visually striking way to articulate the ceremony in terms of some of the more fundamental questions concerning the place of such media in Warlpiri life. The Intitute did not support the idea, and when one of the central performers died shortly thereafter, the community dropped the matter. The tapes took their place on a shelf in the archive that Jupurrurla labeled "not to look." Later, however, the Institute did transfer the Sandall/Peterson film to videotape, and put it into general distribution without, to my knowledge, informing the community that this was being done.

There is no point in isolating any one instance of this failure to address or resolve the problems that the appropriation of Warlpiri images poses. The situation is so general that it proves how fundamental the misunderstandings must be. Alien producers do not know what they take away from the Aborigines whose images, designs, dances, songs, and stories they record. Aborigines are learning to be more careful in these matters. But the conventions of copyright are profoundly different from one context to the other. Perhaps these urgent questions will never be solved: "Who owns that dance now on film?," "Who has the authority to prevent broadcast of that picture of my father who just died?," "How can we make sure women will not see these places we showed to the male film crew?," "Will we see any of the money these people made with our pictures?" . . . Whenever "appropriate" Australian authorities are confronted with such questions, they go straight to the too-hard basket, not only because they are truly difficult questions, but also because they refer to equivocal political positions.

Underlying the problem is not only a failure to specify the processes of reproduction and their place in oral traditions; there is also a contradiction of values regarding the possibilities for Aboriginal futures, and the preferred paths towards these. Many Aborigines do wish to be identified, recognized, and acknowledged in modern media, as well as to become practitioners of their own. They recognize the prestige, the political value, the economic bargaining position that a well-placed story in the national press can provide. They attempt to evaluate the advantages—and what they are told is the necessity—of compromising certain cultural forms to achieve this. But the elements of this exchange, the discrimination between what is fundamental and what is negotiable, resists schematization. On neither side is there a clear sense of what can be given up and what must be kept if Aborigines are to avoid being reprocessed in the great sausage machine of modern mass media. For them, it is the *practices* of cultural reproduction that are essential. If by the next generation the means of representing and reproducing cultural forms are appropriated and lost, then all is destroyed. What remains will just be a few children's stories, place names for use by tourist or housing developments, some boomerangs that don't come back, a Hollywood-manufactured myth of exotica. These will only serve to mask the economic and social oppression of a people who then come into existence primarily in relation to that oppression.

The criteria for Aboriginal media must concern these consequences of recording for cultural reproduction in traditional oral societies. Warlpiri people put it more simply: "Can video make our culture strong? Or will it make us lose our Law?"

The problem about answering this sort of question as straightforwardly as it deserves, is that it usually is asked in deceptive cause/effect terms: What will TV do to Aborigines? The Warlpiri experience resists this formulation. Jupurrurla demonstrates that such questions cannot be answered outside the specific kin-based experiences of their local communities. His productions further demonstrate that television and video are not any one, self-evident thing, a singular cause which can then predict effects. Indeed, Yuendumu's videomakers demonstrate that their television is something wholly unanticipated, and unexplained, by dominant and familiar industrial forms.

Here I want to emphasise *the continuity of modes of cultural production across media,* something that might be too easily over-looked by an ethnocentric focus on content. My researches identify how Jupurrurla and other Warlpiri videomakers have learned ways of using the medium which conform to the basic premises of their tradition in its essential oral form. They demonstrate that this is possible, but also that their efforts are yet vulnerable, easily jeopardized by the invasion of alien and professional media producers.

My work has been subject to criticism for this attention to traditional forms and for encouraging their persistence into modern life. The argument is not meant to be romantic: my intent has been to specify the place of the Law in any struggle by indigenous people for cultural and political autonomy. In the case of Warlpiri television, the mechanisms for achieving this were discovered to lie wholly in the domain of cultural reproduction, in the culture's ability to construct itself, to image itself, through its own eyes as well as the world's.

In the confrontation between Dreamtime and Ourtime, what future is possible? The very terms of such an inquiry have histories that tend to delimit any assured, autonomous future. For example, if it were true that my analysis of Warlpiri TV provided no more than a protectionist agenda, then the charge of romantic indulgence in an idealized past might be justified. I would have failed to escape a "time" that anthropologists call the "ethnographic present"—a fabricated, synchronic moment that, like the Dreamtime, exists in ideological space, not material history. It is implicated in nearly all anthropology, as well as most ethnographic discourse. Certainly, the questions of time that seem essential here cannot be elucidated by constructs of timelessness.

It seems likely that grounding Aborigines in such false, atemporal histories results in projecting them instead into a particular named future whose characteristics are implied by that remarkable word, "Lifestyle." This term now substitutes everywhere for the term culture to indicate the latter's demise in a period of ultra-merchandise. Culture—a learned, inherited tradition—is superseded by a borrowed, or gratuitous model; what your parents and grandparents taught you didn't offer much choice about membership. Lifestyles are, by contrast, assemblages of commodified symbols, operating in concert as packages which can be bought, sold, traded, or lost. The word proves unnervingly durable, serving to describe housing, automobiles, restaurants, clothes, things you wear, things that wear you—most strikingly, both "lifestyle condoms" for men and, for women, sanitary napkins that "fit your lifestyle." Warlpiri people, when projected into this Lifestyle Future, cease to be Warlpiri; they are subsumed as "Aborigines," in an effort to invent them as a sort of special ethnic group able to be inserted into the fragile fantasies of contemporary Australian multiculturalism. Is there no other future for the Warlpiri than as merely another collectivity who have bartered away their history for a "lifestyle?"

I propose an alternative here, and name it the Cultural Future. By this I mean an agenda for cultural maintenance which not only assumes some privileged authority for traditional modes of cultural production, but argues also that the political survival of indigenous people is dependent upon their capacity to continue reproducing these forms.

What I read as the lesson of the Dreaming is that it has always privileged these processes of reproduction over their products, and that this has been the secret of the persistence of Aboriginal cultural identities as well as the basis for their claims to continuity. This analysis confirms Jupurrurla's and Japanangka's claims that TV is a two-edged sword, both a blessing and a curse, a "fire" that has to be fought with fire. The same medium can prove to be the instrument of salvation or destruction. This is why a simple prediction of the medium's effects is so difficult to make. Video and television intrude in the processes of social and cultural reproduction in ways that literate (missionary, bureaucratic, educational) interventions never managed to accomplish. Its potential force is greater than guns, or grog, or even the insidious paternalisms which seek to claim it.

But in a cultural future, *Coniston Story* operates over time to privilege the Japangardi/Japanangka version of that history, to insert it bit by bit into the Dreaming tracks around Crown Creek until the tape itself crumbles and its memory is distributed selectively alongs the paths of local kinship. In this future, when the mourning period for that old Japangardi is passed, his relations will take the Fire Ceremony tape from the "not to look" shelf and review it again, in regard to the presence or absence of recent performances of the ceremony. Audiences at Yuendumu will reinterpret what is on the tape, bring some fellows into the foreground and disattend to others. They might declare this "a proper law tape," and then go on to perform the ceremony exactly the same, but different. I expect, in the highly active interpretative sessions that these attendances have become, there will be much negotiation necessary to resolve apparent contradictions evoked by the recorded history. I expect that a cultural future allows the space and autonomy for this to happen.

In a lifestyle (ethnic, anti-cultural) future, it's not so certain that anyone will be there at Yuendumu to worry about all this. Why should they? After all, the place has only cultural value, lacking any commercial rationale for the lifestyle economy. But if people are to be situated in this future, we can assume that they will be faced with a very different kind of, and participation in, media. Their relation to the forces and modes of cultural reproduction will be quite passive: they will be constituted as an audience, rendered consumers, even though there's not much money to buy anything (the local store is reduced to selling tinned stew and Kung Fu video tapes). But it would be mistaken to claim that the ethnic cultural policy has ignored Aborigines. In fact, they play a major part in the construction of the national, multicultural image; in this scenario, they become niggers. Then they will be regularly on the airways, appearing as well-adjusted families in situation comedies, as models in cosmetic ads, as people who didn't get a "fair go" on *60 Minutes*. Nationally prominent, academically certified Aborigines will discuss Aboriginality on the ABC and commercial stations, filling in the legislated requirements for Australian content. In the lifestyle future, Aborigines can be big media business.

The people at Yuendumu will watch all this on their government-provided, receive-only satellite earth stations; but we can only speculate about what identifications and evaluations they will make. Perhaps the matter will not be inconsequential. Imported programs supplant, but may not so directly intrude on, cultural reproduction. Rather, it is when some archivist wandering through the ABC film library chances on an old undocumented copy of the Peterson Fire Ceremony film, one of the competing versions of the Coniston massacre, or even some old and valuable Baldwin Spencer footage, circa 1929, of Central Australian native dances, that something truly momentous happens. In pursuit of a moment of "primitivism," the tapes go to air, via satellite, to thousands of communities at once, including those of its subjects, their descendants, their relations, their partners in ritual exchange, their children, their women (or men). One more repository guarded by oral secrecy is breached, one more ceremony is rendered worthless, one more possible claim to authenticity is consumed by the voracious appetite of the simulacra for the appearance of reality. At Yuendumu, this already causes fights, verbal and physical, even threatened payback murders, in the hopeless attempt to ascribe blame in the matter, to find within the kin network the one responsible, so that by punishing him or her the tear in the fabric of social reproduction can be repaired. However, the kin links to descendants of Rupert Murdoch or David Hill or Bob Hawke may prove more difficult to trace, and the mechanisms for adjudication impossible to uncover.

A cultural future can only result from political resistance. It will not be founded on any appeal to nostalgia: not nostalgia for a past whose existence will always be obscure and unknown, nor a nostalgia we project into a future conceived only in terms of the convoluted temporalities of our own present. The tenses are difficult to follow here—but in a sense, that is precisely the critical responsibility now before us. Francis Jupurrurla Kelly makes, is making, television at Yuendumu. He intends to continue, and so assure a cultural future for Warlpiri people. His tapes and broadcasts reach forward and backwards through various temporal orders, and attempt somehow to bridge the Dreaming and the historical. This, too, is a struggle which generates Jupurrurla's art.

The only basis for non-Warlpiri interest in their video must recognize these explicitly contemporary contradictions. Channel Four at Yuendumu resists nostalgic sentiment and troubles our desire for a privileged glimpse of otherness. It is we who are rendered other, not its subject. Ultimately, it must be from this compromising position that such work is viewed.

## Notes

The people at Yuendumu were not entirely happy with this text when I brought it for them to review prior to publication. We took out the few offending pictorial images—this wasn't the problem. It was said by some that the pessimism I expressed seemed unwarranted. Certainly, the evidence of continuing motivation and activity at the TV studio was startling. This was late August 1987, and my visit coincided with the installation of a satellite earthstation receiver which introduced the live ABC program schedule to Yuendumu after so many years of waiting, worrying, and preparing.

There still was no license to legitimize the service that Jupurrurla began that week, mixing local programming with the incoming signal (Warlpiri News and documentaries at 6:30 P.M.). Nor had the equipment repaired itself magically since the last visit: signal strength remained unpredictable, and edits were completely unstable. But the community was still passionately involved in making and watching Warlpiri television. This became clear when a battle ensued with the very first day's transmission. Warlpiri News replaced the *EastEnders,* and at least one European resident was incensed. Jupurrurla decided (somewhat unilaterally, it seemed to me) that the service would shut off at 10:30 P.M., so that kids could go to bed and be sure of getting off to school in the morning. No *Rock Arena.* No late movies. It seemed likely there would be a lot of hot negotiating in the coming months.

1. See J. Meggitt, *The Desert People,* Angus & Robertson, 1962, for the classic description of Warlpiri kinship and social organization.

2. P. Loveday and P. Cook, *Aboriginal Arts and Crafts and the Market,* Darwin, Australian National University North Australia Research Unit Monograph, 1983.

3. These cases are explored in detail in E. Michaels, *The Aboriginal Invention of Television, Central Australia, 1982–86,* Canberra, Australian Institute of Aboriginal Studies, 1986.

4. N. Peterson, "Bulawandi: A Central Australian Ceremony for the Resolution of Conflict", in *Australian Aboriginal Anthropology,* ed. R. M. Berndt, Perth, University of Western Australia Press for the Australian Institute of Aboriginal Studies, 1970, pp. 200–215.

# About the Authors

CHRISTOPHER ANDERSON teaches in the Telecommunications Department at Indiana University. He is the author of numerous articles on film and television and of *Hollywood TV: The Studio System in the Fifties*.

IEN ANG teaches at Murdoch University in Australia. Her work is primarily concerned with audience responses to television. She is the author of *Watching "Dallas"* and *Desperately Seeking the Audience*.

DAVID BARKER has taught at Texas Christian University and the University of Missouri. He is currently completing an advanced degree in counseling at Southwest Texas State University and teaching at Concordia College in Austin.

WILLIAM BODDY teaches at Baruch College of the City University of New York. A specialist in the history of media industries, he is the author of *Fifties Television*.

RODNEY BUXTON teaches in the Communication Department at the University of Denver. He has published several articles on the representation of AIDS in fictional television.

RICHARD CAMPBELL teaches in the Communication Department at the University of Michigan. He is the author of *"60 Minutes" and the News: A Mythology for Middle America* and co-author, with Jimmie L. Reeves, of *Cracked Coverage: Television News, the Anti-Cocaine Crusade, and the Reagan Legacy*.

JOHN CAUGHIE teaches in film and television studies at the University of Glasgow, where he is also co-director of the John Logie Baird Centre for Research in Television and Film. He is the author of numerous works on film and television.

CELESTE MICHELLE CONDIT teaches in the Speech Communication Department at the University of Georgia. Her special interests focus on the rhetoric of social movements. She is the author of *Decoding Abortion Rhetoric: Communicating Social Change.*

MICHAEL CURTIN teaches in the Telecommunication Department at Indiana University. His publications deal primarily with the history of documentaries on television.

DANIEL DAYAN is a fellow of the Centre National de la Recherche Scientifique in Paris. He has also taught at the Annenberg School for Communication at the University of Southern California. He writes widely on film and television.

JANE FEUER teaches at the University of Pittsburgh. She is the author of *Hollywood Musicals* and co-editor of *MTM: Quality Television.* She is currently completing a book on television in the Reagan era.

JOHN FISKE teaches at the University of Wisconsin at Madison. His primary interests focus on the roles of television and popular culture in contemporary societies. He is the author of *Television Culture* and *Reading the Popular.*

TODD GITLIN, professor of sociology at the University of California, Berkeley, is the author of *The Whole World Is Watching; Inside Prime Time,* which develops some of the themes of this article; *The Sixties: Years of Hope, Days of Rage;* and many articles on culture, politics, and the media.

HERMAN GRAY teaches at the University of California at Santa Cruz. He is the author of *Producing Jazz* and of numerous articles on mass media and the African-American experience. He is currently at work on a book addressing television and the representation of race.

JAMES HAY teaches in the Speech Communication Department at the University of Illinois at Urbana-Champaign. He is the author of *Popular Film Culture in Fascist Italy: The Passing of the Rex* and articles on television, postmodernism, and Italy.

HAL HIMMELSTEIN teaches at Brooklyn College of the City University of New York. He is the author of *Television Myth and the American Mind* and has worked extensively on media issues in Finland and Russia.

PAUL M. HIRSCH teaches in the Kellog School of Management at Northwestern University. He is the author of many articles on management practices and on mass media organizations.

HENRY JENKINS III teaches at the Massachusetts Institute of Technology. He is the author of numerous articles on film and television and of *Textual Poachers* and *Who Made Pistachio Nuts*.

ELIHU KATZ teaches sociology and communication at Hebrew University in Jerusalem, where he is also scientific director of the Guttman Institute for Applied Social Research. He is also Regents Professor of Communication at the Annenberg School of Communication at the University of Pennsylvania. He is the author of numerous books and articles dealing with mass communication.

DENISE J. KERVIN teaches in the Media Arts Department at the University of Arizona. Her work focuses on women in film and television.

LISA A. LEWIS produces film and television works in Los Angeles. She is the author of *Voicing the Difference: Female Authorship and Music Video* and the editor of *Fan Culture*. She has taught at the University of Arizona and the University of Cincinnati.

JUDINE MAYERLE is chair of the Department of Broadcast and Electronic Communication at Marquette University. She has published a number of articles on television production practices.

EILEEN R. MEEHAN teaches in the Media Arts Department at the University of Arizona. She has written widely on the political economy of popular entertainment.

ERIC MICHAELS was formerly research associate of the Australian Institute for Aboriginal Studies and taught at Griffith University. He is the author of *For a Cultural Future* and of numerous articles on Aboriginal culture and art as well as on mass media.

DAVID MORLEY teaches in the Department of Communication at Goldsmiths' College, University of London. He is co-author of *Nationwide* and the author of *The "Nationwide" Audience* and *Family Television: Cultural Power and Domestic Leisure*.

HORACE NEWCOMB teaches in the Radio-Television-Film Department at the University of Texas at Austin. He writes about television and cultural theory.

JIMMIE L. REEVES teaches in the Communication Department at the University of Michigan. He is the author of numerous articles on television and co-author, with Richard Campbell, of *Cracked Coverage: Television News, the Anti-Cocaine Crusade, and the Reagan Legacy.*

KATHLEEN K. ROWE teaches media studies at the University of Oregon. Her publications focus on feminism and film and television.

MICHAEL K. SAENZ teaches at the University of Iowa. His primary research interests are in the areas of the cultural history and cultural applications of mass media.

LAURIE SCHULZE teaches in the Communication Department at the University of Denver. She has written about female body builders and Madonna as well as about television.

ELLEN SEITER teaches in the Telecommunication Department at Indiana University. She is the author of numerous articles on television and co-author of *Remote Control: Television, Audiences, and Cultural Power* and author of *Sold Separately: Children and Parents in Consumer Culture.*

LYNN SPIGEL teaches in the Cinema Studies Department at the University of Southern California. She is the author of *Make Room for TV* and is a co-editor of the journal *Camera Obscura.*

DAVID THORBURN teaches humanities and is the director of the Program in Cultural Studies at the Massachusetts Institute of Technology. He writes about television and the popular arts.

BERNARD TIMBERG teaches in the Communication Department at Radford University. He is the author of numerous articles about television and of the forthcoming *Titans of Talk.*

MIMI WHITE teaches in the Radio-Television-Film Department at Northwestern University. She is the author of *Tele-Advising: Therapeutic Discourse in American Television* and co-author of *Media Knowledge: Popular Culture, Pedagogy, and Critical Citizenship.*

BETSY WILLIAMS is humanities editor for the University of Texas Press and a doctoral candidate in the Radio-Television-Film Department at the University of Texas at Austin. Her research focuses on changes in the Hollywood film industry in the new media and economic environments.

© Tomorrow-
8-10- TV CXA DL
10- 11- N. Carolina shhpH
+ merge
12:30 - 2:00- NAperoni